Conversion Factors*

Acceleration
$1 \text{ m/s}^2 = 3.2808 \text{ ft/s}^2$

Area
$1 \text{ ft}^2 = 0.0929 \text{ m}^2$
$1 \text{ in.}^2 = 6.4516 \times 10^{-4} \text{ m}^2$

Density
$1 \text{ lbm/ft}^3 = 16.018 \text{ kg/m}^3$
$1 \text{ slug/ft}^3 = 515.3 \text{ kg/m}^3$
$1 \text{ g/cm}^3 = 1.940 \text{ slug/ft}^3$

Energy
$1 \text{ Btu} = 1055.1 \text{ J} = 1055.1 \text{ N} \cdot \text{m}$
$= 778.16 \text{ ft} \cdot \text{lbf}$
$= 0.252 \text{ kcal}$
$1 \text{ J} = 1 \times 10^7 \text{ erg}$

Energy Flow Rate
$1 \text{ Btu/h} = 0.2931 \text{ W}$

Force
$1 \text{ lbf} = 4.448 \text{ N}$
$1 \text{ dyne} = 1 \times 10^{-5} \text{ N}$
$= 2.248 \times 10^{-6} \text{ lbf}$

Length
$1 \text{ in.} = 2.54 \text{ cm}$
$1 \text{ ft} = 0.3048 \text{ m}$
$1 \text{ mi} = 5280 \text{ ft}$
$= 1609.3 \text{ m}$

Mass
$1 \text{ kg} = 2.205 \text{ lbm}$
$= 0.06852 \text{ slugs}$

Mass Flow Rate
$1 \text{ lbm/h} = 0.000126 \text{ kg/s}$

Power
$1 \text{ kW} = 1 \text{ kJ/s}$
$= 3412 \text{ Btu/h}$
$= 1.341 \text{ hp}$
$= 737.57 \text{ ft} \cdot \text{lbf/s}$
$1 \text{ hp} = 550 \text{ ft} \cdot \text{lbf/s}$
$= 2544 \text{ Btu/h}$

Pressure
$1 \text{ kPa} = 1 \times 10^3 \text{ N/m}^2$
$= 0.1450 \text{ lbf/in.}^2$
$= 0.2953 \text{ in. Hg}$
$= 0.75 \text{ cm Hg}$
$= 4.016 \text{ in. H}_2\text{O}$
$= 1 \times 10^{-2} \text{ bar}$
$1 \text{ lbf/ft}^2 = 47.88 \text{ Pa}$
$1 \text{ atm} = 101.325 \text{ kPa}$
$= 14.696 \text{ lbf/in.}^2$

Refrigeration Capacity
$1 \text{ ton} = 3.516 \text{ kW}$
$= 12{,}000 \text{ Btu/h}$

Specific Energy
$1 \text{ kJ/kg} = 0.4299 \text{ Btu/lbm}$
$1 \text{ cal/g} = 4186.8 \text{ J/kg}$

Specific Heat Capacity
$1 \text{ Btu/(lbm} \cdot \text{°F)} = 4186.8 \text{ J/(kg} \cdot \text{K)}$

Temperature
$\text{°F} = 32 + 1.8 \text{ °C}$
$\text{°C} = (\text{°F} - 32)/1.8$
$\text{K} = \text{°C} + 273.15$
$\text{°R} = \text{°F} + 460$
$= 1.8 \text{ K}$

Velocity
$1 \text{ m/s} = 3.281 \text{ ft/s}$
$= 2.237 \text{ mi/h}$
$1 \text{ mi/h} = 1.467 \text{ ft/s}$
$1 \text{ knot} = .5144 \text{ m/s}$
$= 1.151 \text{ mi/h}$

Volume
$1 \text{ ft}^3 = 0.02832 \text{ m}^3$
$= 7.48052 \text{ gal (US)}$
$1 \text{ in.}^3 = 1.6387 \times 10^{-5} \text{ m}^3$
$1 \text{ liter} = 1.057 \text{ quarts}$

Volume Flow Rate
$1 \text{ ft}^3\text{/min} = 0.000472 \text{ m}^3\text{/s}$

Viscosity
DYNAMIC $\quad 1 \text{ lbm/(ft} \cdot \text{s)} = 1.488 \text{ N} \cdot \text{s/m}^2$
KINEMATIC $\quad 1 \text{ ft}^2\text{/s} = 0.0929 \text{ m}^2\text{/s}$

---

*To convert any unit to another, make a fraction using the appropriate equivalency found above with the desired unit on top (numerator). Multiply or divide as required.

Example: To convert 5 hp to kW $\quad 5 \text{ hp} \cdot \left[\dfrac{1 \text{ kW}}{1.341 \text{ hp}} \right] = 3.7286 \text{ kW}$

Classical Thermodynamics

Lynn D. Russell
The University of Alabama in Huntsville

George A. Adebiyi
Mississippi State University

 SAUNDERS COLLEGE PUBLISHING
Harcourt Brace Jovanovich College Publishers

Fort Worth   Philadelphia   San Diego
New York   Orlando   Austin
San Antonio   Toronto   Montreal
London   Sydney   Tokyo

Requests for permission to make copies of any part of the work should be mailed to
Permissions Department, Harcourt Brace Jovanovich, Publishers, 8th Floor, Orlando,
Florida 32887.

Text Typeface: Caledonia
Compositor: Waldman Graphics
Acquisitions Editor: Emily Barrosse
Developmental Editor: Lloyd Black
Managing Editor: Carol Field
Project Editor: Laura Maier
Copy Editor: Kathy Walker
Manager of Art and Design: Carol Bleistine
Art Director: Anne Muldrow
Art Assistant: Caroline McGowan
Text Designer: William Boehm
Cover Designer: Lawrence R. Didona
Text Artwork: GRAFACON, Inc.
Director of EDP: Tim Frelick
Production Manager: Jay Lichty
Marketing Manager: Monica Wilson

Cover Credit: Gabe Palmer/The Stock Market

Printed in the United States of America

CLASSICAL THERMODYNAMICS

ISBN: 0-03-032417-3

Library of Congress Catalog Card Number: 92-050161

2345 039 987654321

This book was written to serve as an undergraduate textbook in classical thermodynamics for engineering students. The text is for students who have had basic calculus (including differentiation and integration), physics, and the fundamentals of chemistry. The text is designed for a two-semester course sequence, or for an introductory one-semester course.

Our purposes in writing this book are (1) to explain fundamental concepts and principles of thermodynamics explicitly and to provide beginning undergraduate students with adequate information for a reasonable understanding of thermodynamics, (2) to provide a more comprehensive treatment of the second law of thermodynamics than is found in most beginning texts and to base this treatment on the most recent research in the field, and (3) to provide software that aids in a physical understanding of the problems while minimizing nonproductive time spent by students in working problems.

The pedagogy for the text is based on approximately 20 years of experience by each of the authors in teaching thermodynamics to a wide variety of students. Important terms and concepts are introduced early (many in Chapter 1), and then fully explained and illustrated later at the appropriate time. These terms and concepts are then used where applicable throughout the remainder of the text in order to reinforce the learning process. As a result of this approach the text provides more in-depth treatment of many difficult concepts, such as exergy, than most other texts. A systematic approach to problem solving is also outlined in Chapter 1 and used throughout the text.

Organization

The book is organized into 15 chapters, the first nine of which cover the fundamentals of thermodynamics, while the remaining six chapters present applications that are common in engineering. The foundation for the entire text is laid in the first two chapters. An overview of thermodynamics is given in Chapter 1 where the importance of thermodynamics and its relationship to our use of energy resources are discussed. Chapters 1 and 2 also provide an overview of the concepts of thermodynamics. Concepts are an essential part of any science, and in the case of thermodynamics, experience has shown that this is an area where students have difficulty. Several examples, drawn from everyday experiences, are used to help students achieve a good understanding of the concepts and also gain an early appreciation of the relevance of thermodynamics to the everyday needs of society. Special emphasis is given throughout the text to physical systems and physical understandings. Fundamental laws and principles are explained with explicit statements, sometimes repeated in different forms, and illustrated with familiar systems in order to assist students in understanding the important principles and concepts. Particular attention is given throughout the text to the definition of system boundaries and to the interactions across these boundaries.

Chapters 3 and 4 deal with the properties of a class of substances known as pure substances. Properties of common substances are provided both in tabular form and on computer disk for use with the text. Four computer programs (for IBM or an IBM-compatible PC) were developed to accompany this text and are contained on the **Thermo·Props™—Thermodynamics Properties Data Finder** disk found at the back of this textbook. The four programs include **STEAM** (for the properties of ice, water, and steam), **R22** (for refrigerant-22 properties), **GAS** (for ideal gases CO_2, CO, O_2, N_2, H_2O, H_2, air, and CH_4 encountered in combustion processes), and **PSY** (for moist air, water, and steam properties needed in the analysis of air conditioning processes). Graphical and tabular outputs are provided by the programs, and the graphical outputs are designed specifically to increase the student's understanding of the system. Problems at the end of the chapter that are more easily solved using the computer programs are identified with a disk symbol next to each problem. These computer programs are not intended to displace the need for students to be able to read tables of properties. In later chapters, however, students are encouraged to use **Thermo·Props**, especially whenever several property values are to be determined, so that the focus can be placed on application of thermodynamic principles to the solution of a problem.

Chapter 5 gives a more detailed treatment of work and heat following the introduction provided in Chapters 1 and 2. This expanded treatment is important to considerations of the first and second laws of thermodynamics. The two laws are often regarded as the pillars of thermodynamics: the first speaks of energy and its conservation (quantity), while the second law deals with the quality aspect of energy. The first law of thermodynamics is discussed in Chapter 6. The exposition of the first law of thermodynamics essentially follows the classical tradition of Poincare and Planck. Starting with a formal statement of the first law of thermodynamics in terms of net heat and net work in a cyclic process, corollaries are established including the important deduction that the thermodynamic concept of energy derives solely from the first law of thermodynamics. Applications of the first law of thermodynamics to nonflow processes (for closed systems) and flow processes (for open systems or control volumes) are explained, and several examples are provided from everyday experiences to demonstrate the importance of the first law analysis to the carrying out of an energy balance whenever a system (closed or open) interacts with its environment.

The second law of thermodynamics occupies a central place in thermodynamics and accordingly is treated in a comprehensive manner consistent with the most recent research in the field. Formal statements of the law are presented and discussed in Chapter 7, followed by a systematic development of the corollaries of the law. Chapter 8 is devoted entirely to consideration of entropy, which is to the second law what energy is to the first law. (Energy and entropy are both derived properties in the sense that their existence can only be inferred as corollaries of the respective laws.) The treatment of the second law in Chapter 8 includes the use of the entropy concept for the evaluation of processes and for the determination of "waste" or "energy degradation" taking place in real processes. In Chapter 9, another second law concept, exergy (or availability) is introduced along with a development of procedures for utilizing the concept in performance evaluation of systems (open or closed). The term *exergy* has been used in preference to *availability* to conform with the international trend.

It should be noted that several applications of thermodynamics in the other sciences are often limited to a first law analysis. However, questions relating to efficiency and evaluation of performance, or the direction of physical and chemical processes, for example, require a second law analysis. There is much contemporary concern about "conservation" and making the available energy resources last longer by using more efficient energy systems. This reality is part of the reason for the prominent place accorded the second law of thermodynamics in this text. After the detailed exposition given to the law in Chapters 7 to 9, the various applications to systems in Chapter 10 to 15 provide ample demonstration of how to carry out a complete thermodynamic analysis based on both the first and second laws of thermodynamics. Second law analyses are illustrated for a broad range of applications, including vapor and gas cycles (Chapter 10), refrigeration cycles (Chapter 11), psychrometrics (Chapter 13), combustion processes (Chapter 14), and chemical equilibria (Chapter 15). The optimum thermodynamic cycle is discussed, and an explanation is given to differentiate *optimum* cycles from *ideal* cycles. The fact that the Carnot cycle is not the optimum cycle for a real heat power plant that receives heat from combustion gases is demonstrated and explained.

Pedagogy

Several special pedagogical features are included in the text. A list of key concepts at the beginning of each chapter introduces the material, while a chapter-end summary of key concepts and ideas provides a handy reference for review. Review questions at the end of each chapter reinforce key concepts and allow students to test their comprehension of material just learned. Significant terms are boldfaced or italicized for emphasis.

Worked examples, followed by exercises, appear throughout the text, so students have models to guide them through new material. The section exercises, which include answers, give students more opportunities to check their progress and build on acquired knowledge as they move through each section and chapter. The problems at the end of each chapter are graded according to three levels of difficulty (low, average, and high), motivating students to challenge their abilities as they progress through the problem sets. Open-ended design problems appear after each problem set in the applications chapters, 10 through 15. Students, given a realistic engineering situation, are asked either to redesign the system to meet specified goals or are asked to analyze the system's capabilities through a series of questions. Also integrated throughout are problems related to safety. A total of 1050 problems and exercises are provided in the text. Answers to selected end-of-chapter problems are given at the back of the book.

Problems, examples, and data are given for both the *Système International d'Unites*

(SI) and the United States Customary System of units (USCS), but the unit systems are not mixed in a specific problem. Although it is apparent that the United States and all other major industrialized countries are moving to SI, the change is not yet complete and it is appropriate for students to develop the capability to deal with both systems. However, if an instructor wishes, he or she can generally use strictly one system or the other throughout.

Package

The following supplements are provided free to adopters.

- **Instructor's Manual with Solutions and Transparency Masters.** Complete solutions to all end-of-chapter problems are provided. In addition, there are 100 transparency masters of selected figures enlarged from the text.
- **Thermo·Props™—Thermodynamics Properties Data Finder.** On disk for IBM-compatible PCs, this software is enclosed with every copy of the text. The four programs presented include **STEAM** (for the properties of ice, water, and steam), **R22** (for refrigerant-22 properties), **GAS** (for ideal gases CO_2, CO, O_2, N_2, H_2O, H_2, air, and CH_4 encountered in combustion processes), and **PSY** (for moist air, water, and steam properties needed in the analysis of air conditioning processes). On-screen graphs and tables offer easy-to-read results, allowing students to concentrate more on problem solving than on time-consuming data searches.

Acknowledgments

We wish to acknowledge the following individuals who assisted in various stages of the review process of the text:

Charles W. Bouchillon, Mississippi State University

C. T. Carley, Mississippi State University

Alan J. Chapman, Rice University

Kenneth D. Kihm, Texas A&M University

Alan A. Kornhauser, Virginia Polytechnic Institute and State University

Robert J. Krane, University of Tennessee, Knoxville

Blaine I. Leidy, University of Pittsburgh

D. C. Look, Jr., University of Missouri—Rolla

John J. McGrath, Michigan State University

Ronald S. Mullisen, California Polytechnic State University, San Luis Obispo

Larry Roe, Virginia Polytechnic Institute and State University

George Tsatsaronis, Tennessee Technological University

Thomas W. Weber, State University of New York at Buffalo

William J. Wepfer, Georgia Institute of Technology

William M. Worek, University of Illinois at Chicago

In addition, we acknowledge the assistance of Thomas D. McCallum, Jr. in the preparation of various figures in the text. Also we appreciate the suggestions and input from students at Mississippi State University who used the text from 1986–1992 while it was in the early stages of development.

The preparation of this text was a 10-year project. We appreciate the continued support of our families during this time. For their encouragement and patience throughout, we dedicate the text to Elaine, Kathy, Brent, Mark, and Jeffrey Russell, and to Iyabo, Nike, and Debo Adebiyi.

Lynn D. Russell
George A. Adebiyi

CONTENTS

NOMENCLATURE

		SI Units	USCS Units
a	Acceleration	m/s^2	ft/s^2
a	Specific Helmholtz function	$J/kg, kJ/kg$	Btu/lbm
$\bar{a}$	Molar specific Helmholtz function	$J/kmol, kJ/kmol$	$Btu/lbmol$
A	Helmholtz function	J, kJ	Btu
A	Surface area	m^2	ft^2
A^*	Surface area per unit mass of system	m^2/kg	ft^2/lbm
AF	Air–fuel ratio	—	—
c	Specific heat	$J/kg \cdot K, kJ/kg \cdot K$	$Btu/lbm \cdot °R$
c_p	Constant pressure specific heat	$J/kg \cdot K, kJ/kg \cdot K$	$Btu/lbm \cdot °R$
$\bar{c}_p$	Molar specific heat at constant pressure	$J/kmol \cdot K, kJ/kmol \cdot K$	$Btu/lbmol \cdot °R$
c_{p0}	Zero pressure specific heat at constant pressure	$J/kg \cdot K, kJ/kg \cdot K$	$Btu/lbm \cdot °R$
$\bar{c}_{p0}$	Zero pressure molar specific heat at constant pressure	$J/kmol \cdot K, kJ/kmol \cdot K$	$Btu/lbmol \cdot °R$
c_v	Constant volume specific heat	$J/kg \cdot K, kJ/kg \cdot K$	$Btu/lbm \cdot °R$
$\bar{c}_v$	Molar specific heat at constant volume	$J/kmol \cdot K, kJ/kmol \cdot K$	$Btu/lbmol \cdot °R$
c_{v0}	Zero pressure specific heat at constant volume	$J/kg \cdot K, kJ/kg \cdot K$	$Btu/lbm \cdot °R$
$\bar{c}_{v0}$	Zero pressure molar specific heat at constant volume	$J/kmol \cdot K, kJ/kmol \cdot K$	$Btu/lbmol \cdot °R$
COP_C	Coefficient of performance for cooling	—	—
COP_H	Coefficient of performance for heating	—	—
E	Total energy of a system	J, kJ	Btu
E	Voltage	V	V
F, f	Force	N	lbf
FA	Fuel–air ratio	—	—
g	Acceleration due to gravity	m/s^2	ft/s^2
g	Specific Gibbs function	$J/kg, kJ/kg$	Btu/lbm

		SI Units	USCS Units
$g°$	Specific Gibbs function at 1 atm pressure	J/kg, kJ/kg	Btu/lbm
$\bar{g}$	Molar specific Gibbs function	J/kmol, kJ/kmol	Btu/lbmol
G	Gibbs function	J, kJ	Btu
g_c	Constant for the USCS force unit = 32.174 $\frac{\text{ft} \cdot \text{lbm}}{\text{lbf} \cdot \text{s}^2}$	—	$\frac{\text{ft} \cdot \text{lbm}}{\text{lbf} \cdot \text{s}^2}$
G	Torque acting on a shaft	N · m	lbf · ft
h	Specific enthalpy	J/kg, kJ/kg	Btu/lbm
$\bar{h}$	Molar specific enthalpy	J/kmol, kJ/kmol	Btu/lbmol
$\bar{h}_i$	Partial molar specific enthalpy for the ith species	J/kmol, kJ/kmol	Btu/lbmol
h_{ma}	Enthalpy of moist air per unit mass of dry air	J/kg da, kJ/kg da	Btu/lbm da
$h°$	Specific enthalpy at 1 atm pressure	J/kg, kJ/kg	Btu/lbm
H	Enthalpy	J, kJ	Btu
$\bar{h}_c°$	Standardized enthalpy of combustion	J/kmol, kJ/kmol	Btu/lbmol
h_c	Convective heat transfer coefficient	W/m^2 · K	Btu/ft^2 · h · °R
h_r	Heat transfer coefficient for radiation	W/m^2 · K	Btu/ft^2 · h · °R
hr (or ω)	Humidity ratio of moist air	—	—
$\bar{h}_R$	Enthalpy of reaction	J/kmol, kJ/kmol	Btu/lbmol
HV	Heating value for a fuel, where LHV is lower heating value and HHV is higher heating value	J/kg, kJ/kg	Btu/lbm
I	Current	amp	amp
k	Thermal conductivity	W/m · K	Btu/ft · h · °R
k	Constant of proportionality in Newton's second law of motion	—	$1/g_c$
k	Proportionality constant for a wire	N/m	lbf/ft
$k \ (= c_p/c_v)$	Ratio of the principal specific heats or specific heat ratio	—	—
K_p	Equilibrium constant	—	—
L	Length dimension (in dimensional analysis)	m	ft
L	Wall thickness	m	ft
m	Mass	kg	lbm
M	Mass dimension (in dimensional analysis)	kg	lbm
M	Molecular weight (or molecular mass)	kg/kmol	lbm/lbmol
m.e.p.	Mean effective pressure	Pa, bar	lbf/in.^{2} (psi)
m.s.	Mixture strength for a fuel–air mixture	—	—
m_a	Mass of dry air	kg da	lbm da
$\dot{m}$	Mass flow rate	kg/s	lbm/s

		SI Units	USCS Units
m_v	Mass of moisture	kg	lbm
n	Polytropic gas index	—	—
N	Number of moles	kmol	lbmol
p	Pressure	Pa, bar	psi, lbf/in.^{2}
p_c	Critical pressure	Pa, bar	psi, lbf/in.^{2}
p_i	Partial pressure of the ith species in a mixture	Pa, bar	psi, lbf/in.^{2}
p_r	Relative pressure	—	—
p_R	Reduced pressure	—	—
q	Heat transfer per unit mass to system	J/kg, kJ/kg	Btu/lbm
Q	Heat transfer to system	J, kJ	Btu
Q_{12}	Heat transfer to system going from state 1 to 2	J, kJ	Btu
$\dot{Q}$	Heat transfer rate to system	W	Btu/h
Q_H	Heat transfer from (or to) a high temperature system	J, kJ	Btu
Q_L	Heat transfer to (or from) a low temperature system	J, kJ	Btu
$\dot{q}_x$	Heat flux (or heat flow rate through a unit area per unit time) in the x-direction	W/m^2	Btu/ft$^2 \cdot$ h
R	Specific gas constant	J/kg $\cdot$ K, kJ/kg $\cdot$ K	Btu/lbm $\cdot$ °R
$\overline{R}$	Universal gas constant	J/kmol $\cdot$ K, kJ/kmol $\cdot$ K	Btu/lbmol $\cdot$ °R
r_c	Cut-off ratio for a Diesel cycle	—	—
rh (or ϕ)	Relative humidity	%	%
r_p	Pressure ratio for a gas turbine cycle (dimensionless)	—	—
R_T	Thermal resistance	K/W	°R $\cdot$ h/Btu
r_v	Compression ratio for a reciprocating IC engine	—	—
r_w	Work ratio (dimensionless)	—	—
s	Specific entropy	J/kg $\cdot$ K, kJ/kg $\cdot$ K	Btu/lbm $\cdot$ °R
$\overline{s}$	Molar specific entropy	J/kmol $\cdot$ K, kJ/kmol $\cdot$ K	Btu/lbmol $\cdot$ °R
$\overline{s}_i$	Partial molar specific entropy for the ith species	J/kmol $\cdot$ K, kJ/kmol $\cdot$ K	Btu/lbmol $\cdot$ °R
$s°$	Specific absolute entropy of a substance at 1 atm pressure	J/kg $\cdot$ K, kJ/kg $\cdot$ K	Btu/lbm $\cdot$ °R
S	Entropy	J/K, kJ/K	Btu/°R
S_{gen}	Entropy generation	J/K, kJ/K	Btu/°R
$\dot{S}_{gen}$	Entropy generation rate	W/K	Btu/h $\cdot$ °R
s.s.c.	Specific steam consumption	kg/kWh	(lbm/h)/hp
SG	Specific gravity (dimensionless)	—	—
SW	Specific weight	N/m^3	lbf/ft^3
t	Time	s	s
T	Temperature	K (°C)	°R (°F)
T^*	Adiabatic saturation temperature	K (°C)	°R (°F)
T_c	Critical temperature	K (°C)	°R (°F)
T_d	Dew point temperature	K (°C)	°R (°F)

		SI Units	USCS Units
T_{db}	Dry bulb temperature	K (°C)	°R (°F)
T_H	Temperature of a high temperature thermal reservoir	K (°C)	°R (°F)
T_L	Temperature of a low temperature thermal reservoir	K (°C)	°R (°F)
T_0	Ambient (or reference environment) temperature	K (°C)	°R (°F)
T_R	Reduced temperature (dimensionless)	—	—
T_{wb}	Wet bulb temperature	K (°C)	°R (°F)
u	Specific internal energy	J/kg, kJ/kg	Btu/lbm
$\bar{u}$	Molar specific internal energy	J/kmol, kJ/kmol	Btu/lbmol
$\bar{u}_c^\circ$	Standardized internal energy of combustion	J/kmol, kJ/kmol	Btu/lbmol
U	Internal energy	J, kJ	Btu
v	Specific volume	m³/kg	ft³/lbm
$\bar{v}$	Molar specific volume	m³/kmol	ft³/lbmol
V	Volume	m³	ft³
V_c	Clearance volume for a reciprocating IC engine	m³	ft³
V_i	Partial volume of the ith species in a mixture	m³	ft³
V_m	Bulk mean velocity	m/s	ft/s
$\lvert \mathbf{V} \rvert$	Velocity (magnitude)	m/s	ft/s
v_r	Relative specific volume (dimensionless)	—	—
w	Work done per unit mass of system	J/kg, kJ/kg	ft · lbf/lbm
W	Work done by system	J, kJ	ft · lbf
W_{12}	Work done by system going from state 1 to state 2	J, kJ	ft · lbf
W_d	Displacement work	J, kJ	ft · lbf
$\dot{W}$	Power, or rate of doing work	W	hp, ft · lbf/s
W_x	Shear (or shaft) work	J, kJ	ft · lbf
$\dot{W}_x$	Shaft power or shear work rate	W	hp, ft · lbf/s
x	Exergy per unit mass	J/kg, kJ/kg	ft · lbf/lbm
$\bar{x}$	Exergy per mole	J/kmol, kJ/kmol	ft · lbf/lbmol
x	Vapor quality (or dryness fraction)	—	—
X	Displacement or extension of wire	m	ft
x_{ma}	Exergy of moist air per unit mass of dry air	J/kg da, kJ/kg da	ft · lbf/lbm da
y_i	Mole fraction for the ith species	—	—
z	Specific property	—	—
Z	Extensive property (in general)	—	—
$\bar{Z}$	Partial molal property	—	—
Z	Compressibility factor (dimensionless)	—	—
Z	System elevation above chosen datum	m	ft

Greek Symbols

α	Coefficient of thermal expansion or volume expansivity
β_T	Coefficient of isothermal compressibility
B_T	Isothermal modulus
γ	Specific weight (N/m^3 or lbf/ft^3)
$\Delta \bar{h}_c$	Enthalpy of combustion (J/kmol, kJ/kmol or Btu/lbmol)
$\Delta \bar{h}_c^\circ$	Standardized enthalpy of combustion (J/kmol, kJ/kmol or Btu/lbmol)
$\Delta \bar{h}_f^\circ$	Enthalpy of formation (J/kmol, kJ/kmol or Btu/lbmol)
$\Delta \bar{h}_R$	Enthalpy of reaction (J/kmol, kJ/kmol or Btu/lbmol)
ε	Efficiency ratio
η	Efficiency
η_c	Carnot cycle efficiency
η_1, η_t	First law efficiency; also thermal efficiency (for heat engines)
η_2	Second-law efficiency
η_{device}	Process efficiency for a device
$\dot{\theta}$	Angular velocity
θ	Angular displacement
μ_h	Isenthalpic Joule–Thomson coefficient
μ	Degree of saturation
$\bar{\mu}_i$	Chemical potential for the ith species in a mixture
ρ	Density
σ	Stefan–Boltzmann constant
σ	Surface tension
ϕ	Relative humidity
χ	Exergy (or availability) (J, kJ or ft · lbf, Btu)
$\dot{\chi}$	Exergy flow rate (W, kW or hp, ft · lbf/s)
ω	Humidity ratio

Subscripts

0,1,2, . . .	States 0, 1, 2, . . .
12	Change or amount during process that takes system from state 1 to state 2
c	Critical state
CS	Control surface
CV	Control volume
e	The condition at exit to a control volume
f	The saturated liquid condition
fg	The phase change from saturated liquid to saturated vapor
g	The saturated vapor condition
i	The condition at inlet to a control volume
i	The saturated ice condition
l	The liquid phase
0	The ambient (or reference environment) condition
p	Constant pressure condition
R	Reduced property, or ratio of property value to that at critical state
s	A saturated pressure or temperature
v	Constant volume condition
v	Vapor condition
+ve	Positive quantity
−ve	Negative quantity

Superscripts

· (overdot)	Time rate of quantity
⁻ (overbar)	Quantity on mole basis
°	Degree on Celsius, Fahrenheit, or Rankine temperature scale
°	Property referenced to the standard reference state

Introduction

1.1 Introduction

Thermodynamics is a fairly modern science. While formal definitions vary, thermodynamics is viewed primarily as the *science of energy*. In other words, thermodynamics can be viewed as the science that deals with energy transformations and the relationships between properties of systems. A *property* is defined formally as any observable characteristic of a system. A *system*, or more precisely, a *closed system*, can be defined as any identifiable collection of matter.

Thermodynamics has much in common with the experimental sciences. It relies entirely on observation and experimental measurements. Data obtained in such endeavors played a crucial role in the eventual formulation of the laws of thermodynamics. The laws of thermodynamics are axioms in the sense that they cannot be proved by demonstration. As long as no observation or measurement, however, is made that contradicts the laws, acceptance of their validity is ensured. Frequently, a student (or an inventor) thinks he or she has found evidence or has come up with an invention that disproves one of the generally accepted laws of thermodynamics. The expression *perpetual motion*

Figure 1.1 A closed-cycle mill proposed by Robert Fludd in 1618 as a source of perpetual power.

machine, for example, was coined and applied to inventions purported to defy either the first or second law of thermodynamics (see Figs. 1.1 and 1.2, which illustrate some classic examples). In several cases, people bought shares in the venture only to discover that the claims of the inventors were fake. Such experiences notwithstanding, any scientific law becomes suspect once a valid observation or experimental measurement is made that is irreconcilable with that law. In this sense, one should realize that the sole basis for thermodynamics is experimental measurements and observation and that no theoretical basis exists for thermodynamics outside of these.

Classical Thermodynamics and Statistical Thermodynamics

Two approaches exist for determining properties of substances (material or matter) which make up systems. In one approach, measurements are made on the large scale and are thus relative to the *macroscopic* behavior of the substance. This approach as-

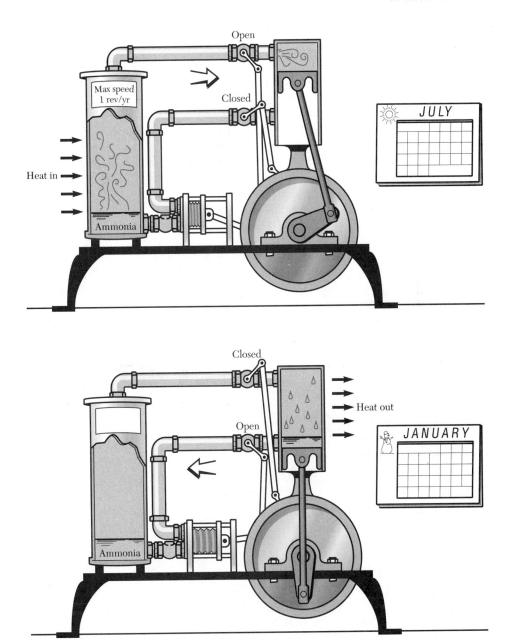

Figure 1.2 John Gamgee's ammonia engine design for limitless production of power from energy extracted from the environment.

sumes that the medium of interest exists as a continuum. The second approach implicitly uses observations on the macroscopic level to postulate behavior at the particle or *microscopic* scale and then uses mathematical calculations on the molecular scale to determine properties from statistical averaging of the behavior of individual particles. *Classical thermodynamics* involves the observation and measurement of properties on a large scale (or macroscopic) basis, while *statistical thermodynamics* aims at the prediction of macroscopic behavior based on molecular (or microscopic) scale events via the appropriate use of mathematics and statistics. Thermodynamics was initially developed by observing the large scale, macroscopic behavior of systems. This "classical" approach will be developed in this text.

History and Expository Traditions

Historically, thermodynamics grew out of an emerging branch of thermal physics in the 18th and 19th centuries, although some of the ideas and considerations involved have a much longer history. The list of individuals whose contributions can be regarded as foundational to modern thermodynamics is long and includes the following: Benjamin Thompson (1753–1814) (later given the title Count Rumford), Sadi Carnot (1796–1832), Rudolph J. Clausius (1822–1888), Lord Kelvin (1824–1907), J. P. Joule (1818–1889), J. Willard Gibbs (1839–1903), and Max Planck (1858–1947). While the development of classical thermodynamics has continued to modern times, much of it is based on the work of these pioneering founders of thermodynamics. From a historical perspective, the 18th century is fairly recent; thus thermodynamics is a relatively modern science.

An excellent historical account of expository traditions in thermodynamics is given by Keenan and Shapiro (1947). They discuss three notable traditions in classical thermodynamics, namely, the *Poincare–Planck structure* (1897–1908), the *Caratheodory structure* (1909), and the *single axiom exposition of Keenan and Hatsopoulos*. (See Hatsopoulos and Keenan 1962, for details.) The Poincare–Planck structure of thermodynamics is the one explained in most contemporary texts on thermodynamics and is the one followed here. Planck's comprehensive treatise was written in 1897, while in 1908, Poincare offered the first logical, complete structure for the exposition of the first law of thermodynamics. Caratheodory was a mathematician, and his exposition is generally acknowledged as being very elegant and mathematically rigorous although lacking in physical intuition and application. In particular, Caratheodory's exposition consciously avoids the use of the concept of heat; instead, the concepts of adiabatic (no heat) process and work were used copiously. The notable feature of the exposition by Keenan and Hatsopoulos is the postulation that all the traditionally stated laws of thermodynamics are, in essence, corollaries of a single axiom, which they termed the *law of stable equilibrium*.

In our view, the Poincare–Planck exposition is relatively simple to understand because it uses concepts and ideas with which most readers are likely to be familiar from experiences with other branches of the sciences, notably physics. It is equally important that valid and accurate conclusions are reached when the principles of thermodynamics, as expounded in this tradition, are applied to most situations encountered in engineering. The approach in this text accordingly follows the Poincare–Planck tradition.

Structure of Thermodynamics

The structure of any scientific discipline includes concepts and laws. *Scientific laws* are postulates believed to be true until contradicted by experimental observation. As stated earlier, the only basis of thermodynamics is the observation of the physical world and the experimental measurements related to this observation. No other theoretical proof exists for thermodynamics. Thus, if a case were observed in nature that was contrary to what is implied by an existing law of thermodynamics, the law would be declared invalid.

Early observations that provided a basis for thermodynamics were generally those on the study of and measurements on a variety of thermal systems, such as steam engines and other work-producing engines that used fuel energy resources. The word *thermodynamics* comes from the Greek words *therme*, which means "heat," and *dynamis*, which means "power." Indeed, thermodynamics was perceived initially as a study of power-producing systems, termed *heat engines*, which used sources that produced heat transfer to the engines. Such applications were studied thoroughly from the turn of the 18th century through much of the 19th century. Today, however, thermodynamics is a much broader science that is relevant to a variety of phenomena encountered in engineering.

The next section introduces several concepts that are needed to arrive at appropriate thermodynamic models for real-life systems and phenomena. Thermodynamic modeling is a necessary first step before one can meaningfully analyze or evaluate processes or phenomena from a thermodynamics standpoint. Section 1.3 provides an overview of the various laws of thermodynamics, after which an introductory survey is given on typical systems and devices of interest in engineering with some explanation on how they can be modeled thermodynamically. A brief perspective then follows on the energy needs of societies now and in the future. Finally, a systematic procedure for solving problems in thermodynamics is presented.

1.2   Basic Concepts and Thermodynamic Modeling

Concepts are an essential part of the development and exposition of any science. They represent the ideas and the language commonly used to express those ideas. Thermodynamics uses many concepts, among which are such basic ones as *system, control volume, surroundings, equilibrium, properties,* and *processes,* to name a few. Still others are consequences of the laws of thermodynamics, such as *energy, entropy,* and *exergy*. Energy is derived from the first law of thermodynamics, while entropy and exergy are derived from the second law. Such concepts are often termed *derived concepts*. The derived concepts cannot be considered in detail until after the pertinent laws have been discussed. At this stage it is useful to introduce the more fundamental concepts and demonstrate their use in the modeling of thermodynamic systems and processes.

The first, and perhaps the most crucial, step in thermodynamic modeling entails what can be termed a process of *abstraction*. In other words, consider with your mind's eye the essential features of whatever is occurring. The operation of a gasoline engine can be used to illustrate the process.

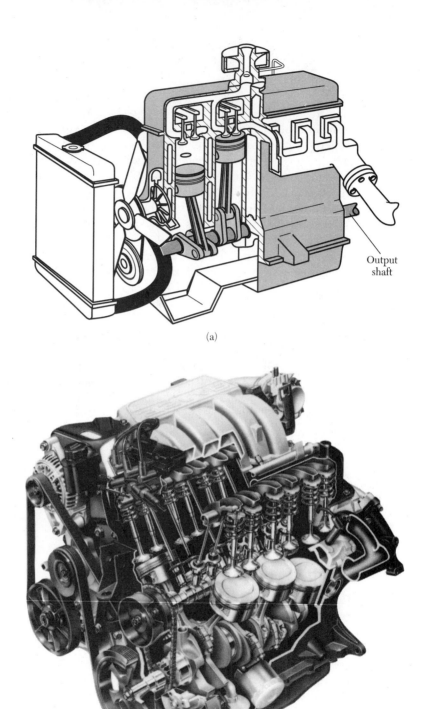

Output
shaft

(a)

(b)

Figure 1.3 (a) A schematic diagram of a gasoline engine. (b) A Chrysler 3.3-liter V-6 engine. (Used with permission from Chrysler Corporation.)

Modeling of a Gasoline Engine

Modeling of the operation of a gasoline engine involves several levels of abstraction. What is going on is easily identified and can be listed as follows:

1. admission of a mixture of air and fuel into the engine
2. burning or combustion of the fuel in air inside the engine cylinder
3. work production resulting from fuel energy released in the combustion process
4. the original mixture entering the engine eventually leaves as an exhaust gas stream

One other notable feature of the typical gasoline engine design is the inclusion of a water-cooling loop to ensure that the engine does not become overheated. The primary function of a gasoline engine, however, is to produce work by burning fuel in air. Therefore, an appropriate model should focus attention not only on the air–fuel mixture entering and the exhaust gas stream leaving the engine but also on the work output resulting from whatever took place inside the engine.

Figure 1.3(a) is a schematic of the gasoline engine showing its principal parts, namely, the carburetor (used for adjusting the ratio of fuel to air in the mixture entering the engine), the engine piston–cylinder (block) unit, the exhaust line from the engine, and a water-cooling loop. Figure 1.3(b) is a pictorial view of a 3.3-liter V-6 engine. In thermodynamics, line sketches and block diagrams suffice to demonstrate models and concepts, as shown in Figure 1.4.

System, Boundary, and Surroundings

A comprehensive and detailed study of the automobile engine is a subject in itself. In thermodynamics, however, attention is often focused on only one aspect of the operation of an engine. For example, the region occupied by the engine unit might be chosen for study. In this case, a *boundary* would be chosen that separates the region from everything else, as shown by dashed lines in Figure 1.4. Such a region in space, selected to provide a focus for the study, is termed a *system*, in this case, an open system or control

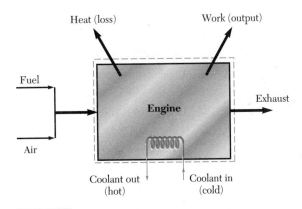

Figure 1.4 A thermodynamic model for the gasoline engine.

volume. Everything else outside the system belongs to the *surroundings* or *environment*. It should be noted that the terms *system* and *control volume* are sometimes used synonymously with the terms *closed system* and *open system*, respectively.

A system is closed if precisely the same collection of matter resides within the boundary of the system throughout the period of observation. For example, the case shown in Figure 1.4 would be a closed system (rather than an open system) if attention were focused on a fixed sample of air–fuel mixture (called the *charge*) rather than on a fixed region in space. For a system so defined, it should be possible for the shape of the boundary to keep changing so that the *same collection of matter* would be confined within the boundary while the interaction occurring between the system and the surroundings is being observed as the engine cycle is carried to completion.

Formal Definitions

A *thermodynamic system* is a region in space, or a fixed collection of matter, enclosed by a real or imaginary boundary. The boundary can be rigid or flexible, and the system can be fixed or moving in space. One of the most important steps in the formulation of any problem in thermodynamics is a *clear and unambiguous definition of the thermodynamic system and its associated system boundary.*

Thermodynamic systems typically have interactions with their surroundings; such interactions involve the transfer or exchange of some commodity across the system boundary. As noted earlier, the *surroundings* are defined as everything external to the system. In practice, only those portions of matter in the surroundings that can be affected by changes occurring within the system are important and therefore need to be considered.

In general, there are three types of systems: closed, open, and isolated. Figure 1.5 gives an example of each type. A closed system, depicted in Figure 1.5(a), has no mass crossing the system boundary. This type of system can have energy transfer (either as heat or work) across the boundary, but no material substance crosses the system boundary.

Figure 1.5(b) is an example of an open system, across whose boundary transfers of energy and matter can occur. An open system is more frequently referred to as a control

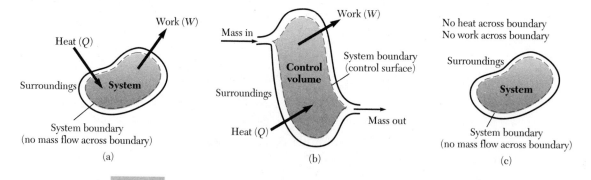

Figure 1.5 A typical representation of (a) a *closed system*, (b) an *open system*, and (c) an *isolated system*.

volume, which is a common concept in fluid mechanics analyses. The boundary of a control volume is called the *control surface*. Mass and energy can flow across the control surface. Whereas an open system can alter its shape, the concept of a control volume has traditionally been limited to a volume of fixed shape and fixed orientation relative to an unaccelerated observer. A simple example to illustrate this distinction is furnished by the inflation of a tire. As air is admitted into the tire, both the shape and size of the tire change. If the inside of the tire is defined as the system, the open system categorization is more appropriate than the control volume description in its widely accepted usage. However, if a rigid container were to be filled with gas, the volume enclosed by the vessel could be described as an open system or a control volume. The control surface will retain its shape and size throughout the process since the container is rigid. If this distinction is of no consequence in a given model, then the terms *control volume* and *open system* can be used interchangeably, as is frequently done in this text.

Figure 1.5(c) illustrates an *isolated system*, which has no interaction whatsoever with its surroundings; that is, neither mass nor energy crosses the system boundary. In general, very few real life applications qualify as isolated systems. The concept, however, is quite valuable for developing certain principles or corollaries relating to the laws of thermodynamics.

Students often experience difficulty deciding whether to use a closed system or an open system for a given problem. This probably accounts for the tendency by many to avoid explicit statements and sketches that clearly indicate the type of system assumed in the model. The following example conveys what one should look for in making the choice.

EXAMPLE 1.1

Suggest appropriate modeling for the following operations (shown in italics) either as a closed system or an open system problem:

(a) *pumping of water* (using a centrifugal pump) from a ground-level tank to an elevated tank

(b) *cooking of food* (held in a covered container)

(c) *generation of steam* (by heating a steady stream of water entering a boiler)

(d) *inflating an automobile tire* with air

Solution:

(a) There is *flow of mass* (water) into and out of the pumping device. If the space corresponding to the inside of the pump is of interest, an open system model would be the appropriate choice. In Figure 1.6(a), the system boundary (or control surface) is shown with a dashed line and is coincident with the inside surface of the pump casing.

(b) Since a *fixed mass* (the food) remains in a covered container throughout the cooking process, the closed system model is the appropriate choice. Figure 1.6(b) shows the inside of the container as the system boundary.

(c) Steam generation results from heating a stream of water as it passes through a

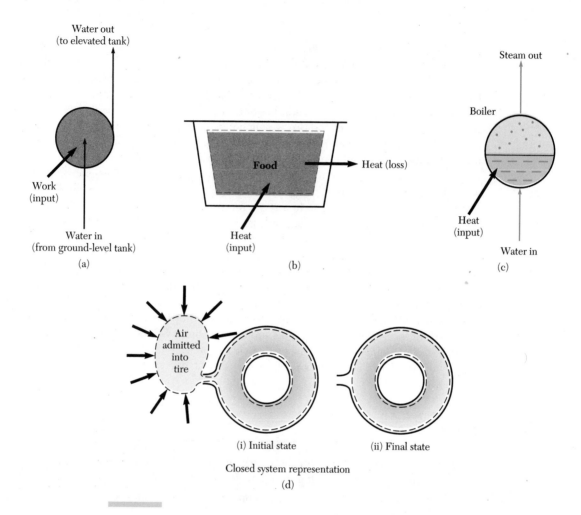

Figure 1.6 Models for (a) a water pumping operation, (b) a food cooking operation, (c) a steam generation operation, and (d) a tire inflation operation.

boiler. The open system model is appropriate here since there is *mass transfer* across the boundary of a fixed region in space. The inside surface of the boiler is the control surface, as shown in Figure 1.6(c).

(d) Both the open and closed system models are appropriate here and are depicted in Figure 1.6(d) for the tire inflation operation. In part (i) of the figure, the mass of air in the tire at any instant defines the system. Since the mass of air in the system changes as the operation proceeds, the model to use is an open system. (As indicated earlier, the volume and shape of such a system changes during the operation and, therefore, a control volume description in the widely understood sense is *not* appropriate.) The closed system model is shown in part (ii) of Figure 1.6(d) and defines the system as the *final mass* of air in the tire. Note that the system boundary initially must include the air in the tire at the beginning and the additional atmospheric air ultimately admitted into the tire as it is inflated.

It is generally impossible to conduct meaningful thermodynamic analyses without first defining the thermodynamic system, the system boundary, and the interactions of the system with its surroundings. In beginning mechanics courses, it is necessary for students to draw a free-body diagram in order to determine the forces acting on a body. Similarly, it is necessary to define the thermodynamic system explicitly in order to analyze thermodynamic behavior. Thus, one should routinely begin solving a problem in thermodynamics with a definition of the system, the system boundary, and the surroundings.

Equilibrium, Property, and State

The three concepts, equilibrium, property, and state, are so closely related that it is difficult to define or discuss one without the others. The following definitions are typical and provide ample demonstration of this fact. A system is in *equilibrium* when no macroscopic changes would occur within the system if it were to be isolated from its surroundings. *Macroscopic changes* refer strictly to changes in *thermodynamic properties*, which are formally defined as any observable characteristics of the system. In thermodynamics, analyses are restricted to properties that are directly or indirectly measurable characteristics of the system. Important thermodynamic properties include pressure, density, specific volume, temperature, and several others that permit the precise definition of the condition, or *state*, in which the thermodynamic system exists.

When a property is measured or computed, the value for the property must be single valued for each state in order for the result to be meaningful. Single valued means that the property must have the same unique value for a specified state. This is true for such properties as mass and volume, which are single valued at each instant for a closed system regardless of whether or not the system is in a state of equilibrium. For most other properties (for example, pressure and temperature), however, a uniquely defined value for the system exists only when the system is in a state of equilibrium. In this case, the thermodynamic property for the system must have the same characteristic value at every location throughout the system. These remarks apply equally to any fixed collection of matter in the system's immediate surroundings since a reciprocal relationship exists between a system and the portions of matter with which it interacts in the surroundings. In other words, without altering the problem, one could choose to regard the original system as surroundings while redefining the former surroundings as the new system. Thus, for the new system, the property for a *fixed collection of matter* must be single valued for each equilibrium state. More detailed discussions of thermodynamic properties are given in Chapter 2.

Once property has been defined, a formal definition of *equilibrium* can be given as the condition in which no tendency exists for any property of the system to change. If a system is in a state of equilibrium, all the thermodynamic properties have uniform values throughout the system, and the only way to produce changes in these properties is through some interaction across the boundary of the system.

Different types of equilibrium exist, such as thermal equilibrium, mechanical equilibrium, and chemical equilibrium. When *thermal equilibrium* has occurred in a closed thermodynamic system, the temperature is the same throughout the system. If a closed system is described as being in thermal equilibrium with its surroundings, this means that the temperature of the closed system is identical to the temperature of its surroundings. When a closed system is in *mechanical equilibrium*, the pressure is uniform throughout the system and no tendency exists for the pressure to change with time. Also, if a system is in mechanical equilibrium with its surroundings, no unbalanced forces are acting in the interior of the system or between the system and its surroundings. A closed thermodynamic system is said to be in *chemical equilibrium* when the chemical composition of the substance in the system is homogeneous throughout and does not change with time. Thus, an evaluation of chemical composition for any macroscopic portion of the system would yield identical results.

In summary, an intimate relationship exists among the thermodynamic concepts of equilibrium, property, and state. While some properties such as mass and volume can be defined for both nonequilibrium and equilibrium states, several others are defined only for the equilibrium states of the system. A state of equilibrium for a system implies that each property has the same value throughout the system and that the properties do not change with time over the period in which the system remains in a state of equilibrium.

The thermodynamic *state* of a system was previously defined as the particular condition in which the system exists. It is identified by specifying the thermodynamic property values for the system. In principle, specification of values for the complete list of thermodynamic properties (such as pressure, temperature, and density) may be necessary to uniquely identify the state of the system. For most substances encountered in engineering, however, it turns out that the values of only a few properties need to be specified to define the state. Such empirical observations are often enshrined in phenomenological laws such as the two-property rule (for the class of substances known as *pure substances*) or the state principle (for simple compressible systems), which are discussed in more detail in Chapter 3. The principal consequence of such laws is that functional relationships exist among at least two or more of the thermodynamic properties. Once the functional relationships are known precisely, the other properties can be determined at any given equilibrium state by using the values for just a few properties. For example, when dealing with properties of a pure substance, two independent thermodynamic properties such as pressure and temperature are generally adequate to specify the thermodynamic state of a simple compressible pure substance. The equations for the relationships among the thermodynamic properties of a substance are referred to as *equations of state*. Chapter 3 considers the equations of state for pure substances in general, while Chapter 4 covers ideal gases in particular.

Process, Path, and Interactions (Work and Heat)

A *process* can be defined simply as that which brings about a change of state of a system. When the thermodynamic state is altered or changed, this is called a *change of state* for the system. The change of state for the system is specified in terms of the initial and final states. Thus, the change of state can be defined by the changes that occur for different thermodynamic properties between the two end states.

A *quasi-equilibrium* or *quasi-static process* is an ideal one in which the deviations from equilibrium are infinitesimal, so that all states that the system passes through during the process can be considered to be equilibrium states. If the system state deviates by more than an infinitesimal amount from equilibrium during a process, it is termed a *nonequilibrium process*. Most *real processes* are nonequilibrium processes, but in some instances, they can be approximated reasonably accurately by *ideal processes*, which are those occurring in a quasi-static manner. In general, detailed and complete thermodynamic analyses can be made only when the processes involved are ideal. Thus, whenever reasonable ideal process approximations of real processes are possible, a complete analysis leading to useful results for actual processes becomes possible. This is a powerful technique frequently employed in classical thermodynamics.

A thermodynamic *path* is defined as the series of states through which the system passes while undergoing a change from one end state to the other. It should be evident that a path is identifiable only when a quasi-equilibrium process is taking place. In other words, a path is the route taken, if identifiable, when a process is executed. If the route cannot be identified, as is the case in nonequilibrium processes, it becomes meaningless to talk of the path for the process. By way of illustration, consider the compression of air in a piston and cylinder arrangement. If the compression is carried out slowly and with the gas temperature constant, the path for the ideal process on a p-V diagram is that given by Boyle's law (where pV is a constant). The path for the ideal process is represented in Figure 1.7 as a solid line corresponding to a change of state for the system from the initial state 1 to the final state 2. If the air compression is carried out quickly, however, the process will almost certainly be nonequilibrium. In this case, only the initial state 1 and the final state (denoted as 3 in Figure 1.7) can be unquestionably indicated on the p-V diagram. Since no identifiable path exists for the process, a dashed line is used, joining states 1 and 3 for the nonequilibrium process.

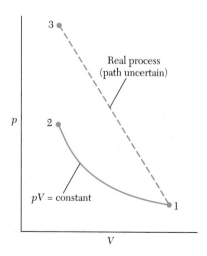

Figure 1.7 Representation of thermodynamic processes on the p-V diagram.

A thermodynamic *cycle* is any thermodynamic process, or set of processes, resulting in a final state for the system that is identical to its initial state. It should be noted that a thermodynamic cycle may comprise either quasi-equilibrium or nonequilibrium processes, subject only to the requirement that the final state for the system must be identical to the initial state. Diagrammatic representations of cyclic processes should be made in the manner shown in Figure 1.7. In other words, a solid line should be used to define the path for a quasi-equilibrium process, while a dashed line should be used for a nonequilibrium process between any two equilibrium states.

In a *nonflow process*, a fixed collection of matter remains within the system boundary throughout. In other words, processes involving closed systems are, by definition, nonflow processes. A *flow process* in contrast, is associated with an open system (or control volume) in which flow of mass across the boundary (or control surface) is permitted.

Certain processes are given particular designations with the prefix *iso-* added. A constant temperature process is one in which the temperature of the system stays constant during the process and is thus referred to as an *isothermal process*. Likewise, a constant pressure process is called an *isobaric process*. Also, the constant volume process is known as an *isochoric process*.

It is important to note that regardless of how a change of state is effected from a specified initial state to a prescribed final state, the change in value of any thermodynamic property is fixed and defined solely by the end states. This is true because properties are characteristics of a system at each equilibrium state, and as such, their values are fixed at any given state regardless of how that state is reached. For this reason, thermodynamic properties are often referred to as *state* or *point functions*.

Other quantities are computed in thermodynamics that are not point functions but instead depend on the process used to cause the change of state. Work and heat are typical of these other quantities, and they are not properties of a system. They are referred to as *interactions*. Interactions involve an exchange, occurring at the system boundary, between the system and its surroundings. Work and heat are both energy interactions between systems and their surroundings. Rigorous thermodynamic definitions of work and heat must be related to the first law of thermodynamics, from which the concept of energy is derived. The more common approach, however, is to take for granted the already familiar notions of work in mechanics and heat in thermal physics as forms of energy flux (or flow of energy). Both of these concepts are defined more precisely later, but here the main concern is that these quantities are not point functions but rather *path functions*, meaning that they are process dependent. (Energy, on the other hand, is a property and is thus a point function.) Interactions are defined in terms of boundary phenomena. For example, if the mechanical work done in a process is to be calculated, the computation must be based entirely on the forces acting at (or through) the system boundary.

Energy

Energy is a central concept in thermodynamics. Indeed, as noted earlier, thermodynamics is commonly viewed as the science of energy. What then is *energy*? The notion of energy is so familiar that it must come as a surprise that no simple definition exists for the concept. Feynman et al. (1963, 4-2) remarked as follows:

It is important to realize that in physics today, we have no knowledge of what energy is. . . . However, there are formulas for calculating some numerical quantity . . . . It is an abstract thing in that it does not tell us the mechanism or the reasons for the various formulas . . . .

The notion of energy as the capacity to do work is popular in mechanics, but this is not a satisfactory definition from a thermodynamic viewpoint. Lindsay (1965, 5) argues that this definition "conveys little unless you have an understanding of what work means. Even in mechanics this is a highly unsatisfactory definition, since it provides no idea of how to measure energy; unless this is made clear, there can be no genuine understanding of the concept." Besides this argument, the observation is made in Section 1.4 that in thermodynamics, the notion of "capacity to do work" is indeed more appropriate to the concept of *exergy*, derived from the second law of thermodynamics, than it is to that of energy, which is derived from the first law of thermodynamics.

It is important to realize that the thermodynamic concept of energy is derived strictly from the first law of thermodynamics. Formal exposition of this law is not provided until Chapter 6, after which a thermodynamic definition of the energy concept is given. At this point, however, some of the valid ideas most readers already have about energy can be reviewed.

Energy can exist in a variety of forms. The gasoline that is used in an automobile possesses a form of energy termed *chemical energy*. Energy resources having stored chemical energy include fossil fuels (coal, oil, and natural gas), wood, and several other combustible substances. When a fuel burns in air, the chemical energy of the fuel is transformed to another form of energy commonly referred to as *heat energy* or *thermal energy*. Neither of these two terms is really appropriate in thermodynamics. Thus, in thermodynamics, the term *internal energy* is used instead to mean the energy that a system possesses by virtue of its thermodynamic state. Although detailed discussion of this form of energy is not provided until Chapter 6, it should be noted here that in thermodynamics, chemical and thermal energy are included in what is termed internal energy. The notion of thermal energy as a form of energy seems to be more popular in thermal physics than in thermodynamics. The following definition has been given for thermal energy by Hudson and Nelson (1990, 132):

The kinetic and potential energies associated with the random motions of atoms and molecules; also called internal energy.

Such definitions of the concept can be very misleading in thermodynamics since internal energy involves more than what is implied here.

Kinetic energy is the energy that a system possesses by virtue of its motion and is a form of mechanical energy. *Gravitational potential energy* is energy that a system has by virtue of its position in a gravitational field. This, too, is a form of mechanical energy. Other forms of energy include nuclear energy, electrical energy, magnetic energy, surface tension energy, and so on. Note that *nuclear energy* is released in certain nuclear reactions that typically involve a loss of mass. In classical thermodynamics, it is generally assumed that the mass and energy conservation laws apply, and therefore nuclear reactions are usually excluded from consideration.

Entropy

Entropy is another concept of considerable importance in thermodynamics. It is a thermodynamic property whose existence is inferred as a corollary of the second law of thermodynamics. *Entropy*(S) is defined in terms of a change between equilibrium states, and is expressed mathematically as $dS = \delta Q/T$ where δQ is the incremental heat addition to the system in a quasi-equilibrium process during which the system temperature is T. The change in entropy between an initial equilibrium state and a final equilibrium state is thus given by the integral of $\delta Q/T$ in any quasi-equilibrium process between the two states. In statistical thermodynamics, entropy is identified as a measure of the state of disorder in the system. Typical applications that require the use of this property include

1. establishing the direction in which physical or chemical processes can occur in nature
2. providing a quantitative measure of the extent of departure of real processes from thermodynamically ideal processes known as *reversible processes*
3. establishing performance limits and evaluating performance efficiencies of actual devices

Students of thermodynamics often encounter difficulty with the entropy concept largely because of the widespread perception that it is a more abstract concept than energy. In reality, entropy is no more abstract a concept than energy since in thermodynamics both are derived from the basic laws of thermodynamics. The pioneers of the science of thermodynamics actually stumbled on the notion of entropy without knowing it before they arrived at the thermodynamic concept of energy as we know it today. In the late 17th century and early 18th century, the *caloric theory* of heat was popular. This theory incorrectly viewed heat as a material substance (called *caloric*) with certain rather unusual properties. For example, Sadi Carnot's famous treatise of 1824 on "the motive power of heat" proposed the idea of a flow of caloric from a high temperature source to a low temperature medium in such a manner that would allow work to be extracted without a diminution in the amount of caloric. It is now widely accepted that a mere substitution of the term *entropy* for *caloric* in Carnot's work would correct the errors in the original presentation.

The entropy concept is discussed more fully in Chapter 8. The main objective in introducing the concept here is to underscore the fact that it is, like energy, a derived concept based on one of the laws of thermodynamics. Also, entropy is a particularly important concept in thermodynamics because such critical questions as direction of reactions and limits of performance of devices, which are unanswerable on the basis of the first law of thermodynamics alone, can be answered using the entropy concept.

EXERCISES

for Section 1.2

1. Use suitable sketches to illustrate each of the following operations (shown in italics) and indicate in each case the appropriate thermodynamic model, whether a closed system or an open system.
   (a) *compression of a steady stream of air* using a rotary compressor
   (b) *mixing* of a stream of ambient air with another stream of air

(c) subjecting a mixture comprising a fixed mass of air and fuel to *compression, ignition, and expansion* processes in a reciprocating piston and cylinder unit of a gasoline engine

2. A body of mass 10 kg falls from rest through a vertical distance of 5 m in the Earth's gravitational field. The acceleration due to gravity (g) can be taken as 9.8 m/s². Assuming negligible air resistance, calculate for the free fall of the body:
   (a) the decrease in potential energy
   (b) the increase in kinetic energy

   Answer: (a) 490 J, (b) 490 J

3. A fork lift raises a load of 250 kg through a vertical distance of 1.5 m. What is the work done in raising the load? (Assume $g = 9.8$ m/s².)

   Answer: 3675 J

4. The energy (E) released in a nuclear reaction can be determined from the famous equation enunciated by Albert Einstein:

$$E = mc^2$$

   where m is the mass deficit and c is the velocity of light. Taking $c = 3 \times 10^8$ m/s, calculate the energy resulting from a mass deficit of 0.001 kg.

   Answer: 9×10^{10} kJ or 1040 megawatt-day

1.3 Fundamental Laws of Thermodynamics

The complete structure of the science of thermodynamics encompasses not only the basic concepts but also the laws of thermodynamics. There are two fundamental laws of thermodynamics involving energy. The *first law of thermodynamics* involves only the quantity aspect of energy and is a statement of the conservation of energy principle, while the *second law of thermodynamics* involves the quality of energy.

Two other laws of thermodynamics have evolved. One involves concepts that should precede the first and second laws, and because a formal statement was not developed until after the first and second laws were developed, this law is called the *zeroth law of thermodynamics*. The final law follows the development of the second law and is referred to as the *third law of thermodynamics*.

An important principle should be mentioned here that is taken for granted in all of classical thermodynamics—*conservation of mass*. In classical thermodynamics, both the conservation of mass and the conservation of energy are presumed to hold. Although Albert Einstein's theory of relativity involves an interchange between energy and mass, the interchange is significant only for velocities of the system that approach the speed of light and for certain other cases such as the conversion of mass to energy in a nuclear reaction. These exceptions are often treated as special topics in thermodynamics and are not discussed in this text. A statement of the conservation of mass appropriate for classical thermodynamics as developed in this text is as follows:

Mass can neither be created nor destroyed, but is conserved for the universe (system plus surroundings) in all processes.

Although the various laws of thermodynamics are repeated and discussed in more detail later, they are briefly stated here to provide some indication of the scope of classical thermodynamics.

Zeroth Law of Thermodynamics   *When two bodies are in thermal equilibrium with a third body, the two bodies are in thermal equilibrium with each other, and all three bodies are said to be at the same temperature.*

First Law of Thermodynamics   *Energy for an isolated system remains constant. When energy is exchanged between a system and its surroundings, the final total energy for the universe (system and its surroundings) equals the initial total energy for the universe (system and its surroundings).*

Figure 1.1 furnishes a classic example of a system that violates the first law of thermodynamics. An appropriate thermodynamic model for the proposed closed-cycle operation of the mill is depicted in Figure 1.8. The mill is represented as a closed system, while the millstones and the corn grinding scheme belong to the surroundings. There is a continuous circulation of water and the system is presumed to be operating in a cycle. Thus, the proposal amounts to producing work at no expense to either the system or the surroundings. This violates the first law of thermodynamics. Robert Fludd proposed the machine in 1618 and it has never worked. While the fact that it never worked was no proof of the first law of thermodynamics (which was promulgated a little over two centuries later), this fact remains part of the evidence for accepting the law. A machine that is supposed to be able to function in violation of the first law is termed a *perpetual motion machine of the first kind (PMM1)*. A major corollary of the first law of thermodynamics is that a PMM1 is impossible.

Second Law of Thermodynamics
Kelvin–Planck Statement   *It is impossible to construct a device that will operate in a cycle and produce no effect other than the raising of a weight and the exchange of heat between the device and a single reservoir.*

The ammonia engine proposed by John Gamgee in the 1880s (Fig. 1.2) is an example of a *perpetual motion machine of the second kind (PMM2)*. The operation of the proposed engine is depicted in Figure 1.9. In effect, energy extracted (as heat) from the ambient air is used for the production of work using a device that operates in a cycle. As with the Fludd engine, no one has been able to get Gamgee's proposed engine to work. Indeed, a major corollary of the second law is that a PMM2 is impossible.

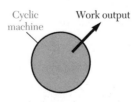

Cyclic machine   Work output

Figure 1.8 A thermodynamic model of Fludd's closed-cycle mill.

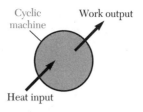

Figure 1.9 A model for the ammonia engine proposed by John Gamgee.

Clausius Statement   *It is impossible to construct a device that operates in a cycle and produces no effect other than the transfer of heat from a low temperature body to a high temperature body.*

If the assertion in the Clausius statement were not true, it should be possible to operate such devices as refrigerators and air conditioners without a supply of energy in the form of electricity or burning gas. We know from intuition and experience that such utopian machines are not possible.

Although not apparent, the two statements of the second law, referring to entirely different situations, are logically equivalent statements of the same principle. In Chapter 7, where the second law of thermodynamics is considered in more detail, deductive logic is used to establish the equivalence of these two statements. The second law of thermodynamics has often been identified as one of the most reaffirmed physical laws known to science.

Third Law of Thermodynamics   *The entropy of a pure substance approaches zero as the temperature of the substance approaches absolute zero temperature.*

Entropy was briefly introduced earlier as a concept derived from the second law of thermodynamics. A more detailed discussion of the concept, as well as the third law of thermodynamics, is given in Chapter 8.

EXAMPLE 1.2

The appliances in the kitchen of a well-insulated apartment include an electric cooker and a microwave oven. The electric cooker is supplied with 1200 W of power and an additional 600 W are supplied to the microwave oven. Estimate the rate of accumulation of energy in the apartment.

Given: An apartment with contents including an electric cooker and a microwave oven. The apartment is well insulated.

Power input ($\dot{E}_{\text{input}}$) = 1200 W (to cooker) + 600 W (to microwave oven)

Find: $\dot{E}_{\text{accumulation}}$ in the apartment. (The dot over the E denotes the time rate.)

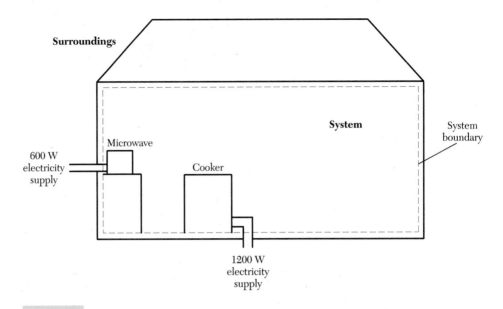

Figure 1.10 First law application to energy flow into an apartment.

Solution: Figure 1.10 provides a definition of the problem. The apartment with its contents is the system; the two appliances supplied with electricity are shown in the figure. Electricity is supplied from the surroundings, and the combined rate of electrical energy flow into the system is 1200 W + 600 W, or 1800 W. The appropriate thermodynamic principle to use is the first law of thermodynamics. Since the system is well insulated, assume that negligible heat loss occurs from the system to the surroundings. Thus, the rate of accumulation of energy in the system must be equal to the net rate at which energy flows into the system. In this case, application of the first law of thermodynamics leads to the estimate of 1800 W for the rate of accumulation of energy in the apartment.

EXAMPLE 1.3

An electric refrigerator is supplied with 120 W of electricity while it cools food at the rate of 300 W. What is the rate of heat rejection from the device assuming that it operates cyclically? If the heat rejection is to the room air, what is the rate of energy input to the air?

Given: Refrigeration device (system) that operates cyclically.

Power input to system ($\dot{E}_{input}$) = 120 W (electrical) + 300 W (cooling rate of food).

Find:

(a) Heat rejection rate ($\dot{Q}_{rejection}$) from system

(b) Rate of energy input to the room air

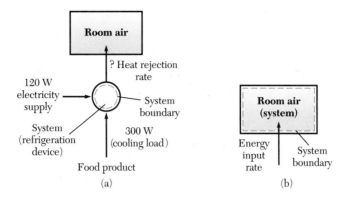

Figure 1.11 First law application to energy flow associated with operation of a refrigerator.

Solution: The refrigeration device is defined as the system in Figure 1.11(a). The total energy input rate to the system is given by

$$\dot{E}_{input} = 120 \text{ W (electrical)} + 300 \text{ W (cooling rate of food)}$$

$$= 420 \text{ W}$$

Since the device operates in cycles, by application of the first law of thermodynamics, the energy output rate from the system must be equal to the energy input rate. Thus, the heat rejection rate from the system is $\dot{Q}_{rejection}$ = 420 W. In Figure 1.11(b), the room air is now defined as the system. The rate of energy transfer to the system is identical to the heat rejection rate from the refrigeration device. Thus, the rate of energy input to the room air is 420 W.

EXERCISES

for Section

1.3

1. 0.01 kg/s of steam is mixed with 1 kg/s of dry air in a humidification process. (*Humidification* means the addition of moisture to air.) What is the mass flow rate of the mixture?

   Answer: 1.01 kg/s

2. Steam supply to an engine comprises a mixture of two streams. The streams prior to mixing have mass flow rates of 0.01 kg/s and 0.1 kg/s, respectively. Upon exit from the engine, the fluid leaves as two streams, one of water at the rate of 0.002 kg/s and the other of steam. What is the mass flow rate of the exit steam?

   Answer: 0.108 kg/s

3. The air conditioning unit of a classroom is operating at full capacity on a hot summer day when 15 additional students enter the room. Assume that each additional student heats up the room at a rate equivalent to that of a 60-W electric bulb. What is the extra load imposed on the air conditioning system due to the arrival of the 15 students? Assume that the room air temperature rises by 1.5°C for

every 100 kJ of heat transferred to it. Estimate the initial rate of increase in the temperature of the room air resulting from the arrival of the additional students.

Answer: 900 W; 0.81°C/min

1.4 Typical Thermodynamic Systems and Processes

As indicated earlier, the primary concern of thermodynamics is energy. Several applications in real life involve the transformation of energy from one form to another. The thermodynamic principles already enumerated are applicable to such phenomena, especially when one is interested in a detailed analysis and evaluation of the processes in relation to the efficiency of energy usage.

Exergy (or Availability) and Quality of Energy

Before discussing a variety of engineering applications of interest in thermodynamics, it is necessary to make a crucial distinction between the various forms of energy with regards to *quality*. Energy has been defined as the capacity to do work. The essential idea implicit in such a definition can be stated more broadly by considering the expanded notion of the *capacity to produce change* in a specified environment. The change desired might be the raising of a weight (which is called work), the heating of a space (called heat), or simply producing a mixed stream of substances at a desired condition from other available streams.

For any of these changes, the transformations of energy taking place can be identified precisely, and in accordance with the first law of thermodynamics, an exact energy balance can be expected. In other words, no more energy can later exist than the sum total of what existed to begin with; neither can there be a destruction of any of the energy that existed at the beginning of the process. From this viewpoint, the first law must be regarded purely as a basis for *energy accounting*.

In contrast, if the *quality* of energy is of concern, one must go beyond the first law of thermodynamics (as discussed later). It is indeed a fact of experience that different amounts of energy may be required to produce the same effect, depending on what form of energy source is being used. For example, a gas-fired heat pump (or air conditioner) uses approximately three times the amount of energy input required by an electric heat pump when the output is the same. Based solely on the first law principle, one could rush to the conclusion that the electric heat pump is the more energy efficient choice. A closer look at this example, however, should reveal that in order to produce one unit of electricity, the energy equivalent of three units of gas must be burned. Thus, the performance of the gas-fired heat pump versus the electric heat pump would, on balance, appear comparable in efficiency.

This example suggests that in evaluating different forms of energy, the quality must be defined in a manner that reflects the capacity or potential to produce change. Energy so defined appears to be what is really intended in the colloquial usage of the term. It is what Gaggioli and Petit (1977) have termed the *real commodity of value*. The terminology variously used in thermodynamics for this energy commodity includes *availability, available energy, exergy,* and *essergy*. This variety of terms exists only in the

English language. In all other languages, the term exergy, with the ending adjusted in each language, is used exclusively. Exergy is the preferred term and is thus used almost exclusively throughout this text.

In Chapters 7 through 9, the quality differentiation between various forms of energy is shown to be derived solely from the second law. Further study of the basic laws of thermodynamics will show that mechanical energy (such as kinetic and potential energy) is of a higher quality than internal energy (which a body possesses by virtue of its thermodynamic state) and that, whereas mechanical energy can be completely converted to internal energy, internal energy cannot be completely converted to mechanical energy in a continuous process. Indeed, this is a corollary of the second law of thermodynamics. From this corollary, an upper limit can be determined for the fraction of the energy of a system (extracted as heat transfer) that can be converted, via a cyclic device, to work (or flow of mechanical energy). In other words, there is a theoretical limit or maximum efficiency for any particular scheme for the conversion of a system's internal energy into work.

The exergy concept is relevant to all thermodynamic processes and should not be thought of as pertaining only to work-producing schemes. As indicated already, it provides a measure of the quality of energy and can, as such, give a true indication of efficiency achieved in any process. Whereas energy as defined by the first law of thermodynamics cannot be used up (it is always conserved for the universe), exergy (or availability) can be consumed, or simply wasted. Several illustrations of this principle are provided later in Chapters 10, 11, 13, and 14, where applications of prime interest in engineering are discussed.

Typical Applications of Interest in Thermodynamics

Heat Engines and Work-Producing Systems

Interest in conversion of heat to work stems from the fact that there are more resources available for energy extraction as heat transfer than there are for direct production of work. Thus, we can burn fuels or collect the solar heat flux (which is quite abundant) to elevate the temperature of a system by heat transfer to it, whereas direct generation of mechanical energy usually requires the use of systems that rely on wind and water movement and are dependent on climate and geographical location. Since more thermal sources of energy exist than mechanical sources, many of the applications of interest in thermodynamics involve the use of heat transfer to a system to produce work. *Heat engines* are cyclic devices to which heat transfer occurs at a high temperature, and as a result, work is produced. The system from which energy flow as heat transfer occurs to a heat engine is referred to as the *heat source*, while the system to which heat transfer from the cyclic device occurs is termed the *heat sink*.

A common application is the *automobile engine*, in which a high temperature condition is produced from the chemical energy stored in the gasoline. The resultant effect is as if a heat source for the engine were produced by burning the fuel in air. Work is produced as a result, and the dispatch of an exhaust gas stream to the environment can be equated to a heat rejection from the engine to the ambient air, which serves as the heat sink for the engine.

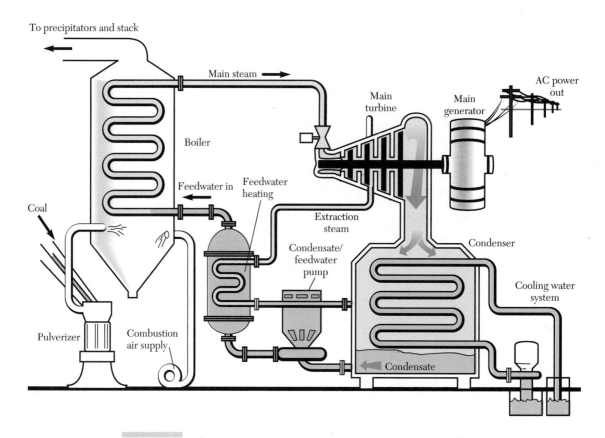

To precipitators and stack

Main steam

AC power out

Main turbine

Main generator

Boiler

Feedwater in Feedwater heating

Extraction steam

Coal

Condensate/ feedwater pump

Condenser

Cooling water system

Pulverizer

Combustion air supply

Condensate

Figure 1.12 A schematic diagram of a coal-fired steam plant.

Another application is the *steam power plant*. A schematic diagram of a conventional steam plant layout is shown in Figure 1.12. In a steam power plant, a fuel (such as coal, oil, or gas) is burned in air to produce heating of water (at high pressure) in a boiler and thus steam is produced at high pressure and temperature. This steam is then expanded through a device, called the *turbine*, to produce mechanical energy, which in turn is used to produce electricity. In this way, the steam power plant utilizes a fuel with stored chemical energy to produce electricity.

A typical *nuclear power plant* is illustrated in Figure 1.13. It does not use a chemical fuel but rather a nuclear fuel in a controlled nuclear reaction to produce the heating needed for production of high pressure steam for power generation. Otherwise, the steam process in the nuclear power plant operates in the same manner as described for the conventional steam power plant to produce mechanical energy, which in turn is used to produce electricity for a variety of end uses.

A *fuel cell* is not a heat engine. It is classified as a direct energy conversion device and is basically a system that converts chemical (stored) energy directly to electricity. The study of such a system is included in thermodynamics, although usually as a special topic in the area of nonequilibrium thermodynamics (for example, see Angrist, 1976,

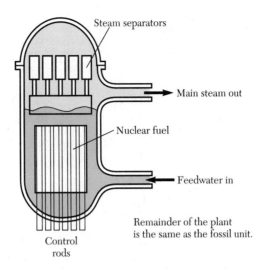

Figure 1.13 A schematic diagram of a typical nuclear power plant.

Chap. 8). The typical automobile battery is another direct conversion device that converts stored chemical energy to electrical energy. It can be operated in reverse to store chemical energy from an input of electrical energy.

A *rocket* or *jet engine* is another engineering system of interest in thermodynamics. In this system, the stored chemical energy of the fuel is released by burning the fuel in a stream of air at high pressure. The high pressure and temperature gas stream is passed through a nozzle device and movement of the rocket is produced by a thrust developed when the gas stream leaves the nozzle. The rocket engine and the jet engine are similar in that both use the nozzle device to generate a thrust. However, a jet engine uses atmospheric air for combustion and only the fuel needed is carried on board, while a rocket engine carries both the fuel and the oxidant (needed for combustion). A typical rocket engine operation is shown in Figure 1.14.

Heat Pumps or Work-Absorbing Systems

Another broad class of applications of interest in thermodynamics involves the use of input energy to produce a cooling or heating effect. The typical *air conditioning system* used in a residence requires an electrical energy input, and its end result is a cooling effect in the residence. A typical air conditioner is shown in Figure 1.15. The system utilizes electrical energy, supplied to an electric motor, which is then transformed to mechanical energy. The electric motor in turn drives a *compressor*, which is used to increase the pressure (compression) of a refrigerant vapor. The refrigerant operates in a cycle to cool the residence (by extracting some of the internal energy of the room air) and to reject the energy extracted to the surroundings (outside the residence). Thus, the device produces a cooling effect by use of input electrical energy. In a similar manner, the typical automobile air conditioning system utilizes mechanical energy di-

Figure 1.14 A typical rocket engine launch of Consort Rocket from White Sands Missile Range, NM, by the University of Alabama in Huntsville. (Photo used with permission.)

rectly from the automobile engine to drive a compressor, which in turn is used to operate a vapor compression system to cool the interior of the automobile.

It is also possible to produce a cooling effect by directly applying heat to a cooling system, such as in the *absorption refrigeration system*. Such systems are in operation in several locations in the United States and are frequently powered by heat sources produced by burning a gas, such as natural gas, or by what is termed *waste heat* from other processes. A common example of this type of system is a propane-fueled refrigerator in a motor home or camper. Figure 1.16 is a schematic diagram of a typical absorption refrigeration system.

Graphical Symbols for Devices and Typical Systems

The previous examples have shown that many systems are in use today that essentially operate to transform energy from one form to another. In some cases, the systems are designed to use energy input as heat transfer to the system, resulting in the production of mechanical or electrical energy. Other systems use electrical or mechanical energy

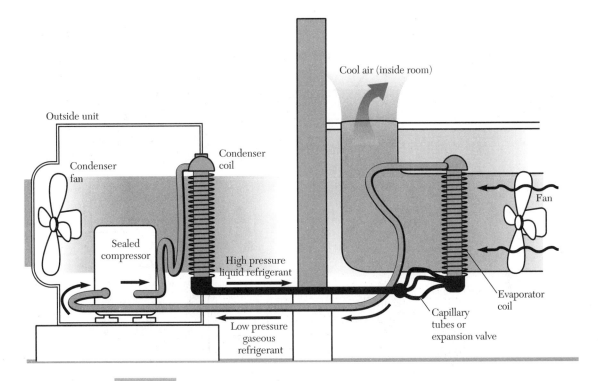

Figure 1.15 A schematic diagram of a typical air conditioner.

as an energy input to produce a heating or cooling effect. A wide range of other applications exist in engineering that involve energy transformations designed to produce particular desired effects. The systems involved may be simple or quite complex, but to facilitate a diagrammatic representation of such a system, certain graphical symbols are commonly used for the range of devices frequently encountered in engineering applications.

Some typical devices used in thermodynamic systems include compressors and pumps, turbines and expanders, evaporators and boilers, condensers, nozzles, and diffusers. A *compressor* is a work-absorbing device that compresses a gas such as air or a refrigerant vapor, resulting in a change from a low pressure to a high pressure state. A *pump* similarly raises the pressure of a fluid; this term is used when the fluid is a liquid. The common graphical symbols for a compressor and a pump are shown in Figures 1.17(a) and (b), respectively. A *turbine*, on the other hand, is a work-producing device through which a gas or vapor stream at high pressure expands to a low pressure while work output is produced. Figure 1.17(c) shows the common symbol for a turbine.

Evaporators, boilers, and condensers are basically heat exchangers. Such devices are designed to transfer energy (as heat transfer) between a heat source (or sink) and a fluid substance passing through the device. An *evaporator* or a *boiler* typically has a liquid flowing into the device where heat is added to the liquid so that it is evaporated or vaporized. Thus, it exits the device as a vapor. In contrast, a *condenser* has a vapor

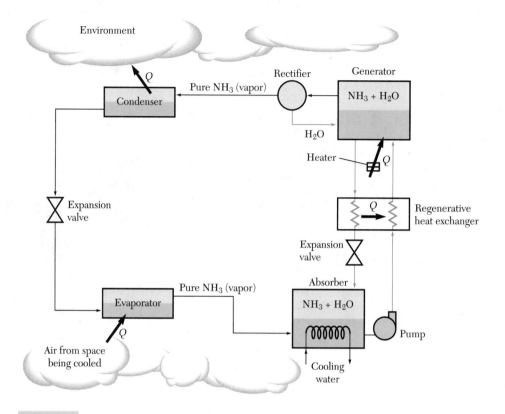

Figure 1.16 A schematic diagram of the major components of a typical absorption refrigeration device.

flowing into the device where extraction of energy (as heat transfer) from the vapor results in the fluid leaving the device as a liquid. In effect, a boiler or an evaporator changes a liquid into a vapor, while a condenser changes a vapor into a liquid. Figures 1.17(d) and (e), respectively, show the common graphical symbols used for a boiler and a condenser.

Nozzles and diffusers are simple devices used with flowing fluids to change their properties. The *subsonic*[1] *nozzle* is a device with a decreasing cross-sectional area that increases the velocity and thereby the kinetic energy of the fluid while decreasing the pressure of the fluid. A *subsonic diffuser*, however, has an increasing cross-sectional area in the direction of fluid flow and thus increases the pressure of the fluid by reducing the fluid velocity and, correspondingly, the kinetic energy of the fluid. Graphical symbols for nozzles and diffusers are illustrated in Figures 1.17(f) and (g), respectively.

Cyclic devices are usually represented by circles in a diagram. Thus, if the device is an engine, a circle marked with an "E," as shown in Figure 1.17(h), is used, whereas

---

[1]*Subsonic* means below the speed of sound in the fluid medium. *Supersonic* (greater than the speed of sound) nozzles require both a converging and a diverging section.

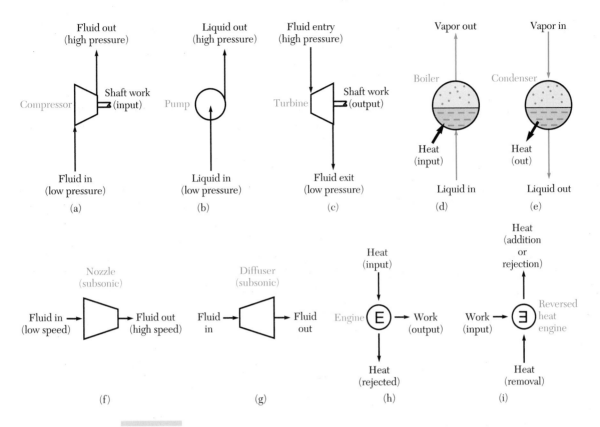

Figure 1.17 Graphical symbols for (a) a compressor, (b) a pump, (c) a turbine, (d) a boiler/evaporator, (e) a condenser, (f) a (subsonic) nozzle, (g) a (subsonic) diffuser, (h) an engine operating cyclically, and (i) a refrigerator.

for a *reversed* heat engine, refrigerator, or heat pump, the circle is marked with a reversed "E" (that is, "Ǝ"), as shown in Figure 1.17(i). As other systems and devices are encountered, the commonly used graphical symbols will be introduced.

1.5 Relationship of Thermodynamics to Energy Needs

Energy occupies a central place in the lives of people everywhere. Energy is needed in the home, for transportation, in industries, in factories and offices, on the farm, and for nearly every activity people do. In short, society's energy needs are phenomenal. We frequently hear comments concerning the "shortage of energy" but this is a misnomer. As previously discussed, energy is conserved, and such comments are actually referring to *exergy* (or "useful energy"), which is a measure of the capacity to produce change.

Most of the primary forms of energy on which we traditionally rely for meeting our energy needs are *capital* or *nonrenewable resources*. Typical of these are *fossil fuels* (coal, oil, and gas), which have generally taken millions of years to form but may easily be used up within a few decades depending on the extent of global usage. Even the *nuclear fuels*, having a high ratio of energy yield to the mass of fuel utilized, may be used up in a matter of centuries. A detailed study of energy in relationship to society's needs involves several considerations, including economics, politics, environmental concerns, and new technology.

Thermodynamics is relevant to the energy needs of society in that it allows us to evaluate the efficiency of processes relative to high quality energy or exergy usage. As noted previously, while the first law of thermodynamics states that energy is conserved, the second law points to a difference in quality between forms of energy. For example, heat supplied at a high temperature level is more valuable than that supplied at a low temperature level. But if the task at hand only requires heat at a low temperature, a supply at a high temperature becomes extremely wasteful. Thus, on the basis of the second law, the conditions can be established on which our use of available energy resources is most prudent for a given application.

The wise and efficient use of this energy is a fundamental concern that can be addressed on the basis of thermodynamic principles. Economic benefits can be derived from the prudent and efficient use of available primary energy resources. The conventional wisdom is that the more energy resources a country uses, the higher the standard of living enjoyed by its citizens. Thus, a fairly close correlation has been found for the United States between its economic growth and its high energy resources usage.

Most of the highly developed economies worldwide use a high level of energy input. *Energy intensity* is defined as the amount of energy input into an economy per unit of economic activity. Evidence shows, however, that a considerable difference exists between some of the most highly developed economies and that of the United States with regard to the energy intensity and economic activity in the country. For example, both Japan and West Germany tend to use approximately one half as much energy as the United States, while sustaining an economic growth that parallels and perhaps exceeds that of the United States (see, for example, Brown et al. 1986). This means that in the United States, there is considerable latitude for improved efficiency in energy usage while still maintaining a high level of economic development and high standard of living. Indeed, such an improvement becomes imperative when one considers the nature of the energy resources that are now being exploited to support economic development in most of the world's energy intensive economies. Conservation of these finite energy resources should become a high priority.

Most of the resources that are currently used to support economic development are the nonrenewable resources, which include primarily oil, natural gas, coal, and uranium, all of which are found in the Earth's crust. All of these nonrenewable resources are finite and will be depleted in a relatively short time if usage rates continually increase. In contrast, the renewable resources such as water flow over a dam, solar energy, and biomass get replenished regularly, thus yielding a certain level of energy input to support economic activities. Unfortunately, the renewable resources currently supply less than one-fifth of the world's energy use, while the nonrenewable resources supply the rest.

A pattern appears to exist concerning the discovery, exploitation, and depletion of nonrenewable or finite resources such as the fossil fuels. Geologist Donald F. Hewett in 1929 analyzed the cycles of metal production that occurred in Europe after the Industrial Revolution and concluded that metal production went through successive stages analogous to those of infancy, adolescence, maturity, and old age. He developed criteria that indicated that the production rate as a function of time tended to follow a bell-shaped curve. Geologist M. King Hubbert between 1956 and 1969 developed studies to extend the analysis of Hewett to oil and coal production. Figure 1.18 shows the general mathematical relationships involved in the cycle for production of these finite resources.

If the production rate is plotted as the ordinate and time as the abscissa, the production rate tends to increase at a very rapid rate to start with, and then when a certain point in time is reached, the production rate peaks. After the peak, a steady decline in production rate occurs until the source is depleted. A typical production rate is shown in Figure 1.18, with the ultimate cumulative quantity of production given by the area under the curve. Using this concept (and recognizing that the total available reserves represent the maximum ultimate cumulative production, coupled with information about historical production rates), Hubbert predicted in 1955 that the peak production rate for oil in the continental United States would occur about 1970. In fact, the peak production did occur in 1970 (see Brown et al. 1986), and the production rate has decreased since that time.

Brown et al. (1986) suggest that world oil production may have peaked in 1979, while Hubbert's analysis predicts that the world oil production will probably peak by the years 2000 to 2010. Regardless, oil resources will decline gradually until all reserves are depleted. According to reliable estimates, it appears that most of the oil to be used for energy needs will be depleted by the year 2060 (Brown et al. 1986).

The use of coal presents many problems different from oil. For example, many air pollution, mining, and handling problems are associated with coal that are often more difficult to deal with than those related to oil. A similar analysis of the supply and demand for coal by Hubbert indicates that coal production will peak around the years 2150 to 2200 and will decline steadily thereafter.

The analytical technique developed by Hubbert could also be applied to natural gas and uranium. The major conclusion from such analyses is that serious long-term considerations exist that should be taken into account when one relies on finite energy

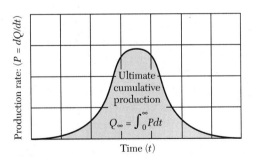

Figure 1.18 Suggested mathematical relationships involved in the cycle of production of a finite energy resource. (Adapted from Hubbert, 1956.)

resources, such as the fossil fuels, for meeting society's energy needs. First, such fuels would be available for use over a much longer period if we were to rely more on renewable resources instead of the present overdependence on the nonrenewable resources. Second, if current conditions persist such that large scale use of finite resources continues, it would be wise for society to strive for the highest efficiency to maximize the benefit derived from use of these nonrenewable resources. In this regard, the principles of thermodynamics can be used to establish how processes can be executed in the most efficient and least wasteful manner.

The energy picture may alter dramatically if new and revolutionary energy technologies should be developed. Currently, a promising example of a source that may prove to be practically limitless is *nuclear fusion*. This is the primary process by which the sun produces the colossal amount of energy that it radiates to all parts of the solar system. Research is currently in progress on development of nuclear fusion systems.

An alternative to the nuclear fusion system for long term energy needs is the nuclear breeder reactor system. Although some of these systems have been operated to date, many problems still exist that relate to their use and to the economics of their operation. Consequently, while breeder reactors offer promise for energy supply for the future, that potential has not yet been realized.

There is much room for improvement at the current time in the use of the finite energy resources. For example, the energy intensity of economic activities in the United States diminished by about 25% from 1973 to 1985 at the same time that a steady economic growth was sustained. This reduction in required energy intensity was achieved through a concerted national effort aimed at using energy resources more efficiently. This national effort was stimulated by the oil embargo that was enforced by OPEC countries in 1973. Thus, it is evident that much can be done to improve efficiency when there is a pressing need to do so. In fact, it is predicted that U.S. energy efficiency will increase by approximately another 20% by the year 2000 (Brown et al. 1986). In the accomplishment of such a goal, thermodynamicists can make a valuable contribution, particularly in helping to identify the major sources of inefficiency and waste in the myriads of applications in engineering that involve energy use.

Examples abound of schemes that have been successfully implemented for achieving improved efficiency of operations. For example, in electric power generation, more complex thermodynamic cycles have been introduced that result in greater efficiency of operation. In another case, waste heat from power generating units, which previously might have been dumped into the environment, is now used for heating of other buildings and facilities in the vicinity of power plants. District heating based on such systems is now widely used in European countries, such as Sweden, and the potential certainly also exists in the United States for improving efficiency in the use of energy resources through similar systems. The situation presents a challenge to future generations of engineers and scientists who will be developing systems that will be operated upon a declining energy resource base. Analysis based on the principles of thermodynamics will almost certainly contribute to meeting this challenge.

Of course, most countries allow their markets to be driven by economics and not thermodynamics, or any other engineering science for that matter. Thus, while thermodynamics may have much relevance to the future of energy and society's energy needs, the opportunities for making a contribution to society may be limited to whatever engineers can do to design more economical and energy-efficient systems that will have a beneficial long-term impact on society.

EXAMPLE 1.4

A well-fed ox can work at a rate equivalent to 200 W. Assume that an ox works at this rate for 5 h each day on a ration equivalent to 11,250 kcal of food energy. What is the ratio of work output to food energy input for the ox? (Assume 1 kcal = 4186.8 J.)

Given: Ox with work output rate of 200 W, works for 5 h

$$\text{Food energy input} = 11{,}250 \text{ kcal}$$

$$(1 \text{ kcal} = 4186.8 \text{ J})$$

Find: The ratio $\dfrac{\text{work output}}{\text{food energy input}}$ for the ox

Solution: Define the ox (and the food ration it consumes) as a system. Such a system is illustrated in Figure 1.19, which indicates a mass interaction in addition to possible heat and work interactions between the system and its surroundings. The food ration consumed by the ox also represents an energy input (associated with mass flow) to the system. The work output is computed as the product of the work output rate and the time. Thus,

$$\text{The daily work output of the ox} = \text{power} \times \text{time}$$

$$= (200 \text{ J/s})(5 \cdot 3600 \text{ s}) = 3.6 \times 10^6 \text{ J}$$

$$\text{Food energy input} = 11{,}250 \cdot 4186.8 \text{ J} = 4.71 \times 10^7 \text{ J}$$

$$\therefore \frac{\text{Work output}}{\text{Food energy input}} = \frac{3.6 \times 10^6 \text{ J}}{4.71 \times 10^7 \text{ J}} = \underline{7.64 \times 10^{-2}}$$

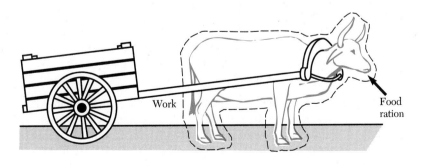

Figure 1.19 Work efficiency of the ox problem considered in Example 1.4.

EXAMPLE 1.5

Assuming a solar constant of 1.39 kW/m² and a mean Earth's radius of 6371 km, estimate the energy intercepted across the diametral plane of the Earth over a period of one year.

Given: Solar constant $= 1.39$ kW/m²
                              Earth's radius $= 6371$ km

Find: Solar energy intercepted in one year by the Earth

Solution: As the planet Earth rotates on its axis, it presents an area equal to that of its diametral plane all the time for the interception of solar radiation. This area is approximately that of a circular disk of radius equal to the Earth's radius of 6371 km given. Thus, the area of the diametral plane of the Earth is

$$\pi(6.371 \times 10^6 \text{ m})^2 = 1.275 \times 10^{14} \text{ m}^2$$

The solar constant given is the extraterrestrial intensity of solar radiation. The energy intercepted across the diametral plane in one year is therefore

$$\left(1.39 \frac{\text{KJ/s}}{\text{m}^2}\right)(3600 \cdot 24 \cdot 365 \text{ s})(1.275 \times 10^{14} \text{ m}^2) = \underline{5.59 \times 10^{21} \text{ kJ}}$$

Note that 1 kJ $= 0.9478$ Btu. One common large scale unit of energy is the Q, which is defined as 10^{18} Btu. The solar energy intercepted across the diametral plane in one year can therefore be expressed approximately as 5300 Q. By contrast, current world energy consumption, which is derived primarily from fossil fuels, is only about 0.5 Q per year.

EXERCISES

**for Section
1.5**

1. One estimate given by Angrist and Hepler (1967) for the energy content of world reserves and resources of fossil fuels is about 130 Q. Assuming future fossil fuel energy usage of about 0.5 Q per year, estimate how long it would take to exhaust the world's supply of this form of energy.

   Answer: 260 years

2. An estimated 6 Q of solar energy flows to the Earth each day. How many days of solar energy inflow to the Earth's surface is equivalent to the estimated 130 Q energy content of the world's fossil fuels?

   Answer: Nearly 22 days

3. Data from Loftness (1984) indicates that an estimated 2.2×10^{15} kcal of energy was used in 1970 in the U.S. food system, whereas the food energy consumed in the same period was 3×10^{14} kcal. Defining efficiency as the ratio of output to input, compute the energy efficiency of food production in the United States in 1970.

   Answer: 0.136

1.6   Systematic Procedure for Solving Thermodynamic Problems

To conclude this introductory chapter, a systematic procedure is presented that can be adopted for the solution of problems in thermodynamics. A quick overview of the entire text is provided as part of this step-by-step procedure.

Step One: Problem Definition

Defining the problem precisely is the most crucial step in attempting to solve any problem. Several elements are involved in arriving at a satisfactory definition of a problem. For problems in thermodynamics, the following steps are suggested as a minimum:

1. **Draw a schematic diagram** of the physical system. As pointed out earlier in Section 1.2, this is a very important step that is comparable to drawing the free-body diagram for a statics problem. The use of line diagrams, aided by the symbols introduced in Section 1.4, will generally be adequate.

2. **Clearly define and show the boundary of the system** (closed or open) for the problem. A most pertinent question to consider is whether you can more appropriately analyze the problem using a closed system model or an open system (or control volume) model. Draw a sketch of the boundary and **label all interactions**, namely, heat, work, and mass (for open systems only), across the boundary. Methods for computing the work or heat interaction are discussed in Chapter 5.

3. **Identify the substance(s) involved** in the problem. Most of the substances you are likely to encounter, as mentioned earlier, belong to the class of pure substances. A comprehensive discussion is given in Chapter 3 on pure substances and how their properties can be determined, usually from tables. In Chapter 2, a more detailed discussion of properties is provided, including the units in which they are expressed—both the SI units (*Le Systeme International d'Unités* or the International System of Units) and the USCS (United States Customary System of units). The section on systems of units also includes a systematic procedure for converting from one system to another.

   Finding values for properties in thermodynamic tables can easily become a tiresome exercise, and in any case, it is not the most essential element of thermodynamics. Accordingly, computer codes have been provided for the thermodynamic properties of the more commonly used substances in engineering applications. These include **STEAM** for the thermodynamic properties of ice, water, and steam (H_2O); **R22** for the properties of refrigerant-22; and **GAS** for the thermodynamic properties of a number of ideal gases including air, methane, oxygen, nitrogen, carbon dioxide, carbon monoxide, and steam or water vapor (at ideal gas conditions).

   Chapter 4 covers the thermodynamic properties of ideal gases. Air behavior at ordinary pressures and temperatures can be approximated as ideal gas behavior. Moist air behavior can also be approximated by ideal gas behavior, although detailed treatment of such mixtures of gases is not given until Chapter 13. The code **PSY** is designed to support a detailed thermodynamic analysis of processes involving moist air. Air conditioning, for example, is a field that requires the determination of several thermodynamic properties of moist air.

Once you have identified the substance(s) involved, you can use available property tables or computer codes for the substance(s) to list the property values, particularly those that may be relevant to the solution of the problem at hand.

4. **List all given or known information** pertaining to the problem.

5. **List what you wish to find**.

Step Two: Outline for Solving the Problem

1. **List all the assumptions** that are reasonable and that may facilitate the solution of the problem. For example,

   Can you assume that the substance is an ideal gas?

   Is the process adiabatic (zero heat interaction), or can you assume that the rate of heat interaction between the system and its surroundings is small enough to be negligible?

   Can you assume no work interaction across the boundary?

2. **Identify the thermodynamic laws or principles that are pertinent** to the solution of the problem. In Section 1.3, a quick overview of the laws of thermodynamics was given. The first law of thermodynamics is considered in detail in Chapter 6. Chapters 7 through 9 are devoted entirely to the second law and its important consequences, including the exergy (or availability) concept and the exergy analysis method. It is often necessary to make use of the general laws of physics in the solution of a given problem. Examples include the conservation of mass principle and Newton's laws of motion. In any case, enough independent relationships must be found to provide a sufficient number of independent equations to solve for all of the unknowns. This fact is well known from mathematics.

3. **Specify symbol, numerical value, and units for solution**. Until you specify the solution in a form that is meaningful to someone else, you have not obtained a satisfactory solution to the problem.

Step Three: Review and Physical Interpretation of Solution

1. **The reasonableness of the solution should be considered**. Errors in algebraic manipulation, for example, or wrong handling of units might lead to numerical results that are incorrect and unreasonable.

2. **Reinterpret the problem based on the physical meaning of the results.**

3. **Consider the practicality of the results.**

1.7 Summary

This chapter has provided an overview of the structure and scope of classical thermodynamics. The structure comprises three principal elements—concepts, laws of thermodynamics, and applications. The fundamental concepts of thermodynamics were introduced in this chapter, with further elaboration to be provided in subsequent chapters. In addition, formal statements of the laws of thermodynamics were given, followed by a synopsis of the relevance of thermodynamics to societal energy needs. The chapter concluded with a systematic procedure for solving problems in thermodynamics.

References

American Petroleum Institute 1990. See U.S. D.O.E./E.I.A.

Angrist, S. W. 1968. "Perpetual Motion Machines." *Scientific American*. Vol. 218. 114–122.

Angrist, S. W. 1976. *Direct Energy Conversion*. 3rd ed. Boston: Allyn and Bacon.

Angrist, S. W. and Hepler, L. G. 1967. *Order and Chaos: Laws of Energy and Entropy*. New York: Basic Books.

Brown, L. R. et al. 1986. *State of the World 1986*. New York: W. W. Norton & Co.

Feynman, R. P., Leighton, R. B., and Sands, M. 1963. *The Feynman Lectures on Physics*. Reading, PA: Addison-Wesley.

Gaggioli, R. A. and Petit, P. J. 1977. "Use the Second Law, First." *Chemtech*, Vol. 7. No. 8. 496–506.

Hatsopoulos, G. N. and Keenan, J. H. 1962. "A Single Axiom for Classical Thermodynamics." *Transactions ASME: Journal of Applied Mechanics*. Vol. 29. 193–199.

Hewett, D. F. 1929. "Cycles in Metal Production." *American Institute of Mining, Metallurgical, and Petroleum Engineers Tech. Pub. Trans.*, Vol. 183. 65–98.

Hubbert, M. K. 1956. "Nuclear Energy and the Fossil Fuels." Presented to the Southwest section of *American Petroleum Institute, Drilling and Production Practice*, San Antonio, TX. 7–25.

Hubbert, M. K. 1962. "Energy Resources." National Academy of Sciences, National Research Council Publication 1000-D.

Hubbert, M. K. 1967. "Degree of Advancement of Petroleum Exploration in United States." *American Association of Petroleum Geologists Bulletin*. Vol. 51. 2207–2227.

Hubbert, M. K. 1969. "Energy Resources." in *Resources and Man: A Study and Recommendations by the Committee on Resources and Man of the Division of Earth Sciences*. National Academy of Sciences, National Research Council. San Francisco: W. H. Freeman. 157–242.

Hudson, A. and Nelson, R. 1990. *University Physics*. Philadelphia: Saunders College Publishing.

Keenan, J. H. and Shapiro, A. H. 1947. "History and Exposition of the Laws of Thermodynamics." *Mechanical Engineering*. Vol. 69. 915–921.

Lindsay, R. B., ed. 1975. *Benchmark Papers on Energy (Vol. 1): Energy: Historical Development of the Concept*. Stroudsburg, PA: Dowden, Hutchinson & Ross.

Loftness, R. L. 1984. *Energy Handbook*. 2nd ed. New York: Van Nostrand Reinhold.

Rankine, W. J. M., *A Manual of the Steam Engine and Other Prime Movers*. 16th ed. Revised by Millar, W. J. 1906. London: Charles Griffin and Co.

Serway, R. A. 1993. *Physics for Scientists and Engineers*. 3rd ed. (updated). Philadelphia: Saunders College Publishing.

U.S. D.O.E./E.I.A. 1990. *U.S. Crude Oil, Natural Gas Liquids, and Natural Gas Reserves. Rest of World Oil and Gas Journal.* "Worldwide Report" issue. (Data supplied by American Petroleum Institute.)

Questions

1. What is thermodynamics? Distinguish clearly between classical and statistical thermodynamics.

2. Explain, in your own words, each of the following concepts.
   (a) closed system, boundary, and surroundings
   (b) open system (or control volume) and control surface
   (c) equilibrium, property, state, and process
   (d) quasi-equilibrium process, cyclic process, and path
   (e) flow and nonflow processes
   (f) isothermal, isobaric, and isochoric processes

3. In which of the following typical operations would it be more appropriate to consider a closed system rather than a control volume?
   (a) filling of an evacuated vessel with gas from a subterranean source
   (b) inflation of a balloon with a gas
   (c) steady flow discharge of hot gases through a jet nozzle
   (d) freezing a given quantity of water

4. Water is heated in a container that is not covered. After some time, the water starts to boil. Which of the following correctly describes the entire process?
   (a) isothermal process
   (b) isobaric process
   (c) isochoric process
   Use a diagram to define the problem. Indicate clearly whether a closed system or a control volume is the more appropriate model. What physical boundary are you assuming for the problem?

5. Which of the following are properties of a system: pressure, temperature, heat, volume, mass, density, and work?

6. Indicate whether the processes listed below are quasi-equilibrium or nonequilibrium processes. Explain your answer in each case.
   (a) Stirring of a fluid using a mechanical agitator. Define the fluid as the system.
   (b) Heating of a metal bar, initially at room temperature, by placing one end of the bar in a furnace. The metal bar is the system.
   (c) Slow expansion of air contained in a piston and cylinder device. The expansion results from a low heating rate from a gas burner. Air is the system in this case.

7. Use suitable diagrams to represent each of the following operations (in italics). Indicate clearly the heat and/or work interactions that may be involved. Also, determine whether a system or a control volume description is the more appropriate one to employ. For the control volume cases, mark on your diagram the direction of flow of the substance in relation to the control volume.
   (a) *Condensation of steam to water* in a flow process.
   (b) *Drying of clothes in a dryer by passing hot air* over the clothes.
   (c) *Freezing of food* in the freezer compartment of a domestic refrigerator.

8. What are the world's principal energy resources? Clearly distinguish between those that are renewable and those that are not. Write a short description of each of the principal sources of energy indicating their relative abundance, costs, and whatever safety or environmental pollution concerns may be involved in their use.

9. Discuss critically the prospects for meeting society's energy needs in the 21st century.

10. It is generally observed that a direct correlation exists between the standard of living and the energy consumption level. Suggest appropriate strategies for maintaining a high standard of living on a reduced level of energy consumption.

11. An inventor offers you an ingenious device that continuously converts heat transfer (Q) from atmospheric air completely to work (W). Use a suitable sketch to represent the operation of the device. Is such a device feasible on the basis of
   (a) the first law of thermodynamics?
   (b) the second law of thermodynamics?

12. A thin sheet of glass is placed on top of a drop of water, as illustrated in Figure Q1.12. On top of the sheet of glass is a thin layer of a volatile liquid, such as alcohol. The volatile liquid evaporates, and the original drop of water is turned into a thin sheet of ice. Does this ice-forming process violate any of the laws of thermodynamics? Explain your answer.

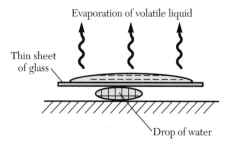

Figure Q1.12 Production of ice using the evaporative cooling effect of a volatile liquid.

Problems

1.1 Assume the following conversion: 1 food Cal = 1 kcal (or 4186.8 J) of energy. The average person needs approximately 2000 to 3000 food Cal per day and an average daily requirement of 2500 Cal may be assumed. If the world population is 5 billion (that is, 5×10^9), determine the total human food energy requirement for one year.
  (a) in kilocalories
  (b) in joules

1.2 Daily solar insolation on the Earth's surface is estimated to average 2×10^{11} J/ha. (1 hectare [ha] equals an area 100 m $\times$ 100 m.) The efficiency of conversion of solar energy to food energy is about 0.1%. The total area of arable and potentially arable land on Earth is 4.24×10^9 ha (approximately 28% of the total land area). If the entire arable land area (including the potentially arable land) were to be under cultivation to meet human food energy needs of 2500 food Cal per person per day, estimate the maximum world population that can be supported.

* **1.3** The daily food intake by the average person in a developed country is the equivalent of 3.5 kWh of electrical energy. (The kilowatt-hour [kWh] is the energy equivalent of working at the rate 1 kW for 1 h.) A human's energy need for sheer survival,

called the *basal* or *resting metabolism* requirement, is approximately equal to that of a 70 W electric light bulb. For manual labor, the food energy requirement (inclusive of the requirement for basal metabolism) is approximately at the rate of 400 W.
  (a) Estimate the maximum hours of manual work that can be done by an average person per day.
  (b) If the work output in 1 h of manual labor is 0.06 kWh, what is the corresponding maximum daily work output for the average person?
  (c) Determine the ratio of maximum work output to the food energy input for the average person.

1.4 The yearly total solar insolation in a particular geographical zone on the Earth's surface is 2×10^{13} Btu/mile^2. Estimate the total surface area (in square miles) of solar collectors required for the production of 1 quad (= 10^{15} Btu) of heat annually, assuming the efficiency of conversion of solar energy to heat is 30%.

1.5 It is estimated that burning (complete fission) of 1 kg of uranium results in a mass deficit of only 0.87 g, which corresponds to 7.83×10^{10} kJ or 906 megawatt-day. (The megawatt-day is the en-

---

*Throughout the book, the least difficult problems are unmarked while the more difficult problems are denoted by one asterisk (*), and the most difficult ones by a double asterisk (**).

ergy equivalent of operating at 1 MW for a 24-h day.) If 32,000 kJ of heat is produced when 1 kg of coal is burnt, determine the mass of coal needed to produce the same heat as the complete fission of 1 kg of uranium.

1.6 The *heating value* (defined as energy released as heat transfer to the surroundings when a unit amount of a fuel is burnt) of bituminous coal can be taken as 26×10^6 Btu/short ton. (1 short ton = 2000 lbm.) Defining 1 quad of energy as 10^{15} Btu, determine the mass (in short tons) of coal required to produce 1 quad of heat transfer.

1.7 In 1982, the total fuel consumed by all the registered vehicles in the United States was about 113×10^9 gal. The heating value of gasoline can be taken as 124,000 Btu/gal. What is the heat transfer equivalent (in quads) of the fuel consumed in 1982?

1.8 In 1973, the gross energy consumption in the United States alone was 75 quads out of an estimated world total of 250 quads. Show this on a pie chart. What percentage of the world's total energy consumption is that of the United States?

* **1.9** It is frequently found that the rate of depletion of nonrenewable resources follows a normal distribution law. One estimate puts the energy content of the world's original supply of crude oil at 8000 quads. One scenario on future use of crude oil assumes that the consumption rate should have peaked in 1990 at about 145 quads for the year. Assuming that the depletion of the world's crude oil reserves follows a normal distribution curve, determine the year by which 99% of the world's original supply will be depleted.

1.10 Table P1.10 gives the estimated proven crude oil reserves for 20 leading nations as of January 1, 1990.
  (a) If the top five nations were all under the domination of one leader, what percentage of the known reserves would he or she control?
  (b) In the year 1989, the United States consumed 5.75×10^9 barrels of oil. If the United States were unable to obtain crude oil from any other nation, how long would the country's crude oil last (at the 1989 rate of consumption)?

TABLE   P1.10

Estimated Proved World Crude Oil Reserves for 20 Leading Nations as of January 1, 1990

Nation	Reserves (in thousands of barrels)
Saudi Arabia	254,959,000
Iraq	100,000,000
Kuwait	94,525,000
Iran	92,860,000
Abu Dhabi	92,205,000
Venezuela	58,504,000
U.S.S.R.°	58,400,000
Mexico	56,365,000
United States	26,501,000
China	24,000,000
Libya	22,800,000
Nigeria	16,000,000
Norway	11,546,204
Algeria	9,200,000
Indonesia	8,200,000
India	7,516,400
Canada	6,133,495
Neutral Zone	5,200,000
Qatar	4,500,000
Egypt	4,500,000
Total for world	1,002,212,623

Source: American Petroleum Institute, 1990.
°A part of the former U.S.S.R. is now called the Commonwealth of Independent States.

* * **1.11** The world oil production and consumption rates can be approximated since 1950 as follows:

Year	Million Barrels per Day
1950	12
1960	20
1970	50
1979	66
1980	60
1990	60

If the consumption of crude oil follows a normal distribution law and if 1979 represents the peak production and consumption rate, in what years will 50%, 60%, 70%, 80%, and 90%, respectively, of the world's original supply be depleted?

1.12 Figure P1.12 illustrates a domestic refrigerator that receives 500 J of electrical energy while producing 1000 J of cooling. With reference to the sketch, indicate the magnitude and direction of the third energy transfer that must have taken place assuming cyclic operation of the refrigeration device. What is the typical medium or environment to (or from) which this third energy transfer normally occurs in real life?

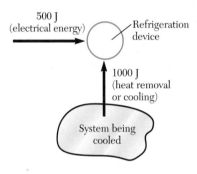

Figure P1.12 A domestic refrigerator.

1.13 Heat pumps are designed to operate with the desired output (space heating energy) being greater in magnitude than the energy input (usually electrical energy) that one must pay for. Figure P1.13

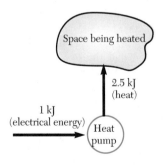

Figure P1.13 A heat pump.

illustrates such a heat pump that receives 1 kJ of electrical energy and delivers 2.5 kJ as heat to the space indicated. Does this violate any of the laws of thermodynamics? Explain your answer. Also indicate the magnitude and direction of the third energy transfer that must occur for cyclic operation of the device.

* **1.14** 0.1 kg of a liquid refrigerant is contained under a laden piston of the piston and cylinder arrangement depicted in Figure P1.14. The pressure and temperature of the ambient air are 100 kPa and 20°C, respectively. The refrigerant is at a pressure of 900 kPa, and its corresponding boiling point at that pressure is 19.6°C. The combined weight of the piston and the load on it is set to exert a pressure of 900 kPa on the refrigerant. The diameter of the piston is 150 mm, and the depth of the liquid refrigerant in the cylinder is 4.7 mm. Subsequently, heat leaks from ambient air to the refrigerant causing it to gradually turn into a vapor (boiling) without a change in temperature or pressure. Assume that the piston is frictionless and that the ambient air pressure decreases by an infinitesimal amount such that the laden piston is raised through a height of 140 mm while boiling of the refrigerant takes place. Determine

(a) the combined weight (in kilonewtons [kN]) of the piston and the load on it;

(b) the work done (in J) by the refrigerant which is defined as the system.

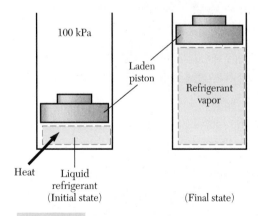

Figure P1.14 Work done when a refrigerant changes from a liquid to a vapor.

Would you expect the energy input to the system as heat leakage to be greater than, less than, or equal to the work done by the system? Why? This process is equivalent to a heat transfer (input) from a single reservoir to a system resulting in work done by the system. Does this violate the second law of thermodynamics? Explain your answer.

The *efficiency* η (or *efficiency ratio*), of a device or a process can be defined as the ratio of the (desired) output to the (required) input. The input is usually construed as what is paid for. This definition can be assumed in the following problems.

* **1.15** The drag force F_D (in N) on an automobile cruising at a speed V (in m/s) can be expressed in terms of the drag coefficient (C_D) as follows:

$$F_D = \left(\frac{1}{2} \rho_{air} \, C_D \, A_{front} \right) V^2 \equiv kV^2$$

where ρ_{air} (in kg/m³) is the air density and A_{front} (in m²) is the frontal (or effective) area over which the drag force acts. By substituting $\rho_{air} = 1.2$ kg/m³, $C_D \approx 1$, and $A_{front} \approx 1$ m², obtain an estimate for k (in kg/m) in this expression for the drag force.

For a particular automobile, assume that the engine output is the sum of the rate of doing work to overcome the resistance due to the drag force

$$\frac{F_D \cdot V}{1000} \text{ kW}$$

and a constant base load of 12 kW. Assume also that the energy efficiency is 24% and remains constant over a wide range of car speeds. The fuel

energy input is estimated at 40 kWh/gal. Determine the fuel consumption (in miles/gal of fuel) for the automobile when it is cruising at
(a) 50 mph (22.3 m/s)
(b) 80 mph (35.8 m/s)

* **1.16** Use current prices for gasoline to estimate the potential savings on a round trip from Starkville, MS, to New York, NY, when the automobile in Problem 1.15 is driven at about 50 mph rather than at 80 mph (for about 2000 mi round trip).

* **1.17** Assume that the performance indicated for the automobile in Problem 1.15 is typical for cars in the United States. What is the likely percentage increase in the annual fuel consumption by automobiles if motorists as a rule drove at a speed of 70 mph instead of 50 mph?

* **1.18** Repeat Problem 1.15 assuming an increased energy efficiency of 30%.

1.19 Professor Rankine (see Rankine, 1906, 84) provided the following data on the capacity of humans to do work:

Work Activity	Resistance Overcome (lbf)	Effective Velocity (ft/s)	Hours per Day
Working pump	13.2	2.5	10
Lifting weights by hand	44	0.55	6

(a) Calculate the work done per day (in ft · lbf) in (i) working a pump and (ii) lifting weights by hand. Given that 737.6 ft · lbf = 1 kJ and that 3600 kJ = 1 kWh, convert your estimates of the work done per day to kilowatt-hours.
(b) Assume that the humans required to work pumps and lift weights are supplied with a food ration of 3000 food Cal per person per day. Convert this to kilowatt-hours. What is the energy efficiency achieved in each of the two work activities?
(c) Suggest a reasonable cost for the food needed daily by a person who works pumps or lifts weights in the manner specified. Use this to estimate the cost of human labor in dollars per kilowatt-hour work output.

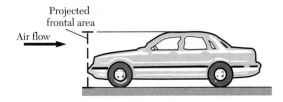
Projected
frontal area
Air flow

Figure P1.15 Air flow over an automobile.

1.20 Professor Rankine (1906, 88) also provided the following information, this time on the work output of horses.

Activity	Resistance Overcome (lbf)	Effective Velocity (ft/s)	Hours per Day
Drawing cart or boat, walking	120	3.6	8
Drawing a gin or mill, walking	100	3	8

(a) Estimate the daily work output in kilowatt-hours for each of the activities listed.

(b) A well-fed horse consumes approximately 10 lbm of grain and 10 lbm of hay daily. Assuming that the horse feed costs $2, what is the cost of the work output of the horse in cents per kilowatt-hour?

(c) Taking 5000 Btu/lbm as the approximate mean value of the energy value obtained by a horse from the feed provided, estimate the work efficiency of the horse based on the previous data.

1.21 A natural gas-fired power plant operates with an overall efficiency of 38%. What is the work output (in kWh) for each 1 million Btu of gas consumed? (Assume 1 kWh ≡ 3413 Btu.) If 1 million Btu of gas costs a power utility $5, estimate the nominal cost in cents per kilowatt-hour of the work output of the plant.

1.22 Heating at the rate of 40,000 Btu/h is the estimated requirement for maintaining a standard U.S. residence at 70°F when the outdoor temperature is 15°F.

(a) Assume that the heating is provided using electric strip heaters. How much electricity (in kWh) must be supplied to meet the demand for heating over a period of 8 h? (Assume 1 kWh ≡ 3413 Btu.) If electricity is sold at 6¢/kWh, determine the cost of heating the residence for the 8-h period.

(b) Suppose gas space heaters are used instead to heat for an 8-h period. Estimate the Btus of gas needed. If gas is sold at $6.50/1 million Btu, what is the cost of heating the house using gas?

(c) An electric heat pump is proposed for meeting the demand for heating the house. The energy efficiency of the heat pump is known as the *coefficient of performance*, which is defined as the ratio of the output (the heat supplied to the space) to the input (the electricity supplied to the heat pump). Assume an energy efficiency of 1.8 in this instance. How much electricity (in kWh) is needed to supply the heat needed over a period of 8 h? How much will this cost assuming a 6¢/kWh cost for electricity?

1.23 The equivalent effect in terms of exergy for the operation described in Problem 1.22 can be taken as approximately 4150 Btu/h. In parts (a) and (c), the electricity input (in kWh) is also the equivalent exergy input. In part (b), however, the Btus of gas needed should be multiplied by a nominal factor of 1.05 to obtain an estimate for the exergy equivalent. Determine the exergy efficiency for each of the options (a), (b), and (c) indicated in Problem 1.22.

1.24 The electricity supply needed for an icemaker is estimated at 0.5 kWh for each 8-lb ice pack. If electricity sells at the rate of 6¢/kWh, what is the energy cost for an 8-lb ice pack? If your local store sells ice at the rate of $1 per 8-lb ice pack, what is the profit margin based on energy cost alone?

1.25 A family of five returned to their air conditioned home after a football game. Assume that each member of the family adds energy as heat transfer to the room air at the rate of 70 W. In addition, various electrical appliances are turned on that draw electricity (a form of work transfer) from the mains at a total rate of 1.8 kW. Define the room air and the electrical appliances as the system and assume that the family members and the mains supply are in the surroundings. Schematically illustrate the system under consideration. Determine the following (assuming 1 kW = 3413 Btu/h):

(a) the additional heating load (in kW) imposed on the system by the family's return home

(b) the additional work output rate (in kW) of the system

(c) the rate of increase in the energy of the system (in Btu/h) due to the additional loads

Thermodynamic Quantities and Units

KEY CONCEPTS

■■■ Macroscopic perspective, continuum, and values of properties at a point

■■■ Extensive and intensive properties, frame of reference for thermodynamic properties, and some thermodynamic properties and their measurement

■■■ Work and heat as interactions, definitions of work in mechanics and thermodynamics, and definitions of heat

■■■ Primary and secondary dimensions; systems of units; SI and USCS units for force, mass, specific weight, specific gravity, energy, power, temperature, and amount of substance; and conversion of units using unity conversion factors

2.1   Introduction

Macroscopic Perspective

As discussed briefly in Chapter 1, the concepts of classical thermodynamics are based on a macroscopic perspective. In this perspective, the nature and behavior of individual particles and their interactions are not studied; instead, the focus is on an overall or large scale view of the behavior of substances. Thus, the macroscopic approach requires no particular theoretical study of the detailed structure of matter on an atomic scale, but instead relies on large scale observation of data concerning the nature, structure, and behavior of substances. The macroscopic approach inevitably results in a loss of some details pertaining to behavior at the microscopic level.

Consistent with the macroscopic perspective is the assumption that systems can be described and modeled as a *continuum*. Implied in this term is the notion that length and volume scales for a system are very large compared to molecular dimensions and that the system contains a large number of molecules. It is generally also assumed that the values of properties at a given point can be defined and that no discontinuity exists between a value at a particular point and a value for the same property of the substance at another point in the immediate region of the first point. This continuum concept is applicable to most conditions encountered in engineering. A notable exception is the case in which the mean free path of a molecule in a dilute gas approaches the order of magnitude of the system dimensions. For example, in typical gas dynamic behavior in outer space and at high atmospheric altitudes, the assumption of a continuum may be incorrect because of the very low density of the gas at these locations. Other approaches outside the scope of this text are needed for such cases.

This chapter covers in detail a comprehensive list of mechanical and thermal quantities of interest in thermodynamics. Thermodynamic properties form the major group of these quantities and are discussed in Section 2.2. Section 2.3 is devoted to further discussion of work and heat. As indicated earlier, neither of these is a property; they are both forms of energy interaction between a system and its surroundings.

No meaningful scientific discussion of thermodynamic quantities can be complete without discussion of the units in which they are commonly measured. Section 2.4 is thus provided to serve as a comprehensive introduction to two principal systems of units that the reader is likely to encounter in engineering practice in the United States—the SI (*Systéme International d'Unites*) and the USCS (United States Customary System). The presentation includes a systematic procedure for conversion from one system of units to any other system through the use of appropriate conversion factors.

2.2 *Thermodynamic Properties*

A thermodynamic property has already been identified as any (macroscopic) observable characteristic of a system. The complete list includes mechanical quantities such as pressure (p) as well as thermal quantities such as temperature (T). Some properties can be measured directly, such as volume (V) and mass (m). For measurement of other properties, indirect methods involving the use of a pertinent physical law or thermodynamic principle must be used. Internal energy (U), temperature (T), and entropy (S) are examples of the second group of properties since they are *derived properties* in the sense that their existence is a consequence of the laws of thermodynamics.

Extensive and Intensive Properties

Thermodynamic properties can be divided into two categories—extensive and intensive. The value of an *extensive property* is directly proportional to the amount of matter present, while an *intensive property* is independent of the mass present. Properties, such as pressure and temperature, that do not depend upon the amount of matter present are intensive properties, while total volume (V) and total mass depend directly on the amount of material substance present and are extensive properties.

In the case of an extensive property (other than mass), one can convert to the corresponding intensive property simply by dividing the total property value for the system by the mass of the system. Thus, a property such as *specific volume* (v), obtained by dividing the total volume by the mass of the (closed) system, is an intensive property and is independent of the amount of the substance present for a given equilibrium state of the system. The nomenclature generally used in this text is a capital letter for the extensive property and a lowercase letter for the corresponding specific property. For example, V is used for total volume (extensive property), while v stands for specific volume (intensive property).

Frame of Reference for Thermodynamic Properties

A thermodynamic property such as temperature, density, or specific volume can be measured and specified without any physical frame of reference, while certain other properties require measurement and specification with respect to some physical frame of reference or coordinate system. Typical properties requiring a reference frame include velocity, elevation, momentum, kinetic energy, and potential energy. Appropriate frames of reference for particular cases are specified as they are encountered in this text.

Thermodynamic Properties and Their Measurement

Until the laws of thermodynamics have been discussed in some detail, only those properties that are either directly measurable or determinable based on the general laws of physics can be presented. Such properties include pressure, density, specific volume, and (empirical) temperature. (Thermodynamic temperature is, strictly, a consequence of the second law of thermodynamics.) If values of a sufficient number of these properties are known for a system, it is possible to determine the particular condition or state in which the thermodynamic system exists. Chapter 1 indicated that, while some properties have a defined value for a system regardless of whether or not an equilibrium condition exists, most properties have a defined value for a system only for equilibrium states. The latter are typically intensive properties that only have values characteristic of the state of the system when such values are the same throughout the system. Formal definitions of some of these properties follow. A brief indication is also provided on how these quantities may be measured.

Pressure

Pressure in thermodynamics is the same as the mechanics concept of pressure and is defined as the normal force per unit area acting on a surface. The term *pressure* is often reserved for fluid media (liquids and gases), while the more general term *normal stress* is usually used when dealing with solids.

For any fluid in the condition termed *static equilibrium*, the pressure at a given point in the fluid is the same in all directions. This condition is also referred to as an *isotropic* condition for the fluid. For a fluid system comprising a vertical column of fluid in a gravitational field, the hydrostatic pressure will vary with depth in the fluid even though a static equilibrium condition exists. Thus, when dealing with a large fluid system, there may be a nonuniformity of pressure due to gravity, and as a consequence, the system will not be in a state of thermodynamic equilibrium even though there is static equilibrium.

Techniques for the measurement of fluid pressure are routinely described in texts on experimental physics. Barometers are typically used for the measurement of the pressure of atmospheric air, while pressure gages (or gauges) are used for the measurement of fluid pressures relative to the surrounding atmospheric (ambient) air pressure. Fluid pressure measured relative to the pressure of ambient air is referred to as the *gage pressure*. The *absolute pressure* is then obtained by adding the gage pressure to the atmospheric pressure.

The USCS unit for pressure is pound force per square inch (lbf/in.^{2}), commonly abbreviated psi. When referring to absolute pressure, lbf/in.^{2} (abs) or psia is used, while lbf/in.^{2} (gage) or psig is employed for gage pressures. Whenever the fluid pressure is below the ambient pressure, the gage pressure is often expressed as a positive quantity with the label "vacuum" added to indicate that the quantity given should be subtracted from the atmospheric pressure to determine the absolute pressure of the fluid.

The SI unit for pressure is the newton per square meter (N/m^2), also called the pascal (Pa), which is equivalent to a force of 1 N (newton) acting on an area of 1 m^2 (that is, 1 Pa $\equiv$ 1 N/m^2). Differentiation between absolute and gage pressures is made simply by appending "(abs)" or "(gage)" after Pa or kPa.

Pressure is also measured in atmospheres. One atmosphere, abbreviated atm, is equal to 101.325 kPa or 14.696 psi and is equivalent to the pressure exerted at the base of a column of mercury that is 760 mm in height. The bar is also used to signify a pressure equal to 100 kPa. This pressure unit is close to the atm and is a commonly used unit in the atmospheric sciences. The bar is also used in several published property tables for water and steam and a host of refrigerants. It is the unit used for the pressure of steam and refrigerant-22 in the property tables provided in Appendix A of this text.

Specific Volume and Density

The *specific volume, v*, of a substance is defined as the volume per unit mass of the substance. If V is the total volume of the system and m is the total mass of the system, then the specific volume is given by

$$v = \frac{V}{m} \tag{2.1}$$

The *density* of the substance, ρ, is defined as the mass per unit volume. Thus, the density is given by

$$\rho = \frac{m}{V} \tag{2.2}$$

As can be seen from these two equations, the density is the reciprocal of the specific volume.

Both the volume occupied by a system and the total mass of the system are relatively easy to measure in most instances. The USCS units for volume and mass are, respectively, the cubic foot (ft^3) and the pound mass, written as lbm. The corresponding units for v and ρ are ft^3/lbm and lbm/ft^3, respectively.

The SI units for volume and mass are m^3 and kg, respectively. Specific volume is then expressed in m^3/kg, while density is measured in kg/m^3. The liter (l or L) is used in the literature, although this is not strictly an SI unit (1 l is 1000 cm^3 or 10^{-3} m^3).

The term *tonne* or *metric ton* is used to designate 1000 kg and is equal to 2205 lbm. The *English* long ton is 2240 lbm while the short ton equals 2000 lbm.

Temperature

Thermodynamic temperature (T) is, strictly speaking, a derived property that is a consequence of the second law of thermodynamics. Detailed discussion of this property is reserved for Chapter 7, which covers the second law. The notion of temperature as the degree of hotness (or coldness) of a body is rooted in our senses and can be considered here. While our senses cannot be relied upon to provide a dependable and objective scale for the measurement of temperature, other systems can, as is found to be the case in empirical thermometry.

Empirical thermometry is based on the zeroth law of thermodynamics, which was introduced in Chapter 1 and can be stated as follows:

> *When two bodies are in thermal equilibrium with a third body, the two bodies are in thermal equilibrium with each other, and all three bodies are said to be at the same temperature.*

An important derived concept from the zeroth law is that of *equality of temperature*. When two bodies at the same temperature are brought into *thermal contact*, there is no discernible thermal effect on either body by the other. In more rigorous expositions of the zeroth law, the nature of the thermal contact is qualified as being through a *diathermic* or *nonadiabatic wall*. An adiabatic wall does not allow discernible thermal effect of one body on the other. Materials called *insulators* are close to the idea of adiabatic media when interposed between two bodies. The diathermic wall, however, is more like the materials classified as conductors.

For two bodies at the same temperature and in thermal contact through a nonadiabatic wall, it is further observed that no change occurs in any other property due to this contact. The two bodies are in *thermal equilibrium* with each other, which is to say that *equality of temperature* is the condition for bodies whose properties do not change when they are brought into thermal contact with each other. This is a corollary of the zeroth law of thermodynamics.

Empirical temperatures are routinely measured in experimental physics using devices called *thermometers*. To measure the temperature of a body, all that is needed is a substance with a property that can be measured and that varies with temperature. Such a substance is called a *thermometric substance*. For the substance to be of use for temperature measurements, the property should vary uniquely with the body's equilibrium temperature. The variation in the property of the substance provides a measure of the temperature on what is clearly an empirical scale. A scale of temperature is simply a formula that prescribes how the thermometric property varies with temperature. In other words, it is a prescription of a scale for determining temperature in terms of thermometric property variation. An absolute scale of temperature is one on which the temperature readings are independent of the substance used.

A good thermometer should have the following desirable characteristics:

1. *Sensitivity* Appreciable changes in the thermometric property should accompany relatively small changes in temperature.

2. *Accuracy* Readings of the thermometer should closely match those obtained on standard scales.

3. *Reproducibility* The thermometer should not give different readings for the same equilibrium temperature.

4. *Fast response* The thermometer should ideally follow temperature changes quickly and should not take long to attain thermal equilibrium with the system whose temperature is being measured.

If the readings obtained on a given thermometer are to be consistent with what other thermometers indicate, calibration relative to a standard is needed. The major categories of temperature measuring devices in use include expansion thermometers, such as the bimetal thermometer (solid expansion), the mercury-in-glass thermometer (liquid expansion), and the gas thermometer (expansion of a gas); resistance thermometers; thermocouples; and pyrometers. Each device has distinct advantages for certain types of application and over particular ranges of temperatures. A standard scale is necessary to integrate and compare these different temperature measuring devices.

Traditional units for temperature (T) that have evolved include degrees Fahrenheit (°F), named after Gabriel Fahrenheit, and degrees Celsius (°C), named after the Swedish astronomer, Anders Celsius. The Fahrenheit unit is used as a USCS unit, while degrees Celsius is regarded as a secondary SI unit. Additional information is provided in Section 2.4 on units for temperature and the conversion relationships between the different systems of units.

EXAMPLE 2.1

The pressure of air in an automobile tire is 30 psia when the atmospheric pressure is 14.7 psia. Convert the pressure to

(a) psig

(b) kPa (gage)

Given:
$$p_{tire} = 30 \text{ psia}, \, p_{atm} = 14.7 \text{ psia}$$
$$(101.325 \text{ kPa} = 1 \text{ atm} = 14.696 \text{ psi})$$

Find: p_{tire} (a) in psig and (b) in kPa (gage)

Solution: The relationship between gage and absolute pressures was given as

$$p \text{ (gage)} = p \text{ (abs)} - p_{atm} \text{ (abs)}$$

Applying the equation, we have

(a) p_{tire} (gage) = 30 psia − 14.7 psia = $\underline{15.3 \text{ psig}}$

(b) p_{tire} (gage) = $15.3 \text{ psig} \left(\dfrac{101.325 \text{ kPa}}{14.696 \text{ psi}} \right)$

$\qquad = \underline{105 \text{ kPa (gage)}}$

EXAMPLE 2.2

Water has a density of 1000 kg/m³ at room temperature. What is the specific volume of the water? Calculate the volume occupied by 2 kg of water at room temperature.

Given: ρ_{water} = 1000 kg/m³, m_{water} = 2 kg

Find: (a) v_{water} and (b) V for 2 kg of water

Solution:

(a) By definition, $v = 1/\rho = 10^{-3}$ m³/kg for water at room temperature.

(b) Again, by definition, $V = mv = \underline{2 \times 10^{-3} \text{ m}^3}$ for the 2 kg of water.

EXERCISES

for Section 2.2

1. A vacuum of 40 kPa (vacuum) is pulled on a fluid in a rigid container using a vacuum pump. The pressure of the ambient air is 100 kPa (abs). What is the absolute pressure of the fluid left in the container that was partially evacuated?

    Answer: 60 kPa (abs)

2. An inflated tire has an air pressure of 16 psig when the atmospheric pressure is 15 psia. What would be the gage pressure of the tire on the atmosphere-free surface of the moon?

    Answer: 31 psig

3. The specific volume of air at standard conditions is approximately 0.84 m³/kg. What is the mass (in kg) of 1.5 m³ of air?

    Answer: 1.79 kg

2.3 Work and Heat Interactions

Work Interaction

Work is defined in mechanics as "the component of force in the direction of displacement" times "the displacement." The USCS unit for work is the ft · lbf, which is the work done when a force of 1 lbf is applied through a distance (in the same direction as the force) of 1 ft. The SI unit for work is the same as for energy and is the joule (J). 1 J of work is done when a force of 1 N is applied through a displacement of 1 m. If, for example, a man lifts a suitcase weighing 200 N through a vertical distance of 1 m, the work done by the man is 200 J.

In thermodynamics, the magnitude of the work done when a force at the system boundary acts through a distance is computed in essentially the same manner as done in mechanics. Thus, in the previous example, the man can be defined as the system, while the suitcase is part of the system's surroundings. As before, work is done by the system (the man) on the surroundings and the magnitude of the work done is 200 J. Two subtle distinctions between the thermodynamic concept of work and the concept in mechanics should, however, be noted. In mechanics, one would say that the work in the example is done by the man, while the suitcase does zero work. In thermodynamics, however, recalling the fact that work is an *interaction*, one would say that the system (the man) does positive work (+ 200 J) and that negative work (− 200 J) is done by the suitcase in the system's surroundings.

A second notable distinction is that the thermodynamic work computed is dependent on how the system is defined. Thus, if the system is defined as the man and the suitcase, the work done becomes zero since no effect is transmitted across the system boundary. Similarly, if the suitcase is defined as the system while the man is now considered part of the system's surroundings, one would say that the system did negative work (− 200 J) and that positive work (+ 200 J) was done by the surroundings.

While detailed treatment of the thermodynamic concept of work is deferred to Chapter 5, it is appropriate at this point to note certain important features of it:

1. Work in thermodynamics is an interaction between a system and its surroundings. As such, it is generally regarded as a boundary phenomenon. A possible exception to this is in the case involving the so-called body forces. These are associated with *force fields*, such as gravitational, electric, and magnetic force fields. The effects due to these force fields can, however, be included in terms for potential energy, as is often done in physics. When approached in this manner, all remaining work interactions should be regarded as boundary phenomena. Thus, computation of work in thermodynamics is ordinarily based on forces acting on the system boundary and the corresponding movements of the boundary taking place.

2. A widely employed sign convention is to consider that a system does *positive work* when the effect external to the system is equivalent to the raising of a weight (in the presence of a gravitational field). Because work is an interaction, if the system does positive work, it follows that the surroundings simultaneously do an equivalent amount of negative work. Conversely, when a system does negative work, the surroundings do an equivalent amount of positive work. Thus, once positive work has been defined for a system, a definition for *negative work* emerges as the work done by the surroundings, such that the algebraic sum of the work done by the system and its surroundings is zero.

3. The thermodynamic definition of work is broader than the mechanics definition in terms of the scalar product of force and displacement. The flow of electricity across a system boundary (which may be stationary), for example, counts as work in thermodynamics. The argument, which is developed more fully in Chapter 5, is that, regardless of the actual end use to which the electricity is put, one could always introduce a hypothetically perfect electric motor to receive the electricity, and use the motor for raising a weight through a distance in the surroundings.

A classic definition for work given by Poincare (see Keenan and Shapiro 1947, 917) can be paraphrased as follows:

Positive work is done by a system on its surroundings during a given process if the system could pass through the same process while the sole effect external to the system was the rise of a weight. The magnitude of the work done is found by counting the number of standard weights that can be raised from one selected level to another. The amount of work flowing from the system is identical to the amount of work flowing into its environment.

Poincare's statements concerning work give rise to a relationship that was indicated earlier between the work done by a system (W_{system}) and that done by its surroundings ($W_{surroundings}$) during a process:

$$W_{system} + W_{surroundings} = 0 \qquad (2.3)$$

A few examples are now provided to explain further the thermodynamic definition of work.

EXAMPLE 2.3

An electric motor-driven crane is used to lift a machine weighing 20 kN through a vertical distance of 2 m. Defining the machine as the system, evaluate

(a) the work done by the surroundings

(b) the work done by the system

Solution:

Figures 2.1(a) and (b) depict the system boundary at the beginning and at the end of the process, respectively. Positive work is done by the surroundings as the motor-driven crane lifts the machine. The magnitude of the work done is computed in terms of the scalar product of the force at the boundary and the displacement of the boundary on which the force acts.

(a) Thus,

$$W_{surroundings} = +20 \text{ kN} \cdot 2 \text{ m}$$

$$= 40 \text{ kJ (kilojoules)}$$

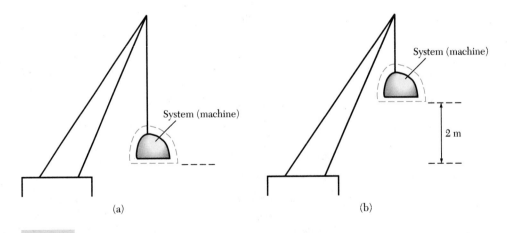

(a) (b)

Figure 2.1 System (the machine) at (a) the start of the process and (b) the end of the process.

(b) From Equation 2.3, we infer

$$W_{system} = -W_{surroundings} = -40 \text{ kJ}$$

EXAMPLE 2.4

Gas at a pressure of 150 kPa (abs) is trapped in a cylinder by a frictionless laden piston, as illustrated in Figure 2.2. The piston cross-sectional area (A) is 0.03 m². Heat transfer occurs to the gas, and as a result, the piston and load are raised through a distance (x) of 0.3 m. The atmospheric air pressure is 100 kPa (abs).

(a) What is the work done by the gas?

(b) If the system is redefined to include the gas and the laden piston, what is the work done (i) by the system and (ii) by the surroundings?

Given: Gas trapped in a cylinder (Figure 2.2)

$$p_{gas} = 150 \text{ kPa (abs)}, \, p_{atm} = 100 \text{ kPa (abs)}$$

$$A_{piston} = 0.03 \text{ m}^2, \, x = 0.3 \text{ m}$$

(a) system is the gas

(b) system is gas + laden piston

Find:

(a) W_{gas}

(b) (i) W_{system}, (ii) $W_{surroundings}$

Solution:

(a) The gas at 150 kPa (abs) is the system. Figures 2.3(a) and (b) depict the system boundary at the initial and final states, respectively, for the process. Work is done by the system across the boundary with the lower face of the piston, which moved

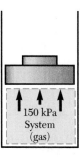

Figure 2.2 Gas expansion problem of Example 2.4.

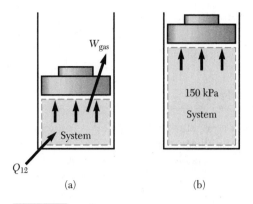

(a)   (b)

Figure 2.3 System boundary at (a) the initial
state and (b) the final state for the gas
expansion problem of Example 2.4.

through a distance of 0.3 m. Thus, assuming that the gas expansion was fully
resisted, the work done by the gas is given by

$$W_{gas} = (150 \text{ kN/m}^2)(0.03 \text{ m}^2)(0.3 \text{ m})$$

$$= 1.35 \text{ kJ}$$

(b) The boundary when the system is the gas and the laden piston combined is
sketched in Figure 2.4. The force opposing the movement of the system boundary
is now that due to atmospheric pressure. Thus,

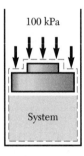

Figure 2.4 Work done by a
system comprising gas and
laden piston in the gas
expansion of Example 2.4.

(i) $W_{\text{system}} = +(100 \text{ kN/m}^2)(0.03 \text{ m}^2)(0.3 \text{ m})$

$\phantom{(i) W_{\text{system}}} = 0.9 \text{ kJ}$

(ii) $W_{\text{surroundings}} = -W_{\text{system}} = -0.9 \text{ kJ}$

Next consider a lump of a substance falling in a gravitational field through an evacuated space, as shown in Figure 2.5. The lump of substance is defined as the system. Whatever forces are acting on the system side of the boundary, there is zero force on the surroundings side since the environment is an evacuated space. Thus, there is no equivalent effect of raising or lowering a weight in an environment that does not offer any resistance to the movement of the system boundary through it. Therefore, the work done by the system is zero. In other words,

$$W_{\text{system}} = 0$$

As a general rule, zero work is done whenever the boundary movement is unresisted on either the system side or the surroundings side. Example 2.4, however, is a case having what is termed *fully resisted* boundary movement. The magnitude of the work done in such cases is easily determined as the scalar product of the boundary forces and the displacement of the system boundary. Further consideration of the work concept is deferred to Chapter 4.

Heat Interaction

Heat is also an interaction between a system and its surroundings. It is not a thermodynamic property, and it is not contained in a substance. Like work, heat is a boundary phenomenon. In other words, heat is an effect of a system on its surroundings (or vice versa) that occurs at the system boundary by virtue of a temperature gradient at the boundary or a difference in temperature between the system and its environment. If a system and its environment are at the same temperature, no heat interaction occurs between them. No heat effect occurs across the system boundary either if the temperature gradient adjacent to the boundary is zero. Such a condition can be achieved by placing an adiabatic wall on either the system side or the surroundings side of the

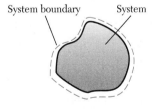

Figure 2.5 A lump of a substance (system) falling through an evacuated space.

boundary. Note that a wall such as an adiabatic wall with a finite thickness should not be regarded as a system boundary because a system boundary is merely a surface with zero thickness. If the system and its environment are a continuum, no discontinuity in temperature occurs at the boundary. In other words, a nonzero temperature gradient is normally required on both sides of the boundary for a heat interaction to take place.

Misgivings are often expressed about attempts to define heat prior to considerations of the first and second laws of thermodynamics. For example, in the classic exposition of thermodynamics provided by Caratheodory (Kestin 1976, 229), an effort was made to exclude consideration of heat altogether as a basic concept. From his treatment of the first law, one can infer heat as a derived quantity (Keenan and Shapiro 1947, 918), which is the sum of work done by the system and the increase in energy of the system.

Poincare, however, defines heat in a phenomenological sense without an explicit reference to energy transfer. Thus, he defines heat simply as "the effect of one system on another by virtue of inequality of temperature" (Keenan and Shapiro 1947, 917). This definition precedes the formal statement of the first law of thermodynamics. Another common definition of heat is "an interaction the sole effect of which external to either system could not have been the rise or fall of a weight" (Hatsopoulos and Keenan 1962, 197). This is equivalent to stating that heat is interaction that is not work.

The SI unit for heat is the joule, which is the same as that for work. The USCS unit for heat is the British thermal unit (Btu), which is different from that for work (see next section).

EXERCISES
for Section
2.3

1. Assume that the crane in Figure 2.1 is used to lift a machine weighing 25 kN through a vertical distance of 1.5 m. Determine the work done by the machine.
   Answer: -37.5 kJ

2. Suppose the crane in Figure 2.1 now *slowly* lowers a machine weighing 25 kN through a vertical distance of 1.5 m. Define the machine as the system. Determine
   (a) W_{system}
   (b) $W_{surroundings}$
   Answer: (a) 37.5 kJ

3. A lump of a substance falls through a vertical distance of 3.5 m while experiencing a drag force of 12 N due to the atmosphere. Define the lump as the system and determine
   (a) W_{system}
   (b) $W_{atmosphere}$
   Answer: (a) 42 J

4. The gas in Example 2.4 is cooled so that its pressure falls slightly below 150 kPa. As a result, the piston and the load on it descend a distance of 0.2 m. Determine
   (a) W_{gas}
   (b) $W_{atmosphere}$
   (c) $W_{piston+load}$.
   Answer: (a) -900 J, (b) 600 J, (c) 300 J

5. A resistor is supplied with electricity, and it rapidly attains a steady-state temperature such that the rate of heat transfer from the resistor to the surroundings is equal to the rate at which electrical energy is supplied to the resistor. If the electrical power supplied is 70 W and the resistor is the system, determine (in W)

(a) $\dot{W}_{system}$, the work output rate of the system

(b) $\dot{W}_{surroundings}$, the work output rate of the surroundings

(c) $\dot{Q}_{system}$, the heat transfer rate to the system

(d) $\dot{Q}_{surroundings}$, the heat transfer rate to the surroundings

Answer: **(a)** -70 W, **(c)** -70 W

2.4   Dimensions and Units

The term *dimension* is used to characterize any measurable physical quantity. Examples of dimensions include mass, length, time, temperature, and electrical current. The term *unit* refers to those arbitrary magnitudes and names assigned as a standard of measurement for the dimensions. Examples of units for the measurement of length include the foot, the mile, and the meter. Similarly, units for measurement of time include the second, the minute, the hour, and the day.

In engineering and scientific applications, mathematical equations are developed to represent certain physical systems. In such applications, each term in an equation must have the same dimensions so that the equation has physical meaning, and computations involving the equation must use the same units for each term. Thus, we must maintain consistency in the treatment of dimensions and units when making calculations concerning physical systems. Furthermore, numbers have no meaning when they are calculated to represent physical quantities unless the numbers are followed by the appropriate units that were used for the calculations. Lord Kelvin is reported to have said, "When you can measure what you are speaking about and express it in numbers, you know something about it; but when you cannot express it in numbers, your knowledge is of a meager and unsatisfactory kind." We might extend his statement to include the fact that unless you specify the units associated with your numbers, you still have not specified a meaningful physical measurement.

Systems of Units

At the present time, two major systems of units are employed in engineering and scientific work. The most common system of units used internationally is the *Systéme International d'Unites* or SI. This system has been adopted as the standard system by all major industrial countries of the world except the United States, and it is widely used in the United States in scientific and several other applications. The SI units evolved from the metric system (which is sometimes called the *absolute metric system*). The other system commonly used in the United States is called the English engineering system or the *U.S. Customary System of Units* or USCS. This system came from the basic English system that had evolved over several centuries. Many engineers in the United States still use this system. SI units are an international standard in most scientific and engineering communications, and the U.S. government passed laws (Metric Conversion Act of 1975 and the Omnibus Trade and Competitiveness Act of 1988) requiring U.S. government agencies to convert, where practical, to the use of the metric system by 1992. However, a working knowledge of USCS units is still necessary because these units are in widespread use in the United States in fields such as refrigeration and air conditioning.

A major weakness of the older and more traditional systems of units is the large number of units for which standards had to be established. In addition, conversions among these units were generally cumbersome. The basic English system, for example, uses several different units for length that are not easily convertible, such as the inch, the foot, the yard, and the mile. Areas are measured not only in terms of the square of length units (for example, the square inch), but also in such units as the acre. Typical units for volume include the gallon (gal) and the pint, in addition to the cube of length units. The establishment and use of these units no doubt arose from practical considerations—that is, they were generally of the appropriate order of magnitude for various measurements actually being made. However, the disadvantage is that one has to work with a long list of related units instead of just a few. Furthermore, the factors to be used for conversion from one unit to another are not easily remembered. For example, it is difficult to remember that 4840 yards^2 equals 1 acre. Thus, converting 6211 yards^2 to acres is a tedious arithmetical exercise without a calculator!

Today, instead of using unrelated units, the trend is to name units for just a few basic quantities or *primary dimensions*. Then other quantities or *secondary dimensions* are defined and measured with reference to the primary dimensions. The most fundamental of these primary dimensions are *length* (L), *mass* (M), and *time* (t). The SI unit for length is the meter (m). Area is a product of two lengths and is accordingly expressed in square meters (m^2). The units for the primary dimensions are referred to as *elementary* or *primary units*, while those for the secondary dimensions are termed *secondary* or *derived units*. The interrelationship between the primary and secondary units is established by making use of the appropriate physical law, such as Newton's second law of motion, which relates force and mass. The list of primary dimensions in any system of units tends to grow as specialized fields are encountered. For example, in thermodynamics, *temperature*, which is unrelated to any of the primary dimensions mentioned earlier, counts as an additional primary dimension. *Charge* is an additional primary dimension in electricity, and so on.

The primary difference between SI and USCS units involves the treatment of the dimensions of mass and force. In SI units, the primary dimensions are *mass* (M), *length* (L), *time* (t), *temperature* (T), *electrical current*, *luminous intensity*, and the *amount* of a substance. In addition to those primary dimensions used in SI, the USCS also uses *force* (F) as a primary dimension. An alternate system of units is the *British gravitational system*, which uses force but not mass as a primary dimension. This system is in widespread use in the aerospace industry in the United States.

A tabulation of the dimensions employed in all of these systems is given in Table 2.1. Tables 2.1 and 2.2 indicate the units used for the basic and derived quantities in the more common system of units. Since the most widely used systems for thermodynamic applications involve either SI units or USCS units, further discussion of units focuses on these two systems.

TABLE   2.1

Systems of Units and Dimensions

Dimension	SI Unit	Centimeter Gram System (cgs) Unit	British Gravitational Unit	USCS Unit
mass (M)	kilogram (kg)	gram (g)	slug°	pound mass (lbm)
length (L)	meter (m)	centimeter (cm)	foot (ft)	foot (ft)
time (t)	second (s)	second (s)	second (s)	second (s)
temperature (T)	kelvin (K)	kelvin (K)	degrees Rankine (°R)	degrees Rankine (°R)
electric current (I)	ampere (A)	ampere (A)	ampere (A)	ampere (A)
amount of substance	mole (mol)	mole (mol)	mole (lbmol)	mole (lbmol)
force (F)	newton° (N)	dyne°	pound force (lbf)	pound force (lbf)

°Called a secondary dimension, not a primary dimension, in this system; units are derived from the primary dimension units.

Force

Newton's second law of motion for a body whose mass is constant can be expressed as

$$F \propto ma \quad \text{or} \quad F = kma \tag{2.4}$$

where k is a constant of proportionality, F is force, m is mass, and a is acceleration. The simplest definition of force is obtained by setting $k = 1$ (without units). This choice

TABLE   2.2

Derived Quantities

Derived Quantity	SI Unit and Symbol	USCS Unit and Symbol
force	newton (1 N = 1 kg · m/s²)	pound force (lbf)°
energy	joule (1 J = 1 N · m)	foot-pound force (ft · lbf) or British thermal unit (Btu)
power	watt (1 W = 1 J/s)	British thermal unit/hour (Btu/h), British thermal unit/second (Btu/s), or horsepower (hp)
pressure	pascal (1 Pa = 1 N/m²) or bar (1 bar = 10^5 N/m²)	atmosphere (1 atm = 14.696 lbf/in.²), or pound-force per square inch (lbf/in.²)
electrical		
potential difference	volt (V)	volt (V)
resistance	ohm (Ω)	ohm (Ω)
capacitance	farad (F)	farad (F)
quantity of charge	coulomb (C)	coulomb (C)

°Primary (not derived) dimension in this system.

results in what is termed the *absolute* unit of force. For force in absolute units, therefore, Equation 2.4 can be rewritten as

$$F = ma \tag{2.5}$$

In SI units, only the absolute unit of force is defined.

The SI unit of mass is the kilogram (kg), that of length is the meter (m), and that of time is the second (s). Force in this system is a secondary or derived dimension. The derived unit of force is the *newton* (N), which is the force necessary to accelerate one kilogram at the rate of one meter per second per second, or

$$1 \text{ N} \equiv 1 \text{ kg} \cdot \text{m/s}^2$$

In the mks (meter/kilogram/second) version of the metric system, the unit for force is also the derived unit, newton. In the cgs (centimeter/gram/second) version, the unit for force is the derived unit, *dyne*, which is defined as the force necessary to accelerate one gram at the rate of one centimeter per second per second, or

$$1 \text{ dyne} \equiv 1 \text{ gm} \cdot \text{cm/s}^2$$

Neither the mks nor the cgs version of the metric system is in widespread use at this time because the earlier versions of the metric system were replaced by the SI units.

In the British gravitational system, force is a primary dimension and mass is a secondary or derived dimension. (The unit of force is an absolute unit in this system.) In this case, the unit of mass is the *slug*, which is defined as the mass (m) that has an acceleration (a) of one foot per second squared (1 ft/s^2) when acted upon by a force (F) of one pound (1 lbf). Thus, Newton's second law of motion (Eq. 2.5) can be rewritten for the unit of mass in this system:

$$m = \frac{F}{a}$$

$$\therefore 1 \text{ slug} = 1 \frac{\text{lbf}}{\text{ft/s}^2} = 1 \frac{\text{lbf} \cdot \text{s}^2}{\text{ft}}$$

In the USCS system, both force and mass are primary dimensions. Furthermore, the constant of proportionality in Equation 2.4 is not unity but $1/g_c$. A unit of force is termed *gravitational* if $k = 1/g_c$, where g_c has a numerical value equal to the acceleration due to gravity at standard conditions on the Earth's surface. The USCS unit of force is thus a gravitational unit. Newton's second law of motion for a body with constant mass can be written in this case as

$$F = \frac{ma}{g_c} \tag{2.6}$$

A force of one pound (1 lbf) is defined as the force necessary to accelerate a one-pound mass (1 lbm) at a rate equal to the standard acceleration of gravity at standard conditions on the Earth's surface, which is 32.174 ft/s^2. With this definition for force, a value for g_c is obtained by transposing Equation 2.6:

$$g_c = \frac{ma}{F} = \frac{(1 \text{ lbm})(32.174 \text{ ft/s}^2)}{1 \text{ lbf}}$$

$$\text{or} \quad g_c = 32.174 \; \frac{\text{ft} \cdot \text{lbm}}{\text{lbf} \cdot \text{s}^2}$$

The dimensions of g_c are

$$g_c = \frac{\text{mass} \times \text{length}}{\text{force} \times (\text{time})^2}$$

The constant g_c can be regarded basically as a factor for converting ma (in $\text{lbm} \cdot \text{ft/s}^2$) to the USCS force unit (lbf).

Weight and Mass

Newton's second law is used to relate mass and weight for a body in a gravitational field. We define the *mass* of a body in terms of the quantity of matter in the body. *Weight*, in contrast, is the force exerted on the body due to the local gravitational field. Thus, the mass of a body does not change when the body is moved from place to place, such as from the Earth to the Moon, while the weight of the body changes considerably according to the magnitude of the local acceleration due to gravity (g). In those systems using an absolute unit of force, Equation (2.5) for Newton's second law of motion can be used to obtain the following expression for the weight of a body:

$$\text{weight} = mg \tag{2.7a}$$

The USCS unit for force is gravitational, and therefore Equation 2.6 must be used for the expression of weight:

$$\text{weight} = \frac{mg}{g_c} \tag{2.7b}$$

The acceleration due to gravity varies with location even on the Earth's surface. On Earth, g is lower at higher elevations than it is at lower elevations. The standard sea level value assumed for g on Earth is

$$g = 32.174 \; \text{ft/s}^2 \; (\text{USCS})$$

$$= 9.807 \; \text{m/s}^2 \; (\text{SI})$$

The variation of weight with location in the Earth's gravitational field is thus directly related to the variation in g since the mass (m) is invariant for a particular body.

EXAMPLE 2.5

The acceleration due to gravity on the Moon's surface is approximately one sixth that for standard conditions on the Earth's surface. If an individual has a mass (m) of 180 lbm, determine the following:

(a) the individual's mass on the Moon

(b) the individual's weight on the Earth at standard conditions

(c) the individual's weight on the Moon

Given:
$$m = 180 \text{ lbm}$$
$$g_{\text{Moon}} = 1/6 \; g_{\text{Earth}}, \; g_{\text{Earth}} = 32.174 \text{ ft/s}^2$$

Find: **(a)** m on Moon, **(b)** weight on Earth, **(c)** weight on Moon

Solution:

(a) The individual's mass is the same everywhere.

$$\therefore m = \underline{180 \text{ lbm}} \text{ on the Moon}$$

(b) In USCS units, Equation 2.7b for weight is the appropriate expression to use:

$$\text{weight} = \frac{mg_{\text{Earth}}}{g_c} = \frac{(180 \text{ lbm})(32.174 \text{ ft/s}^2)}{32.174(\text{ft} \cdot \text{lbm/lbf} \cdot \text{s}^2)}$$

$$= \underline{180 \text{ lbf}} \text{ (on the Earth)}$$

(c) Equation 2.7b can also be used for the weight of the individual on the Moon:

$$\text{weight} = \frac{mg_{\text{Moon}}}{g_c} = \frac{(180 \text{ lbm})\left(\dfrac{32.174}{6} \text{ ft/s}^2\right)}{32.174(\text{ft} \cdot \text{lbm/lbf} \cdot \text{s}^2)}$$

$$= \underline{30 \text{ lbf}} \text{ (on the Moon)}$$

Specific Weight and Specific Gravity

When dealing with a fluid, the term *specific weight* (γ) is sometimes used. This term is defined as weight per unit volume, or

$$\gamma \equiv \frac{\text{weight}}{\text{volume}} = \frac{mg}{V} = \rho g \text{ N/m}^3 \text{ (SI unit)} \qquad (2.8a)$$

$$\gamma \equiv \frac{\text{weight}}{\text{volume}} = \frac{mg/g_c}{V} = \rho \frac{g}{g_c} \text{ lbf/ft}^3 \text{ (USCS unit)} \qquad (2.8b)$$

Another term used with fluids is *specific gravity* (SG), which is defined as the specific weight of the substance divided by the specific weight of liquid water at a standard condition of 1 atm and 4°C:

$$SG \equiv \frac{\gamma}{\gamma_{\text{water}}} = \frac{\rho}{\rho_{\text{water}}} \qquad (2.9)$$

The ratio of specific weights is equal to the ratio of densities for the substance and liquid water. The specific gravity is a dimensionless quantity, and it should be noted that Equation 2.9 applies to both SI units and USCS units as long as the specific weights (or densities) are expressed in the same units.

Energy

Energy in thermodynamics is a derived property related to both work and heat through the first law of thermodynamics. The SI unit of energy is the same as that for work and for heat and is the joule (J), which was defined earlier as the work done when a force of 1 N acts through a displacement of 1 m (in the direction of the force). In other words,

$$1 \text{ joule (J)} \equiv 1 \text{ N} \cdot \text{m} \tag{2.10}$$

There are two separate USCS units for energy. Mechanical forms of energy such as kinetic and potential energy are often measured in the unit for work, which is the ft · lbf. Other forms of energy that were traditionally associated with heating effects are measured in the *British thermal unit*, or Btu. A Btu was initially defined as the amount of energy required to raise the temperature of 1 lbm of water from 59.5°F to 60.5°F. The relationship between the Btu and the ft · lbf has been established from experiments and is given as

$$1 \text{ Btu} = 778.16 \text{ ft} \cdot \text{lbf} \tag{2.11}$$

In working with equations that involve the USCS units for energy and other quantities such as work and heat, special care should be exercised to ensure that the units for the various terms in the equations are consistent.

Power

Power is also a derived quantity and is defined as energy flow per unit time. Most frequently, power refers to *mechanical power*, which is defined as the time rate of doing work, or the work done per unit time. The SI unit of power is the *watt* (W), which is defined as the flow of one joule of energy in one second, or

$$1 \text{ W} \equiv 1 \text{ J/S} \tag{2.12}$$

The USCS units for mechanical power include ft · lbf/s and horsepower (hp). The relationship between these two units is as follows:

$$1 \text{ hp} \equiv 550 \text{ ft} \cdot \text{lbf/s} \tag{2.13}$$

Horsepower is a unit first adopted by James Watt, and it corresponds approximately to the power exerted by a horse in pulling a load. USCS units also exist for defining heating (or cooling) power. These include the Btu/h (sometimes written as Btuh) and the Btu/s, which represent rates of heating or cooling per unit time. The relationship between horsepower and Btu/h or Btu/s is obtained by using Equations 2.11 and 2.13:

$$1 \text{ hp} = 550 \left(\frac{\text{ft} \cdot \text{lbf}}{\text{s}} \right) \left(\frac{1 \text{ Btu}}{778.16 \text{ ft} \cdot \text{lbf}} \right) = 0.707 \text{ Btu/s}$$

$$= 0.707 \left(\frac{\text{Btu}}{\text{s}} \right) \left(\frac{3600 \text{ s}}{1 \text{ h}} \right) = 2545 \text{ Btu/h} \tag{2.14}$$

$$\therefore 1 \text{ hp} = 2545 \text{ Btu/h} = 0.707 \text{ Btu/s}$$

Note that some students get confused about the use of Btuh for energy flow rate, whereas the kWh (kilowatt-hour) in SI refers strictly to energy. The kWh is the energy flow in 1 h corresponding to a power of 1 kW. Likewise, the Wh (watt-hour) is the energy equivalent of a flow of 1 W over a period of 1 h. Btuh, in contrast, is a misnomer and, as indicated earlier, is sometimes used simply as shorthand for Btu/h, which measures energy flow per unit time.

Temperature

Temperature is a primary dimension in all systems of units; however, different units are used in different systems. In SI units, the primary unit of temperature is *kelvin* (K), which is used for absolute temperatures. The *absolute zero temperature* is the lowest temperature that is theoretically attainable and is the temperature at which all macroscopic motion ceases. The Kelvin scale starts at zero at the absolute zero temperature. A derived or secondary SI unit for temperature is *degrees Celsius* (°C), which is related to the Kelvin scale as follows:

$$T\ (°C) = T\ (K) - 273.15 \qquad (2.15)$$

A temperature difference or interval can be expressed in either degrees Celsius or kelvin. A 1°C interval is the same as a 1 K interval:

$$1°C\ (\text{interval}) = 1\ K\ (\text{interval}) \qquad (2.16)$$

Note that the nomenclature here uses K without the ° (degree) symbol in front of it; this approach is consistent with standard SI conventions. Degrees Celsius is sometimes also called degrees Centigrade.

The Celsius numbering system was originally developed by setting a scale between the freezing and boiling points of water at 1 atmosphere (1 atm). A value of zero on the scale was assigned as the freezing point of water (at 1 atm) and a value of 100 was specified for the boiling point of water (also at 1 atm). The interval between the two fixed points was divided into 100 equal increments, with each increment equal to 1°C. The Kelvin numbering scale was defined later with the same increment for one degree, but with zero being set at the absolute zero temperature point. It was found from experimental observation that the absolute zero temperature point is 273.15°C below the freezing point of water at 1 atm (0°C).

Similar systems of units for temperature evolved for USCS units. The *Fahrenheit* (°F) system is defined by setting 32°F as the freezing point of water (at 1 atm), 212°F as the boiling point of water (also at 1 atm), and 180 equal increments or degrees between these two points. The *degrees Rankine* scale, which corresponds to the Kelvin (absolute temperature) scale, was subsequently defined with the same increment for one degree as the Fahrenheit system, but with zero being set at the absolute zero temperature point. Thus, the relationship between the two systems is

$$T\ (°F) = T\ (°R) - 459.67$$

$$1°F\ (\text{interval}) = 1°R\ (\text{interval}) \qquad (2.17)$$

Figure 2.6 shows the relationships among the various temperature scales.

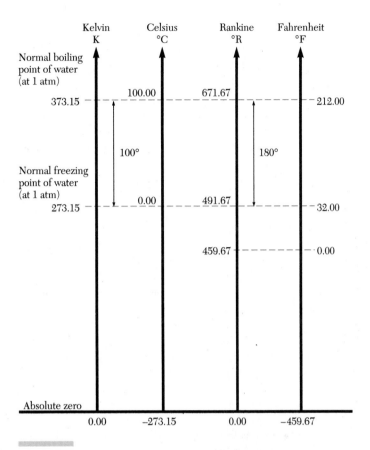

Figure 2.6 A graphical representation of the relationships among different temperature numbering systems.

The interval between the freezing and boiling points of water (at 1 atm) on the Kelvin and Rankine scales is 100 K and 180°R, respectively. Hence, the following equivalence can be inferred between the two scales:

$$1 \text{ K (interval)} = \frac{180}{100} \text{ }^\circ\text{R} = 1.8 \text{ }^\circ\text{R (interval)} \tag{2.18}$$

From Equations 2.16, 2.17, and 2.18, alternate expressions can be derived for the relationships among the different temperature numbering systems. For example,

$$T \text{ (}^\circ\text{C)} = \frac{5}{9} [T \text{ (}^\circ\text{F)} - 32] \tag{2.19}$$

$$T \text{ (}^\circ\text{F)} = \frac{9}{5} T \text{ (}^\circ\text{C)} + 32 \tag{2.20}$$

In general, calculations involving thermodynamic temperatures should use the absolute temperature K or °R unless specified otherwise.

Amount of Substance

The mole is the primary unit for the amount of substance. The *mole* (abbreviated as mol) is formally defined as the amount of substance containing the same number of elementary units (atoms or molecules) as there are atoms in 0.012 kg (or 12 g) of carbon-12. In other words, the mole is the quantity of a substance numerically equal in mass to its gram-molecular weight or molar mass. The *molecular weight* of a substance is the ratio of the mass of one molecule of the substance to that of one-twelfth of an atom of carbon-12. The gram-molecular weight of hydrogen (H_2), for example, is 2 g/mol; while that for nitrogen (N_2) is 28 g/mol. The number of molecules in 1 mol of any substance is 6.022×10^{23} (to four significant figures). This number is called *Avogadro's number*.

The terms "molar mass" and "molecular weight" are sometimes defined and used differently in the literature. The New Encyclopaedia Britannica (1985, 267), for example, defines molecular weight as "the weight in grams of one mole of the substance" and expresses molecular weights in g/mol. This is identical to the definition of molar mass given by Kotz and Purcell (1991, 84) who further stated that molar mass and molecular weight are numerically equal but are expressed in different units, g/mol and atomic mass unit (amu), respectively. (One atomic mass unit equals $\frac{1}{12}$ the mass of a carbon atom with six protons and six neutrons in the nucleus.) A third view (see Waddington 1987, 276) is that "relative molecular mass" replaces the concept of molecular weight. The relative molecular mass is dimensionless and is "the ratio of the average mass per formula unit of the natural nuclidic composition of a substance to $\frac{1}{12}$ of the mass of an atom of nuclide ^{12}C." In this text, "molar mass" and "molecular weight" will be used interchangeably and will be expressed in SI as kg/kmol (or kg/kgmol). The symbol M will be used for the molecular weight (or molar mass) of a substance.

The mole is frequently used as a measure of the amount of substance when dealing with chemical substances and chemical reactions. In chemistry and chemical thermodynamics, SI is the preferred system of units. Thus, the *kilogram-mole*, (kgmol or kmol) can be defined as the SI unit for the amount of substance having a mass numerically equal to the kilogram molecular weight of the substance. The mathematical relationship between the number of moles N (in kmol), the mass m (in kg), and the molecular weight M (in kg/kmol) is thus

$$N = \frac{m}{M} \qquad (2.21)$$

The molecular weight of water (H_2O) is 18 kg/kmol; hence, 1 kmol of water has a mass of 18 kg. Likewise, 1 kmol of carbon dioxide (CO_2) contains 44 kg of the substance. The lbmole (or lbmol) is the USCS unit for the amount of substance. For example, 1 lbmol of H_2 contains 2 lbm of hydrogen.

Conversion of Units

Regardless of whether you work with only one or with more than one system of units, it is often necessary in thermodynamics to convert from one unit to another. For example, a typical expression for the total energy of a system might involve values for the internal energy in Btu, while those for kinetic and potential energy might be in ft · lbf. Both the Btu and the ft · lbf are USCS units. In such a situation, Equation 2.11 must be used for the conversion of one of the units to the other so that all terms in the expression for the total energy are given in consistent units. Likewise, the evaluation of the total energy of a system in SI units might involve an internal energy term measured in kJ, while the kinetic and potential energy terms are measured in J. For consistency of units, the values for the kinetic and potential energy must be divided by 1000 so as to give the total energy in kJ, or else the internal energy term must be multiplied by 1000 to give the total energy in J. In addition to conversions within a given system of units, conversions from one system to another are also required.

Tables 2.1 and 2.2, introduced earlier, provide a summary of the primary and secondary units employed in both the SI and USCS systems. Conversions among the SI units are simple because the units are related by a factor of 1000. Table 2.3 gives a list of commonly used metric multipliers and the prefixes employed. Note that the SI units are strictly those in Table 2.3 that increase (or decrease) by a factor of 10^3. Table 2.4 gives factors for conversion between SI and USCS units.

A systematic technique for conversion from one unit to another is outlined below. The method uses *unity brackets* or *unity conversion factors*. The underlying principles of the method may be summarized as follows:

1. Any physical quantity can be throught of as product of a number and the unit name.
2. Unit names can be treated as if they were algebraic quantities.
3. Relationships between units of the same kind can be expressed by unity brackets or unity conversion factors. In other words, the ratio of equivalent units is numeri-

TABLE 2.3

SI Multipliers and Prefixes		
Multiplier	**Symbol**	**Prefix**
10^{18}	E	eta
10^{15}	P	peta
10^{12}	T	tera
10^{9}	G	giga
10^{6}	M	mega
10^{3}	k	kilo
10^{2}	h	hecto
10^{1}	da	deka
10^{-1}	d	deci
10^{-2}	c	centi
10^{-3}	m	milli
10^{-6}	μ	micro
10^{-9}	n	nano
10^{-12}	p	pico
10^{-15}	f	femto
10^{-18}	a	atto

TABLE 2.4

Conversion Factors

Physical Quantity	Conversion Factor
length	1 ft = 0.3048 m 1 in. = 2.54 cm 1 mi = 1.6093 km
velocity	1 ft/s = 0.3048 m/s 1 mi/h = 0.44703 m/s
area	1 ft^2 = 0.0929 m^2 1 in.^{2} = 6.452 × 10^{-4} m^2
mass	1 lbm = 0.4536 kg = $\dfrac{1}{2.205}$ kg 1 slug = 14.594 kg
mass flow rate	1 lbm/h = 0.000126 kg/s 1 lbm/s = 0.4536 kg/s
volume	1 ft^3 = 7.48052 gal = 0.02832 m^3 = 28.32 l (liter) 1 in.^{3} = 1.6387 × 10^{-5} m^3
volumetric flow rate	1 ft^3/min = 0.000472 m^3/s
density	1 lbm/ft^3 = 16.018 kg/m^3
force	1 lbf = 4.448 N
pressure	1 lbf/in.^{2} = 6894.8 Pa (N/m^2) = $\dfrac{1}{14.50}$ bar 1 lbf/ft^2 = 47.88 Pa (N/m^2) 1 atm = 14.696 lbf/in.^{2} = 101,325 Pa (N/m^2)
temperature	1°R = $\dfrac{1}{1.8}$ K
heat, energy, or work	1 Btu = 1055.1 J = 778.16 ft · lbf = 2.9307 × 10^{-4} kWh 1 cal = 4.1868 J 1 ft · lbf = 1.3558 J 1 hp · h = 2.685 × 10^6 J = 0.7457 kWh
specific internal energy or enthalpy	1 Btu/lbm = 2326.0 J/kg 1 cal/g = 4186.8 J/kg
specific heat	1 Btu/lbm · °F = 4186.8 J/kg · K
heat flow rate	1 Btu/h = 0.2931 W 1 Btu/s = 1055.1 W
power	1 hp = 550 ft · lbf/s = 745.7 W 1 hp = 2544 Btu/h 1 hp = 0.707 Btu/s 1 ft · lbf/s = 1.3558 W 1 Btu/h = 0.2931 W
viscosity, dynamic	1 lbm/ft · s = 1.488 N · s/m^2
viscosity, kinematic	1 ft^2/s = 0.09029 m^2/s 1 ft^2/h = 2.581 × 10^{-5} m^2/s

cally one and is also dimensionless. Thus, multiplying any physical quantity by a unity conversion factor leaves the magnitude and dimensions of the quantity unaltered but makes the desired conversion of units possible based on Principles 1 and 2.

The following examples illustrate the use of the unity conversion factor technique.

EXAMPLE 2.6

Consider an equation for a length, z, that is the sum of two other lengths, x and y. Find z (in inches) given $x = 2$ ft and $y = 3$ in.

Given: $z = x + y$, $x = 2$ ft, $y = 3$ in.

Find: z (in inches)

Solution: The choice of a unity conversion factor is dictated by the need to convert x given in ft to in. Use the equivalence 1 ft = 12 in.

$$z = 2 \text{ ft} \left(\frac{12 \text{ in.}}{1 \text{ ft}} \right) + 3 \text{ in.}$$

$$= 24 \text{ in.} + 3 \text{ in.} = \underline{27 \text{ in.}}$$

EXAMPLE 2.7

A force of 2 N acts on a body to produce an acceleration of 10 cm/s². What is the mass (in kg) of the body?

Given: $F = 2$ N, $a = 10$ cm/s² for a body

Find: m (in kg)

Solution: Newton's second law of motion gives the expression for m:

$$m = \frac{F}{a} = \frac{2 \text{ N}}{10 \text{ cm/s}^2}$$

$$= \frac{2}{10} \frac{\text{N} \cdot \text{s}^2}{\text{cm}} \left(\frac{1 \text{ kg} \cdot \text{m/s}^2}{\text{N}} \right) \left(\frac{100 \text{ cm}}{1 \text{ m}} \right)$$

$$= \underline{20 \text{ kg}}$$

The choice of numerator and denominator in each unity conversion factor is dictated by the units to be eliminated in the expression. Each unity factor is based on the appropriate relationship between units that are equivalent in magnitude and dimension.

EXAMPLE 2.8

The *specific heat* (c) of water (defined as the heat needed to raise the temperature of a unit mass of water by a unit amount) is 1 Btu/(lbm · °R). Convert this to J/(kg · K) using the equivalences for energy and the primary dimensions only.

Given: $c = 1$ Btu/(lbm · °R)(for water)

Find: c[in J/(kg · K)]

Solution: The unity conversion factors are obtained using the equivalences for the primary dimensions given in Table 2.4 and Equation 2.18:

$$c = 1 \frac{\text{Btu}}{\text{lbm} \cdot {}^\circ\text{R}}$$

$$= 1 \frac{\text{Btu}}{\text{lbm} \cdot {}^\circ\text{R}} \left(\frac{1055.1 \text{ J}}{1 \text{ Btu}} \right) \left(\frac{1 \text{ lbm}}{0.4536 \text{ kg}} \right) \left(\frac{1.8 {}^\circ\text{R}}{1 \text{ K}} \right)$$

$$= \underline{4187 \text{ J/kg} \cdot \text{K}}$$

In all engineering calculations, and particularly in thermodynamics, it is good practice to indicate the units alongside each term in an expression. The appropriate unit conversion factor should then be used to reduce all the terms to the same set of units. Furthermore, the units for the final answer obtained must be specified if the result is to have physical meaning.

EXERCISES

for Section 2.4

1. Assume that g at an altitude equal to the radius of the Earth is ¼ that of g on the Earth's surface. If an individual has a mass of 220 lbm, determine the individual's weight (in lbf)
   (a) on the Earth's surface
   (b) at the altitude equal to the radius of the Earth

   Answer: **(b)** 55 lbf

2. The density of water is 62.3 lbm/ft³ at a temperature of 60°F. (Assume $g_{\text{moon}} = 5.18$ ft/s².)
   (a) What is the specific weight (in lbf/ft³) of water on the lunar surface?
   (b) What is the specific gravity of water?

   Answer: **(a)** 10 lbf/ft³

3. The specific gravity of mercury is 13.6. If the density of water is 1000 kg/m³, find the specific weight of mercury at standard conditions on the Earth's surface. (Assume $g = 9.81$ m/s².)

   Answer: 133 kN/m³

4. Convert a weight of 220 lbf to N.

   Answer: 979 N

5. Convert a pressure of 25 psia to kPa (abs).

   Answer: 172 kPa (abs)

6. Convert 6000 Btu/h to kW.

   Answer: 1.76 kW

7. Convert $-40°C$ to °F.

   Answer: $-40°F$

2.5 Summary

This chapter has provided an elaboration of the macroscopic perspective, as well as several other concepts introduced in Chapter 1. The macroscopic perspective assumes that all the systems to be considered can be regarded as a continuum. Thermodynamic properties can be classified as extensive or intensive properties, and several examples of each class were given. Extensive properties, such as mass and volume, depend on the amount of the substance present. Intensive properties, such as pressure and temperature, do not depend on the amount of mass present. Certain other properties such as velocity and elevation must be referenced to a physical coordinate system, while thermodynamic properties do not as a rule require such a frame of reference. The equality of temperature was introduced as a derived concept from the zeroth law of thermodynamics. The zeroth law also provides the basis for empirical thermometry. Work and heat were formally defined. Neither work nor heat is a property of a system; rather, they are forms of interaction between a system and its surroundings. The various concepts and definitions presented in this chapter provide the necessary foundation for subsequent discussions of the laws of thermodynamics and their applications to the analyses of thermodynamic processes.

Dimensions and units for the various quantities to be used throughout this text were introduced in this chapter. While nearly all of the major industrial countries have adopted SI units, both SI and USCS units are still widely used in the United States. In view of this, both systems were discussed here and both will be employed throughout. The need to explicitly specify units and make all mathematical calculations with a consistent set of units was emphasized. The unity conversion factor was defined, and an approach was presented for using this factor to ensure a consistent set of units for all calculations.

References

Caratheodory, C. 1909. "Investigation into the Foundations of Thermodynamics." In Kestin, J. ed. 1976. *Benchmark Papers on Energy (Vol. 5): The Second Law of Thermodynamics*. Stroudsburg, PA: Dowden, Hutchinson & Ross.

Hatsopoulos, G. N. and Keenan, J. H. 1962. "A Single Axiom for Classical Thermodynamics." *Transactions ASME: Journal of Applied Mechanics*. Vol. 29. 193–199.

Keenan, J. H. and Shapiro, A. H. 1947. "History and Exposition of the Laws of Thermodynamics." *Mechanical Engineering*. Vol. 69. 915–921.

Kotz, J. C. and Purcell, K. F. 1991. *Chemistry and Chemical Reactivity*. 2nd ed. Philadelphia: Saunders College Publishing.

The New Encyclopaedia Britannica: Macropaedia. 15th ed. 1985. Vol. 24. Chicago: Encyclopaedia Britannica, Inc.

Waddington, T. C. 1987. In *McGraw-Hill Encyclopedia of Science and Technology*. 6th ed. Vol. 15. 276. New York: McGraw-Hill Book Company.

Questions

1. Define heat and work, and explain why neither is a property.

2. Distinguish carefully between extensive and intensive properties and give four examples of each type.

3. State whether the work done by the specified system (in italics) is positive, negative, or zero in the following examples. Give reasons for your choices.
   (a) A *mass* falls in a vacuum in a gravitational field.
   (b) A *man* walks up a rigid staircase at a constant speed (assume negligible air resistance).
   (c) A *box* is raised by an *agent* from a low level to a higher level in the Earth's gravitational field. Consider the *box* and *agent* separately as systems.

4. A heavy block of copper, which is a good thermal conductor, is dragged along a rough wooden sheet. (Wood is a poor heat conductor.) It is observed that the copper block gets hotter, while the sheet of wood remains at its original temperature. Defining the copper block as a system, indicate whether the heat interaction occurring at the copper–wood interface is positive, zero, or negative. Give reasons for your answer.

5. List the primary (or basic) quantities in the SI units and state the units of each.

6. List the primary quantities in the USCS units and state the units of each.

7. Distinguish between primary and secondary units and illustrate the difference with reference to the SI units for force and mass.

8. Derive from basic principles the unit for g_c in the USCS units.

9. If two objects having differing masses are dropped from the same height in the Earth's gravitational field, which of them will reach the ground first? Why? (See Fig. Q2.9.)

10. An object is weighed at the surfaces of the Earth and the Moon. At which of the surfaces will the object appear heavier and why?

11. Figure Q2.11 represents a simple barometer that comprises a column of liquid in an inverted tube. The space above the liquid column in the tube is a vacuum. Beginning with the definition of pressure and Newton's second law of motion, show that the

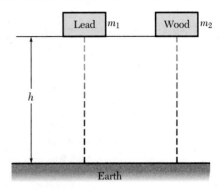

Figure Q2.9 Two masses dropped from same height.

pressure at the base of the liquid column of height z (in ft) is

$$p \ (\text{psia}) = \frac{1}{144} \frac{\rho g}{g_c} z$$

12. A particle of mass m is initially at rest at a height z above ground level. Show that in the free fall of the particle, the sum of the potential and kinetic energies of the particle remains constant.

13. In a chemical reaction, which of these are necessarily conserved?

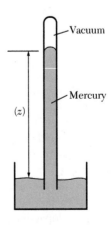

Figure Q2.11 A simple barometer.

(a) mass
(b) number of moles
(c) number of molecules
(d) the atomic species

For those commodities that you believe are not necessarily conserved, give an example that shows such to be the case.

14. Give one example of a process in which work produces a useful heating effect. Give an example in which an undesirable heating effect is a by-product of a work process.

15. Give an example of a device that produces work from a heat input. With the aid of suitable sketches, describe how the device works.

Problems

2.1 Standard atmospheric pressure at sea level is 14.7 lbf/in.^{2} (psia). Convert this to the following:
(a) kPa
(b) bar
(c) in. Hg (mercury) (assume $\rho_{mercury}$ = 850 lbm/ft^3)
(d) ft of water (assume ρ_{water} = 62.4 lbm/ft^3)

2.2 Briefly explain why you cannot use a mercury barometer to measure pressure in a gravity free environment. Also, if a mercury (Hg) barometer reads 30.1 in., express the reading in the following units. (Assume g = 32.2 ft/s^2, and the density of Hg, ρ, is 850 lbm/ft^3.)
(a) lbf/in.^{2}
(b) bar
(c) mm Hg

2.3 What is the total force on a horizontal surface of area 1 hectare (ha) due to an atmospheric pressure of 100 kPa? (Assume 1 ha = 100 m × 100 m.)

2.4 The atmospheric pressure on a certain day is 100 kPa. What is the equivalence of this pressure in terms of a vertical column of mercury in inches (in. Hg)?

2.5 Figure P2.5 illustrates the principal features of a Bourdon gage. The gage is designed to read pressures above the prevailing ambient pressure, in other words, the gage pressure. The pressure of an inflated tire reads 35 psig on a Bourdon gage when the atmospheric pressure is 1 atm. Determine the pressure gage reading for the tire pressure in the atmosphere-free environment of the Moon's surface.

2.6 Figure P2.6 depicts a typical U-tube manometer used for the measurement of pressures relative to atmospheric pressure. A manometer uses water and registers 6 in. of water when an unknown gas

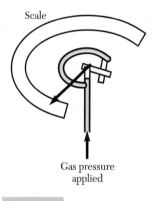

Scale

Gas pressure applied

Figure P2.5 A Bourdon gauge.

pressure is applied on the left side of the U-tube. The atmospheric pressure is 14.7 psia. Determine the applied gas pressure (in psia).

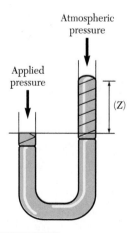

Atmospheric pressure

Applied pressure

(Z)

Figure P2.6 A typical U-tube manometer.

2.7 Express in the following units a difference in level of 9 in. in a manometer using kerosene. (Assume kerosene has a density of 50 lbm/ft^3 and that $g = 32.2$ ft/s^2.)
(a) psi
(b) Pa

2.8 Convert a pressure difference of 1 in. of water to psi.

2.9 A man of mass 75 kg sits on a stool with his entire weight supported over a horizontal surface of 200 mm × 200 mm. What is the average pressure (in Pa) exerted over the horizontal surface?

2.10 The density of water is 1000 kg/m^3. Find the volume occupied by 1 kg of water in
(a) m^3
(b) ft^3

2.11 The density of air at typical atmospheric conditions is about 1 kg/m^3. Compute the volume occupied by 2.5 kg of air in
(a) m^3
(b) ft^3

2.12 The specific gravity of mercury is 13.6. Using a value of 62.3 lbm/ft^3 for the density of water, find the density of mercury in
(a) lbm/ft^3
(b) kg/m^3

2.13 0.5 kg of dry saturated steam, with a specific volume of 1.673 m^3/kg, is contained in a rigid vessel. What is the volume of the vessel?

* **2.14** A rigid vessel of volume 0.1 m^3 contains 1 kg of wet steam, as shown in Figure P2.14. The specific volumes of the liquid and vapor phases are 1.127 × 10^{-3} m^3/kg and 0.1943 m^3/kg, respectively.

Figure P2.14 Wet steam in a rigid vessel.

(a) What is the volume occupied by the vapor phase alone?
(b) What is the fraction, by mass, of liquid water in the system?

2.15 The specific volume of a gas in a container is 0.1 ft^3/lbm, and the volume of the container is 2 ft^3. Find the mass of the gas inside the container.

2.16 3 kg of air is held in a container whose internal dimensions are 0.1 m × 0.2 m × 0.3 m. Find the density of the air.

2.17 An automobile having a mass of 3000 lbm is accelerated horizontally at a constant rate of 5 ft/s^2. What force is required to produce this acceleration?

2.18 A force of 10 N is used to slide a mass of 0.5 kg along a frictionless horizontal surface, as shown in Figure P2.18. What is the acceleration of the mass?

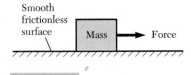

Figure P2.18 Force applied to slide mass along a smooth surface.

2.19 A force of 10 N acts on a body to produce an acceleration of 15 cm/s^2 along a smooth horizontal surface. What is the mass of the body?

2.20 Given that $g_c = 32.174$ ft · lbm/lbf · s^2, estimate the weight in lbf of a 1-ton (short ton) object on the Moon, assuming that the lunar acceleration due to gravity is 5.18 ft/s^2 (1 short ton = 2000 lbm). What is the mass of the object on the surface of the Moon?

2.21 An astronaut weighs 120 lbf on Earth at standard conditions. What would she weigh on the surface of the Moon?

* **2.22** 1 ft^3 of a certain liquid weighs 8 lbf on the surface of the Moon. What is the specific gravity of the liquid?

2.23 What is the density of the liquid in Problem 2.22?

2.24 Newton's law of gravitation can be written as

$$F = G \frac{Mm}{R^2}$$

where $G = 6.67 \times 10^{-11}$ m³/kg · s² is the *gravitational constant*, M and m are the respective masses (in kg) of the two bodies, and R is the distance (in m) between the centers of the two bodies.

(a) The mean radius of the Earth at the equator can be taken as 6.38×10^6 m. Assuming that $g = 9.80$ m/s² on the Earth's surface, estimate the mass of the Earth (in kg).

(b) The mass and mean radius of the Moon are 7.36×10^{22} kg and 1.74×10^6 m, respectively. Estimate the value of g (in ft/s²) at the surface of the Moon.

* **2.25** An astronaut returning from Mars ($g = 12.4$ ft/s²) brings back a set of spring scales that had been calibrated to read mass (in lbm) correctly on the martian surface. The astronaut uses the scales to weigh her new born baby on Earth ($g = 32$ ft/s²) and reads 20.6. What is the mass of the baby in lbm?

* **2.26** Compute the acceleration of gravity on top of a mountain in Colorado that is 5000 ft above sea level.

2.27 Figure P2.27 illustrates a liquid-in-glass thermometer. Calculate the volume at 0°C required in a thermometer to give a 1°C interval a length of 1.5 mm on the stem, the diameter of the bore being 0.24 mm. What would be the volume of this mercury at 100°C? (Assume that the linear expan-

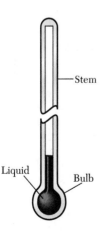

Figure P2.27 A liquid-in-glass thermometer.

sivity of the glass is negligible and that the absolute volumetric expansivity of mercury is 1.81×10^{-4} per K.)

2.28 The temperature T (in °C) on a thermometric scale is defined in terms of a property x by the relationship

$$T = a \log_{10} x + b$$

where a and b are constants. At the ice point, $x = 2.3$, and at the steam point, $x = 6.8$. Evaluate the temperature when $x = 3$.

2.29 Figure P2.29 depicts a chromel–alumel thermocouple. The hot junction is connected via a reference junction to a potentiometer. The reference junction is at 20°C while the potentiometer gives a reading of 3 mV (millivolts). What is the temperature of the hot junction? Assume that

$$T = T_{ref} + 1.2 + 24.1\ e \quad (°C)$$

where T_{ref} is the temperature of the reference junction (in °C) and e is the electromotive force (emf) across the thermocouple junctions (in mV).

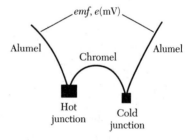

Figure P2.29 A chromel–alumel thermocouple.

2.30 A constant volume gas thermometer has a container filled with nitrogen gas (assumed ideal), as illustrated in Figure P2.30. A pressure gage attached to the container indicates an absolute pressure of 101.4 kPa when the container is at 24°C. The container is then immersed in a hot fluid, and the pressure rises to 135.2 kPa. What is the temperature of the fluid? (For an ideal gas at constant volume, the pressure is directly proportional to the absolute temperature.)

2.31 The resistances of a resistance thermometer measured at 0°C and 100°C were 1.0 and 1.39 ohms,

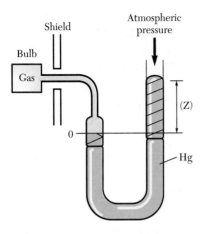

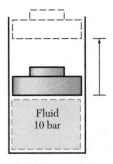

Figure P2.35 Fluid expansion in a frictionless piston and cylinder arrangement.

Figure P2.30 A constant volume gas thermometer. The volume of gas is kept constant by raising or lowering the column of mercury so that plane 0 remains fixed.

respectively. What is the temperature when a resistance of 3.2 ohms is measured if the following are assumed?

(a) a linear scale of temperature

(b) an empirical absolute temperature scale of $\theta = 273.16 \, R/R_t$ K, where R_t is the resistance at 0.01°C, the triple point temperature of water

Which of the temperatures is correct?

2.32 Find the work done by an individual in *slowly* raising 80 lbm of gasoline a vertical distance of 10 ft. Give results in units of ft · lbf.

2.33 Find the work done by an individual in *slowly* lowering 10 kg of water a vertical distance of 8 m.

2.34 An individual slowly raises a body whose mass is 5 kg through a vertical distance of 10 m in a gravitational field for which $g = 9.81$ m/s². Treating the body as a system and assuming that air resistance is negligible, estimate the work done *on* the system. What is the work done *by* the system?

* 2.35 A certain fluid at 10 bar is contained in a cylinder behind a frictionless piston, as illustrated in Figure P2.35. The initial volume is 0.05 m³. Calculate the work done by the fluid when it expands, quasi-statically, at constant pressure to a final volume of 0.08 m³.

* 2.36 Evaluate the work done by the system (in italics) in the following processes:

(a) A *body* of mass 10 kg falls freely through a vertical distance of 30 m in a gravitational field for which $g = 9.81$ m/s². The drag force (full resistance) of the *atmosphere* on the body is 4 N.

(b) A *stirrer* is driven for 10 min by an electric motor exerting a torque of 0.006 N · m and rotating at 900 rpm (revolutions per minute). The system in this instance is the stirrer and the fluid stirred.

(c) A 2-kW *electric kettle* (inclusive of the heating element and water) is turned on for 5 min.

2.37 The Btu is defined as the heat required for raising the temperature of 1 lbm of water 1°F. Determine the heat required for raising the temperature of 5 lbm of water from 25°C to 90°C at standard atmospheric pressure.

2.38 The work done in moving a pile of bricks from one location to another was 50,000 ft · lbf.

(a) What is the magnitude of this work in Btu?

(b) If this amount of energy were used to raise the temperature of 50 lbm of water at 50°F, to what temperature would the water be raised?

2.39 If the work done in moving the bricks in Problem 2.38 were done at a steady rate and accomplished in 5 min, what would be the rate of doing work in the following units?

(a) hp

(b) W

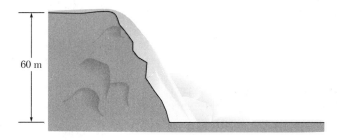

Figure P2.41 Flow of water over Niagara Falls.

2.40 A fluid stream has a velocity of 125 m/s. What is its kinetic energy per unit mass of the fluid in the following units?
(a) J/kg
(b) kJ/kg
(c) ft · lbf/lbm

2.41 Niagara Falls is about 60 m high, as illustrated in Figure P2.41. What is the decrease in potential energy per unit mass of water falling from the top to the bottom of the falls?

∗ **2.42** An astronaut on Earth at standard conditions can jump vertically upward a distance of 2 ft. How high can he jump when he is on the surface of the Moon?

∗ **2.43** A person can broadjump a distance of 15 ft on Earth at standard conditions. How far can that person broadjump on the Moon?

∗ **2.44** One of the Apollo astronauts took a golf ball and golf club to the Moon. If he could hit the ball 300 yd down the fairway on Earth at standard conditions, how far could he hit the ball with the same stroke on the Moon?

2.45 Given 1 ft = 0.3048 m, convert an area of 1 in.² to m².

2.46 Convert a mass of 1 long ton (or 2240 lbm) to kg.

2.47 The normal human body temperature is 37°C. The *Guiness Book of Records* mentions rare cases of human beings surviving minimum and maximum temperatures of 16°C and 44.4°C, respectively. Convert these temperatures to °F.

2.48 The city of Dallol in Ethiopia recorded the highest annual mean temperature (on Earth) of 34.4°C during the period 1960 to 1966. Temperatures as high as 58°C have been recorded in North Africa. In Antarctica temperatures as low as −88°C have

been recorded. Convert these temperatures
(a) to K
(b) to °R
(c) to °F

2.49 The specific heat of air at constant pressure at standard conditions is 1005 J/kg · K. Convert this to Btu/lbm · °R.

2.50 The specific heat of vaporization of water at standard atmospheric pressure is 2268 kJ/kg. Convert this to Btu/lbm using the conversion factors in Table 2.4.

2.51 The specific enthalpy, h, of a system is defined as

$$h = u + pv$$

where u is the specific internal energy, p is the pressure, and v is the specific volume. At a steam pressure of 150 psia, the specific volume is 5.351 ft³/lbm and the specific enthalpy is 1479 Btu/lbm. Calculate the specific internal energy of the steam.

2.52 A new "high temperature" superconductor was reported in 1987 by researchers. This superconductor had negligible resistance to current flow at 95 K. What is this temperature
(a) in °C
(b) in °R
(c) in °F

2.53 Using unity conversion factors, convert the following:
(a) 1 m = _____ in.
(b) 28.2 kPa = _____ lbf/in.²
(c) 10 bar = _____ lbf/in.²
(d) 10 MW (megawatts) = _____ hp
(e) 778 ft · lbf = _____ N · m
(f) 300 lbf = _____ N
(g) 28 cm = _____ in.

(h) 3 slugs = _____ lbm
(i) 2 slugs = _____ kg
(j) 1.4 lbm = _____ g

2.54 One ton of refrigeration is equal to 12,000 Btu/h. Convert 3 tons of refrigeration to kW.

2.55 One kmol of hydrogen gas (H_2) combines with 0.5 kmol of oxygen (O_2) to produce 1 kmol of H_2O.
(a) What is the total mass of the reactants (that is, H_2 and O_2)?
(b) What is the mass of the product of the reaction?

2.56 How many molecules are there in 1 kg of water? If 1 kg of water occupies a volume of 0.001 m³, determine the number of molecules of water per unit volume.

2.57 In Problem 2.55, determine the total number of molecules reacting and the number of molecules in the product of reaction.

2.58 Determine the number of molecules contained in 1 lbmol of any substance.

2.59 Assume that H_2 and O_2 in Problem 2.55 were initially separate and at standard pressure and temperature conditions of 1 atm and 25°C. Their densities at the initial conditions were 0.0824 kg/m³ and 1.31 kg/m³, respectively.
(a) What is the combined volume of the reactants?
(b) The product in Problem 2.55 is cooled to an ambient temperature of 25°C, at which state it exists as a liquid. Assuming that the density of the liquid is 1000 kg/m³, what is the overall change in volume that has occurred?

2.60 A grain of a substance having a molecular weight of 58.5 kg/kmole has a mass of 5×10^{-10} kg. Determine the number of molecules in the grain.

2.61 An appliance is rated as 2.5 hp. Convert this rating to kW.

2.62 A tire is inflated to a pressure of 22 psig at a location on Earth where the atmospheric pressure is 100 kPa (abs).
(a) What is the pressure (in psia) of the air in the tire?
(b) If the tire were transported to the atmosphere free environment of the Moon, what would be the gage pressure of the inflated tire? (Assume that the absolute pressure of air in the tire does not change.)

* 2.63 Gas is trapped in two horizontal cylinders by frictionless pistons, as depicted in Figure P2.63. The cylinders are connected through an expansion valve that is initially closed. Consider the gas in the cylinders (excluding atmospheric air) as the system, and use broken lines to mark the system boundary on the figure. Calculate the magnitude of the force F (in kilonewtons, kN) that must be applied to maintain static equilibrium with the expansion valve closed.

* 2.64 The expansion valve in Figure P2.63 is slightly opened while the force F is applied such that piston B moves a distance 0.2 m to the right while piston C moves through a distance 0.4 m to the right. Assume that the gas pressures on pistons B and C remain constant at 200 and 100 kPa, respectively. Calculate the work done *by* the system.

* 2.65 Convert a pressure difference of 20 in. of water on the surface of the Moon to
(a) lbf/in.²
(b) in. Hg
The density of water is 62.4 lbm/ft³, and the specific gravity of Hg is 13.6. Assume $g_{Moon} = 5.18$ ft/s² and $g_{Earth} = 32.2$ ft/s².

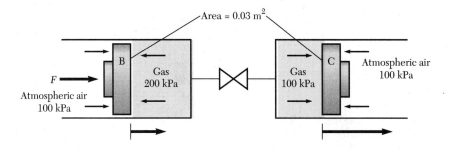

Figure P2.63 Quasi-static compression and expansion of a gas in two horizontal cylinders.

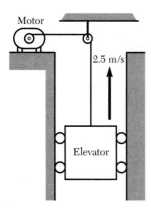

Figure P2.66 Work done
in raising an elevator car.

2.66 The combined weight of the elevator illustrated in
Figure P2.66 and the load it carries is 15 kN. Assume that the elevator travels upward at a constant
speed of 2.5 m/s and that the tension in the cable

is the sum of this weight and a frictional force of
3 kN.

(a) Calculate the work done by the motor in pulling on the cable through a distance of 12 m.
(b) What is the motor's rate of doing work
(in kW)?

2.67 Figure P2.67 illustrates a process for raising the
pressure of air in a rigid vessel from an initial value
of 100 kPa. A frictionless piston is used to force
the air in the cylinder through a one-way valve into
the rigid vessel. The valve opens when the air pressure in the cylinder slightly exceeds 200 kPa. Assume that the air pressure of slightly over 200 kPa
in the cylinder is maintained while the piston ad-
vances 0.2 m to the right, thereby pushing the air
into the rigid vessel. The cross-sectional area of
the piston is 0.03 m². Defining the air in both
the cylinder and the rigid vessel as the system,
determine

(a) the work done by the system
(b) the work done by the surroundings

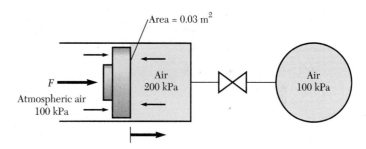

Figure P2.67 Work done in raising the pressure inside a rigid vessel.

Properties of a Pure Substance

3.1 Introduction

Thermodynamic analyses invariably require a knowledge of the properties of the substances being used. Chapters 1 and 2 covered the basic concepts and definitions involved in thermodynamics. Included in these chapters were the related concepts of thermodynamic *equilibrium, state,* and *property*. Chapter 2, in particular, dealt extensively with the thermodynamic concept of property. This chapter begins with a brief general review of the thermodynamic concept of property. Following this, the idea of the *equation of state* is considered, which is essentially the mathematical relationship among properties for a given substance.

The focus here is primarily on the properties of fluid systems, particularly those that belong to the class known as *pure substances*. For pure substances, the *two-property rule* holds—that is, the thermodynamic state of the substance is uniquely defined by specifying only two independent intensive properties. This is fortuitous because it allows formulations for the equation of state that can be expressed for each substance in tables, charts, and even computer codes.

Ideal gases, discussed in Chapter 4, are a subcategory of pure substances. The equations of state for ideal gases take a simple form. Such equations, however, do not generally apply to the entire class of pure substances. The reader is thus cautioned against trying to apply the ideal gas equations to pure substances in general. It is partially to help the reader avoid this pitfall that discussion of the ideal gas equations of state is deferred until later.

3.2 Specification of the Thermodynamic State of Systems

A thermodynamic *property* has been defined as any observable characteristic of an equilibrium system, while the thermodynamic *state* of a system is defined by the condition in which the system exists. If any equilibrium state is to be distinguishable from all others, ordinarily, all the property values for the system at the given equilibrium state would have to be specified. For the thermodynamic state of a system, only the thermodynamic properties of the system need to be considered. In practice, however, one finds that only a few properties need be specified to define the thermodynamic state of many substances uniquely.

A number of thermodynamic properties have already been introduced in the previous chapters. Some of these are *intensive properties*, that is, their magnitude is independent of the quantity of matter within the system. System pressure (p) and temperature (T) are examples of intensive properties. An *extensive property*, in contrast, is dependent on the amount of the substance present. Volume (V) and mass (m) are examples of extensive properties.

Intensive Thermodynamic Properties Normally Used to Specify Thermodynamic State

Several intensive thermodynamic properties could be used to specify the thermodynamic state of a pure substance. Other properties can, in principle, be defined from these properties; however, the discussion at this point is restricted to those properties that are commonly used. These properties can be grouped into two categories, with the first category being those that are easily observed physically and the second category being those that are not.

Easily Observed Properties

Chapter 2 discussed properties that are readily measured. These properties include *pressure, specific volume* (or its reciprocal, *density*), and *empirical temperature*. Laboratory experiments can be defined and instrumentation is readily available that can be used to measure these properties.

Properties Not Easily Observed

Other thermodynamic properties exist that are not easily measured or cannot be measured directly. An example of the former group is *quality* (of wet vapor). Properties that cannot be measured directly are the derived properties, which include *internal energy, enthalpy, thermodynamic temperature, entropy, Gibbs function*, and *Helmholtz function*. These properties are discussed further here.

Derived properties are those whose existence is inferred from the laws of thermodynamics. *Internal energy* (U) is an extensive property derived from the first law of thermodynamics. From a microscopic perspective, the internal energy of a substance is the sum of all the molecular energies associated with the substance. In general, molecules possess translational kinetic energy by virtue of their individual mass and velocity; they also possess vibrational and rotational energies as they rotate and vibrate as a consequence of their motion. Intermolecular forces also exist between molecules. The sum of all molecular energies is the internal energy of the substance. This aggregate of molecular energies cannot be measured directly and must be obtained from other measurable quantities. The interpretation of U in terms of molecular energies is not necessary in classical thermodynamics; it has been included only to provide a physical insight into its meaning. The first law of thermodynamics (considered in detail in Chap. 6) affirms the existence of internal energy (independently of the molecular-scale interpretation) as a property of a system and also provides the basis for the quantitative determination of changes in its value. In other words, in macroscopic thermodynamics, U is a result of the first law, which also provides the formula for determining changes in its value in terms of other quantities that can be measured directly.

Enthalpy (H) is a composite property that is a mathematically defined combination of other properties already defined:

$$H = U + pV \tag{3.1}$$

It is a property that arises in connection with the application of the first law of thermodynamics to flow processes. Since enthalpy involves internal energy, it cannot be measured directly either, and its value must be obtained from other measurable quantities. The term pV is sometimes loosely referred to as *flow work* or *flow energy*. Internal energy, enthalpy, and the pV term have the unit of energy, which in SI is the J or kJ and in USCS is the Btu.

Entropy (S) and *thermodynamic temperature* (T) are both derived properties from the second law of thermodynamics, as is shown in Chapters 7 and 8. The physical notion that most people associate with temperature is that of the degree of hotness (or coldness) of a body. In the case of entropy, however, most students tend to have more difficulty figuring out a physical interpretation. The most frequently offered interpretation is the measure of the *degree of disorder* in a closed system. Thus, for a closed system, a low value of entropy corresponds to a well-ordered condition, while a high value of entropy means a highly disordered state. It should be noted, however, that in classical thermodynamics, such interpretations are not necessary; the existence of entropy and the precise mathematical definition of the property are based entirely on the second law of thermodynamics. In other words, entropy is a result of the macroscopic second law, just as internal energy is a result of the macroscopic first law. Further discussion of these properties must await the more detailed considerations of the first and second laws of thermodynamics in Chapters 6 through 8. The main rationale for introducing the properties at this point is to provide the reader with a complete list of properties that are included in standard property tables for a variety of pure substances. Entropy is an extensive property and is measured in J/K or kJ/K in SI and in Btu/°R in USCS.

Two composite or mathematically defined combinations of properties involve entropy and are frequently encountered in the analysis of chemical systems. They are the *Helmholtz function* (A) and the *Gibbs function* (G), which are defined as

$$A = U - TS \tag{3.2}$$

$$G = H - TS \tag{3.3}$$

Specific Properties

Extensive properties can be reduced to corresponding intensive property form simply by dividing by the "amount" of the substance in the system. Thus, the mass of the system divided by the volume defines the density (ρ), which is an intensive property. For other extensive properties, the mass (m) or the number of moles (N) can serve as the quantity or amount of substance by which to divide the extensive property. Dividing an extensive property (Z) by the mass results in what is known as the *specific property* (z) for the system. If the number of moles is used as the divisor in place of the mass, the result is called the *molar specific property*. The specific property is an intensive property. The following specific properties are defined for the various extensive properties previously given. Appropriate SI and USCS units are given for each.

		SI	USCS	
Specific volume,	$v = \dfrac{V}{m}$	$\dfrac{m^3}{kg}$	$\dfrac{ft^3}{lbm}$	(3.4a)
Molar specific volume,	$\bar{v} = \dfrac{V}{N}$	$\dfrac{m^3}{kgmol}$	$\dfrac{ft^3}{lbmol}$	(3.4b)
Specific internal energy,	$u = \dfrac{U}{m}$	$\dfrac{J}{kg}$	$\dfrac{Btu}{lbm}$	(3.4c)
Molar specific internal energy,	$\bar{u} = \dfrac{U}{N}$	$\dfrac{J}{kgmol}$	$\dfrac{Btu}{lbmol}$	(3.4d)
Specific enthalpy,	$h = \dfrac{H}{m}$	$\dfrac{J}{kg}$	$\dfrac{Btu}{lbm}$	(3.4e)
Molar specific enthalpy,	$\bar{h} = \dfrac{H}{N}$	$\dfrac{J}{kgmol}$	$\dfrac{Btu}{lbmol}$	(3.4f)
Specific entropy,	$s = \dfrac{S}{m}$	$\dfrac{J}{kg \cdot K}$	$\dfrac{Btu}{lbm \cdot °R}$	(3.4g)
Molar specific entropy,	$\bar{s} = \dfrac{S}{N}$	$\dfrac{J}{kgmol \cdot K}$	$\dfrac{Btu}{lbmol \cdot °R}$	(3.4h)
Specific Helmholtz function,	$a = \dfrac{A}{m}$	$\dfrac{J}{kg}$	$\dfrac{Btu}{lbm}$	(3.4i)
Molar specific Helmholtz function,	$\bar{a} = \dfrac{A}{N}$	$\dfrac{J}{kgmol}$	$\dfrac{Btu}{lbmol}$	(3.4j)
Specific Gibbs function,	$g = \dfrac{G}{m}$	$\dfrac{J}{kg}$	$\dfrac{Btu}{lbm}$	(3.4k)
Molar specific Gibbs function,	$\bar{g} = \dfrac{G}{N}$	$\dfrac{J}{kgmol}$	$\dfrac{Btu}{lbmol}$	(3.4l)

The extensive property values are easily calculated from these equations, once you know the intensive property values and the pertinent amount for the substance comprising the system.

Derivative Properties and Other Intensive Properties

In general, the thermodynamic state of a system is specified in terms of a unique set of values for the intensive thermodynamic properties for the system, such as p, T, v, u, h, s, a, and g. Other frequently used intensive properties not yet defined include the *quality* or *dryness fraction* (x) of a wet vapor (or a liquid–vapor mixture), the *constant pressure specific heat* (c_p), and the *constant volume specific heat* (c_v). The last two properties are also referred to as *derivative properties*. Vapor quality (x) is defined only for the wet vapor condition, as shall be explained in a later section. The properties c_p and c_v are defined, respectively, by

$$c_p = \left(\frac{\partial h}{\partial T} \right)_p \tag{3.5}$$

$$c_v = \left(\frac{\partial u}{\partial T} \right)_v \tag{3.6}$$

The ratio of these two specific heats for any particular substance is also a thermodynamic property, the *specific heat ratio* (k), and is given by

$$k = \frac{c_p}{c_v} \tag{3.7}$$

The specific heats and the ratio of specific heats defined by Equations 3.5, 3.6, and 3.7 are especially useful in calculations involving gases, particularly ideal gases. Although k can be reasonably assumed to be a constant in many cases, it is not a constant in general.

3.3   Pure Substances and the Two-Property Rule

As indicated earlier, for many substances of interest in engineering, it is not necessary to know the values for all the system properties to be able to define the equilibrium state of the system. Many of these substances belong to the class named pure substances.

Definitions

Pure substance *A system that has a uniform chemical composition throughout.*

Comments

1. A pure substance must be *homogeneous* in chemical composition. In other words, each portion of the system must have precisely the same chemical constituents combined in the same manner and in identical proportions. For example, if one portion of a system is H_2O while elsewhere in the system the composition is a mixture of H_2 and O_2, not chemically combined but in the proportion 0.5 kmol O_2 to 1 kmol H_2, the system cannot be regarded as a pure substance. It has the same chemical constituents but the chemical *aggregation* (or combination of constituents) is not the same throughout the system. The chemical composition is therefore not homogeneous for the entire system. If, however, each portion of the system comprises H_2O, H_2, and O_2 in identical proportions, the system is homogeneous in composition.

2. A system must also be *invariable* in chemical composition for it to be regarded as a pure substance. This means that the chemical composition should not change with time. For example, a system comprising H_2 and O_2 can be regarded as a pure substance as long as no chemical reaction takes place over the period of interest in a study of the system. If a chemical reaction takes place such that the chemical composition of the system becomes nonhomogeneous over a period of time, the system cannot be regarded as a pure substance for that period. The final products of the reaction can, however, be regarded as a pure substance once the chemical composition again becomes homogeneous throughout the system. The requirements of both a homogeneous and an invariable chemical composition are implied in the definition of a pure substance as one having a *uniform chemical composition* throughout the system.

3. A pure substance does not have to be physically or macroscopically homogeneous. Kline and Koenig (1957) define a *homogeneous system* basically as one having the same intensive properties throughout. For example, a mixture of steam and water in equilibrium will have the same value for nearly all the properties of the system. However, the density of the steam will not be the same as that for the water. Such a system is not physically homogeneous; rather, it is what is called *macroscopically heterogeneous*. Since the system in this example, however, is homogeneous in chemical composition, it is a pure substance. A counter example is a mixture of air and liquid air in equilibrium. While each subsystem (the air and the liquid air) is a pure substance separately, the mixture does not qualify as a pure substance because the relative proportions of the constituents of air are different from those for liquid air.

Phase *A physically homogeneous condition for the substance.*

Elsewhere (see, for example, Kline and Koenig 1957, and Hatsopoulos and Keenan 1962), phase has been defined explicitly as "any part, or parts, of a system in which all intensive properties are identical in both kind and value."

Comments

1. A phase is the same thing as a homogeneous system defined by Kline and Koenig (1957). Thus, steam (or water vapor) is one phase, while water (liquid) is another. This remains the case even when both steam and water (liquid) actually coexist in equilibrium, as long as their densities are not the same.

2. The principal phases in which a substance can exist include solid, liquid, and vapor (or gas). A *solid* is characterized by having a fixed shape and size. A *liquid* takes the shape of the container in which it is placed but has a fixed volume. A *gas*, however, completely fills any vessel in which it is confined. For each phase, the set of intensive properties must be the same throughout the phase.

3. A mixture of substances that is all liquid or solid may not necessarily be a single phase. For example, if immiscible liquids such as oil and water are poured into a container, the oil and water will remain separate and will thus constitute two distinct phases for the system. Also, some substances such as ice and iron may have different physically homogeneous portions, even when each substance is entirely solid. In such cases, each physically homogeneous portion that has uniform intensive properties counts as a phase of the system.

Relationships Among Thermodynamic Properties

For pure substances, the following phenomenological rule is found to be true for relationships among the thermodynamic properties at states of equilibrium:

> *The thermodynamic state of a pure substance is fixed once any two independent intensive thermodynamic properties are specified provided (1) the system is in equilibrium and (2) the effects on the system due to motion, gravity, surface tension (or capillarity), electricity, and magnetism are negligible.*

Some refer to this rule as the *two-property rule* (see, for example, Spalding and Cole 1966). Others refer to it as the *state principle* (Kline and Koenig 1957) or the *state postulate*, although in its original formulation, the state principle applies to a broader class of substances known as *simple systems*. Further consideration of simple systems is deferred to Chapter 13. This principle, as applied to pure substances, will continue to be called the two-property rule.

Comments

1. It should be noted that the thermodynamic state is defined in terms of intensive properties and not in terms of extensive properties nor those properties requiring a physical reference coordinate system. In other words, the thermodynamic coordinates required for the definition of the thermodynamic state are those from the list of intensive thermodynamic properties (such as pressure, specific volume, and temperature) rather than extensive properties (such as total volume) or properties that require a frame of reference (such as velocity and acceleration).

2. The two intensive properties chosen must be *independent* of each other. A typical situation, to be considered more fully later, is the boiling of a liquid at which condition pressure and temperature are no longer independent. For example, water boils at 100°C when the pressure is 1 atm. If the pressure should fall to 80 kPa (at a high altitude location), the boiling point of water drops to 93.5°C. In other words, the boiling point of liquids depends on the pressure, and therefore pressure and temperature cannot in this case be regarded as two independent intensive properties for fixing the state of the pure substance.

3. A further restriction often mentioned is that the two-property rule applies only to *compressible* pure substances. Consider, for example, a solid substance, which is incompressible. The pressure, defined as the normal force on a surface per unit area, is now the same as the normal stress. For such a body, other stresses are usually acting on the surface that may vary with direction. Complete specification of the state of the body must therefore include the magnitude and direction of all the stresses acting on different planes of the body. In such a case, two independent intensive thermodynamic properties will not be adequate to uniquely define the state of the solid body. If the only stresses acting on the body are normal stresses that are the same in all directions and can thus be identified as the pressure, any two independent intensive thermodynamic properties will suffice for a unique specification of the thermodynamic state.

4. The system must be in *equilibrium* for the two-property rule to apply. For very large systems in Earth's gravitational field, for example, the hydrostatic pressure varies with elevation for a fluid system. The system is therefore not in thermodynamic equilibrium, and the two-property rule does not apply. Another example is provided by liquid–vapor mixtures that are separated by a curved interface or phase boundary referred to as a *meniscus*. A difference of pressure occurs between the liquid and vapor phases as a result of surface tension forces. In most cases encountered in engineering, the interface between the liquid and vapor phases is large and flat except at the region of contact with the container, where the meniscus may be significantly curved. The effects of capillary action at the edge of the liquid–vapor interface are often small and negligible. When the effects of capillarity are significant, the pressure will no longer be uniform throughout the system and the two-phase rule can no longer be applied.

5. Perhaps the most important consequence of the two-property rule is the inference that can be drawn concerning the equations of state for a pure substance. If (x, y) are any two independent intensive thermodynamic properties, then a functional relationship of the form $z = z(x, y)$ must exist for every other intensive thermodynamic property z. As indicated previously, for extensive properties Z, the specific properties $z = Z/m$ (where m is the mass) are the appropriate intensive properties in the functional relationship inferred from the two-property rule. The formulation $z = z(x, y)$ is referred to as an *equation of state*. Development of such formulations for various substances is a major research activity in thermodynamics. Empirical data for the substance are used as well as the pertinent laws of thermodynamics in the formulations of equations of state expressing the relationships among thermodynamic properties.

In reality, mathematical equations for the thermodynamic properties of pure substances can be extremely complex, and hence, the usual practice is to tabulate values of z for a range of values of x and y. In Appendix A, for example, Table A.3 gives values of the specific properties v, u, h, and s for a range of values of pressure (p) and temperature (T). Graphical representation in chart form is also common either as *x-y-z* diagrams (in three-dimensional coordinates) or as two-dimensional diagrams known as *projections*. The three-dimensional representation produces contours known as *thermodynamic surfaces* (for example, see Fig. 3.11). This representation is used primarily for qualitative discussion of property relationships for the different phases of a pure substance. It is of little use for quantitative analyses in thermodynamics because of the extreme difficulty in obtaining accurate readings of property values from such diagrams.

The more common charts used are the various projections obtained with one variable plotted against a second one while keeping the third variable constant (for example, see Fig. 3.3). The choice of two independent intensive properties x and y can be made arbitrarily from the long list of intensive properties that have already been introduced. Typical choices of two-dimensional projections in thermodynamics include the *T-s* diagram (for thermodynamic cycles), the *h-s* diagram (for steam power cycles), the *p-h* diagram (for refrigeration cycles), as well as the *T-v*, *p-v*, and *T-p* projections depicting the *pvT* relationships for a pure substance. Further discussions of graphical representations of property relationships are limited at this juncture to the more readily measured properties, which are pressure (p), temperature (T), and specific volume (v) or density (ρ).

3.4 *pvT Relationships for Pure Substances*

The thermodynamic state of a simple compressible pure substance is fixed by specifying any two independent intensive thermodynamic properties. Although one can choose from several intensive thermodynamic properties, the choices here are restricted to those properties that can be readily measured in a laboratory—pressure (p), temperature (T), and specific volume (v) or density (ρ). Other properties are normally obtained by indirect methods that generally involve measurements of different quantities and the use of appropriate thermodynamic relationships. Since density and specific volume are not independent of each other, only one of them can be used at a time; thus either p, v, and T or p, ρ, and T can be used. The most frequent choice is the set p, v, and T. These properties commonly serve as the basis for laboratory measurements and subsequent presentation of the results.

Two-Dimensional (T-v, T-p, and p-v) Projections

Any diagram depicting property relationships for a substance must represent states of equilibrium since thermodynamic properties as a rule have meaning only for a system in equilibrium. In a two-dimensional projection for the representation of the *pvT* relationship for a pure substance, the principal options are as follows:

1. Investigate the variation of T with v (the *T-v* projection) while p takes on a value that is kept constant.

2. Investigate the variation of T with p (the *T-p* projection). In this case, it is found that states corresponding to two phases, coexisting in equilibrium, collapse into a single line since temperature and pressure are dependent properties for the region. The liquid–solid, liquid–vapor, and solid–vapor lines in the diagram demarcate the regions in which the substance exists in one or the other of the principal phases. For this reason, the *T-p* or *p-T* projection is often referred to as a *phase diagram*.

3. Investigate the variation of p with v (the *p-v* projection) while keeping T constant.

These three projections are discussed in subsequent sections in a more comprehensive manner. To begin, the property and phase changes are considered that occur when ice at a below freezing temperature is heated at constant pressure until it becomes steam (or water vapor) at a high temperature. First, however, a few more terms are introduced, while others that were mentioned previously are formally defined in order to make the discussions clearer.

Phase Change Terminology

Vaporization or *boiling* refers to a liquid turning to a vapor. The temperature at which this occurs is the *boiling point*, which is also referred to as the *saturation temperature* (T_s) for the substance. At the saturation temperature, both the liquid phase and the vapor phase of the substance can coexist in equilibrium. The liquid condition is termed the *saturated liquid* state, while that of the vapor is called the *saturated vapor* condition. The letter f is used as a subscript to denote the saturated liquid condition (for example v_f) while g designates the saturated vapor (gas) condition (such as v_g). The pressure of a liquid–vapor mixture in a state of equilibrium is termed the *saturation pressure* (p_s). As indicated earlier, T_s and p_s are dependent properties for a pure substance. The p_s-T_s relationship for water is often determined using the Marcet boiler apparatus.

For a solid changing to a liquid, the term *melting* is used; the temperature at which this occurs is the *melting point*. The change from the liquid phase to the solid phase is referred to as *freezing* or *solidification*. Freezing occurs at the *freezing point*. For a pure substance, the melting point and the freezing point are identical. If a solid phase of a pure substance coexists with a liquid phase in a state of equilibrium, the solid is said to be in the *saturated solid* state, while the liquid is in the *saturated liquid* state. The letter i is used as a subscript for the saturated solid (ice) condition, while f is retained for the saturated liquid condition. When a solid changes directly to a vapor, the term used is *sublimation*. The reverse process of a vapor turning directly into a solid is called *frosting*.

For a pure substance, all three principal phases (solid, liquid, and vapor) can coexist in equilibrium only at a unique condition (for each substance) known as the *triple point*. The corresponding temperature and pressure are referred to as the *triple point temperature* (T_t) and the *triple point pressure* (p_t), respectively. For water, T_t is 0.01°C (32.018°F) while p_t is 0.612 kPa (0.0887 psia). Other important terms are defined throughout whenever they are first mentioned.

T-v Projection for a Pure Substance

Consider a mass of ice contained in a piston–cylinder arrangement such that the ice could be heated while its pressure stayed constant. The mass of the piston is assumed negligible so that the system (defined as the mass of ice) is at ambient pressure, taken as 1 atm. It is further assumed that the heating takes place very slowly such that the system passes through a series of near- or quasi-equilibrium states during the entire process. The initial temperature of the ice is below the melting point. Figure 3.1 depicts the various stages that are observed as the ice is heated at constant pressure and ultimately changed to a vapor at a high temperature.

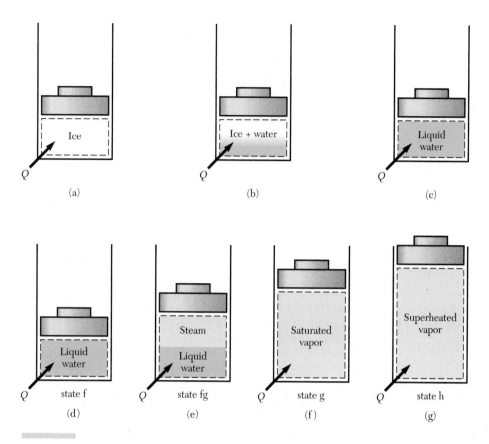

Figure 3.1 Constant pressure heating of an ice–water–steam (solid–liquid–vapor) system. (a) Raising of ice temperature. (b) Phase change at constant temperature. (c) Single phase with temperature rise. f = saturated liquid; fg = saturated liquid and vapor; g = gas.

The first stage represented in Figure 3.1(a) corresponds to the raising of the temperature of the ice from its initial value to the point at which melting is about to commence ($a \rightarrow a'$ in Fig. 3.2(a)). In Figure 3.1(b), the ice is melting and a change of phase occurs from solid (ice) to liquid (water) while the temperature stays constant at the melting point ($a' \rightarrow c$ in Fig. 3.2(a)). The system now comprises a mixture of ice and water at 0°C (32°F) which is the melting point of ice at 1 atm pressure. Physically, a heterogeneous mixture of ice chips in water is observed. When all the ice has melted, the stage represented by Figure 3.1(c) exists. The heat required to completely melt a unit mass of a substance at its melting point is called the *latent* (or *specific*) *heat of fusion*. Latent heat was once thought of as *hidden* heat since it did not produce any perceptible change in temperature. Today, latent heat simply refers to heat that causes a change of phase without altering the temperature of the substance. *Sensible heat*, in contrast, refers to heat that results in a temperature change for the system. The latent heat of fusion of ice at a pressure of 1 atm is approximately 333 kJ/kg (143 Btu/lbm).

Figure 3.1(c) depicts a physically homogeneous phase (liquid water) following the complete melting of the ice. The temperature of the system then rises until the boiling point of water is reached. When the water reaches the boiling point, it is in the condition marked with an f, which is shown in Figure 3.1(d) and was identified earlier as the saturated liquid state. Further heating results in boiling with increasing amounts of the water (liquid) turning into steam (vapor). The boiling process is represented by the sketch marked with an fg, where fg indicates the difference between a property value for the saturated vapor and that for the saturated liquid as shown in Figure 3.2(a). The system is again physically heterogeneous and is characterized by the appearance of a meniscus separating the liquid phase from the vapor phase. The heat required to change a unit mass of a liquid completely to vapor at the boiling point is called the *latent (or specific) heat of vaporization*. The latent heat of steam at 1 atm pressure is 2257 kJ/kg (970 Btu/lbm). When all the liquid has changed into a vapor at the boiling point and the state marked with a g is reached, the substance is said to be at the *saturated vapor* condition. Further heating at constant pressure beyond state g will result in a steady rise in temperature of the vapor. The condition of the steam beyond state g is termed *superheated vapor*, and is indicated by the state h in Figures 3.1(g) and 3.2(a).

For the entire heating process, the trajectory of the intensive properties T and v for the system is represented as the continuous line a-a'-c-f-g-h in Figure 3.2(a). Certain aspects of the behavior indicated for water in the figure are anomalous compared with pure substances in general. Specifically, ice contracts upon melting; for other pure substances, an increase in volume generally accompanies the change from the solid to the liquid phase. Also, for water in the liquid phase, as the temperature increases from the melting point, the specific volume decreases, reaching a minimum at 4°C (39.2°F) and 1 atm. Thereafter, the specific volume increases with temperature. This behavior is peculiar to water. For most other pure substances, the specific volume increases steadily

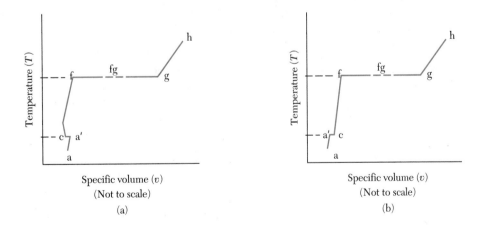

Figure 3.2 Temperature–volume (T-v) diagram for (a) water at 1 atm pressure (expands on freezing) and (b) a typical pure substance at a constant pressure (contracts on freezing).

with a rise in temperature for all the principal phases. Figure 3.2(b) depicts the constant pressure line that is typical of pure substances in general. The constant pressure lines shown include portions that correspond to sensible heating of the solid (a-a'). As briefly indicated earlier, solids may have complex internal stresses when subjected to an external load, and as such, the solid phase of a substance may not behave as a pure substance. For this reason, further discussion of the *T-v* projection is focused on the fluid (liquid or vapor) phases.

Figure 3.3 is the temperature–volume (*T-v*) equilibrium diagram for a typical pure substance when heat is added to change the system from a liquid at state c to a super-heated vapor at state h while the system pressure is kept constant. When the temper-ature of the liquid is below the saturation temperature (or boiling point), T_s, the liquid is called either a *compressed liquid* (in which the pressure is greater than the saturation pressure for the given temperature) or a *subcooled liquid* (in which the temperature is lower than the saturation temperature for the given pressure).

The specific volume of the substance increases substantially as the saturated liquid at state f is completely turned to a saturated vapor at state g. For example, in the case of water, v increases from about 0.001 m^3/kg for saturated water at 100°C to nearly 2 m^3/kg for saturated steam at the same temperature. As mentioned earlier, the process of turning a liquid into a vapor at the saturation temperature is termed *vaporization* or, more loosely, *boiling*. During the process, the fluid is a mixture of liquid and vapor in what is called a *mixed phase* or *wet vapor* condition. For pure substances, the pressure and the temperature are no longer independent properties when the fluid is in the mixed phase region. The pressure and temperature in this region are the *saturation pressure* (p_s) and the *saturation temperature* (T_s), respectively. Constant pressure heating from state g to state h results in increases in both temperature and specific volume, as indi-cated in Figure 3.3. The fluid is now a *superheated vapor*.

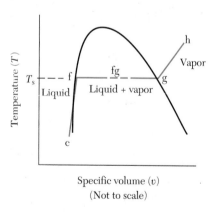

Figure 3.3 Temperature–volume (*T-v*) equilibrium diagram for the fluid phases of a pure substance that is kept at constant pressure.

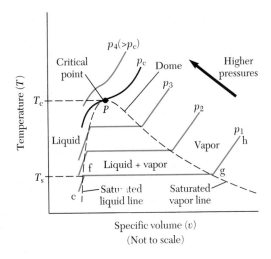

Figure 3.4 Temperature–volume (*T-v*) equilibrium diagram for a pure substance at various pressures.

Figure 3.4 is a typical temperature–volume (*T-v*) equilibrium diagram for pure substances when the heating process is carried out at several different pressures. A *T-v* line at a constant pressure condition is known as an *isobar*. The isobars at higher pressures (and below 217.67 atm for water) are similar to the line c-f-g-h in Figure 3.3 except that the boiling process occurs at a higher temperature as the pressure increases. Also, the change in volume between the fully (or saturated) vapor state and the fully liquid state at the boiling point diminishes as the pressure rises.

The Critical Condition

The isobars continue to show the pattern described until a pressure is reached at which the transition from a saturated liquid condition to a saturated vapor condition occurs instantaneously. The characteristic flat portion of the isobars at lower pressures now reduces to a point P. The isobar through P is now a continuous curve with a point of inflexion at P. Point P is referred to as the *critical point* and the constant pressure line through it is called the *critical isobar*. At pressures higher than that for the critical isobar, it is no longer possible to produce a meniscus (in other words, the condition in which the liquid and vapor phases co-exist in equilibrium can no longer be produced). Rather, in this situation, a change from a "liquid-like" phase to a "gas-like" phase occurs nearly instantaneously at a temperature often referred to as the *pseudocritical temperature* (T_{pc}). At the critical point, the corresponding p, v, and T are called the *critical pressure* (p_c), the *critical specific volume* (v_c), and the *critical temperature* (T_c), respectively.

The line joining the points at which boiling begins is called the *saturated liquid line*, as indicated in Figure 3.4, while the locus of the points for the saturated vapor condition is called the *saturated vapor line*. These two lines meet at the critical point, which for water corresponds to a temperature of 373.976°C (705.16°F) and a pressure of 217.67 atm. The locus of all of the saturated points for both liquid and vapor produces a curve called the *dome* or *saturation envelope*. Thus, for the conditions of Figure 3.4, liquid exists to the left of the dome, a two-phase mixture of liquid and vapor exists under the dome, and superheated vapor exists to the right of the dome. At temperatures above the critical point at the top of the dome, the fluid is said, in the classical view, to be a *gas*. As indicated earlier, a meniscus cannot be produced either by cooling (or heating) at a constant pressure (higher than the p_c) or by compression (or expansion) when the temperature is above T_c. Instead, at pressures and temperatures above the critical point, the liquid changes to a vapor in an imperceptible manner (since no meniscus is produced) and without the process necessarily occurring at a constant temperature.

The p-T Projection

One characteristic feature of the pressure–temperature *p-T* projection is that it delineates the regions occupied by the principal phases of a substance. Figure 3.5 depicts the relationship between pressure and temperature for a pure substance such as water. In the figure, the phase change (from liquid to vapor) occurs at higher and higher temperatures as the pressure is increased, until the critical point is reached. As discussed earlier, the choices for ordinate and abscissa are arbitrary. The vapor pressure curve could be plotted instead with pressure as the ordinate and temperature as the abscissa. However, the temperature is plotted here as the ordinate to be consistent with previous

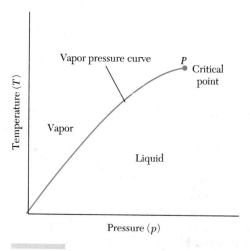

Figure 3.5 Vapor pressure curve for a pure substance such as water.

plots. Vapor pressure curves in many other books are plotted the other way, and this is the approach used when the subject is reconsidered in Chapter 12.

In Figure 3.5, vapor exists to the left of the vapor pressure curve, while liquid exists to the right, for all conditions up to the critical point. The origin for the *T-p* axes in the figure is not zero for either the pressure or the temperature. It should also be noted that water does not exist as a liquid below a certain temperature since it would turn solid (ice) at the freezing point. Figure 3.6 is an expanded version of Figure 3.5. In Figure 3.6, the *T-p* equilibrium diagram is given for a pure substance such as water to include the solid phase in addition to the fluid phases of Figure 3.5. The point at which solid, liquid, and vapor all exist in equilibrium together is designated as the *triple point*. Only one point exists for a pure substance at which such a condition occurs. Table A.22 (SI units) and B.22 (USCS units) in Appendices A and B list the triple points of some common pure substances.

The triple point of Figure 3.6 corresponds to the origin for the *T-p* axes of Figure 3.5. The constant pressure line shown in Figure 3.6 from point a through point h corresponds to the constant pressure heat addition process discussed earlier for Figures 3.1 and 3.2(a). Thus, the solid is changed from state a to state a′ by raising its temperature; then it undergoes a phase change from state a′ to state c. The phase change line crossed is called the *fusion line*, and the heat (per unit mass of substance) required to make this phase change is called the *latent heat of fusion*, as discussed earlier. A continued temperature rise at constant pressure accompanies the change of state from c to f, as shown in Figure 3.6, while the substance remains a liquid. Then heat addition from state f to state g causes a phase change (from liquid to vapor). This heat addition was defined earlier as the latent heat of vaporization, and the line on Figure 3.6 across which

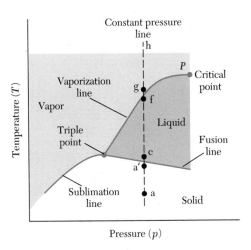

Figure 3.6 Temperature–pressure (*T-p*) equilibrium diagram for a pure substance such as water (including the solid phase region).

such phase change occurs is called the *vaporization line*. Points a′ and c should actually lie on top of each other and on the fusion line, but they are shown separately for clarity. A similar statement applies for points f and g relative to the vaporization line. Note the negative slope of the fusion line for water; the slope is positive for most other pure substances that contract when freezing.

If the pressure is reduced to a value below that of the triple point, the solid will change directly from a solid to a vapor when its temperature is raised. The line that is crossed in the transition from a solid to a vapor is called the *sublimation line*. At the sublimation line, both solid and vapor can exist together in equilibrium.

It is possible to have different phases of solids corresponding to different crystalline structures. Figure 3.7 is a temperature–pressure (T-p) equilibrium diagram depicting different phases in the ice region. The figure corresponds to the region of the solid–liquid interface in Figure 3.6 except that the pressure levels in Figure 3.7 are much higher than those in Figure 3.6. The scale for pressure in Figure 3.7 corresponds roughly to 1000 times the scale for pressure in Figure 3.6. Thus, the fusion line of Figure 3.6 is represented in Figure 3.7 by a short distance near the 0°C point on the curve. In any case, as you can see from Figure 3.7, it is possible to have different phases of solids, which are represented by the Roman numerals I through VII. It is also possible for some of these phases with different crystalline structures to coexist in equilibrium with each other along the phase boundaries of the figure. For certain conditions, it is also possible for most of the phases to coexist with the liquid phase.

This discussion has been restricted to a consideration of water. Most pure substances are known to behave like water in that they can exist in all three phases (solid, liquid,

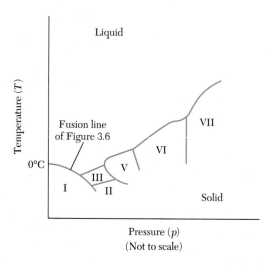

Figure 3.7 Temperature–pressure (T-p) equilibrium diagram for a liquid and a solid region for water showing different solid (ice) phases.

and vapor) and that the three phases can be distinguished from one another by different densities. Furthermore, all substances can be described and characterized by appropriate equilibrium diagrams. Essentially all pure substances behave in a similar manner when changing from a liquid to a vapor, but as indicated earlier, most substances behave differently from water when changing from a liquid to a solid. In particular, water expands upon freezing; this is why ice floats in water. Most other substances, however, tend to contract upon freezing. Figure 3.8 is a T-p equilibrium diagram for a pure substance that contracts upon solidifying such as carbon dioxide.

A comparison of Figures 3.6 and 3.8 shows that all conditions appear similar for water and carbon dioxide except for the slope of the fusion line. For most pure substances, the melting point increases with a rise in pressure, while for water, the melting point of ice decreases with a rise in pressure. In any case, all pure substances can be characterized by equilibrium diagrams in a manner similar to those presented here for water and carbon dioxide.

The p-v Projection

As discussed earlier, it is possible to describe the thermodynamic state by specifying any two independent intensive thermodynamic properties for a simple compressible pure substance. In the previous presentation, temperature was used as the ordinate, but that choice was arbitrary. Figure 3.9 is an alternate presentation with pressure as the ordinate and specific volume as the abscissa. This example is an equilibrium diagram for the substance in the fluid region only. Note in the figure that the liquid is to the left of the

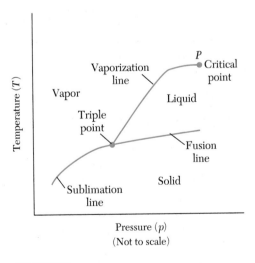

Figure 3.8 Temperature–pressure (T-p) equilibrium diagram for a pure substance such as carbon dioxide that contracts upon solidifying.

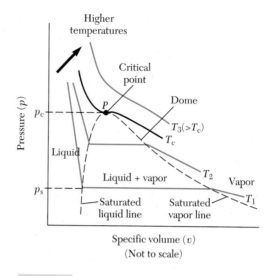

Figure 3.9 Pressure–volume (p-v) equilibrium diagram for a pure substance such as water.

saturated liquid line, vapor is to the right of the dome, and the liquid–vapor mixture is under the dome. This relationship is similar to that in Figure 3.4, where temperature is the ordinate. In place of the constant pressure lines of Figure 3.4, Figure 3.9 has constant temperature lines, which are termed *isotherms*.

Many other choices are available for presentation of relationships among thermodynamic properties for a given pure substance. Some of these other projections are encountered later. The choice is often dictated by the physical situation under consideration and is made to facilitate the analysis of the problem.

Thermodynamic Surfaces

The three primary thermodynamic properties chosen earlier are pressure, specific volume, and temperature. It is possible to present all of these properties in one figure by use of a three-dimensional surface. The main purpose served by a three-dimensional representation of the pvT equation of state is visual and qualitative. The equilibrium states in which the substance can exist are represented by three-dimensional surfaces. Figure 3.10 depicts pvT surfaces for a pure substance such as carbon dioxide that contracts upon solidifying, while Figure 3.11 illustrates similar surfaces for a pure substance such as water that expands upon solidifying. The two-dimensional figures for carbon dioxide and water that were discussed earlier can be viewed as two-dimensional projections from the pvT surfaces in Figures 3.10 and 3.11. All points on the surface represent independent equilibrium states for the substance. Points for a quasi-equilibrium process must also lie on the surface because such a process just passes through equilibrium states. Note that the major differences between Figures 3.10 and 3.11 are in the

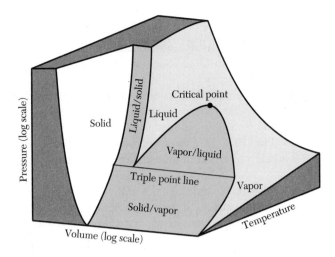

Figure 3.10 *pvT* surface for a pure substance such as carbon dioxide that contracts upon solidifying.

solid–liquid interface region. The atypical behavior of water (which expands upon solidifying) is indicated in Figure 3.11, while the more usual behavior (contracting upon solidifying) of substances is illustrated in Figure 3.10.

To develop such *pvT* surfaces, data that define the relationships among the three properties must be obtained. Such data come from laboratory experimentation and are generally available for most substances used in engineering.

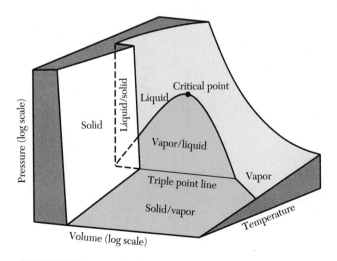

Figure 3.11 *pvT* surface for a pure substance such as water that expands upon solidifying.

3.5   Tables of Thermodynamic Properties

If adequate property data are available, it is conceptually possible to develop an equation of state for a substance. Such an equation relates pvT, or other sets of independent intensive properties xyz, in the form $p = p(v, T)$ or $z = z(x, y)$ for a particular substance in accordance with the two-property rule. Unfortunately, the thermodynamic property relationships for pure substances cannot in general be described with a single simple equation of state except in such simple cases as the ideal gas condition (which is considered in Chap. 4). Therefore, the relationships among thermodynamic properties are frequently presented in tabular form. Basic data for these tables are obtained experimentally, while other data are inferred using mathematical relationships based on the laws of thermodynamics. Tables of thermodynamic properties for several substances are given in Appendix A (SI units) and Appendix B (USCS units). The following discussion concentrates on water–steam, but the approach and general procedures apply to all pure substances.

Figure 3.12 is a temperature–volume (T-v) equilibrium diagram for water showing the regions that are normally included in thermodynamic property tables. The properties vary considerably from region to region. Since different approaches are used to evaluate the properties in different regions, a separate table is normally presented for each such region. Thus, it is common to find data presented in a separate table for the superheated vapor region, for the compressed liquid region, and for the saturated liquid–vapor mixture region. Tables of data for the solid–vapor region are usually only available for water and a few other substances. The following sections examine the tables of data available

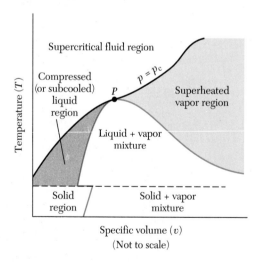

Figure 3.12 Temperature–volume (T-v) equilibrium diagram for water showing regions included in thermodynamic property tables.

for water for the various regions and consider the steps required to find and use such data.

3.5.1 Superheated Vapor Region

In the superheated vapor region, pressure and temperature are independent intensive properties and are often used in property tables for the organization of other intensive properties into columns and rows. The superheated steam properties are given in SI units in Tables A.3 and A.6 of Appendix A. The properties in USCS units are in Tables B.3 and B.6 of Appendix B. For refrigerant-22, the superheated vapor properties are in Tables A.9 and A.10 (SI units), while the values in USCS units are given in Tables B.9 and B.10. The following examples illustrate the use of these tables for finding the properties of a substance existing as a superheated vapor.

EXAMPLE 3.1

Given: H_2O at $p = 10$ bar, $T = 400°C$

Find: v (in m³/kg)

Solution: The first step is to determine the phase(s) in which the substance exists. The specified pressure of 10 bar is below the critical pressure of 220.55 bar for steam, and therefore the state must lie on the constant pressure (10 bar) line illustrated in Figure 3.13. From Table A.2 for saturated water and steam properties (pressure

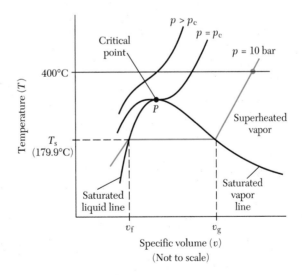

Figure 3.13 Temperature–volume (*T-v*) diagram for water showing the state specified in Example 3.1.

table), the saturation temperature T_s at 10 bar is found to be 179.9°C. The flat portion on the constant pressure line in the T-v diagram (Figure 3.13) is thus marked 179.9°C.

The specified fluid state of 10 bar and 400°C must be the intersection of the $p = 10$ bar line and the horizontal $T = 400$°C line. From Figure 3.13, the intersection is clearly in the superheated steam region, for which property values are given in Table A.3 in Appendix A. Table A.3 shows that $v = 0.307$ m^3/kg.

EXAMPLE 3.2

Given: H_2O at $p = 60$ psia, $v = 10$ ft^3/lbm

Find: **(a)** T (in °F), **(b)** h (specific enthalpy) (in Btu/lbm)

Solution: As in Example 3.1, the saturation temperature, T_s, corresponding to the specified pressure of 60 psia can be found from the table for saturated water and steam (pressure table), which for USCS units is Table B.2. In this case, $T_s = 292.8$°F. In addition, $v_f = 0.01738$ ft^3/lbm and $v_g = 7.174$ ft^3/lbm, from which one can infer that since $v > v_g$, the specified state having $v = 10$ ft^3/lbm must lie in the superheated steam region, as sketched in Figure 3.14. The specified state is the intersection of the $p = 60$ psia line and the $v = 10$ ft^3/lbm ($> v_g$) line.

Superheated properties of steam in USCS units are given in Table B.3. The following abridged portion of the table for $p = 60$ psia serves as a basis for

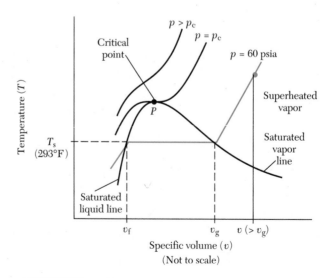

Figure 3.14 Temperature–volume (T-v) diagram for water showing the state specified in Example 3.2.

interpolation, which becomes necessary because no entry exists in the table for the specified v of 10 ft^3/lbm.

T (°F)	v (ft^3/lbm)	h (Btu/lbm)
550	9.915	1307
?	(10)	?
600	10.42	1332

Linear interpolation can be used between the tabulated values to determine T and h at the specified state. Thus,

$$\frac{T - 550}{600 - 550} = \frac{10 - 9.915}{10.42 - 9.915} = \frac{h - 1307}{1332 - 1307}$$

Solving for T and h gives (a) $T = \underline{558°F}$ and   (b) $h = \underline{1311\ \text{Btu/lbm}}$

EXAMPLE 3.3

Given: Refrigerant-22 at $p = 27.5$ bar, $h = 305$ kJ/kg

Find: T (in °C)

Solution: From Table A.8 for saturated refrigerant-22 properties, we find that at $p = 27.5$ bar,

$$T_s = 65.9°C$$

$$h_f = 130.58\ \text{kJ/kg}; \qquad h_g = 260.81\ \text{kJ/kg}$$

The specified h at 27.5 bar is greater than h_g, and thus the fluid is a superheated vapor.

Table A.9 for superheated refrigerant-22 includes property values at $p = 25$ bar and $p = 30$ bar but not at the specified pressure of 27.5 bar. A table can be constructed for $p = 27.5$ bar by interpolation based on the values at $p = 25$ bar and $p = 30$ bar, as follows:

(From Table A.9) $p = 25$ bar		Linear Interpolation for $p = 27.5$ bar		(From Table A.9) $p = 30$ bar	
T (°C)	h (kJ/kg)	T (°C)	h (kJ/kg)	T (°C)	h (kJ/kg)
100	303.10	100	299.93	100	296.76
110	312.46	110	309.68	110	306.89

From the values worked out for $p = 27.5$ bar, the temperature can be determined by further linear interpolation:

$$\frac{T - 100}{110 - 100} = \frac{305 - 299.93}{309.68 - 299.93}$$

Solving for T gives $T = \underline{105°C}$

---

1. Use the appropriate property tables to find the unknown properties in the following. In each case, use the T-v diagram to locate the state and phase(s) of the substance.

   (a) H_2O at $p = 5$ bar, $T = 600°C$; find v

   (b) H_2O at $p = 100$ Pa, $v = 1500$ m^3/kg; find T

   (c) Refrigerant-22 at 20 psia, $T = 115°F$; find h

   (d) Refrigerant-22 at 75 psia, $T = 125°F$; find h

   (e) H_2O at 75 bar and $v = 0.035$ m^3/kg; find T

   Answer: **(a)** 0.804 m^3/kg, **(b)** 51.9°C, **(c)** 123.86 Btu/lbm, **(d)** 123.51 Btu/lbm, **(e)** 377°C

2. The *degrees of superheat* refers to the temperature difference, ΔT, between the actual temperature and the saturation temperature for a superheated vapor. Find T, v, and h for superheated steam at 55 bar with 275°C of superheat.

   Answer: 545°C, 0.06675 m^3/kg, 3533 kJ/kg

3.5.2 Saturated Liquid–Vapor and the Wet Vapor Region

As indicated earlier, pressure and temperature are not independent whenever liquid and vapor coexist in equilibrium. Therefore, an additional independent intensive property besides pressure and temperature must be specified to fix the thermodynamic state in this region. Tables of properties for the region are only for the saturated liquid and the saturated vapor. From these properties, the property values for a mixture of saturated liquid and saturated vapor can be determined once the vapor quality (to be defined shortly) is ascertained. A *pressure table* gives the saturated liquid and vapor properties with the saturation pressure (p_s) as the argument. When the saturation temperature (T_s) is the argument, the table is referred to as a *temperature table*. Tables A.1 and A.2 (SI units) in Appendix A are the temperature and pressure tables, respectively, for the saturated water and steam properties. Their counterparts in USCS units are Tables B.1 and B.2 in Appendix B. For saturated ice and steam properties, Tables A.4 and A.5 (SI units) are the temperature and pressure tables, respectively. The corresponding tables in USCS units are Tables B.4 and B.5. Tables A.7 and A.8 (SI units) are the temperature and pressure tables, respectively, for the properties of the saturated fluid phases of refrigerant-22. Tables B.7 and B.8 are for the same properties in USCS units. As before, the properties presented in these tables are given with the subscript f for properties of the saturated liquid and the subscript g for properties of the saturated vapor (or gas). For saturated ice, the subscript i is used.

Figure 3.15 depicts the liquid–vapor region in the temperature–volume equilibrium diagram for water. Under the dome is the wet vapor region (a mixture of saturated liquid and vapor). The boundary to the left side of the dome is the saturated liquid line, while

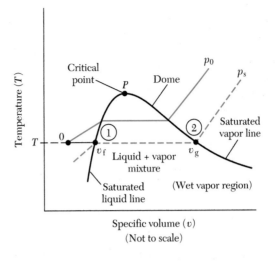

Figure 3.15 Temperature–volume ($T\text{-}v$) equilibrium diagram for water showing the saturated liquid/vapor and the wet vapor regions.

that to the right side of the dome is the saturated vapor boundary. These two lines meet at the critical point at the top of the dome.

One additional thermodynamic property that is defined specifically for a liquid–vapor mixture of a pure substance is the *vapor quality* (or *dryness fraction*), x, which is the ratio of the mass of the vapor (m_g) to the total mass of the mixture (m). It can be expressed mathematically as

$$x = \frac{\text{Mass of vapor}}{\text{Mass of mixture}} = \frac{m_g}{m} = \frac{m_g}{m_f + m_g} \tag{3.8}$$

This property *exists only for two-phase mixtures* (and the saturated phases), and it is frequently used in addition to either pressure or temperature as the second independent property for conditions under the dome. As can be seen from Equation 3.8, x must lie in the range of 0 to 1; the quality is 0 for saturated liquid and 1 for saturated vapor.

Since a mixture in the two-phase region is composed of a saturated liquid in equilibrium with a saturated vapor, any extensive property of the mixture can be determined by adding the contributions due to each component. Thus, the total volume (V) of a mixture of liquid and vapor is the sum of the volume of the saturated liquid component (V_{liquid}) and the saturated vapor component (V_{vapor}):

$$V = V_{\text{liquid}} + V_{\text{vapor}} \tag{3.9}$$

The total volume equals the mass times the specific volume, so Equation 3.9 can also be written as

$$V = mv = m_{\text{liquid}}v_f + m_{\text{vapor}}v_g \tag{3.10}$$

where m and v are the mass and specific volume of the mixture, respectively. Equation 3.10 can be further rewritten as

$$v = \frac{(m - m_{\text{vapor}})}{m} v_{\text{f}} + \frac{m_{\text{vapor}}}{m} v_{\text{g}}$$

$$= \left(1 - \frac{m_{\text{vapor}}}{m}\right) v_{\text{f}} + \frac{m_{\text{vapor}}}{m} v_{\text{g}}$$

(3.11)

Combining Equation 3.11 with 3.8, an equation for the average specific volume of the mixture is obtained:

$$v = (1 - x)v_{\text{f}} + xv_{\text{g}}$$

(3.12)

or

$$v = v_{\text{f}} + x(v_{\text{g}} - v_{\text{f}})$$

(3.13)

The difference between the specific volume of the saturated vapor and that of the saturated liquid is often denoted as v_{fg}, which is given by

$$v_{\text{fg}} = v_{\text{g}} - v_{\text{f}}$$

(3.14)

Thus,

$$v = v_{\text{f}} + xv_{\text{fg}}$$

(3.15)

A similar development that makes use of quality can be made for other intensive thermodynamic properties. With such a development, the following expressions can be obtained for specific internal energy, specific enthalpy, and specific entropy in the two-phase liquid–vapor region:

$$u = u_{\text{f}} + xu_{\text{fg}}$$

(3.16)

$$h = h_{\text{f}} + xh_{\text{fg}}$$

(3.17)

$$s = s_{\text{f}} + xs_{\text{fg}}$$

(3.18)

The tables of properties in the appendices are used in the following worked examples for the wet vapor region.

EXAMPLE 3.4

Given: Refrigerant-22 at $T = 40°C$, $x = 0.5$

Find: v (in m³/kg) and h (in kJ/kg)

Solution: In the fluid region, the steam quality x is defined only for the saturated liquid ($x = 0$), the saturated vapor ($x = 1$), or a mixture of vapor and liquid with x lying between 0 and 1. It is not defined for the superheated vapor region nor for the subcooled or compressed liquid region. Table A.7 gives the saturated refrigerant-22 properties (temperature table) and shows that at 40°C, $v_f = 0.000883$ m³/kg and $v_g = 0.0151$ m³/kg. Thus, v at the specified state can be calculated using Equation 3.15:

$$v = v_f + xv_{fg}$$

$$\therefore v = 0.000883 + 0.5(0.0151 - 0.000883) \text{ m}^3/\text{kg}$$

$$= \underline{0.008 \text{ m}^3/\text{kg}}$$

Following a similar approach, h can be determined as follows:

$$h = h_f + xh_{fg}$$

$$\therefore h = 94.93 + 0.5(261.14 - 94.93) \text{ kJ/kg}$$

$$= \underline{178 \text{ kJ/kg}}$$

EXAMPLE 3.5

Given: H_2O at $p = 30$ psia, $v = 3.0$ ft³/lbm

Find: T (in °F) and h (in Btu/lbm)

Solution: For $p = 30$ psia, from Table B.2 (USCS units), $T_s = 250.4°F$, $v_f = 0.01701$ ft³/lbm, and $v_g = 13.74$ ft³/lbm. A value of $v = 3.0$ ft³/lbm places the fluid state in the wet vapor region under the dome (see Fig. 3.16), and hence the quality x can be determined from Equation 3.15:

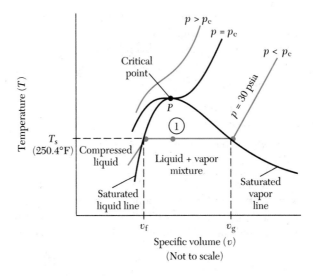

Figure 3.16 Temperature–volume (*T-v*) diagram for water showing the state specified in Example 3.5.

$$x = \frac{v - v_f}{v_{fg}}$$

$$= \frac{3 - 0.01701}{13.74 - 0.01701} = 0.217$$

Thus, for this case, the state is under the dome with $x = 0.217$. T is the saturation temperature, so $T = 250.4°F$.

To determine h, use the following:

$$h = h_f + xh_{fg}$$

$$= 219 + 0.217(1164 - 219) \text{ Btu/lbm}$$

$$= 424 \text{ Btu/lbm}$$

EXERCISES

for Section
3.5.2

1. Use the appropriate property tables to find the following unknown properties. Locate the state and phase(s) of the substance on a T-v diagram in each case.
   (a) H_2O at $p = 1000$ psia, $v = 0.3$ ft³/lbm; find T
   (b) H_2O at $T = 130°F$, $x = 0.9$; find h
   (c) Refrigerant-22 at $p = 3.2$ bar, $h = 102$ kJ/kg; find T, v

   Answer: (a) 545°F, (b) 1015 Btu/lbm, (c) −12.9°C, 0.0244 m³/kg

3.5.3 Compressed or Subcooled Liquid Region

The substance exists as a single phase in this region, and any two independent intensive thermodynamic properties can be used to specify the thermodynamic state for a simple compressible substance in this region.

EXAMPLE 3.6

Given: H_2O at $p = 14.7$ psia, $T = 70°F$

Find: u (specific internal energy) (in Btu/lbm)

Solution: From Table B.1, we find that $T_s = 212°F$ at $p = 14.7$ psia. The specified state is marked in Figure 3.17 as the intersection of the $p = 14.7$ psia line and the $T = 70°F$ line. This corresponds to the compressed liquid region since $T < T_s$. Other property values at the specified state can be read from Table B.3 (USCS units) for compressed water. The value of u is obtained by linear interpolation:

$$\frac{u - 18}{68 - 18} = \frac{70 - 50}{100 - 50}$$

Solving for u gives $u = 38.0$ Btu/lbm.

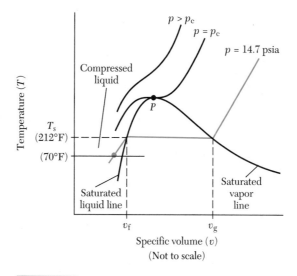

Figure 3.17 Temperature–volume (*T-v*) diagram for water showing the state specified in Example 3.6.

Compressed liquid and compressed solid data are not available for many substances. Fortunately, certain approximations can be made about the variation of properties in this region without a significant loss of accuracy. It has been found from observation that specific volume, internal energy, and entropy are primarily functions of temperature for compressed liquids and solids. Therefore, when data are not available for properties v, u, and s for the compressed liquid region, values are used for the saturated liquid at the same temperature as the compressed liquid state.

Enthalpy (H) was defined earlier by Equations 3.1 and 3.4e, and specific enthalpy (h) can thus be written

$$h = u + pv$$

The state of the compressed liquid at pressure p_0 and temperature T is marked as 0 in Figure 3.15, while that for the saturated liquid at the same temperature is marked as 1. The specific enthalpies at these states are therefore

$$h_0 = u_0 + p_0 v_0 \tag{3.19}$$

$$h_1 = u_1 + p_1 v_1 \tag{3.20}$$

Subtraction gives

$$h_0 = h_1 + (u_0 - u_1) + (p_0 v_0 - p_1 v_1) \tag{3.21}$$

The properties at state 1 can be represented by saturation properties corresponding to a saturation temperature, and since internal energy and specific volume are functions of temperature only, $u_0 = u_1 = u_f$ and $v_0 = v_1 = v_f$. Hence, Equation 3.21 reduces to

$$h_0 = (h_f)_T + (v_f)_T (p_0 - (p_s)_T) \tag{3.22}$$

where $(p_s)_T$ is the saturation pressure at temperature T and h_f and v_f are specific properties for the saturated liquid taken from values at the saturation temperature equal to the actual temperature T of the compressed liquid. Thus, for calculations of enthalpy of a compressed liquid, the enthalpy (h_f) at the saturated liquid state for the thermodynamic state temperature is used and a correction term for the pressure differences is added, as given by the last term in Equation 3.22.

EXAMPLE 3.7

Given: Compressed H_2O in state 1 at $p = 15$ MPa ($= 150$ bars), $T = 60°C$

Find: Obtain values for v, u, s, and h for state 1 either from the compressed liquid table for water or from the computer code **STEAM** (see Section 3.7). Then obtain values for v, u, and s for the saturated liquid at 60°C and for h for the compressed liquid as previously discussed (state 2). Compute the percentage error resulting from this approximation.

Solution: State 1 is the true value from the code **STEAM**. The properties v, u, and s for state 2 are taken from Table A.1 in Appendix A at the saturated liquid state for $T = 60°C$. The value for h_2 (h at state 2) is calculated from Equation 3.22 as

$$h_2 = h_0 = h_f + v_f(p_0 - p_s)$$

$$= 251.0 \text{ kJ/kg} + 0.00102 \text{ m}^3/\text{kg} (15,000 - 20 \text{ kPa})$$

$$= 266.3 \text{ kJ/kg}$$

	State 1	State 2	% Error for State 2
v (m³/kg)	0.00101	0.00102	0.99
u (kJ/kg)	248.6	251.0	0.97
s (kJ/kg · K)	0.823	0.831	0.97
h (kJ/kg)	263.8	266.3	0.95

The error is calculated by

$$\% \text{ error} = \left(\frac{\text{estimated value} - \text{true value}}{\text{true value}} \right) 100$$

Note that all values have an error of less than 1%, even for this very high pressure. If we had used the saturation value for enthalpy, the error would have been -4.85%.

EXAMPLE 3.8

Given: Refrigerant-22 at $p = 400$ psia, $T = 100°F$

Find: v (in ft³/lbm), s (in Btu/lbm · °R), h (in Btu/lbm)

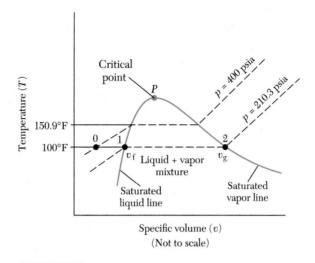

Figure 3.18 Temperature–volume (T-v) diagram for refrigerant-22 for Example 3.8.

Solution: Table B.8 for the saturated refrigerant-22 properties (pressure table) shows that $T_s = 150.9°F$. The T-v diagram (Fig. 3.18) indicates that the point of intersection of the $p = 400$ psia line and the $T = 100°F$ ($<T_s$) line is in the compressed liquid region. Thus, v and s can be given to a good approximation by the corresponding saturated liquid properties at a saturation temperature equal to 100°F. Table B.7 shows that at the saturation temperature of 100°F, $v_f = 0.014$ ft³/lbm and $s_f = 0.0801$ Btu/lbm · °R. Therefore, v and s at the specified state are as follows:

$$v = 0.014 \text{ ft}^3/\text{lbm}$$

$$s = 0.0801 \text{ Btu/lbm} \cdot °R$$

For h at 400 psia and 100°F, Equation 3.22 can be used:

$$h = (h_f)_{(T_s = 100°F)} + (v_f)_{(T_s = 100°F)}[p - (p_s)_{(T_s = 100°F)}]$$

Table B.7 gives h_f and p_s at the saturation temperature of 100°F, from which h can be computed using the previous equation:

$$h = 39.56 \frac{\text{Btu}}{\text{lbm}} + 0.014 \frac{\text{ft}^3}{\text{lbm}}(400 - 210.3)\frac{\text{lbf}}{\text{in.}^2}$$

$$= 39.56 \frac{\text{Btu}}{\text{lbm}} + 2.656 \frac{\text{ft}^3 \cdot \text{lbf}}{\text{lbm} \cdot \text{in.}^2}\left(\frac{144 \text{ in.}^2}{1 \text{ ft}^2}\right)\left(\frac{1 \text{ Btu}}{778.16 \text{ ft} \cdot \text{lbf}}\right)$$

$$= 40.1 \text{ Btu/lbm}$$

1. Find v and h for H_2O at $p = 150$ psia and $T = 120°F$.

   Answer: 0.0162 ft^3/lbm, 88 Btu/lbm

2. Find v and h for refrigerant-22 at 150 psia and 30°F.

   Answer: 0.0124 ft^3/lbm, 19.1 Btu/lbm

3. Find u and h for H_2O at 30 bar and 10°C.

   Answer: 41.8 kJ/kg, 44.8 kJ/kg

4. The *degrees of subcooling* refers to the temperature difference between the saturation temperature and the actual temperature of a subcooled liquid. For refrigerant-22 at 20 bar with 20°C of subcooling, find T, v, and h.

   Answer: 31.3°C, 8.55×10^{-4} m^3/kg, 83.8 kJ/kg

3.5.4 Saturated Solid–Vapor Mixture Region

Two independent intensive thermodynamic properties are necessary to specify properties in the saturated solid–vapor mixture region for a simple compressible pure substance. Since pressure and temperature are not independent, one additional property must be used in addition to these two to independently specify the thermodynamic state uniquely. Thus, the calculations involved in computing thermodynamic properties for this region are similar to those for the saturated liquid–vapor mixture region. Vapor quality is normally used in the computations, and the subscript g indicates the saturated vapor state. The subscript i is normally used for the saturated solid state, and the quality for this two-phase mixture is defined as the mass of vapor divided by the total mass of the mixture. Thus, the procedure here generally follows the one previously given in Section 3.5.2 for a liquid–vapor mixture.

EXAMPLE 3.9

Given: H_2O at $T = -20°C$, $v = 1000$ m^3/kg

Find: p (in Pa), h (in kJ/kg)

Solution: The temperature $T = -20°C$ indicated is below the saturated liquid–vapor temperatures, and thus Table A.4 must be consulted for the ice–steam region instead. From Table A.4 for saturated ice and steam, the saturation pressure p_s is found to be 103.3 Pa at the specified temperature. Also, $v_i = 1.087 \times 10^{-3}$ m^3/kg and $v_g = 1131$ m^3/kg at a saturation temperature of $-20°C$. The specified v lies between v_i and v_g, indicating that the H_2O is an ice–steam mixture, as illustrated in Figure 3.19. Thus, from Equation 3.15 with v_i in place of v_f

$$x = \frac{v - v_i}{v_g - v_i} = \frac{1000 - 0.001}{1131 - 0.001} = 0.88$$

and applying Equation 3.17 with h_i in place of h_f

$$h = h_i + x(h_g - h_i)$$

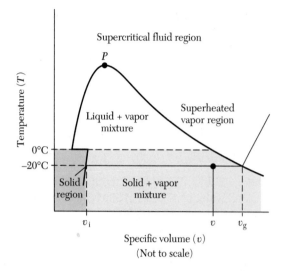

Figure 3.19 Temperature–volume (T-v) diagram for water (including the ice–steam region) showing the state specified in Example 3.9.

$$= -374.1 + 0.88[2464 - (-374.1)] \text{ kJ/kg}$$

$$= 2123 \text{ kJ/kg}$$

EXERCISES
for Section
3.5.4

1. Find T, v, and h for H_2O at 60 Pa and $x = 0.9$. Locate the state on a T-v diagram.
   Answer: $-25.5°C$, 1.71 m^3/kg, 2170 kJ/kg

2. Find p and u for H_2O at $-25°C$ and $v = 1200$ m^3/kg. Locate the state and phase(s) of the substance on a T-v diagram.
   Answer: 63.3 Pa, 1423 kJ/kg

3. Find p and v for H_2O at $-55°F$ and $h = 985$ Btu/lbm. Locate the state and phase(s) of the substance on a T-v diagram.
   Answer: 7.01×10^{-4} psia, 3.29×10^5 ft^3/lbm

3.5.5 Compressed Solid Region

The compressed solid region is treated in a fashion similar to the compressed liquid region. For a solid, provided the pressure (the normal stress) is the same in all directions, two independent intensive thermodynamic properties are sufficient to specify the thermodynamic state. A compressed solid is normally regarded as incompressible, and the properties specific volume (v), internal energy (u), and entropy (s) are assumed to be functions of temperature only. Thus, values for v, u, and s can be obtained by use of the saturated solid properties at the same temperature. In this case, Equation 3.22 can be used to compute the enthalpy for the solid using the enthalpy at the saturation solid state at the same temperature as the compressed solid. If stresses exist inside the body that are not the same in different directions, two properties do not define the thermodynamic state and Equation 3.22 cannot be used for the computation of enthalpy.

EXAMPLE 3.10

Given: H_2O at $T = -30°C$, $p = 100$ Pa

Find: v (in m^3/kg), s (in $kJ/kg \cdot K$), h (in kJ/kg)

Solution: Table A.5 for saturated ice and steam (pressure table) shows that $T_s = -20.3°C$ when $p = 100$ Pa. Since the specified T is below T_s, the H_2O is thus compressed ice (see Fig. 3.20).

At $-30°C$, the saturation pressure p_s is found from Table A.4 to be 38.02 Pa. The saturated solid properties v_i and s_i at $T = -30°C$ can be equated to v and s, respectively, at the specified state. The specific enthalpy can be determined using Equation 3.22 with h_i and v_i at $T = -30°C$ substituted for h_f and v_f, respectively, in the equation. Therefore,

$$v = 0.001086 \text{ m}^3/\text{kg}$$

$$s = -1.452 \text{ kJ/kg} \cdot \text{K}$$

$$h = h_i + v_i(p - p_s)$$

$$= -393.2 + 0.001086 \text{ m}^3/\text{kg} (0.1 - 0.038 \text{ kPa})$$

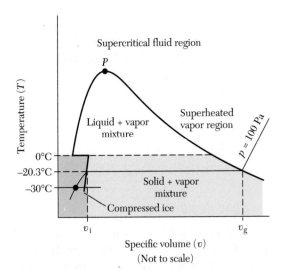

Figure 3.20 Temperature–volume $(T\text{-}v)$ diagram for water (including the ice–steam region) showing the state specified in Example 3.10.

Hence, $h = -393.2$ kJ/kg (to 1 decimal place). A negligible difference between the saturation value and the corrected value for enthalpy exists in this case. Note that the enthalpy value is negative because a datum state was set with a value of enthalpy of liquid water equal to zero at the triple point.

EXERCISES

for Section 3.5.5

1. Find v and h for H_2O at $p = 1$ atm and $T = -90°C$.
   Answer: 1.08×10^{-3} m^3/kg, -493 kJ/kg

2. Find v, u, and h for H_2O at $p = 1$ atm and $T = -65°F$.
   Answer: 0.0173 ft^3/lbm, -187 Btu/lbm, -187 Btu/lbm

3.6 Systematic Procedure for Reading Property Tables

Now that several examples have been provided on the use of property tables, the following is offered as a general guide to reading thermodynamic property tables for pure substances.

Step 1. Sketch a phase diagram (for example, the T-v diagram). To illustrate this, the fluid phases for a pure substance are shown in Figure 3.21. This T-v diagram shows the critical isobaric line, a constant pressure line at $p < p_c$, and a constant pressure line at $p > p_c$.

Step 2. Use the specified properties to identify the phase region in which your system lies on the phase diagram. The system pressure is often specified. If this is the

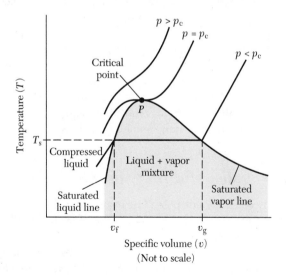

Figure 3.21 Temperature–volume (T-v) diagram for a typical pure substance.

case, you can identify the phase region in the following manner. (a) First check whether $p < p_c$ or $p > p_c$. If $p > p_c$, the wet vapor region is ruled out. Use the property table for the supercritical pressure region in this case. Note that the property tables in the appendix do not extend to the supercritical pressure region. Several tables are available for steam that cover a wider range of conditions than those in this book. Examples include the NBS/NRC Steam Tables by Haar et al. (1984) in SI units and the Steam Tables by Keenan et al. (1969) in English units. (b) Next, if $p < p_c$, then use the saturated fluid property table to find the saturation temperature (T_s) that corresponds to the specified pressure (p). When $p < p_c$, use the second property (Φ) value given to decide on which side of the saturation envelope the fluid state lies. The possibilities (excluding the possibility that Φ is density) are

- compressed liquid when $\Phi < \Phi_f$
- saturated liquid when $\Phi = \Phi_f$
- wet vapor when $\Phi_f < \Phi < \Phi_g$
- saturated vapor when $\Phi = \Phi_g$
- superheated vapor when $\Phi > \Phi_g$

Values of Φ_f and Φ_g can be read from the saturated fluid property table along with the saturation temperature.

Step 3. Look up the appropriate table of properties based on the conclusions reached in Steps 1 and 2 (Sections 3.5.1 through 3.5.3 explain how this can be done for the fluid phases).

Similar procedures as outlined here can be used if T is known rather than p. If both T and p are known, this procedure can be followed; however, the thermodynamic state is not uniquely specified by T and p if the substance exists as a liquid–vapor mixture. In this case, an additional property must be specified. The quality x is often used in such a case. When neither T nor p is known, it is difficult to determine the other state properties and a trial-and-error procedure is usually needed.

It is often necessary to interpolate between tabulated values to determine the property values at a specified state. In general, linear interpolation yields reasonably accurate results. For entropy, however, the pattern of variation is often logarithmic, in which case a linear interpolation may introduce significant errors. By contrast, the computer codes make use of accurately defined equations of state and thus avoid this difficulty.

3.7   *Computer Routines for Thermodynamic Properties*

It is generally not possible to provide a simple equation of state that can be used to compute thermodynamic properties for all the different phases in which a pure substance might exist. However, computer codes have been developed that can be used with any IBM-PC compatible computer for a wide range of calculations for thermodynamic properties. These codes are based on published thermodynamic data that were obtained from laboratory experimentation and observation. The tabulations provided in Appendices A (SI units) and B (USCS units) for ice, water, and steam and for refrigerant-22 were obtained from the computer codes **STEAM** and **R22**, respectively. Further information on these computer codes, including a step-by-step procedure on their use, is furnished in Appendix D.

EXERCISES
for Section
3.6

1. Verify your answers to Exercise 1 in Section 3.5.1 using the appropriate property code.

2. Verify your answers to Exercise 1 in Section 3.5.2 using the appropriate property code.

3. Verify your answers to Exercises 1 through 4 in Section 3.5.3 using the appropriate code.

4. Verify your answers to Exercises 1 through 3 in Section 3.5.4 using the **STEAM** code.

5. Verify your answers to Exercises 1 and 2 in Section 3.5.5 using the **STEAM** code.

3.8 Summary

The properties of a pure substance were discussed in this chapter. First, it was noted that, in principle, unique characterization of an equilibrium state for a general thermodynamic system might require specification of the values of all the thermodynamic properties of the system. However, many substances of interest in engineering belong to categories for which the values of only a few properties need to be specified. The specification of the values of only two independent intensive thermodynamic properties is adequate for fixing the equilibrium state of pure substances. The postulate that it takes two independent intensive thermodynamic properties to define the thermodynamic state of a simple compressible pure substance is called the *two-property rule*. Thermodynamic property values for particular substances are generally obtained from experimental observation and are available for many pure substances.

The *phases* of a pure substance were discussed along with pvT relationships for such substances. Much attention was devoted here to fluid (liquid and vapor) systems, with only brief attention given to solids and the various possible phases of solids. Several different phase equilibrium diagrams were presented showing the characteristics for water, a widely used pure substance. Thermodynamic surfaces were presented as a way to represent three different thermodynamic properties in one diagram. Tables of thermodynamic properties were discussed, with several different examples given to illustrate their use.

Finally, computer codes developed for support of this book were mentioned. These codes allow for computation of thermodynamic properties and for presentation of thermodynamic states by use of any IBM-PC compatible computer. Either the tables or the computer routines can be used to solve problems that are presented throughout the remainder of this book. However, several problems are marked with ▣; these problems either require the use of a computer code or can be solved more easily with the aid of one of the codes.

▣ indicates that the use of computer code is required or advised.

References

Bejan, A. 1988. *Advanced Engineering Thermodynamics*. New York: John Wiley & Sons.

Gibbs, J. W. 1928. *The Collected Works of J. Willard Gibbs. Vol. I. Thermodynamics*. New Haven, CT: Yale University Press. (Reprinted 1948.)

Haar, L., Gallagher, J. S., and Kell, G. S. 1984. *NBS/NRC Steam Tables: Thermodynamic and Transport Properties and Computer Programs for Vapor and Liquid States of Water in SI Units*. Washington, D.C.: Hemisphere.

Hatsopoulos, G. N. and Keenan, J. H. 1962. "A Single Axiom for Classical Thermodynamics." *Transactions ASME: Journal of Applied Mechanics*. Vol. 29. 193–199.

Haywood, R. W. 1980. *Equilibrium Thermodynamics for Engineers and Scientists*. New York: John Wiley & Sons.

Keenan, J. H., Keyes, F. G., Hill, P. G., and Moore, J. G. 1969. *Steam Tables: Thermodynamic Properties of Water Including Vapor, Liquid, and Solid Phases (English Units)*. New York: John Wiley & Sons.

Kline, S. J. and Koenig, F. O. 1957. "The State Principle: Some General Aspects of the Relationships Among the Properties of Systems." *Journal of Applied Mechanics*. Vol. 24, 29–34.

Spalding, D. B. and Cole, E. H. 1966. *Engineering Thermodynamics*. 2nd ed. London, U.K.: Edward Arnold (Publishers) Ltd.

Questions

1. A pure substance is a system that satisfies the following tests: (a) the same chemical species are present throughout the system, (b) the chemical species are chemically combined the same way and throughout, and (c) the chemical composition and aggregation are the same throughout the time period of interest. Use these tests to determine whether or not the substances listed here are pure substances. Check each column as appropriate.

System	Test (a)	Test (b)	Test (c)	Pure Substance?
Water–steam mixture				
Air				
Liquid air				
Air–liquid air mixture				
Hydrogen–nitrogen mixture				
Air–water mixture				
Automobile engine exhaust				
Air–fuel vapor undergoing combustion				
Air–fuel vapor (not reacting)				
Air–water vapor mixture				

2. State the two-property rule for a pure substance. What are the important restrictions on the applicability of the rule?

3. Indicate which of the following properties are *intensive* and which are *extensive*. Give both the SI unit and the USCS unit for each property.
   (a) pressure   (f) quality
   (b) temperature   (g) specific heat
   (c) volume   (h) specific internal energy
   (d) enthalpy   (i) Gibbs function
   (e) specific entropy (j) Helmholtz function

4. Explain what is meant by the *phase* of a substance. Use the phase diagram for water to explain the terms often used for describing changes of phase for a pure substance.

5. Use suitable diagrams to define or explain the following terms for a pure substance.
   (a) triple point
   (b) critical point
   (c) saturation envelope (or dome)

6. Water has a maximum density at 4°C when the pressure is 1 atm. In the neighborhood of this state, water has the same density at two different temperatures. Does the two-property rule apply to water under these conditions? Explain your answer.

7. With the aid of suitable diagrams, describe the physical and phase changes you would expect when compressed water is heated at a constant pressure (p) until it becomes a vapor (or gas). The description should cover three cases:
   (a) $p < p_c$
   (b) $p = p_c$
   (c) $p > p_c$

8. An ice–steam mixture is confined to a rigid vessel and then heated until it becomes a dry saturated vapor. Assuming the process took place in a quasi-equilibrium manner, show the path on the following diagrams for the ice–water–steam system.
   (a) T-v projection
   (b) p-v diagram
   (c) T-p projection

9. Use the appropriate data in the tables provided in Appendix A to produce the following projections (to scale). In each case, plot the dome and clearly indicate the different phase regions.
   (a) $\ln p$ versus h diagram for refrigerant-22
   (b) T versus h diagram for H_2O (inclusive of the ice region)
   (c) h versus s diagram for H_2O

10. Figure Q3.10 illustrates the T-v diagram for a pure substance. Points 1 and 2 are marked on a constant pressure $(< p_c)$ line. Temperature T_1 is below the critical temperature T_c, while T_2 is above the critical temperature.
    (a) Draw lines showing isothermal compression of the superheated vapor at states 1 and 2 until the fluid has turned liquid in each case.
    (b) If the fluid is compressed in a transparent container, what physical changes do you expect to see in (i) the isothermal compression to the liquid state from state 1 and (ii) the isothermal compression to the liquid state from state 2?

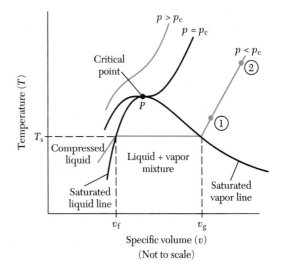

Figure Q3.10 Temperature–volume $(T$-$v)$ diagram for isothermal compression of a pure substance at subcritical and supercritical temperatures.

Problems

3.1 A number of states for water are listed below. For each state, use the tables of properties given in Appendices A and B to determine (i) the condition of the substance (whether it is superheated steam, subcooled or compressed water, wet steam, etc.) and (ii) the phase(s) present. Mark each of these

states on a T-v diagram that also shows the dome and the appropriate constant pressure line.
(a) $p = 1$ atm, $T = 27°C$
(b) $p = 0.05$ bar, $s = 7.368$ kJ/kg · K
(c) $p = 103.5$ Pa, $T = -20°C$
(d) $p = 4$ MPa, $v = 0.0988$ m³/kg
(e) $T = 100°C$, $x = 0.8$
(f) $T = 100°F$, $p = 14.7$ psia
(g) $x = 0.6$, $T = 500°F$
(h) $p = 200$ psia, $v = 2$ cu ft/lbm
(i) $p = 800$ psia, $T = 1000°F$
(j) $T = 1000°F$, $p = 2000$ psia

3.2 For each of the states listed in Problem 3.1, find
(a) the specific enthalpy
(b) the specific internal energy

3.3 Repeat Problem 3.2 using the computer code **STEAM** and compare these answers with your solutions to the problem.

3.4 5 kg of water is heated from 20°C to 90°C at a constant pressure of 1 atm. Estimate the increase in enthalpy
(a) using the steam table
(b) assuming that water has an average specific heat at constant pressure c_p of 4185 J/kg · K
(c) using the computer code **STEAM**

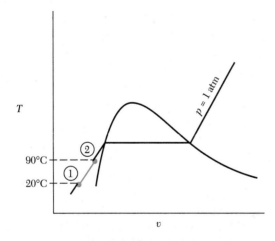

Figure P3.4 Constant pressure heating of water.

3.5 2 kg of steam at 1 atm is heated from 120°C to 200°C. Estimate the increase in enthalpy
(a) using the steam table

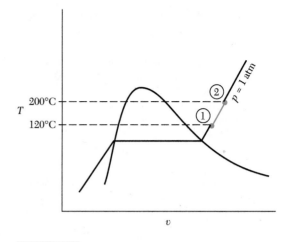

Figure P3.5 Constant pressure heating of steam.

(b) assuming that steam at 1 atm has an average specific heat at constant pressure c_p of 1.86 kJ/kg · K
(c) using the computer code **STEAM**

3.6 1 kg of ice at 0°C is mixed with 5 kg of water at 90°C. The final enthalpy of the mixture is the sum of the initial enthalpies of the ice and the water. What is the final temperature attained? Verify your specific enthalpy values using the **STEAM** code.

Figure P3.6 Mixing of ice and liquid water.

3.7 Dry saturated steam at 20 psia is admitted into the cylinder of the piston-and-cylinder arrangement of a Cornish beam engine. The cylinder has a bore (inside diameter) of 42 in. and a stroke of 8 ft.

Calculate the mass of steam admitted into the cylinder at each complete stroke of the engine.

3.8 Steam at a pressure of 100 psia and a temperature of 400°F is compressed in a piston-and-cylinder arrangement to a pressure of 250 psia. The decrease in specific enthalpy of the steam is found to be 120 Btu/lbm. Find the final temperature of the steam. If the steam is wet in the final condition, also determine its quality.

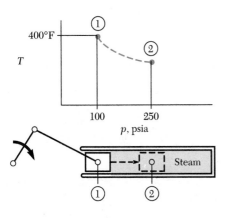

Figure P3.8 Steam compressed in a cylinder from state 1 to state 2.

3.9 Use steam tables to determine the unknown properties in the following:
(a) $p = 1$ bar, $v = 2.41$ m³/kg; $T = $ _____ °C
(b) $p = 1$ MPa, $T = 120°C$; $v = $ _____ m³/kg
(c) $T = 100°C$, $x = 0.6$; $p = $ _____ kPa, $v = $ _____ m³/kg
(d) $p = 6$ MPa, $T = 475°C$; $v = $ _____ m³/kg
(e) $p = 150$ bar, $v = 0.005$ m³/kg; $T = $ _____ °C
(f) $p = 1$ psia, $v = 631.1$ ft³/lbm; $T = $ _____ °F
(g) $p = 30$ psia, $T = 150°F$; $v = $ _____ ft³/lbm
(h) $T = 100°F$, $x = 0.8$; $p = $ _____ psia, $v = $ _____ ft³/lbm
(i) $p = 50$ psia, $T = 425°F$; $v = $ _____ ft³/lbm
(j) $p = 200$ psia, $v = 1.5$ ft³/lbm; $T = $ _____ °F

3.10 Verify the correctness of your solution to Problem 3.9 using the computer code **STEAM**.

* 3.11 A rigid vessel of volume 0.01 m³ contains 90% (by volume) water and 10% steam at a pressure of 2 bar. Determine
(a) the temperature of the water
(b) the total mass of the water and steam in the vessel

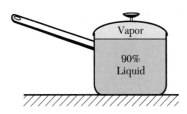

Figure P3.11 Kettle filled with liquid and vapor H_2O.

* 3.12 If the container in Problem 3.11 now has 10% (by volume) water and 90% steam at a pressure of 3 bar, determine
(a) the temperature of the system
(b) the total mass of the system

* 3.13 Wet steam at 250°C is contained in a vessel of volume 0.025 m³. The mass of the liquid water present is 4 kg. Evaluate the following for the system:
(a) the pressure
(b) the mass
(c) the quality
(d) the enthalpy

* 3.14 A rigid vessel contains 1 kg wet steam at a pressure of 150 kPa. When the system is heated, it eventually becomes dry saturated steam at 40 bar. Determine
(a) the quality of the wet steam initially in the vessel
(b) the volume of the vessel
(c) the increase in internal energy of the steam as it is heated from its initial state to the dry saturated vapor state

3.15 Adiabatic throttling in steady flow is an isenthalpic (or constant enthalpy) process. A steady flow of refrigerant-22 is throttled adiabatically from an initial pressure of 15 bar and a temperature of 35°C to the evaporator pressure of 2.019 bar. Use

(i) the refrigerant-22 property tables and (ii) the computer code **R22** to determine
(a) the final temperature of the refrigerant
(b) the quality of the refrigerant at the end of the throttling process
(c) the increase in entropy per unit mass of the fluid

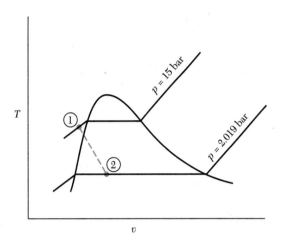

Figure P3.15 Adiabatic throttling process for refrigerant-22.

* **3.16** 1 kg of saturated liquid refrigerant-22 is confined in a cylinder beneath a weighted piston. If the temperature is 30°C, what is the pressure of the

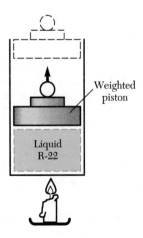

Figure P3.16 Constant pressure heating of refrigerant-22.

fluid? If the fluid is heated at constant pressure until the volumes of liquid and vapor are equal, determine the increase in enthalpy produced. Use the refrigerant-22 property tables first and then verify the property values using the computer code **R22**.

3.17 1 kg of superheated refrigerant-22 is cooled at a constant pressure of 20 bar from an initial temperature of 80°C to a final temperature of 40°C. Using (i) the property tables and (ii) the code **R22**, determine
(a) the change in entropy
(b) the change in enthalpy

3.18 2 lbm of wet refrigerant-22 vapor, having a quality of 0.3, is at a temperature of $-20°F$.
(a) What is the pressure of the fluid?
(b) If the fluid is heated at constant pressure until it all becomes a saturated vapor, what is the increase in enthalpy? Compare this with the product of the increase in specific entropy and the absolute temperature of the fluid.

3.19 The properties for superheated steam at 10 bar and 400°C are given in Table A.3 of Appendix A. At this state,

$$v = 0.3066 \text{ m}^3/\text{kg}$$
$$h = 3264 \text{ kJ/kg}$$
$$s = 7.465 \text{ kJ/kg} \cdot \text{K}$$

For 3 kg of steam at the specified state, compute
(a) the internal energy
(b) the Helmholtz function
(c) the Gibbs function

3.20 For 2 kg of steam at 60 bar and 500°C, compute the following (using the appropriate table in Appendix A for the data on steam at this state):
(a) the internal energy
(b) the Helmholtz function
(c) the Gibbs function

3.21 For 1 lbm of steam at 100 psia and 700°F, determine
(a) the internal energy
(b) the Helmholtz function
(c) the Gibbs function

3.22 For 1 lbm of refrigerant-22 at 300 psia and 200°F, determine
(a) the internal energy
(b) the Helmholtz function
(c) the Gibbs function

* 3.23 The kilogram molecular weight M of H_2O is 18 kg/kmol. Steam is maintained in a container at 100 bar and 500°C. For this system, determine its

   (a) specific volume
   (b) molar specific volume
   (c) specific internal energy
   (d) molar specific internal energy
   (e) specific enthalpy
   (f) molar specific enthalpy
   (g) specific entropy
   (h) molar specific entropy
   (i) specific Helmholtz function
   (j) molar specific Helmholtz function
   (k) specific Gibbs function
   (l) molar specific Gibbs function

* 3.24 Determine the properties listed in (a) through (l) of Problem 3.23 for steam at $p = 100$ psia and $T = 700°F$.

* 3.25 Determine the properties listed in (a) through (l) of Problem 3.23 for dry saturated refrigerant-22 vapor at $T = 25°C$. $M = 86.49$ kg/kmol for refrigerant-22 vapor.

* 3.26 Determine the properties listed in (a) through (l) of Problem 3.23 for saturated liquid refrigerant-22 at $T = 100°F$.

* 3.27 Equation 3.5 gives the definition of constant pressure specific heat, c_p. Using property values from the steam tables at $p = 1$ psia and at $T = 850°F$ and 950°F, estimate the value of c_p for steam at $T = 900°F$. Compare your result with the value for c_p given in the ideal gas tables for steam at $T = 900°F$ (1360°R).

* 3.28 Equation 3.6 gives the definition of constant volume specific heat, c_v. Using property values from the steam tables at $v = 809.8$ ft³/lbm and at $T = 850°F$ and 950°F, estimate the value of c_v for steam at $T = 900°F$. (You may wish to use the computer code supplied with this book instead of the steam tables for this problem.)

* 3.29 Equation 3.5 gives the definition of constant pressure specific heat, c_p. Using property values from the steam tables at $p = 0.0061$ bar and at $T = 200°C$ and 300°C, estimate the value of c_p for steam at $T = 250°C$. Compare your result with the value for c_p given in the ideal gas tables (Table A.19) for steam at $T = 250°C$ (523 K).

* 3.30 Equation 3.6 gives the definition of constant volume specific heat, c_v. Using property values from the steam tables (or from the computer code) at

$v = 48.3$ m³/kg and at $T = 200°C$ and 300°C, estimate the value of c_v for steam at $T = 250°C$.

* 3.31 A two-phase mixture of saturated water and steam consists of 1 kg of vapor and 0.5 kg of liquid at $p = 100$ bar. Compute the quality x for this mixture. Also compute v, h, and s for the mixture.

* 3.32 A two-phase mixture of saturated water and steam consists of 0.33 lbm of vapor and 0.15 lbm of liquid at $p = 10$ psia. Compute x, v, h, and s for the mixture.

* 3.33 A two-phase mixture of refrigerant-22 liquid and vapor consists of 0.1 kg of vapor and 0.06 kg of liquid at $T = 0°C$. Compute x, v, h, and s for the mixture.

* 3.34 A two-phase mixture of refrigerant-22 liquid and vapor consists of 1 kg of vapor and 1 kg of liquid at $T = -40°F$. Compute x, v, h, and s for the mixture.

* * 3.35 Superheated refrigerant-22 exists at $T = 50°F$ and $h = 110$ Btu/lbm. What is the pressure of the substance at this state? You may use the computer code **R22** for this.

* 3.36 Superheated steam exists at $T = 70°F$ and low pressure. What is the limiting value of the specific enthalpy for the substance at this temperature?

* 3.37 Superheated steam exists at $p = 15$ psia and $h = 1155$ Btu/lbm. What is the density of steam at this state?

* 3.38 Wet refrigerant-22 vapor exists at $T = 20°C$ and $h = 200$ kJ/kg. What is the density of the substance at this state?

* 3.39 Compressed liquid H_2O is at $p = 1000$ psia and $T = 50°F$. Obtain values for v, u, s, and h either from the compressed liquid table for H_2O or from the computer code **STEAM**. Then calculate these values from saturated liquid properties at $T = 50°F$ by using the procedures outlined in Section 3.5.3. Determine the percentage error for the calculated values.

* 3.40 Compressed liquid H_2O is at $p = 100$ bar and $T = 30°C$. Obtain values for v, u, s, and h for the substance at this state from the computer code **STEAM**. Then calculate these values from saturated liquid properties at $T = 30°C$ by using the procedures outlined in Section 3.5.3. Determine the percentage error for the calculated values.

* 3.41 Compressed refrigerant-22 liquid is at $p =$

100 bar and $T = 0°C$. Determine v, s, and h for this state.

* **3.42** Compressed refrigerant-22 liquid is at $p = 500$ psia and $T = 0°F$. Determine v, s, and h for this state.

3.43 The substance H_2O exists in equilibrium at $T = -30°C$ and $v = 2000$ m³/kg. Determine p and h at this state.

3.44 The substance H_2O exists in equilibrium at $T = -10°C$ and $v = 300$ m³/kg. Determine p, u, and h at this state.

3.45 The substance H_2O exists in equilibrium at $T = -40°F$ and $p = 1$ atm. Determine v, u, and h at this state. Describe the substance at this state.

3.46 The substance H_2O exists in equilibrium at $T = -20°C$ and $p = 1$ atm. Determine v, u, and h at this state. Describe the substance at this state.

* **3.47** Refrigerant-22 exists in equilibrium at $T = -40°C$ and $p = 26$ bar. Determine v, h, u, and s for this state.

* **3.48** Refrigerant-22 exists in equilibrium at $T = -40°F$ and $p = 20$ atm. Determine v, h, u, and s for this state.

* **3.49** A rigid vessel contains 0.5 kg of wet steam at $p = 200$ kPa. When the system is heated, it eventually reaches the critical point as shown in Figure P3.49. Name one intensive property that stays constant throughout the process. Determine
(a) the quality of the wet steam at the initial state
(b) the volume of the vessel
(c) the increase in internal energy of the steam

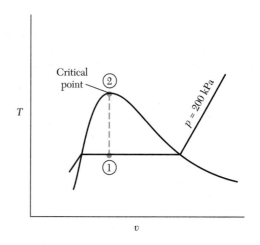

Figure P3.49 Constant volume heating of wet steam.

* **3.50 (a)** The initial state of a fixed mass of steam is $p = 5$ bar and $T = 250°C$. At the final state, $p = 1$ bar.
  (i) If the specific volume is the same at the initial and final states, determine, using property tables, the quality of steam at the final state.
  (ii) Assuming instead that the specific enthalpy remains constant, determine, using property tables, the temperature of the system in the final state.
(b) Check your results in part **(a)** using the computer code **STEAM**.

Ideal Gas and Real Gas

■■■ Ideal gas pvT relationships, Avogadro's law, and conditions for ideal gas approximation of real gases

■■■ Internal energy and enthalpy of ideal gases, principal specific heats and zero pressure specific heats, perfect gases, ideal gas tables, and the computer code **GAS** for gas properties

■■■ Principle of corresponding states, compressibility factor, reduced properties, and compressibility charts

■■■ Real gas pvT equations of state, van der Waals equation, Beattie–Bridgeman equation, Redlich–Kwong equation, the virial form of equation of state in powers of density, and the virial form of equation of state in powers of pressure

4.1 Definition of Ideal Gas

No simple equation is available to relate the thermodynamic properties of a pure substance for all phases of the substance. As discussed in Chapter 3, an equation to relate the thermodynamic properties at different thermodynamic states is called an *equation of state*. Although no such equation with general applicability is available, relatively simple equations of state do exist that can be used for pure substances in certain regions, such as for a single phase of a substance. Such an equation does exist for gases at low pressures with temperatures that are high in comparison to the critical temperature. A gas in this state is called an *ideal gas* or a *perfect gas*. Although these terms are used interchangeably by some authors, the use of the term *perfect gas* is restricted here to a specific category of ideal gas in the manner defined in Section 4.4.

An *ideal gas* is defined as a gas that obeys the following pvT equation of state:

$$pv = RT \qquad (4.1)$$

where R has a fixed value for any particular gas and is called the *specific gas constant* for the gas. The SI unit for R is kJ/(kg · K), while the USCS unit is Btu/(lbm · °R). This definition is adequate, and it can be shown that for gases that obey the $pv = RT$ law, the internal energy is a function of temperature only. The proof that the internal energy of a gas, which obeys Equation 4.1, depends only on temperature is usually based on a Maxwell relation and on a mathematical theorem that is considered in Chapter 12 (see also Haywood 1980, 278–280, for a formal proof). Since the internal energy of an ideal gas depends only on temperature, the enthalpy (h) and the specific heats (c_v and c_p) are also functions of temperature only.

Historically, the ideal gas pvT equation was discovered through experimental observation on gases at relatively low pressure and high temperature conditions. The principle is also validated from the kinetic theory of gas model in which assumptions are made that are equivalent to the low pressure and high temperature conditions observed in ideal gas behavior. The ideal gas model assumes that the gas molecules are infinitesimally small and exert no force on one another. For high pressure gases, the intermolecular distances are very small and a force of repulsion exists among the molecules. As the gas temperature is lowered, the intermolecular distances are again reduced, leading to significant intermolecular forces. Equation 4.1 is the defining equation for ideal gas behavior, and any gaseous system that obeys the ideal gas pvT equation is an ideal gas.

Boyle's law and Charles's law are special cases of the ideal gas pvT equation; these laws preceded the ideal gas law and led to its development. Boyle's law relates pressure and specific volume when temperature is held constant. Robert Boyle found that this relationship is given by

$$p_1 v_1 = p_2 v_2 = pv = \text{constant}_1 \tag{4.2}$$

Charles's law can be expressed as follows when the volume is held constant:

$$\frac{T_1}{p_1} = \frac{T_2}{p_2} = \frac{T}{p} = \text{constant}_2 \tag{4.3}$$

and by the following equation when the pressure is held constant:

$$\frac{T_1}{v_1} = \frac{T_2}{v_2} = \frac{T}{v} = \text{constant}_3 \tag{4.4}$$

A combination of Boyle's law and Charles's law leads to the following equation:

$$\frac{p_1 v_1}{T_1} = \frac{p_2 v_2}{T_2} = \frac{pv}{T} = \text{constant}_4 = R \tag{4.5}$$

where constant$_4$ in this case is R, the specific gas constant for the gas.

The specific volume (v) is defined as the total volume (V) divided by the total mass (m), or

$$v = \frac{V}{m} \tag{4.6}$$

The ideal gas pvT equation can therefore also be written in terms of the total volume as

$$pV = mRT \tag{4.7}$$

Let's now define the volume per mole of a substance as

$$\bar{v} = \frac{V}{N} \tag{4.8}$$

where N is the amount (in mol or kmol) of the substance. An overbar on a quantity indicates that the substance is on a molar basis (and not on a mass basis). The mol and kmol (or kgmol) were defined earlier in Section 2.4. Thus, the SI unit of kmol (instead of kg) is used for the amount of the substance on a molar basis. This approach is convenient and useful because 1 mol of any gas occupies the same volume as 1 mol of any other gas if the pressure and temperature are the same for the gases. This conclusion follows from Avogadro's law, which states that equal volumes of different gases at the same temperature and pressure contain the same number of molecules. Also, 1 kmol of any gas contains 6.022×10^{26} molecules (Avogadro's number), and 1 kmol of any substance contains a mass equal to the molecular (or molar) mass M of the substance. From this knowledge, one can compute the amount of a substance (in kmol) as follows:

$$N = \frac{m}{M} \tag{4.9}$$

where M is the molecular weight (or molar mass) in kg/kmol for the substance. If Equations 4.7, 4.8, and 4.9 are combined, the ideal gas law can be written as

$$p\bar{v} = MRT \tag{4.10}$$

From experimental observation it has been found that the quantity MR is a universal constant. This constant is called the *universal gas constant* and is defined by

$$\bar{R} = MR \tag{4.11}$$

The universal gas constant is a physical constant whose value is given by

$$\begin{aligned}
\bar{R} &= 8314.4 \text{ J/kmol} \cdot \text{K} \\
&= 8.3144 \text{ kJ/kmol} \cdot \text{K} \\
&= 1545 \text{ ft} \cdot \text{lbf/lbmol} \cdot \text{°R} \\
&= 1.986 \text{ Btu/lbmol} \cdot \text{°R}
\end{aligned} \tag{4.12}$$

Combining Equations 4.10 and 4.11 gives the following as another expression for the ideal gas pvT equation:

$$p\bar{v} = \bar{R}T \tag{4.13}$$

Combining Equations 4.8 and 4.13 leads to an alternate expression:

$$pV = N\bar{R}T \tag{4.14}$$

It is important to note that the ideal gas pvT equation merely represents an approximation for the behavior of real gases. The following section compares ideal gas with real gas to determine more precisely the regions of applicability of this ideal gas equation of state.

4.2   Comparison of Ideal Gas with Real Gas

The ideal gas equation of state was first developed from experimental observations. Consideration of molecular behavior and comparison of calculated results and laboratory data over a wide range of conditions have allowed the appropriate region to be established for use of the ideal gas equation. Figure 4.1 shows the general region of applicability of the ideal gas assumption for a pure substance. The region indicated on the temperature–volume (T-v) diagram in Figure 4.1(a) corresponds to pressures much lower than the critical pressure and to very high values of specific volume (or very low density). Furthermore, the ideal gas model is generally a more accurate description of a real gas when the gas temperature is above the critical temperature rather than below it. This is illustrated in Figure 4.1(b), which gives the temperature–pressure (T-p) diagram for a typical pure substance. The ideal gas region in Figure 4.1(b) corresponds to high temperatures and low pressures, but also extends to low temperatures, relative to the critical point, when the pressure approaches zero.

In summary, from a molecular perspective, the ideal gas model is valid when the intermolecular forces in the gas are small and negligible and when the mean free path between gas molecules is high. From a macroscopic perspective, based on a study of the vast experimental data available for several substances, the following criteria define the conditions for the ideal gas model to be appropriate.

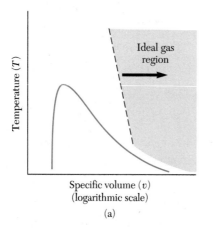
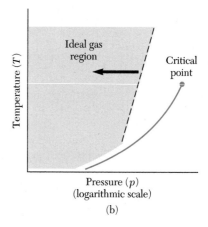

Figure 4.1 (a) Equilibrium temperature–volume (T-v) diagram for a pure substance showing region of applicability of ideal gas assumption. (b) Equilibrium temperature–pressure (T-p) diagram for a pure substance showing region of applicability of ideal gas assumption.

1. The pressure approaches zero (that is, the pressure, p, should be much less than the critical pressure, p_c for the substance).
2. The temperature is greater than twice the critical temperature (that is, $T > 2T_c$).
3. The molecular weight of the substance is low.

Of the three criteria listed, the single most important criterion for an ideal gas is the condition that *the pressure is low* (or approaches zero) *relative to the critical pressure*. For monatomic and diatomic gases, the ideal gas model is generally good up to 10 to 20 atm for temperatures at room temperature and above.

A particular case of considerable interest in engineering is atmospheric air. The primary constituents of atmospheric air are nitrogen and oxygen, both of which have critical pressures much greater than 1 atm and critical temperatures much lower than typical ambient temperature. Thus, it is found that the ideal gas equation of state is valid for air at pressures as high as 25 atm and temperatures of 25°C and higher. For pressures less than or equal to 1 atm, the equation applies at lower temperatures down to about − 100°C. The ideal gas equation of state can be applied to air over this wide range of conditions with an error of less than 1%.

The situation is not as favorable when considering water vapor (or steam). The critical pressure of water is close to 218 atm, while the critical temperature is 647 K. While steam at pressures and temperatures close to ambient are not far from the ideal gas approximation (in spite of the low temperature), at higher pressures approaching the critical pressure, the ideal gas model is inappropriate. The following examples for water vapor illustrate quantitatively the region in which steam behavior can be reasonably regarded as ideal gas behavior.

EXAMPLE 4.1

Given: Water vapor at the following states:

State 1. $p = 0.010$ MPa, $T = 100°C$
State 2. $p = 0.10$ MPa, $T = 100°C$
State 3. $p = 0.5$ MPa, $T = 500°C$
State 4. $p = 1.00$ MPa, $T = 600°C$
State 5. $p = 10.0$ MPa, $T = 600°C$
State 6. $p = 15.0$ MPa, $T = 600°C$

Find:
(a) Tabulate the properties p, v, and T for the thermodynamic states indicated.
(b) Compute the quantity pv/RT for each of these states.
(c) Explain the significance of the results in part (b) concerning ideal gas behavior of water vapor.

Solution: For water vapor, the ideal gas constant (R) is obtained from Equation 4.11:

$$R = \frac{\overline{R}}{M} = \frac{8.314 \text{ kJ/kmol} \cdot \text{K}}{18 \text{ kg/kmol}} = 0.4619 \text{ kJ/kg} \cdot \text{K}$$

Values for the specific volume (v) for steam at the stated conditions are obtained from Table A.3 in Appendix A.

(a)

State	p (kPa)	v (m³/kg)	T (K)	pv/RT
1	10	17.2	373	0.998
2	100	1.70	373	0.987
3	500	0.711	773	0.996
4	1000	0.401	873	0.994
5	10000	0.0383	873	0.950
6	15000	0.0248	873	0.923

(b) Since $pv/RT = 1$ for ideal gas, one can see from the last column that increasing the pressure reduces the agreement of the ideal gas equation with the actual properties. Furthermore, comparison of states 2 and 3 shows that increasing the temperature increases the agreement of the ideal gas equation with the actual properties.

EXERCISES

for Section 4.2

1. The molecular weight (or molar mass) of air is 28.96 lbm/lbmol. Use Equation 4.11 to calculate the specific gas constant for air
   (a) in ft · lbf/lbm · °R
   (b) in Btu/lbm · °R

   Answer: (a) 53.3 ft · lbf/lbm · °R, (b) 0.0686 Btu/lbm · °R

2. Assume a molecular weight of 18.015 for H_2O. Calculate R for H_2O
   (a) in ft · lbf/lbm · °R
   (b) in Btu/lbm · °R

   Answer: (a) 85.8 ft · lbf/lbm · °R, (b) 0.110 Btu/lbm · °R

3. The total volume of air in a house is 12,000 ft³. If the air pressure and temperature are 14.5 psia and 78°F, respectively, determine
   (a) the mass (in lbm) of air in the house
   (b) the number of moles (in lbmol) of the air in the house

   Answer: (a) 874 lbm, (b) 30.1 lbmol

4. Given $M_{air} = 28.96$, what is the specific gas constant for air?

   Answer: 0.287 kJ/kg · K

5. Atmospheric air is at 101 kPa (abs) and 25°C. Determine
   (a) the density of the air
   (b) the molar specific volume of the air

   Answer: (a) 1.18 kg/m³, (b) 24.5 m³/kmol

6. Use steam tables to determine (a) v for H_2O at 1000 kPa (abs) and 455 K. Calculate (b) $\bar{v}$ and (c) $p\bar{v}/T$. Is it reasonable to regard the steam as an ideal gas?

   Answer: (a) 0.196 m³/kg, (b) 3.52 m³/kmol, (c) 7.74 kJ/kmol · K

4.3 Internal Energy and Enthalpy of Ideal Gas

As stated in Section 4.1, the specific internal energy of an ideal gas is a function of temperature only. This result was demonstrated by Joule in a classic experiment of thermodynamics and can be expressed mathematically as follows:

$$u = u(T) \text{ only} \quad \text{(for ideal gas)} \qquad (4.15)$$

Since specific enthalpy is defined by

$$h = u + pv \qquad (4.16)$$

and since

$$pv = RT \qquad (4.1)$$

a combination of these two equations leads to the following expression for the specific enthalpy of an ideal gas:

$$h = u + RT \qquad (4.17)$$

where R is a constant and u is a function of temperature only. Therefore, h is also a function of temperature only. Mathematically,

$$h = h(T) \text{ only} \quad \text{(for ideal gas)} \qquad (4.18)$$

Thus, the specific internal energy and the specific enthalpy of an ideal gas are *both functions of temperature only*. The expression in terms of molar quantities, corresponding to Equation 4.17 is

$$\bar{h} = \bar{u} + \bar{R}T \qquad (4.19)$$

4.4 Specific Heats of Ideal Gas

The constant pressure specific heat (c_p) and the constant volume specific heat (c_v) are thermodynamic properties that were defined earlier as

$$c_p = \left(\frac{\partial h}{\partial T}\right)_p \qquad (3.5)$$

$$c_v = \left(\frac{\partial u}{\partial T}\right)_v \qquad (3.6)$$

Since specific enthalpy and specific internal energy are functions of temperature only for an ideal gas, these expressions become, for an ideal gas,

$$c_p = \frac{dh}{dT} \qquad (4.20)$$

$$c_v = \frac{du}{dT} \qquad (4.21)$$

or

$$dh = c_p \, dT \tag{4.22}$$

$$du = c_v \, dT \tag{4.23}$$

Regardless of the processes involved, Equations 4.22 and 4.23 can be used for the computation of changes in enthalpy and internal energy for any ideal gas. The equations are important and are frequently used in analyses involving the application of the first law of thermodynamics to gaseous systems.

A simple relationship exists between c_p and c_v for an ideal gas. Forming the differential of Equation 4.17 gives

$$dh = du + R \, dT \tag{4.24}$$

Equations 4.22, 4.23, and 4.24 can be combined to obtain the following:

$$c_p \, dT = c_v \, dT + R \, dT \tag{4.25}$$

or

$$c_p - c_v = R \tag{4.26}$$

Equation 4.26 for an ideal gas is a very useful relationship because if any two of the three values c_p, c_v, and R are known, the third can be determined from the other two.

Combining Equation 3.7 for the specific heat ratio ($k = c_p/c_v$) with Equation 4.26 leads to the following alternate expressions for c_p and c_v:

$$c_p = \frac{kR}{k - 1} \tag{4.27}$$

$$c_v = \frac{R}{k - 1} \tag{4.28}$$

Thus, the specific gas constant and the specific heat ratio of an ideal gas are adequate to determine both c_p and c_v for the gas.

It has been found from the kinetic theory of gases that c_p for an ideal monatomic gas such as argon or neon is given by

$$c_p = \frac{5}{2}R \tag{4.29}$$

Combining Equations 4.26 and 4.29 leads to the following equation for the ideal monatomic gas c_v:

$$c_v = \frac{3}{2}R \tag{4.30}$$

Perfect Gases

The term *perfect gas* is used here for a gas that obeys Equation 4.1 and whose specific heats are constant. For any case where c_p and c_v can be considered as constants, Equations 4.22 and 4.23 can be integrated to obtain the following expressions for changes in enthalpy and internal energy for the gas:

$$h_2 - h_1 = c_p(T_2 - T_1) \tag{4.31}$$

$$u_2 - u_1 = c_v(T_2 - T_1) \tag{4.32}$$

Zero Pressure Specific Heats

As indicated earlier, real gases exhibit ideal gas behavior at very low pressures. For these conditions, one can conclude that the specific heats are functions of temperature only. Thus, results of measurements made at very low pressures as a function of temperature for the specific heats of the gases can be used to approximate those for the gas at zero pressure. The theoretical limit at the zero pressure condition is referred to as the *zero pressure specific heat* corresponding to the ideal gas behavior of the substance. Figure 4.2 shows results from laboratory measurements for constant pressure specific heat (c_{p0}) for gases at low pressure. The subscript 0 (zero) on the specific heat indicates zero pressure and generally means that the measurements were made at low enough pres-

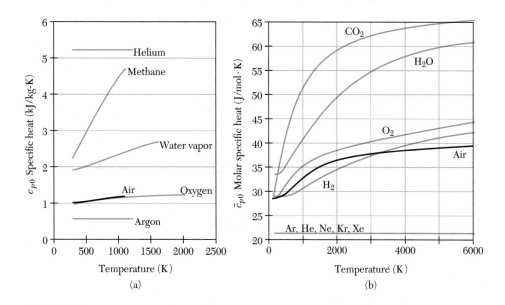

Figure 4.2 Constant pressure specific heats for a number of gases at zero pressure (charts developed from JANAF Thermochemical Data, Dow Chemical Company, Midland, Michigan).

sures for ideal gas behavior to be exhibited by the gas. Figure 4.2 shows that some gases have a constant specific heat, while the specific heat for others varies considerably with temperature.

To obtain enthalpy and internal energy changes for the gases in the ideal gas region where the specific heats are not constants, mathematical expressions can be developed for the specific heats from the data. The following expressions can then be used to obtain the specific enthalpy and internal energy changes, respectively:

$$h_2 - h_1 = \int_{T_1}^{T_2} c_{po}(T)dT \tag{4.33}$$

and

$$u_2 - u_1 = \int_{T_1}^{T_2} c_{vo}(T)dT \tag{4.34}$$

Extensive data are available for specific heats of gases at low pressures. A sample of such data is given in Table A.21 and B.21 in the appendices. The trends in Figure 4.2 show that over a small range of temperatures, a constant value for specific heats can be used. For a larger temperature range, a linear variation of specific heats with temperature can be assumed. For a very wide temperature range, however, a more precise curve fit will often be required.

The previous equations are applicable to ideal gas systems only. For real gases at high pressures or at conditions outside the ideal gas region, other approaches are needed to correctly determine the thermodynamic properties.

EXAMPLE 4.2

Calculate the molar specific heat at constant pressure for argon. The molecular weight (M) of argon is 39.94. Also determine c_p and c_v for argon. The temperature of a mass of argon gas is raised from 25°C to 150°C. Calculate the increase in the specific enthalpy of the gas.

Given: Argon gas, $M = 39.94$, $T_1 = 25°C$, $T_2 = 150°C$

Find: (a) $\bar{c}_p$, (b) c_p, (c) c_v, (d) $\Delta h = h_2 - h_1$

Solution:

(a) Argon is a monatomic gas. Assume that it behaves as an ideal gas. Hence, Equation 4.29 applies:

$$c_p = \frac{5}{2}R$$

The molar specific heat at constant pressure ($\bar{c}_p$) can be shown from its basic definition to be related to c_p according to the following:

$$\bar{c}_p = Mc_p$$

Thus,

$$\bar{c}_p = \frac{5}{2}MR = \frac{5}{2}\bar{R}$$

Substituting for the universal gas constant in this expression gives

$$\bar{c}_p = 20.8 \text{ kJ/kmol} \cdot \text{K}$$

(b)

$$c_p = \bar{c}_p/M$$

$$= 0.520 \text{ kJ/kg} \cdot \text{K}$$

(c) R for argon can be calculated by using Equation 4.11:

$$R = \frac{\bar{R}}{M} = 0.208 \text{ kJ/kg} \cdot \text{K}$$

Equation 4.26 gives

$$c_v = c_p - R$$

$$= 0.312 \text{ kJ/kg} \cdot \text{K}$$

(d) Equation 4.31 can be used to determine the increase in specific enthalpy:

$$h_2 - h_1 = c_p(T_2 - T_1) = 65 \text{ kJ/kg}$$

EXERCISES

for Section 4.4

1. Specific heat at constant pressure (c_p) for air at typical ambient temperatures can be taken as 1.005 kJ/kg · K. Convert this to $\bar{c}_p$ (in kJ/kmol · K). Determine $\bar{c}_v$, c_v, and k for air. ($M_{air} = 28.96$).

   Answer: 29.1 kJ/kmol · K, 20.8 kJ/kmol · K, 0.718 kJ/kg · K, $k = 1.4$

2. Ideal gas steam has $\bar{c}_p = 33.6$ kJ/kmol · K at 25°C. Determine the ideal gas $\bar{c}_v$ and k for steam at this temperature.

   Answer: 25.3 kJ/kmol · K, $k = 1.33$

3. Use the steam table to determine the increase in specific enthalpy of low pressure steam at 1 psia when its temperature is raised from 200°F to 500°F. Compare your estimate with the value obtained when the low pressure steam is treated as an ideal gas with $c_p = 0.459$ Btu/lbm · °R.

   Answer: 138 Btu/lbm

4. Use the steam table to determine the increase in specific enthalpy of steam at 150 psia whose temperature is raised from 400°F to 600°F. Compare your estimate with the value obtained assuming ideal gas behavior with c_p taken as 0.474 Btu/lbm · °R.

   Answer: 106 Btu/lbm

4.5   *Ideal Gas Tables and Computer Routines*

Measurements have been made for many gases at low pressures to obtain specific heat data as a function of temperature only. Such data have been used here to develop appropriate correlations for both the computer code **GAS** and the ideal gas (zero pressure) Tables A.11 through A.20 (SI units) in Appendix A and Tables B.11 through B.20 (USCS units) in Appendix B. The computer code **GAS** applies to the ideal gas properties of the following gases: air, methane, oxygen, nitrogen, carbon dioxide, carbon monoxide, and water vapor. The code is supplied with this book. Further details on the code are provided in Appendix D.

The primary data used for each ideal gas is the constant pressure specific heat, (c_{p0}) as a function of temperature. Appropriate curve-fitting techniques were employed to generate the functional relationships for $c_{p0} = c_{p0}(T)$, which were then used with Equation 4.33 to produce equations for the specific enthalpy of the gas. Other equations in Chapter 8 have been used for the computation of the entropy of each ideal gas at a standard pressure of 1 atm. Unlike enthalpy, the specific entropy for an ideal gas depends not only on temperature but also on pressure or volume.

4.6   *Principle of Corresponding States and Compressibility Charts*

Several different approaches exist for determining the thermodynamic properties of real gases. Real gases behave as ideal gases only in certain regions and, more generally, are not accurately modeled by the ideal gas pvT equation. It has already been observed that pressure makes a significant difference in the behavior of gases and that all gases tend to approach the behavior of ideal gas only as they approach zero pressure. One approach for dealing with real gases that evolved from this knowledge involves the use of a *compressibility factor* (Z), which is defined by

$$Z = \frac{pv}{RT} \qquad (4.35)$$

If Equation 4.35 is compared with Equation 4.1, one notes that the compressibility factor Z = 1 for an ideal gas. How then does one find the value of Z for a real gas? The answer is, from experimental measurements.

Figure 4.3 is a compressibility chart for nitrogen, which is a typical real gas. From examination of the figure, you can see that the compressibility factor Z approaches unity as the pressure approaches zero in accordance with the earlier assertion. Also observe from the figure that the value of Z significantly deviates from unity at high pressures and low temperatures.

If a compressibility chart is available for the gas under consideration, it can be used along with Equation 4.35 as the pvT equation of state for the gas. Unfortunately, such charts are not available for many gases, but the principle of corresponding states allows one to develop a generalized compressibility chart that can be used for approximate calculations for all gases.

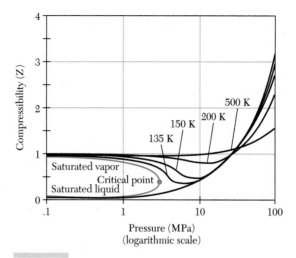

Figure 4.3 Compressibility chart for nitrogen (developed from selected data obtained from Thermodynamic Properties of Nitrogen, National Standard Reference Data Service of the U.S.S.R., Hemisphere Publishing Corp., 1987).

From experimental observations, it has been found that all gases exhibit similar trends in their compressibility behavior. Furthermore, it has been found that nondimensional pressures and temperatures can be used to normalize the properties of all gases with respect to the critical state. Thus, a single generalized compressibility chart can be developed that applies to all gases. A *reduced property* is defined as the ratio of the property to the value of the property at the critical state. Thus, the *reduced pressure* is defined as

$$p_R \equiv \frac{p}{p_c} \tag{4.36}$$

and the *reduced temperature* as

$$T_R \equiv \frac{T}{T_c} \tag{4.37}$$

According to the *principle of corresponding states*, the compressibility factor for any gas is a function of only p_R and T_R. Thus,

$$Z = Z(p_R, T_R) \tag{4.38}$$

The principle of corresponding states is thus an assertion, based on empirical studies, that the compressibility factors for most gases are identical when the gases have the same reduced pressure and temperature. It is clearly an approximation, but this observation allows preparation of one generalized compressibility chart that can be used in conjunction with Equation 4.35 to determine the thermodynamic pvT relationships for all gases. Such a generalized compressibility chart is presented in Figure C.1 in Appendix C. (Note that an iterative solution procedure is necessary for Z if v is specified instead of T or p.)

EXAMPLE 4.3

Given: Nitrogen at $p = 100$ atm and $T = 300$ K. The critical properties of nitrogen are $p_c = 3390$ kPa and $T_c = 126.2$ K.

Find: Specific volume v by use of Equation 4.38 and the compressibility chart. Compare with results from ideal gas equation.

Solution:

$$T_R = \frac{T}{T_c} = \frac{300 \text{ K}}{126.2 \text{ K}} = 2.38$$

$$p_R = \frac{p}{p_c} = \frac{(100)(101.325) \text{ kPa}}{3390 \text{ kPa}} = 2.99$$

From the generalized compressibility chart, $Z = 1.01$. R is computed using Equation 4.11 and substituted in Equation 4.35 along with Z, p, and T to determine v.

$$R = \frac{\overline{R}}{M} = \frac{8.314 \text{ kJ/kmol} \cdot \text{K}}{28 \text{ kg/kmol}} = 0.297 \text{ kJ/kg} \cdot \text{K}$$

$$\therefore v = \frac{ZRT}{p} = \frac{(1.01)(0.297 \text{ kJ/kg} \cdot \text{K})(300 \text{ K})}{(100)(101.325) \text{ kPa}}$$

$$= 8.88 \times 10^{-3} \frac{\text{kN} \cdot \text{m} \cdot \text{K}}{\text{kg} \cdot \text{K}} \frac{m^2}{\text{kN}}$$

$$= \underline{8.88 \times 10^{-3} \text{ m}^3/\text{kg}}$$

Since $Z = 1.01$, the calculated v using Equation 4.35 is 1% more than the value obtained from the ideal gas equation.

EXAMPLE 4.4

Given: Nitrogen at $p = 100$ atm and T = 145 K.

Find: Specific volume v and compare results with those in Example 4.3.

Solution:

$$T_R = \frac{T}{T_c} = \frac{145 \text{ K}}{126.2 \text{ K}} = 1.15$$

$$p_R = \frac{p}{p_c} = \frac{(100)(101.325) \text{ kPa}}{3390 \text{ kPa}} = 2.99$$

From the generalized compressibility chart, $Z = 0.52$. Using R for nitrogen from Example 4.3 and employing Equation 4.35 gives

$$v = \frac{ZRT}{p} = \frac{(0.52)(0.297 \text{ kJ/kg} \cdot \text{K})(145 \text{ K})}{(100)(101.325) \text{ kPa}}$$

$$= 2.21 \times 10^{-3} \text{ m}^3/\text{kg}$$

For Example 4.3 (with $T > 2\,T_c$), v obtained by using the compressibility chart was approximately equal to the ideal gas value. For Example 4.4 (with $T = 1.15T_c$), v varies by a factor of two from the ideal gas result.

Examination of the generalized compressibility chart in Figure C.1 in Appendix C shows that in some regions, the gases tend to behave essentially as an ideal gas, while in other regions, significant variations from ideal gas behavior occur. If you refer back to Example 4.1, you will see that the factor pv/RT, which was tabulated in that example, corresponds to the compressibility factor Z. Thus, the computations of Example 4.1 indicate that the water vapor behaves more nearly like an ideal gas at low pressures and high temperatures.

While compressibility charts are useful for some applications, a single mathematical equation of state is much preferred whenever you are doing repetitive calculations, especially when using the computer. Some real gas equations of state that have been developed are discussed in the next section.

EXERCISES

for Section 4.6

1. Consider high pressure steam at 150 bar and 400°C. Use the steam table to read v and then compute the compressibility (Z). For steam, assume $p_c = 220.5$ bar and $T_c = 374$°C. Compute p_R and T_R and use these to obtain an estimate for Z from the generalized compressibility chart. How well do the two estimates compare?

   Answer: $Z = 0.757$

2. Refrigerant-22 has $p_c = 4986$ kPa and $T_c = 369.28$ K. Use Table A.8 in Appendix A to obtain v_g and T_s for dry, saturated refrigerant-22 vapor at $p = 30$ bar. The molecular weight of refrigerant-22 is 86.48. Compute the compressibility (Z) for the refrigerant-22 vapor. Calculate p_R and T_R and obtain an estimate for Z using the generalized compressibility chart. How well do the two estimates compare?

   Answer: $Z = 0.621$.

4.7   Real Gas Equations of State

Several pvT equations of state have been proposed for real gases. Some apply with great accuracy only to a particular region, while others apply with less accuracy to a broader range. Each of the equations is simply a correlation obtained using available experimental data. Correlations obtained from such data for particular gases do not necessarily fit other gases. This is probably the reason why an accurate generalized equation of state does not exist that is applicable to all regions and to all real gases. A few of the equations that have been developed are presented in this section. The original references for these equations are van der Waals (1899), Beattie and Bridgeman (1927 and 1928), and Redlich and Kwong (1949).

4.7.1   Van der Waals Equation

One of the oldest equations of state for real gases was presented by van der Waals in 1899:

$$\left(p + \frac{a}{v^2}\right)(v - b) = RT \tag{4.39}$$

or

$$\left(p + \frac{\bar{a}}{\bar{v}^2}\right)(\bar{v} - \bar{b}) = \bar{R}T \tag{4.39a}$$

where the constants a, b, $\bar{a}$, and $\bar{b}$ are related to the critical state properties according to the following:

$$a = \frac{27}{64}\frac{R^2 T_c^2}{p_c}, \qquad b = \frac{1}{8}\frac{RT_c}{p_c} \tag{4.40}$$

and

$$\bar{a} = \frac{27}{64}\frac{\bar{R}^2 T_c^2}{p_c}, \qquad \bar{b} = \frac{1}{8}\frac{\bar{R}T_c}{p_c} \tag{4.40a}$$

Note that Equation 4.39 is similar to the ideal gas equation. The constant b is included to correct for volume occupied by the gas molecules, while the term a/v^2 is included to account for intermolecular forces of attraction and repulsion. Table 4.1 gives, for several gases, numerical values for constants appearing in the equation.

TABLE   4.1

Van der Waals Equation of State Constants*				
Substance	$\bar{a}$ $kPa\left(\dfrac{m^3}{kmol}\right)^2$	$\bar{b}$ $\left(\dfrac{m^3}{kmol}\right)$	a $kPa\left(\dfrac{m^3}{kg}\right)^2$	b $\left(\dfrac{m^3}{kg} \times 10^3\right)$
Air	135.86	0.0365	0.1620	1.260
Carbon dioxide	366.81	0.0429	0.1894	0.975
Carbon monoxide	147.66	0.0396	0.1882	1.414
Helium	3.518	0.0239	0.2194	5.971
Methane	229.01	0.0428	0.8898	2.668
Nitrogen	137.0	0.0387	0.1746	1.381
Oxygen	137.84	0.0318	0.1346	0.994
Refrigerant-12	1049	0.0971	0.0718	0.803
Refrigerant-22	797.4	0.0769	0.1066	0.889
Water vapor	553.7	0.0305	1.7062	1.693

*Data calculated from critical properties.

The van der Waals equation is not highly accurate over a wide range of conditions. If constants a and b are modified to vary slightly with temperature, the resulting equation would become more accurate. Note that the equation is explicit in temperature and pressure, but implicit in volume. Therefore, it is not easy to use when the volume is not known. Also, multiple solutions exist for volume, and one must choose the correct one from physical considerations. The form of the van der Waals equation is still used for much ongoing research on thermodynamic fundamental equations, and it is widely referenced in current literature.

EXAMPLE 4.5
Use of van der Waals Equation

Given: Nitrogen at $T = 145$ K and $v = 0.0022$ m³/kg (same state as that of Example 4.4).

Find: Calculate p using van der Waals equation and compare with p in Example 4.4.

Solution: Using Equations 4.39 and 4.40 with T_c and p_c given in Example 4.3 gives

$$a = \frac{27}{64}\frac{R^2 T_c^2}{p_c} = \frac{27}{64}\frac{(0.297\text{ kJ/kg} \cdot \text{K})^2 (126.2\text{ K})^2}{3390\text{ kPa}}$$

$$= 0.175\text{ kPa} \cdot (\text{m}^3/\text{kg})^2$$

$$b = \frac{1}{8}\frac{RT_c}{p_c} = \frac{1}{8}\frac{(0.297\text{ kJ/kg} \cdot \text{K})(126.2\text{ K})}{3390\text{ kPa}}$$

$$= 0.00138\text{ m}^3/\text{kg}$$

(a and b from Table 4.1 are 0.1746 kPa (m³/kg)² and 1.381×10^{-3} m³/kg respectively.)

$$\therefore p = \frac{RT}{v - b} - \frac{a}{v^2}$$

$$= \frac{(0.297\text{ kJ/kg} \cdot \text{K})(145\text{ K})}{(0.0022 - 0.00138)(\text{m}^3/\text{kg})} - \frac{0.175\text{ kPa} \cdot (\text{m}^3/\text{kg})^2}{(0.0022\text{ m}^3/\text{kg})^2}$$

$$= 16{,}361\text{ kPa} = \underline{161\text{ atm}}$$

Thus, the result is 61% higher than the correct value for p for nitrogen at this state.

4.7.2 Beattie–Bridgeman Equation

Another of the well-known equations of state is the Beattie–Bridgeman equation, which was developed in 1928. This equation can be written as follows

$$p = \frac{RT}{v^2}\left(1 - \frac{c}{vT^3}\right)(v + B) - \frac{A}{v^2} \tag{4.41}$$

where

$$A = A_0(1 - a/v), \quad B = B_0(1 - b/v)$$

TABLE 4.2

Beattie–Bridgeman Equation of State Constants*

Gas	A_0 (N · m⁴/kg²)	a (m³/kg)	B_0 (m³/kg)	b (m³/kg)	c (m³ · K³/kg)
He	136.789	1.496×10^{-2}	3.50×10^{-3}	0.0	10.0
Ne	52.768	1.087×10^{-3}	1.0198×10^{-3}	0.0	50,000
Ar	82.1067	5.833×10^{-4}	9.8497×10^{-4}	0.0	1500.88
H_2	4926.758	-2.5107×10^{-3}	1.04×10^{-2}	-2.1628×10^{-2}	250.07
N_2	173.566	9.3411×10^{-4}	1.8011×10^{-3}	-2.4664×10^{-4}	1499.14
O_2	147.545	8.0063×10^{-4}	1.445×10^{-3}	1.315×10^{-4}	1500.0
Air	157.161	6.6669×10^{-4}	1.592×10^{-3}	-3.8013×10^{-5}	1498.41
CO_2	262.027	1.621×10^{-3}	2.3809×10^{-3}	1.6443×10^{-3}	15,000.0
CH_4	897.739	1.1571×10^{-3}	3.4852×10^{-3}	-9.8997×10^{-4}	8003.34
$(C_2H_5)_2O$	577.549	1.6774×10^{-3}	6.135×10^{-3}	1.6137×10^{-3}	4499.37

*Constants from Beattie and Bridgeman 1928.

p is in Pa when R is in J/kg · K, v is in m³/kg, T is in K and A_0 is in N · m⁴/kg². a, b, and B_0 have the same units as v while c must be expressed in m³ · K³/kg.

Equation 4.41 is essentially an empirical curve fit of data, and values for the constants for the Beattie–Bridgeman equation are given in Table 4.2. The equation is reasonably accurate when values of the specific volume are greater than the critical specific volume. For many applications, the pressure and temperature are known and the specific volume must be calculated. Since the equation is implicit in v, Equation 4.41 must be solved by iteration in such cases.

EXAMPLE 4.6
Use of Beattie–Bridgeman Equation

Given: Nitrogen at $T = 145$ K and $v = 0.0022$ m³/kg (same as Example 4.5).

Find: Calculate p using the Beattie–Bridgeman equation and compare with p in Example 4.4.

Solution: Equation 4.41 and the constants given in Table 4.2 give the following:

$$A = A_0(1 - a/v) = 173.566 \left(1 - \frac{9.3411 \times 10^{-4}}{0.0022} \right)$$

$$= 99.87 \text{ N} \cdot \text{m}^4/\text{kg}^2$$

$$B = B_0(1 - b/v) = 1.8011 \times 10^{-3} \left(1 - \frac{(-2.4664 \times 10^{-4})}{0.0022} \right)$$

$$= 2.003 \times 10^{-3} \text{ m}^3/\text{kg}$$

$$p = \frac{RT}{v^2} \left(1 - \frac{c}{vT^3} \right)(v + B) - \frac{A}{v^2}$$

$$= \frac{(297)(145)}{0.0022^2}\left(1 - \frac{1499.14}{(0.0022)(145^3)}\right)(0.0022 + 0.002003) - \frac{99.87}{0.0022^2}$$

$$= 8.404 \times 10^6 \text{ Pa}$$

$$= \underline{82.9 \text{ atm}}$$

In this case, the Beattie–Bridgeman equation gives an estimate that is very different from that obtained using the van der Waals equation. The estimate from the Beattie–Bridgeman equation is, however, closer to that obtained using compressibility data.

4.7.3 Redlich–Kwong Equation

The Redlich–Kwong equation is another equation of state. It was proposed by Redlich and Kwong in 1949 and is given by

$$p = \frac{RT}{v - b} - \frac{a}{T^{1/2}v(v + b)} \tag{4.42}$$

or

$$p = \frac{\overline{R}T}{\overline{v} - \overline{b}} - \frac{\overline{a}}{\overline{v}(\overline{v} + \overline{b})T^{1/2}} \tag{4.42a}$$

where the constants a and b are given by

$$a = 0.4275\frac{R^2 T_c^{2.5}}{p_c}, \quad b = 0.0867\frac{RT_c}{p_c} \tag{4.43}$$

TABLE 4.3

Redlich–Kwong Equation of State Constants*

Gas	$\overline{a}$ $kPa\left(\dfrac{m^3}{kmol}\right)^2 K^{0.5}$	$\overline{b}$ $\left(\dfrac{m^3}{kmol}\right)$	a $\left(kPa\left(\dfrac{m^3}{kg}\right)^2 K^{0.5}\right)$	b $\left(\dfrac{m^3}{kg} \times 10^3\right)$
Air	1584	0.02533	1.8892	0.8747
Carbon dioxide	6454	0.0297	3.332	0.674
Carbon monoxide	1725	0.02743	2.1988	0.9793
Helium	8.168	0.01655	0.5097	4.130
Methane	3204	0.02967	12.449	1.849
Nitrogen	1560	0.0268	1.988	0.958
Oxygen	1735	0.0221	1.695	0.6907
Refrigerant-12	20,860	0.0674	1.427	0.557
Refrigerant-22	15,526	0.0533	2.076	0.6163
Water vapor	14,270	0.0212	43.98	1.174

*Data calculated from critical properties.

$\bar{a}$ and b are given by

$$\bar{a} = 0.4275 \, \bar{R}^2 T_c^{2.5}/p_c, \qquad \bar{b} = 0.0867 \, \bar{R} T_c/p_c \qquad (4.43a)$$

A great advantage of this equation is that it has only two constants. Furthermore, it has generally been found to produce good estimates at high pressures with temperatures above the critical value. Note that the mathematical form of this equation is similar to the van der Waals equation. Values for the constants are given in Table 4.3. Consistency of units is obtained with p in kPa, v and b in m³/kg, T in K, R in kJ/kg · K and a in kPa(m³/kg)²(K)⁰·⁵. For the molar version, Equation 4.42a, $\bar{R}$ = 8.3144 kJ/kmol · K, $\bar{v}$ and $\bar{b}$ are in m³/kmol and $\bar{a}$ is in kPa (m³/kmol)²(K)⁰·⁵.

EXAMPLE 4.7
Use of the Redlich–Kwong Equation

Given: Nitrogen at $T = 145$ K and $v = 0.0022$ m³/kg (same state as previous examples).

Find: Calculate p using the Redlich–Kwong equation and compare with p calculated in the previous examples.

Solution: Using Equations 4.42 and 4.43 gives

$$a = 0.4275 \frac{R^2 T_c^{2.5}}{p_c}$$

$$= 0.4275 \frac{(0.297 \text{ kJ/kg} \cdot \text{K})^2 (126.2 \text{ K})^{2.5}}{3390 \text{ kPa}}$$

$$= 1.99 \text{ kPa}(\text{K}^{0.5})(\text{m}^3/\text{kg})^2$$

$$b = 0.0867 \frac{R T_c}{p_c}$$

$$= 0.0867 \frac{(0.297 \text{ kJ/kg} \cdot \text{K})(126.2 \text{ K})}{3390 \text{ kPa}}$$

$$= 9.59 \times 10^{-4} \text{ m}^3/\text{kg}$$

(Table 4.3 shows $a = 1.988$ kPa(m³/kg)²K⁰·⁵ and $b = 9.58 \times 10^{-4}$ m³/kg.)

$$p = \frac{RT}{v - b} - \frac{a}{T^{1/2} v (v + b)}$$

$$= \frac{(0.297 \text{ kJ/kg} \cdot \text{K})(145 \text{ K})}{(0.0022 - 0.00096) \text{ m}^3/\text{kg}} -$$

$$\frac{1.99 \text{ kPa}(\text{K}^{0.5})(\text{m}^3/\text{kg})^2}{(145 \text{ K})^{1/2}(0.0022 \text{ m}^3/\text{kg})(0.0022 + 0.00096) \text{ m}^3/\text{kg}}$$

$$= 10958 \text{ kPa} = \underline{108 \text{ atm}}$$

Thus, this equation gives a result that differs only 8% from the compressibility data, which is considerably more accurate than the results obtained in Examples 4.5 and 4.6.

4.7.4 Virial Form of the Equation of State

A number of successful empirical equations of state for real gases can be shown to be equivalent to the following:

$$Z = \frac{pv}{RT} = 1 + \frac{b(T)}{v} + \frac{c(T)}{v^2} + \cdots \tag{4.44}$$

This is known as the *virial form* of the equation of state, with $b(T)$, $c(T)$, ... referred to as the second, third, and so on *virial coefficients* in the infinite series expansion in powers of density (or $1/v$). This form is well established on the basis of statistical mechanics (see, for example, Haar and Shenker 1971).

An alternative virial form of the equation of state is in an infinite series expansion in powers of pressure:

$$Z = \frac{pv}{RT} = 1 + b'(T)p + c'(T)p^2 + \cdots \tag{4.45}$$

where $b'(T)$, $c'(T)$, ... are the corresponding second, third, and so on virial coefficients. This version is of considerable interest in engineering because the compressibility factor here is expressed in terms of the frequently measured quantities pressure and temperature. While this form is convenient in engineering analysis, in general, one has to include terms to much higher powers of p than is required in the series expansion in powers of $1/v$.

The Keyes equation (see Keenan and Keyes 1936) for water is in terms of powers of pressure and has the form:

$$v = \frac{RT}{p} + b + cp + ep^3 + gp^{12} \tag{4.46}$$

where b, c, e, and g are functions of temperature only. This equation is in excellent agreement with experimental data for specific volumes greater than about three times the critical volume.

As stated in Chapter 3, empirical equations and computer routines have been developed by the authors to compute the thermodynamic properties of steam. Tabulated results for these equations are given in various tables in the appendices. Samples of the equations of state for steam for temperatures above the critical temperature are as follows:

1. Subcritical pressures (10 to 220 bar) and supercritical temperatures:

   For the temperature range 500°C to 900°C, the compressibility factor (Z) is given by

$$Z = \frac{pv}{RT} = 1 + \sum_{I=1}^{3} \sum_{J=1}^{7} A_{IJ} p_R^I T_R^{2-J} \tag{4.47}$$

TABLE   4.4

Coefficients for the Virial Form of the Equation of State (Eqs. 4.47 and 4.48)

Coefficient	$500°C \le T \le 900°C$ $(1.195 \le T_R \le 1.813)$ $10 \text{ bar} \le p \le 220 \text{ bar}$ $(0.045 \le p_R \le 0.998)$	$600°C \le T \le 1000°C$ $(1.35 \le T_R \le 1.97)$ $240 \text{ bar} \le p \le 500 \text{ bar}$ $(1.088 \le p_R \le 2.27)$
A_{11}	0.00156755	0.019687
A_{12}	0.0173113	−0.0496664
A_{13}	0.0143244	0.0866681
A_{14}	−0.0839084	−0.137661
A_{15}	−0.0799404	0.0840803
A_{16}	−0.102522	−0.332846
A_{17}	−0.103064	—
A_{21}	−0.0363106	−0.00946782
A_{22}	0.0755601	−0.00892537
A_{23}	0.041979	0.0949427
A_{24}	−0.0514225	−0.0489785
A_{25}	−0.023028	0.0246788
A_{26}	−0.122061	−0.144831
A_{27}	0.0340672	—
A_{31}	0.0638299	0.00867791
A_{32}	−0.155143	−0.0324624
A_{33}	−0.0123554	0.07008
A_{34}	0.0675587	−0.169507
A_{35}	0.0991729	0.216617
A_{36}	0.311772	−0.0739871
A_{37}	−0.403706	—

The coefficients A_{IJ} are given in Table 4.4. The maximum deviation between v given by Equation 4.47 and the tabulation in the *NBS/NRC Steam Tables* (Haar et al. 1984) is only about 1% in the specified range of pressure and temperature.

2. Supercritical pressure (240 to 500 bar) and supercritical temperatures:
   For the temperature range 600°C to 1000°C, Z is given by

$$Z = \frac{pv}{RT} = 1 + \sum_{I=1}^{3} \sum_{J=1}^{6} A_{IJ}p_R^I T_R^{2-J} \tag{4.48}$$

The coefficients A_{IJ} for this case are also given in Table 4.4. The maximum deviation (relative to tabulated values) in the computed values of v using Equation 4.48 is only about 0.5%.

It should be apparent from considering these equations that the development of equations of state is rather difficult, but adequate equations of state are available for many substances covering a wide range of conditions. Computer codes for calculating properties are undoubtedly useful for solving problems in thermodynamics.

1. Consider high pressure steam at 400°C and $v = 0.01567$ m³/kg. Obtain estimates of the pressure using
   (a) the van der Waals equation
   (b) the Redlich–Kwong equation

   The actual pressure is 150 bar. What is the percentage error in each of these estimates?

   Answer: (a) 153 bar, (b) 148 bar

2. Nitrogen gas at 300 K has a specific volume (v) of 6.09×10^{-3} m³/kg. Obtain estimates of the pressure using
   (a) the van der Waals equation
   (b) the Beattie–Bridgeman equation
   (c) the Redlich–Kwong equation

   The actual pressure is 150 bar. What is the percentage error in each of these estimates?

   Answer: (a) 142 bar, (b) 148 bar, (c) 147 bar

4.8   Summary

This chapter considered ideal gases and real gases. An *ideal gas* is defined as one that obeys the ideal gas pvT equation of state. Although no gas is totally "ideal," all gases can be modeled by the ideal gas equation for some regions. In general, the gases that most nearly approach ideal gas behavior are those at pressures approaching zero with temperatures significantly higher than the critical temperature.

Specific internal energy, enthalpy, and specific heats of an ideal gas are all functions of temperature only. This fact allows the development of ideal gas tables for many different gases. The computer code **GAS**, which is included with the text, contains the ideal gas properties of several gases.

The *principle of corresponding states* implies that all gases behave in the same general manner and further postulates that the compressibility factor for all gases is a function of only the reduced temperature and pressure for the gas. This principle has been used to develop generalized compressibility charts that can be employed to determine the thermodynamic state for all gases. One such empirical compressibility chart is presented in Appendix C.

Examination of the applicability of the ideal gas equation shows that most regions for real gases cannot be accurately modeled by the ideal gas equation. Many different and often complex equations of state have been developed historically in attempts to model real gas behavior realistically. A number of these equations were presented to illustrate the general approach taken. In general, all of these equations reduce to the ideal gas equation at low pressures and relatively high temperatures. It should be noted that, while the ideal gas equations are simple and easy to use, the limitations in the range of conditions over which they apply should not be ignored. For more general applicability, one must use a more complex real gas equation that is based on experimental data for the gas.

References

Beattie, J. A. and Bridgeman, O. C. 1927. "A New Equation of State for Fluids. I. Application to Gaseous Ethyl Ether and Carbon Dioxide." *Journal of the American Chemical Society.* Vol. 49, Part 2. 1665–1667.

Beattie, J. A. and Bridgeman, O. C. 1928. "A New Equation of State for Fluids." *Proceedings of the American Academy of Arts and Science.* Vol. 63, No. 5. 229–308.

Haar, L. and Shenker, S. H. 1971. "Equation of State for Dense Gases." *J. Chem. Physics.* Vol. 55, No. 10. 4951–4958.

Haar, L., Gallagher, J. S., and Kell, G. S. 1984. *NBS/NRC Steam Tables.* New York: Hemisphere.

Haywood, R. W. 1980. *Equilibrium Thermodynamics for Engineers and Scientists.* New York: John Wiley & Sons.

Keenan, J. H. and Keyes, F. G. 1936. *Thermodynamic Properties of Steam Including Data for the Liquid and Solid Phases.* New York: John Wiley & Sons.

Redlich, O. and Kwong, J. N. S. 1949. "On the Thermodynamics of Solutions. V. An Equation of State: Fugacities of Gaseous Solutions." *Chemical Reviews.* Vol. 44, No. 1. 233–244.

van der Waals, J. D. 1899. *Die Continuitat des Gasformigen und Flussigen Zustandes.* Leipzig: Barth.

Questions

1. Define an ideal gas. Give the approximate conditions for the gaseous phase of a pure substance to behave as an ideal gas.

2. What is a perfect gas? Give two examples.

3. Which of the following are unlikely to behave as an ideal gas at a typical ambient condition of 1 atm and 25°C: oxygen, hydrogen, nitrogen, carbon dioxide, refrigerant-22, or H_2O?

4. Write down the relationship among c_v, c_p, and R for an ideal gas. Which of these can vary with temperature and which one does not?

5. Define k, the ratio of principal specific heats. Does k for ideal gases in general vary with temperature?

6. Real gases approach ideal gas behavior under certain conditions of pressure and temperature. Which of the two conditions is the more critical?

7. What is the generalized compressibility chart? Discuss its importance in the derivation of the pvT characteristics for a real gas whose critical parameters are known but very little else.

8. Write down the virial forms of the equation of state in powers of density and pressure, respectively. Discuss the relative merits of using one or the other of these equations.

9. Assume that the pvT equation of state for a real gas can be approximated by the following:

$$\bar{v} = \frac{\overline{R}T}{p} + b(T)$$

If $b(T)$ is a positive quantity that is small compared with $\bar{v}$, sketch the paths for two isothermal compression processes (at different temperatures) on a p—$\bar{v}$ diagram.

10. Use the appropriate steam table in Appendix A to obtain values of h for superheated steam at a temperature of 300°C in the range of pressures from 1 to 70 bar. Plot h versus p (to scale) and comment on the dependence or independence of the enthalpy for the gas on pressure.

Problems

4.1 Evaluate $p\bar{v}/\bar{R}T$ for each of the gas conditions specified in the following table. From this, determine whether or not the gas state corresponds to the ideal gas condition.

Gas	p (kPa)	v (m³/kg)	T (K)
H_2O	101.3	3.651	800
H_2O	5,000	0.0394	537
H_2O	15,000	0.0285	973
H_2O	5	29.8	323
NH_3	69	2.588	365
R-22	100	0.215	232
CO_2	101.3	1.322	700
Air	101.3	0.85	300
He	101.3	5.126	250
Ar	101.3	0.616	300

4.2 If both the volume and the pressure of an ideal gas are doubled (Fig. P4.2), what is the ratio of the absolute temperatures?

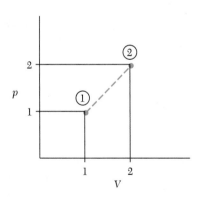

Figure P4.2 Change of state for an ideal gas.

4.3 The temperature of a closed cylinder containing air at 25°C and a pressure of 760 mm Hg is raised to 100°C (Fig. P4.3). What is the new pressure?

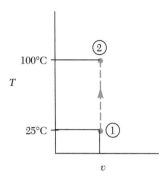

Figure P4.3 Change of state for air.

4.4 Assume that the volume of an inflated automobile tire does not change significantly when the temperature changes. If the initial pressure in the tire is 240 kPa (gage) at a temperature of 10°C, calculate the new pressure when the temperature is 35°C. (The ambient atmospheric pressure can be taken as 100 kPa.)

4.5 An automobile tire has a volume of 0.0165 m³ when inflated to a pressure of 165 kPa (gage) at 0°C (Fig. P4.5). What is the mass of the air in the tire? What is the new pressure when the temperature rises to 2°C and the volume of the tire is 0.0168 m³? Assume the atmospheric pressure is 100 kPa.

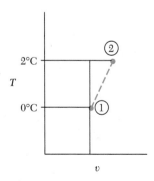

Figure P4.5 Change of state for air in a tire.

4.6 Use the relationship between the specific gas constant and the universal gas constant to determine the specific gas constant for the following gases:
(a) carbon monoxide
(b) oxygen
(c) nitrogen
(d) mercury vapor
(e) hydrogen

4.7 (a) A room measures 5 m × 6 m × 4 m. What is the mass of air contained in the room if the temperature is 20°C and the pressure is 1 atm?
(b) Gas at 0.1 MPa and 2 m³ volume is compressed to a volume of 0.5 m³ at constant temperature. If the gas behaves as an ideal gas, what is the new pressure?

4.8 A 5-l container holds 0.005 kg of an ideal gas when the pressure is 750 mm Hg and the temperature is 50°C (Fig. P4.8). What will be the pressure if 0.006 kg of this gas is confined in a 2-l container at 0°C?

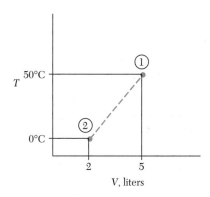

Figure P4.8 Change of state for an ideal gas.

4.9 (a) A boiler drum of volume 2 m³ holds steam at a pressure of 1.5 bar and a temperature of 400°C. Estimate the mass of the steam in the drum:
(i) assuming that steam obeys the ideal gas equation
(ii) using steam property tables
(b) Repeat Part (a) for dry, saturated steam at 50 bar.

4.10 Oxygen at 200 bar is stored in a steel vessel at 27°C (Fig. P4.10). The capacity of the vessel is 0.05 m³. Assuming that oxygen is a perfect gas, calculate the mass of oxygen stored in the vessel. The vessel is protected against excessive pressure by a fusible plug that will melt if the temperature rises to a certain level. At what temperature must the plug melt to limit the pressure in the vessel to 250 bar?

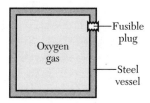

Figure P4.10 Storage of oxygen.

4.11 Enclosed in a container is 10 kg of carbon dioxide gas at a temperature of 100°C and a pressure of 1 bar. Compute the volume of the container.

4.12 Enclosed in a container is 3 lbm of carbon monoxide gas at a temperature of 150°F and a pressure of 20 psia. Determine the volume of the container.

4.13 Nitrogen gas at a temperature of 25°C and a pressure of 0.5 bar is enclosed in a cylinder whose internal volume is 1 m³. Determine the mass of the gas in the cylinder.

4.14 Pure oxygen gas at 70°F and 8 psia is enclosed in a space station module whose volume is 6000 ft³. Determine the mass of the oxygen gas in the module.

* **4.15** An unknown gas is enclosed in a cylinder having a volume of 1 m³. The mass of the gas is determined to be 1 kg, and its pressure and temperature are 10^5 N/m² and 529.23 K, respectively. Determine the molecular weight (molar mass) of the gas.

4.16 An ideal gas is enclosed in a balloon. If its pressure is doubled while its temperature remains constant, how does the volume change?

4.17 An ideal gas is enclosed in a rigid cylinder. Heat is added to the gas until its pressure doubles. What is its final temperature if its initial temperature is 70°F?

4.18 An ideal gas is maintained at constant pressure in a balloon. If heat is removed from the gas until its temperature is reduced from 100°F to 0°F, how does the volume of the gas change?

4.19 An ideal gas is initially at an equilibrium state with a temperature of 40°F, a pressure of 1 atm, and a volume of 1 ft³. If its temperature is raised to 600°R and its pressure is raised to 2 atm, what is its final volume?

* **4.20** Enclosed in a rigid sphere is 5 kmol of an ideal gas at 30°C and 2 bar. What is the radius of the sphere? How many molecules of the gas are in the sphere?

* **4.21** The ideal gas in the sphere of Problem 4.20 is replaced with 2.5 kmol of nitrogen gas at 30°C. What is the pressure of the nitrogen gas in the sphere? How many molecules of nitrogen are in the sphere?

* **4.22** Water vapor is in an equilibrium state at 100 psia and 1000°F. Using the specific volume value from the steam tables as the "correct" value, how much deviation from this value would occur if you used the ideal gas equation at this state? If the temperature of the water vapor is lowered while the pressure remains constant, what is the lowest temperature to which the gas can be lowered with the ideal gas equation still remaining accurate with the error margin not exceeding ±2%?

* **4.23** Water vapor is in an equilibrium state at 0.1 bar and 400°C. Using the specific volume value from the steam tables as the correct value, how much deviation from this value would occur if you used the ideal gas equation at this state? If the pressure of the water vapor is increased while the temperature remains constant, what is the highest pressure to which the gas can be raised with the ideal gas equation still remaining accurate with the error margin not exceeding ±2%?

* **4.24** Water vapor exists as dry, saturated steam at 100 psia. If the pressure remained constant, the ideal gas equation could be used for superheated steam to reasonably approximate values of specific volume at what lowest temperature? Assume that $Z = pv/RT$ must lie between 0.99 and 1.01 for ideal gas behavior.

* **4.25** Water vapor exists as dry saturated steam at 150°C. If the temperature remained constant, the ideal gas equation could be used for superheated steam to reasonably approximate values of specific volume at what highest pressure? (Assume that $Z = pv/RT$ must lie between 0.99 and 1.01 for ideal gas behavior.)

* **4.26** A quantity of a certain perfect gas is compressed from an initial state of 0.085 m³ and 100 kPa to a final state of 0.034 m³ and 390 kPa (Fig. P4.26). The specific gas constant is 296 J/kg · K. The observed temperature rise is 146 K. Calculate the mass of gas present.

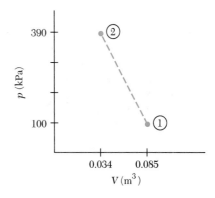

Figure P4.26 Change of state for a perfect gas.

* **4.27** Refrigerant-22 has a critical pressure of 4988 kPa and a critical temperature of 369.30 K. If the specific volume of the vapor is found to be 0.019 m³/kg at a temperature of 30°C, estimate the pressure
  (a) assuming that the vapor obeys the ideal gas equation
  (b) using the van der Waals equation
  (c) from refrigerant-22 property tables

* **4.28** An air receiver of volume 1 m³ contains air at a pressure of 100 atm and a temperature of 300 K. The pseudocritical parameters of air can be taken as $p_c = 37.25$ atm and $T_c = 132.4$ K.
  (a) Use the ideal gas equation to estimate the mass of air in the receiver.
  (b) Use the Redlich–Kwong equation to estimate the specific volume and then the mass of air in the receiver. (An iterative procedure is needed for the determination of v from the Redlich–Kwong equation.)

* **4.29** The critical pressure and temperature of water are 22.05 MPa and 374°C, respectively. Use the virial form of the equation of state given in Section 4.7.4 to determine the compressibility and specific volume of steam at 20 MPa and 700°C.

4.30 When 1 kg of a certain gas is heated at constant pressure from 30°C to 110°C, the heat required is 1136 kJ. When 1 kg of the same gas is heated at constant volume between the same temperatures, the heat required is 808 kJ. Calculate c_v, c_p, k, and R, and the molecular weight (M) of the gas.

4.31 In an experiment, the value of k (the ratio of the principal specific heats) for carbon dioxide was found to be 1.3. Assuming that carbon dioxide is a perfect gas, calculate the gas constant (R) and the principal specific heats $(c_p$ and $c_v)$.

4.32 A fixed mass of air is compressed from an initial pressure of 100 kPa and volume of 2 m³ to a final pressure of 400 kPa and volume of 0.5 m³ (Fig. P4.32). What is the increase in internal energy?

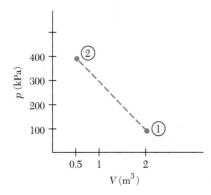

Figure P4.32 Compression of air.

4.33 The specific heat at constant pressure for argon is 0.5203 kJ/kg · K. Show that (c_p/R) for argon is, to a good approximation, equal to 2.5. The increase in enthalpy of a fixed mass of argon in a process is 11 kJ. If the mass of the gas is 0.5 kg, calculate the rise in temperature of the gas.

** **4.34** A mass of 1 kg of a gas expands from an initial volume V_1 to a final volume V_2 $(>V_1)$ in a quasi-static isothermal process (Fig. P4.34). Assume the van der Waals equation of state and show that the work done is

$$w_{12} = RT \ln\left(\frac{V_2 - b}{V_1 - b}\right) + a\left(\frac{1}{V_2} - \frac{1}{V_1}\right)$$

If the gas were ideal, show that the same process would produce more work than w_{12} provided $T < (27/8)\, T_c$ and assuming that both V_1 and V_2 are much greater than b.

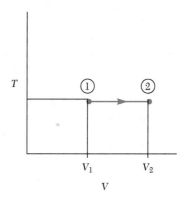

Figure P4.34 Isothermal expansion of a gas.

** **4.35** The molar specific heat of methane gas in the temperature range 275 K to 1150 K is given by

$$\frac{\bar{c}_p}{R}\left(\frac{T}{T_c}\right)^{0.5} = \sum_{i=1}^{5} A_i\left(\frac{T}{T_c}\right)^{i-1}$$

where A_i indicates coefficients with the following values:

Coefficient	Value
A_1	3.880095
A_2	− 1.452249
A_3	1.951505
A_4	− 0.2897473
A_5	0.01528889

T is the absolute temperature, and T_c, the critical temperature for methane, is 190.7 K. Given that the reference molar enthalpy of methane at the reference temperature of 25°C is − 74,848 kJ/kmol, determine the molar enthalpy at 1000 K using the above specific heat correlation for methane. Check your answer using the **GAS** computer code.

4.36 The temperature of argon gas at atmospheric pressure is raised from 100°C to 200°C. How much did the internal energy of the gas increase during this process?

4.37 The enthalpy of neon gas increased by 10 kJ/kg during a process. The gas was initially at 25°C and 1 atm. What was its final temperature?

* **4.38** Nitrogen gas is at 150 K and 68 atm. Find the specific volume of the gas at this state by using the compressibility chart for nitrogen. Compare the result with that obtained from the ideal gas equation.

* **4.39** Oxygen gas is at 200 K and 200 atm. Find the specific volume of the gas at this state by using the generalized compressibility chart. Compare the result with that obtained from the ideal gas equation.

* **4.40** Water vapor is in an equilibrium state at $T = 150°C$ and $v = 1.94 \text{ m}^3/\text{kg}$. Compute the pressure of the water vapor at this state by using the van der Waals equation, and compare the result with the value given in the steam tables. Also, compare it with the value obtained from the ideal gas equation for this state.

* * **4.41** Water vapor is in an equilibrium state of 10^6 N/m^2 and 250°C. Compute the specific volume of the water vapor at this state by using the van der Waals equation and compare the result with the value given in the steam tables.

* * **4.42** Refrigerant-22 is in an equilibrium state at $9 \times 10^5 \text{ N/m}^2$ and 100°C. Compute the specific volume of the substance at this state by using the van der Waals equation.

* **4.43** Air is in an equilibrium state at $T = 500°C$ and $v = 2.03 \text{ m}^3/\text{kg}$. Compute the pressure of the air at this state by using the Beattie–Bridgeman equation, and compare your result with the value obtained from the ideal gas equation.

* **4.44** Nitrogen gas is in an equilibrium state at $T = 200 \text{ K}$ and $v = 0.001 \text{ m}^3/\text{kg}$. Compute the pressure at this state and compare results using
  (a) the van der Waals equation
  (b) the Beattie–Bridgeman equation
  (c) the ideal gas equation

* **4.45** Water vapor is in an equilibrium state at $T = 500°C$ and $v = 2.85 \text{ m}^3/\text{kg}$. Compute the pressure of the water vapor at this state by using the Redlich–Kwong equation, and compare your result with the value given in the steam tables.

* **4.46** Refrigerant-22 is in an equilibrium state at $9 \times 10^5 \text{ N/m}^2$ and 100°C. Compute the specific volume of the substance at this state by using the Redlich–Kwong equation.

* **4.47** Nitrogen gas is in an equilibrium state at $T = 300 \text{ K}$ and $v = 0.001 \text{ m}^3/\text{kg}$. Compute the pressure at this state by using the Redlich–Kwong equation, and compare your result with that from the ideal gas equation.

* * **4.48** Compute the compressibility factor by using Equation 4.47 for water vapor at 150 bar and 800°C. Compare your result with the value of Z calculated from p, v, and T at this state from the steam tables or the computer code **STEAM**.

* * **4.49** Using Equation 4.33 and appropriate data and/or equations, compute the enthalpy change when 1 kmol of carbon monoxide is raised from 1000 K to 3000 K. Compare your result with the value obtained from the ideal gas tables for this process. (c_{p0} data for carbon monoxide as a function of temperature are provided in Appendix A.)

* * **4.50** Using Equation 4.34 and appropriate data and/or equations, compute the internal energy change when 1 kmol of air is raised from 300 K to 1100 K. (Zero pressure specific heat data as a function of temperature are provided for air in Appendix A.)

Processes, Work, and Heat

5.1 Introduction

This chapter deals with processes, work, and heat. Both equilibrium and nonequilibrium processes are considered. The discussion focuses on the definitions of work and heat and gives several examples of each. These concepts are essential to understand the first law of thermodynamics, which is presented in Chapter 6.

Prior to considering processes, recall that a *thermodynamic system* is defined as a region in space enclosed by a physical or an imaginary boundary. In general, systems of interest are those that include matter inside the system boundary. Recall that such a thermodynamic system is said to be in *equilibrium* when there is no tendency for any property of the system to change. The thermodynamic *state* of a system is defined by the condition in which the system exists. Such a state is specified by an adequate number of thermodynamic properties for the system (two for a pure, simple, compressible system). When a thermodynamic state is changed, this is called a *change of state* for the system. Now let's consider thermodynamic processes in greater detail.

5.2   Processes

The thermodynamic concept for a process was introduced and discussed in Chapter 1. A *process* is that which brings about a change of state of a system. A *thermodynamic cycle* is a process (or series of processes) that results in the final state of the system being identical with the initial state. A distinction exists between quasi-equilibrium and nonequilibrium processes. A *quasi-equilibrium process* is one in which the system passes through an infinite series of equilibrium states from its initial state to its final state. The system remains infinitesimally close to a condition of equilibrium throughout the process. Such a process would have to occur at an extremely slow rate for the system to remain in a quasi-equilibrium state throughout. Any *real process*, in contrast, does not occur slowly and involves factors, such as friction and temperature gradients, that cause the system to depart from equilibrium. Thus, real processes are usually *nonequilibrium processes* since the system will in general pass through nonequilibrium states.

When calculating the work output from a work-producing device or machine, the actual work output is less than that calculated if quasi-equilibrium processes are assumed. If, instead, the work input required to produce a certain effect (such as the compression of a gas) is determined, the actual work required is greater than that computed assuming quasi-equilibrium processes. From such observations, some amount of *work is said to be lost* in nonequilibrium processes (in comparison to work required to produce the desired effect via quasi-equilibrium processes).

A recognizable or identifiable *thermodynamic path* exists only for quasi-equilibrium processes. This path is the series of equilibrium states through which the system passes while undergoing a change from one equilibrium state to another. Nonequilibrium thermodynamic processes occur without the system going through a series of equilibrium states, and therefore no discernible path exists for such processes. The processes can, however, be defined by specifying the initial and final states of the system in addition to heat and/or work interaction involved. The following section further illustrates these points.

Constant Volume Heating Process

Consider a fixed mass of air contained in a rigid vessel of fixed volume, as shown in Figure 5.1. A gas heater supplies heat to the air. The temperature of the layer of air at the bottom of the vessel will rise first, and as a result, this layer will become less dense than the rest of the air in the vessel. Our first observation, therefore, is that during the heating process, density (or specific volume) and temperature will not have uniform values throughout the mass of the air. Furthermore, if the heating process is carried out in a gravitational environment, at some point in the process, convection currents are likely to occur with the hot and less dense air rising to the top of the vessel while the colder and denser air at the top of the vessel falls to the bottom. Clearly, the pressure of air in the vessel will be nonuniform as well.

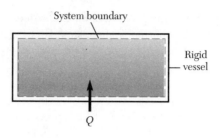

Figure 5.1 Constant volume heating process.

Suppose that the heating process was terminated and an ample thermal insulation was applied to the vessel. After a sufficiently long time, the air in the vessel will attain thermal and mechanical equilibrium and the temperature and pressure will be uniform throughout the system (air). This describes a typical, real process that produces a change of state of a system from an initial (equilibrium) state 1 to a final (equilibrium) state 2. Both states 1 and 2 are equilibrium states for which the values of all the thermodynamic properties can, in principle, be ascertained. Now let's depict what has transpired graphically using the T-v coordinates. Since both the volume (V) and the mass (m) for the system stay constant, v_1 must be equal to v_2 ($= V/m$). T_1 is the initial system temperature, while T_2 ($> T_1$) is the final temperature attained. We can therefore represent equilibrium states 1 and 2 on the T-v diagram (Fig. 5.2) as the points 1 and 2 with coordinates (v_1, T_1) and (v_2, T_2), respectively.

An earlier discussion noted that neither v ($= 1/\rho$) nor T is likely to have a uniform value for the system during the heating process. Thus, between states 1 and 2 in Figure 5.2, no path can be ascribed that correctly shows the thermodynamic states that the system passes through since such states are nonequilibrium states. Therefore, it is surmised that a path does not exist, and a broken or dashed line between points 1 and 2 on Figure 5.2 indicates that the system had its state changed from initial state 1 to final state 2 along an undefinable path.

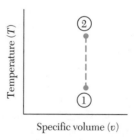

Figure 5.2 Representation of a real (constant v) process on the T-v diagram.

It is not difficult to imagine how the heating process just described could have been accomplished in a quasi-equilibrium manner. If the heating rate is kept very low, there would always be sufficient time for the system to attain equilibrium during the process. The specific volume (v) would be uniform throughout the system and would have a constant value (V/m) for the entire process. The temperature (T) would rise gradually but very slowly, such that at each instant, it would have a uniform value throughout the system. This case would thus have an identifiable path that is a solid line between points 1 and 2, shown in Figure 5.3.

Although real processes as a rule are nonequilibrium processes, in several instances, they can be quite close to the quasi-equilibrium ideal. Thus, if a real process is executed slowly or if the system volume is small enough, the system may not depart significantly from equilibrium, and it may be reasonable to approximate the process as an ideal, quasi-equilibrium process. This is important in classical thermodynamics because detailed analyses involving the application of the laws of thermodynamics can, in general, be made only for quasi-equilibrium processes. Thus, if real processes are approximated by ideal processes, the results obtained for the ideal processes can be applied to the real processes through the use of appropriately defined process efficiencies. These process efficiencies are typically determined from experimental measurements.

In summary, a thermodynamic path is definable only for quasi-equilibrium processes. The path for a quasi-equilibrium process between any two equilibrium states 1 and 2 of a system can be represented by a solid line on a phase or thermodynamic property diagram, whereas for a nonequilibrium process, only broken lines (or dashes) are used since the system does not pass through equilibrium states. Also, in order to adequately define a nonequilibrium process, one must have information concerning either the work or the heat or both.

Chapters 1 and 2 repeatedly pointed out that thermodynamic properties (which can be specified at a point) are regarded as *point functions*, while certain other quantities such as work and heat (whose values are process dependent) are not point functions but *path functions*. All thermodynamic properties (such as pressure, temperature, internal energy, and entropy) are point functions, while heat and work are path functions. This

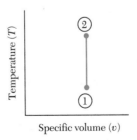

Figure 5.3 Representation of a quasi-equilibrium (constant v) process on the T-v diagram.

text follows the popular convention of using an *exact differential* (for example, dz) to describe a small change in a point function, while the *inexact differential* (for example, δQ) is used to represent a small increment in a path function.

Chapter 1 also distinguished between *nonflow* and *flow processes*. In the former, a fixed collection of matter remains within the (closed) system boundary throughout the processes or change of state, while in flow processes, flow of matter across the (open) system boundary is permitted. Whether a process is nonflow or flow therefore depends on whether the problem is formulated as a closed system or as an open system. Both types of processes can involve either work or heat or both. Only closed system processes are considered in the rest of this chapter. Work and heat in relation to an open system are discussed extensively in the presentation of the first law of thermodynamics in Chapter 6.

5.3   Work

Work and energy are intimately connected. Energy is inferred to be a property of a system as a consequence of the first law of thermodynamics. Discussion of the relationship between work and energy should strictly come after formal exposition of the first law, which is given in Chapter 6. At this juncture, however, it is important to stress that while energy is a property, work (which is flow of energy) is *not*. For example, you do not look at a very active toddler and say "he has a lot of work in him!" Rather, one would say of a very active child, "he is full of energy." Likewise, while the amount of money in a bank account (property of the account) bears a relationship to the amount of money withdrawn or deposited (flow of property of the account), the withdrawal or deposit is never considered a property of the account. In other words, you cannot tell whether the account is in the red or in the black from looking at the amount written on a check. This differentiation between work transfer (a nonproperty) and energy (a property) is important in thermodynamics and will be reiterated from time to time in this and subsequent chapters. In this section, definitions of work in mechanics and in thermodynamics will be discussed, and later, expressions will be derived for calculating the work done in various categories of nonflow processes.

5.3.1   Definitions of Work

Work Done as a Result of Mechanical Action

The definitions of work in mechanics and thermodynamics were discussed briefly in Chapter 2. In both cases, *work* is computed as a *force times displacement* effect. In thermodynamics, however, work is an *interaction*, and the magnitude of the work done must be determined based on forces acting at the system boundary and the accompanying displacement of the boundary. This is represented quantitatively by considering the work done when a system is bodily moved from an initial position 1 to a final position 2. If **f** is the boundary force (vector) exerted by the surroundings in opposition to the movement of the system boundary, the work done by the system can be determined using the following equation:

$$\delta W = - \sum_{\substack{\text{system} \\ \text{boundary}}} \mathbf{f} \cdot d\mathbf{x}$$

and   (5.1)

$$W_{12} = \sum_{\text{point 1}}^{\text{point 2}} \delta W$$

where $d\mathbf{x}$ is an infinitesimal displacement vector for the patch of the system boundary on which the force acts. The negative sign is included because the force vector $\mathbf{f}$ is defined in the opposing sense to the boundary displacement, whereas the work done by the system is deemed positive when the effect on the surroundings is equivalent to the raising of a weight (in the presence of a gravitational field). A typical application of Equation 5.1 is illustrated in Figure 5.4 in which an individual is pulling on a rope and moving from point 1 to point 2 along a horizontal surface. During the process, the individual pulls the rope and raises the weight (A) from position 1a to position 2a.

A free-body diagram is used in mechanics to indicate the forces acting on an object (Fig. 5.5). To use Equation 5.1 to compute the work done on the weight while moving it from position 1a to 2a, one must determine the force acting during the movement and must also define the distance (in the direction of the force) through which the weight is moved. A similar procedure to that just described for mechanics is followed in thermodynamics. However, you must first explicitly define the system and the system boundary. In this example, positive work is done by the individual, who could therefore be defined as the system. Once you have defined the system and the system boundary, you can determine the work interaction by identifying the forces acting at the boundary and the displacements taking place.

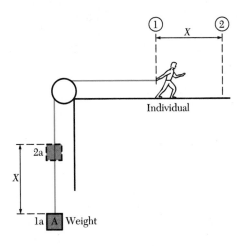

Figure 5.4 Work done by an individual pulling on a rope.

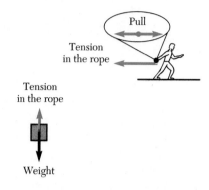

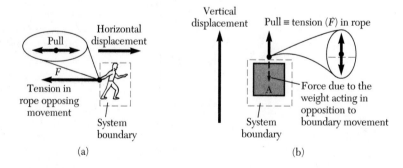

Figure 5.5 Free-body diagram for forces acting on the individual and the weight raised in Figure 5.4.

In this example, let's regard the individual as the system. Figure 5.6(a) indicates a boundary resistance, F, to the pull exerted by the individual on the rope. Thus, if $\mathbf{i}$ denotes a unit vector to the right in the horizontal direction, $\mathbf{f} = -F\mathbf{i}$. If X is the distance moved in the process, $\mathbf{x} = X\mathbf{i}$. Applying Equation 5.1 gives

$$W_{\text{individual}} = W_{\text{system}} = -(-F\mathbf{i}) \cdot (X\mathbf{i}) = +FX$$

The work done by the weight is determined from

$$W_{\text{weight}} = W_{\text{surroundings}} = -W_{\text{system}} = -FX$$

Note that to keep the analysis simple, the rope is assumed to be inelastic and the pulley is assumed to be frictionless so that the tension in the rope is uniform throughout and is equal to F. It is also assumed that the boundary movement is fully resisted; in other words, F is only infinitesimally greater than the weight of body A.

Figure 5.6 Work done by (a) the individual and (b) the weight in the process depicted in Figure 5.4.

Suppose the weight is defined as the system instead. The forces at the system boundary would then be as shown in Figure 5.6(b). In this case, work is done on the system while positive work is done by the surroundings; hence,

$$W_{surroundings} = +FX \, (= W_{individual})$$

Therefore,

$$W_{weight} = W_{system} = -W_{surroundings} = -FX$$

Note that work is not a property, and it is not defined at either the initial or the final state of the system. Rather, it is only defined for the process that occurs while moving the weight from position 1a to position 2a. How much work is done thus depends on the manner in which the process is executed.

Equation 5.1 can be expressed in a more general form by considering the force vector $\mathbf{f}$ as that force acting over a patch of the system boundary in opposition to the movement of that patch of the boundary, while dx is the infinitesimal displacement vector for the patch on which the force acts. The force can be resolved into two components, $\mathbf{f}_n$ and $\mathbf{f}_s$, respectively, in the directions normal to and tangential with the boundary. Likewise, the displacement vector can be resolved into dx_n and dx_s along the normal and tangential directions, respectively. The product $\mathbf{f}_n \cdot dx_n$ defines what is commonly referred to as the *displacement work*, while $\mathbf{f}_s \cdot dx_s$ is the *shear work* component. Summation of the products $\mathbf{f}_n \cdot dx_n$ over the entire system boundary gives the incremental displacement work, δW_d, that is done. Similarly, the incremental shear work done, δW_x, is the sum total of the products $\mathbf{f}_s \cdot dx_s$ for the entire system boundary. Thus,

$$\delta W_d = -\sum_{\substack{system \\ boundary}} \mathbf{f}_n \cdot dx_n \tag{5.1a}$$

As indicated earlier, the negative sign is included because the force vector is defined as that resisting the boundary movement. The displacement vector is assumed to be in the direction outward from the system, and the work indicated is positive work done by the system. The incremental shear work is given by

$$\delta W_x = -\sum_{\substack{system \\ boundary}} \mathbf{f}_s \cdot dx_s \tag{5.1b}$$

The shear work is sometimes produced as output from the system by use of a shaft and is also called the *shaft work*.

These equations can be used in virtually all cases of work resulting from forces at the system boundary acting through a distance. One notable exception is encountered when the system boundary coincides with the surface of discontinuity between two bodies that are in relative motion to each other. The classic example of this is the sliding of a block over a rough surface. If the surface of contact between the block and the body (over which the block slides) is taken as part of the system boundary, the force $\mathbf{f}$ will be the frictional force, but there is an ambiguity as to what the displacement is. The ambiguity arises because, for the block that is moved, both the force and displacement are nonzero, whereas the displacement is zero for the body (assumed stationary) over which the block slides. The decision is arbitrary if either the nonzero displacement of the block or the zero displacement of the body is used to determine the work done from Equation 5.1. This difficulty can be circumvented by redefining the system boundary to avoid the surface of discontinuity. An alternative approach requires the application of the first law of thermodynamics. (The interested reader should refer to Bridgman 1941, 47–56 or to Spalding and Cole 1966, 107–109, for a more complete discussion of this problem.)

Formal Definition of Work in Thermodynamics

In thermodynamics one considers that work is done only on or by a thermodynamic system, which must be defined before one can determine whether or not work is being done. Work interaction can take several forms, including mechanical work, electrical work, or magnetic work. The previous discussion of work focused on mechanical work. Chapter 2 introduced a formal definition of *work* by Poincare. The definition is adequate for the thermodynamic work concept in the most inclusive sense and can be stated as follows:

> *Positive work is done by a system on its surroundings during a given process if the system could pass through the same process while the sole effect external to the system was the rise of a weight (in the presence of a gravitational field).*

Negative work is defined as the work done by the surroundings when the system does positive work. Equation 2.3 expressed this relationship as follows:

$$W_{system} + W_{surroundings} = 0 \tag{2.3}$$

In other words, the amount of work flowing *from* the system is identical to the amount flowing *to* its environment. Work is designated by the symbol W, and W_{12} means that this is the work done by a system in going from state 1 to state 2. A sign convention is implied in this definition of work. This widely employed convention in thermodynamics regards W as *positive* if the effect external to the system is equivalent to a rise in weight (in a gravitational field). This text follows this convention, so that work done by the system on the surroundings is positive, while work done on the system by the surroundings is negative for the system. From this thermodynamic definition of work, you can see that work involves action across the system boundary and that the work done is measured in terms of an effect external to the system equivalent to raising a weight in the presence of a gravitational field.

Returning to the system of Figure 5.4, you can also compute the work based on the formal thermodynamic definition of work. In such cases, you would draw the boundary for the system, as indicated in Figures 5.6(a) and (b), and then determine whether or not you could show that the interaction between the system and the surroundings, from state 1 to state 2, could be equivalent to the raising of a weight in a gravitational field (for either the system or the surroundings).

Figures 5.7(a) and (b), corresponding to the systems defined in Figures 5.6(a) and (b), respectively, show the work interaction between each system and its surroundings in terms of the equivalent raising of a weight, Thus, in Figure 5.7(a), the work done by the system (the individual) is equivalent to raising a weight (F) through a distance (X). For the system comprising the weight, the work interaction is equivalent to the raising of a weight by the surroundings, as shown on Figure 5.7(b).

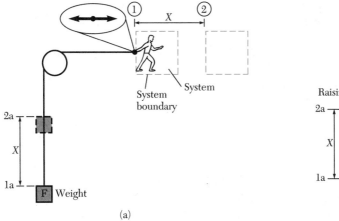

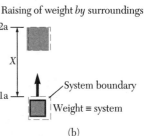

Figure 5.7 (a) Work done by the system (the individual) in raising a weight, which is equivalent to the work done by the individual in Figure 5.4. (b) Work done by the surroundings (the individual) in raising a weight through a distance.

For a particular process, the definition of the system boundary can determine whether or not work is being done. Figures 5.8(a) and (b) show how the definition of the system boundary can alter the work done by a system. In Figure 5.8(a), a gasoline engine is being used to turn a shaft, which in turn raises a weight. With the system boundary as shown in the figure, it is evident that work is done in the process because the effect *external* to the system can be described as the rise of a weight. In Figure 5.8(b), however, the system boundary now envelopes the weight being raised by the shaft, and therefore no work is done by the system since there is no effect external to the system. Instead, the relocation of the weight is occurring within the system boundary. This reemphasizes the importance of a clear definition of the system boundary for the proper evaluation of work interaction in a specified process.

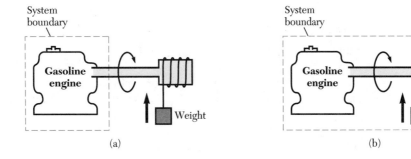

Figure 5.8 (a) Work done by a system (gasoline engine) in raising a weight. (b) Zero work resulting from redefinition of the system boundary.

As Chapter 2 indicated, the units used for work come from the definition of work as force times distance. Thus, the USCS unit for work is the ft · lbf. One ft · lbf is equivalent to 1 lbf (pound force) acting through a distance of 1 ft. Similarly, in SI units, 1 J (joule) is defined as the equivalent of 1 N (newton) acting through a distance of 1 m (1 kJ is 1000 J).

Power is defined as a time rate of doing work. The symbol $\dot{W}$ is used for power, which is defined as

$$\dot{W} = \frac{\delta W}{dt} \tag{5.2}$$

This type of derivative with respect to time is used in this text for both work and heat. In other words, an inexact differential is used with heat and work, both of which are process dependent and not point functions. The rate of change with time for point functions is written herein using the exact differential notation, as indicated earlier. The watt (W) is a derived unit in SI units that is used for power and is equivalent to 1 J/s. Horsepower (hp) is frequently used as an USCS unit of power. It is defined as 1 hp = 550 ft · lbf/s. (For conversion to SI units, the equivalence is 1 hp = 745.66 W.) In some cases the work per unit mass is of interest. In such cases, the symbol w ($= W/m$) is used, and w has units of J/kg or kJ/kg in SI and ft · lbf/lbm in USCS.

The following sections consider several examples of mechanical (displacement and shaft) work as well as other forms of work. These other forms of thermodynamic work may not involve an actual force acting through a displacement, but the fact that the interaction is work can be established in each case by showing that the sole effect external to the system (or surroundings) is equivalent to the rise of a weight.

5.3.2   Mechanical Displacement Work

Equation 5.1a is used to compute mechanical displacement work when an actual force is acting on the boundary through a distance. Common examples include displacement work due to pressure at the system boundary, stretched wire work, and surface tension work. In calculating the displacement work, both the force at the boundary and the displacement are in a direction normal to the surface.

Work at the Boundary Due to Pressure

An example of work frequently encountered in thermodynamics is work involving the motion of a system boundary under the action of pressure forces. Figure 5.9 depicts an example of work done by a system (a gas) on a piston via the movement of the system boundary from position 1 to position 2. In this case, the system volume increases and the piston is moved outward. If the force resisting the gas expansion at the moving boundary can be defined, the work done during the process can be computed by Equation 5.1 or 5.1a for force acting through a distance. In this example, positive work is done by the system, and therefore computation of the work done must be based on the force in the surroundings that opposes movement of the system boundary. The approach is similar to that of the previous examples. In all cases, look at the effects *external* to

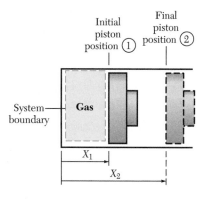

Figure 5.9 Work done by a system
(gas) in moving against a piston from
position 1 to position 2.

the system (or surroundings) and compute the resulting work (or equivalent work) done
on the surroundings (or system).

This is especially important when dealing with work done by a system when the
boundary movement is only *partially resisted*. In such cases, the forces on either side
of the system boundary are not equal in magnitude. A correct determination of the
positive work done must be based on the equivalent raising of a weight effect produced
in the action against the lesser force that is partially resisting the movement of the system
boundary. In contrast, when the system boundary movement is *fully resisted*, the magnitude of the opposing force is the same as the actual force in the direction of movement
of the boundary. Furthermore, for a simple compressible substance, if the process
should take place in a quasi-static or quasi-equilibrium manner, the pressure at the
system boundary is the same as the uniform pressure throughout the system at each
instant. In such cases, the work done can be evaluated from knowledge of the changes
in the thermodynamic state of the system. This section considers the determination of
displacement work when the movement of the system boundary is either partially resisted or fully resisted.

EXAMPLE 5.1
Partially Resisted Expansion of a System

A gas is held at a pressure of 0.2 MPa under a frictionless leakproof piston of cross-
sectional area 0.01 m^2, as shown in Figure 5.10. The ambient pressure is 0.1 MPa,
and the gas is held at the pressure indicated by a pair of studs riveted to the cylinder
so as to project over the top face of the piston. The studs are instantaneously
sheared off, and the piston moves up a distance of 150 mm before finally coming to
rest. Assume that the piston has a negligible mass. Ignoring the work required to
shear the studs, determine the work done by the gas held in the cylinder.

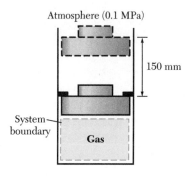

Atmosphere (0.1 MPa)

150 mm

System—
boundary

Gas

Figure 5.10 Partially resisted
expansion of a gas.

Given: The system of Figure 5.10. A partially resisted expansion of the gas takes place against the ambient pressure of 0.1 MPa. Piston mass is negligible, and frictionless piston movement can be assumed. For the system boundary that moves, $A = 0.01$ m^2 and $X = 0.15$ m.

Find: W_{gas}

Solution: The system is defined as the gas held under the piston in Figure 5.10. The interface between the gas and the lower face of the piston is the only portion of the system boundary for which neither the resisting force nor the boundary movement is zero. At this interface, the resisting force f is

$$f = \frac{\text{Ambient}}{\text{pressure}} \times \text{area} = (100 \text{ kPa})(0.01 \text{ m}^2) = 1 \text{ kN}$$

Therefore, Equation 5.1a for mechanical displacement work gives

$$W_{gas} = f_n X_n = (1 \text{ kN})(0.15 \text{ m}) = \underline{150 \text{ J}}$$

It is important to note that the opposing force, rather than the force exerted by the gas on the moving boundary, has been used for calculating the work done on the surroundings. This is the correct measure of the effect equivalent to the rise of a weight produced in the surroundings.

EXAMPLE 5.2
Fully Resisted Expansion of a System

A fixed mass of air is trapped in a vertical cylinder under a frictionless leakproof piston A of cross-sectional area 0.01 m^2, as shown in Figure 5.11. The cylinder is connected via a pipe and a valve to a horizontal cylinder that also holds air trapped by another frictionless leakproof piston B of cross-sectional area 0.005 m^2. The upper

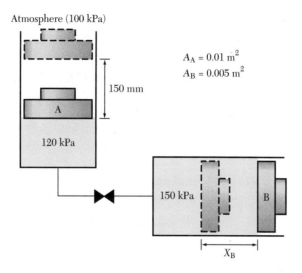

Atmosphere (100 kPa)

$A_A = 0.01$ m^2
$A_B = 0.005$ m^2

150 mm

A

120 kPa

150 kPa

B

X_B

Figure 5.11 Fully resisted expansion of a system (air).

face of piston A is exposed to the atmosphere at a pressure of 100 kPa. The weight of the piston is such that the air trapped in the vertical cylinder is at a pressure of 120 kPa. The connecting valve is opened slightly, while a horizontal force is applied to piston B to maintain the air pressure in the horizontal cylinder at 150 kPa. At the same time, piston A moves up *slowly* through a distance of 150 mm from its initial position. Calculate the work done

(a) by piston A at its upper face

(b) by the atmosphere at its interface with piston A

(c) by piston A

(d) by the system comprising the air trapped in both the vertical and horizontal cylinders

Assume that the pressure of the air in the vertical cylinder remains constant at approximately 120 kPa throughout the process. Neglect any changes in the density of air in the process.

Given: The arrangement in Figure 5.11 with air trapped between two pistons A and B in a vertical and a horizontal cylinder. Piston A moves a distance of 150 mm. $A_A = 0.01$ m^2 and $A_B = 0.005$ m^2. Atmospheric pressure $= 100$ kPa. The pressure of air under piston A $= 120$ kPa. The pressure of air in the horizontal cylinder $= 150$ kPa.

Find:

(a) $W_{\text{piston A}}$ at its upper surface

(b) $W_{\text{atmosphere}}$ at interface with piston A

(c) $W_{\text{piston A}}$

(d) $W_{\text{air mass}}$

Solution:

(a) Piston A is the system, as indicated in Figure 5.12(a). The force opposing the movement of the upper surface of piston A is

$$f_{top} = (100 \text{ kN/m}^2)(0.01 \text{ m}^2) = 1 \text{ kN}$$

The distance moved = 0.15 m. Applying Equation (5.1a) gives

Work done by piston A (system) at its upper face = $(1 \text{ kN})(0.15 \text{ m})$

$$= \underline{150 \text{ J}}$$

(b) In Figure 5.12(a), the atmosphere is part of the surroundings. For the interaction across the upper surface of piston A

$$W_{surroundings} = -W_{system}$$

$$\therefore W_{atmosphere} = \underline{-150 \text{ J}} \text{ (at the interface with piston A)}$$

(c) The total work done by the system (piston A) in Figure 5.12(a) must include work done at the lower face of the piston. Work is done *on* the piston at the lower face, the magnitude of which from Equation 5.1a is

$$(120 \text{ kPa})(0.01 \text{ m}^2)(0.15 \text{ m}) = 180 \text{ J}$$

Note that the force acting on the lower face of the piston is assumed to be fully resisted by the piston. This is reasonable since the process occurs slowly.

$$\therefore W_{piston\ A} = +150 \text{ J} + (-180 \text{ J})$$

$$= \underline{-30 \text{ J}}$$

In other words, 30 J of work is done *on* piston A.

(d) The system is depicted in Figure 5.12(b) and is defined as the air trapped in the vertical and horizontal cylinders. The system does positive work at the interface

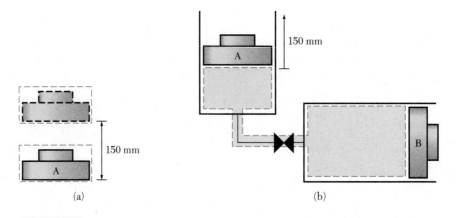

150 mm

150 mm

A

B

A

(a)   (b)

Figure 5.12 (a) Work done by piston A (the system) in Figure 5.11. (b) Work done by the air (the system) in Figure 5.11.

with piston A, while work is done on the system by piston B. If changes in the density of the air are negligible, the increase in the volume of air in the vertical cylinder is equal to the decrease in the volume of air in the horizontal cylinder. Thus, denoting the displacement of piston B by X_B, it follows that $\Delta V_B = \Delta V_A$, or

$$X_B(0.005 \text{ m}^2) = (0.01 \text{ m}^2)(0.15 \text{ m})$$

Hence,

$$X_B = 0.3 \text{ m}$$

Since the process occurs slowly, one can assume that the system boundary movement is fully resisted. The pressure acting on the boundary with piston A is 120 kPa, while that on piston B is 150 kPa. Applying Equation 5.1a thus gives

$$W_{\text{system}} = +(120 \text{ kPa})(0.01 \text{ m}^2)(0.15 \text{ m}) + [-(150 \text{ kPa})(0.005 \text{ m}^2)(0.3 \text{ m})]$$

$$= +180 \text{ J} - 225 \text{ J}$$

$$= \underline{-45 \text{ J}}$$

The work done *on* the system is 45 J.

Computation of the work done in each case in Example 5.2 is equivalent to a product of the pressure and a volume. It should be noted, however, that the volume involved is *not* a change in volume for the system. For example, in the case of piston A, the system volume does not change. Also a negligible change in volume was assumed for the air trapped in the cylinders. The volume used in calculating the work done under the action of a constant pressure force is sometimes referred to as the *swept volume* and is measured by the volume through which the patch of system boundary "sweeps."

A second point to note in Example 5.2 is that for the system comprising the air trapped in the cylinders, the system boundary movement is fully resisted, but the process is *not* quasi-equilibrium. Pressure, for example, is nonuniform for the system, and therefore the system is not in an equilibrium state. However, the work done can be calculated since the forces acting in opposition to the movement of the system boundary are known.

Work Done at Boundary by a Pure, Simple, Compressible Substance During a Quasi-Equilibrium Process

Several cases are encountered in thermodynamics that involve volume changes for a pure, simple, compressible substance under conditions that can be approximated by quasi-equilibrium processes. Typically, the processes occur sufficiently slowly so that at each instant in time, the system is close to an equilibrium condition. Since the quasi-equilibrium assumption implies that a state of mechanical equilibrium exists, it follows that the pressure of the system must be balanced by the external pressure at the system boundary during the process. In other words, the system boundary movement must be *fully resisted* if the process is quasi-equilibrium.

The magnitude of the work done due to pressure for a quasi-equilibrium process can be derived simply by reference to Figure 5.9. Consider an infinitesimal movement of the piston through a distance dX. From Equation 5.1a, the work done is

$$\delta W = f \, dX \tag{5.3}$$

The resisting force f in this case is the same in magnitude as the force exerted by the system on the surface of the piston, which is the product of the pressure times the piston area:

$$f = pA \tag{5.4}$$

Since the product of area times dX gives an infinitesimal volume change for the system, Equation 5.3 for the work done becomes

$$\delta W = pA \, dX = p \, dV \tag{5.5}$$

where $A \, dX = dV$. For the case illustrated by Figure 5.9, work is being done by the system on the surroundings, and the work from state 1 to state 2 can be computed by integrating the expression for work increment in Equation 5.5. Thus, on the premise that the process is quasi-equilibrium, the final expression for the work done by the system is

$$W_{12} = \sum_{\text{state } 1}^{\text{state } 2} \delta W = \int_{\text{state } 1}^{\text{state } 2} p \, dV \tag{5.6}$$

Since $\int p \, dV$ equals the area under the p-V curve for the process (Fig. 5.13), the work for such a process can be obtained by evaluating the area. One could measure the pressures and the volumes during the process to obtain the pressure–volume diagram, or one could compute the relationship using an appropriate equation of state for the substance during the process. Note from Figure 5.13 that the area under the curve

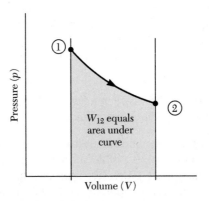

Figure 5.13 Pressure–volume (p-V) diagram for the process in Figure 5.9 (assumed quasi-equilibrium).

clearly depends on the path, and thus the work depends on the path. Also, as noted earlier, unless the process is quasi-equilibrium, a "path" does not exist at all. Therefore, $\int p \, dV$ can be used to compute work *only* for a quasi-equilibrium process with p expressible as a function of V. Furthermore, it is not possible to relate W_{12} to $(W_2 - W_1)$ because no such thing as W_1 or W_2 exists.

EXAMPLE 5.3
Work Done During Quasi-Equilibrium Compression of a Gas

A piston compresses a gas in a cylinder during a quasi-equilibrium process that is the reverse of that shown in Figure 5.9. The pressure in the cylinder varies according to the relationship $pV^{1.4} = $ constant. The initial pressure in the cylinder is 101,325 N/m^2, and the initial volume of the cylinder is 0.01 m^3. Compute the work in compressing the gas to a final volume of 0.005 m^3.

Given: Gas compressed during quasi-equilibrium process, with $pV^{1.4} = C$. The system is the gas.

$$p_1 = 101{,}325 \text{ N/m}^2, \; V_1 = 0.01 \text{ m}^3, \; V_2 = 0.005 \text{ m}^3$$

Find: W_{12}

Solution: From Equation 5.6,

$$W_{12} = \int_{\text{state } 1}^{\text{state } 2} p \, dV$$

$$= \int_{V_1}^{V_2} \frac{C}{V^{1.4}} \, dV = C \left(\frac{V^{-0.4}}{-0.4} \right)_{V_1}^{V_2}$$

$$= \frac{C}{-0.4} \left(\frac{1}{V_2^{0.4}} - \frac{1}{V_1^{0.4}} \right)$$

and

$$C = p_1 V_1^{1.4} = (101{,}325)(0.01)^{1.4} = 160.6$$

Hence,

$$W_{12} = \underline{-809 \text{ J}}$$

Thus, 809 J of work is done on the gas during the process.

Constant Volume Process

The work done by the force of pressure at the system boundary can be computed using Equation 5.6 for quasi-equilibrium processes. Thus, for the case in which the volume does not change, dV would equal zero, and hence,

$$W_{12} = \int_{\text{state } 1}^{\text{state } 2} p \, dV = 0 \quad \text{(constant volume process)}$$

This result is valid for the nonequilibrium constant volume process also, provided the system volume is fixed and no movement of the system boundary takes place. Thus, in the absence of shaft work, the work done by a closed system (of any substance) in a constant volume process is zero.

Constant Pressure Process

If the pressure remains constant during a process, p in Equation 5.6 can be taken outside the integral sign, and the following result can be obtained for the constant pressure process:

$$W_{12} = \int_{V_1}^{V_2} p \, dV = p \int_{V_1}^{V_2} dV$$

$$= p(V_2 - V_1) \quad \text{(constant pressure process)} \tag{5.7}$$

This result applies to any constant pressure, quasi-equilibrium process for a closed system comprising any substance.

Ideal Gas at Constant Temperature

As discussed in Chapter 4, the ideal gas pvT equation (Eq. 4.7) can be written in the following form:

$$pV = mRT \tag{4.7}$$

If the temperature of the gas remains constant and if no mass crosses the system boundary so that the system mass m is constant, then Equation 4.7 reduces to

$$pV = \text{Constant} = C \tag{5.8}$$

for the case of constant temperature ideal gas. The constant at state 1 is

$$C = p_1 V_1 \tag{5.9}$$

Solving for the pressure in Equation 5.8 gives

$$p = \frac{C}{V} = \frac{p_1 V_1}{V} \tag{5.10}$$

Now Equation 5.10 can be combined with Equation 5.6 to obtain the following expression for the work done during the quasi-equilibrium process at a constant temperature:

$$W_{12} = p_1 V_1 \int_{V_1}^{V_2} \frac{dV}{V} = p_1 V_1 \ln\left(\frac{V_2}{V_1}\right) \quad \begin{array}{l}\text{(constant temperature}\\ \text{ideal gas process)}\end{array} \tag{5.11}$$

If V_2 is greater than V_1, the work done during this constant temperature process with an ideal gas is positive, and work is done by the system on the surroundings in such a case.

Ideal Gas with a Polytropic Process

A quasi-equilibrium polytropic process for an ideal gas can be described by

$$pV^n = \text{constant} \tag{5.12}$$

where n is a constant for the particular process and is referred to as the *polytropic index* or *polytropic exponent*. Following the same procedures as previously outlined for the constant temperature process, Equations 5.12 and 5.6 can be combined to obtain an expression for work explicitly in terms of V. Combining the two equations and integrating gives

$$W_{12} = \frac{p_2 V_2 - p_1 V_1}{1 - n} = \frac{mR(T_2 - T_1)}{1 - n} \tag{5.13}$$

for a quasi-equilibrium polytropic process provided $n \neq 1$. Note that the result is not defined for the case of $n = 1$, but a solution is given by Equation 5.11 for this constant temperature case. For the case of $n = 0$, note from Equation 5.12 that p is constant and the result in Equation 5.13 reduces to the result in Equation 5.7 for the constant pressure process with $n = 0$. If $n \to \infty$, Equation 5.12 corresponds to the constant volume case, and the work given by Equation 5.13 becomes zero, which is the result obtained earlier for the constant volume process.

Stretched Wire Work

Figure 5.14 shows a wire (the thermodynamic system) that is stretched from length L_1 to length L_2. The work done can be computed by applying Equation 5.1a for displacement work:

$$W_{12} = -\int_{L_1}^{L_2} f \, dL \tag{5.14}$$

where f is the applied force, which is equal in magnitude to the tension in the wire (opposing the stretching of the wire), and dL is the incremental change in length of the wire as it is stretched. The tension in the wire increases as the length of the wire increases, and the applied force must be at least infinitesimally greater than the tension in the wire. (The negative sign is included in front of the integral because work is done

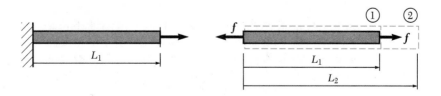

Figure 5.14 Work done in stretching a wire (the system) from L_1 to L_2.

on the system during the stretching of the wire). If the wire obeys Hooke's law, the tension in the wire is proportional to the extension (X) of the wire from its unstretched state; that is, $f = kX$. Therefore, the work done by the system is

$$W_{12} = -\int_{X_1}^{X_2} kX\, dX = -\tfrac{1}{2}k(X_2^2 - X_1^2)$$

where   (5.15)

$$X = L - L_0$$

and where L_0 is the length of the wire when it is unstretched and k is a proportionality constant. Note that stress–strain relationships could be considered here to develop a more specific expression for the energy stored by stretching the wire. However, the main objective at this point is the work interaction involved.

EXAMPLE 5.4
Stretched Wire Work

A 5-m-long wire is stretched as shown in Figure 5.14. The applied force is proportional to the displacement, and the proportionality constant for the wire is $k = 7850$ N/m. Compute the work done in stretching the wire 1 cm.

Given: Wire stretching as shown in Figure 5.14 with $f = kX$, where $X = L - 5$ m.

$$L_1 = 5 \text{ m}, L_2 = 5.01 \text{ m}, k = 7850 \text{ N/m}$$

Find: W_{12}

Solution: $X_1 = 0$ and $X_2 = 0.01$ m. Therefore, from Equation 5.15,

$$W_{12} = -\tfrac{1}{2}k(X_2^2 - X_1^2)$$

$$= -\tfrac{1}{2}\left(7850\,\frac{N}{m}\right)(0.01^2 \text{ m}^2) = \underline{-0.393 \text{ J}}$$

Thus, 0.393 J of work is done on the system during the process.

Surface Tension Work

Surface tension work is a form of mechanical work. Surface tension effects are usually important only for liquids in special situations that include capillary phenomena (such as the sap rising in a tree or kerosene rising in the wick of a lamp), the behavior of small quantities of liquid (such as raindrops or water on a duck's back), and liquid behavior in a container in a microgravity field (such as liquid in an orbiting space station). While surface tension effects are generally small and negligible in comparison to several other effects for earth-based systems, they can dominate the behavior of liquids in the micro-gravity environment of outer space.

Surface tension (σ) between a liquid and vapor (or gas) can be defined as the work per unit area required (in a quasi-equilibrium process) to increase the surface area of the liquid. In other words,

$$\delta W_\sigma = -\sigma \, dA \tag{5.16}$$

where σ is the surface tension and δW_σ is the work required to increase the surface area by an increment dA. The minus sign is used in front of δW_σ because input work is required to the liquid system to increase its surface area of contact with the gas. The units of surface tension are N $\cdot$ m/m^2, or N/m, and ft $\cdot$ lbf/ft^2, or lbf/ft. Note that the units of surface tension represent force $\times$ distance $\div$ area, which are equivalent to force $\div$ length. Thus, the surface tension is also defined as the force per unit length exerted in the plane of the liquid surface in opposition to the stretching of the surface. Minimum work is done to increase the surface area of a liquid when the process is quasi-equilibrium. Assuming the surface tension (σ) is constant, the work done by the system in a quasi-equilibrium process is given by

$$W_{12} = \sum \delta W_\sigma = -\sigma \int_{\text{state } 1}^{\text{state } 2} dA \tag{5.17}$$

The surface tension of a liquid can be determined by measuring the force necessary to pull an inverted U-shaped wire slowly upward through the surface of the liquid (Hausmann and Slack 1948). Figure 5.15 illustrates a simple technique for the measurement of surface tension. The force required to pull the wire through the surface is the sum of the weight of the wire and the force F due to surface tension. The total surface tension pulling downward on the wire is $2\sigma L$ since a liquid surface exists on each side of the wire. Balancing of the forces gives $F = 2\sigma L$. Thus, the surface tension (σ) can be calculated by knowing F and L. Typical values of surface tension for some fluids at 20°C are given in Table 5.1 (from Hausmann and Slack 1948). Surface tension also exists between two immiscible liquids in contact with each other. This *interfacial tension* is approximately given by the difference between their individual surface tensions when each is in contact with air (Knudsen and Katz 1958). Thus, from the data of

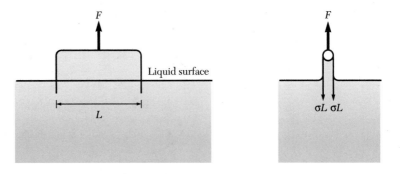

Figure 5.15 A technique for the measurement of liquid surface tension.

TABLE   5.1

Typical Values of Surface Tension for Some Fluids at 20°C*

Liquid	In Contact with	Surface Tension (N/m)
Benzene	Air	29×10^{-3}
Glycerine	Air	63×10^{-3}
Mercury	Air	470×10^{-3}
Mercury	Water	392×10^{-3}
Olive oil	Air	35×10^{-3}
Olive oil	Water	19×10^{-3}
Water	Air	75×10^{-3}

*Data taken from Hausmann and Slack (1948).

Table 5.1, the interfacial surface tension between mercury and olive oil, for example, should be about 435×10^{-3} N/m.

EXAMPLE 5.5
Surface Tension Work

Show that the minimum work required to produce a spherical soap bubble of diameter d is $2\pi\sigma d^2$.

Given: The system is a thin liquid film bounded by two spherical liquid surfaces, as illustrated in Figure 5.16. The soap bubble of diameter d is formed in a quasi-equilibrium process.

Find: Minimum work required, W.

Solution: Minimum work is done when the process is a quasi-equilibrium process. Equation 5.17 can therefore be used, and if the surface tension σ is assumed constant, it follows that

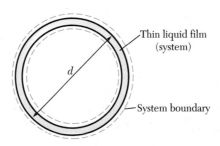

Figure 5.16 Quasi-equilibrium work in producing a spherical soap bubble.

$$W_{12} = \sum \delta W_\sigma = -\sigma \int_{\text{state } 1}^{\text{state } 2} dA$$

$$= -\sigma 2(\pi d^2) - 0 = -2\pi\sigma d^2$$

The initial surface area of the system is assumed to be zero since the soap bubble was drawn from within a liquid mass. Also, two liquid surfaces are involved, and provided that the liquid film is thin, the surface areas are each given by πd^2.

EXERCISES

for Section
5.3.2

1. A crane is used to lift a piece of equipment through a vertical distance of 5 ft. The equipment weighs 3500 lbf. Define the equipment as the system and determine
   (a) W_{system}
   (b) $W_{\text{surroundings}}$
   Answer: (b) 17,500 ft · lbf

2. The crane in Exercise 1 is used for lowering a weight of 4000 lbf through a vertical distance of 5 ft. Define the weight as the system and determine
   (a) W_{system}
   (b) $W_{\text{surroundings}}$
   Answer: (a) 20,000 ft · lbf

3. Steam at a pressure of 5 bar pushes on a piston, as illustrated in the accompanying figure. The cross-sectional area of the piston is 0.07 m². What is the force (in kN) exerted by the steam on the piston?
   Answer: 35 kN

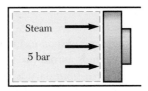

Steam pushing on a piston.

4. The piston in the previous figure is pushed to the right a distance of 0.2 m as a result of the constant steam pressure of 5 bar acting. Assume that the movement of the piston is fully resisted. What is the work done by the steam at its boundary with the piston?
   Answer: 7 kJ

5. A cable having k (spring constant) = 2000 kN/m is used for lifting a 10-kN load. If the cable is initially unstretched, calculate the work done on the cable when it is stretched to the point at which it just lifts the load off the ground.
   Answer: 25 J

6. Repeat Exercise 5 for a cable with k = 1000 kN/m.
   Answer: 50 J

7. A gas is compressed from a state of 100 kPa and 0.0069 m^3 to a volume of 0.00115 m^3 in a piston and cylinder arrangement, as illustrated in the accompanying figure. The compression process is quasi-equilibrium. Calculate the work done on the gas if the p-V compression law is
   (a) $pV^{1.33}$ = constant
   (b) $pV^{1.1}$ = constant
   (c) pV = constant

   Answer (a) 1.69 kJ, (b) 1.35 kJ, (c) 1.24 kJ

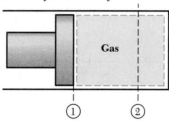

Gas compression in a piston and cylinder arrangement.

8. Repeat Exercise 7 for gas compression from the state of 14.5 lbf/in.^{2} and 0.24 ft^3 to a volume of 0.04 ft^3.

   Answer: (a) 1225 ft · lbf, (b) 983 ft · lbf, (c) 898 ft · lbf

9. What is the work done in producing a soap bubble of diameter 50 mm? Estimate the additional work needed to increase the diameter to 100 mm under isothermal conditions. The surface tension of soapy water can be taken as 0.025 N/m.

   Answer: 3.93 × 10^{-4} J, 1.18 × 10^{-3} J

5.3.3 Mechanical Shaft Work

Figure 5.8(a) illustrates a common example of work done by a system on the surroundings via a shaft that extends across the system boundary. If the plane of intersection of the shaft with the system boundary is considered, work is done via the shear forces acting through an angular displacement. As with mechanical displacement work, the rotation of the shaft in the surroundings may be either partially resisted or fully resisted. Thus, in the example of Figure 5.8(a), the shaft could rotate without any weight attached to it. Unless the rotation of the shaft is partially or fully resisted in the surroundings, no effect is transmitted that could be deemed equivalent to the raising of a weight. In other words, for unresisted rotation of the shaft, the shaft work done is zero. If, however, **G** is the torque in the surroundings acting in opposition to the rotation of the shaft while $d\boldsymbol{\theta}$ is the infinitesimal angular displacement vector, the shaft work done by the system can be calculated as

$$\delta W = -\mathbf{G} \cdot d\boldsymbol{\theta}$$

and

$$W_{12} = -\sum_{\text{state }1}^{\text{state }2} \mathbf{G} \cdot d\boldsymbol{\theta}$$

(5.18)

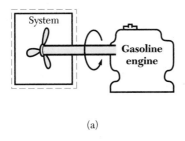
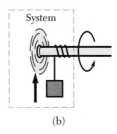

(a)   (b)

Figure 5.17 Paddle wheel work done on a system via the rotating shaft from a gasoline engine.

It is easily seen that the effect transmitted is equivalent to the raising of a weight, and this effect is given in magnitude by Equation 5.18. Mechanical shaft work is encountered in the operation of several thermodynamic systems, including gasoline engines, electric motors, compressors, turbines, pumps, and a number of other devices.

Figure 5.17 shows another example of shaft work. In Figure 5.17(a), a gasoline engine external to the system is turning a shaft that penetrates the system boundary causing a paddle wheel inside the system to rotate. The system includes the fluid that is stirred (or agitated) and the paddle wheel. The resisting torque that the fluid exerts on the paddle wheel can be replaced by one due to a weight, as illustrated in Figure 5.17(b). Positive shaft work is done by the surroundings in this case, and its magnitude must be calculated based on the torque exerted by the fluid on the shaft via the paddle wheel. By the sign convention used in this book, the shaft work done by the system is negative since work is being done on the system. It should be emphasized that no work is done by a rotating shaft unless it penetrates the system boundary.

EXAMPLE 5.6
Mechanical Shaft Work

The shaft on a gasoline engine such as that shown in Figure 5.8(a) turns at 3600 rpm (revolutions per minute). The torque transmitted through the shaft is 15 lbf · ft. Compute the rate of doing work by the engine in raising a weight.

Given: Gasoline engine, angular rotation of shaft = 3600 rpm.

$$\mathbf{G} = -15\,\mathbf{k}\ \text{lbf} \cdot \text{ft}, \qquad \dot{\boldsymbol{\theta}} = 3600\,\mathbf{k}\ \text{rpm}$$

where **k** represents a unit vector parallel to the shaft.

Find: Power output, $\dot{W}$

Solution: The gasoline engine is considered to be the system. Equation 5.18 applies, with the torque opposing the rotation of the shaft being $\mathbf{G} = -15\mathbf{k}$ lbf · ft (constant). Thus,

$$\frac{\delta W}{dt} = -\mathbf{G} \cdot \dot{\boldsymbol{\theta}}$$

$$= (15 \text{ lbf} \cdot \text{ft})\left(3600 \ \frac{\text{rev}}{\text{min}}\right)\left(\frac{2\pi \ \text{rad}}{1 \ \text{rev}}\right)\left(\frac{1 \ \text{min}}{60 \ \text{s}}\right)$$

$$= 5655 \text{ ft} \cdot \text{lbf/s}\left(\frac{1 \ \text{hp}}{550 \ \text{ft} \cdot \text{lbf/s}}\right)$$

$$= \underline{10.3 \text{ hp}}$$

This is the power output from the engine.

EXERCISES

for Section 5.3.3

1. The electric motor used for an elevator delivers 100 hp at a motor speed of 1000 rpm. What is the torque delivered by the machine?

   Answer: 525 ft · lbf

2. The torque output from an automobile engine shaft is 100 ft · lbf at 3600 rpm. What is the horsepower delivered?

   Answer: 75.4 hp

3. An electric motor delivers 1.5 kW to the compressor of a room air conditioner. If it operates at 3500 rpm, determine the torque applied to the compressor shaft.

   Answer: 4.09 N · m

5.3.4 Electrical Work

The flow of electricity across a system boundary is regarded in thermodynamics as a form of work. That electricity is a form of work can be demonstrated with the aid of Figures 5.18(a) and (b). In Figure 5.18(a), a battery (the system) supplies electricity to a resistor in the surroundings. The resistor is replaced in Figure 5.18(b) by an electric motor that receives the current from the battery and causes the shaft to rotate, thereby raising a weight. The change from the actual situation in Figure 5.18(a) to the alternate situation in Figure 5.18(b) shows that the system can pass through the same process while the sole effect external to the system is the raising of a weight. Therefore, positive work is done by a system when electricity flows from it to the surroundings. This form

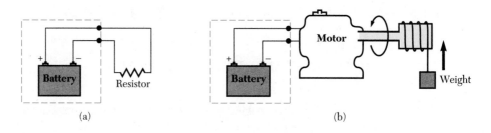

(a)   (b)

Figure 5.18(a) Electrical work done by a battery supplying electricity to a resistor. **(b)** Equivalent shaft work process in place of the process in Figure 5.18(a).

of work is termed *electrical work* and it is classified as shaft work since it is not accompanied by an actual physical displacement of the system boundary. If E is the voltage output of the battery and I the current, the work done is given by

$$\delta W = EI \, dt$$

or (5.19)

$$W_{12} = \int_{time_1}^{time_2} EI \, dt$$

where t denotes time.

Figure 5.19 gives another illustration of how the system for the example in Figure 5.18(b) might be defined. In this case the system boundary encloses only the motor. Consequently, there is work input into the system at the system boundary due to the electrical output of the battery, and there is work output at another part of the system boundary due to the rotating shaft that is used to raise the weight. The magnitude of the electrical work input into the system in Figure 5.19 is the same as the magnitude of the electrical work output in Figure 5.18(b). In both cases, Equation 5.19 can be used to compute the magnitude of electrical work.

EXAMPLE 5.7
Electrical Work

Compute the rate of work output from the battery in Figure 5.18(a) if the terminal voltage (E) is 12 v and the current (I) is 5 amp.

Given: The system is the battery with $E = 12$ v and $I = 5$ amp.

Find: $\delta W/dt$

Solution: From Equation 5.19,

$$\delta W/dt = EI = (12)(5) = \underline{60 \text{ W}}$$

which is the rate of work output from the battery.

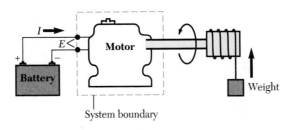

Figure 5.19 Electrical work into a motor (the system) and shaft work out of the system.

1. The electrical power supplied to the motor in Exercise 1 in Section 5.3.3 is 90 kW. The motor in turn delivers 100 hp to the hoist arrangement for the elevator. Define the electric motor as the system and calculate $\dot{W}_{\text{system}}$.

    Answer: -15.4 kW

2. A low consumption solar refrigerator uses a 12-v d.c. battery for the supply of 60 W of electrical power to the motor that drives the compressor of the refrigeration device. What is the current supplied to the motor?

    Answer: 5 amp

3. The electric motor of a ceiling fan draws 100 W of electrical power from the mains. Define the room air, the fan, and its motor as the system and determine $\dot{W}_{\text{system}}$.

    Answer: -100 W

4. Suppose the shaft power output of the fan motor in Exercise 3 is 95 W and the fan speed is 300 rpm. Calculate the following:
    (a) the torque output of the fan motor
    (b) the rate of work done by the fan motor alone
    (c) the rate of work done by the system comprising the room air and the ceiling fan (but excluding the fan motor)

    Answer: **(a)** 3.02 N $\cdot$ m, **(b)** -5 W, **(c)** -95 W

5.3.5 Other Types of Work

Many other types of work are encountered in various engineering applications. Such work includes that produced by a *magnetic field* or an *electrical field*. When more than one type of work interaction occurs between the system and the surroundings, the total work done by the system is obtained by algebraically summing all the different work interactions computed. Thus, in general,

$$\delta W = - \sum_{\substack{\text{system} \\ \text{boundary}}} \mathbf{f}_n \cdot d\mathbf{x}_n - \sum_{\substack{\text{system} \\ \text{boundary}}} \mathbf{f}_s \cdot d\mathbf{x}_s$$

$$- \sigma dA + EI\, dt + \text{field work} + \text{all other work} \qquad (5.20)$$

In all cases, you must first define the system and the system boundary in order to compute the work done. For an interaction to count as work, it must be possible to re-enact the process and be able to produce the raising of a weight as the sole effect external to the system (or the surroundings) doing positive work. Furthermore, work does not exist unless an action across the system boundary is involved. Chapter 6 gives additional examples and considerations of work involving the first law of thermodynamics.

5.4 Heat

A formal definition of *heat* credited to Poincare was given in Chapter 2 as the effect of one system on another by virtue of inequality of temperature. Another definition can be paraphrased as follows: heat is energy interaction that is not work.

The nature of heat was much misunderstood for a long time prior to Joule's work in the 1840s. An early (but erroneous) view that was commonly accepted was that heat was a material substance called "caloric." This caloric was described by John Dalton in 1808 as ". . . an elastic fluid of great subtility, the particles of which repel one another, but are attracted by all other bodies." Later, Joule conducted experiments showing that an equivalent heating effect could be produced via a work interaction. Such results led to the conclusion that heat, like work, is purely a form of energy interaction between a system and its surroundings. Chapter 6 considers Joule's experiment. Although the caloric theory eventually fell into disrepute, various terms used in the theory, such as calorie and calorimetry, have remained to the present day. This point is further illustrated by the definition of such properties as specific heat in terms of the heat required to raise the temperature of a unit mass of the substance by a unit degree of temperature. The remainder of this section discusses the definition of heat in greater detail and concludes with a comparison of heat with work.

5.4.1   Definition of Heat

At this juncture, an adequate definition of *heat* is as follows: the effect across a system boundary by virtue of inequality of temperature between the system and its surroundings. No explicit reference is made to energy in this definition since the thermodynamic definition of energy is derived from the first law of thermodynamics, which is yet to be discussed. Heat is defined only for a process; it is not a property. It can only exist for interactions across the system boundary. The symbol Q (or Q_{12}) is used to designate heat, and the inexact differential δQ is used to indicate an infinitesimal heat interaction. Thus, the incremental and the total heat transfer to a system during a process are given by

$$\delta Q = -\sum_{\substack{\text{system} \\ \text{boundary}}} (\mathbf{q} \cdot d\mathbf{A}_s)\, dt$$

and

$$Q_{12} = \sum_{(\text{state } 1)}^{(\text{state } 2)} \delta Q$$

(5.21)

where $\mathbf{q}$ is the heat flux vector (or heat flow per unit area) and $(\mathbf{q} \cdot d\mathbf{A}_s)$ denotes the heat flow rate through a patch or an infinitesimal area (dA_s) of the system boundary. ($d\mathbf{A}_s$ is a vector of magnitude dA_s and in the direction normal to the patch of the system boundary.) dt is the infinitesimal time period for the heat flow. Sometimes the heat interaction per unit mass of the system is of interest. In this case, the symbol q is used for the heat transfer to a system per unit mass of the system, which is computed as

$$q = \frac{Q}{m}$$

(5.22)

When the time rate of heat interaction is of interest, an overdot above the symbol is used to indicate the time rate with respect to the heat interaction across the system boundary. Thus, the heat rate to a system is

$$\dot{Q} = \frac{\delta Q}{dt} \qquad (5.23)$$

and

$$\dot{q} = \frac{\delta q}{dt} \qquad (5.24)$$

Heat into the system is designated as positive ($+$) and heat out of the system as negative ($-$). A system does not *hold* heat; heat is an effect *across* the system boundary. As with work, the magnitude of heat flowing *into* a system is identical to the amount flowing *from* its environment. In other words,

$$Q_{\text{system}} + Q_{\text{surroundings}} = 0 \qquad (5.25)$$

When no heat flow occurs across the system boundary in a process, the term *adiabatic* is used to indicate that no heat interaction has taken place. Thus, we speak of an *adiabatic process* as one for which $\dot{Q} = 0$ and $Q = 0$.

5.4.2   Heat from a Phenomenological Perspective

Computation of the magnitude of heat flow is made using certain phenomenological laws. Detailed treatment of *rates* at which the interaction occurs is in the domain of *heat transfer*, which is an engineering science. In classical thermodynamics, the rates at which processes occur are not often of major concern. Instead, the common assumption is that processes take place at a sufficiently slow pace so that they can be treated as approximately quasi-equilibrium processes. For completeness, however, the following elementary introduction to heat transfer is provided to enable the reader to make estimates of the magnitude of heat flow in much the same way that the work being done can be computed based on the principles of mechanics.

The principal modes by which heat transfer can occur are conduction, convection, and radiation. In *heat conduction*, the interaction occurs through a physical medium in which no movement of macroscopic parts takes place. This is the mode of interaction that occurs in solid media as well as in the stationary layers of fluids adjacent to solid surfaces. *Convection* refers to a mode of interaction in which the effect is transmitted via actual physical movement of macroscopic parts of the intervening medium. Convection typically occurs in fluid media. If the motion of the fluid is artificially induced, the term *forced convection* is used. In *free convection*, however, the heating process produces a temperature and density gradient in the fluid and fluid motion is induced by the action of gravity.

Radiation is the third principal mode by which heat interaction can occur. In furnaces and combustion chambers where temperatures are very high, radiation is usually the predominant mode. Unlike conduction and convection, radiation does not require a material medium. Thus, heat transmission from the Sun at a temperature of several million degrees to the Earth is by radiation through about 93 million miles of empty space. *Thermal radiation* is electromagnetic radiation and therefore has the characteristics of electromagnetic waves (or photons). All electromagnetic waves travel at the speed of light, which is approximately 3×10^8 m/s in a vacuum. They differ only by

TABLE 5.2

Values of Thermal Conductivity for Some Common Materials* Used in Engineering

Material	Thermal Conductivity k (W/m · K)	Thermal Conductivity k (Btu/h · ft · °R)
Aluminum, pure (at 20°C or 68°F)	222	128
Copper, pure (at room temperature)	393	227
Steel (hot-worked)	47	27
Glass, pyrex (at 93°C or 200°F)	1.02	0.59
Concrete, sand, and gravel (at 24°C or 75°F)	1.82	1.05
Fiberglass with asphalt coating (board) (at 38°C or 100°F)	0.04	0.023
Air (at 300 K or 540°R and 1 atm)	0.0262	0.0151

*Data extracted from *Marks' Standard Handbook for Mechanical Engineers. 9th ed.* (1987).

their wavelengths, and the wavelength region generally associated with thermal radiation is 0.1 to 100 μm (1 μm, or micron = 10^{-6} m). The phenomenological laws governing each of the principal modes of heat interaction are briefly presented and discussed here.

Heat Conduction

The basic law of heat conduction is *Fourier's law*. Like Ohm's law for electric resistors and the laws of thermodynamics, Fourier's law is based on empirical observation. Fourier's law of conduction can be stated for one-dimensional conduction as follows:

$$\dot{q}_x = -k \frac{dT}{dx} \tag{5.26}$$

$\dot{q}_x$ is the *heat flux*, or flow through a unit area per unit time, in the x-direction. The coefficient k is the *thermal conductivity* of the physical medium. It is a property of each material, and it generally varies with the temperature of the substance. The SI unit for k is W/(m · K), while the USCS unit often used is Btu/(h · ft · °R). Values of thermal conductivity are given in Table 5.2 for some common materials used in engineering.

EXAMPLE 5.7
Thermal Conduction

A glass window is 5 mm thick and has an area of 1 m². The two faces of the window are maintained at temperatures of 20°C and −10°C. If the thermal conductivity of the glass is 1 W/(m · K), calculate the heat transmission rate through the glass.

Given: A sheet of glass with surface temperatures as indicated in Figure 5.20. $T_1 = 20°C$, $T_2 = -10°C$, $L = 0.005$ m, $A = 1$ m². For glass, $k = 1$ W/m · K.

Find: Heat transmission rate, $\dot{Q}$

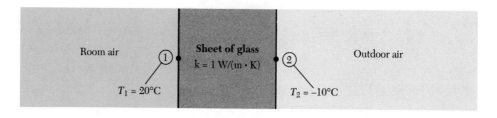

Figure 5.20 Heat flow by conduction through a sheet of glass.

Solution: The situation described could be the heat loss in winter from a room to the outside through a glass sheet that is used for the window. The room air can be defined as the system so that the surface of the glass at 20°C becomes part of the system boundary. Application of Fourier's law of conduction then provides a quantitative determination of the heat interaction rate (at the glass–room air interface) between the system and its surroundings. Equation 5.26 can be written as follows in the present case:

$$\dot{Q}_{loss} = \frac{kA}{L}(T_1 - T_2)$$

$$= \frac{(1 \ W/m \cdot K)(1 \ m^2)}{(0.005 \ m)} [20°C - (-10°C)]$$

$$= 6 \ kW$$

The analogy of Fourier's law with Ohm's law is striking. Equation 5.26 can be re-written in the following manner:

$$\text{Fourier's law:} \quad \dot{Q} = \frac{\text{temperature drop } (\Delta T)}{\left(\dfrac{L}{kA}\right)} \tag{5.27}$$

This can be compared with Ohm's law:

$$\text{Ohm's law:} \quad \text{Current } (I) = \frac{\text{voltage drop } (\Delta V)}{\text{resistance } (R)} \tag{5.28}$$

From this analogy, a *thermal resistance* (R_T) can be defined such that

$$R_T = \frac{\Delta T}{\dot{Q}} = \frac{L}{kA} \tag{5.29}$$

The result in Equation 5.29 is extremely useful for dealing with heat conduction through composite structures. As in electrical engineering, if the conduction paths are in series, the overall thermal resistance is computed as the sum of the individual thermal resistances. For a pair of thermal resistances R_{T1} and R_{T2} in parallel, the equivalent thermal resistance R_T is obtained by

$$\frac{1}{R_T} = \frac{1}{R_{T1}} + \frac{1}{R_{T2}} \tag{5.30}$$

Once the thermal resistance is determined, Equation 5.29 can be used for calculating the heat transmission rate through the physical medium.

Convection

The basic law for calculating the heat transmission rate in the convection mode is analogous to *Newton's law of cooling*. The approach uses a *heat transfer coefficient* (h_c), which is defined as

$$h_c = \frac{\dot{Q}}{A(T_w - T_f)} \tag{5.31}$$

where A is the surface area through which the heat transmission occurs, while T_w and T_f are the solid surface and the fluid medium temperatures, respectively. The units for h_c are $W/(m^2 \cdot K)$ and $Btu/h \cdot ft^2 \cdot °R$. The heat transfer coefficient is analogous to the electrical conductance, and an equivalent thermal resistance for the fluid medium can be defined as $1/(h_c A)$. Table 5.3 gives some typical values of h_c.

EXAMPLE 5.8
Thermal Convection

The glass sheet in Example 5.7 is interposed between room air at a temperature of 20°C and outside air at -10°C. The heat transfer coefficient from room air to the glass is 15 $W/(m^2 \cdot K)$ while that for convection between the glass surface and the outside air is 20 $W/(m^2 \cdot K)$. Determine the heat loss rate from the room air through the glass.

TABLE   5.3

Typical Values* for Convective Heat Transfer Coefficient (h_c)	
Phenomenon	**Typical Range of h_c** $(W/m^2 \cdot K)$ or $(Btu/h \cdot ft^2 \cdot °F)$
Forced convection (with water)	(250 to 12,000) or (50 to 2000)
Boiling water (natural convection)	(600 to 50,000) or (100 to 9000)
Forced convection (with air)	(30 to 600) or (5 to 100)
Free convection (in air)	(5 to 60) or (1 to 10)

*Estimates were taken from Mikheyev (undated).

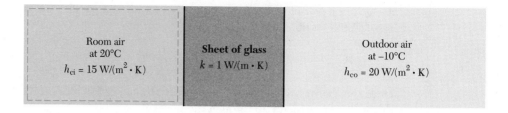

Figure 5.21 Heat transfer by conduction and convection from room air (the system) to outside air.

Given: Heat interaction between room air and outside air for the situation defined in Figure 5.21.

Find: The heat loss rate from the room air.

Solution: The room air can be defined as the system. The glass surface exposed to the room air is then part of the system boundary. The heat transmission path now comprises the convective layer (the room air) adjacent to the glass, the conduction path (through the glass), and the convective layer on the outside adjacent to the glass. The paths are in series; consequently, the total thermal resistance (R_T) is obtained by adding the individual thermal resistances:

$$R_T = \frac{1}{h_{ci}A} + \frac{L}{kA} + \frac{1}{h_{co}A}$$

$$= \frac{1}{(15 \text{ W/m}^2 \cdot \text{K})(1 \text{ m}^2)} + \frac{0.005 \text{ m}}{(1 \text{ W/m} \cdot \text{K})(1 \text{ m}^2)} + \frac{1}{(20 \text{ W/m}^2 \cdot \text{K})(1 \text{ m}^2)}$$

$$= 0.1217 \text{ K/W}$$

From this, the heat loss rate can be determined:

$$\dot{Q}_{loss} = \frac{\Delta T}{R_T} = \frac{(T_{room\ air} - T_{outside\ air})}{R_T}$$

$$= \frac{(20°C) - (-10°C)}{0.1217 \text{ K/W}} = \underline{247 \text{ W}}$$

Thermal Radiation

The fundamental law for thermal radiation is the *Stefan–Boltzmann law*, which in its simplest form can be written as

$$\frac{\dot{Q}_{radiation}}{A_{surface}} = \sigma T_w^4 \qquad (5.32)$$

where T_w is the temperature of the emitting surface (or wall). Equation 5.32 gives the *total emissive power* of what is termed a *blackbody*. The Stefan–Boltzmann constant is

σ, which has a numerical value of 5.67×10^{-8} W/m$^2 \cdot$ K^4. For radiant heat interaction between surfaces at different temperatures, the expression for heat transmission involves the fourth power of the temperatures, in addition to factors that depend on the geometry, relative orientation, and the nature or characteristics of the surfaces. The resulting formula is complex and is typically of the form

$$\dot{Q}_{\text{radiation}} = C(T_A^4 - T_B^4) \tag{5.33}$$

where the heat flow is from surface A to surface B with temperature T_A assumed to be higher than T_B. The constant C combines the area and the previously mentioned factors on geometry, relative orientation, and radiative characteristics of the surfaces. For moderate temperature differences between the surfaces, the following approximate linearized expression, which is based on an equivalent heat transfer coefficient for radiation h_r, can be used:

$$\dot{Q}_{\text{radiation}} = h_r A(T_{w1} - T_{w2}) \tag{5.34}$$

where the subscripts w1 and w2 refer to the two surfaces involved.

All heat interaction processes are accompanied to some extent by thermal radiation. Heat transmission by radiation is usually negligible in solid and liquid media. In gases, however, radiation can be significant, especially when high surface temperatures and high temperature differences are involved.

EXERCISES

for Section 5.4.2

1. A $\frac{3}{4}$-in.-thick layer of plywood ($k = 0.069$ Btu/h $\cdot$ ft $\cdot$ °R) is used for a ceiling over a room having a floor area of 190 ft^2. On a hot day, the top surface of the plywood facing the attic is at 105°F, while the lower surface facing the room air is at 78°F. Estimate the heat transfer rate to the room air from the attic.

   Answer: 5660 Btu/h

2. Suppose the ceiling in Exercise 1 is replaced with a composite slab comprising a 4-in.-thick layer of fiberglass ($k = 0.022$ Btu/h $\cdot$ ft $\cdot$ °R) sandwiched between two $\frac{3}{4}$-in.-thick plywood sheets ($k = 0.069$ Btu/h $\cdot$ ft $\cdot$ °R). What is the heat transfer rate to the room air from the attic assuming the same plywood surface temperatures of 105°F and 78°F?

   Answer: 302 Btu/h

3. The exposed surface area of an electronic device is 100 mm^2. To ensure that the surface does not exceed 50°C when the ambient air is at 35°C, heat must be removed at the rate of 0.6 W. Determine the required heat transfer coefficient.

   Answer: 400 W/m$^2 \cdot$ K

5.4.3   Comparison of Heat with Work

When comparing heat with work, the previous discussions show that both are inexact differentials, both are only defined for a process, and both are only defined for action across the system boundary. Systems do not possess heat or work, and neither of these quantities is a property of a system.

As Chapter 6 shows, Joule's experiment demonstrated that the same heating effect can be produced via either a heat or a work interaction. Both have dimensions of force times distance. Thus, we can use the J, the N · m, or the ft · lbf as appropriate units for either heat or work. Because heat was earlier thought to be different from work, thermal units were developed for heat in both the English and the earlier metric systems. The thermal unit developed in the English system was the British thermal unit (Btu). This unit was originally defined as the amount of heat required to raise 1 lbm of water from 59.5°F to 60.5°F. Similarly, in the metric system, the *calorie* (cal) was defined to be the amount of heat required to raise 1 g of water from 14.5°C to 15.5°C. Once it was recognized that both heat and work interactions are fundamentally forms of energy interaction, it became appropriate to use the same units for both. Many traces, however, can be seen of the earlier ideas concerning the peculiarity of heat in our modern usage of Btu's and calories.

5.5 Summary

This chapter defined a process as that which produces a change of state for a system. It is possible to conceive of a quasi-equilibrium process that could occur such that the system passes through an infinite number of equilibrium states in moving from one state to another state. Real processes are nonequilibrium processes that do not occur in this manner. However, the concept of a quasi-equilibrium process allows work and heat to be computed for a process with reasonable accuracy for many situations.

Both work and heat can be defined as effects transmitted across a system boundary. Work is equivalent to a force acting at the system boundary through a distance, and the action at that system boundary is equivalent to the raising of a weight in the presence of a gravitational field. This chapter considered many different forms of work and showed how one might quantify such work for the interactions between a system and its surroundings.

Heat is defined as an effect across a system boundary by virtue of inequality of temperature between the system and its surroundings. Both heat and work are only defined for a process. They are not properties of a system, and a small change in either is given by an inexact differential of the quantity. Both heat and work are flows (or effects transmitted) across the system boundary, and the same units can be used for computations of both quantities. The sign convention used here is that heat into the system is positive, while work out of the system is positive.

It is useful to note that work is a more "valuable" form of energy flow than heat. Work can always be substituted for heat to produce a desired change of state for a system, but the reverse is not usually possible. Further discussion of the relative qualities of heat and work is given later when the second law of thermodynamics is considered.

References

Bridgman, P. W. 1941. *The Nature of Thermodynamics*. Cambridge, MA: Harvard University Press. (Reprint by Harper & Brothers, New York, 1961.)

Hausmann, E. and Slack, E. P. 1948. *Physics*. 3rd ed. Princeton, NJ: D. Van Nostrand Co.

Knudsen, J. G. and Katz, D. L. 1958. *Fluid Dynamics and Heat Transfer*. New York: McGraw-Hill.

Marks, L. S., Avallone, E. A. and Baumeister III, T. Eds. 1987. *Marks' Standard Handbook for Mechanical Engineers*. 9th ed. New York: McGraw-Hill, Inc.

Mikheyev, M. (undated) *Fundamentals of Heat Transfer*. Moscow: Peace Publishers.

Spalding, D. B. and Cole, E. H. 1966. *Engineering Thermodynamics*. 2nd ed. London: Edward Arnold.

Questions

1. What is the work done in the following processes?
   (a) constant volume heating of a fluid held in a rigid container
   (b) unresisted expansion of a fluid

2. Figures Q5.2(a) and (b) depict the partially resisted expansion of a gas contained in a cylinder under a weightless piston. The piston is held in position using a pair of studs that are designed to shear off when the gas pressure reaches a predetermined value p_{limit}. The value p_{limit} is higher than the atmospheric pressure p_{atm}. The gas is heated and eventually the gas pressure reaches p_{limit}. At this stage, the studs holding down the piston are sheared off while the piston shoots up until it finally settles to the position indicated in Figure Q5.2(b). For the gas expansion, assume an initial pressure and volume of p_{limit} and V_1, respectively, and a final pressure and volume of p_{atm} and V_2. Show the process on a p-V diagram. What is your best estimate of the work done by the gas?

3. Figure Q5.3 illustrates a ceiling fan used in the lounge of an apartment. The fan blades do work on the room air by causing movement of the air. Would you consider the movement of the surface of the blades against the air to be fully resisted or only partially resisted?

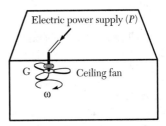

Figure Q5.3 Schematic diagram of the operation of a ceiling fan in the lounge of an apartment.

4. In question 3, assume that the torque G exerted by the fan motor and the rotational speed ω of the fan are known. Also assume that the electrical power supply (P) to the fan motor is known.
   (a) Would you expect the power computed from the torque G and the speed ω to be greater than, equal to, or less than P? Explain your answer.
   (b) If the effect of the operation of the fan on the room air is to be studied, suggest an appropriate choice of system boundary that would ensure a reliable estimate of work input to the system. What is the work input to the system you have chosen?

5. Figure Q5.5 depicts a stirrer driven by an electric motor. The stirrer is in an evacuated rigid vessel.

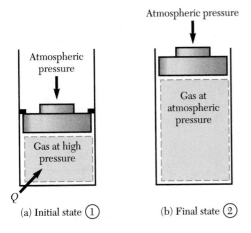

(a) Initial state ①   (b) Final state ②

Figure Q5.2 The partially resisted expansion of a gas contained in a cylinder under a weightless piston in the (a) initial state and (b) final state.

When electricity is supplied to the motor, the stirrer races at first and eventually rotates at a high speed determined by the friction in the motor–stirrer drive. What is the work done by each of the following systems?

(a) the vacuum
(b) the vacuum and stirrer
(c) the electric motor and stirrer

Sketch the system boundary for each of these cases.

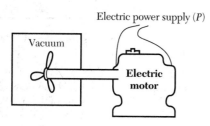

Figure Q5.5 A stirrer driven by an electric motor.

6. Figure Q5.6 illustrates the temperature distribution for a steady-state heat conduction through a wall having a thermal conductivity k_w. The thickness of the wall is L, and the temperatures on the inside and outside surfaces of the wall are T_{wi} and T_{wo}, respectively. If the room air is defined as the system, show an appropriate boundary surface relative to the wall in Figure Q5.6. Give an expression for the heat transfer rate to the system through the wall in terms of the quantities indicated.

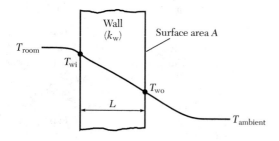

Figure Q5.6 Steady-state temperature distribution for heat transfer between room air and outdoor air.

7. What is the work done in raising a load in a microgravity environment? Suggest an appropriate definition for work in a microgravity environment.

8. Warmth is provided to an individual in her study by an electric heater. It is extremely cold outside, and the heating provided by the heater is just sufficient to maintain the study at a reasonable comfort level. If the system boundary is as defined in Figure Q5.8, identify and clearly mark on the diagram the work and/or heat transfer across the boundary.

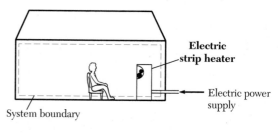

Figure Q5.8 Comfort heating of an individual's study using an electric heater.

9. By what mechanisms is warmth provided to the individual in Figure Q5.8 from the electric heater? Which particular mechanism, if any, is likely to be dominant? Does the warming provided by the heater count as heat relative to the system boundary indicated in Figure Q5.8?

10. In Figure Q5.10, an electric motor is supplied with electricity, resulting in the motor raising itself using the hoist arrangement shown. Identify all the work interactions taking place for a system defined as the electric motor.

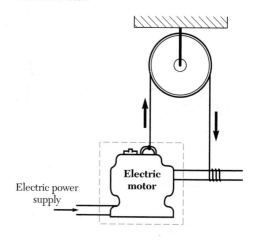

Figure Q5.10 An electric motor raising itself with a hoisting apparatus.

11. Figure Q5.11 illustrates the heating of water using an electric heating coil. The coil is supplied with electrical power, as a result of which the heating

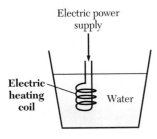

Figure Q5.11 Heating of water using an electric heating coil.

element attains and remains at a very high temperature. Characterize the interaction occurring across the system boundary as heat or work in each of the following cases:
  (a) the system is the water and the electric heating coil
  (b) the system is the electric heating coil
  (c) the system is the water

12. What are the principal mechanisms for heat transfer?

13. It is necessary to keep a set of microcomputer chips from being overheated by cooling at the rate of a few watts per chip. Suggest appropriate means for accomplishing this.

14. Barefoot water skiers say that their feet get warm when skiing on extremely smooth water, but they do not get warm when skiing on relatively rough water. Explain why.

Problems

5.1 Figure P5.1 depicts a man lifting a load of 150 N with the aid of a simple pulley. The man slowly lifts the load through a distance of 0.6 m. The pulley can be assumed frictionless and the rope is inelastic.
  (a) If the system is defined as everything inside the enclosure—the man, the pulley, the load, and all—what is the work done by the system?
  (b) Suppose instead that the man alone is the system. What is the work done by the system?
  (c) If the load is defined as the system, what is the work done?
Draw appropriate sketches clearly indicating the system boundary in each case.

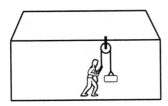

Figure P5.1 Lifting of a load by the man in Problem 5.1.

5.2 An individual slowly lowers a weight of 100 N through a vertical distance of 10 m. Which does

positive work: the weight or the individual? Explain your answer. What is the magnitude of the work done?

5.3 An individual slowly pushes a load of 20 kg of gravel a distance of 3 m along a rough road bed. The friction coefficient is 0.6 and g can be taken as approximately 10 m/s^2. What is the minimum force that the individual must apply? Calculate the work done by the individual in moving the load.

5.4 A mass of 1 kg of water at 27°C is contained in a sealed, rigid container of volume 0.01 m^3. The initial pressure is 1 atm. The water temperature is raised to 100°C by using a gas heater. Calculate the work done by the water in the container.

5.5 An automobile has a mass of 1500 kg and has an engine that develops 20 kW when traveling at a speed of 60 km/h. Neglecting losses, determine the resistance to motion in kN.

5.6 An individual pushes an automobile 2 km with a constant force of 300 N. Compute the work done
  (a) by the individual on the automobile
  (b) by the automobile on the individual

5.7 If the person in Problem 5.6 moves the automobile at a constant speed of 1 km/h, what is the power exerted by the individual in pushing the automobile?

* **5.8** Two horses are used to pull a wagon loaded with wood up a mountain trail 2 mi to a cabin.
  (a) If the horses exert a constant force of 200 lbf on the wagon, how much work is done by the horses in pulling the wagon?
  (b) If this work is done in 45 min, what is the rate of doing work in hp?
  (c) What is the total work done by the wagon on the horses during the process?

5.9 Figure P5.9 shows a vertical cylinder closed by a frictionless, leakproof piston of cross-sectional area 0.01 m². The cylinder is connected via a pipe and a valve to a large air tank (or receiver) containing air at a pressure of 200 kPa. The upper face of the piston is exposed to ambient air at a pressure of 100 kPa. The total weight of the laden piston is 250 N. What is the pressure of the air (in the cylinder) under the piston at equilibrium?

If the valve is opened so as to admit the high pressure air in the receiver into the cylinder and cause the piston to move up slowly through a distance of 100 mm from its initial position, calculate the work done by the following:
  (a) the piston at its upper face
  (b) the atmosphere
  (c) the piston
  (d) the system comprising the air in the receiver and under the piston in the cylinder

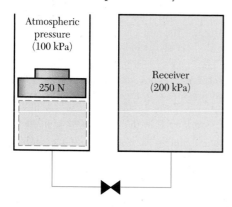

Figure P5.9 Work done in the piston and cylinder device of Problem 5.9.

5.10 In the induction stroke of a reciprocating piston and cylinder device, ambient air at 100 kPa is admitted into the cylinder of the device, as shown in Figure P5.10. The air pressure in the cylinder is 25 kPa below that of ambient air and remains at this level throughout the induction process. The cross-sectional area of the piston is 6000 mm² and the stroke is 60 mm. A volume of 3×10^{-4} m³ (at 100 kPa) of ambient air is admitted into the cylinder in each complete induction stroke. What is the work done by the air finally admitted into the cylinder in each induction stroke?

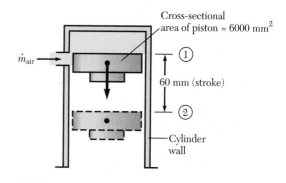

Figure P5.10 Induction stroke for a piston and cylinder device.

5.11 A volume of 2.5×10^{-3} m³ (at 100 kPa) of fresh air from the atmosphere is admitted into a previously evacuated rigid container. The ambient pressure is 100 kPa. Assuming that filling of the container occurs slowly, calculate the work done by the following:
  (a) the atmosphere (excluding the air going into the container)
  (b) the air admitted into the rigid vessel

5.12 A man uses a spring to lift a body of weight 50 N off the ground. The spring constant is 500 N/m. Determine the work done by the man in pulling the spring
  (a) to the stage when the load just leaves the ground surface
  (b) after raising the load through a distance of 1 m

* **5.13** A rope initially 50 ft long is stretched until it is 51 ft long. The applied force is proportional to the displacement, and the proportionality constant for the rope is $k = 20$ lbf/ft.
  (a) Compute the work done on the rope in stretching it.
  (b) If this work was done in 10 sec, what was the average rate of doing work in hp?

* 5.14 A 10-m-long wire is stretched a distance of 3 cm. The tension in the wire varies according to the expression $f = 1.5(L^2 - 100)$ N, where L is the length of the wire in meters. Compute the work done *by* the wire.

* 5.15 An individual slowly lowers a weight of 200 N onto a compression spring. During this operation, the spring is shortened by 5 mm. What is the work done by the following?
(a) the compression spring
(b) the weight
(c) the individual

5.16 What is the minimum work required to increase the surface area of liquid glycerine by 20 m² when the liquid is in contact with air?

5.17 What is the minimum work required to increase the surface area of liquid mercury by 10 m² when the mercury is in contact with olive oil?

5.18 An inverted U-shaped wire is pulled upward slowly through the surface of a liquid. The distance between the ends of the wire is 3 cm. The mass of the wire is 5 g. The upward force is 5340 dynes. What is the surface tension of the liquid? (Assume $g = 981$ cm/s².)

* 5.19 Steam at 1 MPa and 250°C flows through a steam main at a low velocity, as illustrated in Figure P5.19. Two rigid containers, each of capacity 0.03 m³, are connected to this main by pipes that are closed by valves. Container A is initially evacuated, while container B already holds steam at 0.5 MPa and 250°C. Both containers have heat-conducting walls and are immersed in a constant temperature bath at 250°C. The valve at the inlet to each container is opened until each chamber holds steam at 1 MPa and 250°C. Calculate the following:
(a) the mass of steam initially in container B
(b) the mass of steam finally in each of the containers
(c) the volume of steam at 1 MPa and 250°C admitted (or *swept*) into each of the containers
(d) the work done by the system comprising the steam finally in container A
(e) the work done by the steam finally in container B

* 5.20 An automobile tire of volume 0.02 m³ is to be inflated from a large air tank (or *receiver*). The air in the tank is at 500 kPa (abs) and 300 K. Assume that the tire is relatively inelastic so that its volume remains constant. Initially, the air in the tire is at the atmospheric pressure of 100 kPa and a temperature of 300 K. What is the mass of air in the tire? The valve connecting the tire to the air receiver is opened so as to admit air into the tire slowly until the pressure in the tire rises to 300 kPa. Determine the total mass of air now in the tire. (Assume that the temperature of air remains at 300 K.) If the air pressure and temperature in the receiver remain approximately constant during the inflation of the tire, calculate
(a) the work done by the system comprising the air in the receiver and the tire
(b) the work done by the air finally in the tire

* 5.21 A single-stage reciprocating compressor operates on an open cycle. The processes in each cycle are as follows:
1→2: Induction of air in a constant pressure process with $p_1 = 100$ kPa $= p_2$, $V_1 \approx 0$, and $V_2 = 0.0035$ m³.
2→3: The mass of air induced is compressed according to the polytropic law $pV^{1.25} = $ constant, with $p_3 = 600$ kPa.
3→4: Discharge of the compressed air at 600 kPa to an air receiver with $p_4 = 600$ kPa and $V_4 \approx 0$.

Assume that all the processes are *quasi-static* (that is, one that occurs *slowly* so that fully resisted movement of the system boundary can be assumed). Sketch the processes on a p-V diagram. Determine the work done by the (induced) air on

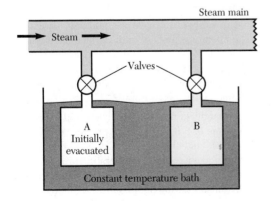

Figure P5.19 Filling of rigid containers from a steam main as described in Problem 5.19.

the piston (at the induced air–piston interface) in each of the processes 1→2, 2→3, and 3→4. Determine the work input per cycle required for compressing the air.

*5.22 An idealized open cycle for a single-stage, reciprocating air motor comprises the following quasi-static processes:

1→2: Air supply at a supply pressure of 600 kPa such that $V_1 \approx 0$, $p_1 = 600$ kPa $= p_2$ and $V_2 = 1.8 \times 10^{-4}$ m^3.

2→3: Polytropic expansion, with an index of 1.3, to $V_3 = 3.6 \times 10^{-4}$ m^3.

3→4: Constant volume process to a pressure of 100 kPa.

4→5: Exhaustion of the air at a constant pressure of 100 kPa such that $p_5 = 100$ kPa $= p_4$ and $V_5 \approx 0$.

The supply air temperature is 300 K. Show the processes on a *p-V* diagram. Calculate

(a) the mass of air supplied per cycle

(b) the air temperature at the end of the expansion process

(c) the net work output per cycle of the air motor

*5.23 An idealized steam engine (open) cycle comprises the following quasi-static processes:

1→2: Admission of dry, saturated steam at 1 MPa into the cylinder of the reciprocating steam engine at a constant pressure with $V_1 \approx 0$, $p_1 = 1$ MPa $= p_2$ and $V_2 = 0.005$ m^3. Steam supply to the cylinder is cut off at state 2.

2→3: Following the cut off of steam supply at state 2, the steam in the cylinder expands according to $pV =$ constant, to the maximum cylinder volume of 0.02 m^3 ($= V_3$).

3→4: Instantaneous blow down at constant volume ($V_4 = V_3$) takes place, resulting in a drop in pressure to $p_4 = 105$ kPa.

4→5: The steam is exhausted at 105 kPa ($= p_5$) against the back pressure until V_5 approaches V_1 ($\approx$0).

Sketch the processes on a *p-V* diagram. Determine

(a) the mass of steam supplied to the engine per cycle

(b) the net work output, per cycle, of the steam engine

*5.24 A piston compresses a gas in a cylinder during a quasi-equilibrium process that is the reverse of that shown in Figure 5.9. The pressure in the cyl-

inder varies according to the relation $pV^{1.3} =$ constant. The initial pressure in the cylinder is 2 atm, and the initial volume of the cylinder is 100 in.^{3}. Compute the work (in ft · lbf) done on the system in compressing the gas to a final volume of 40 in.^{3}.

5.25 A piston moves from position 1 to 2, as shown in Figure 5.9. During the process, the pressure of the gas inside the cylinder (that is, inside the system boundary) decreases according to a linear relationship with volume from p_1 to p_2, while the volume increases from V_1 to V_2. Develop an expression for the work done *by the piston* on the gas during the process. (Assume the process is quasi-equilibrium.)

5.26 A quantity of 10 lbm of air is enclosed in a rigid container at 200 psia. Heat is added to the air until the pressure increases to 300 psia. What is the work done on the system (the air) during the process?

5.27 A quantity of air at the atmospheric pressure of 1 atm is heated during a constant pressure process so that the volume of the air increases from 1 m^3 to 3 m^3. Compute the work done by the air during the process.

5.28 A quantity of nitrogen gas at 5 psia is cooled during a constant pressure process until its volume decreases from 1 ft^3 to 0.3 ft^3. Compute the work done by the gas during the process.

5.29 A quantity of air is initially at 10^6 Pa with a volume of 0.5 m^3. If the temperature of the air remains constant while the volume of the air slowly increases to 1 m^3, what is the work done by the gas during the process?

5.30 A quantity of nitrogen gas undergoes a constant temperature process until its final state is $p_2 = 10^6$ Pa, $V_2 = 3$ m^3, and $T_2 = 100$°C. If the gas initially occupied a volume of 2 m^3, how much work was done by the gas during the process (assumed quasi-equilibrium)?

5.31 A quantity of oxygen gas is contained initially in a volume of 2 ft^3. The initial pressure of the gas is 15 psia and the temperature is 500°R. The gas undergoes a constant temperature process until its final pressure is 30 psia. What was the work done by the gas during the process? (Assume the process was quasi-equilibrium.)

*5.32 A quantity of neon gas undergoes a polytropic process described by $pv^{1.3} =$ constant. The initial pressure and volume are 10^5 Pa and 2 m^3, respectively, while the final pressure is 10^6 Pa. Compute the work done by the gas during the process.

* **5.33** Air undergoes a polytropic process described by pV^n = constant. The initial pressure and volume are 15 psia and 4 ft^3, respectively. The final pressure is 21 psia and the final volume is 3 ft^3. Compute the work done by the air during the process.

* **5.34** Argon undergoes a polytropic process described by pV^n = constant. The initial pressure is 10^5 N/m^2 and the initial volume is 2 m^3. The final pressure is 2.5 $\times$ 10^5 N/m^2 and the final volume is 1 m^3. Compute the work done by the gas during the process.

** **5.35** An unknown gas undergoes a polytropic process. The gas is initially enclosed in a volume of 10 m^3 with a pressure of 10^5 N/m^2. At the final state, the gas has a volume of 5 m^3. If 750 kJ of work was done on the gas during the process, what was the polytropic exponent (index) for the process?

5.36 Develop an expression for the work done per unit mass for a gas during a polytropic process in terms of only the following parameters: T_1, T_2, R, and n.

5.37 An electric mixer rotates a stirrer a total of 1000 rad (radians) while mixing a milk shake.
   (a) If the torque required to stir the mix is 10 N $\cdot$ m, what is the work done by the mixer during the process?
   (b) If the process was completed in 5 min, what was the rate of doing work in W?

5.38 A gasoline engine is used to stir concrete. The shaft of the engine turns at the rate of 1 rpm, and the torque transmitted through the shaft is 100,000 lbf $\cdot$ ft. Compute the rate of doing work (in stirring concrete) by the engine in hp.

5.39 An automobile engine delivers 150 hp through the output shaft from the engine. If the shaft is turning at a rate of 2000 rpm, what is the torque transmitted through the output shaft in lbf $\cdot$ ft?

5.40 A 12-V automobile battery is used only to provide lighting for the vehicle. During a period when the lights are using 4 amp, what is the rate of work output from the battery?

* **5.41** Power for an electric car is provided by a bank of 20 12-V batteries. If the batteries are wired in parallel and the output current of each battery is 9 amp, what is the power available for the car in W? In hp?

5.42 A 120-V electric heater is rated 1.5 kW. What is the current drawn by the heater?

5.43 Determine the thermal resistance of a wall that allows a heat transfer of 300 Btu/h $\cdot$ ft^2 with a temperature difference of 60°F.

* **5.44** An industrial oven wall is composed of 10 in. of fireclay brick (inside), 4 in. of insulating brick, and 6 in. of masonry brick (outside), as shown in Figure P5.44. The thermal conductivities of the bricks are

$$k = 0.6 \text{ Btu/h} \cdot \text{ft} \cdot °\text{F (fireclay)}$$

$$k = 0.08 \text{ Btu/h} \cdot \text{ft} \cdot °\text{F (insulating)}$$

$$k = 0.4 \text{ Btu/h} \cdot \text{ft} \cdot °\text{F (masonry)}$$

The inner and outer surface temperatures are

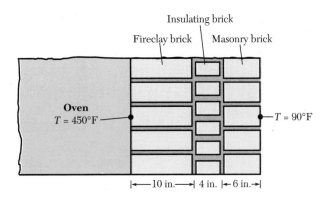

Figure P5.44 Heat transfer through an oven wall.

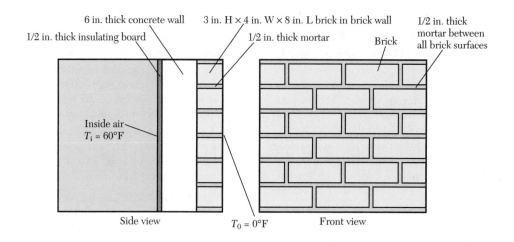

Figure P5.45 Heat transfer through a composite wall.

450°F and 90°F, respectively. Assuming that the resistances of the mortar joints are the same as for the adjacent bricks, determine the temperatures of the inner surfaces.

* **5.45** A masonry wall (Fig. P5.45) is composed of 3-inch high by 4-inch wide by 8-inch long brick ($k = 0.4$ Btu/h · ft · °F) with $\frac{1}{2}$-in.-thick mortar joints ($k = 0.7$ Btu/h · ft · °F) on the outside, a 6-in.-thick concrete wall ($k = 0.4$ Btu/h · ft · °F) in the middle, and a $\frac{1}{2}$-in.-thick insulating board ($k = 0.04$ Btu/h · ft · °F) on the inside. The outside surface temperature is 0°F and the inside surface temperature is 60°F. Compute the heat flux (in Btu/h · ft^2) through the wall. (Use thermal resistances. Note that there is parallel flow in the outer layer.)

* **5.46** Now assume that the outside and inside surface temperatures for the wall in Problem P5.45 are unknown, but that the outside air temperature is 10°F while the inside air temperature is 68°F. Assuming that the convective heat transfer coefficients for the outside and inside are 1.5 Btu/h · ft^2 · °F and 2.5 Btu/h · ft^2 · °F, respectively, compute the heat flux (in Btu/h · ft^2) through the wall for this situation.

* **5.47** Certain electronic components are dissipating heat at the rate of 0.5 W. The components are

enclosed in a cubic container of aluminum ($k = 128$ Btu/h · ft · °F), and the thickness of the aluminum is $\frac{1}{8}$ in., as shown in Figure P5.47. The outer dimensions of the container are 1 in. × 1 in. × 1 in. If the outside air temperature is 70°F and the outside convective heat transfer coefficient is 1.2 Btu/h · ft^2 · °F, what is the inside surface temperature of the aluminum container? What is the most significant resistance to the heat transfer? (The cooling of electronic components is often a severe problem in microelectronics because of the difficulty of getting the heat transferred away from the chips.)

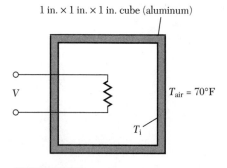

Figure P5.47 Heat dissipation from electronic components.

The First Law of Thermodynamics

6.1 Introduction

This chapter considers the first law of thermodynamics in detail. As stated earlier, this law is the principle of conservation of energy. The model for this discussion assumes (1) that the system is a simple compressible pure substance and (2) that both the initial and final states of a closed system are equilibrium states. In other words, the process can be nonequilibrium or quasi-equilibrium as long as the initial and final states of the system are equilibrium states. Consideration of the first law involves careful observation of the interactions across the system boundary between the system and its surroundings.

6.2 Producing a Heating Effect by Doing Work

Chapter 5 covered work and heat. From 1840 to 1848, an Englishman named James Prescott Joule conducted his famous experiments that established a relationship between work and heat and paved the way to the formulation of the first law of thermodynamics. Joule demonstrated that work could be used to produce a rise in temperature of a system

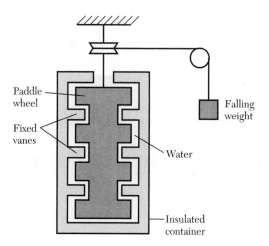

Figure 6.1 Joule's experiment demonstrating that work can be used to produce the same effect as heat.

in the same way that heat can be used. More importantly, however, Joule established that the same magnitude of work input is involved, by different work processes, in producing the same change of state for a system under *adiabatic* conditions.

Figure 6.1 illustrates the paddle wheel apparatus that Joule used for some of his experiments. A weight, lowered at a constant velocity, was used to turn a paddle wheel and thereby do work on the system. The temperature of the water increased as a result of the paddle wheel motion. From measurements of the temperature rise of the water, Joule was able to establish the fact that a fixed amount of work on the system was needed to produce a specified change of state for the system. Joule performed additional experiments using other work processes to produce an increase in the temperature of a system. The major conclusion of his studies was that *the same amount of work was needed to produce the same heating effect, regardless of the actual work process*. Thus, it was found that work of magnitude 778 ft · lbf produces a heating effect equivalent to 1 Btu. This quantitative equivalence between heat and work has been more precisely defined in USCS units such that 1 Btu is equal to exactly 778.169 ft · lbf. The Btu so defined is termed the *international British thermal unit*. In the United States there is widespread use of the conversion 1 Btu = 778.16 ft · lbf that was introduced in Chapter 2. This conversion will continue to be used in this text.

Joule's experiment is repeatable and demonstrates the fact that work can always be substituted for heat to obtain a desired heating effect. In the particular case illustrated in Figure 6.1, the lowering of the weight causes work to be done on the system via the paddle wheel. The action of friction, as the paddle wheel turns the water against the resistance due to the fixed vanes, results in a rise in temperature of the water (the system). The temperature rise is always directly proportional to the amount of work done on the system via the lowering of the weight. The same effect can be produced by heat interaction of an appropriate magnitude. The next section presents the classical formulations of the first law of thermodynamics and points out their relationship to Joule's findings.

6.3   First Law of Thermodynamics for a Closed System

A variety of formal statements of the first law of thermodynamics are given in the literature. Any could serve as the *primary statement* of the law, and then, using logic, the others could be shown to be *equivalent statements* or *corollaries* of the primary statement. Regardless of how the first law of thermodynamics is stated, one important consequence of the law is the *derived concept of energy*. In other words, strictly speaking, there is no definition for energy in thermodynamics until the first law of thermodynamics is considered. Thus, in the classical statements of the first law principle, the basic concepts work and heat are used without being defined in reference to energy. A few of these statements are given here.

Statements of the First Law

Planck's Formulation

The starting point for Planck's statement (see Planck 1927, 40–47) of the first law of thermodynamics is the empirical observation that a perpetual motion machine is impossible:

> . . . *it is impossible to construct an engine which will operate in a cycle and produce continuous work, or kinetic energy, from nothing.*

An engine or machine that can produce work continuously from nothing is termed a *perpetual motion machine of the first kind (PMM1)*. Fludd's closed-cycle mill, which was mentioned in Chapter 1, is a classic example of a PMM1. The inventor's claim was that the device operating in a cycle would produce work continuously at no cost to the surroundings. The first law of thermodynamics unequivocally proclaims that a PMM1 is impossible.

Planck's formal statement of the first law of thermodynamics is as follows:

> . . . *the algebraic sum of the mechanical equivalents of the external effects produced outside the system, when it passes from the given to the normal state, is independent of the manner of the transformation.*

Joule's famous experiments are cited as evidence in support of this principle. The "mechanical equivalents of the external effects produced outside the system" would appear to correspond to what was described in Chapters 2 and 5 as the "raising" of a weight, in other words, *work interaction*.

Poincare's Statement

Poincare's statement dates back to 1908 and uses the basic concepts of heat, work, and cyclic processes (Keenan and Shapiro 1947, 917). Both heat and work are defined in this formulation as interactions between a system and its surroundings, but with no explicit reference to energy transfer. Thus, for example, heat is defined as "the effect of one system on another by virtue of inequality of temperature." The formal statement of the first law principle given by Poincare is as follows:

In a cyclic process, the net heat is proportional to the net work.

Poincare's statement can be related to the version of Joule's experiment depicted in Figure 6.1. After using work to raise the temperature of the water in the apparatus, an appropriate amount of heat rejection to the surroundings can be used to restore the system to its original temperature. The combination of the work process followed by the heat process restores the system to its original state and therefore constitutes a thermodynamic cycle. Poincare's statement of the first law of thermodynamics thus asserts that the heat to the surroundings must be equal to the work done on the system.

The cyclic integral ($\oint$) will be used as the symbol to indicate the summation of a quantity around the cycle. Thus, the first law for the cycle can be written as

$$\oint \delta Q = \oint \delta W \tag{6.1a}$$

A corollary of this principle is that a property E exists for the system such that

$$dE = \delta Q - \delta W \tag{6.1b}$$

The proof of the corollary is presented later in this section. This text has adopted the Poincare statement as the primary statement of the first law of thermodynamics.

Caratheodory's Formulation

Caratheodory used work and the adiabatic wall concepts in his statement of the first law principle in 1909 (Keenan and Shapiro 1947, 918):

The work done by a system which is surrounded by an adiabatic wall is fixed by the end states of the system and is independent of the nature of the process.

This too is a formulation that is consistent with the findings of Joule in his famous experiments. As indicated earlier, Joule found that the same amount of work was needed in an adiabatic process for producing a specified change of state for a system, regardless of the precise nature of the work process.

The immediate corollary of Caratheodory's statement of the first law is that a property E can be defined for a system such that

$$dE = -\delta W_{\text{adiabatic}} \tag{6.2a}$$

or

$$E_2 - E_1 = -(W_{12})_{\text{adiabatic}} \tag{6.2b}$$

The subscript adiabatic is a reminder that the process must be adiabatic.

First Law and Energy

Equation 6.1b was given as a corollary of the first law in connection with the Poincare statement, but no proof was presented. A proof is now given based on that statement of the first law of thermodynamics.

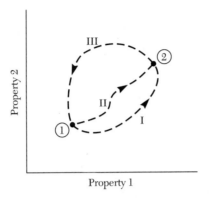

Figure 6.2 A schematic illustration of processes (I, II, and III) between two equilibrium states 1 and 2.

Let I and II designate any two arbitrary processes by which a system at an equilibrium state 1 is transformed to another equilibrium state 2. A process III is required to restore the system to the initial state 1. The situation described is illustrated on the phase diagram of Figure 6.2, with the processes I, II, and III shown as dashed lines since they generally do not have to be quasi-equilibrium processes. The pairs of processes (I and III) and (II and III) are each cyclic and, therefore Equation 6.1a can be applied to each pair in turn. In other words, for the cyclic process pair (I and III),

$$\oint (\delta Q - \delta W) = 0 = \int_{\text{state } 1}^{\text{state } 2} (\delta Q - \delta W)_{\text{I}} + \int_{\text{state } 2}^{\text{state } 1} (\delta Q - \delta W)_{\text{III}} \quad (6.3a)$$

Similarly, for the pair (II and III),

$$\oint (\delta Q - \delta W) = 0 = \int_{\text{state } 1}^{\text{state } 2} (\delta Q - \delta W)_{\text{II}} + \int_{\text{state } 2}^{\text{state } 1} (\delta Q - \delta W)_{\text{III}} \quad (6.3b)$$

Comparing Equation 6.3a with 6.3b, one concludes that

$$\int_{\text{state } 1}^{\text{state } 2} (\delta Q - \delta W)_{\text{I}} = \int_{\text{state } 1}^{\text{state } 2} (\delta Q - \delta W)_{\text{II}} \quad (6.3c)$$

The processes I and II were chosen arbitrarily, but note that the value of the integral $\int_{\text{state } 1}^{\text{state } 2} (\delta Q - \delta W)$ is determined solely by the end states and does not depend on the process. *A thermodynamic quantity whose value is determined only by the end states is a thermodynamic property.* The integral of $(\delta Q - \delta W)$ for any process transforming the system from state 1 to state 2 thus defines a change in a thermodynamic property. This property is referred to as *energy* and is designated E. It also follows that the incremental change in E (dE) is equal to the quantity $(\delta Q - \delta W)$ as given earlier by Equation 6.1b.

For the entire process 1→2, Q_{12} can be used to designate the integral $\int_{\text{state } 1}^{\text{state } 2} \delta Q$ and W_{12} for the integral $\int_{\text{state } 1}^{\text{state } 2} \delta W$. The increase in energy of the system is then

$$E_2 - E_1 = \Delta E = Q_{12} - W_{12} \tag{6.4}$$

Equation 6.4 is generally referred to as the *nonflow energy equation (NFEE)*, which is an expression for the first law of thermodynamics applied to nonflow processes (with a fixed mass in the system).

The intimate connection that exists among heat, work, and energy is evident in Equation 6.4. The SI units are the same for the three quantities. Both heat and work can now be seen as *modes of energy interaction*. It is important to remember, however, that while energy is a property of a system, heat and work are not. Heat and work only exist as interactions between a system and its surroundings.

Datum State for Energy

The first law only provides a basis for the measurement of *changes* in energy of a system and not the absolute value of energy. Thus, to compile a table of properties for a substance that includes energy, one would carry out experiments involving arbitrary amounts of heat and/or work and compute the resulting changes in E using Equation 6.4. The choice of the reference state relative to which changes in E are measured must be made explicit, however. For most applications, the changes in energy between states is of much greater interest than the absolute amount of energy at each state. Thus, the choice of the reference state could be arbitrary and is usually based on convenience. For example, the energy values in tables of properties for water are usually those relative to the triple point of water as the reference state. For some common refrigerants, a reference condition of saturated liquid at $-40°C$ is frequently used instead.

Conservation of Energy Principle

By applying the NFEE to changes in energy for both a closed system and its surroundings, one can demonstrate that the conservation of energy principle is a corollary or an alternative statement of the first law of thermodynamics. Heat and work are interactions, and therefore Q_{12} into the system will correspond to $-Q_{12}$ into the surroundings. Likewise, W_{12} from the system corresponds to $-W_{12}$ from the surroundings. Thus, applying Equation 6.4 to the energy changes in the (closed) system and its surroundings gives

$$(\Delta E)_{\text{system}} = Q_{12} - W_{12} \tag{6.5a}$$

and

$$(\Delta E)_{\text{surroundings}} = -Q_{12} - (-W_{12}) \tag{6.5b}$$

Adding both sides of Equations 6.5a and 6.5b gives

$$(\Delta E)_{\text{universe}} = (\Delta E)_{\text{system}} + (\Delta E)_{\text{surroundings}} = 0 \tag{6.5}$$

This is the *conservation of energy principle*. It affirms that the total energy of the universe is constant. Energy cannot be created, neither can it be destroyed. Equation 6.5

also means that if an increase occurs in the energy of a closed system, a simultaneous decrease (of the same magnitude) must occur in the energy of the surroundings. As previously noted, heat and work are simply different modes by which energy is exchanged between a system and its surroundings. If a system is isolated from its surroundings, both heat and work are zero and Equation 6.4 implies that the energy of the *isolated system* is conserved since ΔE for the isolated system is equal to zero.

Total System Energy (E)

The *total energy* of a closed system includes what is termed *internal energy* (U) as well as *kinetic energy* (KE) and *potential energy* (PE). The internal energy of a system is defined as the energy that the system has by virtue of its thermodynamic state. Potential energy usually means gravitational potential energy. In physics, the gravitational field is recognized as only one of several force fields that are described as *conservative*. Other examples of conservative force fields are electrical fields and magnetic fields. Potential energy due to these other sources must be included in the expression for the total energy when these fields are of importance. The symbol E represents the total energy of the system, which in all cases of interest in this text is given by

$$E = U + KE + PE \qquad (6.6)$$

The kinetic energy of the system must be measured relative to a specified physical coordinate system. The potential energy requires the definition of the pertinent force field and a specification of the position of the system relative to a physical coordinate system. The *specific internal energy, u* ($= U/m$), is measured with respect to the datum state defined for the substance. In the evaluation of the total energy, E, the same reference coordinates must be used for all thermodynamic states involved. Thus, if a system is transformed from an initial state 1 to a final state 2, the change in total energy is given by

$$\Delta E = (U_2 + KE_2 + PE_2) - (U_1 + KE_1 + PE_1) \qquad (6.7)$$

where both U_1 and U_2 are measured relative to the same reference state. The changes in kinetic energy and potential energy are given by

$$\Delta(KE) = KE_2 - KE_1 = \frac{1}{2}m(|\mathbf{V_2}|^2 - |\mathbf{V_1}|^2) \qquad (6.8)$$

and

$$\Delta(PE) = PE_2 - PE_1 = mg(Z_2 - Z_1) \qquad (6.9)$$

where $\mathbf{V}$ represents the velocity of the system and Z represents its elevation above the chosen datum plane. The measurements corresponding to the initial and final states must be made relative to the same physical coordinate system.

If the expressions given by Equations 6.6 through 6.9 are now substituted into Equation 6.4, the following general form for the nonflow energy equation is obtained:

$$Q_{12} - W_{12} = U_2 - U_1 + m\left(\frac{|\mathbf{V_2}|^2}{2} - \frac{|\mathbf{V_1}|^2}{2}\right) + mg(Z_2 - Z_1) \qquad (6.10)$$

Particular care is needed with the manipulation of the units for internal energy, kinetic energy, and potential energy in the expressions for total energy. The SI unit for U is the joule (J), which is consistent with the units for both kinetic energy and potential energy provided the mass is measured in kg, the velocity in m/s, the elevation in m, and the acceleration due to gravity in m/s^2. However, the USCS unit for internal energy (U) is the Btu, while that for kinetic energy and potential energy is lbm $\cdot$ ft^2/s^2. Thus, the changes in kinetic energy and potential energy computed using Equations 6.8 and 6.9 should be divided by g_c ($= 32.174$ ft $\cdot$ lbm/lbf $\cdot$ s^2) first to convert to the more usual mechanical energy USCS unit, which is the ft $\cdot$ lbf. Thereafter, the unit equivalence 1 Btu $= 778.16$ ft $\cdot$ lbf can be used for converting all the energy values to a consistent set of units.

First Law for a Closed System for Time-Varying Conditions

Classical thermodynamics is generally not concerned with temporal (time) variations of properties for a system since many of these properties are defined for the system only when there is a thermodynamic equilibrium condition. One exception to this, however, is the total energy of a closed system. This is an extensive property and has a fixed value at any given instant in time, regardless of whether the instantaneous condition of the system is an equilibrium or a nonequilibrium state. This can be shown to be true by considering what happens if the process were interrupted at any stage and the system were completely isolated from its surroundings. From Equation 6.1b, it is evident that $dE = 0$ when there is neither a heat interaction nor a work interaction between the system and its surroundings. In other words, the total energy of a closed system does not change unless heat or work interaction occurs with the surroundings.

By forming a differential with respect to time for Equation 6.1b, the time rate of increase of the total energy of the system is obtained in terms of the difference between the heat input rate to the system and the work output rate from the system:

$$\frac{\delta Q}{dt} - \frac{\delta W}{dt} = \frac{dE}{dt}$$

or

$$(6.11)$$

$$\dot{Q} - \dot{W} = \frac{dE}{dt}$$

where $\dot{Q}$ and $\dot{W}$ represent differentials with respect to time of the path-dependent quantities heat and work, respectively. In words, Equation 6.11 for the conservation of energy principle for time-varying conditions is as follows:

$$\begin{pmatrix} \text{Net rate of} \\ \text{heat into} \\ \text{system across} \\ \text{the boundary} \\ \text{of the system} \end{pmatrix} - \begin{pmatrix} \text{Net rate of} \\ \text{work out of} \\ \text{system across} \\ \text{the boundary} \\ \text{of the system} \end{pmatrix} = \begin{pmatrix} \text{Net rate of} \\ \text{energy} \\ \text{accumulation} \\ \text{within the} \\ \text{system} \end{pmatrix} \qquad (6.11a)$$

where the term *rate* means "time rate." Substituting the expressions for energy (E)

presented in Equations 6.6 through 6.9 into Equation 6.11 and integrating for a change from state 1 to state 2 gives

$$Q_{12} - W_{12} = U_2 - U_1 + m\left(\frac{|\mathbf{V}_2|^2}{2} - \frac{|\mathbf{V}_1|^2}{2}\right) + mg(Z_2 - Z_1)$$

which is the same as Equation 6.10 obtained earlier.

Several examples involving the application of the first law of thermodynamics to nonflow processes are given in the following section. These examples illustrate different types of applications of the law when the system has a fixed mass. Applications involving an open system are considered in Section 6.5.

6.4 *Examples of First Law Applied to a Closed System*

Figure 6.3 represents a closed system that undergoes a change of state from an initial equilibrium state 1 to a final equilibrium state 2. Q_{12} and W_{12} are, respectively, the heat and work exchanged between the system and its surroundings. Both heat and work transfers are taken in positive directions, as shown in the figure, which is in keeping with the more popular convention mentioned earlier.

EXAMPLE 6.1
Change of State Involving Work and Heat

A closed system such as that shown in Figure 6.3 undergoes a change of state. During the process, 10 kJ of heat transfer occurs into the system, which does 3 kJ of work on its surroundings. Find the increase in energy of the system.

Given:
$$Q_{12} = + 10 \text{ kJ (heat into system)}$$
$$W_{12} = +3 \text{ kJ (work out of system)}$$

Find: ΔE

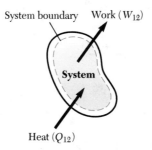

Figure 6.3 Representation of a closed system that undergoes a change of state from state 1 to state 2.

Solution: Simply apply the nonflow energy equation (NFEE), Equation (6.4), as follows:

$$E_2 - E_1 = Q_{12} - W_{12}$$
$$= 10 \text{ kJ} - 3 \text{ kJ}$$
$$= \underline{7 \text{ kJ}}$$

Thus, 7 kJ of energy is the increase in energy of the system resulting from this process.

EXAMPLE 6.2
Closed System Cycle (Constant Volume Processes)

A closed system in a container, such as that shown in Figure 6.1, undergoes a process that takes it from state 1 to state 2. The container is rigid and so the volume of the system is unchanged in the process. The only work interaction between the system and the surroundings is via the paddle wheel shown. The work done on the system via the paddle wheel is 220 ft · lbf, and the heat interaction can be assumed negligible.

(a) Determine the increase in internal energy (in Btu) of the system.

(b) What is the heat interaction needed to return the system to its original state?

Given:

(a) $W_{12} = -220 \text{ ft} \cdot \text{lbf}$

(b) $Q_{12} = 0$

Process 2→3 such that $U_3 = U_1$ ($W_{23} = 0$)

Find:

(a) $U_2 - U_1$ (in Btu)

(b) Q_{23}

Solution: The NFEE (Eq. 6.10) applies. Since changes in kinetic energy and potential energy are not involved for the system, the NFEE can be written as

$$U_2 - U_1 = Q_{12} - W_{12}$$

(a) For the process 1→2,

$$U_2 - U_1 = 0 - (-220 \text{ ft} \cdot \text{lbf}) \left(\frac{1 \text{ Btu}}{778.16 \text{ ft} \cdot \text{lbf}} \right)$$

$$= \underline{+0.283 \text{ Btu}}$$

(b) For the process 2→3,

$$Q_{23} = W_{23} + (U_3 - U_2)$$
$$= W_{23} + (U_1 - U_2) \quad \text{(since } U_3 = U_1\text{)}$$
$$= 0 + (-0.283 \text{ Btu})$$
$$= \underline{-0.283 \text{ Btu}}$$

The combined process 1→2→3 is a thermodynamic cycle since states 3 and 1 are identical. Therefore, from Equation 6.1a,

$$Q_{12} + Q_{23} = W_{12} + W_{23}$$

Since $Q_{12} = 0$ and $W_{23} = 0$, it follows that $Q_{23} = W_{12} = -0.283$ Btu.

EXAMPLE 6.3
Constant Pressure Process

A closed system, such as that shown in Figure 6.3, undergoes a constant pressure process from state 1 to state 2. The volume of the system is doubled from 1 m³ to 2 m³ while 200 kJ of heat is added to the system. If no work is done except for the boundary displacement work and the pressure remains constant at 101.3 kPa (1 atm) throughout the process, what is the change in internal energy of the system? Develop an expression for the heat added in terms of properties of the system assuming a quasi-equilibrium process.

Given:

$$p = \text{constant} = 101.3 \text{ kPa}$$

$$V_1 = 1 \text{ m}^3, \ V_2 = 2 \text{ m}^3$$

$$Q_{12} = 200 \text{ kJ}$$

Find:

(a) ΔU

(b) an expression for Q_{12} in terms of thermodynamics properties for a quasi-equilibrium process

Solution:

(a) The general NFEE (Eq. 6.10) applies. If negligible changes occur in kinetic and potential energy, Equation 6.10 gives

$$U_2 - U_1 = Q_{12} - W_{12}$$

where $Q_{12} (= 200 \text{ kJ})$ is given. The work done is calculated using Equation 5.7 for the displacement work in a constant pressure quasi-equilibrium process:

$$W_{12} = p(V_2 - V_1)$$

$$= (101.3 \text{ kPa})(2 - 1)\text{m}^3 = 101.3 \text{ kJ}$$

Thus,

$$U_2 - U_1 = (200 - 101.3) \text{ kJ}$$

$$= \underline{98.7 \text{ kJ}}$$

(b) If $\Delta(KE)$ and $\Delta(PE)$ are negligible and other forms of work (such as shaft work and electrical work) are zero for the process, Equation 6.10 (the general NFEE) becomes

$$Q_{12} = (U_2 - U_1) + W_{12}$$

Equation 5.7 for the displacement work done in a quasi-equilibrium process and the fact that $p_1 = p_2 = p$ give the following for Q_{12}:

$$Q_{12} = (U_2 - U_1) + p(V_2 - V_1)$$
$$= (U_2 + p_2 V_2) - (U_1 + p_1 V_1)$$

Equation 3.1 defines the combination of properties $(U + pV)$ as the total enthalpy (H) of the system. Thus, for the quasi-equilibrium constant pressure process,

$$Q_{12} = (H_2 - H_1)$$

EXAMPLE 6.4
Heterogeneous System (Constant Pressure Process)

A mass of 10 kg of ice at $-10°C$ is added to 250 kg of hot tea at 95°C. Both the ice and tea are at the ambient pressure of 1 atm. What is the eventual temperature of the mixture? Assume that the container is well insulated and that the thermodynamic properties of the hot tea are approximately the same as those for pure water.

Given: System: (10 kg ice at $-10°C$) + (250 kg hot tea at 95°C). Assume (i) container is well insulated and (ii) properties of tea are approximately the same as for water

Find: Final temperature for the mixture

Solution: The system as defined is heterogeneous since it has discrete parts that are initially in different states of equilibrium. Figures 6.4(a) and (b) depict initial state 1 and final state 2 of the system, respectively.
    The process is nonflow, and hence, the NFEE applies. Thus, by neglecting changes in the kinetic and potential energy terms in Equation 6.10, it follows that

$$Q_{12} - W_{12} = U_2 - U_1$$

Furthermore, since the mixing process occurs at a constant pressure, $p(= p_1 = p_2)$, one can substitute $W_{12} = p(V_2 - V_1)$ into the energy equation. Thus, as in Example 6.3,

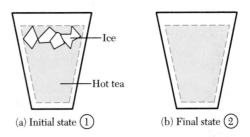

(a) Initial state ①   (b) Final state ②

Figure 6.4 A heterogeneous system of ice and hot tea undergoing a constant pressure mixing process from (a) initial state 1 to (b) final state 2.

$$Q_{12} = (U_2 + pV_2) - (U_1 + pV_1)$$
$$= (U_2 + p_2V_2) - (U_1 + p_1V_1)$$
$$= H_2 - H_1$$

The container is well insulated (given). Therefore, $Q_{12} \approx 0$ and so $H_2 = H_1$. Now,

$$H_1 = m_{ice}h_{ice} + m_{hot\ tea}h_{hot\ tea}$$

Either the property tables (A-4 for ice and A-1 for liquid water in Appendix A) or the **STEAM** code can be used to determine the specific enthalpies for the ice and the hot tea. The initial enthalpy for the system is thus

$$H_1 = (10\ kg)(-354.1\ kJ/kg) + (250\ kg)(398\ kJ/kg) = 95,959\ kJ$$

At the final state, the system becomes homogeneous with

$$H2 = m_{system}h2 = H1$$

Thus,

$$h_2 = \frac{95,959\ kJ}{260\ kg} = 369.1\ kJ/kg$$

Now two independent intensive properties (h_2, p_2) are known for the final state of the system, and hence either the property tables for water or the **STEAM** code can be used for the determination of any other property. (It is reasonable in problems of this type to assume a system pressure of 1 atm if the ambient pressure is not given.) The alternative to the use of a property table or the computer code in this case is to use the mean specific heat (at constant pressure) for liquid water, which is 4.2 kJ/kg · K, and to determine the final temperature of the mixture from

$$h_2 = (c_p)_{water}T_2(^\circ C) = 369.1\ kJ/kg$$

$$\therefore T_2 = \frac{369.1}{4.2} = 87.9^\circ C$$

Note that this equation assumes that the reference temperature for the specific enthalpy of water given in the tables is 0°C, although the actual temperature is in fact 0.01°C. It is also assumed that the constant pressure specific heat for a liquid remains constant over a fair range of temperatures.

The final temperature of the mixture is 88°C.

EXAMPLE 6.5
Free Expansion of an Ideal Gas into a Vacuum

An ideal gas at pressure p_1 and temperature T_1 is enclosed in a rigid insulated container, such as that shown in Figure 6.5. The gas is separated from a vacuum by a membrane, as shown in the figure. The membrane is ruptured and the gas expands freely into the vacuum. Neglecting the work done in rupturing the membrane, determine the final temperature of the gas after it has reached equilibrium in the entire volume of the container.

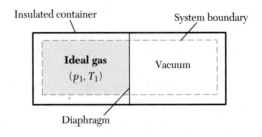

Figure 6.5 Free expansion of an ideal gas into a vacuum.

Given: Ideal gas at p_1, T_1 expands freely into a vacuum (not a quasi-equilibrium process).

$$Q_{12} = 0 \text{ (insulated container, assume adiabatic)}$$

Find: T_2

Solution: The simplest way to solve the problem is to define the system as comprising the gas and the evacuated space in the insulated container. The system boundary is then the inside surface of the container, as shown on Figure 6.5.

There is no movement of the system boundary and so $W_{12} = 0$. Since the container is insulated, you can also assume $Q_{12} = 0$. Applying the NFEE (Eq. 6.4) thus gives

$$E_2 - E_1 = Q_{12} - W_{12}$$
$$= 0 - 0 = 0$$

Changes in the KE and PE terms in the expression for ΔE are zero, and hence,

$$E_2 - E_1 = U_2 - U_1 = 0$$
$$U_2 = U_1$$

The internal energy for the system is identical to that for the gas since the vacuum does not contribute anything to the total energy. For an ideal gas, the specific internal energy is a function only of temperature, as was indicated in Chapter 4. Therefore, $T_2 = T_1$.

Note: If the gas were a real gas, U_2 would still equal U_1, but some difference in temperature would exist between state 1 and state 2 due to the thermodynamic state change associated with the pressure change. For a real gas, u is no longer a function of temperature only but rather a function of any two independent intensive properties, such as the temperature and pressure (or specific volume).

EXAMPLE 6.6
Constant Temperature Process for an Ideal Gas

A closed system, such as that shown in Figure 6.3, undergoes a quasi-equilibrium constant temperature process from state 1 to state 2. The system is an ideal gas that is initially at 200 kPa and 300 K. The system volume doubles during the process from 2 m^3 to 4 m^3. Compute (**a**) the work done and (**b**) the heat added during the process.

Given: $T = T_1 = T_2 = $ constant for the process

$$p_1 = 200 \text{ kPa}, \; T_1 = 300 \text{ K}$$

$$V_1 = 2 \text{ m}^3, \; V_2 = 4 \text{ m}^3$$

Find: (**a**) W_{12}, (**b**) Q_{12}

Solution:

(**a**) The expansion work for an ideal gas system undergoing a constant temperature quasi-equilibrium process is given by Equation 5.11:

$$W_{12} = p_1 V_1 \ln \frac{V_2}{V_1}$$

$$= (200 \text{ kPa})(2 \text{ m}^3) \ln \frac{4}{2}$$

$$= 277 \text{ kJ}$$

(**b**) The NFEE (Eq. 6.10) applies:

$$E_2 - E_1 = Q_{12} - W_{12}$$

Since the gas is an ideal gas and $T_2 = T_1$, it follows that $U_2 = U_1$. Also, if changes in the KE and PE for the system are negligible, E_2 must be equal to E_1. Therefore,

$$Q_{12} = W_{12}$$

$$= 277 \text{ kJ}$$

EXAMPLE 6.7
Polytropic Process for an Ideal Gas

A closed system, such as that shown in Figure 6.3, undergoes a quasi-equilibrium polytropic process from state 1 to state 2. The system is air (ideal gas), which is initially at 100 kPa and 300 K. The system volume is reduced from 2 m^3 to 1 m^3 during the process, and the polytropic index n equals 1.3 for the process. Assume a constant value of $c_v = 0.718 \text{ kJ/(kg} \cdot \text{K)}$ for air during the process. Compute

(**a**) the pressure and temperature at state 2

(**b**) the work done by the system during the process

(**c**) the heat added to the system during the process

Given: Ideal gas (air), quasi-equilibrium polytropic process, with $n = 1.3$

$$p_1 = 100 \text{ kPa}, \; T_1 = 300 \text{ K}, \; V_1 = 2 \text{ m}^3, \; V_2 = 1 \text{ m}^3$$

$$c_v = 0.718 \text{ kJ/(kg} \cdot \text{K)} \text{ for air}$$

Note: $R = 287 \text{ (N} \cdot \text{m)/(kg} \cdot \text{K)}$ or $\text{J/(kg} \cdot \text{K)}$ for air.

Find: **(a)** p_2, T_2, **(b)** W_{12}, **(c)** Q_{12}

Solution:

(a) The polytropic process is defined by Equation 5.12, where pV^n = constant and n is the polytropic index. Thus, $p_1V_1^n = p_2V_2^n$, and hence,

$$p_2 = p_1\left(\frac{V_1}{V_2}\right)^n = (100 \text{ kPa})\left(\frac{2}{1}\right)^{1.3} = 246.2 \text{ kPa}$$

The final pressure is thus 246 kPa.

The final temperature can be determined from the ideal gas equation (Eq. 4.7), which is $p_2V_2 = m_2RT_2$, and $m_2 = m_1 = p_1V_1/(RT_1)$. Thus,

$$m_2 = \frac{(10^5 \text{ N/m}^2)(2 \text{ m}^3)}{[287 \text{ N} \cdot \text{m}/(\text{kg} \cdot \text{K})](300 \text{ K})} = 2.323 \text{ kg}$$

and hence,

$$T_2 = \frac{p_2V_2}{m_2R}$$

$$= \frac{(246{,}200 \text{ N/m}^2)(1 \text{ m}^3)}{2.323 \text{ kg}[287 \text{ N} \cdot \text{m}/(\text{kg} \cdot \text{K})]} = 369.3 \text{ K}$$

The final temperature of the gas is thus 369 K.

(b) Equation 5.13 gives the displacement work done in a quasi-equilibrium, polytropic process:

$$W_{12} = \frac{p_2V_2 - p_1V_1}{1 - n}$$

$$= \frac{(246.2 \text{ kN/m}^2)(1 \text{ m}^3) - (100 \text{ kN/m}^2)(2 \text{ m}^3)}{1 - 1.3}$$

$$= -154 \text{ kJ}$$

Thus, 154 kJ of work is done *on* the system.

(c) Changes in KE and PE are negligible, and hence, the NFEE can be written as

$$Q_{12} - W_{12} = (U_2 - U_1)$$

Also, for an ideal gas $U_2 - U_1 = mc_v(T_2 - T_1)$ if c_v is constant. Therefore,

$$Q_{12} = mc_v(T_2 - T_1) + W_{12}$$

$$= (2.323 \text{ kg})(0.718 \text{ kJ/kg} \cdot \text{K})(369.3 \text{ K} - 300 \text{ K}) + (-154 \text{ kJ})$$

$$= -38.4 \text{ kJ}$$

Thus, 38.4 kJ of heat is rejected from the system to the surroundings.

EXAMPLE 6.8
Application for Time-Varying Conditions

A closed system, such as that shown in Figure 6.3, undergoes a process from state 1 to state 2. During the process, the heat to the system is at a rate of 5 kJ/s, while work is done by the system at a rate of 2 kJ/s. If the process lasts for 1 hr, compute

(a) the rate of energy accumulation in the system during the process

(b) the total energy increase of the system during the change from state 1 to state 2

Given: $\dot{Q} = 5$ kJ/s, $\dot{W} = 2$ kJ/s

Find: (a) dE/dt for process, (b) $E_2 - E_1$

Solution:
(a) Equation 6.11 for the transient or time-varying condition can be used in this case:

$$\dot{Q} - \dot{W} = \frac{dE}{dt}$$

$$\therefore \frac{dE}{dt} = 5 \text{ kJ/s} - 2 \text{ kJ/s}$$

$$= \underline{3 \text{ kW}}$$

(b) $E_2 - E_1 = \int_{\text{time}_1}^{\text{time}_2} \left(\frac{dE}{dt}\right) dt$

where dE/dt is constant and equal to 3 kJ/s. Therefore,

$$E_2 - E_1 = (3 \text{ kJ/s})(3600 \text{ s})$$

$$= \underline{10{,}800 \text{ kJ}}$$

Exercises for Section 6.4

1. Heat transfer (Q_{12}) occurs to a fluid of mass m whose volume remains constant, as illustrated in the accompanying figure. The initial pressure of the fluid is 21 bar.

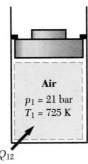

Constant volume heating of a fluid.

Assume that the fluid is air ($c_v = 0.718$ kJ/kg · K) and that $T_1 = 725$ K and $T_2 = 1100$ K. Determine in kJ/kg

(a) w_{12} (or W_{12}/m)

(b) $\Delta U_{12}/m$

(c) Q_{12}/m

Answer: (b) 269 kJ/kg, (c) 269 kJ/kg

2. Heat transfer (Q_{12}) occurs to a fluid of mass m whose volume increases at a constant pressure of 21 bar. Assume that the fluid is air ($c_v = 0.718$ kJ/kg · K and $R = 0.287$ kJ/kg · K) and that $T_1 = 725$ K and $T_2 = 1100$ K. Assume that the fluid expansion is fully resisted. Determine in kJ/kg

(a) w_{12} (or W_{12}/m)

(b) Δu (or $\Delta U/m$)

(c) q_{12} (or Q_{12}/m)

Answer: (a) 108 kJ/kg, (b) 269 kJ/kg, (c) 377 kJ/kg

3. Suppose the air in Exercise 2 actually pushes against a load that is only equal to a boundary pressure of 15 bar resisting the expansion of the air. Define the air as the system and assume that T_1, T_2, and the volume change that occurs are the same as in Exercise 2. Determine in kJ/kg

(a) w_{12}

(b) Δu

(c) q_{12}

Answer: (a) 76.9 kJ/kg, (b) 269 kJ/kg, (c) 346 kJ/kg

4. A well-insulated rigid vessel of volume 0.2 m^3 is divided into two equal compartments by a thin diaphragm, as illustrated in the accompanying figure. One of the compartments is a vacuum. The other is a fluid at a pressure of 200 kPa and a temperature of 125°C. The diaphragm is ruptured and the fluid fills the entire volume of the vessel. Assume that the fluid is an ideal gas. After you have determined Q_{12}, W_{12}, and ΔU for the process, calculate the final temperature of the gas.

Answer: 125°C

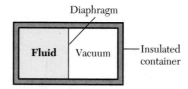

Diaphragm

Fluid   Vacuum   —Insulated
container

Unresisted expansion of a fluid in
an adiabatic enclosure.

5. Assume that the fluid in Exercise 4 is water vapor at a pressure of 200 kPa and temperature of 125°C prior to the thin diaphragm being ruptured. The final pressure in the vessel is found to be 1 bar. After you have determined Q_{12}, W_{12}, and ΔU for the process, calculate the final temperature of the water vapor.

Answer: 120°C

6. The resistance wire for a heating element is supplied with electrical power at the rate of 60 W/in. of wire length. The surface area of the wire is 1.01×10^{-3} m^2 per inch of wire length. At steady state, the heat transfer coefficient for heat loss from the surface of the resistance wire is 65 W/m$^2 \cdot$ K. The temperature of the environment is kept at 20°C. Define a 1-in. length of resistance wire as the system. At the steady state, determine

   (a) $\dot{W}_{system}$

   (b) $\left(\dfrac{dE}{dt}\right)_{system}$

   (c) $\dot{Q}_{system}$

   (d) the surface temperature T_s of the wire

   Answer: (d) 934°C

6.5   First Law of Thermodynamics for a Control Volume (Open System)

Introduction

The first law of thermodynamics as formulated in the previous section was for a closed system. Most of the problems encountered in engineering, however, are those involving a flow of mass across a system boundary. It is therefore also necessary to formulate the first law for an open system across whose boundary a flow of mass is permitted.

In dealing with open systems, the *control volume* concept is used. The control volume is traditionally regarded as fixed in space and of fixed volume, but it allows mass to flow across its boundary in addition to heat and work interactions across the control surface. Chapter 1 introduced the concept of a control volume more broadly as a volume of fixed shape and of fixed orientation relative to an unaccelerated observer. In principle, the control volume could thus be defined less restrictively as an arbitrary volume in space that can move, rotate, expand (or contract) with time so long as it retains a fixed shape and orientation relative to an unaccelerated observer.

One possible advantage in defining a control volume as one having a *fixed volume* in addition to being of *fixed shape* and at a *fixed position* is that the boundary displacement work component of the external work done is eliminated. In any case, the control volume must be defined in relation to some physical coordinate system that can either be stationary or in uniform motion (that is, no acceleration). Once the control volume and control surface are defined, one can focus attention on the interactions across the boundary of the control volume. As discussed in Section 6.3, it is not necessary for the mass within the control volume to be in equilibrium in order to evaluate the instantaneous m and E for a system. This is because both quantities are extensive properties that have defined values whether or not a state of equilibrium exists.

The following sections first examine the conservation of mass for the control volume and then proceed to derive the general flow energy equation (FEE) via the application of the first law of thermodynamics to the flow process.

Conservation of Mass for Control Volume

Figure 6.6 defines the conditions for mass flow in and out of a control volume. A general statement for the conservation of mass for the control volume can be written as follows:

$$\begin{pmatrix} \text{Mass rate} \\ \text{into} \\ \text{control volume} \\ \text{across} \\ \text{boundary} \end{pmatrix} - \begin{pmatrix} \text{Mass rate} \\ \text{out of} \\ \text{control volume} \\ \text{across} \\ \text{boundary} \end{pmatrix} = \begin{pmatrix} \text{Time rate of} \\ \text{change of mass} \\ \text{accumulated} \\ \text{within} \\ \text{control volume} \end{pmatrix} \qquad (6.12)$$

This mass balance can be expressed mathematically for the control volume in Figure 6.6 as

$$\dot{m}_i - \dot{m}_e = \frac{m_{t+\Delta t} - m_t}{\Delta t} \qquad (6.13)$$

where $\dot{m}_i$ is the mass flow rate across the boundary into the control volume, $\dot{m}_e$ is the mass flow rate across the boundary out of the control volume, m_t is the mass of matter within the control volume at time t, and $m_{t+\Delta t}$ is the mass of matter within the control volume at $t + \Delta t$. If the limit is taken as Δt approaches zero, Equation 6.13 becomes

$$\dot{m}_i - \dot{m}_e = \frac{dm_{\text{CV}}}{dt} \qquad (6.14)$$

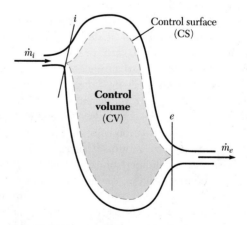

Figure 6.6 Conservation of mass principle applied to a control volume.

where m_{CV} designates the mass of matter within the control volume. If there is more than one inlet and more than one exit, Equation 6.14 can be generalized to include any additional number of entry lines and exit lines by summation of the mass flow rates:

$$\sum_{\text{inlet}} \dot{m}_i - \sum_{\text{exit}} \dot{m}_e = \frac{dm_{CV}}{dt} \qquad (6.15)$$

Equation 6.15 is the *conservation of mass principle* applied to a control volume, and it is often referred to as the *continuity equation*. The two terms involving mass flow across the control surface can be consolidated and incorporated into a surface integral that defines the mass flux across the control surface. Such an approach is frequently followed in fluid mechanics. Note that if both terms involving mass flow across the boundary are zero, the right-hand term will also be zero. The mass within the control volume is thus constant, and the problem reduces to that of a closed system, which was discussed earlier.

It is common practice when dealing with fluid flow to define and use a *bulk mean velocity*, V_m, such that the mass flow rate is given by

$$\dot{m} = \int_{A_{CS}} \rho \mathbf{V} \cdot \mathbf{dA} = \bar{\rho} V_m A_{CS} \qquad (6.16)$$

where A_{CS} is the cross-sectional area of the portion of the control surface across which the fluid with a mean density, $\bar{\rho}$, flows.

General Form of First Law of Thermodynamics for Control Volume

Application of the first law of thermodynamics to open system processes (which are called *flow processes*) leads to the equation often termed the *flow energy equation* (*FEE*). A common procedure for deriving the FEE is to consider the unsteady process for the closed system shown in Figures 6.7(a) and (b). In the actual flow process, only a single mass stream into the control volume and a single mass stream out of the volume are assumed. The instantaneous mass flow rates are, respectively, $\dot{m}_i$ and $\dot{m}_e$. In most cases, however, multiple streams occur into and out of the control volume, but this case is limited to the single stream case to keep the derivation of the flow energy equation simple.

The boundary of the closed system in Figure 6.7(a) is defined to include all the matter in the control volume at time t as well as the small elemental mass Δm_i, which would enter the control volume in the ensuing incremental time period Δt. At time $t + \Delta t$, the system boundary will thus take the new shape shown in Figure 6.7(b). The closed system now comprises all the mass in the control volume and the small mass Δm_e, which has moved into the exit line.

Recall that Equation 6.11 for the first law of thermodynamics when applied to unsteady, nonflow processes does not require instantaneous equilibrium states for the system. The equation can therefore be applied to the unsteady process in the present case, in which the closed system shown changes from the nonequilibrium condition of

Figure 6.7(a) to the nonequilibrium condition of Figure 6.7(b). However, *local equilibrium* conditions must be assumed for the small masses Δm_i and Δm_e entering and leaving the control volume, respectively, during the time period Δt for the process. This assumption enables one to express values of the extensive properties for the discrete lumps in terms of their masses and the corresponding instantaneous intensive properties, assumed uniform throughout each lump. By making Δt sufficiently small, and provided that the entry or exit cross-sectional areas are not too large, the assumption of local equilibrium is reasonable in all but exceptional cases in which sharp discontinuities exist in the flow (due to a shock wave, for example). In such exceptional cases, the problem is circumvented by using control volumes large enough to ensure that the region of discontinuity remains entirely within the control volume.

With reference to Figures 6.7(a) and (b), Equation 6.1b can be applied to the unsteady process by determining the appropriate expressions for δQ, δW, and dE. The heat δQ to the system in the time period Δt is

$$\delta Q = \dot{Q}\,\Delta t \tag{6.17}$$

The work done by the system, δW, is the combination of a shaft work component, δW_x, and the displacement work, δW_d. By assuming a control volume of fixed shape, size, and position, the displacement work is only that due to the boundary movements at the entry and exit lines as the discrete lumps are moved into and out of the control volume. The shaft work is given by

$$\delta W_x = \dot{W}_x\,\Delta t \tag{6.18}$$

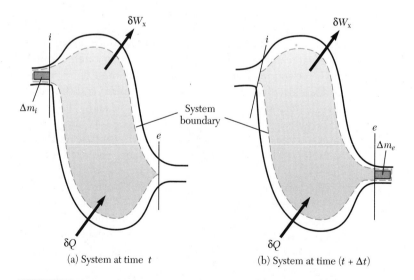

(a) System at time t (b) System at time $(t + \Delta t)$

Figure 6.7 The first law of thermodynamics applied to a control volume by considering the unsteady process for a closed system. (a) System at time t. (b) System at time $(t + \Delta t)$.

For the displacement work component, the changes in volume (or swept volumes) ΔV_i and ΔV_e at the entry and exit lines, respectively, are used to obtain the following

$$\delta W_d = p_e\, \Delta V_e - p_i\, \Delta V_i \tag{6.19}$$

Therefore,

$$\begin{aligned}
\delta W &= \delta W_x + \delta W_d \\
&= \dot{W}_x\, \Delta t + p_e\, \Delta V_e - p_i\, \Delta V_i
\end{aligned} \tag{6.20}$$

Since ΔV_i and ΔV_e are related to Δm_i and Δm_e via the respective specific volumes (v) at the entry and exit conditions, then

$$\Delta V_i = (\Delta m_i)v_i \tag{6.21}$$

and

$$\Delta V_e = (\Delta m_e)v_e \tag{6.22}$$

Substituting for ΔV_i and ΔV_e in Equation 6.20 gives

$$\delta W = \dot{W}_x\, \Delta t + (\Delta m_e)p_e v_e - (\Delta m_i)p_i v_i \tag{6.23}$$

The expression for dE is obtained as the difference between the total energy for the closed system in the states corresponding to Figures 6.7(b) and (a). Equations 6.7 through 6.9 are used to derive the following for dE:

$$\begin{aligned}
dE = {} & \left[(E_{CV})_{t+\Delta t} + \Delta m_e \left(u_e + \frac{(V_m)_e^2}{2} + gZ_e \right) \right] \\
& - \left[(E_{CV})_t + \Delta m_i \left(u_i + \frac{(V_m)_i^2}{2} + gZ_i \right) \right]
\end{aligned} \tag{6.24}$$

where E_{CV} is the instantaneous total energy of the entire contents of the control volume. $u(=U/m)$ denotes the specific internal energy.

The final step is to apply Equation 6.1b:

$$\delta Q - \delta W = dE \tag{6.1b}$$

Substituting Equations 6.17, 6.23, and 6.24 into Equation 6.1b and rearranging terms gives

$$\begin{aligned}
\dot{Q} - \dot{W}_x = {} & \frac{dE_{CV}}{dt} + \dot{m}_e \left[(u_e + p_e v_e) + \frac{(V_m)_e^2}{2} + gZ_e \right] \\
& - \dot{m}_i \left[(u_i + p_i v_i) + \frac{(V_m)_i^2}{2} + gZ_i \right]
\end{aligned} \tag{6.25}$$

The term $(u + pv)$ was defined in Chapter 3 as the specific enthalpy. This combination of the properties u, p, and v arises in the application of the first law of thermodynamics to flow processes, and the use of enthalpy for the combination is a choice dictated largely by mathematical convenience. Using enthalpy in Equation 6.25 gives the following for the *single* (*inlet* and *outlet*) *stream version* of the flow energy equation (FEE):

FEE (single stream version)

$$\dot{Q} - \dot{W}_x = \frac{dE_{CV}}{dt} + \dot{m}_e \left[h_e + \frac{(V_m)_e^2}{2} + gZ_e \right]$$

$$- \dot{m}_i \left[h_i + \frac{(V_m)_i^2}{2} + gZ_i \right] \tag{6.26}$$

The instantaneous total energy E_{CV} of the mass in the control volume is given by the Equation 6.6 and can be expressed as

$$E_{CV} = \left(U_{CV} + \frac{1}{2} m_{CV} (V_{CV})^2 + m_{CV} \, gZ_{CV} \right) \tag{6.26a}$$

The result for the flow energy equation can be generalized to cover the *multistream version* as well. Thus, for multiple streams into and out of the control volume,

FEE (multiple streams version)

$$\dot{Q} - \dot{W}_x = \frac{dE_{CV}}{dt} + \sum_{\substack{\text{all} \\ \text{exit streams}}} \dot{m}_e \left[h_e + \frac{(V_m)_e^2}{2} + gZ_e \right]$$

$$- \sum_{\substack{\text{all} \\ \text{entry streams}}} \dot{m}_i \left[h_i + \frac{(V_m)_i^2}{2} + gZ_i \right] \tag{6.27}$$

Recall that the previous derivation used a rather restrictive definition of control volume as one of fixed volume, shape, and position. If the volume is not fixed, an external boundary displacement work rate term should be added to the shaft power term, W_x, in the flow energy equations derived.

Steady-State Steady-Flow Process

A common application of the first law for a control volume is to the *steady-state steady-flow process*. This process is typically encountered in the operation of such devices as turbines, compressors, pumps, nozzles, diffusers, and several others. The term *steady state* means that no variation with time occurs for any property or condition at any point in the control volume. Although no variations with time occur within the control volume, spatial variations are allowed, but the control volume does not move relative to the coordinate frame. In addition, the mass flow rates of matter into or out of the control volume stay constant. The rates of heat and work interactions across the control surface also remain constant for the duration of the process.

The following simplifications result from these features of the steady-state steady-flow process.

1. No accumulation of matter occurs in the control volume for steady-state operation. In other words, the mass accumulation term dm_{CV}/dt in Equations 6.14 and 6.15 vanishes while the total mass flow rate into the control volume becomes equal to

the total mass flow rate out of the control volume. Equations 6.14 and 6.15 thus reduce to

$$\dot{m}_i = \dot{m}_e = \dot{m} \quad \text{(a constant)} \tag{6.28a}$$

and

$$\sum_{\substack{\text{all} \\ \text{entry streams}}} \dot{m}_i = \sum_{\substack{\text{all} \\ \text{exit streams}}} \dot{m}_e \tag{6.28b}$$

2. The energy accumulation rate for the control volume must be zero for the steady-state steady-flow process. Thus, the dE_{CV}/dt term in Equations 6.26 and 6.27 becomes zero. Equation 6.28a can thus be used, along with the fact that dE_{CV}/dt is zero, to obtain the following for the single (inlet and outlet) stream version of the flow energy equation:

SFEE (single stream version)

$$\dot{Q} - \dot{W}_x = \dot{m}\left[h_e + \frac{(V_m)_e^2}{2} + gZ_e\right]$$
$$- \dot{m}\left[h_i + \frac{(V_m)_i^2}{2} + gZ_i\right] \tag{6.29}$$

Equation 6.29 is often referred to as the single (inlet and outlet) stream version of the *steady-flow energy equation (SFEE)*. The term $\dot{m}$ is the flow rate, which in this case is constant and equal to either the mass flow rate of the inlet stream or the mass flow rate of stream leaving the control volume. Likewise, the multiple stream version of the steady-flow energy equation (SFEE) is obtained from Equations 6.27 and 6.28b and from the fact that dE_{CV}/dt is zero:

SFEE (multiple stream version)

$$\dot{Q} - \dot{W}_x = \sum_{\substack{\text{all} \\ \text{exit streams}}} \dot{m}_e\left[h_e + \frac{(V_m)_e^2}{2} + gZ_e\right]$$
$$- \sum_{\substack{\text{all} \\ \text{entry streams}}} \dot{m}_i\left[h_i + \frac{(V_m)_i^2}{2} + gZ_i\right] \tag{6.30}$$

As indicated earlier, extreme caution should be exercised to ensure the consistency of units whenever the energy equations are used. Whenever the velocity is in m/s, the kinetic energy (per unit mass) term in Equations 6.29 and 6.30 is in J/kg. Likewise, if Z is in m (meters) and g is in m/s², the potential energy (per unit mass) term is in J/kg. If the velocity is measured in ft/s, Z in ft, and g in ft/s², the terms for kinetic energy and potential energy per unit mass must be divided by g_c if the result in USCS units is in ft · lbf/lbm. The specific enthalpy, however, is likely to be given in Btu/lbm in USCS units. Thus, suitable conversion factors must be used to ensure consistency of units for the terms in the energy equation. An appropriate version for Equation 6.29 in USCS units can be derived as follows:

$$\dot{m}\left(\frac{V_m^2}{2}\right) \quad \text{has units of} \quad \frac{\text{lbm}}{\text{s}}\left(\frac{\text{ft}^2}{\text{s}^2}\right) = (\text{lbm} \cdot \text{ft/s}^2)\frac{\text{ft}}{\text{s}}$$

To convert lbm · ft/s² to lbf, divide by g_c since 1 lbf = $(1 \text{ lbm} \cdot \text{ft/s}^2)/g_c$, or g_c = 32.174 lbm · ft/(lbf · s²). An identical conversion is needed for the term involving potential energy. Thus, Equation 6.29 for the SFEE can be written as follows when USCS units are involved:

$$\left(\dot{Q}\frac{\text{Btu}}{\text{s}}\right)\left[\frac{778.16 \text{ ft} \cdot \text{lbf}}{1 \text{ Btu}}\right] - \left(\dot{W}_x\frac{\text{ft} \cdot \text{lbf}}{\text{s}}\right)$$

$$= \left(\dot{m}\frac{\text{lbm}}{\text{s}}\right)\left[\left(\Delta h\frac{\text{Btu}}{\text{lbm}}\right)\left(\frac{778.16 \text{ ft} \cdot \text{lbf}}{1 \text{ Btu}}\right)\right. \qquad (6.29a)$$

$$\left. + \left(\frac{\Delta V_m^2}{2g_c}\frac{\text{ft} \cdot \text{lbf}}{\text{lbm}}\right) + \left(\frac{\Delta(gZ)}{g_c}\frac{\text{ft} \cdot \text{lbf}}{\text{lbm}}\right)\right]$$

In Equation 6.29a, consistent units of ft · lbf/s have been achieved among the individual terms.

6.6 Examples of First Law Applied to a Control Volume

This section presents several examples that apply the first law to a control volume (open system). Figure 6.8 represents a control volume with a single mass flow into and out of it. The heat input rate is $\dot{Q}$ and the shaft work output rate is $\dot{W}_x$.

EXAMPLE 6.9
Flow of an Ideal Gas Through a Control Volume

Air is flowing through a control volume, such as that shown in Figure 6.8. The conditions for air are such that it can be considered as an ideal gas during the process. The state of the air entering the control volume at position i is 2 atm and 300 K, while the state of the air exiting at plane e is 1 atm and 400 K. No shaft work occurs, and the changes in kinetic energy and potential energy between the entry and exit conditions for the air stream are negligible. If the mass flow into the control volume is 100 kg/s and the mass flow leaving the control volume is 100 kg/s, find the net rate of heat transfer to the control volume for the steady-state steady-flow process.

Given: Air with $\dot{m}_i = \dot{m}_e = 100$ kg/s, $\dot{W}_x = 0$, $\Delta(\text{KE}) \approx 0$, $\Delta(\text{PE}) \approx 0$

$$p_i = 2(101.325) \text{ kPa}, \ p_e = 101.325 \text{ kPa}$$

$$T_i = 300 \text{ K}, \ T_e = 400 \text{ K}$$

Find: $\delta Q/dt$

Solution: Equation 6.29 for steady-state steady-flow processes applies in this situation.

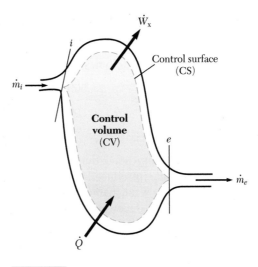

Figure 6.8 A control volume with a single mass flow in and a single mass flow out.

The enthalpy values of air at the stated conditions can be obtained either from the table of properties for air (Table A-11 in Appendix A) or by using the **GAS** code. The code shows that

$$h_i = 2 \text{ kJ/kg} \quad \text{at } T_i = 300 \text{ K}$$

$$h_e = 103 \text{ kJ/kg} \quad \text{at } T_e = 400 \text{ K}$$

Therefore,

$$\dot{Q} = \dot{W}_x + \dot{m}[(h_e - h_i) + \Delta(\text{KE}) + \Delta(\text{PE})]$$

$$\approx \dot{m}(h_e - h_i) = (100 \text{ kg/s})(103 - 2 \text{ kJ/kg})$$

$$= \underline{1.01 \times 10^4 \text{ kW}}$$

Alternatively, $h_e - h_i$ can be determined more simply, and without loss of accuracy, by using Equation 4.31 for a perfect gas:

$$h_e - h_i = c_p(T_e - T_i)$$

where c_p for air can be taken as 1.005 kJ/kg · K. This gives the same result as previously obtained.

EXAMPLE 6.10
Fluid Flow Through a Turbine

Steam expands through a turbine in a steady-state steady-flow adiabatic process. The mass flow rate of the steam is 3.0 lbm/s. The entering state of the steam is 500 psia and 1000°F, while the exiting state is 100 psia and 560°F. Negligible changes in kinetic and potential energies occur through the device. Find the power output (rate of doing work) for the turbine.

Given: Steam with $\dot{m} = 3.0$ lbm/s, $\dot{Q} = 0$, $\Delta(KE) \approx 0$, $\Delta(PE) \approx 0$

$$p_i = 500 \text{ psia}, \ p_e = 100 \text{ psia}$$

$$T_i = 1000°F, \ T_e = 560°F$$

Find: $\delta W_x / dt$

Solution: Figure 6.9 represents the flow process through a turbine using the graphical symbol for the turbine introduced in Chapter 1. The SFEE, Equation 6.29 or 6.29a, applies. If the process is adiabatic and changes in kinetic and potential energies are negligible, the SFEE reduces to

$$\dot{W}_x = \dot{m}(h_i - h_e)$$

The enthalpy of steam at entry or exit conditions can be determined either from the steam tables (Table B-3 in Appendix B) or by using the **STEAM** code. The computer code shows that

$$h_i = 1521 \text{ Btu/lbm}, \quad h_e = 1309 \text{ Btu/lbm}$$

Therefore,

$$\dot{W}_x = (3 \text{ lbm/s})(1521 - 1309 \text{ Btu/lbm})\left(\frac{1 \text{ hp}}{0.707 \text{ Btu/s}}\right)$$

$$= \underline{900 \text{ hp}}$$

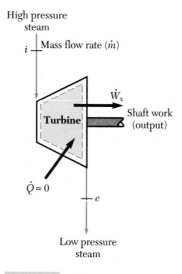

Figure 6.9 Steam expansion through an adiabatic turbine.

EXAMPLE 6.11
Fluid Flow Through a Pump

Water flows through a pump with a mass flow rate of 3 lbm/s. The entering state of the water is 100 psia and 70°F, while the exiting state is 500 psia and 70°F. Negligible changes in potential and kinetic energies occur during the process, and the steady-state steady-flow process is adiabatic. Find the power input required by the pump.

Given: Water with $\dot{m} = 3$ lbm/s, $\dot{Q} = 0$, $\Delta(\text{KE}) \approx 0$, $\Delta(\text{PE}) \approx 0$

$$p_i = 100 \text{ psia}, p_e = 500 \text{ psia}$$

$$T_i = 70°F, T_e = 70°F$$

Find: $\delta W_x/dt$

Solution: The control surface is drawn around the pump, as indicated in Figure 6.10. You could use either the compressed water table (Table B-3 in Appendix B) or the **STEAM** code. The specific enthalpies h_i and h_e from the computer code are

$$h_i = 38.3 \text{ Btu/lbm}, \quad h_e = 39.4 \text{ Btu/lbm}$$

Since the process is adiabatic and changes in kinetic and potential energies are negligible, application of the reduced form of the SFEE yields the following for the power input to the pump:

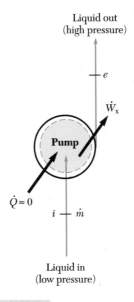

Liquid out
(high pressure)

e

$\dot{W}_x$

Pump

$\dot{Q} \approx 0$

i $\dot{m}$

Liquid in
(low pressure)

Figure 6.10 Water flow through a pump.

$$-\dot{W}_x = \dot{m}(h_e - h_i)$$

$$= (3 \text{ lbm/s})(39.4 - 38.3 \text{ Btu/lbm})\left(\frac{1 \text{ hp}}{0.707 \text{ Btu/s}}\right)$$

$$= 4.67 \text{ hp}$$

Note in Example 6.11 that the difference between h_i and h_e is small. A reliable estimate of the difference can be obtained using Equation 3.22:

$$h_e - h_i \approx v_f(p_e - p_i)$$

$$= \left(0.01605 \frac{\text{ft}^3}{\text{lbm}}\right)\left(500 - 100 \frac{\text{lbf}}{\text{in.}^2}\right)\left(\frac{144 \text{ in.}^2}{1 \text{ ft}^2}\right)$$

$$= 924.5 \frac{\text{ft} \cdot \text{lbf}}{\text{lbm}}$$

Thus,

$$-\dot{W}_x = \left(3 \frac{\text{lbm}}{\text{s}}\right)\left(924.5 \frac{\text{ft} \cdot \text{lbf}}{\text{lbm}}\right)\left(\frac{1 \text{ hp}}{550 \text{ ft} \cdot \text{lbf/s}}\right)$$

$$= 5.04 \text{ hp}$$

This alternative approach is encouraged because the property values from the tables and the computer code are rounded off.

EXAMPLE 6.12
Fluid Flow Through a Compressor

Refrigerant-22 flows through a compressor in a steady-state steady-flow process. The process is adiabatic, and negligible changes in potential and kinetic energies occur through the device. The mass flow rate of the refrigerant is 0.3 lbm/s. The entering state is 100 psia and 55°F, and the exiting state is 400 psia and 200°F. Find the power input to the compressor.

Given: Refrigerant-22 with $\dot{m} = 0.3$ lbm/s, $\dot{Q} = 0$, $\Delta KE = \Delta PE = 0$

$$p_i = 100 \text{ psia}, p_e = 400 \text{ psia}$$

$$T_i = 55°F, T_e = 200°F$$

Find: $\delta W_x/dt$

Solution: The process is represented in Figure 6.11, which shows the control surface inside the casing of the compressor. Applying Equation 6.29 to the process and simplifying gives

$$-\dot{W}_x = \dot{m}(h_e - h_i)$$

The enthalpies for the fluid can be obtained either from Table B-9 for superheated

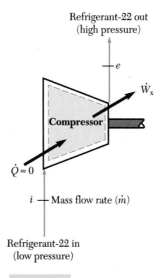

Figure 6.11 Compression of a
steady stream of refrigerant-22
vapor.

refrigerant-22 or by using the **R22** computer code. The code gives the following
values:

$$h_i = 109.77 \text{ Btu/lbm}, \quad h_e = 126.03 \text{ Btu/lbm}$$

Hence, the power input to the compressor is

$$-\dot{W}_x = \dot{m}(h_e - h_i)$$

$$= (0.3 \text{ lbm/s})\left(126.03 - 109.77 \frac{\text{Btu}}{\text{lbm}}\right)\left(\frac{778.16 \text{ ft} \cdot \text{lbf}}{1 \text{ Btu}}\right)\left(\frac{1 \text{ hp}}{550 \text{ ft} \cdot \text{lbf/s}}\right)$$

$$= \underline{6.9 \text{ hp}}$$

EXAMPLE 6.13
Flow Through a Subsonic Nozzle

A subsonic nozzle expands air from a temperature of 190°F and a pressure of 145 psia
to 150°F and 110 psia. The process is adiabatic. If the inlet velocity is small, determine
the exit velocity of the air. For air, assume $c_p = 0.24$ Btu/lbm · °R.

Given: Air flowing through a subsonic nozzle with $\dot{Q} = 0$, $(V_m)_i \approx 0$

$$p_i = 145 \text{ psia}, \ p_e = 110 \text{ psia}$$

$$T_i = 190°F, \ T_e = 150°F$$

$$c_p = 0.24 \text{ Btu/lbm} \cdot °R$$

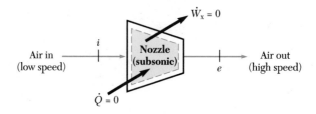

Figure 6.12 Adiabatic flow of air through a subsonic nozzle.

Find: $(V_m)_e$

Solution: Figure 6.12 depicts the adiabatic nozzle. Using Equation 6.29a with the information provided gives the following simplified form:

$$0 = \left(h_e - h_i \frac{\text{Btu}}{\text{lbm}} \right) \left(\frac{778.16 \text{ ft} \cdot \text{lbf}}{1 \text{ Btu}} \right) + \frac{(V_m)_e^2}{2g_c} \frac{\text{ft} \cdot \text{lbf}}{\text{lbm}}$$

If the air is treated as a perfect gas, then $\Delta h = c_p(T_e - T_i)$. Therefore,

$$(V_m)_e^2 = 2(32.174)(0.24)(190 - 150)(778.16) \text{ ft}^2/\text{s}^2$$

$$= 4.807 \times 10^5 \text{ ft}^2/\text{s}^2$$

Hence,

$$(V_m)_e = \underline{693 \text{ ft/s}}$$

EXAMPLE 6.14
Filling an Evacuated Rigid Tank

Steam at a pressure of 300 psia and a temperature of 500°F is flowing in a pipe, as shown in Figure 6.13. An evacuated rigid tank is connected to the steam pipeline through a valve, as shown in the figure. At a certain time, the valve is opened and steam flows into the tank until the pressure in the tank reaches that in the pipeline. Then the valve is closed. The process of filling the tank is adiabatic, and changes in kinetic and potential energies are negligible. Determine the final temperature of the steam in the tank after equilibrium is reached. Note that this is a nonsteady situation.

Given: Steam at 500°F and 300 psia flowing into an evacuated tank (Fig. 6.13)

Find: Final temperature T_2 of steam in tank

Solution: The control surface for the tank is indicated in Figure 6.13. Equation 6.26, which applies to the process, takes a much simpler form as a result of the following assumptions:

1. $\dot{Q} = 0$, since the process is adiabatic
2. $\dot{W}_x = 0$, since shaft work is not mentioned

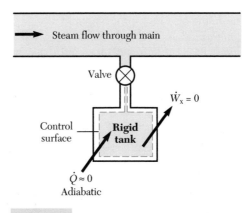

Figure 6.13 Flow of steam from a steam main into a rigid container.

3. $\dot{m}_e = 0$ since the mass flow occurs only *into* the control volume

4. kinetic and potential energy terms can be neglected

The flow energy equation (FEE) thus reduces to the following simple form:

$$0 = \left(\frac{dE_{CV}}{dt} \frac{\text{Btu}}{\text{s}} \right) - \left(\dot{m}_i \frac{\text{lbm}}{\text{s}} \right) \left(h_i \frac{\text{Btu}}{\text{lbm}} \right)$$

The subscript i represents the inlet stream at the state at which it enters the control volume. In this case, the entry condition is 500°F and 300 psia, at which the remainder of the steam flows through the pipe. Integration of the reduced equation with respect to time yields the following:

$$(E_{CV})_{\text{final}} - (E_{CV})_{\text{initial}} = m_i h_i$$

where m_i is the total mass of steam admitted into the rigid tank by the end of the process. The total energy of mass in the control volume can be determined at the beginning and end of the process as follows:

$$(E_{CV})_{\text{initial}} = 0, \quad \text{since the tank was evacuated initially}$$

$$(E_{CV})_{\text{final}} = m_i u_{\text{final}}$$

Substituting these in the reduced equation and dividing through by m_i gives this result:

$$u_{\text{final}} = h_i$$

For steam at 300 psia and 500°F, $h = 1257$ Btu/lbm. Steam property values can be obtained either from the tables in Appendix B or by using the **STEAM** code. The final state of the steam in the tank is at a pressure of 300 psia and a specific internal energy, u_{final}, of 1257 Btu/lbm. The steam code gives $T = 731°F$ for the steam in the tank at the end of the process.

The filling of a tank process could also be analyzed as a nonflow process, in which case one would apply the nonflow energy equation (NFEE) instead of the flow energy equation (FEE). Figures 6.14(a) and (b) depict the (closed) system boundary at the beginning and final states when the problem is formulated as a nonflow process. The system comprises the steam (of mass m_i) that finally enters the tank plus the evacuated space in the tank. The terms in the NFEE are determined as follows:

$$Q_{12} = 0, \quad \text{since the process is adiabatic}$$

$$W_{12} = -p_i V_{\text{swept}} = -p_i(m_i v_i)$$

where p_i is the pressure of the steam in the pipe, while v_i is the specific volume. The volume swept, V_{swept}, is the volume of steam that was swept into the tank under the action of the pressure of the steam flowing inside the pipe.

$$\Delta U = m_i(u_{\text{final}} - u_i)$$

Substituting in the NFEE (Eq. 6.4), neglecting the kinetic and potential energy terms, and dividing through by m_i leads to the same result as obtained earlier, that is,

$$u_{\text{final}} = u_i + p_i v_i = h_i$$

A substantial increase in the temperature of the steam is indicated as a result of the flow of steam into a previously evacuated tank when the tank is well insulated. At first, this must appear surprising. A physical insight into why the temperature rise occurs can be gained by looking again at the analysis of the process as a nonflow process. The rise in temperature can be attributed to the work done on the system by the surroundings, which is $p_i v_i$ per unit mass of steam admitted to the tank. The work transfer to the system produces an increase in internal energy that is manifested by the substantial increase observed in the temperature of the steam. From the flow process perspective, the steam flowing into the tank transports its internal energy in addition to the $p_i v_i$, which some would refer to as *flow work*.

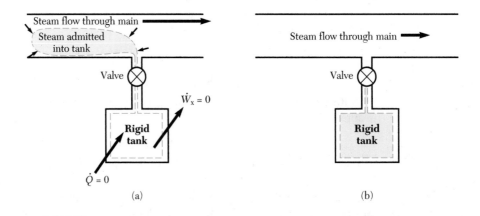

Figure 6.14 System boundary at the (a) initial state and (b) final state for Example 6.14 on the filling of an evacuated rigid tank.

EXAMPLE 6.15
Filling a Rigid Tank

Steam at 300 psia and 500°F is flowing in a pipe, as shown in Figure 6.13. A rigid tank containing steam at 500°F and 20 psia is connected to the steam pipeline through a valve, as shown in the figure. At a certain time, the valve is opened, and steam flows into the tank until the pressure in the tank reaches that in the pipeline. Then the valve is closed. The process is adiabatic, and changes in kinetic and potential energies are negligible.

(a) Determine the final temperature of the steam in the tank after equilibrium is reached.

(b) Compare the result with that from Example 6.14.

Given: Steam at 500°F and 300 psia flowing into a tank that is initially filled with steam at 500°F and 20 psia

Find:

(a) Final temperature T_2 of steam in tank

(b) Compare result with T_2 from Example 6.14

Solution: Following the same procedure outlined in Example 6.14, simplify Equation 6.26 to

$$m_i h_i = m_2 u_2 - m_1 u_1 \tag{1}$$

where m_i is the mass of steam admitted from the pipe into the tank; m_1 and m_2 denote the mass of steam in the tank at the beginning and end of the process, respectively; and u_1 and u_2 are the specific internal energy of the steam in the tank at the initial and final states.

From the conservation of mass principle, the following relationship must exist between the masses defined:

$$m_i + m_1 = m_2 \tag{2}$$

Also, if V denotes the volume of the tank, it follows from the definition of specific volume that

$$m_1 = V/v_1 \quad \text{and} \quad m_2 = V/v_2 \tag{3}$$

where v_1 and v_2 are the specific volume of the steam in the tank at the initial and final states, respectively. Combining Equations 2 and 3 gives

$$m_i = V(1/v_2 - 1/v_1) \tag{4}$$

Now, combining Equations 1, 3, and 4 gives

$$V(1/v_2 - 1/v_1)h_i = Vu_2/v_2 - Vu_1/v_1 \tag{5}$$

The volume V of the tank cancels out, and the reduced equation further simplifies to

$$(1/v_2)(h_i - u_2) = (1/v_1)(h_i - u_1) \tag{6}$$

Note that the only unknowns in Equation 6 are u_2 and v_2, and these must be uniquely related to each other for the steam at the final state of 300 psia and T_2. An iterative procedure is needed in this case, and use of the **STEAM** code is advised.

$$h_i = 1257 \text{ Btu/lbm}, \quad \text{at 300 psia and 500°F}$$

At state 1, $p_1 = 20$ psia and $T_1 = 500$°F. Thus, $v_1 = 28.5$ ft^3/lbm and $u_1 = 1181$ Btu/lbm.

(a) Using $p_2 = 300$ psia along with the previous values, and a trial-and-error process with Equation 6, gives $T_2 = 716$°F.

(b) When steam is initially in the tank, the final temperature of the steam in the tank (716°F) is not as high as it was in Example 6.14 when the tank was initially evacuated (731°F). Since an initial mass of steam exists in the tank in the present example, the additional amount that is admitted to bring up the steam pressure in the tank to 300 psia is less than in Example 6.14. Thus, the smaller amount of flow work involved in the present case produces a correspondingly smaller increase in internal energy of the steam finally in the tank as compared to Example 6.14.

Note that if the volume of the tank were known, m_1, m_2, and m_i could be calculated from Equations 2, 3, and 4.

EXAMPLE 6.16
The Throttling Process

A fluid stream is throttled whenever some type of restriction is placed in the flow. A significant pressure drop is likely to occur without any work being produced and with negligible changes in kinetic and potential energy. The throttling process usually takes place under adiabatic condition. Throttling can be achieved by use of a valve (which is partially closed), a porous plug, or any other means that restrict the fluid flow thereby producing a drop in pressure in the flow.

Figure 6.15 shows the throttling process for a capillary tube, which is often used in air conditioning applications. In this case, the fluid flows from a larger diameter tube through a smaller diameter tube. The flow through the smaller diameter tube involves

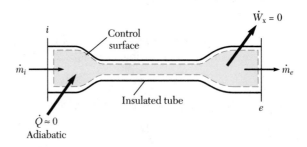

Figure 6.15 The throttling process through a capillary tube.

a significant pressure drop. The pipe is insulated from the surroundings, and therefore the process can be considered as adiabatic.

If the flow in the capillary tube of Figure 6.15 is steady flow and if the entering mass equals the exiting mass, develop an expression for the change in enthalpy between plane i and plane e.

Given: System shown in Figure 6.15 with $\dot{m}_i = \dot{m}_e$, $\dot{Q} = 0$ (insulated), $\dot{W}_x = 0$ (since no shaft work is involved), $\Delta(KE)$ and $\Delta(PE)$ are negligible

Find: $h_e - h_i$

Solution: Apply Equation 6.29 for the fluid flow through the control volume. In this case, the equation reduces to

$$h_e - h_i = 0$$

Thus, the enthalpy of the fluid at the exit plane is identical to the enthalpy of the fluid at the inlet plane. Note that this does not mean that the enthalpy stays constant during the process; it only means that the enthalpy at state e is equal to that at state i for the process. The process is termed a *constant enthalpy* or *isenthalpic process* and is sometimes called the *Joule–Thomson process*. It should be noted that a steady-flow throttling is analogous to a free expansion process for a closed system; that is, $h_e = h_i$ for the adiabatic throttling flow process, while $u_2 = u_1$ for the adiabatic free (or unresisted) expansion process (nonflow).

EXERCISES

for Section 6.6

1. Water flows through a tube at a rate of 15 lbm/min while heat is added to it at a rate of 15,000 Btu/h. If the water entered at 70°F, determine the exit temperature. Assume the pressure remains constant at 14.7 psia throughout.

   Answer: 86.5°F

2. Repeat Exercise 1 but assume a heat addition rate of 150,000 Btu/h instead. What is the exit state of the water?

   Answer: 212°F, $x = 0.025$

3. Ambient air at 1 atm and 15°C flows at the rate 0.05 kg/s over an electric strip heater, as illustrated in the accompanying figure. The power supply to the heater is

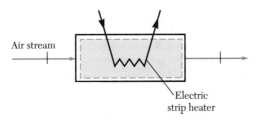

Raising of temperature of an air stream using an electric strip heater.

1.8 kW. The control volume includes the electric strip heater, and the enclosure through which the air passes is well-insulated thermally. Determine

(a) $\dot{Q}$

(b) $\dot{W}$

(c) the air exit temperature.

c_p for the air can be taken as 1.005 kJ/kg · K.

Answer: (c) 50.8°C

4. A flow at the rate of 600 lbm/h of hot air at 1 atm and 1500°F mixes adiabatically with 120 lbm/h of ambient air at 1 atm and 70°F. The mean specific heat at constant pressure for air in this temperature range can be taken as 0.26 Btu/lbm · °R. Determine for the mixed air stream

(a) the mass flow rate in lbm/h

(b) the temperature in °F

Answer: (b) 1260°F

6.7   Summary

The chapter began with a brief account of Joule's famous experiments, which demonstrated (1) that work could be used to produce the same effect on a system as heat and (2) that the same amount of work was needed to produce the same heating effect regardless of the actual work process. Various statements of the first law of thermodynamics were presented and discussed in relation to the observations in Joule's experiments. These different statements can be shown to be equivalent statements of the law by choosing any one of them as the primary statement and then proving that the other statements are logical consequences, or corollaries, of the primary statement. Poincare's statement of the first law of thermodynamics is used in this book as the primary statement from which certain corollaries have been established. The corollaries include (1) energy as a derived concept from the first law of thermodynamics and (2) the conservation of energy principle.

Further consideration of the first law of thermodynamics leads to the energy equation, which defines a relationship among heat, work, and energy. Thus, both heat and work are seen as forms of energy interaction or transfer between a system and its surroundings. Application of the first law of thermodynamics to the closed system or non-flow processes results in the nonflow energy equation (NFEE), written as Equation 6.10 in its most comprehensive form. The first law of thermodynamics only provides a basis for determination of changes in the energy of a system and not the absolute value of energy. As such, a datum state must be defined relative to which tabulated values of the energy of a given substance are expressed. Several worked examples were provided on the use of the NFEE, including applications to the transient behavior or time-varying conditions for a closed system.

In considering open systems, the conservation of mass principle was introduced and discussed. Equations 6.14 and 6.15 are statements of the principle for the single (inlet and outlet) stream and the multiple stream cases, respectively. The first law of thermodynamics was also applied to flow processes that occur when a system is open. The energy equation was derived for a control volume, defined as an open system of fixed volume, shape, and position. Derivation of the equation was circumscribed by the assumption of local equilibrium condition for entering and exiting masses. For exceptional cases in which sharp discontinuities exist in the flow, one should use control volumes that are large enough to ensure that the region of discontinuity remains entirely within the control volume. Equations 6.26 and 6.27 are the flow energy equations (FEE) that emerge. For steady-state steady-flow conditions, the corresponding equations are called the steady-flow energy equations (SFEE) and are Equations 6.29 and 6.30 for the single stream and multiple stream cases, respectively. An appropriate version of the SFEE in USCS units for the single stream case is Equation 6.29a. Several examples were also considered involving the application of the first law of thermodynamics to a control volume.

References

Keenan, J. H. and Shapiro, A. H. 1947. "History and Exposition of the Laws of Thermodynamics." *Mechanical Engineering*, Vol. 69, 915–921.

Kliewer, K. L. 1987. "Perpetual Motion." *McGraw-Hill Encyclopedia of Science and Technology*, 6th ed., Vol. 13, 232–234.

Planck, M. 1927. *Treatise on Thermodynamics*. (Translated by A. Ogg from the 7th German edition.) New York: Dover Publications Inc.

Questions

1. An "inventor" claims that he has developed an energy machine for a car that produces more energy than it uses, thus creating an unlimited source of power. The car is supposedly powered by a series of batteries connected to the energy machine. Would you buy shares in this venture? Why or why not?

2. Figure Q6.2 illustrates what has been described as the *hydrostatic paradox* (see Kliewer 1987), which was first proposed by Abbe de la Roque in 1686. The claim was that water would flow continuously because its weight in the large vessel exceeds that in the narrow tube segment. Show that if the hydrostatic machine could work, it would be a perpetual motion machine of the first kind.

3. Figure Q6.3 represents a *magnetic machine*, which is supposed to produce perpetual motion of a ball. The magnet pulls the ball up the ramp and the ball is supposed to drop through the hole near the magnet and then roll down the second inclined plane until it is back at the base of the first ramp only to

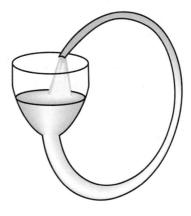

Figure Q6.2 The hydrostatic paradox.

be pulled up the ramp again. Show that if such a machine could work, it would be a perpetual motion machine of the first kind. Do you think it could work? Why or why not?

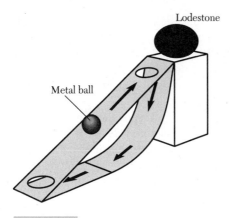

Figure Q6.3 The magnetic machine.

4. A wooden block is hauled at a constant speed over a metallic slab, as illustrated in Figure Q6.4. As a result of the frictional resistance between the two surfaces, the metallic slab gets warmer as the block is hauled over it. Assume that the wooden block is perfectly insulating thermally, and as a result, its temperature does not change. Also assume that heat loss is negligible from the metallic slab to the ambient air and to the ground on which it is placed. By considering the wooden block and the metallic slab in turn as the system, indicate whether the quantities Q, W, and ΔU are positive, zero, or negative.

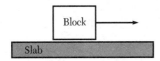

Figure Q6.4 The hauling of a block over a slab.

5. Reconsider Question 4 but this time assume that the block is metallic while the slab is wooden.

6. A steady current I flows through a resistor, as illustrated in Figure Q6.6. After a short time, the temperature of the resistor stays constant. The voltage across the resistor is V. Define the resistor as the system and indicate the magnitude and direction of $\dot{Q}$, $\dot{W}_x$, and dU/dt when the temperature of the resistor has reached the steady value.

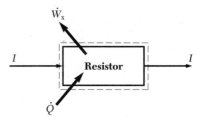

Figure Q6.6 An electrical resistor at steady state.

7. Write the nonflow energy equation (NFEE) and identify each term in it. Indicate the correct SI unit for each term in the equation.

8. Write the appropriate version of the NFEE for USCS units. Indicate the USCS unit for each term and the conversion factor to use so as to ensure consistency of units for all the terms in the equation.

9. Write the steady-flow energy equation (SFEE) for a single entry stream and a single exit stream. Identify each term in the equation and indicate the SI unit for each term.

10. Write the single stream version for the SFEE appropriate for quantities measured in USCS units. Simplify the equation to give an expression for the exit velocity of a fluid passing through an adiabatic nozzle. You can assume that the entry velocity of the fluid is small and negligible.

11. Figure Q6.11 illustrates the velocity profile for a viscous fluid flowing through a pipe. The fluid velocity is zero at the wall of the pipe, while a nonzero

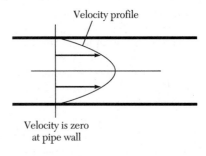

Figure Q6.11 Velocity profile for the flow of a viscous fluid in a pipe.

wall shear stress (τ_w) acts on the fluid in contact with the pipe wall surface. If the inside surface of the pipe is defined as the control surface, what is the rate of shear work done by the fluid?

12. When is it correct to use $\Delta u = c_v \, \Delta T$ for a pure substance? Note that c_v is the specific heat at constant volume.

13. Use property data for steam in either Appendix A or B to obtain the *T-h* diagram for a pure substance. Clearly indicate on your diagram a typical constant

pressure line at pressures below the critical pressure for the substance. When is it reasonable to use $\Delta h = c_p \, \Delta T$ for a pure substance? Note that c_p is the specific heat at constant pressure.

14. What is the value of c_p for a pure substance in the wet vapor region?

15. One corollary of the first law of thermodynamics is that the total energy of the universe remains constant. What, then, is meant by the expression "the world's energy supply is being used up rapidly"?

Problems

6.1 A thermodynamic cycle for a closed system consists of processes I and III, as shown in Figure 6.2. During process I from state 1 to state 2, the work done by the system is 20 kJ, while the heat added to the system is 50 kJ. During process III from state 2 back to state 1, the heat rejected by the system is 12 kJ.
 (a) What is the work done by the system during process III?
 (b) What is the net work done by the system during the cycle?

6.2 A thermodynamic cycle for a closed system consists of processes I and III, as shown in Figure 6.2. During process I from state 1 to state 2, the heat rejected by the system is 500 Btu. During process III from state 2 back to state 1, the heat added to the system is 1200 Btu, and the work done by the system during process III is 500 Btu.
 (a) What is the work done by the system during process I?
 (b) What is the net work done by the system during the cycle?

6.3 During a process from state 1 to state 2, the heat rejected by a closed system is 60 kJ, while the work added to the system is 75 kJ. What is the change in total energy of the system during the process?

6.4 A thermodynamic cycle for a closed system consists of five processes. The heat added to the system during the processes I through V are 12 Btu, 16 Btu, -5 Btu, 0 Btu, and -10 Btu, respectively. The work done by the system during the first four processes are 4 Btu, 8 Btu, 5 Btu, and -4 Btu.
 (a) What is the work done by the system during process V?

 (b) What is the net work done by the system during the cycle?
 (c) What is the net work done by the surroundings during the cycle?
 (d) What is the total energy change of the universe during the cycle?

* **6.5** An automobile with a mass of 3000 lbm is initially at rest on the top of a mountain at an elevation of 5000 ft above sea level. One hour later the automobile is traveling along a road that is 2000 ft above sea level at a speed of 50 mi/h. No work was done on or by the automobile during the process, but 300 Btu of heat was rejected by the automobile during the process.
 (a) What was the change in total energy of the automobile during the process?
 (b) What was the change in internal energy of the automobile during the process?
 (c) What was the change in total energy of the surroundings during the process?

6.6 A heat input of 42 kJ is made to a gas held in a rigid container. Determine
 (a) the work done by the gas
 (b) the increase in internal energy of the gas

6.7 Water of mass 2 kg is held at a constant volume in a container while a heat transfer of 10 kJ occurs to it slowly via a flame. The container is not well insulated, and as a result, 2 kJ of heat is lost to the surroundings over the same period. Taking the specific heat of water as 4.2 kJ/kg · K, determine the temperature increase of the water.

6.8 The shear work done on a system having a thermal capacity (defined as mass × specific heat) of 20 J/K is the equivalent of a force of 200 N acting through a distance of 2 m. If the system is well insulated thermally, determine the temperature increase produced.

6.9 A 500 kg mass is attached to a pulley and a rotating paddle submerged in water. The mass drops 2.5 m under the influence of gravity, with $g = 10$ m/s^2. Find the heat (in kJ) needed to subsequently return the water to its original temperature.

6.10 Water flows over a waterfall that is 60 m high. Assuming that all of the potential energy of the water at the top of the fall is converted to an increase in internal energy at the bottom of the fall, determine the corresponding increase in water temperature. Take $g = 10$ m/s^2 and $(c_p)_{water} = 4.2$ kJ/kg · K.

6.11 A gas is enclosed in a container fitted with a piston of cross-sectional area 0.1 m^2, as shown in Figure P6.11. The pressure of the gas is maintained constant at 8 kPa, while the gas is heated slowly and the piston moves outward a distance of 40 mm. Heat input to the gas during this process is 42 J. What is the increase in internal energy of the gas?

6.12 The volume of a system changes slowly from 0.5 m^3 to 0.6 m^3 at a constant pressure of 250 kPa. At the same time, the internal energy of the system decreases by 20 kJ. Calculate the magnitude and direction of the heat transfer.

6.13 Fluid at 1 MPa contained in one half of a rigid chamber expands adiabatically to fill the entire volume of the chamber, which is 2 m^3. Initially, the other half of the chamber is a vacuum. What is the change in internal energy of the fluid?

6.14 Calculate the heat transfer required to convert 2 kg of water at 40°C and 6 bar to dry saturated steam at a pressure of 6 bar in a quasi-equilibrium, nonflow isobaric process. What is the work done?

6.15 A mass of 1 tonne (= 1000 kg) of ice at −20°C is to be produced by cooling water at 27°C and 100 kPa in a constant pressure nonflow process. Determine the amount of cooling load in kWh (1 kWh = 3600 kJ).

6.16 A mass of 2 kg of air at 600 K is compressed isothermally from 100 kPa to 4 MPa. Assuming the process is quasi-equilibrium, determine
(a) ΔU
(b) W_{12}
(c) Q_{12}

6.17 A rigid vessel contains 2 kg of wet steam with a quality $x = 0.5$. The temperature of the steam is raised from 250°C to 500°C by heating, as shown in Figure P6.17. Determine
(a) ΔU
(b) Q_{12}

6.18 A sealed, rigid 5-m^3 air tank is solar heated, which raises the air temperature from 27°C to 77°C. The initial air pressure, p_1, is 100 kPa (abs). What is the heat transfer to the air in the tank?

6.19 A mass of 5 kg of −5°C ice is added to 50 kg of hot tea at 90°C. What is the eventual temperature of the mixture? Assume the container is well insulated and that the thermodynamic properties of the hot tea are approximately the same as those for water.

6.20 In an experiment to determine the specific heat of water, a drum holding 0.02 kg of water is driven against a brake that applies a braking torque of 0.306 N · m. The thermal capacity of the container (the drum) is 90 J/K. After the drum is turned

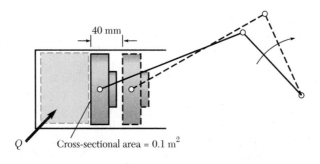

Figure P6.11 Expansion of gas at constant pressure.

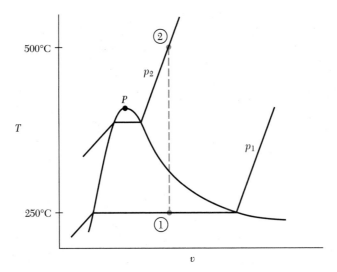

Figure P6.17 Constant volume heating of wet steam.

through 110 revolutions, the rise in temperature of the water and drum was found to be 1.23°C. Determine

(a) the work done on the system comprising the drum and the water

(b) the specific heat of the water

6.21 0.5 kg of air at 100 kPa and 300 K is compressed in a quasi-equilibrium process according to the polytropic law pV^n = constant. The final pressure is 1 MPa. For (i) n = 1.4 and (ii) n = 1, determine

(a) W_{12}

(b) Q_{12}

6.22 A 5000-kg object moving at 20 m/s rams into a plunger in 0.012 m³ of water and is eventually brought to rest. What is the maximum possible rise in temperature of the water?

6.23 A perfectly insulated bucket holds a quantity of water that has its temperature raised using an electric coil. The coil, which is supplied with 300 kJ of electricity, delivers 300 kJ of energy to the water.

(a) Defining the coil and the water contained in the bucket as the system, calculate Q_{12}, W_{12}, and ΔU. Assume that the coil and the bucket have a negligible thermal capacity.

(b) If the system is now only the water in the bucket, determine Q_{12}, W_{12}, and ΔU for the system.

6.24 (a) 50 J of work is done on a system in a rigid container while the internal energy increases by 10 J. What is the heat transfer to the system?

(b) A rigid, well-insulated vessel contains 2 kg of a perfect gas having c_v = 0.715 kJ/kg · K. A stirring paddle is inserted into the vessel, and the work done on the system via the paddle is 50 J. The initial temperature of the gas is 300 K. Find the final temperature of the gas.

6.25 A mass of 1000 kg of fish at 25°C is to be cooled to −20°C. The freezing point of the fish is −2.2°C, and the specific heat of the fish above and below the freezing point can be taken as 3.2 and 1.7 kJ/kg · K, respectively. The specific heat of fusion of the fish is 235 kJ/kg. Determine the cooling load.

6.26 A mass of 3000 kg of fresh beef is to be chilled to 7°C from 39°C in a chilling cooler. The specific heat of beef can be taken as 3140 J/kg · K. The amount of cooling needed for the beef is referred to as the *product load*. Compute the product load (in kWh). If the cooling is carried out over a period of 24 h, what is the average cooling rate in kW?

* **6.27** A tire has a volume of 0.03 m³ and initially contains atmospheric air at 100 kPa and 300 K. The tire *slowly* admits air from a large compressed air storage tank (or receiver) at 1 MPa and 300 K. The final pressure and temperature of the air in the tire are 200 kPa (gage) and 300 K, respectively. Assume that the volume of the tire remains constant. Defining the mass of air finally in the tire as the system, determine for the system

(a) the total mass
(b) W_{12}
(c) Q_{12}

* **6.28** An evacuated rigid bottle of volume 1 m³ has its valve opened such that the bottle slowly fills up to the ambient air condition of 100 kPa and 300 K. Calculate

(a) the mass of air finally in the bottle
(b) the work done by the atmosphere
(c) Q_{12} for the system defined as the air finally in the bottle

* **6.29** Figure P6.29 depicts an ideal engine cycle that comprises nonflow processes and uses air as the working fluid. All processes are quasi-equilibrium and are as follows:

1→2: adiabatic compression according to the law $pV^k = $ constant ($k = 1.4$ for air)
2→3: heating of the gas at a constant volume (that is, $V_3 = V_2$)
3→4: adiabatic expansion according to the law $pV^k = $ constant
4→1: cooling of air at a constant volume (that is, $V_4 = V_1$)

In a particular case, $p_1 = 100$ kPa, $V_1 = 0.001$ m³, and $T_1 = 300$ K. If $V_2 = 6 \times 10^{-5}$ m³ and $T_3 = 1650$ K, calculate

(a) p_2, T_2, p_3, p_4, and T_4

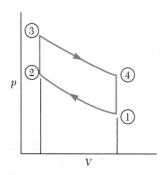

Figure P6.29 Ideal engine cycle.

(b) Q, W, and ΔU for each of the processes involved

* **6.30** A magnetic field is used to launch (from the surface of the Earth) a missile whose mass is 1200 kg. If the initial velocity of the missile is 300 m/s upward, what is the maximum height the missile will go if no heat is transferred and no work is done during the process? Assume $g = 9.81$ m/s².

* **6.31** A projectile is launched vertically upward from the surface of the Moon with an initial velocity of 20 mi/h. If no heat is transferred and no work is done during the process, what is the maximum height the projectile will go? Assume $g = 5.18$ ft/s² on the surface of the Moon.

* **6.32** If heat is being added to a closed system at a rate of 10 Btu/s and work is being done by the system at a rate of 10 hp, what is the net rate of energy accumulation within the system?

6.33 An electric car is depleting its stored electrical energy at a rate of 10 kW, while it loses heat to the surroundings at a rate of 500 J/s. What is the net rate of work output from the car?

* **6.34** A closed system undergoes a process from state 1 to state 2. During the process, the potential energy of the system increases by 200 ft · lbf, while its kinetic energy decreases by 32 ft · lbf. The internal energy of the system increases by 5 Btu, while 20 Btu of heat is added to the system. What is the work done by the system during the process in ft · lbf?

6.35 A closed system in a rigid insulated container, such as that shown in Figure 6.1, undergoes a process from state 1 to state 2.

(a) If 200 kJ of paddle wheel work is done on the system during the process, what is the change in internal energy of the system during the process?
(b) What heat interaction is needed to return the system to its original state?

* **6.36** A closed system consisting of air at 15 psia undergoes a constant pressure process from state 1 to state 2. During the process, the volume changes from 10 ft³ to 6 ft³, while 30 Btu of heat is added to the system and 10 Btu of paddle wheel work is done on the system.

(a) What is the change in internal energy of the system during the process?
(b) Develop an expression for the heat added in

terms of the shaft work and properties of the system. Assume that the volume change is fully resisted.

6.37 A closed system consisting of 1 kg of steam at 1 bar and 200°C undergoes a constant pressure process to 300°C. How much heat was added to the system during the process?

* 6.38 A mass of 10 lbm of water exists at atmospheric pressure and 70°F in an insulated thermos bottle. If 4 lbm of ice at 20°F is added to the water inside the thermos bottle, what is the eventual temperature of the mixture?

* 6.39 A mass of 5 kg of steam at 1 atm and 200°C is mixed with 10 kg of liquid water at 1 atm and 20°C in a well-insulated container. What is the eventual temperature of the mixture?

* * 6.40 A mass of 10 kg of steam at 70 bar and 400°C is enclosed in a rigid insulated container, such as that shown in Figure 6.5. The steam is separated from a vacuum by a membrane, as shown in the figure. The membrane is ruptured, and the steam expands freely into the vacuum until the total volume of the steam is doubled. Neglecting the work done in rupturing the membrane, determine the final temperature of the steam after it has reached equilibrium in the entire volume of the container.

* 6.41 A mass of 10 kg of air at 1 bar and 100°C is enclosed in a rigid insulated container, such as that shown in Figure 6.5. The air is separated from a vacuum by a membrane, as shown in the figure. The membrane is ruptured, and the air expands freely into the vacuum until the total volume of the air is tripled. Neglecting the work done in rupturing the membrane, determine the final temperature of the air after it has reached equilibrium in the entire volume of the container.

* 6.42 A closed system consisting of 5 kg of an ideal gas (with $R = 0.287$ kJ/kg · K) initially at 150 kPa and 400°C undergoes a quasi-equilibrium constant temperature process from state 1 to state 2. If 200 kJ of heat transfer occurs to the system during the process, what is the final volume of the system?

* 6.43 A closed system consisting of 3 lbm of ideal gas (with $R = 53.35$ ft · lbf/lbm · °R) initially at 2 atm and 150°F undergoes a quasi-equilibrium constant temperature process from state 1 to state 2. If the heat added to the system during the process is 10 Btu, what is the final volume of the system?

* 6.44 A closed system undergoes a quasi-equilibrium polytropic process from state 1 to state 2. The system is nitrogen initially at 500 kPa and 350 K. The system volume is increased from 3 m^3 to 5 m^3 during the process, and the polytropic index n equals 1.4 for the process. For nitrogen, assume $c_{v0} = 0.743$ kJ/kg · K and $R = 0.297$ kJ/kg · K. Compute
(a) the pressure and temperature at state 2
(b) the work done by the system during the process
(c) the heat added to the system during the process

* 6.45 A closed system undergoes a quasi-equilibrium polytropic process from state 1 to state 2. The system is air initially at 1 atm and 90°F. The system volume is reduced from 5 ft^3 to 3 ft^3 during the process, and the polytropic index n equals 1.35 for the process. For air, assume $c_{v0} = 0.1715$ Btu/lbm · °R and $R = 53.3$ ft · lbf/lbm · °R. Compute
(a) the temperature (in °F) at state 2
(b) the work done (in ft · lbf) by the system during the process
(c) the heat added (in Btu) to the system during the process

6.46 A closed system undergoes a process from state 1 to state 2. During the process, work is done on the system at a rate of 10 kJ/s, while heat is added to the system at a rate of 8 kJ/s. If the process lasts for 20 min, compute
(a) the rate of energy accumulation (in kW) in the system during the process
(b) the total energy increase (in MJ) of the system during the change from state 1 to state 2

6.47 A closed system undergoes a process from state 1 to state 2. During the process, heat is added to the system at the rate of 50 Btu/h, while work is done by the system at the rate of 20,000 ft · lbf/h. If the process lasts for 32 min, compute
(a) the rate of energy accumulation (in Btu/h) in the system during the process
(b) the total energy increase (in Btu) of the system during the change from state 1 to state 2

6.48 Gasoline is added to an automobile tank at the rate of 20 lbm/min, while the fuel is being removed from the tank at the rate of 10 lbm/h. What is the time rate of change of mass (in lbm/min) of the gasoline that is accumulating within the tank?

6.49 A large 10,000-gal storage tank is used to supply nonleaded gasoline to automobiles through five different pumps. At 7:30 A.M., the gas pumps are being used to supply automobiles at the rates of 10 kg/min, 8 kg/min, 6 kg/min, 7 kg/min, and 12 kg/min, respectively, while the storage tank is being filled at the rate of 1000 kg/min. What is the time rate of change of gasoline accumulating within the storage tank at this instant?

6.50 Water at 70°F is flowing through a 2-in.-diameter circular pipe at the rate of 30 lbm/min. What is the bulk mean velocity (in ft/min) of the fluid?

∗ 6.51 Nitrogen gas (with $R = 0.297$ kJ/kg · K) at 200 kPa and 27°C is flowing through a 20-cm diameter circular duct at the rate of 0.6 kg/s. What is the bulk mean velocity (in m/s) of the gas?

∗ 6.52 Water at 1 atm and 27°C is flowing with a bulk mean velocity of 5 m/s into a large tank at the rate of 50 kg/s. Paddle wheel work is being done in the tank at a rate of 20 kJ/s, while heat is being lost from the tank at a rate of 4 kJ/s. An exit stream of water at 30°C is leaving the tank with a bulk mean velocity of 3 m/s and a flow rate of 20 kg/s. The elevation of the entrance stream is the same as that for the exit stream.
  (a) What is the rate of accumulation of mass within the tank?
  (b) What is the time rate of change of total energy inside the tank (control volume)?

∗ 6.53 Air (with $c_p = 1.02$ kJ/kg · K) is flowing through a control volume, such as that shown in Figure 6.8. The conditions for the air are such that it can be considered an ideal gas during the process. The state of the air entering the control volume at position 1 is 3 atm and 100°C, while the state of the air exiting at plane 2 is 2 atm and 150°C. Paddle wheel work input to the control volume is at a rate of 100 kJ/s. The changes in kinetic and potential energies between the entry and exit conditions for the air stream are negligible. If the mass flow into the control volume is 180 kg/s and the mass flow leaving the control volume is 180 kg/s, find the net rate of heat transfer to the control volume for this steady-state steady-flow process.

∗ 6.54 Oxygen is flowing through a control volume, such as that shown in Figure 6.8. The conditions are such that the oxygen can be considered an ideal gas during the process. The state of the gas entering the control volume at position 1 is 30 psia and 70°F, while the state of the gas exiting at plane 2 is 20 psia and 200°F. No shaft work is done, and the changes in kinetic and potential energies between the entry and exit conditions are negligible for the gas stream. If the mass flow into the control volume is 500 lbm/s and the mass flow leaving the control volume is 500 lbm/s, find the net rate of heat transfer to the control volume for this steady-state steady-flow process.

∗ 6.55 Steam expands through a turbine that is enclosed by a control volume, such as that shown in Figure 6.9. The mass flow rate of the steam is 20 lbm/s. The entering state of the steam is 400 psia and 800°F, while the exiting state is 50 psia and 350°F. Negligible changes in kinetic and potential energies occur through the device, and the steady-state steady-flow process is adiabatic. Find the power output (rate of doing work), in hp, for the turbine.

∗ 6.56 Steam expands through a turbine that is enclosed by a control volume, such as that shown in Figure 6.9. The mass flow rate of the steam is 7 kg/s. The entering state of the steam is 50 bar and 800°C, while the exiting state is 0.05 bar and 50°C. Changes in kinetic and potential energies are negligible through the device, and the steady-state steady-flow process is adiabatic. Find the power output (rate of doing work) for the turbine.

∗ 6.57 Water flows through a pump with a mass flow rate of 10 lbm/s. The entering state of the water is 30 psia and 100°F, while the exiting state is 600 psia and 100°F. Changes in potential and kinetic energies are negligible during the process, and the steady-state steady-flow process is adiabatic. Find the power input (in hp) required by the pump.

∗ 6.58 Water flows through a pump with a mass flow rate of 10 kg/s. The entering state of the water is 1 bar and 30°C, while the exiting state is 40 bar and 40°C. Heat is added to the water at the rate of 421 kJ/s, and changes in potential and kinetic energies are negligible during the process. Find the power input (in kW) required by the pump.

∗ 6.59 Refrigerant-22 is flowing through a compressor in a steady-state steady-flow process. The process is adiabatic, and changes in potential and kinetic energies are negligible through the device. The mass flow rate of refrigerant-22 is 30 lbm/min. The entering state is 50 psia and 60°F, and the exiting state is 150 psia and 160°F. Find the power input (in hp) to the compressor.

∗ **6.60** Refrigerant-22 is flowing through a compressor in a steady-state steady-flow process. The process is adiabatic, and changes in potential and kinetic energies are negligible through the device. The mass flow rate of refrigerant-22 is 10 kg/s. The entering state is 2.5 bar and 40°C, and the exiting state is 12 bar and 120°C. Find the power input (in kW) to the compressor.

∗ **6.61** Refrigerant-22 is flowing through a nozzle in a steady-state steady-flow process. The process is adiabatic, and no work is done during the process. Changes in kinetic and potential energies are negligible during the process. The refrigerant enters the nozzle as saturated liquid at 135°F and exits at 90°F. Determine the pressure of the refrigerant as it exits the nozzle and the pressure drop through the nozzle.

∗ **6.62** Water is flowing through a nozzle in a steady-state steady-flow process. The process is adiabatic, and no work is done during the process. Changes in kinetic and potential energies are negligible during the process. The water enters the nozzle as saturated liquid at 80°C and exits at 25°C. Determine the pressure of the fluid as it exits the nozzle and the pressure drop through the nozzle.

∗ **6.63** Steam at a pressure of 400 psia and a temperature of 600°F is flowing in a pipe, as shown in Figure 6.13. An evacuated rigid tank is connected to the steam pipeline through a valve, as shown in the figure. At a certain time, the valve is opened and steam flows into the tank until the pressure in the tank reaches that in the pipeline. Then the valve is closed. The process of filling the tank is adiabatic, and the steam velocity in the pipeline is low. Determine the final temperature of the steam in the tank after equilibrium is reached.

∗ **6.64** Steam at a pressure of 100 bar and a temperature of 400°C is flowing in a pipe, as shown in Figure 6.13. An evacuated rigid tank is connected to the steam pipeline through a valve, as shown in the figure. At a certain time, the valve is opened and steam flows into the tank until the pressure in the tank reaches that in the pipeline. Then the valve is closed. The process of filling the tank is adiabatic, and the steam velocity in the pipeline is low. Determine the final temperature of the steam in the tank after equilibrium is reached.

∗∗ **6.65** Steam at 100 bar and 400°C is flowing in a pipe, as shown in Figure 6.13. A rigid tank containing steam at 400°C and 10 bar is connected to the steam pipeline through a valve, as shown in the figure. At a certain time, the valve is opened and steam flows into the tank until the pressure in the tank reaches that in the pipeline. Then the valve is closed. The process is adiabatic, and the steam velocity in the pipeline is low. Determine the final temperature of the steam in the tank after equilibrium is reached.

∗∗ **6.66** Steam at 400 psia and 600°F is flowing in a pipe, as shown in Figure 6.13. A rigid tank containing steam at 600°F and 40 psia is connected to the steam pipeline through a valve, as shown in the figure. At a certain time, the valve is opened and steam flows into the tank until the pressure in the tank reaches that in the pipeline. Then the valve is closed. The process is adiabatic, and the steam velocity in the pipeline is low. Determine the final temperature of the steam in the tank after equilibrium is reached.

∗ **6.67** Refrigerant-22 exits a condenser and flows through a throttling valve. The substance enters the throttling valve as a saturated liquid at 110°F and exits at a temperature of 40°F as represented in Figure P6.67. Determine
  (a) the pressure drop (in psi) through the valve
  (b) the change in specific volume of the substance (in ft³/lbm) between the inlet and exit of the valve

∗ **6.68** Refrigerant-22 exits a condenser and flows through a throttling valve. The substance enters the throttling valve as a saturated liquid at 50°C and exits at a temperature of 10°C. Determine
  (a) the pressure drop through the valve
  (b) the change in specific volume of the substance between the inlet and exit of the valve

∗ **6.69** Water exits a condenser and flows through a throttling valve. The substance enters the throttling valve as a saturated liquid at 120°F and exits at a temperature of 40°F. Determine
  (a) the pressure drop through the valve
  (b) the change in specific volume of the substance between the inlet and exit of the valve

∗ **6.70** Water exits a condenser and flows through a throttling valve. The substance enters the throttling valve as a saturated liquid at 60°C, and exits at a temperature of 20°C. Determine
  (a) the pressure drop through the valve

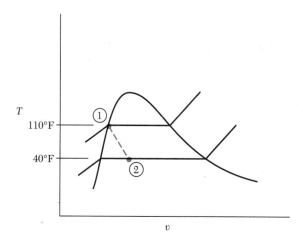

Figure P6.67 Flow through a throttling valve.

(b) the change in specific volume of the substance between the inlet and exit of the valve

6.71 Methane gas enclosed in a balloon undergoes a complete cycle consisting of two processes. During the first process, the gas expands and the internal energy of the gas decreases by 500 kJ. The gas is compressed during the second process, and 100 kJ of heat is transferred from the system. How much work was done by the gas during the compression process?

* **6.72** Steam enters a pipeline from three sources, all at 50 bar, as shown in Figure P6.72. The first stream has a quality $x = 0.6$ (see Chap. 3) and a mass flow rate of 50 kg/s. The second stream is at 300°C with a mass flow rate of 40 kg/s. The third stream is at 600°C with a mass flow rate of 20 kg/s. The pipeline is well insulated such that no heat is lost to

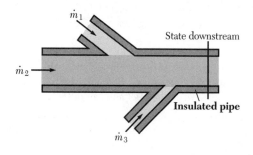

Figure P6.72 Steam flow through a pipeline.

the surroundings during the mixing processes. If the pressure in the pipeline remains constant at 50 bar, what is the temperature of the mixture in the pipeline downstream from the mixing processes? If the mixture is a wet vapor, determine its quality also.

* **6.73** An automobile battery is fully charged and stored at 35°C. During a period of storage, the battery slowly discharges while losing 500 kJ of heat to the surroundings. The battery is then recharged to its initial state with 5 kWh of electrical work. What is the heat transfer to the battery during the charging process?

* **6.74** A closed cylinder fitted with a piston contains air initially at 1 atm and 70°F in a volume of 1 ft³. Then the piston compresses the air until its pressure is 5 atm and its temperature is 250°F. During this compression process, the work done on the air is 5 Btu. What is the heat transfer to the gas during the compression process?

The ideal operation of a four-stroke gasoline engine can be conceptually represented as the cycle depicted in Figures P6.75(a) and (b), which are p-v and T-v diagrams, respectively. All the processes are deemed nonflow and are assumed to be quasi-equilibrium. Each process is defined more precisely as follows:

$1 \rightarrow 2$ is an adiabatic compression process. The ratio of maximum volume (v_1) to minimum volume (v_2) is defined as the compression ratio (r_v). The proc-

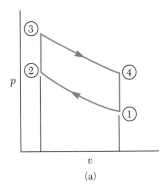 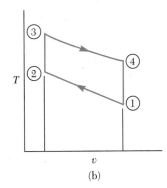

Figure P6.75 The OTTO cycle on (a) p-v coordinates and (b) T-v coordinates for the ideal operation of a four-stroke gasoline engine.

ess path is $pv^k = \text{constant}_1$, where k is the ratio of the principal specific heats for the working fluid, which is taken as air.

$2 \rightarrow 3$ is a constant volume heating process. In other words, $v = \text{constant} = v_2 = v_3$.

$3 \rightarrow 4$ is an adiabatic expansion process terminating with $v_4 = v_1$. The process path is $pv^k = \text{constant}_2$.

$4 \rightarrow 1$ is a constant volume heat rejection (or cooling) process that completes the cycle.

When air is taken as the working fluid for the engine, the cycle is termed an *air-standard cycle*, which is named the *OTTO cycle*. Problems P6.75 to P6.80 that follow cover the application of the first law of thermodynamics to the analysis of the ideal cycle illustrated in Figures P6.75(a) and (b) and as just described.

* **6.75** Use w to denote the work done per unit mass by the system (the working fluid) and q for the heat transfer to the system per unit mass of the system. By applying the NFEE to the processes $1 \rightarrow 2$, $2 \rightarrow 3$, $3 \rightarrow 4$, and $4 \rightarrow 1$ in turn, show that

$$w_{\text{net}} = q_{\text{net}}$$

for the cycle.

* **6.76** By eliminating v between the process path equation $pv^k = \text{constant}$ and the ideal gas pvT equation, show that for a quasi-equilibrium adiabatic expansion or compression process involving an ideal gas,

$$Tv^{(k-1)} = \text{constant}$$

Then show that the expression for w_{net} obtained

in Problem 6.75 can be rewritten as

$$w_{\text{net}} = c_v T_1 (1 - r_v^{k-1}) + c_v T_3 \left(1 - \frac{1}{r_v^{k-1}} \right)$$

* **6.77** The positive heat transfer (input) to the system occurs during the process $2 \rightarrow 3$. Show that this input, q_{input}, can be written as

$$q_{\text{input}} = c_v (T_3 - T_1 r_v^{k-1})$$

* **6.78** For an engine cycle, the *thermal efficiency*, η, is defined as

$$\eta = \left(\frac{w_{\text{net}}}{q_{\text{input}}} \right) 100\%$$

Use the results in Problems 6.76 and 6.77 to obtain the following expression for the OTTO cycle efficiency:

$$\eta = \left(1 - \frac{1}{r_v^{k-1}} \right) 100\%$$

6.79 Assume the following parameters for the design of a gasoline engine:

$$r_v = 9, \quad T_1 = 300 \text{ K}, \quad T_3 = 1100 \text{ K}$$
$$k = 1.4 \quad \text{and} \quad c_v = 0.718 \text{ kJ/kg} \cdot \text{K}$$

Determine

(a) w_{net}

(b) q_{input}

(c) η

* **6.80** A more realistic model for the gasoline engine assumes that neither the adiabatic compression $(1 \rightarrow 2)$ nor the adiabatic expansion $(3 \rightarrow 4)$ is a

quasi-equilibrium process. For this revised model, assume that $T_1 = 300$ K, $T_2 = 825$ K, $T_3 = 1100$ K, and $T_4 = 520$ K. Apply the NFEE to the processes 1→2 (adiabatic compression), 2→3 (constant volume heating), 3→4 (adiabatic expansion), and 4→1 (constant volume cooling) in turn, and hence, determine the following:

(a) w_{net}

(b) q_{input}

(c) η (as first defined in Problem 6.78)

The Second Law of Thermodynamics

7.1 Introduction

Chapter 6 considered the first law of thermodynamics and observed that Joule's experiment showed an equivalence between heat and work. This equivalence is quantitative and means only that heat and work are alternate methods of transferring energy. No indication was given of the relative values of heat and work. Consideration of several different applications of heat and work, however, shows that *work is more valuable than heat*. Many more feats can be accomplished with work. For example, holding a candle under a block of steel will not make the block rise in a gravitational field, but an applied force acting through a distance upward will. Work can be converted 100% to heat, but heat cannot be continuously converted 100% to work. This distinction is a dissymmetry of nature. The first law does not address it, but the distinction definitely exists.

A microscopic perspective might provide an insight into this dissymmetry of nature. In general, molecular actions involving work are ordered, while molecular actions involving heat are disordered. Molecular actions involving work are coherent, while molecular actions involving heat are incoherent. High quality energy is undispersed energy, as in a lump of coal, while low quality energy is dispersed, such as that stored in the water of the oceans. High quality energy generally requires that the atoms storing the energy be more highly structured, as in the lump of coal, while low quality energy generally allows less structuring of the atoms. Energy tends to disperse, to spread, and to lose coherence with time. Thus, high quality energy tends to disperse with time and move toward a state of higher chaos and less coherence for the atoms. The first law does not provide any clue as to the direction for a process. However, nature shows that there is a direction for all processes involving work and heat. For example, one observes that hot water in a bucket cools with time; it does not spontaneously heat. Furthermore, a ball rolls down a hill to the bottom; it does not spontaneously return to the top. Also, gasoline is used in an automobile engine to drive up a hill; the gasoline in the tank cannot be replenished by coasting back down the hill. Thus, there is a preferred direction involved in all natural actions involving heat, work, and energy. The second law of thermodynamics addresses this direction.

Before investigating the second law of thermodynamics, let's review the basis of logic that allows the statement and use of such laws. Aristotle (384–322 b.c.) developed the concepts and form for modern logic. Although Aristotle concentrated on deductive logic, he also laid the groundwork that eventually led to inductive logic. Scientists and engineers generally use both inductive and deductive logic to reach conclusions. While conclusions and statements established by inductive reasoning cannot be proven, the method has been found to be extremely valuable to the advancement of science and engineering. Basically, inductive logic is used to formulate certain general statements concerning natural behavior based on experimental evidence and observation. Many cases in nature concerning an area of interest are observed, and a law or principle is formulated that accounts for the observed behavior. Thus, *inductive logic* is a process of reasoning from particulars to the general, from a part to a whole, from individual cases to universal truths. Even after a general law is established, it cannot be proved. One simply observes that no behavior in nature has been found that is contrary to the law. If even a single contrary behavior were found, either the law would be abandoned or it would be restated in a new form to account for the contrary behavior.

Once natural processes have been observed adequately and a law has been established by inductive logic, the law can then be used with deductive logic to examine individual cases and to establish corollaries of the general law that apply to specific subsets or cases. Thus, *deductive logic* entails reasoning from the general to a particular, from the universal to the individual, or from a given premise to the necessary conclusion. Deductive logic can be used to establish a corollary of a general law. Such a *corollary* is a consequence or additional inference from the known proposition. It is something that logically follows from the established proposition.

Section 7.3 states the second law of thermodynamics that was arrived at by inductive logic. Sections 7.4 through 7.7 then illustrate how you can apply deductive logic to reach a number of important conclusions after you have accepted the second law from inductive logic. In general, all of the laws of thermodynamics and most of the scientific laws such as Newton's laws were obtained from inductive logic. They are accepted as valid, and then conclusions are drawn related to specific situations and circumstances through the method of deductive logic. Before formally stating the second law of thermodynamics, Section 7.2 briefly considers certain concepts that have particular relevance to the statement of the second law.

7.2 Reservoirs, Heat Engines, and Refrigerators

Discussions of the second law of thermodynamics often refer to the concepts of reservoirs, heat engines, and refrigerators. These concepts are considered in some detail here.

Thermal Reservoirs and Heat Engines

A *thermal reservoir* is a heat source or heat sink that remains at a constant temperature, regardless of the energy interaction. Note that a system or device that is supplying heat is a *source*, while a *sink* is a system or device to which heat is rejected. No matter how much heat is added to or removed from the reservoir, it remains at a constant temperature throughout the process, with only an infinitesimal change in its temperature. Ocean water and atmospheric air can be considered as close examples of thermal reservoirs in nature. Figure 7.1 shows two constant temperature reservoirs. One of the reservoirs is a high temperature reservoir at temperature T_H, which is available to supply heat Q_H to the system. The high temperature reservoir remains at T_H throughout the process, no matter what the temperature of the system is. Also shown in Figure 7.1 is a low temperature reservoir that remains at T_L while receiving heat Q_L from the system. Observation of nature shows that the system must be at a temperature lower than T_H and higher than T_L for the heat transfers to occur in the directions indicated. It should be noted, however, that a heat source does not always have to be a high temperature reservoir (for example, as in refrigeration cycles). Also, in general, heat sources and heat sinks are not necessarily thermal reservoirs. A heat source (or sink) is described as *finite* if its temperature changes when heat is extracted from it (or added to it).

A *heat engine* is a device that can operate continuously to produce work while receiving heat from a high temperature source and rejecting heat to a low temperature sink. Figure 7.2 is a schematic diagram of a heat engine. Figure 7.3 illustrates a simple

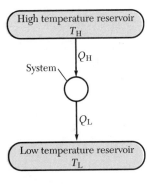

Figure 7.1 Constant temperature reservoirs.

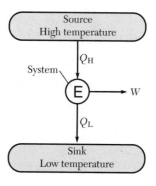

Figure 7.2 Schematic diagram of a heat engine.

steam power plant, which is a typical type of heat engine. In the steam power plant, heat is added to water in a boiler to produce steam. The steam is then expanded through a turbine that produces a positive work output. Some of the work output of the turbine is used to drive a pump. The net output from the turbine (total turbine output work minus the pump work) is available as the work output of the engine. After the steam passes through the turbine to produce work, it goes to a condenser where it is condensed as a result of heat rejection from the device to a sink. After the steam is condensed, it

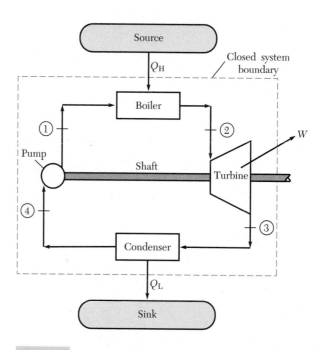

Figure 7.3 Schematic diagram of a steam power plant.

is pumped as water to a high pressure and returned to the boiler to complete the cyclic process. Thus, this device receives heat from a source at high temperature and rejects heat to a sink at low temperature while producing a net work output.

An *automobile engine* is another example of a heat engine. The burning of gasoline with air is equivalent to a heat flow into a fluid from a high temperature source. As a consequence, work is produced to turn the wheels of the automobile. Heat rejection to the surroundings takes place through the radiator and the exhaust gases. Unlike the steam power plant, the automobile engine does not constitute a complete thermo-dynamic cycle since the exhaust gas never returns to its original state of fuel–air mixture. Indeed, the internal combustion engine follows a *mechanical cycle*, not a thermodynamic cycle. Nevertheless, to facilitate a thermodynamic analysis of the automobile engine, the cycle is often approximated as a thermodynamic cycle with a few assumptions and sim-plifications, as Chapter 10 will show.

It is often important to determine the efficiency of devices and systems. In general, the *efficiency* of any device or system is defined as follows:

$$\text{Efficiency} = \frac{\text{desired output}}{\text{required input}} \tag{7.1}$$

For a heat engine, the desired output is work, while the required input is heat that is provided from the high temperature source. Thus, the efficiency (η_1) of a heat engine can be expressed as follows:

$$\eta_1 = \left(\frac{W}{Q_H}\right) 100\% \tag{7.2}$$

This is strictly a *first law efficiency* and is not based on any evaluation of quality of the forms of energy flow. The first law efficiency for a heat engine (η_1) is also referred to as the *thermal efficiency*.

This chapter abandons the previous sign convention used for work and heat and instead designates the direction of work and heat each time in accordance with the scheme under consideration. Thus, the nomenclature Q_H for the heat engine of Figure 7.3 indicates that the heat flow is into the system in accordance with the diagram, while Q_L indicates that the heat is out of the system.

A first law analysis of the heat engine cycle in Figure 7.3 gives the following equation:

$$W = Q_H - Q_L \tag{7.3}$$

Combining Equations 7.2 and 7.3 gives the following for the first law thermal efficiency of the heat engine:

$$\eta_1 = \left(\frac{Q_H - Q_L}{Q_H}\right) 100\% = \left(1 - \frac{Q_L}{Q_H}\right) 100\% \tag{7.4}$$

Reversed Heat Engines

Two other devices of interest are the refrigerator and the heat pump, which are similar devices that operate as a *reversed heat engine*. Such a device, illustrated in Figure 7.4,

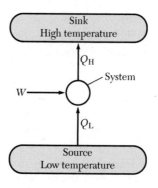

Figure 7.4 Schematic diagram of a refrigerator.

receives heat from a low temperature reservoir (source) and rejects heat to a high temperature reservoir (sink). In typical applications, when the device is used for cooling of the source, it is referred to as a *refrigerator* or *air conditioner*, and when it is used for heating the sink (or for both heating and cooling), it is often called a *heat pump*.

Figure 7.5 shows a typical component diagram of a refrigeration cycle that can be used to produce a cooling effect. In this cycle, a *refrigerant* fluid is circulated around a

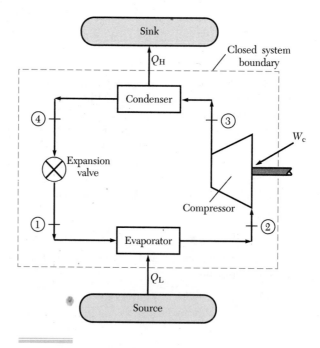

Figure 7.5 Schematic diagram of a typical refrigeration cycle.

loop. The refrigerant is evaporated (or vaporized) at a low temperature in the evaporator. During this process, heat is removed from the medium to be cooled to produce a cooling effect in that medium (low temperature heat source). After the refrigerant is evaporated, it is compressed as a gas by a compressor to the higher temperature and pressure. This compression process requires work input to the compressor. The gas leaving the compressor is condensed to a liquid in the condenser. During this condensation process, heat is rejected, usually to the ambient surroundings, at a high temperature (the sink). After the refrigerant leaves the condenser, it is expanded through an expansion or throttle valve to a low temperature and pressure in the evaporator, from where the cycle is repeated.

Upper Limits on First Law Efficiency

The early 19th century work of a French engineer, Sadi Carnot (1796–1832), is widely acknowledged as the foundation for the eventual formulation of the second law of thermodynamics. Carnot engaged in extensive speculation as to the highest possible efficiency that could theoretically be attained by a heat engine operating between a given pair of thermal reservoirs, one at a high temperature (T_H) and the other at a low temperature (T_L). Typical thermal efficiencies for practical heat engines that have been developed to date vary from about 0.5% for the Newcomen steam engine (1750) to about 35% to 40% for the large modern steam power plants. Evidently, limits for the first law efficiencies of heat engines are significantly below the 100% mark. The question that arises is whether the limit is set by practical factors alone or whether, in fact, theoretical limits are set by nature. In his treatise, Carnot (1824) established a theoretical limit, less than 100%, for the thermal efficiency of heat engines, a conclusion that can now be shown by deductive logic to be a corollary or consequence of the second law of thermodynamics.

Whereas a heat engine is designed for producing work from heat input, a reversed heat engine produces heating or cooling from a work input. The first law efficiencies for practical reversed heat engines typically turn out to be greater than 100%. From first law considerations alone, one cannot figure out the theoretical upper limit for the first law efficiency of reversed heat engines. As seen later, such limits can be established, however, as a corollary of the second law of thermodynamics.

The first law efficiency of a refrigeration or heat pump cycle is usually called the *coefficient of performance* of the device. For the refrigeration or heat pump cycle, the desired effect can be either a cooling effect (refrigeration or air conditioning), which involves heat flow from the source to the device, or it can be a heating effect (heat pump), which involves heat flow from the device to a sink at the higher temperature. The *coefficient of performance for cooling* is designated as COP_C, while the *coefficient of performance for heating* is COP_H. Thus, COP_C is given by

$$COP_C = \frac{\text{desired effect}}{\text{required input}} = \frac{Q_L}{W} \tag{7.5}$$

while COP_H is given by

$$COP_H = \frac{\text{desired effect}}{\text{required input}} = \frac{Q_H}{W} \qquad (7.6)$$

Applying the first law to the arrangement depicted in Figure 7.5 gives the following expressions for COP_C and COP_H:

$$COP_C = \frac{Q_L}{Q_H - Q_L} \qquad (7.7)$$

and

$$COP_H = \frac{Q_H}{Q_H - Q_L} \qquad (7.8)$$

Algebraic manipulation of Equations 7.7 and 7.8 reveals that

$$COP_H - COP_C = 1 \qquad (7.9)$$

Refrigerators and heat pumps are considered in more detail later.

7.3 Statements of the Second Law

As indicated earlier in Chapter 1, there are two classical statements of the second law of thermodynamics:

Kelvin–Planck statement *It is impossible to construct a device that will operate in a cycle and produce no effect other than the raising of a weight (in the presence of a gravitational field) and the exchange of heat with a single reservoir.*

Clausius statement *It is impossible to construct a device that operates in a cycle and produces no effect other than the transfer of heat from a low temperature body to a high temperature body.*

Several observations can be made concerning these statements. The Kelvin–Planck statement concerns a device that exchanges heat with a single reservoir while raising a weight, and the raising of a weight in a gravitational field is equivalent to doing work. The statement implies that *it is impossible to construct a device that will operate continuously to produce work while exchanging heat with a single thermal reservoir* (source or sink). One can conclude from this that it is impossible to construct a device that will *continuously* convert heat totally into work. The statement therefore implies that work is of more value than heat because work can always be converted completely and continuously to heat.

While the Kelvin–Planck statement is of particular relevance to a heat engine, the Clausius statement relates more directly to a reversed heat engine. Both the Kelvin–Planck statement and the Clausius statement were arrived at independently by inductive logic after considering experimental evidence and observations of nature. The two statements are equivalent. We could use either as the primary statement of the second law of thermodynamics and then establish the other as a corollary of the primary statement. This approach is similar to that of Chapter 6 for demonstrating the equivalence of different statements of the first law of thermodynamics.

Several propositions about the behavior of thermodynamic systems, called corollaries, have been deduced from the laws of thermodynamics. A method known as *reductio ad absurdum* is often used to prove these corollaries. In this method of proof, the contrary of the proposition is assumed to be true. If one can then show that the logical consequence of the assumption contradicts the primary law, the assumption becomes "absurd" and thus untenable. The original proposition therefore stands. This method is illustrated in the following paragraphs. The Kelvin–Planck statement (which was originally established by inductive logic) is regarded here as the primary statement of the second law of thermodynamics. The Clausius statement will be shown to naturally follow as a corollary of the Kelvin–Planck statement.

Consider the system shown in Figure 7.6. This device violates the Clausius statement. It operates in a cycle and produces no effect except the transfer of heat (Q_L) from a low temperature reservoir to a high temperature reservoir. Now consider the heat engine shown in Figure 7.7 (which does not violate either the Kelvin–Planck or the Clausius statement). This heat engine is operating in a cycle to produce net work, while receiving heat (Q_H) from a high temperature reservoir and rejecting heat (Q_L) to a low temperature reservoir. In this case, $Q_H = Q_L + W$, and the Q_L in Figure 7.7 is set equal to the Q_L in Figure 7.6.

Let's combine the device of Figure 7.6 with the heat engine of Figure 7.7 into the composite arrangement shown in Figure 7.8. In this case, the low temperature reservoir is redundant since the heat flow Q_L from the engine can pass directly to the refrigeration device.

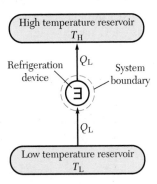

Figure 7.6 A hypothetical device that violates the Clausius statement of the second law.

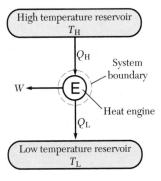

Figure 7.7 A heat engine operating in a cycle while producing work and rejecting heat to a low temperature reservoir.

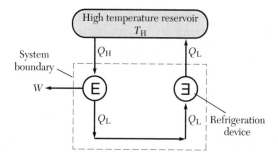

Figure 7.8 Combined heat engine and the hypothetical refrigeration device violating the Kelvin–
Planck statement of the second law.

Thus, the system in Figure 7.8 operates in a cycle and produces net work while exchanging heat with a single reservoir. This is a direct violation of the Kelvin–Planck statement. Thus, the assumption of a device that violates the Clausius statement has led to a contradiction of the Kelvin–Planck statement, which is considered as the primary law. Consequently, it is proved that the Clausius statement is a corollary of the Kelvin–Planck statement. Note that by following a similar procedure, it could be shown that the Kelvin–Planck statement is a corollary of the Clausius statement. Thus, the two statements are equivalent. As stated earlier, either of the two statements can be considered as the primary law. However, in view of the traditional strong interest in heat engines in thermodynamics, the Kelvin–Planck statement has been chosen as the primary law in this text since it deals directly with the conversion of heat into work.

EXERCISES

for Section 7.3

1. A volume of 0.005 m³ of air at 300 K and 200 kPa is contained in a cylinder under a frictionless laden piston. Heat transfer occurs to the air until its volume doubles, while the pressure remains constant at 200 kPa throughout. Assume that the expansion of the system (the air) is fully resisted. Calculate
   (a) W_{12}
   (b) Q_{12}
   Answer: **(a)** 1 kJ, **(b)** 3.5 kJ

2. Consider the reverse of the process in Exercise 1. The air at 600 K and 200 kPa is now cooled such that the volume of the air is halved from an initial volume of 0.01 m³. Assume that the process is quasi-equilibrium and determine
   (a) the work done *on* the system (the air)
   (b) the heat transfer *to* the surroundings
   Answer: **(a)** 1 kJ, **(b)** 3.5 kJ

3. A system comprising 0.01 kg of saturated water at 2 bar is heated until all the water is turned into saturated steam at 2 bar. The expansion of the system (H_2O) is fully resisted. Determine

(a) W_{12}

(b) ΔU

(c) Q_{12}

Answer: (a) 1.77 kJ, (b) 20.2 kJ, (c) 22 kJ

4. Repeat Exercise 3 for saturated water at 60 bar that is heated until it all turns into saturated steam at 60 bar.

   Answer: (a) 1.87 kJ, (b) 13.8 kJ, (c) 15.7 kJ

5. An ideal gas of mass m that is initially at a pressure p_1 and a volume V_1 expands isothermally according to the law $pV =$ constant. Heat (Q_{12}) is transferred to the gas, while work (W_{12}) is done by the system (the gas). Assume that the process is quasi-equilibrium and obtain the expressions for Q_{12} and W_{12} in terms of V_1, V_2, m, R, and the temperature (T) of the gas. Does the result contradict the second law of thermodynamics?

7.4 *Perpetual Motion Machines*

Perpetual motion machines were discussed in Chapters 1 and 6. They are hypothetical devices that can be shown to violate the laws of thermodynamics. Chapter 6 considered a perpetual motion machine of the first kind (PMM1), a device that produces continuous work from nothing in violation of the first law of thermodynamics. One corollary of the first law of thermodynamics is that a PMM1 is impossible. A *perpetual motion machine of the second kind (PMM2)* is a device that can produce net work continuously while exchanging heat with a single reservoir. That is, the device can operate in a cycle while converting all heat input (from a single thermal reservoir) into work output for the device. John Gamgee's ammonia engine of the 1880s, mentioned in Chapter 1, was designed for limitless production of power from energy extracted from the environment (a single thermal reservoir). Such a device is in clear violation of the second law of thermodynamics and must be deemed impossible. The hybrid engine and refrigeration device depicted in Figure 7.8 is another example of a perpetual motion machine of the second kind.

7.5 *Reversibility and Irreversibility*

As discussed in Sections 7.3 and 7.4, the second law of thermodynamics states that no device can be developed that *continuously* converts all heat input to work output. However, this formal statement did not provide information concerning how much of the heat input could be converted to work. This issue is obviously of importance. It is particularly important to know what fraction of input heat could be converted to work for an ideal device. The answer to the question turns out to be a corollary of the second law. The device that gives the maximum output is an ideal device. An *ideal device* is one in which all processes for the device and for the surroundings are reversible. A thermodynamic process is said to be *reversible* if at any time during the process, both the system and the surroundings can be returned to their initial states without any evidence of any change remaining in the system or its surroundings. An ideal device is thus one that can undergo reversible processes. Conceptually, such an ideal device could be a piston that is moved very slowly by adding heat at an infinitesimal rate to a gas inside the piston cylinder through an infinitesimal temperature difference and in the absence of friction in the device. Such a process could be reversed by heat removal from the gas such that all elements of the system returned to the original state without any change in either the system or its surroundings. Other processes that are sometimes regarded as approximating reversible processes include elastic deformation of a solid, magnetization effects, electrical currents flowing in superconducting materials (which exhibit almost zero resistance to current flow), and various other phenomena involving extremely slow movements of mechanical devices in the absence of friction.

An *internally reversible process* is visualized as one in which the matter inside the system can be returned to its original state without any change from its original state, even though some change is left in the surroundings. Similarly, an *externally reversible process* is visualized as one in which the surroundings are returned to their original state but a change is left within the system. An example of an internally reversible process is the vaporization of water at constant pressure by the addition of heat while the temperature of the system (the boiling water) stays constant. Heat removal of the same magnitude will return the system to its original state. It is assumed that the phase changes for the system occurred at constant temperature and pressure regardless of whether the heat was added to the system (or heat was removed from the system) with a finite or an infinitesimal temperature difference between the system and its surroundings. Consequently, such a system phase change at constant temperature and pressure could be envisioned as an internally reversible process. In this case, the changes for the system have occurred in such a manner that the system can be returned to its original state without leaving any change within the system, even though changes could remain in the surroundings. When changes remain in the surroundings, the process is said to be *externally irreversible*. A process is said to be *completely reversible* if it can be interrupted at any stage, and means can be found to restore both the system and its surroundings to their respective initial states without leaving a trace in either the system or its surroundings.

If changes are left in either the system or the surroundings after the process occurs, the process is called an *irreversible process*. Irreversible effects arise whenever dissipation of any kind of energy occurs within the system or the surroundings or when heat transfer takes place across a nonzero temperature gradient or in the absence of any type of equilibrium during the process. Thus, irreversibilities are caused by phenomena such as friction, unrestrained expansion, inelastic deformation of solids, mixing of different fluids, viscous flow of fluids, or any finite departure from equilibrium during a process.

Real processes are irreversible because they do not occur reversibly for both the system and its surroundings. The necessary conditions for a reversible process are the same as those for the quasi-equilibrium process discussed previously in Chapter 5. For the completely reversible or quasi-equilibrium process, both a system and its surroundings must pass through a series of equilibrium states, and therefore the changes in state must occur very slowly. In addition, the interactions across the system boundary must occur with infinitesimally small differences between the relevant system properties and those for the surroundings. Devices are not built to operate at extremely slow rates, neither do the interactions across the system boundary occur with infinitesimally small differences between the pertinent system properties and those for the surroundings. However, an ideal device that operates with all processes reversible can be visualized to help determine the maximum possible efficiency that can be obtained when the device is operated either as a heat engine or as a reversed heat engine. The concept of reversibility is a powerful one that allows an upper limit for the efficiency of performance of heat engines and refrigerators to be established. Such a concept allowed Sadi Carnot to develop the Carnot cycle, which turned out to be an important foundation for the eventual formulation of the second law of thermodynamics.

7.6 Carnot Cycle

The *Carnot cycle* is a hypothetical cycle developed by Carnot for either a heat engine or a reversed heat engine. All the processes involved in the Carnot cycle are reversible, thereby providing the best possible device that one could construct. Results from the cycle analysis can be used to determine the maximum efficiency of performance possible for either a heat engine or a reversed heat engine.

The steam power plant shown in Figure 7.3 can serve as a model for the discussion of the important features of a typical *Carnot cycle heat engine*. The working fluid of the heat engine is initially a saturated liquid or wet vapor at state 1. It circulates around the cycle from point 1 to point 2 through a boiler (or evaporator), then from point 2 to point 3 through a turbine, from point 3 to point 4 through a condenser, and finally back to point 1 through a pump or compressor. The thermodynamic states for the working fluid of the Carnot heat engine cycle are shown in Figure 7.9. (For a steam or vapor cycle all the states 1 through 4 will be located under the dome in the *T-v* diagram.) There are four basic processes in the Carnot cycle, and each is completely reversible. The processes are listed as follows, with the numbers corresponding to the states in Figure 7.9 and the positions in the schematic diagram in Figure 7.3:

Process 1→2. A *reversible isothermal process* in which heat flows from a high temperature source to the working fluid, which is at a constant temperature that is only infinitesimally below that of the source.

Process 2→3. A *reversible adiabatic process* during which the working fluid expands through a turbine to produce a positive work output.

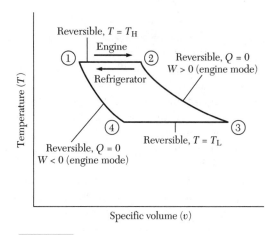

Figure 7.9 Thermodynamic states for the working fluid in a Carnot cycle. (Refer back to Fig. 7.3.)

Process 3→4. A *reversible isothermal process* in which heat flows to a low temperature sink from the working fluid (during its passage through the condenser). Both the working fluid and the sink are at a constant temperature, which are different only by an infinitesimal amount.

Process 4→1. A *reversible adiabatic process* during which the temperature of the working fluid is raised back to the temperature level of the high temperature source. Work is supplied to the fluid stream to compress it to the higher pressure at state 1.

A component diagram for a *reversed Carnot cycle* device is shown in Figure 7.10. This device is similar to the typical refrigeration cycle shown in Figure 7.5 except that the expansion valve in Figure 7.5 is replaced by the turbine and shaft in the Carnot device of Figure 7.10. Also, all processes in the Carnot device in Figure 7.10 are reversible. In this case, the device could serve either as a refrigerator or as a heat pump with heat removal from a low temperature reservoir (source) and heat rejection to a high temperature reservoir (sink). To operate the device, work input is required. All processes are reversible, and both of the heat transfer processes occur through an infinitesimal temperature difference. The thermodynamic state diagram in Figure 7.9 illustrates the reversed cycle for the device in Figure 7.10. In the case of the heat engine, the working fluid is flowing around the cycle from 1→2→3→4→1, while in the reversed Carnot cycle device, the working fluid flows in the order of 1→4→3→2→1. The basic processes for the reversed Carnot cycle device are as follows:

Process 1→4. *Reversible adiabatic expansion* of the working fluid during which the temperature of the working fluid decreases from T_H to T_L. Work is

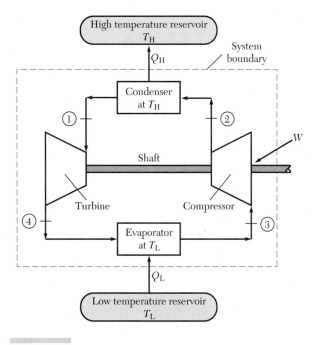

Figure 7.10 Reversed Carnot cycle device.

done by the turbine during the process, and this work is transmitted via the shaft to supplement the work input to the compressor.

Process 4→3. A *reversible isothermal process* during which heat is transferred through an infinitesimal temperature difference from the low temperature reservoir to the working fluid in the evaporator.

Process 3→2. *Reversible adiabatic compression* of the working fluid during which the temperature of the working fluid is raised from T_L to T_H. Work is supplied to the compressor from the turbine and from an external source for this compression process.

Process 2→1. A *reversible isothermal process* during which heat is rejected from the working fluid in the condenser to the high temperature reservoir at T_H.

Note that the temperature of the condenser is T_H and the temperature of the evaporator is T_L for the reversible heat transfer processes with infinitesimal temperature differences. Many different Carnot cycle devices can be visualized. All possess the necessary elements to produce the thermodynamic states shown in Figure 7.9 for the working fluid. Furthermore, all are reversible and are thus idealized processes. Such a Carnot cycle device provides the maximum possible thermal efficiency for a heat engine or the maximum possible coefficient of performance for a refrigeration device provided that the heat source and sink are both thermal reservoirs. Furthermore, the Carnot cycle efficiency is the thermal efficiency of any reversible heat engine operating between T_L

and T_H and has a value that is independent of the working fluid employed. The following section considers proofs of these statements.

7.7 Some Corollaries of the Second Law

As previously discussed, the Kelvin–Planck statement (established from experimental evidence and observation by inductive logic) of the second law is regarded here as the primary law. This section considers a number of corollaries that are logical deductions from the second law. Some of these corollaries are proved here by the *reductio ad absurdum* method of logic, which was outlined earlier. After a corollary is established, it assumes the same validity as the accepted statement of the second law, and any disproof of the corollary is in effect a disproof of the primary law.

Now let's look at some of the corollaries of the second law that are firmly established. These corollaries include those concerned with the characteristics of reversible heat engines and reversed heat engines operating between only two reservoirs, the thermodynamic scale of temperature, and the inequality of Clausius or Clausius inequality, which is especially relevant to the case of engines operating between a series of reservoirs. (The terms "inequality of Clausius" and "Clausius inequality" are used interchangeably in the literature, although use of the former has a longer history.) The inequality of Clausius is used in the next chapter to prove the existence of entropy as a derived property of a system based on the second law of thermodynamics.

7.7.1 Carnot Cycle Engines Operating Between Two Reservoirs

Corollary 1 *No heat engine can be constructed to operate between only two heat reservoirs with a higher efficiency than a reversible heat engine operating between the same two reservoirs.*

The proof of this corollary can be developed as follows. Assume the converse of this corollary to be true, and let I in Figure 7.11 be the irreversible engine operating between the two reservoirs while R is the reversible engine operating between the same two reservoirs. As shown in Figure 7.11(a), both engines receive Q_H from the high temperature reservoir. The work produced by the irreversible engine is W_I, while the work produced by the reversible engine is W_R. By the converse assumption, W_I is greater than W_R. The reversible engine can be operated as a reversed engine with only the Q_L, Q_H, and W_R reversed in direction. Now reverse engine R so that it receives Q_L from the low temperature reservoir and rejects Q_H at temperature T_H. This heat (Q_H) can be supplied from engine R to engine I as shown in Figure 7.11(b), thereby eliminating the need for the high temperature reservoir at T_H. Now the system shown in Figure 7.11(b) can produce net work because W_I is greater than W_R by the assumption made earlier. Thus, the system produces net work while exchanging heat with a single temperature reservoir. Such a system is a perpetual motion machine of the second kind and is a violation of the Kelvin–Planck statement of the second law of thermodynamics. Therefore, the initial converse assumption is not valid, and the corollary must be true.

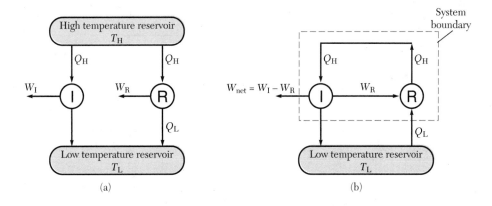

Figure 7.11 (a) Reversible (R) and irreversible (I) engines operating between two reservoirs. (b) Assumes $W_I > W_R$, leading to PMM2.

Another corollary relating to the Carnot cycle can be stated as follows:

Corollary 2 *All reversible engines operating between the same two reservoirs have the same efficiency.*

The proof of this corollary is similar to that of Corollary 1. Consider the two engines in Figure 7.11. Assume that engine I is also reversible so that two reversible heat engines R_1 and R_2 operate between the two reservoirs. Following the previous arguments shows that $\eta_{R_1} \leq \eta_{R_2}$ and that $\eta_{R_2} \leq \eta_{R_1}$. Therefore, it is concluded that $\eta_{R_1} = \eta_{R_2}$ unless the second law is violated. Thus, the corollary must be true.

7.7.2 Corollary Concerning Thermodynamic Temperature Scale

One of the most important results of Carnot's studies is the previous statement of Corollary 2 that *all reversible heat engines have the same thermal efficiency when operating between the same two reservoirs*. This corollary indicates that the thermal efficiency of a reversible heat engine does not depend on the nature of the heat engine or the working fluid. The thermal efficiency only depends on the temperatures of the two reservoirs. This fact allows one to establish an *absolute thermodynamic temperature scale* that is independent of the properties of the substance used for the measurement of temperature and is based only on the characteristics of reversible heat engines and constant temperature reservoirs. The corollary involving the thermodynamic temperature scale is stated as follows:

Corollary 3 *A temperature scale can be defined that is independent of the properties of the substance used for measurement of temperature.*

Proof of this corollary can be demonstrated as follows. Consider the reversible heat engines shown in Figure 7.12. The reversible heat engine R_1 receives heat Q_H from the high temperature reservoir at T_H and produces net work W_1 while rejecting heat Q_L to

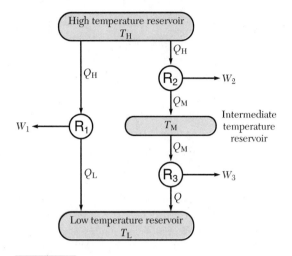

Figure 7.12 Reversible heat engines (R_1, R_2, and R_3) used to demonstrate the thermodynamic temperature scale.

the low temperature reservoir at T_L. The reversible heat engine R_2 also receives heat Q_H from the high temperature reservoir at T_H while producing net work W_2 and rejecting heat Q_M to the intermediate temperature reservoir at T_M. The reversible heat engine R_3 is sized so that it receives heat Q_M from the intermediate temperature reservoir at T_M and produces net work W_3 while rejecting heat Q to the low temperature reservoir at T_L.

The thermal efficiency of the reversible heat engine R_1 can be obtained from Equation 7.4 by first law analysis of the engine. Also, from Corollary 2, the efficiency of engine R_1 can be expressed as a function of the temperatures of the high and low temperature reservoirs. Thus, the thermal efficiency of engine R_1 is as follows:

$$(\eta_1)_{R_1} = 1 - \frac{Q_L}{Q_H} = f(T_L, T_H) \tag{7.10}$$

Similarly, for engines R_2 and R_3,

$$(\eta_1)_{R_2} = 1 - \frac{Q_M}{Q_H} = f(T_M, T_H) \tag{7.11}$$

and

$$(\eta_1)_{R_3} = 1 - \frac{Q}{Q_M} = f(T_L, T_M) \tag{7.12}$$

Engines R_2 and R_3 can be considered as equivalent to a single reversible engine operating between the high temperature reservoir T_H and the low temperature reservoir at T_L since Q_M (which is rejected from engine R_2) can be taken directly into engine R_3, thereby eliminating the intermediate temperature reservoir at T_M. Now the efficiency of this combined engine must be identical to that of engine R_1 based on Corollary 2. Therefore, work $W_1 = W_2 + W_3$, and a first law analysis of the combined system in comparison with engine R_1 leads to the conclusion that Q and Q_L are equal. Thus, Equation 7.12 can be expressed as

$$(\eta_1)_{R_3} = 1 - \frac{Q_L}{Q_M} = f(T_L, T_M) \tag{7.12a}$$

To simplify the functional expressions, the following functional relationships for the ratio of heat terms in Equations 7.10, 7.11, and 7.12a can be written:

$$\frac{Q_L}{Q_H} = \phi(T_L, T_H) \tag{7.13}$$

$$\frac{Q_M}{Q_H} = \phi(T_M, T_H) \tag{7.14}$$

$$\frac{Q_L}{Q_M} = \phi(T_L, T_M) \tag{7.15}$$

Now, since

$$\frac{Q_L}{Q_H} = \left(\frac{Q_M}{Q_H}\right)\left(\frac{Q_L}{Q_M}\right),$$

the functional relationships can be expressed as

$$\phi(T_L, T_H) = \phi(T_M, T_H)\,\phi(T_L, T_M) \tag{7.16}$$

Since the intermediate temperature T_M was arbitrarily chosen and does not appear on the left-hand side, it must also cancel on the right-hand side. To simplify the relationships, express the two functional relationships on the right-hand side as

$$\phi(T_M, T_H) = \frac{\psi(T_M)}{\psi(T_H)} \tag{7.17a}$$

and

$$\phi(T_L, T_M) = \frac{\psi(T_L)}{\psi(T_M)} \tag{7.17b}$$

Combining Equations 7.13, 7.16, 7.17a, and 7.17b gives

$$\frac{Q_L}{Q_H} = \phi(T_L, T_H) = \frac{\psi(T_M)}{\psi(T_H)}\frac{\psi(T_L)}{\psi(T_M)}$$

Therefore,

$$\frac{Q_L}{Q_H} = \frac{\psi(T_L)}{\psi(T_H)} \tag{7.18}$$

It has been found appropriate to select the ψ functions of T_L and T_H such that

$$\frac{Q_L}{Q_H} = \frac{T_L}{T_H} \tag{7.19}$$

One reason for the particular choice of a linear function is because temperature defined in this manner is fortuitously the same as the (absolute) ideal gas temperature.

Equation 7.19 is universally accepted as the definition of the *thermodynamic temperature scale*. It is also called the *absolute thermodynamic Kelvin temperature scale* after Lord Kelvin who first proposed it. In establishing the Kelvin temperature scale, the reference temperature T_i of a heat reservoir at the triple point of water was assigned the value 273.16 K. Note that the triple point for water is 0.01 K above the ice point. The ice point was defined in Chapter 3 as the melting point of water at a pressure of 1 atm. If a reversible heat engine is operated between a reservoir at the triple point of water and another reservoir at any temperature T, then T can be determined by

$$T = 273.16\frac{Q}{Q_i} \tag{7.20}$$

where Q_i is the (finite) heat exchanged with the heat reservoir at 273.16 K and Q is the heat exchanged with the reservoir at temperature T.

Note from Equation 7.20 that Q approaches zero as T approaches zero; thus, $Q = 0$ when $T = 0$. In other words, a heat engine operating between T_i and $T = 0$ would produce net work while rejecting zero heat to the low temperature reservoir at $T = 0$. Such a device is impossible according to the Kelvin–Planck statement. Therefore, the inescapable conclusion is that it is not possible to reach absolute zero temperature with any system or device.

As discussed in Chapter 2, increments of temperature for the Kelvin temperature were established by using 100 equal increments between the boiling point and ice point of water at 1 atm. This unit increment is the same as 1/273.16 of the triple point temperature of water on the Kelvin scale. While in practice, temperature measurements via the use of reversible heat engines are hardly feasible, it can be shown that the temperature scale thus developed leads to numerical values for temperatures that are identical with those measured by using a constant volume ideal gas thermometer. Other temperature measuring devices in use have empirically defined scales that closely match the results that would be obtained on the thermodynamic temperature scale.

7.7.3   Efficiency of Carnot Devices

Now that the thermodynamic temperature scale has been established, the efficiency for reversible heat engines and reversed heat engines can be expressed in terms of thermodynamic temperatures. Thus, Equations 7.4 and 7.19 can be combined to obtain the following expression for the efficiency of a reversible Carnot heat engine:

$$\eta_c = 1 - \frac{T_L}{T_H} \tag{7.21}$$

Similarly, combining Equations 7.19 with Equations 7.7 and 7.8 leads to the following expressions for the coefficients of performance of refrigeration or heat pump devices operating reversibly between T_L and T_H:

$$(COP_C)_{rev} = \frac{T_L}{T_H - T_L} \qquad (7.22a)$$

and

$$(COP_H)_{rev} = \frac{T_H}{T_H - T_L} \qquad (7.22b)$$

The subscript "rev" is added as a reminder that these coefficients of performance are for the reversible devices only. These expressions for the maximum possible efficiency of a heat engine and the maximum coefficients of performance of refrigeration or heat pump systems are now in terms of temperature only. Upper bounds can now be set to the best possible performance of such devices. Thus, the second law of thermodynamics leads to specific numerical results that allow one to define the best possible performance of a heat engine or refrigeration or heat pump device when the heat source and sink are thermal reservoirs.

EXERCISES

for Section 7.7.3

1. Given a heat source at 1100 K and a heat sink at 300 K, calculate the maximum possible thermal efficiency for a heat engine operating between the heat source and the sink.

   Answer: 72.7%

2. Repeat Exercise 1 for a heat source at 1100°F and a heat sink at 80°F.

   Answer: 65.4%

3. A heat pump is to be designed to operate between an outdoor temperature of 20°F and an indoor temperature of 75°F. What is the maximum efficiency COP_H possible?

   Answer: 9.73

4. Repeat Exercise 3 for source and sink temperatures of -20°C and 20°C, respectively.

   Answer: 7.33

7.7.4 Inequality of Clausius

In many real cycles, heat is received and rejected during processes that involve a continuous change in the temperature of the working fluid. Such a cycle can be considered reversible if the sources and sinks are assumed to consist of an infinite number of reservoirs that differ infinitesimally from one another in temperature. At any instant of time during the heating or cooling processes, heat is exchanged between the system and the source or sink with only an infinitesimal difference in temperature between the reservoir and the fluid in the system. This is equivalent to considering a heat engine that is operating between many different reservoirs, as shown in Figure 7.13(a). A corollary of the second law known as the *Inequality of Clausius* (or Clausius inequality) deals with such a heat engine operating between many different reservoirs. This corollary can be stated as follows:

Corollary 4 *Whenever a system undergoes a complete cycle, the cyclic integral of $\delta Q/T$ around the cycle is less than or equal to zero.*

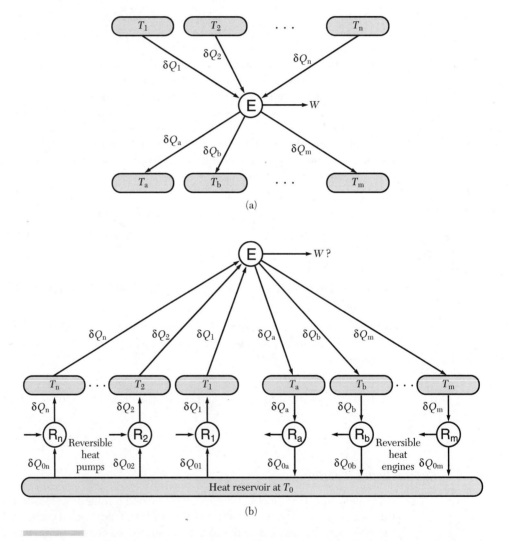

(a)

(b)

Figure 7.13 (a) Heat engine (E) cycle with several constant temperature sources and sinks. (b) An equivalent operation to the cyclic engine in part (a) such that heat is exchanged with one heat reservoir only.

This corollary is a direct consequence of the second law and can be proved as follows. Consider the heat engine operating with many different reservoirs, as shown in Figure 7.13(a). This heat engine can be either reversible or irreversible. Heat flows of magnitudes δQ_1, δQ_2, δQ_3, ... δQ_n occur to the engine (the system) from sources at temperatures T_1, T_2, T_3 ... T_n, while heat rejections of magnitudes δQ_a, δQ_b, δQ_c, ... δQ_m occur from the system to sinks at temperatures T_a, T_b, T_c, ... T_m. The system undergoes complete cycles of operation. In Figure 7.13(b), the source reservoirs have been replaced by a series of reversible heat pumps that receive δQ_{01}, δQ_{02}, δQ_{03}, ... δQ_{0n} from a single heat reservoir at T_0 while delivering the heat flows δQ_1, δQ_2, δQ_3, ... δQ_n to the heat engine as before in Figure 7.13(a). Likewise, the sink reservoirs of Figure 7.13(a) are replaced in Figure 7.13(b) by a series of reversible heat engines that receive δQ_a, δQ_b, δQ_c, ... δQ_m while rejecting δQ_{0a}, δQ_{0b}, δQ_{0c}, ... δQ_{0m} to the single heat reservoir at T_0. The temperature T_0 can be arbitrarily chosen as long as it is lower than that for any of the reservoirs in Figure 7.13(a).

Now consider the reversible heat pumps in Figure 7.13(b). In accordance with Equation 7.19, the relationship between each of the δQ_1, δQ_2, δQ_3, ... δQ_n from the reversible heat pumps and the corresponding δQ_{01}, δQ_{02}, δQ_{03}, ... δQ_{0n} from the single heat reservoir is as follows:

$$\frac{\delta Q_{0i}}{\delta Q_i} = \frac{T_0}{T_i} \tag{7.23}$$

Similarly for the reversible heat engines, the heat flows δQ_a, δQ_b, δQ_c, ... δQ_m and δQ_{0a}, δQ_{0b}, δQ_{0c}, ... δQ_{0m} are in the following manner:

$$\frac{\delta Q_{0j}}{\delta Q_j} = \frac{T_0}{T_j} \tag{7.24}$$

The j in Equation 7.24 denotes the a, b, c, ... m in the expressions for heat flow.

Now consider the original engine together with all the reversible engines and heat pumps in Figure 7.13(b) as a single complex system. This complex system undergoes a cycle and returns to its initial state after one cycle. The net heat received by this system during the cycle is given by

$$Q_0 = \sum_{i=1}^{n} \delta Q_{0i} - \sum_{\substack{\text{all } j \\ \text{a, b,...} m}} \delta Q_{0j} \tag{7.25}$$

Consideration of the first law of thermodynamics for the cycle shows that

$$W = \sum \delta W = \sum \delta Q = Q_0 \tag{7.26}$$

Since the system is undergoing a cycle and exchanging heat with only a single temperature reservoir, the Kelvin–Planck statement of the second law shows that the summation of all the work for the cycle must be less than or equal to zero. That is, the Kelvin–Planck statement asserts that positive work output cannot occur from the device while exchanging heat with a single reservoir. Thus, the work output for this device must be less than or equal to zero. Therefore, from Equation 7.26, Q_0 must be less than or equal to zero, and we can rewrite Equation 7.25 as

$$\sum_{i=1}^{n} \delta Q_{0i} - \sum_{\substack{\text{all } j \\ \text{a, b,...}m}} \delta Q_{0j} \leq 0 \tag{7.27}$$

Combining Equations 7.23, 7.24, and 7.27 gives

$$\sum_{i=1}^{n} \frac{\delta Q_i}{T_i} T_0 - \sum_{\substack{\text{all } j \\ \text{a, b,...}m}} \frac{\delta Q_j}{T_j} T_0 \leq 0 \tag{7.28}$$

Dividing through by the constant temperature T_0 and reverting to the original sign convention whereby Q is positive for input heat and negative for reject heat, the following general expression is obtained for the cycle:

$$\sum_{\text{all } i} \frac{\delta Q_i}{T_i} \leq 0 \tag{7.29}$$

Recall that T_i in Equation 7.29 is the temperature of the reservoir. Referring now to the original system, the system boundary can be chosen such that T_i is also the temperature at the patch of the boundary through which δQ_i flows into the system. Thus, if all such transfers across the system boundary are considered, Equation 7.29 can be written as follows:

$$\oint \frac{\delta Q}{T_{\text{boundary}}} \leq 0 \tag{7.30}$$

where δQ is the heat flow to the system through a patch of the system boundary at temperature T_{boundary}. This equation is called the inequality of Clausius. The inequality was developed here directly from the Kelvin–Planck statement of the second law. The relationship applies to systems operating in a cycle regardless of whether the processes are reversible or irreversible. Note that when the processes are irreversible, no unique temperature exists for the system, although temperatures at various portions of the system boundary will be well defined. Hence, Equation 7.30 is equally applicable to cycles executed in a reversible or an irreversible manner.

Now consider the application of the inequality of Clausius to a reversible heat engine cycle. If the heat engine of Figure 7.13(b) is reversed, all the Q_i's will change sign, giving the following inequality:

$$\sum_{\text{all } i} \frac{-\delta Q_i}{T_i} \leq 0 \quad \text{or} \quad \sum_{\text{all } i} \frac{\delta Q_i}{T_i} \geq 0 \tag{7.31}$$

Thus, for a reversible cycle, $\Sigma_i \, \delta Q_i/T_i$ must be less than or equal to zero according to Equation (7.29), while $\Sigma_i \, \delta Q_i/T_i$ must be greater than or equal to zero according to Equation 7.31. The only way both of these expressions can be true is for

$$\sum_{\text{all } i} \frac{\delta Q_i}{T_i} = 0 \quad \text{for reversible cycle} \tag{7.32}$$

or in more general terms,

$$\oint \frac{\delta Q}{T} = 0 \quad \text{for reversible cycle} \tag{7.33}$$

Thus, the inequality of Clausius is given by the inequality of Equation 7.30 for any cycle, whether reversible or irreversible, while Equation 7.33 applies only to the reversible cycle. For the reversible case, T is the temperature of the system that must be uniform at each instant and is the same as the system boundary temperature. Equations 7.32 and 7.33 are at times referred to as the *equality of Clausius*. These results are used in Chapter 8 to prove the existence of a property called *entropy* as a consequence of the second law of thermodynamics. Entropy is an important property in thermodynamics, and it is most frequently used in analyses requiring application of the second law of thermodynamics.

EXAMPLE 7.1
Application of Inequality of Clausius to a System with More than Two Reservoirs

Given: A heat engine receives half of its heat at a temperature of 1000 K and the rest at 500 K while rejecting heat to a sink at 300 K. What is the maximum possible thermal efficiency of this heat engine?

Find: η_1

Solution: Figure 7.14 is a schematic of the system just described. The maximum possible efficiency is for a reversible heat engine operating as shown. Therefore, Equation 7.32 applies to this situation. For the reversible heat transfer processes,

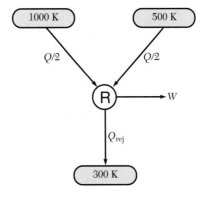

Figure 7.14 Schematic diagram of a reversible heat engine (R) that operates with more than two heat reservoirs.

$$\sum_{i=1}^{3} \frac{\delta Q_i}{T_i} = \frac{Q/2}{1000 \text{ K}} + \frac{Q/2}{500 \text{ K}} - \frac{Q_{\text{rej}}}{300 \text{ K}} = 0$$

Solving for Q_{rej} gives $Q_{\text{rej}} = 0.45Q$. Now the thermal efficiency can be calculated using Equation 7.1.

$$\eta_1 = \frac{W_{\text{output}}}{Q_{\text{input}}} = \frac{Q_{\text{input}} - Q_{\text{rej}}}{Q_{\text{input}}}$$

$$= \frac{Q/2 + Q/2 - 0.45Q}{Q} = 0.55 \text{ or } 55\%$$

Note that the inequality of Clausius provides a test for the reversibility of a cycle. If $\Sigma_i \, \delta Q_i/T_i$ is equal to zero, the cycle is reversible, while if $\Sigma_i \, \delta Q_i/T_i$ is less than zero, it is irreversible. If $\Sigma_i \, \delta Q_i/T_i$ is greater than zero, the cycle is impossible because it violates the Kelvin–Planck statement of the second law of thermodynamics.

EXERCISES

for Section 7.7.4

1. It is claimed that a heat pump that delivers 2 Btu/s to room air at 75°F when the outdoor air is at 20°F requires $\frac{1}{4}$ hp of electrical power supply. Use the inequality of Clausius to determine whether or not this claim can be true.

2. A reversible cyclic device operates in the manner indicated in the accompanying figure. Show that at steady-state, $\dot{Q}_1 = \frac{4}{7}\dot{Q}$ and $\dot{Q}_2 = \frac{3}{7}\dot{Q}$.

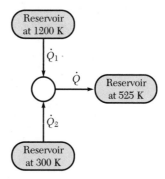

7.8 Summary

This chapter considered the second law of thermodynamics. It reviewed the methods of logic that led to the development of the second law and indicated that the second law of thermodynamics was formulated through inductive logic by observation of experimental evidence. Although the law cannot be proven, it is acceptable until it is disproved. The chapter also used deductive logic to state and illustrate several corollaries or consequences of the second law.

The primary statement of the second law used in this book is the Kelvin–Planck statement. This statement is based on the observation that work is more valuable than heat and that heat cannot be converted continuously 100% to work with any system or device. Although the first law of thermodynamics shows that heat and work are equivalent means of transferring energy, the second law of thermodynamics further demonstrates that a dissymmetry exists in nature—that work is more valuable than heat.

The chapter considered heat engines and refrigeration or heat pump devices and a number of logical consequences of the second law. The best possible performance is obtained with a system that operates with all processes reversible. The Carnot heat engine is one example of the ideal heat engine operating between two thermal reservoirs with all processes reversible. All reversible heat engines operating between constant temperature reservoirs at T_H and T_L have the same thermal efficiency as the Carnot cycle efficiency. After establishing the thermodynamic temperature scale, it becomes possible to compute the maximum possible efficiency of a Carnot heat engine or of a reversed Carnot heat engine from knowledge of thermodynamic temperature levels alone. The inequality of Clausius helps deal with systems that exchange heat with more than two reservoirs. The second law and its corollaries thus enable computation of the maximum possible efficiency for the conversion of heat into work by a heat engine, and the analysis of cycles, which may involve both heat and work. Chapter 8 considers another corollary that will lead to the property called entropy, which allows quantification of second law analysis for processes and systems. This property will provide information for analyzing the behavior of thermal systems in more detail.

References

Carnot, S. 1824. *Reflections on the Motive Power of Fire*. Revised edition edited by Mendoza, E. 1960. New York: Dover Publications.

Shapiro, G. 1979. *Physics Without Math: A Descriptive Introduction*. Englewood Cliffs, NJ: Prentice-Hall, Inc.

Questions

1. Figures Q7.1(a) and (b) on page 276 illustrate the popular toy known as the "dipping" or "drinking" bird (see Shapiro 1979, for example, for a description of how the toy works). Should the toy be regarded as a heat engine or as a perpetual motion machine? Give reasons for your answer.

2. Suggest a few examples in which a limited quantity of work can be produced by a device that interacts with only one thermal reservoir. Do these contradict the second law of thermodynamics?

3. Can you think of any two naturally occurring thermal reservoirs at different temperatures in a given location on Earth? What are the typical temperatures encountered, and what is the maximum possible thermal efficiency for a heat engine operating between the reservoirs?

4. In Figure Q7.4 on page 276, two blocks of metals, one hot and the other cold, are separated using an adiabatic wall. Subsequently, the adiabatic wall is removed and the blocks are placed in direct contact. What happens to the temperatures of the metal blocks? Is the ensuing process reversible or irreversible? Why?

5. What is a thermal reservoir? Give examples of systems in nature that may be regarded as approximations to the thermal reservoir.

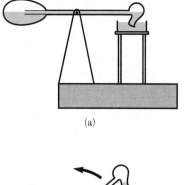

(a)

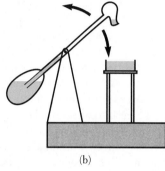

(b)

Figure Q7.1 Dipping or drinking bird in operation.

6. Distinguish carefully between reversible and irreversible processes. Give two examples of real processes that can reasonably be regarded as close to reversible processes.

7. For a heat engine designed to operate on a space station, the "outer space" (deep space) at a temperature approaching absolute zero could conceiv-

ably serve as heat sink. Suggest reasons why such an idea might not work in practice.

8. The domestic refrigerator is used for the initial cooling of foods and for maintaining the foods at a low temperature. How would you categorize the foods as a heat source—as a thermal reservoir or a finite heat source? What about the heat sink? Explain your answer in each case.

9. External combustion heat engines typically use a high temperature stream of combustion products as a heat source. Heat extraction from the stream of hot gases results in a lowering of the temperature of the heat source. What kind of heat source is this? A thermal reservoir or a finite heat source? Does the Carnot cycle correctly express the upper limit of thermal efficiency in this situation? Why?

10. Heat is supplied to a system of large thermal ca-

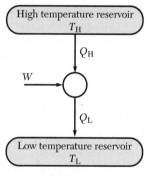

Figure Q7.10 Use of work to increase heat transfer to a low temperature reservoir.

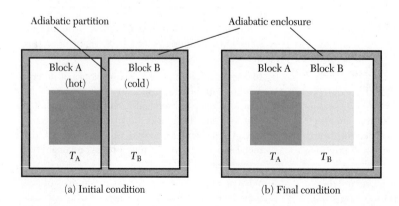

Figure Q7.4 A hot metal block placed in contact with a cold metal block.

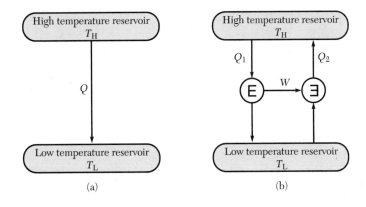

Figure Q7.11 (a) Direct heat transfer between two thermal reservoirs, (b) Proposed scheme for restoration of the two reservoirs to their original conditions.

pacity that remains at low temperature T_L. The cyclic device for this process is illustrated in Figure Q7.10. Heat Q_H is drawn from the ambient air at T_H ($> T_L$), and in addition, work W is supplied to the cyclic device. If both Q_L and W are positive in the directions indicated, show that the process must be irreversible.

11. Figure Q7.11(a) illustrates heat transfer Q from a thermal reservoir at T_H to a second thermal reservoir at T_L where $T_H > T_L$. The use of reversible

devices is proposed, as indicated in Figure Q7.11(b), for restoring both thermal reservoirs to their original conditions. Establish the relationship between Q_1, Q_2, and Q by applying the first law of thermodynamics to the process that is intended to restore the thermal reservoirs to their original states. Use the second law of thermodynamics to show that the proposed scheme cannot work if $T_H - T_L$ is finite. Was the original process reversible or irreversible?

Problems

7.1. Define the thermal (or first law) efficiency of a heat engine. Determine the first law efficiency of the following devices or systems. (Assume that coal has a heating value of 33,500 kJ/kg.)

(a) A set of Newcomen steam engines produces 2.08 kWh of work while consuming 37.5 kg of coal.

(b) A coal-fired power station generates 2000 MW of power. Coal is burnt at a rate of 8.1×10^5 kg/h.

(c) An oil engine for a submarine develops 1200 kW. Its rate of fuel consumption is 5 kg/minute. The heating value of the fuel can be taken as 43,500 kJ/kg.

7.2. Define the first law efficiencies COP_C and COP_H for a reversed heat engine. Calculate COP_C and COP_H for the following:

(a) $\dot{Q}_c = 350$ W, $\dot{W} = 110$ W

(b) $\dot{Q}_c = 240$ W, $\dot{E}_{in} = 300$ W

(c) $\dot{Q}_H = 14$ kW, $\dot{W} = 4$ kW

(d) $\dot{Q}_c = 1.5$ kW, $\dot{Q}_H = 3$ kW

7.3. Show that for a reversed heat engine, $COP_H = COP_C + 1$.

7.4. During a heat engine cycle, 1000 Btu of heat is added to the system (the engine) while 700 Btu of heat is rejected by the system to the surroundings.

(a) What is the work done (in ft · lbf) by the heat engine?

(b) What is the first law efficiency for the heat engine cycle?

7.5. A heat engine operating in a cycle produces

250 kJ of work while rejecting 600 kJ of heat to the surroundings.
(a) What is the heat added (in kJ) to the engine during the cycle?
(b) What is the first law (or thermal) efficiency for the heat engine cycle?

7.6. A heat pump operates in a cycle to heat a building, as shown in Figure P7.6. A quantity of 36,000 Btu is supplied as heat to the building, while 24,000 Btu of heat is input to the heat pump from the surroundings.
(a) What is the work input (in ft · lbf) to the heat pump during this time period?
(b) What is the coefficient of performance for heating (COP_H) for the heat pump?
(c) What is the first law efficiency for the heat pump?

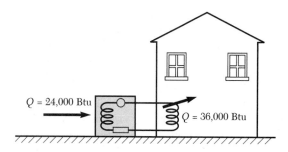

Figure P7.6 Heating of a building by a heat pump.

7.7. How much refrigeration (in kWh) is required to freeze 1 tonne (= 1000 kg) of fish? The fish enters the plant at 25°C and is to be taken down to −20°C. Assume the freezing point of fish is −2.2°C and that the specific heat of fusion is 235 kJ/kg. The specific heat capacities for the fish at temperatures above and below the freezing point are 3.2 and 1.7 kJ/kg · K, respectively. If a refrigeration device with a COP_C equal to 1.5 were used, calculate the minimum electrical energy input (in kWh) to the compressor of the device.

7.8. An engine working on the Carnot cycle has maximum and minimum temperatures of 1227°C and 327°C. Determine its efficiency. Also determine the heat transfer (in kJ/h) from the high temperature medium when the power output of the engine is 25 kW.

7.9. A reversible Carnot heat engine operates between 1000 K and 300 K.
(a) What is the thermal efficiency of the engine?
(b) If 500 kJ of heat is added to the engine at 1000 K, what are the work output and the heat rejected?

7.10. A reversible Carnot heat engine operates between 1200°F and 100°F.
(a) What is the thermal efficiency of the engine?
(b) If 1600 Btu of heat is rejected by the engine, what are the work output (in ft · lbf) and the heat input (in Btu)?

7.11. A reversible heat engine receives 900 Btu of heat from a high temperature reservoir while rejecting 300 Btu to a low temperature reservoir, as shown in Figure P7.11.
(a) What is the thermal efficiency of this heat engine?
(b) How does the efficiency of this engine compare with that of the Carnot heat engine?
(c) What is the work output (in ft · lbf)?
(d) If the low temperature reservoir is at 140°F, what is the temperature of the high temperature reservoir (in °F)?

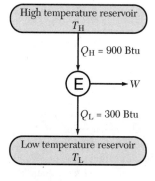

Figure P7.11 Operation of a reversible heat engine.

7.12. A refrigerator works reversibly between the temperatures of −23°C and 37°C. Determine its COP_C. Determine the cooling load (in kW) if the work input is at the rate of 5 kW.

* **7.13.** A reversible Carnot device operates as a heat pump. It absorbs heat from the surroundings at

0°F and rejects heat to the inside of a building at 80°F.

(a) What is the coefficient of performance for heating (COP_H) for this device?

(b) If the work input to the device is 10 hp, what is the rate of heat supplied to the building (in Btu/h)?

(c) Plot COP_H for values of the temperature of the surroundings from 0°F to -100°F.

7.14. The reactor of a nuclear power generating station consumes 0.05 tonnes per day of uranium-235 while the net power output is 150 MW. The thermal energy yielded when 1 tonne of uranium fuel is burned is approximately 12,000 megawatt/day. Calculate the first law efficiency achieved.

7.15. In thermal power plants, metallurgical considerations generally dictate the maximum allowable cycle temperature. For many conventional plants, this limit is about 900 K. If the sink is at 300 K, what is the maximum first law (or thermal) efficiency attainable?

7.16. A boiler furnace at a temperature of 1500°C is the heat source for a power plant that rejects heat to a sink at 30°C. What is the absolute maximum thermal efficiency that can be achieved? If the power cycle temperature is not to exceed 800 K, what is the maximum first law efficiency?

7.17. The average human has a maximum steady power output of about 70 W. If he or she expends muscular energy at this rate for 4 h on food intake having an energy value of 2000 food Cal (or kcal), calculate the first law efficiency achieved.

7.18. Power is to be generated from a geothermal heat source (hot springs) providing energy at a temperature of 90°C. The lowest temperature for heat rejection is 27°C. What is the maximum possible thermal efficiency?

7.19. An inventor claims to have designed a heat engine having an efficiency of 38% when using the exhaust gas from an engine, at a temperature of 145°C, as the heat source and atmospheric air at 27°C as the sink. Examine the validity of the claim.

* 7.20. Two reversible heat engines operate in series between a source at 800 K and a sink at 300 K. If the engines have equal efficiencies and the heat transfer from the first engine to the second is 400 kJ (Fig. P7.20), calculate:

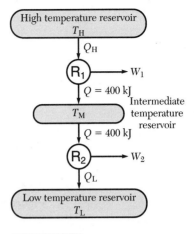

Figure P7.20 Operation of two reversible heat engines in series.

(a) the temperature at which heat transfer occurs to the second engine

(b) the heat transfer from the source at 800 K to the first heat engine

(c) the work done by each engine

7.21. An inventor says that he has invented an engine that operates by taking heat from the air and rejecting heat to the water in the ocean.

(a) Assuming the air is at 70°F and the water is at 60°F, what is the maximum efficiency of his engine?

(b) What is your assessment of the potential value of such an engine for use on ships?

* 7.22. An inventor has developed a combined heat pump and heat engine system (illustrated in Fig. P7.22) that operates as follows. The heat pump absorbs heat from the surroundings at 50°F and provides the heat output from the heat pump as heat input to a heat engine at 150°F. The heat engine then operates to produce work while rejecting heat to the surroundings at 50°F. What is the maximum first law efficiency of this combined system?

* 7.23. A Carnot heat engine operates between 800 K and 400 K. An engineer proposes to increase the thermal efficiency of the engine by raising the high temperature source to 1000 K. However, the engineer's supervisor is concerned about metallurgical limits of the materials and suggests that

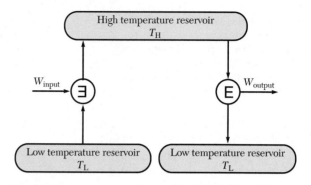

Figure P7.22 Combined heat pump and heat engine system.

she look at increasing the efficiency by reducing the temperature of the heat sink. To what temperature must the sink temperature be reduced to produce the same thermal efficiency as that produced by raising the source temperature to 1000 K?

* **7.24.** An inventor proposes a heat engine that operates in a cycle. The working fluid (system) receives 1300 kJ/kg at a system boundary temperature of 311°C while rejecting 220 kJ/kg at 100°C to the surroundings. If no other heat interactions occur during a cycle, determine whether or not the cycle satisfies the inequality of Clausius.

* **7.25.** A proposed heat engine has the following heat interactions: process I, 500 Btu input at 120°F; process II, 800 Btu input at 180°F; process III, 1200 Btu input at 220°F; process IV, 1000 Btu input at 700°F; and process V, 2500 Btu rejected at 80°F. Does this engine satisfy the inequality of Clausius?

* **7.26.** A heat engine (shown in Fig. P7.26) receives one-third of its heat at 700 K, one-third at 800 K, and one-third at 900 K, while rejecting heat to a sink at 400 K. What is the maximum possible thermal efficiency of this heat engine?

* **7.27.** A heat engine receives its heat as follows: 25% of heat input at 1000 K, 25% of heat input at 900 K, 25% of heat input at 800 K, and the remaining 25% of heat input at 700 K. One-half of the rejected heat from the engine is at 500 K, while the other half is at 300 K. What is the maximum possible thermal efficiency of this heat engine?

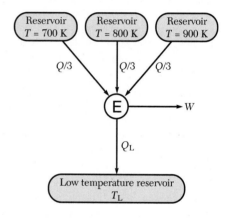

Figure P7.26 Operation of a heat engine with different temperatures of sources.

* **7.28.** A heat engine receives heat input as follows: 600 Btu at 1400 K, 500 Btu at 1200 K, 400 Btu at 1000 K, 200 Btu at 800 K, and 100 Btu at 600 K. It rejects heat to the surroundings at 300 K. What is the maximum possible thermal efficiency of this heat engine?

** **7.29.** A Carnot cycle heat engine is operating on a space station. Heat (Q_H) is supplied to the engine from a constant temperature reservoir at T_H, while heat (Q_L) is rejected from the engine by a space radiator, as shown in Figure P7.29. The space radiator rejects heat by thermal radiation to deep space,

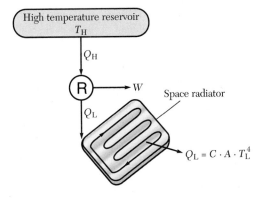

Figure P7.29 Heat rejection by thermal radiation from a Carnot heat engine operating on a space station.

and this rejected heat is proportional to the area of the space radiator (A) and to temperature to the fourth power. Thus, $Q_L = C \cdot A \cdot T_L^4$, where C is a constant. If the work of the engine remains constant while Q_H is supplied at a constant value of T_H, then Q_L must be fixed. Also, an optimum value for T_L exists to minimize the area of the space radiator. Find this optimum value for T_L expressed in terms of T_H.

∗ ∗ 7.30. A perfect gas is used as the working fluid in a heat engine. The engine cycle is illustrated in Figure P7.30 on T-p coordinates. All the processes are steady-flow processes. The heat input to the working fluid occurs at a constant pressure, while a continuous change occurs in the temperature of the working fluid from T_2 to T_3. The heat rejected from the working fluid also occurs at a constant pressure, while a continuous change occurs in the temperature of the working fluid from T_4 to T_1. No other heat interactions occur during the cycle. Work is done by the fluid in a steady-flow process between states 3 and 4, and work is done on the gas between states 1 and 2. Both of these processes are reversible and adiabatic. The instantaneous heat transfer rate to (or from) the gas that changes the gas temperature by an amount dT during the constant pressure processes can be determined from $\delta Q / dt = \dot{m} c_p \, dT$.

  (a) Assuming that $T_4 = T_2$, use the inequality of Clausius to develop an expression for the maximum possible thermal efficiency of this

heat engine in terms of only the maximum temperature T_3 and the minimum temperature T_1 of the cycle. Assume total reversibility for the processes.

  (b) Compare your result from part (a) with the thermal efficiency of a Carnot heat engine operating between a high temperature reservoir at T_3 and a low temperature reservoir at T_1. Make this comparison by plotting the thermal efficiency as a function of the ratio T_3 / T_1 for the engine described in part (a) and for the Carnot engine over a range of values for T_3 / T_1 from 1 to 5.

  (c) Is there a difference in cycle thermal efficiency for the case when all the heat supplied is at a constant temperature and all the heat rejected is at a constant temperature versus the case when heat is supplied and rejected at continuously changing temperatures? A heat engine cycle similar to this is analyzed in Chapter 10. The cycle is called a *Brayton cycle* (but note that T_4 does not necessarily equal T_2 for the Brayton cycle).

∗ ∗ 7.31. Assume that the Clausius statement of the second law of thermodynamics is the primary statement and show that the Kelvin–Planck statement is a corollary of the Clausius statement. Be very precise with your logic, your assumptions, and your statements.

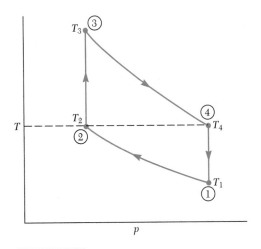

Figure P7.30 The engine cycle for Problem P7.30 shown on T-p coordinates.

****7.32** Beginning with the Kelvin–Planck statement of the second law, prove that it is not possible to reduce the temperature of a system to absolute zero.

****7.33.** Beginning with the Kelvin–Planck statement of the second law, prove that the efficiency of any reversible engine operating between two thermal reservoirs is independent of the working fluid in the engine and depends only on the temperature of the reservoirs.

Entropy

8.1 Introduction

Chapter 7 considered the second law of thermodynamics for systems that undergo cyclic
processes. However, most thermodynamic applications involve noncyclic processes. The
results from the last chapter will be used in this chapter to establish, as a consequence
of the second law of thermodynamics, the existence of an extensive property called
entropy. Entropy is a *derived concept* in the same way that internal energy for a system
is a derived concept. The existence of entropy turns out to be especially useful in the
analysis of noncyclic processes. Section 8.2 presents a formal proof of the existence of
entropy as a derived property based on the second law of thermodynamics. Thereafter,
various applications involving the use of the concept are discussed.

8.2   Entropy as a Property

To demonstrate the existence of entropy as a property, let's begin with a corollary that follows from the second law of thermodynamics:

Corollary 5—*A property of a closed system exists such that a change in its value for any reversible process undergone by the system between state 1 and state 2 is equal to $\int \delta Q/T$.*

This property can be illustrated as follows. Consider Figure 8.1, in which three reversible process paths are shown between states 1 and 2 on the T-v diagram. Note that the T-v diagram is chosen arbitrarily and is only used to show the hypothetical equilibrium states 1 and 2 and the thermodynamic paths for the reversible processes. Consider a thermodynamic cycle that begins at state 1 and proceeds along path I to state 2 and then returns from state 2 to state 1 along path III. Both paths I and III are for *reversible processes*. The two reversible processes constitute a *reversible cycle*, so the equality case in the general inequality of Clausius principle can be applied to this cycle to obtain

$$\oint \frac{\delta Q}{T} = 0 = \int_{\text{state } 1}^{\text{state } 2} \left(\frac{\delta Q}{T}\right)_{\text{I,rev}} + \int_{\text{state } 2}^{\text{state } 1} \left(\frac{\delta Q}{T}\right)_{\text{III,rev}} \tag{8.1}$$

Note that this follows the normal sign convention for δQ; in other words, heat into the system is positive and heat out of the system is negative. Also, since the processes are reversible, T is the temperature of the system and T is uniform throughout the system at each instant. Now consider another reversible cycle that begins at state 1 and follows path II to state 2 and then follows path III back to state 1. The Clausius equality can be again applied to this cycle to obtain

$$\oint \frac{\delta Q}{T} = 0 = \int_{\text{state } 1}^{\text{state } 2} \left(\frac{\delta Q}{T}\right)_{\text{II,rev}} + \int_{\text{state } 2}^{\text{state } 1} \left(\frac{\delta Q}{T}\right)_{\text{III,rev}} \tag{8.2}$$

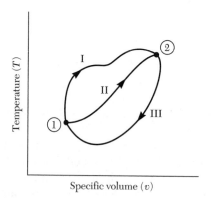

Figure 8.1 Thermodynamic processes used to demonstrate that entropy is a property.

Subtracting the second equation from the first gives

$$\int_{\text{state }1}^{\text{state }2} \left(\frac{\delta Q}{T}\right)_{\text{I,rev}} = \int_{\text{state }1}^{\text{state }2} \left(\frac{\delta Q}{T}\right)_{\text{II,rev}} \tag{8.3}$$

In both of these cases, the special case for the inequality of Clausius where the cyclic integral $\oint \delta Q/T$ equals zero applies because these are totally reversible cycles. Since paths I and II were chosen arbitrarily, one can conclude that the integral of $\delta Q/T$ for a reversible process is a function whose value is determined only by the end states and not by the path. A thermodynamic quantity whose value is determined only by the end states was identified earlier as a thermodynamic property. Thus, the integral of $\delta Q/T$ for a reversible process defines a change in a thermodynamic property. This property is called *entropy* and is normally designated by S.

Since entropy is a thermodynamic property, a small change in entropy is designated by the exact differential dS. This quantity can be expressed as follows:

$$dS = \left(\frac{\delta Q}{T}\right)_{\text{rev}} \tag{8.4}$$

The change in entropy for a system as it undergoes a change of state from 1 to 2 can be obtained by integrating Equation 8.4 along any thermodynamic path for a reversible process to yield

$$S_2 - S_1 = \int_{\text{state }1}^{\text{state }2} \left(\frac{\delta Q}{T}\right)_{\text{rev}} \tag{8.5}$$

To integrate $\delta Q/T$, the mathematical relationship between δQ and T must be known for the reversible process chosen. However, the major conclusion at this time is that *entropy is a property.*

Although Equation 8.5 was developed for a reversible process, the change in entropy between state 1 and state 2 must be the same whether the process is reversible or irreversible, in view of the fact that entropy is a property. The previous definition of the incremental change of entropy (dS) for a system in terms of $\delta Q/T$ for a reversible process, however, provides a theoretical basis for the computation of changes of the property between specified equilibrium states. For an *irreversible process*, the mathematical relationship between δQ and T, if any, does not provide any basis to determine the property change except that the amount of change of the entropy (S) must be greater than $\int \delta Q/T$ as will be shown later in Section 8.8.

The preceding discussion has proved that entropy is a direct consequence of the second law of thermodynamics. Chapter 6 showed that internal energy is a direct consequence of the first law of thermodynamics. As for internal energy, one is normally concerned only with *changes* of entropy. The entropy at any arbitrary reference state can be made equal to zero, and then the entropy for any other state can be found by evaluating the integral of $\delta Q/T$ for any reversible process from the reference state to this other state. Since real processes are irreversible, the values for entropy cannot be found from direct measurements of Q and T for these processes. Instead, entropy is expressed as a function of other thermodynamic properties that can be measured. Section 8.4 considers such relationships later. Meanwhile, the similarity and the major differences between the properties of internal energy and entropy are summarized here in relation to the respective laws of thermodynamics from which they derive.

The first law of thermodynamics led to a definition of internal energy as a property, while the second law led to the definition of entropy as a property. Chapter 6 showed that one could evaluate the change in internal energy from a knowledge of work and heat for any process joining the two states, without regard to whether the process was reversible or irreversible. On the other hand, a change in entropy can be evaluated from a knowledge of heat and temperature only for a reversible process joining the two states. Entropy is a property, and its change between two states is independent of the process joining the two states. The change in entropy, however, is equal to the integral of $\delta Q/T$ only for a reversible process. (Section 8.8 considers the value of the integral of $\delta Q/T$ for an irreversible process.)

8.3   Entropy and the Third Law of Thermodynamics

The third law of thermodynamics can be stated as follows:

The entropy of a pure substance in thermodynamic equilibrium approaches zero as the temperature of the substance approaches absolute zero.

Since the third law sets a value for entropy that approaches zero as the temperature approaches absolute zero, it would be logical to use this value as the reference state in property tables. However, that is not the precedent that has been established. For example, the entropy in most thermodynamic steam tables is referenced to the triple point of water, with a value of zero for saturated liquid at the triple point of water. The reference state for the entropy of most gases that are used in combustion processes is the *standard reference state* of 1 atm and 25°C (77°F). In this case, however, the value of entropy assigned is such that the entropy for the substance is zero at the absolute zero temperature. Other substances use different reference points for a zero value of entropy. In general, such inconsistent approaches do not create any difficulty since we are normally interested only in the changes of the entropy for a particular substance rather than the absolute value. Thus, the arbitrary reference point leads to no numerical differences for calculations between any two specified states.

Although an absolute zero temperature cannot be achieved in accordance with the discussion in Chapter 7, a temperature of approximately 10^{-8} K had been reached prior to 1984 (Atkins, 1984). At this temperature, no macroscopic motion occurs and all materials are highly structured. From this, one can surmise that entropy provides a measure of structure, of chaos, and of disorder. The value of entropy increases as the temperature increases; chaos increases while structure diminishes. *Entropy* is sometimes defined from a microscopic perspective as the measure of disorder within the system. Thus, high values of entropy indicate high levels of disorder, while low values indicate low levels of disorder.

8.4   The Combined First and Second Law

An important equation can be developed by combining the first law of thermodynamics with the expression for entropy, which came from the second law of thermodynamics.

Consider a simple compressible substance that undergoes a reversible process. For this substance the first law can be written as

$$\delta Q - \delta W = dE \tag{8.6}$$

Now assume (1) that the total energy of the system, E, is internal energy only and (2) that the work done by the system in the (quasi-equilibrium) reversible process can be expressed as $p\,dV$ work only. With these assumptions, the first law can be expressed as

$$\delta Q_{\text{rev}} - p\,dV = dU \tag{8.7}$$

Furthermore, since the process is reversible, the heat term, δQ, can be expressed in terms of T and dS based on Equation 8.4:

$$\delta Q_{\text{rev}} = T\,dS \tag{8.8}$$

As noted earlier, this equation is a consequence of the second law of thermodynamics. Combining Equations 8.7 and 8.8 gives

$$T\,dS = dU + p\,dV \tag{8.9}$$

This equation now relates thermodynamic properties only and thus can be used to determine thermodynamic property changes between specified thermodynamic states. It applies for both reversible and irreversible processes (subject only to the restrictions indicated later) because the properties of a substance depend only on the state. This equation is sometimes called the *combined first and second law* of thermodynamics. It is also referred to as the *Gibbs equation* or as a $T\,dS$ *equation*. It is an extremely important relationship that is used extensively in the thermodynamics of pure substances.

The combined first and second law can also be developed in terms of enthalpy as follows. Enthalpy was defined earlier as

$$H = U + pV \tag{8.10}$$

Therefore,

$$dH = dU + p\,dV + V\,dp \tag{8.11}$$

Combining Equations 8.9 and 8.11 gives

$$T\,dS = dH - V\,dp \tag{8.12}$$

Both Equations 8.9 and 8.12 are extremely useful relationships in the thermodynamics of pure substances. Certain restrictions, however, are implied in the derivation of these $T\,dS$ equations, which should be noted:

1. The derivation was for a *closed system*. Thus, when dealing with open systems, the entropy of masses flowing into and out of the control volume must be factored into the expression for the entropy changes of the open system. Further consideration of entropy changes for the open system is made later in Section 8.9.

2. By assuming that the system is a pure substance, those substances whose chemical composition could change while a process is taking place have been excluded. In

other words, the $T\,dS$ equations apply to closed systems having a fixed chemical composition. It is known, for example, that they do not apply when irreversible chemical reactions (or changes in chemical composition) take place.

Equations 8.9 and 8.12 can also be written on a unit mass basis as follows:

$$T\,ds = du + p\,dv \tag{8.13}$$

and

$$T\,ds = dh - v\,dp \tag{8.14}$$

The equivalent form of the equations on a mole basis are as follows:

$$T\,d\bar{s} = d\bar{u} + p\,d\bar{v} \tag{8.15}$$

and

$$T\,d\bar{s} = d\bar{h} - \bar{v}\,dp \tag{8.16}$$

The above *intensive T ds equations* are applicable whether dealing with closed or open systems since only the intensive properties are involved. They are frequently used for developing explicit forms for the equations of state for pure substances, as explained in Section 8.5 along with some examples.

8.5 Entropy Change of a Pure Substance

Since entropy is a thermodynamic property, it can be specified for every equilibrium state of any pure substance. Entropy followed directly from the second law of thermodynamics in much the same manner as the thermodynamic temperature scale. Furthermore, measurement of entropy changes for a substance is similar to the measurement of temperature changes for a substance. That is, neither property is normally measured directly, but rather they are inferred from the measurement of other properties (or quantities). Thus, when measuring temperature using empirical thermometry, the rise of mercury in a glass container is observed or the difference in pressure in a constant volume gas thermometer is measured. Similarly, values for entropy changes are normally obtained by measurements of other quantities. The following paragraphs illustrate two methods for measuring entropy changes.

Consider the system (a mixture of liquid and vapor) shown in Figure 8.2. Electrical energy is supplied to a resistance heating element that is in good thermal contact with the container holding the two-phase mixture. The process is carried out slowly and can be considered (for the system) a quasi-equilibrium or internally reversible process. The liquid vaporizes, and the laden frictionless piston rises slowly as more vapor is formed so that the process is also a constant pressure process. Thus, the vaporization process is both a constant temperature and a constant pressure process. For the closed system, therefore, the entropy change as a result of the process can be computed from Equation 8.5 as

$$S_2 - S_1 = \frac{1}{T}\int_{\text{state }1}^{\text{state }2} (\delta Q)_{\text{rev}} = \frac{(Q_{12})_{\text{rev}}}{T} \tag{8.17}$$

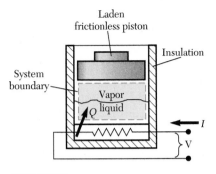

Figure 8.2 A method for the measurement of the change in entropy for a reversible process.

where the temperature has been taken outside the integral because it remained constant throughout the process. Thus, from measurement of the temperature and the heat input (from the resistance heating element) for the process, the entropy change for the system can be determined. In the surroundings, the transformation of the electrical energy input to the resistance heating element, to heat flow to the two-phase mixture, is irreversible. The two-phase mixture (the system), however, passes through a series of quasi-equilibrium states with both the temperature and pressure staying constant. The process is *internally reversible*, thus justifying the use of Equation 8.5 to determine the change in entropy of the system. Note that only internal reversibility, as opposed to complete reversibility, is required for Equation 8.17 to be true.

Now assume that the heat addition in Figure 8.2 is not made slowly, but that the vaporization of the liquid still takes place at a constant pressure. The initial and final temperatures are the same in this case, too, although at intermediate stages the temperature may be nonuniform in the system. In this case, however, the entropy change can be determined from Equation 8.12 as

$$dS = \frac{dH}{T} - \frac{V}{T}\,dp$$

$$= \frac{dH}{T} \quad \text{since } p = \text{constant}$$

(8.18)

Equation 8.18 is an equation of state. Thus, for any process between initial state 1 and final state 2 such that $p_2 = p_1$ and $T_2 = T_1$, Equation 8.18 can be integrated assuming constant p and constant T yielding

$$S_2 - S_1 = \frac{H_2 - H_1}{T}$$

(8.19)

In this case, the entropy change can be obtained directly from knowledge of the enthalpy change for the process. However, the enthalpy change can be obtained from a first law analysis as $Q = (H_2 - H_1)$ only if the change in volume of the system is fully resisted and the heat input (Q) to the substance during the process occurred at a constant pressure.

Since entropy is an important parameter involved in second law analysis of many systems, entropy magnitudes are available for most pure substances used in thermodynamic applications. Entropy values for some of these substances are provided in the tables of Appendices A and B and are calculated by the computer codes provided with this book. As noted in a previous chapter, entropy varies primarily with temperature and has units of kJ (or Btu) per degree. (As can be seen from Equation 8.19, entropy has the units of kJ/K or Btu/°R.) Specific entropy, defined as entropy per unit mass, is expressed by the symbol s. The units for specific entropy are kJ/(kg · K) or Btu/(lbm · °R).

Thermodynamic analyses often present property information graphically. Up to this point, this book has concentrated primarily on use of the T-v diagram. Subsequent sections of this and later chapters will use the T-s and h-s diagrams quite often. Figures 8.3 and 8.4 depict typical T-s and h-s diagrams, respectively, for a pure substance. The h-s diagram is also called the *Mollier diagram*, after Richard Mollier (1863–1935), who introduced its use. The Mollier diagram is especially useful for applications such as power generation by steam.

As indicated in Chapter 3, the entropy values in the two-phase region can be calculated from Equation 3.18 using the same procedure as that for the values of enthalpy and internal energy in that region. In cases where tabulated data are not readily available for values of entropy of a substance (for example, a solid, compressed liquid, or ideal gas), the combined first and second law can be used to compute values, as outlined in the following sections.

Solids and Incompressible Liquids

The specific volume of a solid or an incompressible liquid is essentially constant. Thus, Equation 8.13 can be simplified for these systems by equating dv to zero. The entropy change for a solid or an incompressible liquid is thus given by

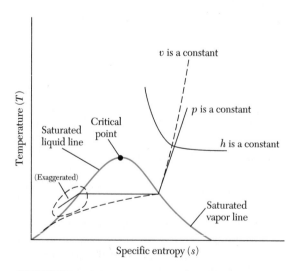

Figure 8.3 Typical T-s diagram for a pure substance.

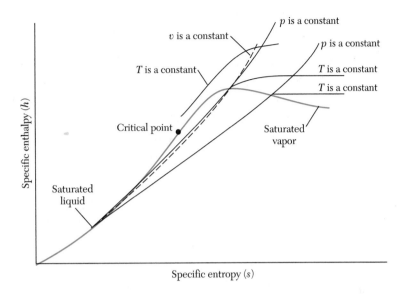

Figure 8.4 Typical *h-s* diagram for a pure substance.

$$ds \approx \frac{1}{T} \, du \qquad (8.20)$$

Since c_p is approximately equal to c_v ($= c$) for a solid or incompressible liquid, $du = c \, dT$ can be substituted in Equation 8.20 to obtain the entropy change between two states:

$$ds = \frac{c}{T} \, dT$$

Integrating gives

$$s_2 - s_1 = \int_{T_1}^{T_2} \frac{c}{T} \, dT \qquad (8.21)$$

If the variation of specific heat c with temperature is known, the integration of the expression on the right-hand side of Equation 8.21 can be completed. Frequently, c is constant, in which case the integral is $c \ln(T_2/T_1)$.

Entropy Change for an Ideal Gas

The entropy change for an ideal gas can be found from the ideal gas equation of state and the combined first and second law. One approach is to use Equation 8.13 combined with Equation 4.23 and the ideal gas pvT equation of state (Eq. 4.1) to obtain

$$ds = c_{v0} \frac{dT}{T} + R \frac{dv}{v} \qquad (8.22)$$

where c_{v0} is the specific heat at constant volume for an ideal gas and R is the specific gas constant. Similarly, Equations 8.14, 4.22, and 4.1 can be combined to obtain

$$ds = c_{p0} \frac{dT}{T} - R \frac{dp}{p} \qquad (8.23)$$

where c_{p0} is the specific heat at constant pressure for an ideal gas. If data are available for specific heat, Equations 8.22 and 8.23 can be integrated to obtain the entropy change for an ideal gas between two equilibrium states. When the specific heat can be considered constant for the process, the results obtained are as follows:

$$s_2 - s_1 = c_{v0} \ln \frac{T_2}{T_1} + R \ln \frac{v_2}{v_1} \qquad (8.24)$$

and

$$s_2 - s_1 = c_{p0} \ln \frac{T_2}{T_1} - R \ln \frac{p_2}{p_1} \qquad (8.25)$$

Note the similarity of these equations to the results for solids or liquids with $\Delta v = 0$ and $\Delta p = 0$.

In general, the values for specific heat for gases vary with temperature. Thus, a more accurate computation for entropy change would involve the use of either Equation 8.22 or 8.23 and the appropriate relationships for the specific heats as functions of temperature. The values given in gas tables normally take into account the specific heat variations with temperature. Such tables are usually provided for a system pressure of 1 atm and relative to a reference temperature (T_{ref}). Thus, both sides of Equation 8.23 can be integrated to obtain

$$s_2 - s_1 = \int_{T_1}^{T_2} \frac{c_{p0}}{T} dT - R \ln \frac{p_2}{p_1} \qquad (8.26)$$

Now let's define the value of the remaining integral in Equation 8.26 between the reference temperature T_{ref} and any other temperature as

$$s^0(T) = \int_{T_{ref}}^{T} \frac{c_{p0}}{T} dT \qquad (8.27)$$

where $s^0(T)$ is termed the *standardized specific entropy* at the standard pressure $p = 1$ atm and any temperature T. Now,

$$s^0(T_2) - s^0(T_1) = \int_{T_{ref}}^{T_2} \frac{c_{p0}}{T} dT - \int_{T_{ref}}^{T_1} \frac{c_{p0}}{T} dT = \int_{T_1}^{T_2} \frac{c_{p0}}{T} dT \qquad (8.28)$$

Thus, Equation 8.26 can be rewritten as

$$s_2 - s_1 = s^0(T_2) - s^0(T_1) - R \ln \frac{p_2}{p_1} \qquad (8.29)$$

The function given in Equation 8.27 is tabulated in gas tables with the variations of c_{p0} with temperature taken into account. Using this function, Equation 8.29 can be used to compute entropy changes for ideal gases with better accuracy than that given by Equation 8.24 or 8.25 because the variation of c_{po} with temperature has been accounted for.

Note that a significant difference exists between the property values given in gas tables and those in tables for the gaseous phases of pure substances, such as steam and refrigerants. The gas tables are based on ideal gas relationships with properties listed (at a standard pressure of 1 atm) as functions of temperature only, while the properties in the steam tables are presented in terms of two independent variables (such as temperature and pressure). Also note that the reference temperature normally chosen for the standardized specific entropy $s^0(T)$ is absolute zero. Since the entropy of a pure substance approaches zero as the temperature approaches absolute zero, the absolute specific entropy (s) of an ideal gas at any temperature T and pressure p (in atm) can be calculated using the following reduced version of Equation 8.29

$$s = s^0(T) - R \ln p \qquad (8.29a)$$

These equations, used for the ideal gas tables, are valid only in the region where the substance behaves as an ideal gas.

EXERCISES

for Section 8.5

1. The specific heat of water at 1 atm and room temperature can be taken as 4200 J/kg · K. Use Equation 8.21 to calculate $s_2 - s_1$ for water when $T_1 = 20°C$ and $T_2 = 50°C$. Compare your result with that obtained from the steam tables.

   Answer: 0.409 kJ/kg · K

2. A mass of 0.5 kg of air at $p_1 = 100$ kPa and $T_1 = 300$ K is heated at a constant volume until it attains a temperature $T_2 = 500$ K. Assume $c_p = 1.005$ kJ/kg · K and $R = 0.2871$ kJ/kg · K for air. For the air, determine
   (a) p_2 (in kPa)
   (b) ΔS (in kJ/K)

   Answer: (a) 167 kPa, (b) 0.183 kJ/K

3. Use Equation 8.29 and the appropriate table in Appendix A to determine $s_2 - s_1$ for the air in Exercise 2. Hence, compute ΔS for the air.

8.6 Isentropic Process

If a process is both adiabatic and reversible, $\delta Q \,(=\delta Q_{rev}) = 0$, and from Equation 8.4, which defines entropy change, it is concluded that $dS = 0$ for the process. In other words, entropy stays constant for an adiabatic and reversible process. Such a process is called an *isentropic process*. Thus, an isentropic process is defined as the *reversible, adiabatic process* for which entropy stays constant throughout.

The isentropic process is the standard against which adiabatic processes are judged. This is true when dealing with both nonflow processes for a closed system and flow processes for an open system. For example, if a fixed mass of a substance is compressed adiabatically, the least amount of work is required when the compression process is isentropic. Conversely, the maximum work output is obtained from the adiabatic expansion of a system (closed or open) when the process is isentropic, since the process is reversible. Section 8.8 further considers this principle in a broader context for closed systems, and Section 8.9 for open systems. The following examples consider devices that operate adiabatically. The illustrations include both *flow* and *nonflow processes*. As indicated earlier in Section 8.4, the use of *specific entropy* (which is an intensive property) rather than total entropy enables one to analyze problems involving closed and open systems in a similar fashion. Thus, for a reversible adiabatic process involving a fixed mass of a substance, the specific entropy (s_2) at the final state of the system must be the same as that (s_1) at the initial state. Likewise, for a reversible single stream steady flow of mass through an adiabatic device, the specific entropy of the substance at the exit (s_e) is the same as that at the inlet condition (s_i). This condition frequently serves as a valuable linkage between a known thermodynamic state of a substance and another state that is to be determined.

EXAMPLE 8.1
Isentropic Process for a Turbine

Steam enters a turbine at 500°C and 10 bar. This steam undergoes an isentropic process through the turbine and exits at a pressure of 1 bar. Determine the temperature at the exit state and compute the turbine output work per unit mass of steam.

Given:

$$T_i = 500°C \quad p_i$$

$$= 10 \text{ bar}, \quad p_e = 1 \text{ bar}$$

Isentropic flow through turbine

Find: T_e and w

Solution: For isentropic flow, $s_e = s_i$. The value of s_i $(= 7.762 \text{ kJ/kg} \cdot \text{K})$ is obtained from either the steam tables or the **STEAM** code, and hence, s_e is known. Two independent intensive properties are known for the exit condition of the steam. (Note that if you decide to continue with steam tables to determine other properties of steam at the exit condition, you will have to interpolate between the tabulated property values for superheated steam at 1 bar. A linear interpolation is not accurate for entropy; the variation of entropy with temperature is closer to a logarithmic function. Also, excessive effort looking up property values in tables can easily obscure or divert attention from the thermodynamic principles under consideration. Thus, the computer codes are used here as the first choice.)

The **STEAM** code gives the following values for the specific enthalpy of the steam at the entry and exit conditions:

$$h_i = 3478.5 \text{ kJ/kg}, \qquad h_e = 2841.7 \text{ kJ/kg}$$

Application of the first law of thermodynamics to the steady-flow process (Eq. 6.29) gives

$$w = \frac{\dot{W}_x}{\dot{m}} = h_i - h_e$$

$$= 3487.5 \text{ kJ/kg} - 2841.7 \text{ kJ/kg} = \underline{637 \text{ kJ/kg}}$$

This reduced form of the steady-flow energy equation (SFEE) assumes that changes in kinetic energy and potential energy are negligible. It was also assumed that the process is adiabatic and thus the heat flow rate is zero.

EXAMPLE 8.2
Work Done in Isentropic Expansion of Steam

A mass of 0.5 kg of steam at 5.5 bar and a temperature of 200°C expands reversibly and adiabatically to a final pressure of 1.4 bar. Calculate the work done by the steam.

Given: System: 0.5 kg of steam
Process: isentropic expansion ($Q_{12} = 0$)

State 1	State 2
$p_1 = 5.5$ bar	$p_2 = 1.4$ bar
$T_1 = 200$°C	$(s_2 = s_1)$

Find: W_{12}

Solution: The system comprises the 0.5 kg of steam. First, apply the nonflow energy equation (NFEE), Equation 6.10:

$$Q_{12} - W_{12} = U_2 - U_1$$

It is assumed that changes in kinetic and potential energy are negligible. Since $Q_{12} = 0$, the equation simplifies to

$$W_{12} = m(u_1 - u_2)$$

Next, use the **STEAM** code to obtain the steam properties (including u_1) at state 1:

$$u_1 = 2640.4 \text{ kJ/kg}$$

$$s_1 = 7.010 \text{ kJ/kg} \cdot \text{K}$$

The process is isentropic and therefore $s_2 = s_1 = 7.010$ kJ/kg · K. Use the **STEAM** code again to obtain the steam properties at state 2 (with $p_2 = 1.4$ bar and $s_2 = 7.010$ kJ/kg · K). $u_2 = 2433.7$ kJ/kg.

$$W_{12} = 0.5 \ (2640.4 - 2433.7) \text{ kJ}$$

$$= \underline{103 \text{ kJ}}$$

Isentropic Relationships for an Ideal Gas

For an isentropic process connecting an initial state 1 of a system to a final state 2, the specific entropy is unchanged. Thus, for an isentropic process involving an ideal gas with constant specific heats, Equations 8.24 and 8.25 can be reduced to the following explicit T-v and T-p forms:

$$\frac{T_2}{T_1} = \left(\frac{v_1}{v_2}\right)^{R/c_{v0}} = \left(\frac{v_1}{v_2}\right)^{(k-1)} \tag{8.30}$$

and

$$\frac{T_2}{T_1} = \left(\frac{p_2}{p_1}\right)^{R/c_{p0}} = \left(\frac{p_2}{p_1}\right)^{(k-1)/k} \tag{8.31}$$

where k is the specific heat ratio. Equations 8.30 and 8.31 are derived from Equations 8.24 and 8.25, respectively, by putting $s_2 = s_1$ and making use of the logarithmic property $a \ln b = \ln b^a$ when $b > 0$. The relationships $R = c_{p0} - c_{v0}$ and $k = c_{p0}/c_{v0}$ have been used to further simplify the expressions. Equations 8.30 and 8.31 are in terms of intensive properties and can be applied to either the *isentropic nonflow process* or the *isentropic flow process* involving an ideal gas with constant specific heats.

Equations 8.30 and 8.31 can be combined to yield the following equation in terms of pressure and the specific volume only:

$$\frac{p_2}{p_1} = \left(\frac{v_1}{v_2}\right)^k \tag{8.32}$$

which can also be written as

$$p_1 v_1^k = p_2 v_2^k \tag{8.33}$$

or in general terms as

$$pv^k = \text{constant} \tag{8.34}$$

Comparing Equation 8.34 with Equation 5.12, which defines the polytropic process for an ideal gas, shows that the value of the exponent is equal to k. Thus, the polytropic index n is equal to k for the isentropic process involving the ideal gas with constant specific heats. The displacement work done by a fixed mass of an ideal gas in an isentropic process is calculated by substituting $n = k$ into Equation 5.13 (from Chap. 5). This gives

$$W_{12} = \frac{p_2 V_2 - p_1 V_1}{1 - k} = \frac{mR(T_2 - T_1)}{1 - k} = mc_{v0}(T_1 - T_2)$$

The last step in the equation for W_{12} results from making the substitutions $R = c_{p0} - c_{v0}$ and $k = c_{p0}/c_{v0}$.

Equations 8.30 through 8.34 were developed for the ideal gas undergoing an isentropic process on the assumption that the specific heats of the gas are constant. As mentioned earlier, the gas tables account more accurately than these equations for the effects due to the temperature dependence of the specific heats.

Use of Gas Tables for Isentropic Processes

The gas tables provide several properties as a function of temperature. For example, Tables A.11 to A.20 in Appendix A (SI units) and corresponding Tables B.11 to B.20 in Appendix B (USCS units) give values of specific enthalpy (h) and the standardized specific entropy $s^0(T)$ as a function of temperature. The tabulations for air and methane in particular include two additional quantities: the *relative pressure* (p_r) and the *relative specific volume* (v_r). Both of these quantities are defined shortly and their use is demonstrated to facilitate the analysis of isentropic processes involving ideal gases.

Substituting $s_2 = s_1$ in Equation 8.29 for an isentropic process gives

$$\ln \frac{p_2}{p_1} = \frac{s^0(T_2) - s^0(T_1)}{R} \tag{8.35}$$

Thus, the pressure ratio between two states connected by an isentropic process depends only on the values of $s^0(T_2)$ and $s^0(T_1)$ and therefore on temperature only. Equation 8.35 provides the basis for the use of the tabulated pressure ratio values in Tables A.11 and B.11 for air when the process is isentropic. For instance, a third state at temperature T_0 and pressure p^* could be considered such that the specific entropy is the same as that for states 1 and 2. Thus, the following equations, which are of the same form as Equation 8.35, can be written:

$$\ln \frac{p_1}{p^*} = \frac{s^0(T_1) - s^0(T_0)}{R} \tag{8.35a}$$

and

$$\ln \frac{p_2}{p^*} = \frac{s^0(T_2) - s^0(T_0)}{R} \tag{8.35b}$$

or more generally,

$$\ln \frac{p}{p^*} = \frac{s^0(T) - s^0(T_0)}{R} \tag{8.35c}$$

In Equation 8.35c, p and T denote the pressure and temperature at any other state at which the specific entropy for the ideal gas is the same as that at states 1 and 2. Note that the choice of p^* is not arbitrary but rather must be such that at the state with temperature T_0 and pressure p^*, the specific entropy is also equal to that at states 1 and 2. Thus, the *relative pressure* p_r is defined as

$$p_r = \frac{p}{p^*} \tag{8.36a}$$

For a given isentropic process, p^* is fixed. Thus, the following relationship can be established between the relative pressures and the actual pressures for the states connected by the isentropic process:

$$\frac{p_{r2}}{p_{r1}} = \frac{p_2}{p_1} \tag{8.36b}$$

Equation 8.35c can be rewritten in the following form using Equation 8.36a for the relative pressure:

$$p_r = \exp \frac{s^0(T) - s^0(T_0)}{R} \qquad (8.37)$$

The relative pressures indicated in Tables A.11 and A.13 for air and methane, respectively, were calculated using Equation 8.37 with T_0 arbitrarily set equal to 298 K (or 25°C). Thus, if the gas temperature is given, the relative pressure can be read off the table. If the actual pressure of the gas is also known, the reference pressure p^* can be determined, thus permitting the evaluation of the relative pressure for some other state with the same specific entropy as the original gas state, but at a different pressure. If the relative pressure at the second state is known, the gas table can be used to determine the gas temperature at this state. An illustration of this procedure is provided in Example 8.3.

In some cases, it is preferable to work with volume ratios instead of pressure ratios for isentropic processes. Using the ideal gas equation (Eq. 4.1) to solve for the specific volume ratios between two states gives

$$\frac{v_2}{v_1} = \frac{p_1}{p_2} \frac{T_2}{T_1} \qquad (8.38)$$

Since the pressure ratio is a function of temperature only, it follows from Equation 8.38 that the specific volume ratio is also a function of temperature only. Thus, a relative specific volume can be defined and used with the gas tables in the same way that relative pressure is used. The *relative specific volume* is defined as

$$v_r = \frac{v}{v^*} \qquad (8.39a)$$

where v^* is the gas specific volume when the pressure is p^* and the temperature is T_0. This leads to the following relationship between the relative specific volumes at states 1 and 2 and the actual specific volumes:

$$\frac{v_{r2}}{v_{r1}} = \frac{v_2}{v_1} \qquad (8.39b)$$

Combining Equations 8.37, 8.38, and 8.39 leads to the following equation that was used for the computation of the relative specific volumes given in Tables A.11 and A.13:

$$v_r = \frac{T}{T_0} \exp\left(-\frac{s^0(T) - s^0(T_0)}{R}\right) \qquad (8.40)$$

The relative specific volume values in the table can also be used for the determination of gas states in an isentropic process.

Note that the choice of T_0 is arbitrary, and therefore the values indicated for p_r and v_r in an ideal gas table may differ from one source to another. This should not pose any problem as long as the same table is used for an isentropic process involving a given ideal gas. Also note that you should not use either v_r or p_r as the actual v or p. Furthermore, v_r and p_r are dimensionless variables that have little (if any) utility except in isentropic processes involving ideal gases.

The examples that follow illustrate the use of the gas tables.

EXAMPLE 8.3
Use of Relative Pressure Ratio with Gas Tables

Air at 830 K and 5 bar is enclosed in a cylinder. The air expands isentropically against a piston, and the pressure of the air is reduced to 1 bar. Using the ideal gas tables for air, find the temperature of the air at the end of the isentropic process and compute the work done per unit mass by the gas during the process. Compare your results with those obtained by assuming constant specific heat values for air.

Given: Air at $T_1 = 830$ K and $p_1 = 5$ bar; $p_2 = 1$ bar
Isentropic air expansion (against piston)

Find: T_2 and w_{12} **(a)** using ideal gas tables, **(b)** assuming constant specific heat

Solution:

(a) *Solution using gas tables*

The ideal gas properties for air are given in Table A.11 in Appendix A, which gives the following interpolated value for h_1:

$$\frac{h_1 - 524}{579 - 524} = \frac{830 - 800}{850 - 800}$$

$$\therefore h_1 = 557 \text{ kJ/kg}$$

Linear interpolation was used for obtaining this estimate of h_1.

For the isentropic expansion of an ideal gas, Equation 8.35 applies. Table A.11 can be used in the same manner as just indicated. Thus, using linear interpolation, the relative pressure p_{r1} is estimated as 39.8 at the indicated temperature of 830 K. The corresponding p_{r2} for the isentropic process is obtained using Equation 8.36b:

$$p_{r2} = \frac{p_2}{p_1} p_{r1}$$

$$= \frac{1}{5}(39.8) = 7.96$$

Referring to Table A.11 again, use a linear interpolation to obtain $T_2 = 534.6$ K and $h_2 = 241$ kJ/kg. The first part of the problem is thus solved, with the final temperature of the gas being 535 K (or 262°C).

The work done by the system in the isentropic process is determined by applying the nonflow energy equation (NFEE) (Eq. 6.10) to the nonflow process. In this case, assuming that changes in kinetic energy and potential energy are negligible, the NFEE reduces to

$$w = \frac{W_{12}}{m} = u_1 - u_2$$

$$= (h_1 - p_1 v_1) - (h_2 - p_2 v_2)$$

$$= (h_1 - RT_1) - (h_2 - RT_2)$$

For air, $R = 0.2871$ kJ/kg $\cdot$ K. Substituting in the previous expression thus leads to

$$w = (557 - 241) - 0.2871(830 - 534.6 \text{ kJ/kg})$$

$$= 231 \text{ kJ/kg}$$

(b) *Solution using constant specific heat equations*

Equation 8.31 can be used for the determination of T_2 when the process is isentropic and constant specific heat can be assumed. For air, $k = 1.4$. Therefore,

$$\frac{T_2}{T_1} = \left(\frac{p_2}{p_1}\right)^{(k-1)/k} = \left(\frac{1}{5}\right)^{0.4/1.4} = 0.631$$

Hence, $T_2 = 524$ K. Note that the difference between this estimate of T_2 and that obtained using the gas tables is less than 2%.

The value of w is determined using the following reduced form of the NFEE:

$$w = u_1 - u_2$$

$$= c_v(T_1 - T_2)$$

For air in the temperature range of 524 K to 830 K, c_p can be taken as 1.07 kJ/kg $\cdot$ K. This estimate is obtained from the c_p values for air given in Table A.11. Thus, the specific heat at constant volume $c_v = c_p - R = 0.783$ kJ/kg $\cdot$ K.

$$\therefore w = 0.783(830 - 534.6 \text{ kJ/kg})$$

$$= 231 \text{ kJ/kg}$$

Evidently in this instance, the loss in accuracy is not significant when constant specific heat values are assumed.

EXAMPLE 8.4
Use of Relative Specific Volume Ratio with Gas Tables

Air at 830 K is contained in a cylinder with an initial volume of 0.1 m³. The air expands isentropically against a piston so that the volume increases to 0.322 m³. Use the ideal gas property tables for air to find the temperature of the air at the end of the isentropic process, and compute the work done (per unit mass of air) during the process. Compare the results obtained with those when constant specific heat values are assumed for the air.

Given: Air at $T_1 = 830$ K, $V_1 = 0.1$ m³ expanding isentropically to $V_2 = 0.322$ m³

Find: T_2 and w (a) using ideal gas tables, (b) assuming constant specific heat

Solution:

(a) *Solution using gas tables*

From Table A.11 for the ideal gas properties for air, determine v_{r1} and h_1 by linear interpolation as follows:

$$\frac{v_{r1} - 0.0762}{0.0641 - 0.0762} = \frac{830 - 800}{850 - 800} = \frac{h_1 - 524}{579 - 524}$$

$$\therefore v_{r1} = 0.06894$$

$$h_1 = 557 \text{ kJ/kg}$$

Using Equation (8.39b) determine v_{r2}:

$$v_{r2} = v_{r1} \frac{v_2}{v_1} = 0.222$$

The estimate obtained for v_{r2} can be used with Table A.11 to obtain the following interpolated values for T_2 and h_2:

$$T_2 = 540.8 \text{ K}$$

$$h_2 = 247.4 \text{ kJ/kg}$$

The final temperature is thus 541 K.

As in Example 8.3, the work done per unit mass can be calculated using

$$w = h_1 - h_2 - R(T_1 - T_2)$$

$$= (557 - 247.4 \text{ kJ/kg}) - (0.2871 \text{ kJ/kg} \cdot \text{K})(830 - 540.8 \text{ K})$$

$$= 226.6 \text{ kJ/kg}$$

Thus, the work done is 226.6 kJ/kg.

(b) *Solution using constant specific heat equations*

Equation 8.30 can be used to calculate T_2 when constant specific heats are assumed:

$$T_2 = T_1 \left(\frac{V_1}{V_2}\right)^{k-1} = 830 \text{ K} \left(\frac{0.1}{0.322}\right)^{0.4} = 520 \text{ K}$$

This estimate is nearly 4% less than that determined using the ideal gas tables for air. Note, however, that in the temperature range of 520 K to 830 K, the average values for c_p and c_v are 1.07 kJ/kg · K and 0.783 kJ/kg · K, respectively. Using these gives $k = 1.37$ instead of the value of 1.4 assumed in the application of Equation 8.30 to this problem. If $k = 1.37$ is used, the calculated value for T_2 is 540.7 K, which is very close to the value determined with the aid of the ideal gas tables.

The work per unit mass can be calculated using the following reduced form of the NFEE:

$$w = u_1 - u_2 = c_v(T_1 - T_2)$$

$$= 0.783 \text{ kJ/kg} \cdot \text{K}(830 - 540.7 \text{ K})$$

$$= \underline{226.5 \text{ kJ/kg}}$$

While this estimate is nearly identical to that obtained using the ideal gas tables, note that if the specific heat values assumed are those for air at room temperature, the discrepancy could be significant. Thus, the initial estimate for T_2 obtained by assuming $k = 1.4$ was nearly 4% below that obtained using the ideal gas tables. The discrepancy is less in the estimate for w when c_v is taken as 0.718 kJ/kg $\cdot$ K and $T_2 = 520$ K is used. The estimated value of w is about 223 kJ/kg, which is close to that determined using the ideal gas tables.

Isentropic Process for Solid or Incompressible Fluid

The entropy change for a solid or an incompressible fluid was given by Equation 8.20 as

$$ds \approx \frac{du}{T} \qquad (8.20)$$

Since $ds = 0$ for an isentropic process, it follows from Equation 8.20 that $du \approx 0$. Furthermore, since $du = c \, dT$, it follows that $dT \approx 0$. Thus, an isentropic process for a solid or an incompressible liquid is essentially an isothermal process also.

By considering Equation 8.14, one can deduce the following equation for the change in specific enthalpy when the process is isentropic:

$$dh = v \, dp \qquad (8.41)$$

Since the specific volume does not change significantly for an incompressible fluid or solid, Equation 8.41 can be integrated to give

$$h_2 - h_1 = v(p_2 - p_1) \qquad (8.42)$$

This equation is particularly useful for computing work done in the pumping of liquids.

EXAMPLE 8.5
Calculation of Work for a Pump

A centrifugal pump is used to increase the pressure of water at 70°F from the atmospheric pressure of 1 atm to 50 psia pressure. Compute the work done per lbm of water for isentropic flow through the pump.

Given: Water at $T_i = 70°F$ and $p_i = 1$ atm is pumped to $p_e = 50$ psia
Process is isentropic

Find: Work (w) done per lbm of water

Solution: For a steady-flow process involving a single mass flow, the SFEE (Eq. 6.29) can be applied. If changes in kinetic energy and potential energy are assumed negligible, the equation reduces to

$$w = \frac{\dot{W}_x}{\dot{m}} = h_i - h_e$$

Equation 8.42 can be used for a liquid to compute the change in specific enthalpy between two states having the same specific entropy. Thus, the expression for w can be written as

$$w = h_i - h_e = v(p_i - p_e)$$

For water at 70°F, $v = 0.01605$ ft³/lbm (from Table B.1 in Appendix B). Thus,

$$w = v(p_i - p_e)$$

$$= \left(0.01605 \ \frac{\text{ft}^3}{\text{lbm}}\right)\left(14.7 - 50 \ \frac{\text{lbf}}{\text{in.}^2}\right)\left(\frac{144 \ \text{in.}^2}{1 \ \text{ft}^2}\right)$$

$$= -81.6 \ \text{ft} \cdot \text{lbf/lbm}$$

A special note should be made concerning the isentropic process for liquids. Equation 8.42 was developed from Equation 8.14 without any stipulation except that the process is isentropic. As seen earlier, if the substance is totally incompressible, then the process is also isothermal. However, for adiabatic compression of a liquid such as water, some compressibility effects can be observed and the substance is not totally incompressible. Thus, an adiabatic process that increases the pressure of a liquid flowing through a pump also leads to a slight increase in temperature. This increase is usually very small, generally less than 1°C of temperature rise for a pressure increase of several atmospheres. In any case, only a slight increase in temperature will occur for the adiabatic compression of liquids.

EXERCISES

for Section 8.6

1. Steam at 50 bar and 500°C expands isentropically through a turbine to a pressure of 0.05 bar. Use steam tables to determine the steam quality at the exit to the turbine. Cross check your result with the **STEAM** code.

   Answer: 0.821

2. Wet steam at 0.05 bar is to be compressed isentropically such that its final condition is saturated water at 50 bar. What is the quality of the wet steam at the initial state?

   Answer: 0.309

3. Use the appropriate $T \, ds$ equation to show that for the isentropic expansion or compression of an ideal gas,

$$Tv^{k-1} = \text{constant}$$

Air at 300 K is compressed isentropically to $\frac{1}{5}$ of its initial volume. Determine the final temperature of the air. Assume $k = 1.4$ for air.

Answer: 571 K

4. Air at 6 bar and 1100 K expands isentropically to a pressure of 1 bar. Use Equation 8.31 to determine the final temperature of the air. Assume $k = 1.4$.

Answer: 659 K

5. Repeat Exercise 4 assuming $c_p = 1.117$ kJ/kg · K and $R = 0.2871$ kJ/kg · K for air.

Answer: 694 K

6. Use the ideal gas table in Appendix A to determine the final temperature of the air in Exercise 4.

Answer: 693 K

7. Starting with the appropriate $T\,ds$ equation, show that for the isentropic compression of a liquid (or solid),

$$h_2 - h_1 \approx v_1(p_2 - p_1)$$

A steady stream of saturated water at 0.05 bar is compressed isentropically to 50 bar using a feed pump. Determine the work input to the pump per unit mass of water flowing.

Answer: 5.02 kJ/kg

8.7 Carnot Cycle T-S Diagram

The typical temperature–specific entropy (T-s) diagram for pure substances was briefly discussed earlier and illustrated in Figure 8.3. This diagram is particularly useful in the analysis of thermodynamic cycles. Since the Carnot cycle consists of two isothermal processes and two reversible adiabatic processes, the representation of the cycle on the temperature–entropy (T-S) diagram is a rectangle, as shown in Figure 8.5(a). The process from state 1 to 2 in the Carnot cycle is constant temperature heat addition ($\Delta S > 0$), while the process from state 2 to 3 is a reversible adiabatic or constant entropy process. The process from state 3 to 4 is constant temperature heat removal ($\Delta S < 0$), while the process from state 4 to 1 is a reversible adiabatic or isentropic process.

For a reversible cycle, the net work done and the cycle efficiency can be determined from the T-S diagram. For a closed system, one can rewrite Equation 8.4 for the incremental heat flow as

$$\delta Q_{\text{rev}} = T\,dS \tag{8.43}$$

Thus, for the Carnot cycle shown in Figure 8.5(a),

$$Q_{12} = T_{\text{H}} \int_{S_1}^{S_2} dS \tag{8.44}$$

and

$$Q_{34} = T_{\text{L}} \int_{S_3}^{S_4} dS \tag{8.45}$$

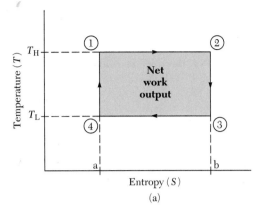

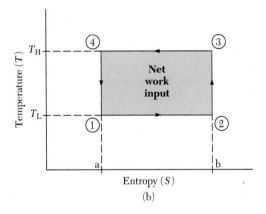

Figure 8.5(a) The Carnot heat engine cycle on the T-S diagram. **(b)** The reversed Carnot cycle on the T-S diagram.

Integrating these expressions gives

$$Q_{12} = T_{\mathrm{H}}(S_2 - S_1) \qquad (8.46)$$

and

$$Q_{34} = T_{\mathrm{L}}(S_4 - S_3) \qquad (8.47)$$

The processes between states 2 and 3 and between 4 and 1 are both isentropic, which means that $S_3 = S_2$ and $S_4 = S_1$. Therefore, the heat transfer for the process $3{\rightarrow}4$ can be written as

$$Q_{34} = -T_{\mathrm{L}}(S_2 - S_1) \qquad (8.48)$$

First law analysis demonstrates that the net heat transfer for each cycle is equal to the net work done. Therefore, the net work for the Carnot cycle can be computed as

$$W = Q_{12} + Q_{34} \tag{8.49}$$
$$= (T_H - T_L)(S_2 - S_1)$$

Furthermore, the thermal efficiency for the cycle is

$$\eta_1 = \frac{W}{Q_{input}} = \frac{W}{Q_{12}}$$

or

$$\eta_1 = \frac{(T_H - T_L)(S_2 - S_1)}{T_H(S_2 - S_1)} \tag{8.50}$$

which reduces to

$$\eta_1 = 1 - \frac{T_L}{T_H} \tag{8.51}$$

This expression for efficiency is identical to that for the thermal efficiency of the Carnot engine obtained earlier in Chapter 7.

Another interesting feature of the *T-S* diagram is that heat and work for the Carnot cycle (or any other ideal cycle) can be represented by areas in the diagram. From observation of Equations 8.44 and 8.46, it must be evident that the area a–1–2–b–a in Figure 8.5(a) represents the heat added to the system during the cycle. Similarly, consideration of Equations 8.45 and 8.47 shows that the area a–4–3–b–a represents the magnitude of the heat removed from the system during the cycle. Equation 8.49 demonstrates that the difference between these two areas represents the work done. Thus, the area 1–2–3–4–1 corresponds to the work done by the system during the cycle.

It has now been demonstrated that the *T-S* diagram provides a useful representation for a cycle. Such diagrams are often used for the representation and analysis of ideal cycles generally.

If the Carnot heat engine cycle just discussed is reversed, the resulting cycle represents a refrigeration or heat pump cycle. Such a cycle is shown in Figure 8.5(b). Following a development similar to that just outlined, the cooling effect or heat input during the cycle can be shown equal to the area a–b–2–1–a in Figure 8.5(b). The heat rejected during the cycle is equal to the area a–b–3–4–a, and the work input to the system during the cycle equals the area 1–2–3–4–1. Furthermore, from an analysis similar to that previously outlined, the COPs for the cycle are given by

$$(COP_C)_{rev} = T_L/(T_H - T_L) \tag{8.52}$$

$$(COP_H)_{rev} = T_H/(T_H - T_L) \tag{8.53}$$

which are identical to the relationships for COP given in Chapter 7. If the process from state 1 to 2 in Figure 8.5(b) represents an internally reversible heat transfer process in which a phase change is occurring for the fluid at constant pressure and constant temperature, then the entropy change for the process can be computed as

$$S_2 - S_1 = \int_{\text{state 1}}^{\text{state 2}} \frac{1}{T} (\delta Q)_{\text{rev}}$$

$$= \frac{1}{T} \int_{\text{state 1}}^{\text{state 2}} (\delta Q)_{\text{rev}} \tag{8.54}$$

$$= \frac{Q_{12}}{T}$$

Since the heat transfer for a constant pressure closed system process can be determined from a first law analysis to be equal to the enthalpy difference for the process, Equation 8.54 can be expressed as

$$S_2 - S_1 = (H_2 - H_1)/T \tag{8.55}$$

Comparison of tabulated data for the two-phase region for a fluid such as refrigerant-22 or water verifies this relationship. Also, for the phase change process just discussed, $s_{fg} = h_{fg}/T_{sat}$, where fg indicates the change from saturated liquid to saturated vapor.

EXERCISES

for Section 8.7

In these exercises, assume that the Carnot cycle in Figure 8.5(a) is made up of nonflow processes and that the ideal engine uses a mixture of saturated water and saturated steam as the working fluid. State 1 is the saturated liquid state at a pressure of 50 bar, while state 2 is the dry, saturated vapor state at the same pressure. States 3 and 4 correspond to a condenser pressure of 0.05 bar.

1. Determine, per unit mass of the working fluid (in kJ/kg),
   (a) q_{12}
   (b) w_{12}
   (*Hint:* Apply the NFEE to determine w_{12}.)
   Answer: (a) 1,640 kJ/kg, (b) 191 kJ/kg

2. Compute
   (a) q_{23} (in kJ/kg)
   (b) w_{23} (in kJ/kg)
   Answer: (b) 875 kJ/kg

3. Consider process 3→4 and compute
   (a) q_{34} (in kJ/kg)
   (b) w_{34} (in kJ/kg)
   Answer: (a) −934 kJ/kg, (b) −54.4 kJ/kg

4. Calculate, per unit mass of the working fluid,
   (a) q_{41} (in kJ/kg)
   (b) w_{41} (in kJ/kg)
   Answer: (b) −306 kJ/kg

5. Assume that the mass of the working fluid is 0.5 kg. Calculate
   (a) the net work output per cycle, W_{net} (in kJ)
   (b) the heat input per cycle, Q_{input} (in kJ)

(c) the cycle efficiency, η (in %)

(d) the *work ratio*, r_w, defined as the ratio of net work output to the positive work output in each cycle

Answer: (a) 353 kJ, (b) 820 kJ, (c) 43%, (d) 0.662

8.8 The Principle of Increase in Entropy for a Closed System

The Basic Principle

This section considers how irreversibilities affect the change in entropy for a closed system undergoing real processes. Returning to Figure 8.1, assume that process I is irreversible while III is reversible. Applying the inequality of Clausius for the complete cyclic processes I and III gives

$$\int_{\text{state }1}^{\text{state }2} \left(\frac{\delta Q}{T_{\text{boundary}}}\right)_{\text{I,irrev}} + \int_{\text{state }2}^{\text{state }1} \left(\frac{\delta Q}{T}\right)_{\text{III,rev}} \leq 0 \tag{8.56}$$

or

$$\int_{\text{state }1}^{\text{state }2} \left(\frac{\delta Q}{T_{\text{boundary}}}\right)_{\text{I,irrev}} + \int_{S_2}^{S_1} dS \leq 0 \tag{8.57}$$

Thus,

$$S_2 - S_1 \geq \int_{\text{state }1}^{\text{state }2} \left(\frac{\delta Q}{T_{\text{boundary}}}\right)_{\text{I,irrev}} \tag{8.58}$$

Equation 8.58 is often recast in the following differential form:

$$dS \geq \frac{\delta Q}{T_{\text{boundary}}} \tag{8.59}$$

The equality sign in Equation 8.59 applies only when the process is reversible, in which case S is unambiguously the entropy for the system at an equilibrium or near-equilibrium condition. There is some controversy[1], however, concerning the validity of the inequality in Equation 8.59 for irreversible processes. Notwithstanding such an objection regarding the applicability of this inequality to real processes, its validity is widely assumed and its use is generally found to lead to correct conclusions provided the system under consideration departs only slightly from equilibrium conditions during the process. Therefore, development of the principle of increase in entropy in the differential form can be continued.

---

[1]The point of controversy pertains to whether or not it is possible to assign a unique entropy to nonequilibrium states of a system. Kestin (1976) reproduces a 1970 paper by Meixner that addresses this issue. The principal assertion of the paper is that a unique nonequilibrium entropy is nonexistent and indeterminate. Thus, the inequality in Equation 8.59 is objected to for nonreversible processes on the grounds that S is uniquely defined only for equilibrium conditions. No such objection is made against Equation 8.58 since the entropies are for equilibrium states 1 and 2.

The inequality of Equation 8.59 can be written more explicitly as follows:

$$dS > \frac{\delta Q}{T_{\text{boundary}}} \qquad dS = \frac{\delta Q}{T} \qquad dS < \frac{\delta Q}{T_{\text{boundary}}} \tag{8.60}$$

Irreversible   Reversible   Impossible

Note that for the irreversible case, T_{boundary} has been written as a reminder that the temperature in the inequality of Clausius refers strictly to the instantaneous temperature at the patch of the system boundary through which δQ flows to the system.

Let's now apply Equation 8.59 to both the system and the surroundings for a process. Assume that the heat source in the surroundings is instantaneously at T_0. For a continuum, the instantaneous temperature on either side of the system boundary is the same. Thus, by a judicious choice of the boundary, the temperature on the system side will also be T_0 regardless of whether the process is reversible or irreversible. Applying Equation 8.59 to the system and the surroundings in turn, the following is obtained for the process taking place over a small time interval:

$$dS_{\text{system}} \geq \frac{\delta Q}{T_0} \tag{8.61}$$

and

$$dS_{\text{surr}} \geq \frac{(-\delta Q)}{T_0} \tag{8.62}$$

Note that a positive δQ to the system corresponds to a negative δQ to the surroundings, and vice-versa.

Adding the entropy changes for the system and the surroundings gives the total increase of entropy for the universe in the incremental process. The inequalities of Equations 8.61 and 8.62 thus give

$$dS_{\text{universe}} = dS_{\text{system}} + dS_{\text{surr}} \geq \delta Q(1/T_0 - 1/T_0)$$

The right-hand side of this inequality is zero, and hence it can be written in the following final form:

$$dS_{\text{universe}} \geq 0 \tag{8.63}$$

In Equation 8.63, the equality applies only to a reversible process, while the inequality applies to real or irreversible processes. This result is often referred to as the *principle of increase of entropy*. It states that the total change in entropy for a system and its surroundings (or the total change of entropy for the universe) for any process is always greater than or equal to zero.

If no interaction occurs between the system and its surroundings such that $\delta Q = 0$, then the inequality of Equation 8.61 reduces to the following for an isolated system:

$$dS_{\text{isolated system}} \geq 0 \tag{8.64}$$

where the equality applies only to reversible processes and the inequality applies to irreversible processes. This result shows that entropy always increases for an irreversible process in an isolated system.

EXAMPLE 8.6

Computation of Entropy Increase for the Universe During a Process

Steam at 200°F is condensed at a constant pressure from saturated vapor (g) to saturated liquid (f). During the process, the heat of condensation is rejected to the surroundings, which are at 70°F. Compute the entropy change of the universe per lbm of steam condensed during the process.

Given: Steam at $T_1 = 200°F$ and $x_1 = 1$, condensed such that $T_2 = 200°F$ and $x_2 = 0$; surroundings at $T_0 = 70°F$ (530°R)
Heat of condensation rejected to surroundings

Find: $\Delta S_{universe}$ per lbm of steam condensed

Solution: Consider the initial mass of steam as the system, with m as the mass of the system in lbm.

First law analysis

For each lbm of saturated steam condensed, let q_0 be the heat rejected to the surroundings. For a constant pressure process, the reduced form of the NFEE is obtained by combining Equation 5.7 for the work done with Equation 6.10. Also, assume that changes in kinetic energy and potential energy are negligible. Thus,

$$\frac{Q_{12}}{m} = h_2 - h_1$$

Therefore, in this case,

$$-q_0 = h_f - h_g = -h_{fg}$$

At 200°F, $h_{fg} = 978$ Btu/lbm (from Table B.1 in Appendix B). Therefore, $q_0 = 978$ Btu/lbm.

Second law analysis

Define the system boundary such that the temperature is constant and only infinitesimally different from that of the surroundings in those regions where heat is rejected to the surroundings. For the system, the increase in entropy is given by

$$\Delta S_{system} = S_2 - S_1 = m(s_f - s_g)$$

For saturated water and saturated vapor at 200°F, $s_f = 0.2940$ Btu/lbm · °R and $s_g = 1.7761$ Btu/lbm · °R (from Table B.1 in Appendix B). Thus, $\Delta S = -1.4821m$ Btu/°R for the system, where m is the mass of the steam condensed (in lbm).

The heat addition to the surroundings takes place at a constant temperature T_0, which is also the temperature of the surroundings. The process is internally reversible (for the surroundings). Thus, using Equation 8.4 for entropy change gives

$$\Delta S_{surr} = \frac{mq_0}{T_0}$$

$$= \frac{978m \text{ Btu}}{530°R} = 1.845m \text{ Btu/°R}$$

The increase in entropy for the universe is obtained as the sum of the increases in entropy for the system and the surroundings:

$$\Delta S_{universe} = \Delta S_{system} + \Delta S_{surr}$$

$$= -1.4821m + 1.845m = 0.363m \text{ Btu/°R}$$

Thus, ΔS for the universe (per lbm of steam condensed) is 0.363 Btu/°R.

Entropy Generation

Since the total entropy always increases for an irreversible process, the term *entropy generation* can be introduced to specify the amount of entropy generated or produced inside the system due to the irreversibility of the process. Equation 8.59 can be written as follows:

$$dS - \frac{\delta Q}{T_{boundary}} \geq 0 \tag{8.59}$$

As stated in Equation 8.60, the equality applies only when the process is internally reversible for the system, while the inequality applies for all real (internally irreversible) processes. Therefore, a *nonnegative* term δS_{gen} can be added that provides a measure of the entropy generation within the system due to irreversibilities during the process. In this case, Equation 8.59 can be rewritten as

$$dS = \frac{\delta Q}{T_{boundary}} + \delta S_{gen} \tag{8.65}$$

The first term on the right-hand side of the equation represents the entropy change due to heat exchange with the surroundings. This term is often identified as an *entropy flux* that accompanies the flow of energy as heat δQ. It has been suggested that in Carnot's treatise of 1824, the much used "caloric" concept would be correct if it were interpreted as the entropy flux rather than as a flow of energy as heat. Also note that while no flow of entropy is associated with work, heat flow is always accompanied by a flow of entropy. This can be used as a basis for distinguishing between work and heat. Both heat and work are forms of energy flow across the boundary of a system. *Energy transfer as heat is always accompanied by a flow of entropy, while the flow of energy as work is not.*

The second term on the right-hand side of Equation 8.65 (δS_{gen}) represents the incremental amount of entropy generated inside the system during the process. The factors causing these internal irreversibilities are the same phenomena discussed earlier, including friction, energy flow through a nonzero temperature gradient, viscous dissipation, uncontrolled expansion, and several other phenomena. It turns out that all of these factors reduce the system's capacity to do work. Note that the entropy generation term is expressed as an inexact differential. The entropy generation is not a property because it depends on the process. In other words, it is a path function in the same sense that heat and work are also path functions. Thus, the specific process determines how much entropy is generated. Furthermore, the entropy generation provides a quantitative measure of irreversibility for the process.

A nondifferential form of Equation 8.65 can be developed from Equation 8.58:

$$S_2 - S_1 = \int_{\text{state } 1}^{\text{state } 2} \frac{\delta Q}{T_{\text{boundary}}} + S_{\text{gen}} \qquad (8.65a)$$

The first term on the right-hand side, again represents the entropy flux associated with the flow of energy as heat across the system boundary. S_{gen} is the total entropy generation inside the system for the irreversible process 1→2. The entropy generation is non-negative and is zero only for internally reversible processes.

Entropy Generation, Lost Work, and Irreversibility

Figure 8.6(a) illustrates a typical real process for a system that results in a change of state for the system from state 1 to state 2. The same changes of state occur for the system and the heat source shown in Figure 8.6(b) as in part (a) except that the arrangement in part (b) uses reversible cyclic devices to achieve complete reversibility of all the processes. In both cases, an ambient environment that can be regarded as a thermal reservoir at a constant temperature T_0 is assumed in the system's surroundings. Parts (a) and (b) each show two system boundaries, one for the original system and the other for an enlarged system that includes the heat source. The boundary for the enlarged system is judiciously chosen such that heat interaction with the ambient environment occurs at a boundary temperature equal to T_0. Considering the enlarged system of part (a) and applying the first law yields

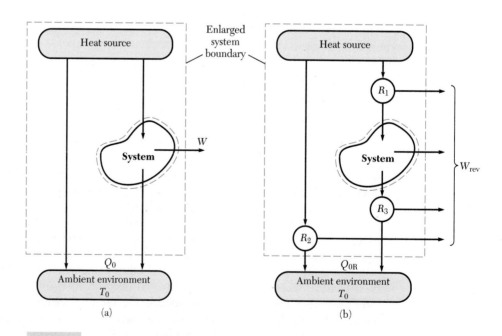

Figure 8.6 Work output for (a) a real process, and (b) completely reversible processes, in producing a change of state of a system.

$$-Q_0 - W = \Delta U_{system} + \Delta U_{source} \tag{8.66a}$$

where W is the actual work output from the enlarged system, Q_0 is the total heat transfer from the enlarged system to the ambient environment, and ΔU_{system} and ΔU_{source} are the changes in internal energy of the original system and the heat source, respectively. Next, the second law can be applied to determine the entropy changes for the enlarged system of part (a) using Equation 8.65a:

$$\Delta S_{system} + \Delta S_{source} = \frac{-Q_0}{T_0} + S_{gen} \tag{8.66b}$$

ΔS_{system} and ΔS_{source} are the changes in entropy for the original system and the heat source, respectively. S_{gen} is the total entropy generation in the enlarged system of part (a). Eliminating Q_0 between Equations 8.66a and 8.66b yields the following expression for the actual work (W):

$$W = T_0(\Delta S_{system} + \Delta S_{source})$$
$$- (\Delta U_{system} + \Delta U_{source}) - T_0 S_{gen} \tag{8.66}$$

The enlarged system in Figure 8.6(b) operates with complete reversibility with Q_{0R} equal to Q_0 in part (a) and both heat transfer processes being reversible. The changes of state produced in the enlarged system of part (b) are the same as for the enlarged system in part (a). (Note that the changes of state for the cyclic devices added in part (b) are zero because the devices are operated cyclically.) Thus, following the same procedure that led to Equation 8.66 for the real process of part (a), the result of applying the first and second laws to the reversible processes for the enlarged system of part (b) is

$$W_{rev} = T_0 (\Delta S_{system} + \Delta S_{source})$$
$$- (\Delta U_{system} + \Delta U_{source}) \tag{8.67}$$

Subtracting the actual work (W) given by Equation 8.66 from the reversible work (W_{rev}) leads to the following

$$(W_{rev} - W) = T_0 S_{gen} \tag{8.68}$$

where T_0 represents the temperature of the ambient environment and S_{gen} is the total entropy generation (or entropy production) during the real process. Note that since the heat transfer Q_0 to the ambient environment in Figure 8.6(a) occurs reversibly, zero entropy generation takes place internally to the ambient environment. Therefore, S_{gen} also represents the total entropy generation for the universe.

From Equation 8.68, the term *entropy generation* is directly related to what is often called *lost work* due to irreversibility of real processes. For a process that is designed to produce net positive work from a system, the actual work output (W) will fall short of the maximum possible (W_{rev}) by an amount equal to ($T_0 S_{gen}$), which is therefore regarded as lost work. For a completely reversible process, $S_{gen} = 0$ and $W = W_{rev}$.

Let's suppose instead that a process requires net work input to a system. From Equation 8.68, the actual work input ($-W$) must exceed the minimum required ($-W_{rev}$) by an amount equal to ($T_0 S_{gen}$). Several other processes are neither designed for "work production" nor for "work absorption." Equation 8.68 is still relevant and it provides a measure of loss in work capability equal to $(T_0\, S_{gen})$ that results from the irreversibility of the process. Thus, the term $(T_0\, S_{gen})$ is also referred to as the *irreversibility* (I) for an actual process.

Methods are available in the field of irreversible thermodynamics for computing the entropy generation in processes for several different systems. However, such analyses are beyond the scope of this text, and their use often focuses on specialized applications in engineering. The most important point at this time is to recognize that irreversibilities always exist for any real process; these irreversibilities lead to entropy generation and thus to lost work for any process. Thus, the reversible process always produces the maximum amount of work for a work-producing system and requires the minimum amount of work for a work-absorbing system.

1. A lump of iron at 800 K is quenched in a large mass of water at 300 K. The mass of the iron is 0.5 kg, and the mean specific heat of iron can be taken as 560 J/kg · K. Calculate
   (a) the heat transfer from the iron to the pool of water
   (b) the increase in entropy ΔS_{iron} of the lump of iron
   (c) the increase in entropy ΔS_{water} of the pool of water
   What is the $\Delta S_{universe}$?
   Answer: (a) 140 kJ, (b) -275 J/K, (c) 467 J/K

2. A volume of 0.1 m^3 of air (mass m_A) at 200 kPa and 300 K is separated by a thin diaphragm from another 0.1 m^3 of air (mass m_B) at 100 kPa and 300 K inside a rigid, well-insulated vessel. The diaphragm is ruptured so that the two air masses mix and attain a uniform pressure (p_f) and temperature (T_f). Assuming $R = 0.2871$ kJ/kg · K and $c_p = 1.005$ kJ/kg · K for the air, determine
   (a) m_A and m_B
   (b) T_f and p_f for the mixture
   (c) the entropy change for the system comprising the initial two air masses
   Answer: (a) 0.232 kg, 0.116 kg; (b) 300 K, 150 kPa; (c) 5.66 J/K

3. A quantity of 42 kJ of shaft work transferred to 1 kg of water in a well-insulated container produces an increase in temperature of the water from 300 K to 310 K. It is suggested that a spontaneous decrease in temperature of the water from 310 K to 300 K could occur leading to a shaft work transfer of 42 kJ to the surroundings. Calculate the change in entropy for the universe resulting from this hypothetical process, and hence, determine whether or not the suggested process could ever occur. Assume a specific heat of 4.2 kJ/kg · K for the water.
   Answer: -0.138 kJ/K

8.9 The Principle of Increase in Entropy for a Control Volume

The Basic Principle

The previous section considered the principle of increase in entropy for a closed system. This principle can be extended to an open system or a control volume using a method similar to that used in Chapter 6 for the application of the first law of thermodynamics to a control volume. Figures 8.7(a) and (b) depict a closed system at two instants in time, t and $(t + \Delta t)$. In Figure 8.7(a), the closed system comprises all the mass in the control volume as well as the incremental mass Δm_i that is admitted into the control volume in the interval t to $(t + \Delta t)$. The closed system in Figure 8.7(b) includes all the mass in the control volume at time $(t + \Delta t)$ in addition to the incremental mass Δm_e that leaves the control volume in the interval t to $(t + \Delta t)$.

Referring to Figure 8.7(a), the total entropy of the system at time t can be written as

$$S^{(t)} = S_{CV}^{(t)} + (\Delta m_i)s_i \tag{8.69}$$

where $S_{CV}^{(t)}$ is the instantaneous entropy of the mass contained in the control volume at time t. Similarly, at the slightly later time of $(t + \Delta t)$ illustrated in Figure 8.7(b), the total entropy of the system is

$$S^{(t+\Delta t)} = S_{CV}^{(t+\Delta t)} + (\Delta m_e)s_e \tag{8.70}$$

Subtracting Equation 8.69 from 8.70 gives

$$
\begin{aligned}
dS &= S^{(t+\Delta t)} - S^{(t)} \\
&= S_{CV}^{(t+\Delta t)} - S_{CV}^{(t)} + (\Delta m_e)s_e - (\Delta m_i)s_i
\end{aligned}
\tag{8.71}
$$

By substituting this expression for dS in Equation 8.59, which expresses the principle of increase of entropy for a closed system, the corresponding equation for the control volume is obtained:

$$S_{CV}^{(t+\Delta t)} - S_{CV}^{(t)} + (\Delta m_e)s_e - (\Delta m_i)s_i \geq \sum_{\substack{\text{control} \\ \text{surface}}} \frac{\delta Q}{T_{\text{boundary}}} \tag{8.72}$$

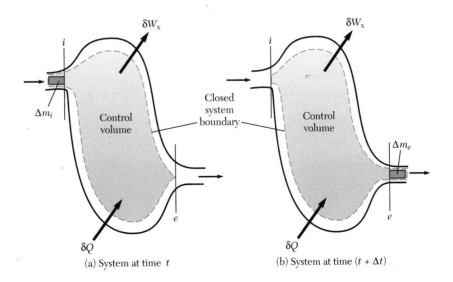

(a) System at time t (b) System at time $(t + \Delta t)$

Figure 8.7 The entropy balance equation for the control volume. (a) System at time t. (b) System at time $(t + \Delta t)$.

Dividing through by Δt leads to

$$\frac{\Delta S_{CV}}{\Delta t} + \left(\frac{\Delta m_e}{\Delta t}\right) s_e - \left(\frac{\Delta m_i}{\Delta t}\right) s_i \geq \sum_{\substack{\text{control} \\ \text{surface}}} \frac{1}{T_{\text{boundary}}} \frac{\delta Q}{\Delta t} \tag{8.73}$$

Taking limits as $\Delta t \to 0$ finally gives

$$\frac{dS_{CV}}{dt} + \dot{m}_e s_e - \dot{m}_i s_i \geq \sum_{\substack{\text{control} \\ \text{surface}}} \frac{\dot{Q}}{T_{\text{boundary}}} \tag{8.74}$$

This inequality is the *principle of increase of entropy for a control volume.* In words, it can be expressed as follows:

$$\begin{pmatrix} \text{Rate of} \\ \text{increase} \\ \text{of entropy} \\ \text{for mass} \\ \text{within CV} \end{pmatrix} \geq \begin{pmatrix} \text{Rate of flow} \\ \text{of entropy to CV} \\ \text{due to the} \\ \text{flow of heat} \end{pmatrix} + \begin{pmatrix} \text{Rate of} \\ \text{entropy} \\ \text{addition} \\ \text{due to} \\ \text{mass flow} \\ \text{into CV} \end{pmatrix} - \begin{pmatrix} \text{Rate of} \\ \text{entropy} \\ \text{depletion} \\ \text{due to} \\ \text{mass flow} \\ \text{out of CV} \end{pmatrix}$$

The expression has been derived for the *single mass flow* into and out of the control volume. Extension of the inequality to the *multiple stream* case yields the following:

$$\frac{dS_{CV}}{dt} + \sum_{\substack{\text{exit} \\ \text{streams}}} \dot{m}_e s_e - \sum_{\substack{\text{inlet} \\ \text{streams}}} \dot{m}_i s_i \geq \sum_{\substack{\text{control} \\ \text{surface}}} \frac{\dot{Q}}{T_{\text{boundary}}} \tag{8.75}$$

As indicated in Section 8.8 when discussing the criteria given in the expressions in Equation 8.60, the equality in Equations 8.74 and 8.75 applies for reversible processes only, while the inequality applies for the irreversible processes.

Equations 8.74 and 8.75 can also be expressed without the inequality by use of the entropy generation term in a manner similar to that given earlier for the closed system. Thus, invoking Equation 8.65 instead of Equation 8.59 and substituting for dS from Equation 8.71 leads to

$$\frac{dS_{CV}}{dt} + \dot{m}_e s_e - \dot{m}_i s_i = \sum_{\substack{\text{control} \\ \text{surface}}} \frac{\dot{Q}}{T_{\text{boundary}}} + \dot{S}_{\text{gen,CV}} \tag{8.76}$$

where

$$\dot{S}_{\text{gen,CV}} = \frac{\delta S_{\text{gen,CV}}}{dt} \tag{8.77}$$

and $\dot{S}_{\text{gen,CV}}$ is a nonnegative quantity and is the *entropy generation rate* in the control volume due to irreversibilities. For multiple streams into and out of the control volume, the equation corresponding to Equation 8.76 is

$$\frac{dS_{CV}}{dt} + \sum_{\substack{\text{exit} \\ \text{streams}}} \dot{m}_e s_e - \sum_{\substack{\text{inlet} \\ \text{streams}}} \dot{m}_i s_i = \sum_{\substack{\text{control} \\ \text{surface}}} \frac{\dot{Q}}{T_{\text{boundary}}} + \dot{S}_{\text{gen,CV}} \tag{8.78}$$

As discussed in Section 8.8, the entropy generation term (that is, the second term on the right-hand side of Eq. 8.78) provides a measure of the difference between the reversible process and the irreversible process. Furthermore, the entropy generation term is positive for any irreversible process and zero for a reversible process. As in the case of a closed system, the rate of entropy generation term provides a measure of the *lost power* due to the irreversibility of the process.

Steady-State Steady-Flow Process

The steady-state steady-flow process is one in which the mass flow rates and the properties within the control volume do not vary with time. When such is the case, dS_{CV}/dt becomes zero and Equation 8.78 reduces to

$$\sum_{\substack{\text{exit} \\ \text{streams}}} \dot{m}_e s_e - \sum_{\substack{\text{inlet} \\ \text{streams}}} \dot{m}_i s_i = \sum_{\substack{\text{control} \\ \text{surface}}} \frac{\dot{Q}}{T_{\text{boundary}}} + \dot{S}_{\text{gen,CV}} \qquad (8.79)$$

If only one mass flow enters with uniform properties and one mass flow exits with uniform properties, Equation 8.79 reduces to

$$\dot{m}(s_e - s_i) = \sum_{\substack{\text{control} \\ \text{surface}}} \frac{\dot{Q}}{T_{\text{boundary}}} + \dot{S}_{\text{gen,CV}} \qquad (8.80)$$

where the mass flow entering is equal to mass flow exiting for the steady-state steady-flow process.

Equations 8.79 and 8.80 provide definitive criteria based on the second law of thermodynamics as to whether a specified steady-state steady-flow process is possible or impossible. For a given process, either Equation 8.79 or 8.80 can be used to determine the entropy generation rate. If the entropy generation rate is zero, then the process is reversible. A positive entropy generation rate means that the process is possible but that it is irreversible. If a negative entropy generation rate is found, the process is impossible. Typical applications of these criteria are given later as worked examples.

For an adiabatic process with no heat exchange with the surroundings, Equations 8.79 and 8.80, respectively, reduce to

$$\sum_{\substack{\text{exit} \\ \text{streams}}} \dot{m}_e s_e - \sum_{\substack{\text{inlet} \\ \text{streams}}} \dot{m}_i s_i = \dot{S}_{\text{gen,CV}} \geq 0 \qquad (8.81a)$$

and

$$\dot{m}(s_e - s_i) = \dot{S}_{\text{gen,CV}} \geq 0 \qquad (8.81b)$$

Reversible Steady-State Steady-Flow Process

In a reversible process, the entropy generation term is zero. Thus, Equation 8.80 reduces to the following for a reversible, steady-state steady-flow process:

$$\dot{m}(s_e - s_i) = \sum_{\substack{\text{control} \\ \text{surface}}} \frac{\dot{Q}}{T_{\text{boundary}}} \tag{8.82}$$

If the process is also adiabatic, the equation further reduces to

$$s_e = s_i \tag{8.83}$$

The reversible adiabatic process was identified earlier for the closed system as an isentropic process. Equation 8.83 confirms that a reversible adiabatic flow process is also isentropic. The implied restrictions in Equation 8.83 are (1) that the mass flowing is a pure substance and (2) that only one mass flow occurs into and out of the control volume. Worked examples involving the use of Equation 8.83 were given in Section 8.6 when the isentropic process was discussed.

The Bernoulli Equation

Let's consider the steady-state steady-flow of a fluid through a *stream tube* or a *duct*. If the fluid flow is treated as frictionless, the shaft or shear work can be regarded as zero. Equation 6.29 is the version of the first law of thermodynamics that applies to this situation and can be rewritten as

$$\frac{\dot{Q}}{\dot{m}} = (h_e - h_i) + \left(\frac{(V_m)_e^2}{2} - \frac{(V_m)_i^2}{2} \right) + g(Z_e - Z_i) \tag{8.84}$$

Reversible adiabatic process

If the process is adiabatic, Equation 8.84 reduces to the following:

$$0 = (h_e - h_i) + \left(\frac{(V_m)_e^2}{2} - \frac{(V_m)_i^2}{2} \right) + g(Z_e - Z_i) \tag{8.85}$$

Assuming that the process is also reversible, Equation 8.83 shows that the specific entropy stays constant. Putting $ds = 0$ in Equation 8.14 and integrating gives

$$h_e - h_i = \int_{p_i}^{p_e} v \, dp \tag{8.86}$$

For an incompressible fluid, Equation 8.86 can be further simplified by performing the integration with v assumed constant:

$$h_e - h_i = v(p_e - p_i) \tag{8.87}$$

Substitution of the expression for $(h_e - h_i)$ in Equation 8.85 leads to

$$\frac{p_e}{\rho} + \frac{(V_m)_e^2}{2} + gZ_e = \frac{p_i}{\rho} + \frac{(V_m)_i^2}{2} + gZ_i \tag{8.88}$$

Equation 8.88 is the very familiar *Bernoulli equation* of fluid mechanics. It applies to the reversible, adiabatic process defined above.

Reversible isothermal process

Suppose you assume instead that the flow process is reversible and isothermal. Equation 8.82 can be applied, leading to the following equation:

$$T(s_e - s_i) = \frac{\dot{Q}}{\dot{m}} \qquad (8.89)$$

where T is constant and is the fluid temperature. By integrating the $T\,ds$ equation (Eq. 8.14) for an incompressible fluid and assuming a constant temperature T, the following equation is also obtained:

$$T(s_e - s_i) = h_e - h_i - v(p_e - p_i) \qquad (8.90)$$

Combining Equations 8.89 and 8.90 and substituting for the heat flow per unit mass of fluid in Equation 8.84 leads to

$$0 = v(p_e - p_i) + \left(\frac{(V_m)_e^2}{2} - \frac{(V_m)_i^2}{2} \right) + g(Z_e - Z_i) \qquad (8.91)$$

Equation 8.91 is identical to Equation 8.88, which has been identified as the Bernoulli equation. Thus, Equation 8.88 applies to both a reversible adiabatic process and a reversible isothermal process for an incompressible fluid. Any reversible process can be constructed from a large number of combinations of reversible adiabatic and reversible isothermal processes. Therefore, it is concluded that Equation 8.88 applies to any reversible, steady-state steady-flow process with zero work for an incompressible fluid.

Equation 8.88 has widespread application to such processes in the fluid mechanics field. It is an unusual equation because it can also be developed from Newton's second law of motion. Normally, identical equations cannot be obtained from independent laws for a process. However, this is a case in which independent physical laws lead to the same equation for the set of assumptions that are specified.

Note from Equation 6.29 that for the case of a reversible adiabatic process with negligible changes in kinetic and potential energy, the work can be computed as

$$\frac{\dot{W}}{\dot{m}} = w = h_i - h_e \qquad (8.92)$$

Combining Equations 8.87 and 8.92 also gives the equation for computation of pump work for an incompressible fluid, which was developed in Example 8.5.

Shaft Work in Quasi-Equilibrium Processes

Consider a steady-state steady-flow of a pure substance through a device. Assume that a single mass flow occurs into the device and a single mass flow out of the device. The objective is to obtain an expression for the shaft work done in a reversible flow process. Conceptually, the situation can be viewed as in Figure 8.8, which depicts the steady-flow process as equivalent to a steady flow through a succession of miniature control volumes each of which instantaneously holds a fixed mass at a near-equilibrium condition. The entry and exit conditions for a typical miniature control volume are indicated

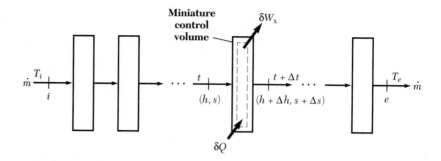

Figure 8.8 A conceptual representation for shaft work done in a reversible flow process for a control volume.

in Figure 8.8. In a short time interval between t and $(t + \Delta t)$, the heat input and the shaft power output are δQ and δW_x, respectively. The control volume is small enough to justify the assumption that the temperature, T, at the control surface is effectively equal to that of the fluid mass in the control volume. Also, it is assumed that the material stream enters the miniature control volume with specific enthalpy and entropy of h and s, respectively, while leaving with corresponding values of $(h + \Delta h)$ and $(s + \Delta s)$.

Thus, at steady state, applying Equation 8.80 to the miniature control volume gives

$$\delta S_{gen,CV} + \frac{\delta Q}{T} + (\dot{m}\,\Delta t)[s - (s + \Delta s)] = 0 \tag{8.93}$$

If the steady-flow energy Equation 6.29 is applied under the assumption that changes in kinetic energy and potential energy for the substance are negligible, it follows that

$$\delta Q - \delta W_x = (\dot{m}\,\Delta t)\,\Delta h \tag{8.94}$$

Thus, δQ can be eliminated between Equations 8.93 and 8.94 to obtain

$$\frac{\delta W_x}{\Delta t} = \dot{m}(T\,\Delta s - \Delta h) - T\left(\frac{\delta S_{gen,CV}}{\Delta t}\right) \tag{8.95}$$

The term $T\,\Delta s - \Delta h$ can be replaced by $-v\,\Delta p$ using Equation 8.14 derived earlier. Therefore, taking the limit as Δt approaches zero, the shaft power output is obtained from the miniature control volume as

$$\frac{\delta W_x}{dt} = \dot{m}(-v\,dp) - T\left(\frac{\delta S_{gen,CV}}{dt}\right) \tag{8.96}$$

If the process is reversible, then $\delta S_{gen,CV}/dt$ must be zero, and hence, the shaft power output from the miniature control volume is given from Equation 8.96 by

$$\left(\frac{\delta W_x}{dt}\right)_{rev} = -\dot{m}v\,dp \tag{8.97}$$

The specific shaft work output, $(\delta w)_{rev}$, for the reversible case (from each miniature control volume) is thus

$$\delta w_{rev} = \frac{\left(\dfrac{\delta W_x}{dt}\right)_{rev}}{\dot{m}} = -v \, dp \tag{8.98}$$

Thus, in a steady-flow reversible process with the fluid stream entering the control volume at state 1 and leaving at state 2, the specific shaft work output $(w)_{rev}$ is given by the summation of $(\delta w)_{rev}$ in Equation 8.98:

$$w_{rev} = \sum \delta w_{rev} = -\int_{p_1}^{p_2} v \, dp \tag{8.99}$$

It should be noted that Equation 8.99 for reversible shaft work is valid for both compressible and incompressible flows. The previously derived expression for pump work for incompressible flows given in Example 8.5 is one specific example of the case.

Shaft Work in Reversible Processes Involving an Ideal Gas in a Control Volume

The ideal gas pvT equation is $pv = RT$. This equation, together with the process path equation, will be used for the derivation of the shaft work in reversible flow processes using Equation 8.99.

Isothermal Processes

If the temperature (T) is constant, the following can be calculated from Equation 8.99 after substituting RT/p for v for an ideal gas:

$$w_{rev} = -RT \ln \frac{p_2}{p_1} = RT \ln \frac{p_1}{p_2} \tag{8.100}$$

Polytropic Processes

If $pv^n = \lambda^n$ (a constant) and the process is quasi-equilibrium, w_{rev} can be found from Equation 8.99 as

$$w_{rev} = -\lambda \int_{p_1}^{p_2} \frac{dp}{p^{1/n}} = -\lambda \left[\frac{p^{1-1/n}}{1 - 1/n} \right]_{p_1}^{p_2} \tag{8.101}$$

$$= \frac{n}{n - 1} (p_1 v_1 - p_2 v_2)$$

Using $pv = RT$ finally allows expression of Equation 8.101 as

$$w_{rev} = \frac{nR}{n - 1}(T_1 - T_2) \tag{8.102}$$

These expressions can be used to compute reversible shaft work for the various idealized processes.

Total Entropy Change for the Universe for a Process Involving Control Volume

The *total entropy change*, or the entropy change of the universe, for a process involving a control volume is equal to the entropy change of the control volume plus the entropy change of the surroundings. Consider Figure 8.9, which illustrates the total entropy change of the universe for a process involving a control volume.

In this case, both mass and heat transfer occur between the control volume and the surroundings. The total entropy change for the control volume was given by Equation 8.78. The mass flow into the control volume is out of the surroundings, the mass flow out of the control volume is into the surroundings, and the heat transfer into the control volume is out of the surroundings. Thus, an equation similar to Equation 8.78 can be written for the surroundings as

$$\frac{dS_{surr}}{dt} - \underset{\substack{\text{inlet} \\ \text{streams}}}{\sum} \dot{m}_e s_e + \underset{\substack{\text{exit} \\ \text{streams}}}{\sum} \dot{m}_i s_i = \underset{\substack{\text{control} \\ \text{surface}}}{\sum} \frac{(-\dot{Q})}{T_0} + \dot{S}_{gen,surr} \qquad (8.103)$$

The designations *inlet* and *exit* in Equation 8.103 refer to streams entering and leaving the *surroundings*. A heat exchange with the surroundings has been assumed and is considered to be at a constant temperature T_0. This implies that the control surface is defined such that the temperature $T_{boundary}$ in Equation 8.78 is also constant for the control volume and is equal to T_0. Adding the total entropy change for the control volume and the surroundings, which are given by Equations 8.78 and 8.103, leads to

$$\frac{dS_{total}}{dt} = \frac{dS_{CV}}{dt} + \frac{dS_{surr}}{dt}$$
$$= \dot{S}_{gen,CV} + \dot{S}_{gen,surr} \qquad (8.104)$$

As indicated earlier, the entropy generation rate is always a nonnegative quantity. Consequently, the expression for the total entropy change of the universe can be written as

$$\frac{dS_{total}}{dt} = \dot{S}_{gen,CV} + \dot{S}_{gen,surr} \geq 0 \qquad (8.105)$$

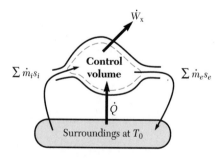

Figure 8.9 The total entropy change for the universe comprising the surroundings at a constant temperature and a control volume.

Thus, the total entropy, or the entropy of the universe, always *increases* for any real process regardless of whether the system is closed or open. The product $T_0 \cdot (\dot{S}_{gen,CV} + \dot{S}_{gen,surr})$ gives the lost power due to irreversibilities.

If there is no entropy generation in the surroundings during the process, then Equation 8.103 can be written for some interval of time as

$$(S_2 - S_1)_{surr} = \sum_{\substack{inlet \\ streams}} m_e s_e - \sum_{\substack{exit \\ streams}} m_i s_i - \frac{Q}{T_0} \qquad (8.106)$$

where Q now represents the total heat flow from the surroundings to the control volume.

EXAMPLE 8.7
Emptying a Tank

The rigid tank that was shown in Figure 6.13 is initially filled with steam at $p = 5$ bar and $T = 600°C$. Steam at $400°C$ and 2 bar flows in the pipeline to which the tank is connected via a valve. The valve is opened slightly, and steam flows slowly from the tank until the pressure in the tank decreases to 2 bar. The process is adiabatic.

(a) Show that if steam discharge from the tank occurred slowly, then the specific entropy of the fluid in the tank would remain constant.

(b) Determine the final temperature of the steam remaining in the tank.

Given: Open system—steam in a rigid tank at $p_1 = 5$ bar, $T_1 = 600°C$
Process—steam discharged slowly and adiabatically ($\dot{Q} = 0$) from rigid tank until $p_2 = 2$ bar in the tank

Find:
(a) Prove that $s = $ constant in the tank
(b) T_2 for the steam in the tank

Solution: Define the steam in the tank as the open system (or control volume) and the inside surface of the tank as the control surface (see Figure 6.13).

(a) The key assumption is that the steam discharge from the tank occurred sufficiently slowly to ensure that the fluid in the tank passed through a series of equilibrium states. In other words, $\dot{S}_{gen,CV} = 0$ and $s_e = s_{CV}$ may be assumed since the instantaneous property values must be uniform throughout the control volume at each equilibrium state. Also, $\dot{Q} = 0$ since the process is adiabatic. Thus, Equation 8.76 applied to the transient steam discharge process reduces to

$$\frac{dS_{CV}}{dt} + \dot{m}_e s_{CV} = 0 \qquad (1)$$

Use the mass balance Equation 6.14 to obtain

$$-\dot{m}_e = \frac{dm_{CV}}{dt} \qquad (2)$$

Finally, use $S_{CV} = m_{CV}\, s_{CV}$ to obtain

$$\frac{dS_{CV}}{dt} = m_{CV}\frac{ds_{CV}}{dt} + s_{CV}\frac{dm_{CV}}{dt} \qquad (3)$$

Combining Equations (1) through (3) yields

$$\frac{ds_{CV}}{dt} = 0$$

and, hence, s_{CV} = constant for the steam in the control volume. (Note that while the entire process, including the mixing of two streams of steam at different states, was highly irreversible, the changes in the control volume took place reversibly and adiabatically with the result that s remained constant throughout the process for the steam in the tank.)

(b) Use the **STEAM** code to obtain s_i = 8.352 kJ/kg · K for steam initially in the rigid tank at 5 bar and 600°C. The final state for the steam left in the tank is p_2 = 2 bar and $s_2 = s_1$ = 8.352 kJ/kg · K. Using the **STEAM** code again gives T_2 = 443.4°C. Thus, the final temperature of steam in the tank is 443°C.

EXAMPLE 8.8
An Inventor's Claim Examined

An inventor claims she has a device that uses 10 kg/s of geothermal water at 95°C and 1 atm pressure and produces a mechanical shaft power of 400 kW. The geothermal water stream leaves the device at 1 atm pressure and 30°C. The device rejects heat continuously to the ambient air at 1 atm and 25°C. Determine the heat rejection rate from the device to the surroundings. Is the inventor's claim plausible?

Given: The hypothetical device is illustrated in Figure 8.10. The fluid is water, and the conditions at inlet and exit are as follows:

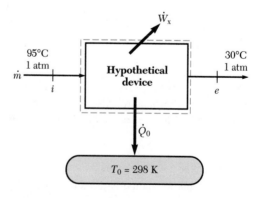

Figure 8.10 The hypothetical device of Example 8.8.

$$p_i = 1 \text{ atm} = p_e; \; T_i = 95°C, \; T_e = 30°C; \; T_{surr} = T_0 = 25°C$$

$$\dot{m} = 10 \text{ kg/s}$$

Find:

(a) Heat rejection rate to the surroundings

(b) $\dot{S}_{gen,CV}$ and hence if the claim is plausible

Solution: In Figure 8.10, the control surface is defined so as to enclose the hypothetical device. $\dot{Q}_0$ denotes the heat rejection rate from the control volume to the surroundings at T_0.

(a) *First law analysis*

The SFEE (Eq. 6.29) can be applied to this case. Thus, if changes in kinetic energy and potential energy are assumed negligible,

$$(-\dot{Q}_0) - \dot{W}_x = \dot{m}(h_e - h_i)$$

The **STEAM** code gives $h_i = 398.1$ kJ/kg and $h_e = 125.8$ kJ/kg. Thus,

$$\dot{Q}_0 = \dot{m}(h_i - h_e) - \dot{W}_x$$

$$= (10 \text{ kg/s})(398.1 - 125.8 \text{ kJ/kg}) - 400 \text{ kW} = \underline{2323 \text{ kW}}$$

The heat rejection rate is 2323 kW.

(b) *Second law analysis*

The entropy generation rate in the control volume is computed using Equation 8.80:

$$\dot{S}_{gen,CV} = \dot{m}(s_e - s_i) - \frac{-\dot{Q}_0}{T_0}$$

The use of T_0 for $T_{boundary}$ is justified since the control surface can be chosen such that the boundary temperature is that of ambient air, which is at T_0. The **STEAM** code gives $s_i = 1.25$ kJ/kg · K and $s_e = 0.437$ kJ/kg · K. Substituting in the previous equation for the entropy generation rate leads to

$$\dot{S}_{gen,CV} = (10 \text{ kg/s})(0.437 - 1.25 \text{ kJ/kg} \cdot K) + \frac{2323 \text{ kW}}{298 \text{ K}}$$

$$= -0.335 \text{ kW/K (impossible!)}$$

The entropy generation rate in the control volume cannot be negative. Hence, the hypothetical device is impossible.

EXAMPLE 8.9
The Vortex Tube

A hypothetical device is supplied with 2 kg/s of air at 400 kPa and 300 K. Two separate streams of air leave the device. Each stream is at an ambient pressure of

100 kPa, and the mass flow rate is the same for both streams. The inventor claims that one of the exit streams is at 330 K while the other is at 270 K. The ambient environment is at 300 K. Determine whether such a device is possible.

Given: The hypothetical device is illustrated in Figure 8.11.

$$p_1 = 400 \text{ kPa}, \ p_2 = 100 \text{ kPa}, \ p_3 = 100 \text{ kPa}, \ p_0 = 100 \text{ kPa}$$

$$T_1 = 300 \text{ K}, \ T_2 = 330 \text{ K}, \ T_3 = 270 \text{ K}, \ T_0 = 300 \text{ K}$$

Find: Is the device possible?

Solution: This is a control volume problem. The control surface is indicated in Figure 8.11. No shaft work is involved, but heat interaction can occur with the surroundings. In this case, application of the multiple stream version of the SFEE (Eq. 6.30) indicates a zero heat transfer rate.

Second law analysis

The entropy generation rate for the control volume is calculated using Equation 8.81a:

$$\dot{S}_{\text{gen,CV}} = \sum_{\substack{\text{exit} \\ \text{streams}}} \dot{m}_e s_e - \sum_{\substack{\text{inlet} \\ \text{streams}}} \dot{m}_i s_i$$

$$= \dot{m}_2 s_2 + \dot{m}_3 s_3 - \dot{m}_1 s_1$$

$$= \dot{m}_2 (s_2 - s_1) + \dot{m}_3 (s_3 - s_1)$$

The last step in this equation follows from mass balance considerations. In other words, the supply mass flow rate must be equal to the sum of the mass flow rates of the two exit streams. The changes in specific entropy terms can be determined using Equation 8.25 and assuming $c_{p0} = 1.005 \text{ kJ/kg} \cdot \text{K}$ and $R = 0.2871 \text{ kJ/kg} \cdot \text{K}$ for air. Thus,

$$s_2 - s_1 = c_{p0} \ln \frac{T_2}{T_1} - R \ln \frac{p_2}{p_1}$$

$$= (1.005 \text{ kJ/kg} \cdot \text{K}) \ln \frac{330 \text{ K}}{300 \text{ K}} - (0.2871 \text{ kJ/kg} \cdot \text{K}) \ln \frac{100 \text{ kPa}}{400 \text{ kPa}}$$

$$= 0.494 \text{ kJ/kg} \cdot \text{K}$$

Likewise,

$$s_3 - s_1 = c_{p0} \ln \frac{T_3}{T_1} - R \ln \frac{p_3}{p_1}$$

$$= (1.005 \text{ kJ/kg} \cdot \text{K}) \ln \frac{270 \text{ K}}{300 \text{ K}} - (0.2871 \text{ kJ/kg} \cdot \text{K}) \ln \frac{100 \text{ kPa}}{400 \text{ kPa}}$$

$$= 0.292 \text{ kJ/kg} \cdot \text{K}$$

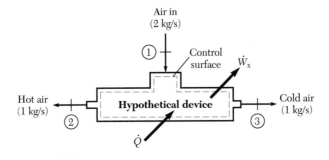

Figure 8.11 The hypothetical device of Example 8.9.

Hence, the entropy generation rate is 0.786 kW/K. Since this is definitely nonnegative, the device is possible. Such devices actually exist and are called *vortex tubes*. Although they typically have low efficiencies, their suitability for certain applications cannot be ignored. Typical applications include rapid cooling of soldered parts, electronic component cooling, cooling of machine operations, and several other industrial spot cooling operations. The vortex tube is essentially a passive device with no moving mechanical parts. It is therefore relatively maintenance free and durable.

EXERCISES

for Section 8.9

1. Water is heated from 25°C to 90°C as it flows at a rate of 0.5 kg/s through a tube that is immersed in a hot bath at 100°C. Define the tube through which the water flows as the control volume and calculate
   (a) $\dot{Q}$, in kW
   (b) $\dot{m}_{water}(s_e - s_i)$, in kW/K
   (c) $\dfrac{dS_{oil\ bath}}{dt}$, in kW/K
   (d) $\dfrac{dS_{universe}}{dt}$, in kW/K
   (e) the lost power, in kW (assume $T_0 = 298$ K)
   Answer: (a) 136 kW, (b) 0.413 kW/K, (c) -0.364 kW/K, (d) 0.0485 kW/K, (e) 14.5 kW

2. Assume that in Exercise 1, the specified temperature rise for the water is produced in an adiabatic process using an electric strip heater. Define the control volume to include the strip heater as well as the adiabatic enclosure through which the water flows. Supply of electricity is from outside the control volume. Calculate
   (a) $\dot{W}_x$, in kW
   (b) $\dot{m}_{water}(s_e - s_i)$, in kW/K
   (c) $\dfrac{dS_{universe}}{dt}$, in kW/K
   (d) the lost power, in kW (assume $T_0 = 298$ K)
   Answer: (a) -136 kW, (b) 0.413 kW/K, (c) 0.413 kW/K, (d) 123 kW

3. A steady stream of air is compressed reversibly and isothermally from a pressure of 1 bar to 5 bar. The temperature of the air is 300 K. Determine the work transfer to the surroundings, per unit mass of air, in the compression process. R for air is 0.2871 kJ/kg · K.

   Answer: -139 kJ/kg

4. A steady stream of air is compressed isentropically from a pressure of 1 bar to 5 bar. The air temperature at the entry to the compressor is 300 K. Determine
   (a) the exit temperature of the air
   (b) the work transfer to the surroundings, per unit mass of air, in the compression process

   Answer: **(a)** 475 K, **(b)** -176 kJ/kg

8.10 Efficiency of Devices

The previous chapter dealt with cycles and specified the first law efficiency for a cyclic device as

$$\text{Efficiency } (\eta_1) = \frac{\text{desired output}}{\text{required input}} \qquad (7.1)$$

The present chapter has been considering processes in general, including noncyclic processes. Of interest here are a variety of devices, such as turbines, compressors, pumps, nozzles, and diffusers, that undergo noncyclic processes and typically deliver a desired output. An efficiency for any output device undergoing a process is specified as

$$\frac{\text{Process efficiency}}{\text{of output device}} = \frac{\text{actual output of device}}{\text{ideal output of device}} \qquad (8.107)$$

Conversely, if the device requires a work or heat input to accomplish a specified task, the process efficiency for the device is specified as

$$\frac{\text{Process efficiency}}{\text{of input device}} = \frac{\text{ideal input required by device}}{\text{actual input to device}} \qquad (8.108)$$

For all devices, the best possible performance is achieved for a reversible process with no entropy generation and no losses. In the case of *adiabatic devices*, the processes have to be reversible and adiabatic, in other words, *isentropic*. For *isothermal devices*, the best possible performance is a *reversible isothermal process* for the device. Now let's consider some specific devices.

Turbines

A *turbine* is a device that expands a fluid from a high pressure to a low pressure while producing work. Most turbines are designed for adiabatic flow through the device. Figure 8.12 shows a typical process for adiabatic flow through a turbine. The ideal process is an isentropic process from state 1 to state 2s, while the actual process goes from state

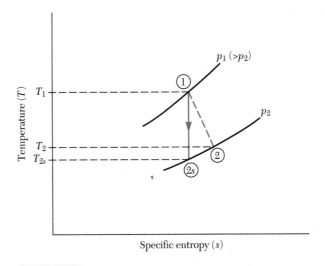

Figure 8.12 *T-s* diagram for flow through an adiabatic turbine.

1 to state 2 with an associated increase in entropy due to entropy generation caused by irreversibilities. In this case, the process efficiency can be specified as

$$\eta_{\substack{\text{adiabatic} \\ \text{turbine}}} = \frac{w_a}{w_s} \tag{8.109}$$

where w_a represents the actual specific work output of the device and w_s represents the ideal specific work output of the device with isentropic flow assumed through the device. This efficiency is also referred to as the *isentropic efficiency* for the turbine. For adiabatic flow through the turbine with negligible changes in kinetic and potential energy, the efficiency of the device can also be expressed in terms of enthalpy as

$$\eta_{\substack{\text{adiabatic} \\ \text{turbine}}} = \frac{h_1 - h_2}{h_1 - h_{2s}} \tag{8.110}$$

Compressors

A *compressor* is a device that requires work input to compress a gas from a low pressure to a higher pressure. Most compressors are designed for adiabatic flow through the device. Figure 8.13 shows a typical process for adiabatic flow through a compressor. In this case, the ideal process is the reversible adiabatic process that goes from state 1 to state 2s, while the actual process involves an increase in entropy and goes from state 1 to state 2. Since the compressor is a work input (or work-absorbing) device, Equation 8.108 can be expressed mathematically for the compressor as

$$\eta_{\substack{\text{adiabatic} \\ \text{compressor}}} = \frac{w_s}{w_a} \tag{8.111}$$

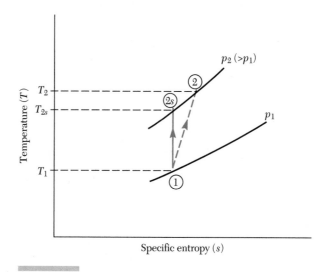

Figure 8.13 T-s diagram for flow through an adiabatic compressor.

This is also referred to as the *isentropic efficiency* of the compressor. With negligible changes in kinetic and potential energy through the compressor, the efficiency of the adiabatic compressor can also be expressed in terms of enthalpy as

$$\eta_{\substack{\text{adiabatic}\\\text{compressor}}} = \frac{h_{2s} - h_1}{h_2 - h_1} \tag{8.112}$$

The gas flowing through the compressor can be cooled by fins or other means to approximate isothermal flow through the compressor. The ideal process for such a flow is the reversible isothermal process shown in Figure 8.14. Compared to this ideal process, the efficiency for the actual flow through the isothermal compressor is defined by

$$\eta_{\substack{\text{isothermal}\\\text{compressor}}} = \frac{w_T}{w_a} \tag{8.113}$$

where w_T represents the work for the reversible isothermal process between states 1 and 2 for the device.

Pumps

The term *pump* is generally used for any device that increases the pressure of a liquid, while the term *compressor* is used for a device that increases the pressure of a gas. Both devices operate to increase the pressure of the fluid flowing through the device. Thus, with the same assumptions as just outlined for the compressor, equations comparable to Equations 8.111, 8.112, and 8.113 can be written for pumps. For adiabatic flow through a pump, the efficiency is given by

$$\eta_{\substack{\text{adiabatic}\\\text{pump}}} = \frac{w_s}{w_a} \tag{8.114}$$

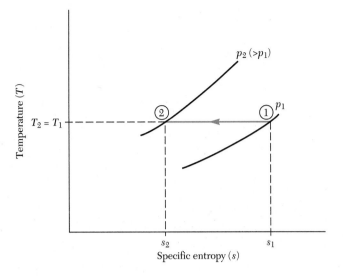

Figure 8.14 *T-s* diagram for flow through an isothermal compressor.

which can be specified in terms of enthalpy for negligible changes in kinetic and potential energy through the device as

$$\eta_{\substack{\text{adiabatic} \\ \text{pump}}} = \frac{h_{2s} - h_1}{h_2 - h_1} \tag{8.115}$$

Efficiency for isothermal flow through the pump is given by

$$\eta_{\substack{\text{isothermal} \\ \text{pump}}} = \frac{w_T}{w_a} \tag{8.116}$$

As discussed earlier, the reversible adiabatic flow of an incompressible liquid through a pump is also isothermal flow. Therefore, w_T equals w_s for this case. For many pumping processes involving incompressible liquids, the known quantities are the pressures at the beginning and end states, and the isentropic efficiency of the device. In such cases, Equation 8.42 can be used in conjunction with Equation 8.115 to determine the end state of the fluid for the pumping process.

Nozzles for Subsonic Flow

A *nozzle* is a device that expands a fluid flowing at subsonic speed from a high pressure to a lower pressure while increasing the kinetic energy of the fluid flowing through the nozzle. A schematic diagram of such a device is shown in Figure 8.15(a). The flow through a nozzle can be approximated by adiabatic flow. Thus, the ideal process for a nozzle is the reversible adiabatic or the isentropic process. Flow through a nozzle is similar to that given in Figure 8.12 for adiabatic flow through a turbine except that no work is extracted from the fluid flowing through the nozzle. The ideal process is the isentropic process, while the actual process involves an increase in entropy for the

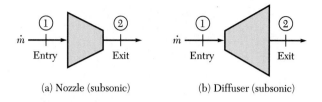

Figure 8.15 Schematic diagrams of (a) a subsonic nozzle and (b) a subsonic diffuser.

process. The efficiency for the nozzle compares the actual exit kinetic energy to the exit kinetic energy that would have been achieved for an isentropic process. The inlet state and the exhaust pressures are the same for both processes. Thus, the nozzle efficiency is

$$\eta_{nozzle} = \frac{(KE)_a}{(KE)_s} = \frac{\frac{1}{2}(V_m)_2^2}{\frac{1}{2}(V_m)_{2s}^2}$$

Diffusers for Subsonic Flow

While a nozzle is used to increase a fluid's kinetic energy by decreasing its pressure, a *diffuser* increases the fluid's pressure by decreasing its kinetic energy. Figure 8.15(b) shows a schematic diagram of a diffuser. While the cross-sectional area decreases in the direction of the flow for the nozzle, the cross-sectional area increases in the direction of the flow for the diffuser. In both cases, the ideal flow is isentropic.

The efficiency of the diffuser (assuming negligible exit velocity) is defined as

$$\eta_{diffuser} = \frac{\Delta h_s}{\frac{1}{2}(V_m)_1^2} \tag{8.118}$$

where Δh_s represents the enthalpy increase through the diffuser for isentropic flow.

Thus, it is apparent that the efficiency of all devices for a process is a comparison of the actual performance with the ideal performance for such a device. In general, the ideal performance for most devices of the type just considered is obtained with a reversible adiabatic or isentropic process.

EXERCISES

for Section 8.10

1. Steam at 750 psia and 950°F expands through an adiabatic turbine in a steady-flow process. The exit pressure of the steam is 1 psia. If the isentropic efficiency of the turbine is 0.9, determine
   (a) the vapor quality at the exit to the turbine
   (b) the work output of the turbine, in Btu/lbm of steam
   Answer: (a) 0.866, (b) 498 Btu/lbm

2. A steady flow of air at 1 atm and 300 K is compressed adiabatically to a pressure of 8 atm. The isentropic efficiency of the compressor is 0.85. Calculate
   (a) the temperature of the air at exit to the compressor
   (b) the work input to the compressor, per unit mass of air
   Answer: (a) 586 K, (b) 288 kJ/kg

8.11   *Summary*

This chapter showed that a property called *entropy* could be defined as a consequence of the second law. The property is particularly useful and relevant to performance evaluation of systems and processes. Entropy can be specified for any thermodynamic state of a pure substance and is one of the thermodynamic coordinates that can be used to define the thermodynamic state of a system.

Isentropic or constant entropy processes are frequently encountered as part of many ideal cycles of interest in engineering. Furthermore, all real processes result in an increase in entropy of the universe; that is, the total entropy of the combined system and surroundings always increases for real processes. This increase in entropy is caused by the irreversibilities (such as friction and dissipation) that are observed in nature for real processes. For devices that operate isentropically in the ideal case, a process or isentropic efficiency has been defined that provides a measure of the extent to which the actual performance of the device comes close to that of the ideal case.

For any process, heat produces a transfer of both energy and entropy across the system boundary, while work produces only energy transfer. Another significant finding was that the work produced by any device is always less than the ideal work that could be produced by the device in the absence of irreversibilities. The difference between the work produced for a reversible process and the actual work is called the *lost work* and can be expressed in terms of the entropy generation for the process. For all real processes the entropy of the universe increases, and this increase in entropy represents a deterioration of the quality of energy and is closely related to the concept of lost work. When all processes are reversible, however, no change occurs in the entropy of the universe. One purpose of thermodynamic design analyses is to devise thermodynamic systems to optimize the utilization of energy and thereby minimize the increase in entropy for thermodynamic processes.

References

Atkins, P. W. 1984. *The Second Law*. New York: Scientific American Library.

Kestin, J. ed. 1976. *The Second Law of Thermodynamics*. Benchmark Papers on Energy, Vol. 5. Stroudsburg, PA: Dowden, Hutchinson, and Ross, Inc.

Meixner, J. "On the Foundation of Thermodynamics of Processes," *in* Kestin, J. ed. 1976. 313–323.

Questions

1. Which one of the following statements about entropy is correct?
   - (a) Entropy is defined as $dS = \delta Q/T$ regardless of whether the process is reversible or nonreversible.
   - (b) The entropy of a closed system cannot decrease.
   - (c) The entropy of the universe cannot decrease.

2. A reversible Carnot engine operates between thermal reservoirs at source and sink temperatures of T_H and T_L, respectively. Show this cycle on *T-s* coordinates.

3. Illustrate on a *T-s* diagram a reversed Carnot cycle for source and sink temperatures of T_L and T_H, respectively.

4. A student has analyzed the Carnot cycle of Figure 8.5(a) for an ideal gas and advised you that the efficiency is zero because (1) ideal gas enthalpy is a function of temperature only; (2) the heat transferred to the ideal gas from state 1 to 2 is zero because $q = (h_2 - h_1)$ and $h_2 = h_1$ (both functions of temperature only), and similarly, the heat transferred from state 3 to 4 is also zero; and (3) the process work from state 2 to 3 is $-W_{2\text{-}3} = \Delta h = f(T_2, T_3)$ only, and similarly, for process $4\rightarrow1$, $-W_{4\text{-}1} = \Delta h = f(T_4, T_1)$ only. Therefore, since $T_1 = T_2$ and $T_3 = T_4$, $W_{2\text{-}3} + W_{4\text{-}1} = 0$. Now he says that since the net work for the cycle is zero, the efficiency is also zero for the Carnot cycle for an ideal gas. Do you agree with his conclusions? If not, how can you show the fallacy of his arguments?

5. A fixed mass of an ideal gas is held in a rigid container. If heat is added to the gas, will the entropy of the gas increase, decrease, or stay constant? Explain your answer.

6. A few blocks of ice are added to hot water contained in a well-insulated vessel. Will the total entropy of the mixture be higher or lower than the sum of the initial entropy of the ice and the water?

7. Define *entropy*. Suggest an experimental method for the determination of s_{if} for ice at its melting point, where $s_{if} = s_f - s_i$.

8. Write down the version of the intensive $T\,ds$ equation that involves specific enthalpy. What are the restrictions, if any, on the applicability of this equation?

9. Show the following processes for an ideal gas on a T-s diagram:
   (a) isobaric heating process at a pressure p_0 from a temperature T_0 to a final temperature T_1
   (b) isochoric heating process from the same initial state as in part (a) to a final temperature T_1
   (c) isentropic compression from the same initial state as in part (a) to a final temperature of T_1

10. Write down the expression for lost work in terms of entropy generation in a nonflow process. Can the entropy generation in a process ever be a negative quantity?

11. Sketch the h-s diagram for a pure substance such as water. Using the intensive $T\,ds$ equation as a guide, show a constant pressure $(p < p_c)$ line on the diagram to traverse the compressed liquid, wet vapor, and superheated vapor regions, respectively.

Problems

8.1. A heat transfer of 15 kJ occurs to a thermal reservoir at 300 K. What is the increase in entropy of the reservoir?

8.2. During a certain reversible process for a closed system, 500 kJ was added as heat to the system at a constant temperature of 500 K. What was the entropy change for the system during this process?

8.3. During a certain reversible process for a closed system, the entropy of the system increased by 2 Btu/°R. The process occurred at a constant temperature of 80°F. What was the heat input to the system during the process?

8.4. A mass of 2 kg of ice at 0°C is completely melted at 0°C. What is the heat transfer (in kJ) to the ice? Also determine the increase in entropy of the system.

8.5. A mass of 5 kg of water at 20°C is turned completely to ice at 0°C. Determine
   (a) the heat transfer (in kJ) to the system
   (b) the increase in entropy of the system

8.6. A mass of 5 kg of water at 1 atm pressure is heated from 20°C to 50°C. Determine
   (a) the heat transfer, in kJ, to the water
   (b) the change in entropy of the water

8.7. Assume the heat transfer to the water in Problem 8.6 is from a thermal reservoir at 400 K. Determine
   (a) the increase in entropy (in kJ/K) of the thermal reservoir
   (b) the change in entropy of the universe
   Is the process reversible or irreversible? Why?

8.8. The specific heat of copper at room temperature is 385 J/kg · K. If a block of copper of mass 2 kg is heated from 15°C to 35°C, what is the increase in entropy of the system? (*Hint:* Solve this problem working from the appropriate form of the $T\,ds$ equation.)

8.9. Compute the entropy change for 1 lbm of copper during a process in which the temperature of the copper is raised from 100°F to 200°F. (Assume that the specific heat of the copper during this process is 0.092 Btu/lbm · °R.)

8.10. Compute the entropy change for 10 kg of gold during a process in which the temperature of the gold is raised from 40°C to 250°C. (Assume that the specific heat of the gold is 0.13 kJ/kg · K.)

8.11. Air of 2 kg mass at 300 K is heated at a constant volume to a temperature of 400 K. Calculate
   (a) the heat transfer to the system
   (b) the increase in entropy of the system
   If the process did not occur at constant volume, but the final volume equals the initial volume, would the increase in entropy of the system remain the same as before?

8.12. The initial and final states of 5 kg of air are given by $p_1 = 100$ kPa, $T_1 = 300$ K and $p_2 = 500$ kPa, $T_2 = 300$ K. Calculate $S_2 - S_1$ for the system.

8.13. A mass of 5 kg of air is heated at a constant pressure from 300 K to 400 K. What is the resulting increase in entropy of the system?

8.14. Air is initially at 300 K in a volume of 1 m³. The air behaves as an ideal gas and undergoes an isentropic process (Fig. P8.14). Its final temperature is 400 K. What is its final volume?

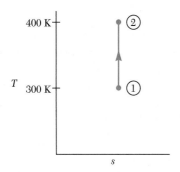

Figure P8.14 Isentropic process for air.

8.15. Air behaves as an ideal gas and undergoes an isentropic process. The volume of the air doubles during the process. If the initial temperature of the air is 500°R, what is the final temperature?

8.16. During a certain process with air, the temperature doubles and the volume doubles. Compute the entropy change per unit mass of air for the process.

8.17. During a certain process with nitrogen gas, the temperature is raised from 50°C to 200°C while the pressure triples. Compute the entropy change per kg of gas for the process.

* 8.18. Nitrogen gas undergoes a process in which its temperature is raised from 300 K to 500 K, while its pressure is increased from 1 atm to 2 atm. With the aid of the gas tables, compute the entropy change per kg for the gas.

* 8.19. One kmol of oxygen is initially at 1 atm and 400 K. The gas undergoes a process during which its temperature is raised to 700 K while its entropy increases by 10 kJ/(kmol · K). Using the gas tables, determine the final pressure of the gas.

* 8.20. A mass of 10 lbm of carbon dioxide gas is initially at 1000°R and 20 psia. The gas undergoes a process during which its temperature is raised to 2000°R while its entropy increases by 2 Btu/°R. Using the gas tables, determine the final pressure of the gas.

8.21. Determine the entropy (in kJ/K) of 5 kg of steam at 2 MPa and 500°C. If the steam is expanded isentropically to a pressure of 5 kPa, find the quality of the steam at the end state. What is the change in entropy?

* 8.22. Steam enters a turbine at 600°C and 80 bar. This steam undergoes an isentropic process through the turbine and exits at a pressure of 3 bar (Fig. P8.22). Determine the temperature at the exit state and compute the work done (per kg of steam) for the process.

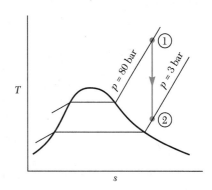

Figure P8.22 Isentropic process for flow through a turbine.

* 8.23. Steam enters a turbine at 1000°F and 600 psia. This steam undergoes an isentropic process through the turbine and exits at a pressure of 5 psia. Determine the temperature at the exit state and compute the work done (per lbm of steam) for the process.

* **8.24.** Refrigerant-22 enters a compressor as a saturated vapor at 0°C. The gas is compressed during an isentropic process through the compressor and exits at 40°C (Fig. P8.24). Determine the pressure at the exit state and compute the work input to the compressor (per kg of gas) for the process. (*Hint:* Use the R22 code and guess T_s at the exit state. Iterate until the correct T_s is found to give $s_e = s_i$ and $T_e = 40°C$.)

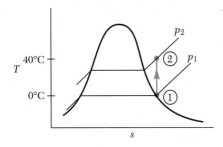

Figure P8.24 Isentropic flow through a compressor.

* **8.25.** Refrigerant-22 enters a compressor as a saturated vapor at 40°F. The gas is compressed during an isentropic process through the compressor and exits at 110°F. Determine the pressure at the exit state and compute the work input to the compressor (per lbm of gas) for the process.

8.26. The rigid tank shown in Figure 6.13 is filled initially with steam at 10 bar and 500°C. Steam at 300°C and 1 bar flows in the pipeline to which the tank is connected via a valve. The valve is opened slightly, and steam flows slowly from the tank until the pressure in the tank decreases to 1 bar. Assume that the tank is well insulated and that the specific entropy of the fluid in the tank stays constant throughout. Determine the final temperature of the steam in the tank.

8.27. The rigid tank shown in Figure 6.13 is filled initially with steam at 200 psia and 800°F. Steam at 20 psia and 300°F flows in the pipeline to which the tank is connected via a valve. The valve is opened slightly, and steam flows slowly from the tank until the pressure in the tank decreases to

20 psia. Assume that the tank is well insulated and that the specific entropy of the fluid in the tank stays constant throughout. Determine the final temperature of the steam in the tank.

8.28. Air behaves as an ideal gas while undergoing an isentropic process from 10 bar and 300 K to 5 bar. Determine
(a) T_2
(b) the work done per unit mass of the air.

8.29. Carbon dioxide behaves as an ideal gas while undergoing an isentropic process from 20 psia and 400°F to 10 psia. Determine
(a) T_2
(b) the work done per unit mass of the gas. Assume $k = 1.24$ and $c_{v0} = 0.188$ Btu/lbm · °R for carbon dioxide.

* **8.30.** Air at 1 atm and 600°R undergoes an isentropic process to 1200°R.
(a) By use of the gas tables, determine the pressure of the gas at the end of the process.
(b) Compute the work done (per unit mass) by the gas during the process.

* **8.31.** Methane at 5 atm and 650 K undergoes an isentropic process to 500 K.
(a) By use of the gas tables determine the pressure of the gas at the end of the process.
(b) Compute the work done (per unit mass) by the gas during the process.

* **8.32.** Air at 800 K and 6 bar is enclosed in a cylinder. The gas expands isentropically against a piston until its pressure is reduced to 2 bar. By use of the gas tables
(a) find the temperature of the gas at the end of the process
(b) compute the work done (per unit mass) by the air during the process.

* **8.33.** Methane gas at 1000°R and 90 psia is enclosed in a cylinder. The gas expands isentropically against a piston until its pressure is reduced to 20 psia. By use of the gas tables, find the temperature of the gas at the end of the process and compute the work done (per unit mass) by the gas during the process.

* **8.34.** Air at 1700°R is contained in a cylinder with an initial volume of 0.01 ft³. The air expands isentropically against a piston until the volume increases to 0.04 ft³. By use of the gas tables, find the temperature of the air at the end of the is-

entropic process and compute the work done (per unit mass) by the air during the process.

* **8.35.** Methane gas at 1000 K is contained in a cylinder with an initial volume of 0.005 m³. The gas expands isentropically against a piston until the volume increases to 0.02 m³. By use of the gas tables, find the temperature of the gas at the end of the isentropic process and compute the work done (per unit mass) by the gas during the process.

8.36. A centrifugal pump is used to increase the pressure of water at 300 K from 1 bar to 10 bar. Define the pump as control volume and compute the work done (per kg of water) by the control volume for isentropic flow through the pump.

8.37. A centrifugal pump is used to increase the pressure of water at 100°F from 20 psia to 100 psia. Define the pump as control volume and compute the work done (per lbm of water) by the control volume for isentropic flow through the pump.

8.38. A heat transfer of 10 kJ occurs from a thermal reservoir A at 127°C to a second thermal reservoir B at 27°C. Calculate
  (a) the increase in entropy of system A
  (b) the increase in entropy of system B
  (c) the total change in entropy (in kJ/K) of the universe
  Is the process reversible or irreversible?

* **8.39.** A Carnot heat engine is operating between 900 K and 300 K. If the value of entropy of the working fluid at the end of the heat input process is 100 kJ/K and the entropy at the end of the heat rejection process is 50 kJ/K, what is the work output for a cycle? Plot the cycle on a *T-S* diagram and show the work output on the diagram.

* **8.40.** A Carnot heat engine is operating between 1000°R and 500°R. The work output per cycle is 1500 Btu, and the entropy of the working fluid at the beginning of the heat input process is 900 Btu/°R. What is the entropy of the working fluid at the beginning of the heat rejection process?

* **8.41.** A reversed Carnot heat engine is used for refrigeration. The high temperature reservoir is at 60°C, and the low temperature reservoir is at −20°C. The entropy of the working fluid at the beginning of the heat rejection process is 15 kJ/K, and the entropy at the beginning of the heat input

process is 5 kJ/K. What is the input work required for the cycle? Plot the cycle on a *T-S* diagram and show the input work on the diagram.

* **8.42.** A reversed Carnot heat engine is used for refrigeration. The high temperature reservoir is at 110°F, and the low temperature reservoir is at 40°F. The entropy values at the beginning and end of the heat rejection process are 10 Btu/°R and 5 Btu/°R, respectively.
  (a) What is the cooling effect during the cycle?
  (b) What is the coefficient of performance for cooling for the cycle?

* **8.43.** Figure P8.43 shows a thermodynamic model of a complete absorption refrigeration system. T_L is the *evaporator* temperature, T_H the *condenser/absorber* temperature, and T_G the *generator* temperature in an actual typical device. Regarding the heat sources and sinks as thermal reservoirs, use either the inequality of Clausius theorem or the principle of increase of entropy to show that

$$\text{COP}_C = \frac{Q_C}{Q_G} \le \frac{T_L(T_G - T_H)}{T_G(T_H - T_L)}$$

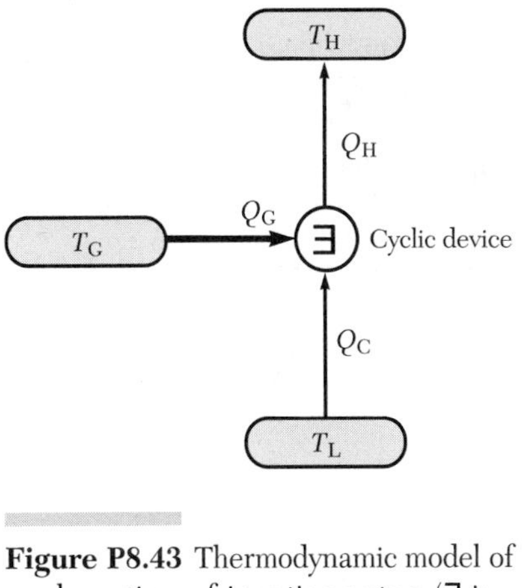

Figure P8.43 Thermodynamic model of an absorption refrigeration system (∃ is reversed engine).

8.44. A typical design of a solar heated absorption refrigeration system might have $T_L = 273$ K, $T_H = 300$ K, and $T_G = 340$ K. Using the expression in Problem 8.43, calculate the maximum possible value of COP_C.

8.45. An inventor claims to have designed a heat engine that produces 40 kJ of work while receiving 100 kJ of heat from a source at 147°C. The heat sink is at 27°C. Calculate

(a) the increase in entropy of the source
(b) the increase in entropy of the sink
(c) the total increase in entropy of the universe

Based on your results (or other information), indicate whether or not the inventor's claim is plausible.

8.46. Figure P8.46 shows a rigid, well-insulated container that is partitioned in half using a thin diaphragm. One half contains an ideal gas with a pressure of 5 bar, a volume of 0.05 m³, and a temperature of 300 K. The other half is a vacuum. If the diaphragm is now ruptured, determine

(a) the final temperature of the gas
(b) the increase in entropy of the system
(c) the lost work, assuming that the temperature of the surroundings is 300 K

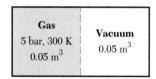

Figure P8.46 Unresisted gas expansion.

8.47. Using Equation 8.55, compute s_{fg} at 500°R for H$_2$O and refrigerant-22 and compare your results with tabulated data in the appendix.

8.48. Using Equation 8.55, compute s_{fg} at 25°C for H$_2$O and refrigerant-22 and compare your results with tabulated data in the appendix.

* **8.49.** Steam at 500 K is condensed at constant pressure from saturated vapor to saturated liquid, and the heat of condensation is rejected to the surroundings, which are at 25°C. Compute the entropy change of the universe per kg of steam condensed during the process.

* **8.50.** Steam at 1 atm is condensed at constant pressure from saturated vapor to saturated liquid, and the heat of condensation is rejected to the surroundings, which are at 40°F. Compute the entropy change of the universe per lbm of steam condensed during the process.

* **8.51.** Refrigerant-22 at 10°C is evaporated at constant pressure from saturated liquid to saturated vapor. During the process, the heat of vaporization is absorbed from the surroundings, which are at 30°C. Compute the entropy change of the universe per kg of refrigerant-22 evaporated during the process.

* **8.52.** Adiabatic throttling is an isenthalpic process. Saturated liquid refrigerant-22 is adiabatically throttled between a condenser pressure of 20 bar and an evaporator pressure of 2 bar in a steady-flow process. Determine the lost power assuming a refrigerant mass flow rate of 6.4 kg/h and an ambient temperature of 300 K.

* **8.53.** A steady-state steady-flow process occurs in a heat exchanger, as shown in Figure P8.53. Steam en-

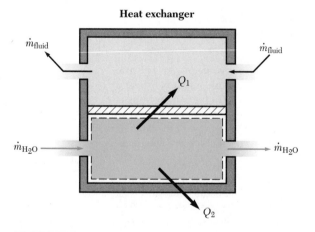

Figure P8.53 Flow through a heat exchanger.

ters the heat exchanger as a saturated vapor at 300°F and exits as a liquid at 100°F. The flow rate of the H_2O is 10 lbm/h. Two heat interactions occur with the H_2O in the heat exchanger. Heat at the rate of 10,000 Btu/h is being removed from the H_2O by another fluid that is at a constant temperature 90°F, while heat is also being lost to the surroundings at 70°F. Determine the rate of entropy generation for the control volume (the heat exchanger) for this steady-state steady-flow process.

* **8.54.** Steam flows through a turbine during an adiabatic, steady-state steady-flow process. The mass flow rate of the steam is 10 kg/s. The steam enters at 450°C and 80 bar and exits at 150°C and 1 bar. Determine the rate of entropy generation for this process.

* **8.55.** Steam flows through a turbine during a *reversible* adiabatic, steady-state steady-flow process. The mass flow rate of the steam is 10 kg/s. The steam enters at 450°C and 80 bar and exits at 1 bar.
   (a) What is the rate of entropy generation for this process?
   (b) What is the exit temperature?
   (c) What is the work done during this process?

* **8.56.** Water at 70°F flows through a device (the control volume) during a steady-state steady-flow process. The velocity of the water at the inlet is 10 ft/s, while the velocity at the exit is 20 ft/s. The elevation of the inlet is 50 ft above the exit. The pressure at the inlet is 40 psia. Assume the process is reversible and adiabatic with the shear work output rate equal to zero. What is the pressure of the water at the exit of the control volume?

* **8.57.** Air flows through a control volume in a reversible, steady-state steady-flow isothermal process. The air enters the control volume at 100°C and 5 atm and exits at 2 atm. Compute the reversible shaft work per kg of air for this process.

* **8.58.** Air flows through a device (the control volume) in a reversible, steady-state steady-flow polytropic process. The air enters the control volume at 200°F and 30 psia and exits at 100°F. The polytropic index equals 1.3 for the process. Compute the reversible shaft work per lbm of air for this process.

* **8.59.** During a steady-state steady-flow process, two streams of refrigerant-22 enter a control volume and one stream exits. One entering stream is a

saturated liquid at 20°C with a mass flow rate of 0.5 kg/s. The second entering stream is a saturated liquid at −10°C with a mass flow rate of 0.1 kg/s. A single stream of refrigerant exits the control volume as a saturated vapor at 25°C. Determine the heat input to the control volume from the surroundings. The surroundings are at 25°C. Compute the rate of entropy generation for the control volume for this process.

* **8.60.** Steam enters an adiabatic turbine at 400°C and 30 bar and exits at 150°C and 1 bar. Determine the adiabatic turbine efficiency for the process.

* **8.61.** Steam enters an adiabatic turbine at 700°F and 150 psia and exits at 200°F and 10 psia. Determine the adiabatic turbine efficiency for the process.

* **8.62.** Refrigerant-22 enters a compressor as saturated vapor at 0°C and exits at 10 bar and 60°C. Determine the adiabatic compressor efficiency for the process.

* **8.63.** Refrigerant-22 enters a compressor at 70 psia and 40°F and exits at 150 psia and 140°F. Determine the adiabatic compressor efficiency for the process.

* **8.64.** Air enters a compressor at 1 atm and 300 K and flows at a steady rate through the compressor.
   (a) Case 1. The air is compressed during a reversible adiabatic (isentropic) process until the pressure is 10 atm. Compute the work done per kg of air for this isentropic process.
   (b) Case 2. Now the air is compressed during a reversible isothermal process until the pressure is 10 atm. Compute the work done per kg of air for this isothermal process and compare your result with the work required in the isentropic process in case 1.

* **8.65.** It is desirable to compress steam from a state of 80 psia and 500°F to 400 psia in a steady-flow process. Determine the difference in work required between a reversible isentropic process and a reversible isothermal process for this compression.

* **8.66.** The pressure of a stream of water at 70°F is increased from 1 atm to 10 atm.
   (a) If the adiabatic efficiency of the pump is 0.85, what is the work required per lbm of water for this process?

(b) How much work is required if the process is reversible and isothermal?

* **8.67.** Steam at 500°F and 100 psia flows through a nozzle. The exit velocity for isentropic flow through the nozzle is 3 ft/s. If the nozzle efficiency is 0.9, what is the actual exit velocity for the flow?

* **8.68.** Warm air at 35°C is cooled sensibly to 20°C at a constant pressure of 1 atm by passing it over the evaporator coils of a refrigeration device, as shown in Figure P8.68. The mass flow rate of the air is 3 kg/s. The ambient air temperature is 35°C, while the evaporator temperature can be taken as constant and equal to −5°C. Calculate

(a) the heat transfer rate in kW from the air stream to the evaporator coils

(b) the rate of increase in the entropy of the universe

(c) the lost power resulting from irreversibilities in the cooling process

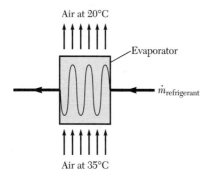

Air at 20°C

Evaporator

$\dot{m}_{\text{refrigerant}}$

Air at 35°C

Figure P8.68 Air flow over evaporator coils.

* **8.69.** A mass of 0.5 kg of ice at 0°C is added to 2 kg of hot tea at 90°C. Assuming that the thermodynamic properties of tea are approximately the same as those for pure water, estimate the final temperature of the mixture. The mixture is thermally insulated from the ambient surroundings, whose temperature is 30°C. Calculate the increase in entropy of the universe and, hence, the lost work in the process.

8.70. A 1.5-kg lump of iron at 1500 K is cooled rapidly to 300 K in a large pool of water at 300 K. If $c_p = 680$ J/kg · K for the iron, calculate

(a) the heat transfer from the iron to the pool of water

(b) the increase in entropy of the iron

(c) the increase in entropy of the pool of water

(d) the lost work, assuming the ambient surroundings temperature is 300 K

* **8.71.** An ideal heat engine uses air as the working fluid and operates on an ideal cycle that comprises the following reversible, steady-flow processes:

(i) a polytropic compression process, according to the law $pV^{1.3} = $ constant, from $p_1 = 100$ kPa and $T_1 = 300$ K to $T_2 = 600$ K

(ii) isothermal expansion such that $p_3 = 500$ kPa

(iii) a polytropic expansion process, according to the law $pV^{1.3} = $ constant, to $T_4 = 300$ K

(iv) isothermal compression to the initial state $p_1 = 100$ kPa and $T_1 = 300$ K

Show the processes on a T-s diagram. Calculate

(a) p_2 and p_4

(b) q_{12}, q_{23}, q_{34}, and q_{41}

(c) w_{12}, w_{23}, w_{34}, and w_{41}

(d) the first law efficiency of the heat engine

* **8.72.** Refrigerant-22 is compressed from 4 bar and 0°C to 15 bar. If the compression is adiabatic but has a process (isentropic) efficiency of 85%, determine

(a) the final temperature

(b) the work done (in kJ/kg refrigerant), assuming a steady-flow process

(c) the lost work (in kJ/kg refrigerant) if the ambient temperature is 300 K

* **8.73.** Air at 10 bar and 500°C expands through an adiabatic turbine to a pressure of 1 bar in a steady-flow process. The isentropic efficiency of the adiabatic turbine is 85%. Calculate

(a) the air temperature at exit to the turbine

(b) the actual specific work output of the turbine

(c) the lost work (in kJ/kg air) if the ambient temperature is 300 K

* **8.74.** Air expands adiabatically from a volume of 1 m³ to a volume of 2 m³ by flowing into an evacuated chamber. The initial pressure and temperature are 10 bar and 300 K, respectively.

(a) What is the change in internal energy of the fluid?

(b) Determine the increase in entropy (in kJ/K) of the air in the process.

Is the process reversible? Give reasons for your answer.

8.75. Air expands reversibly and adiabatically in a non-flow process from an initial pressure, volume, and temperature of 10 bar, 1 m³, and 300 K, respectively, to a final volume of 2 m³. Calculate
(a) the final air temperature and pressure
(b) the increase in entropy of the air
(c) the work done by the air during the process

* **8.76.** A reversible refrigeration device operating between 0°C and 27°C is used for producing ice from water initially at 0°C. Calculate the minimum work required in the manufacture of 1000 kg of ice from water initially at 0°C.

* **8.77.** A cyclic heat engine operates between a finite heat source and an infinite sink at T_0. During the operation of the heat engine, the source temperature decreases from T to T_0 while heat transfer Q occurs from the source to the engine. By applying the principle of increase of entropy, show that the (shaft) work output of the engine, W, must satisfy

$$W \leq Q - T_0(S - S_0)$$

where S and S_0 refer to the finite heat source.

* **8.78.** A steady stream of a fluid, initially at a pressure p_0 and a temperature T_0, is compressed by a device that can exchange heat with the ambient surroundings at a temperature T_0. The fluid leaves the device at a higher pressure p and at a temperature T. The heat transfer rate from the compression device to the ambient surroundings is $\dot{Q}_0$, while the mechanical power to the device is $\dot{W}_x$. Show that

$$\dot{W}_x \geq \dot{m}[(h - h_0) - T_0(s - s_0)]$$

where $\dot{m}$ is the mass flow rate of the fluid, h_0 and s_0 are the specific enthalpy and entropy of the fluid at the entry to the device, and h and s are the corresponding values at the exit of the device.

* **8.79.** If the fluid in Problem 8.78 is an ideal gas and the compression process occurs such that $T = T_0$, show that

$$\dot{W}_x \geq \dot{m}RT_0 \ln \frac{p}{p_0}$$

where R is the specific gas constant for the fluid.

* **8.80.** Assume that the fluid in Problem 8.78 is an incompressible liquid and that the exit temperature equals T_0. Show that

$$\dot{W}_x \geq \frac{\dot{m}}{\rho}(p - p_0)$$

where ρ is the mean density of the liquid.

* **8.81.** Figure P8.81 depicts a scheme for power production using air stored at a high pressure in a large receiver of volume V. Initially, the air receiver holds a mass of air m_s at a high pressure p_s and at a temperature equal to the ambient temperature T_0. An air motor is connected to the air receiver such that (shaft) work (W_x) is produced, while the pressure of air in the receiver eventually falls to the ambient pressure p_0. The air receiver and motor exchange heat (Q_0) with the ambient environment such that the final temperature of the air left in the receiver is T_0. The air exits the motor at a pressure p_0 and a temperature T_0. Show that

$$W_x \leq m_s RT_0 \left(\frac{p_0}{p_s} - 1 + \ln \frac{p_s}{p_0} \right)$$

where R is the specific gas constant for air.

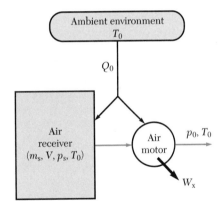

Figure P8.81 Work production from compressed air.

8.82. Develop an expression for the combined first and second law of thermodynamics as an extension of Equation 8.9 for a closed system for the case in which the work done for the process includes $p \, dV$ work, mechanical work, shaft work, stretched wire work, surface tension work, and electrical work.

8.83. Develop an expression similar to Equation 5.11

for the work by a closed system for an ideal gas with a constant temperature process in terms of p_1, p_2, and V_1 only.

8.84. Develop an expression similar to Equation 5.13 for the work by a closed system for an ideal gas with a polytropic process $(n \neq 1)$ in terms of n, p_1, p_2, and V_1 only.

*   **8.85.** Develop an expression for the change in enthalpy

of a closed system for an ideal gas with isentropic process in terms of p_1, p_2, v_1, and k only.

8.86. Combine the pvT equation of state for an ideal gas (Eq. 4.1) with Equation 8.35 to obtain an expression for $\ln (v_2/v_1)$. From this result, you can conclude that the volume ratio for an isentropic process for a closed system is dependent only on what variables?

Thermodynamic Availability

■■■ Exergy or availability as a second law concept; exergy as maximum net (or shaft) work potential; reference environment and the dead state; dissipation of exergy, entropy production, and irreversibility of processes

■■■ Exergy equations, the nonflow availability equation (NFAE), the steady-flow availability equation (SFAE), and second law efficiency

■■■ Exergy in nonflow processes, exergy of a closed system, exergy of ideal gas systems, exergy for solids and liquids, exergy and work, exergy and mechanical energy, exergy associated with heat flow, and exergy consumption and entropy generation

■■■ Exergy in steady-flow processes, exergy of flow of mass, exergy flow rate associated with work and heat transfer rates, exergy dissipation and entropy generation (the Gouy–Stodola lost work theorem), and alternate expressions for exergy of flow of mass

■■■ Exergy and optimum thermodynamic cycles, complete reversibility criterion for optimum thermodynamic efficiency, unlimited and finite heat sources and sinks, optimum engine cycle when heat source is finite, and optimum refrigeration cycle when heat source is finite

9.1 Introduction

Thermodynamic availability is a concept that, until recently, was largely ignored in most expositions of thermodynamics. The concept derives from the second law of thermodynamics and is closely related to reversibility and entropy, which were discussed in Chapters 7 and 8. Two conclusions from the consideration of the second law of thermodynamics in this book are (1) that *work is a higher quality form of energy transfer than heat* and (2) that whenever a system interacts with its surroundings, the *optimum thermodynamic efficiency is achieved when all processes are reversible*. Both of these are embraced in the availability concept, as will be shown later when formal definitions are given for the concept.

Many of the problems encountered in thermodynamics that require the application of the second law of thermodynamics can be solved either by using entropy alone or by using the *availability* concept. Both approaches should be viewed as complementary, although in individual cases, one may be more suitable than the other. Consider, for example, the problem of calculating the maximum power obtainable from a stream of geothermal water. The stream of hot water could serve as the heat source for a heat engine, but it should be noted that as heat flows from the geothermal water to the working fluid of the engine, the temperature of the heat source will also fall. The Carnot cycle efficiency does not provide the correct measure of ideal efficiency in such a circumstance since the heat source is not a constant temperature reservoir. Instead, one must consider a *reversible* engine that interacts with the heat source and sink in a reversible manner. The problem can be solved by making use of the relationship between entropy changes and the heat flow in a reversible process. More simply, however, the maximum possible power output can be determined by calculating the availability (or exergy) of the geothermal water stream. The appropriate expression to use is one of several that will be presented in this chapter.

Historical Notes and Terminology

Historically, the ideas that are central to availability are widely attributed to Josiah W. Gibbs and James C. Maxwell, whose contributions date back more than a century. A brief historical sketch of the development of the availability concept has been given by Haywood (1974a, 1974b). The highlights are summarized here.

1871—Maxwell first used the term *available energy*, although he credited Kelvin with the original ideas.

1873—Gibbs was the first to provide an analytical basis for determining the available energy in terms of *maximum possible net (or shaft) work* (in nonflow processes).

1875—Maxwell's fourth edition of his *Theory of Heat* included the first and "extremely simple and straightforward" analysis in terms of *maximum possible gross work* (in nonflow processes).

1889—Gouy's expression for *availability* in nonflow processes was published. His derivation was much simpler than that of Gibbs.

1898—Stodola independently derived an expression for maximum possible gross work in a nonflow process and extended his work to include (1) the case of steady-flow processes and (2) the relationship between loss of gross work potential and entropy creation.

1932—Keenan presented Gibbs' results in "simpler and more practical terms," including giving the name *availability* to the maximum possible net (or shaft) work.

1956—Rant (1956) first proposed the use of the term *exergy* as a new word for *work capability* used previously by Bosnjakovic (1938).

From this brief summary, it is evident that the concept of availability has a long history, and thus it is surprising that not much use has been made of the idea in the real world of engineering. It is noteworthy that in recent times, renewed interest in the idea has always coincided with periods of energy crises, such as the energy crisis of the 1970s. The pattern seems to be that whenever there is concern with the efficiency of energy usage, interest in the availability concept heightens. Then when there is a perception that energy is cheap and abundant, interest in availability analysis wanes dramatically.

The historical sketch given also indicates that other terms are frequently used instead of availability. At present, there is no universal agreement in the English language on the preferred terminology. The international trend is to use *exergy*. The term *exergy* is widely employed in Europe and Japan and is also the preferred term recommended in recent international conferences on second law analysis. Here in the United States, the term *availability* is still widely used. This book uses *exergy*, although occasional cross-references to *availability* are given for the benefit of those who may be more familiar with this term.

Exergy as Maximum Work Potential

Chapter 1 gave a definition of exergy (or availability) as *the capacity to produce change in a specified environment*. Following consideration of the first and second laws of thermodynamics, exergy can now be defined in a more precise fashion. To set the stage, consider a closed system that can interact with a reference environment in the surroundings. Conceptually, a *reference environment* is a system that remains essentially in a state of complete thermodynamic equilibrium despite finite flows of energy (heat and work) and mass between it and the system. Ambient (or atmospheric) air can be considered as a reference environment since its pressure and temperature are unlikely to change by a significant amount due to a heat or work interaction with a given (finite) system. Other large systems that can be considered as reference environments include lakes, the ocean, and the Earth's crust, as long as their intensive thermodynamic state remains constant over the period of interest.

The situation described is illustrated in Figure 9.1, which depicts a closed system having a heat flow and a boundary displacement work interaction between it and the

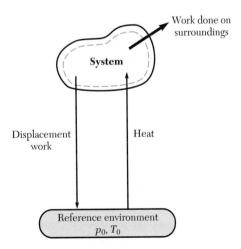

Figure 9.1 Work production by a closed system that interacts with a reference environment.

reference environment. While the interaction takes place, a (useful) work output occurs to the surroundings. It is assumed that the temperature (T_0) and pressure (p_0) of the reference environment are uniform throughout and remain constant for the duration of the processes taking place.

> **Exergy (of a closed system)** *The maximum (useful) work that can be extracted from the system as it interacts with its reference environment.*

Comments

1. The total work output of a system is termed the gross work output (W_{gross}). A part of this is the displacement work (W_d) associated with system boundary displacement against the pressure (p_0) of the reference environment. The net or useful work output (W_{net}) is the difference $(W_{gross} - W_d)$. The net work is also referred to as shaft work (see Haywood 1974a, for example) because the net work can be converted completely to shaft work by pushing on a frictionless piston, which is anchored to a shaft via a crank and a connecting rod as in a typical reciprocating piston and cylinder arrangement.

2. Heat flow occurs between the system and the reference environment only as long as the system temperature (T) is not the same as that of the reference environment (T_0). When T becomes equal to T_0, a state of *thermal equilibrium* is said to exist between the system and the reference environment. At this stage, the heat flow becomes zero.

3. A state of *mechanical equilibrium* is reached between the system and the reference environment when the system pressure (p) becomes equal to p_0. At this condition, the system cannot produce work via a volume change against the pressure exerted by the reference environment.

4. When a system attains thermal and mechanical equilibrium with the reference environment, the system is said to have arrived at the *thermal–mechanical dead state*. At the dead state, the system becomes incapable of producing work in the surroundings by a heat or (boundary displacement) work interaction with the reference environment.

5. From previous considerations of the second law of thermodynamics, it can be concluded that the maximum work possible is obtained when all processes are reversible. Thus, the exergy of the closed system is obtained by calculating the work output when the system is brought to the dead state as a consequence of processes that are all reversible. This principle is used later to develop an expression for the exergy of a closed system.

6. The idea of measuring exergy in terms of how much work can be produced is reasonable and most appropriate in consequence of the second law of thermodynamics. While mechanical energy and work can always be converted completely to other forms of energy, such as internal energy (or the flow of energy as heat), the reverse is not always possible, at least not on a continuous basis. In other words, *mechanical energy* (or flow of energy as work) *represents the highest quality of energy* that is possible. Thus, it can be reasonably used as the most appropriate "currency" or measure of energy when considering both quality and quantity.

Exergy and Reversibility of Processes

It is necessary to point out that considerations of availability are important not only when dealing with work-producing (or work-absorbing) devices or systems but also when one is interested in the evaluation of the efficiency of any process whatsoever. By way of illustration, consider a nonflow process that takes a closed system from an initial state 1 to a final state 2 during which the heat and work exchanged between the system and the surroundings are Q_{12} and W_{12}, respectively.

The first law of thermodynamics can be applied to the system and surroundings separately to obtain

$$Q_{12} - W_{12} = (\Delta E_{12})_{\text{system}} \tag{9.1}$$

and

$$-Q_{12} + W_{12} = (\Delta E_{12})_{\text{surr}} \tag{9.2}$$

Adding both sides of Equations 9.1 and 9.2 gives

$$(\Delta E_{12})_{\text{universe}} = (\Delta E_{12})_{\text{system}} + (\Delta E_{12})_{\text{surr}} = 0 \tag{9.3}$$

Thus, on the basis of the first law alone, it is concluded that *energy is conserved* (for a system and its surroundings) *regardless of the processes involved* and that reversible and irreversible processes cannot be distinguished.

The situation is radically altered when it is viewed from a second law perspective. The inequality of Clausius theorem in terms of entropy changes (Eq. 8.58) can be written for the system and the surroundings as

$$(S_2 - S_1)_{\text{system}} \geq \int_{\text{state 1}}^{\text{state 2}} \frac{\delta Q}{T_{\text{boundary}}} \tag{9.4}$$

while

$$(S_2 - S_1)_{\text{surr}} \geq \int_{\text{state 1}}^{\text{state 2}} \frac{(-\delta Q)}{T_{\text{boundary}}} \tag{9.5}$$

Adding gives

$$(\Delta S_{12})_{\text{universe}} = (\Delta S_{12})_{\text{system}} + (\Delta S_{12})_{\text{surr}} \geq 0 \tag{9.6}$$

In words, the inequality of Equation 9.6 expresses the second law principle that *entropy can only remain constant or else increase for the universe*. The equality in Equation 9.6 applies only when all the processes occur reversibly; otherwise, the entropy for the universe must increase.

Chapter 8 pointed out that *entropy production* is synonymous with what is termed *lost work*. In other words, a loss of something of value occurs in real processes, and this loss manifests itself as entropy production. The something of value that is lost is not energy since, by Equation 9.3, the energy of the universe is conserved. The loss can only be viewed as a reduced capacity to produce change; in other words, there is a loss or dissipation of exergy due to the irreversibility of the process. In reversible processes, exergy is conserved, while in real processes, some of it is irretrievably consumed. Some processes are indeed more wasteful than others in this regard. It is only in reversible processes that availability is conserved. Thus, the aim in availability analyses is to establish in a given situation what the best possible performance is so as to have a yardstick against which the performance of actual systems or real processes can be compared and judged.

Exergy Equations and the Second Law Efficiency

This chapter deals with exergy (or availability) analysis for both nonflow and flow processes. In either case, the concern is with what happens to a supply of exergy when real processes occur.

Nonflow Processes and the Nonflow Availability[1] Equation (NFAE)

When dealing specifically with a nonflow process, one should be able to identify the exergy supply, χ_{input}, to the process. For example, if work is done on a system in a process, the magnitude of the net work done on the system is equivalent to an exergy supply to the system. Likewise, since electricity counts as shaft work, the electrical work done on a system represents an exergy flow to the system and can be regarded as the exergy input to a process. Similarly, one can identify the exergy product $\chi_{product}$, which is defined as the desired outcome of the process expressed in terms of exergy or exergy flow. If, for example, the objective is to produce ice from a mass of water initially at the prevailing ambient temperature and pressure, the exergy product is measured simply as the exergy of the final product, which in this case is the ice produced. Section 9.2 develops expressions for the exergy of pure substances as well as the flow of exergy associated with either heat flow or work. The latter is particularly simple since the net mechanical work translates directly to exergy.

The exergy product cannot exceed the exergy input to a process unless a component of the total exergy supply has been left out erroneously. In fact, in real processes, exergy is partially consumed due to irreversibilities associated with the process. The exergy consumed is denoted by $\chi_{consumed}$. This term is frequently called *exergy destruction* by researchers in the area. It is also possible to have an exergy product that is not used, such as the exergy associated with the temperature rise of a mass of air (to be stored in an air receiver) when it is compressed. In this example, the desired product is air at a higher pressure but at room temperature; the exergy associated with the accompanying rise in temperature can therefore be regarded as unutilized or unused exergy, χ_{unused}. In typical applications, heat transfer occurs to the surroundings over a period of time, and the air in the receiver eventually attains ambient temperature. Whatever exergy was associated with the rise in temperature during the compression process would thereby be dissipated. In the general case, the *nonflow availability equation (NFAE)* can thus be written as

$$\chi_{input} = \chi_{product} + \chi_{consumed} + \chi_{unused} \tag{9.7}$$

---

[1]Use of the term *availability* rather than *exergy* is preferred here so as not to confuse the acronym for the exergy equation with NFEE (for the nonflow energy equation).

Evidently, when all the processes are reversible, $\chi_{consumed}$ is zero. If, at the same time, there is no unused exergy, it follows that the (useful) exergy product is equal to the exergy input to the process. In all other processes, the exergy product must be less than exergy input.

Second Law Efficiency (η_2)

A second law efficiency, η_2, for a process can be defined as

$$\eta_2 = \frac{\text{exergy product}}{\text{exergy input}} \cdot 100\%$$

$$= \frac{\chi_{product}}{\chi_{input}} \cdot 100\%$$

(9.8)

Clearly, η_2 must always be less than 100%; it attains the value of 100% only when all processes involved are reversible (and the entire exergy product is useful). Thus, η_2 provides us with a realistic comparison between what is achieved in an actual situation and the best that could possibly be achieved when a system changes from a given state to any other.

Flow Processes and the Steady-Flow Availability Equation (SFAE)

When dealing with flow processes, exergy flows are associated with mass flows as well as with flows of energy either as heat or as work across the control surface. Thus, in each application, one should be able to identify the exergy input rate, $\dot{\chi}_{input}$, to the process and the resulting (useful) exergy production rate, $\dot{\chi}_{product}$. For steady-flow processes, the corresponding *steady-flow availability equation (SFAE)* has basically the same form as Equation 9.7 for nonflow processes:

$$\dot{\chi}_{input} = \dot{\chi}_{product} + \dot{\chi}_{consumed} + \dot{\chi}_{unused}$$

(9.9)

The $\dot{\chi}_{consumed}$ and $\dot{\chi}_{unused}$ in this case, respectively, denote the exergy consumption rate and the component of the exergy production rate that is unused.

As with nonflow processes, the best that could possibly be attained is for all processes to be reversible, in which case the exergy consumption rate is zero. If all the exergy product is useful, the exergy production rate is equal to the input rate. The second law efficiency, η_2, is defined for the case involving flow processes as

$$\eta_2 = \frac{\text{exergy production rate}}{\text{exergy input rate}} \cdot 100\%$$

$$= \frac{\dot{\chi}_{product}}{\dot{\chi}_{input}} \cdot 100\%$$

(9.10)

Thus, $\eta_2 \leq 100\%$ follows from the second law of thermodynamics.

Once equations are derived for computing the exergy associated with flows of mass and the flow of energy (as heat or work) across the control surface, Equation 9.9 can readily be used as a basis for the availability analysis of a steady-flow process.

First Law and Second Law Efficiencies

It is important to remember that it is the second law efficiency that furnishes the proper measure of performance and not the first law efficiency. Chapter 7 defined the first law efficiency, η_1, as

$$\eta_1 = \frac{\text{desired (energy) output}}{\text{required (energy) input}} \cdot 100\% \tag{7.1}$$

In some cases, this definition is found to give efficiencies that are higher than 100%. For example, the first law efficiency defined for a heat pump for heating is the coefficient of performance (COP_H), given by

$$COP_H = \frac{\begin{array}{c}\text{heat delivered to the high}\\\text{temperature medium } (Q_H)\end{array}}{\begin{array}{c}\text{energy input for operating}\\\text{the device } (W)\end{array}} \tag{7.6}$$

The heat pump actually draws heat (Q_L) from a lower temperature medium or environment, usually ambient air, in addition to the energy input as work (W) to the device. Thus,

$$Q_H = W + Q_L$$

or

$$Q_H > W \tag{9.11}$$

Thus, the first law efficiency, η_1, which is given by

$$\eta_1 = \frac{Q_H}{W} \cdot 100\% \tag{9.12}$$

will, in this instance, be greater than 100%. For a typical vapor compression unit, η_1 can be as high as 400%!

On the basis of the first law alone, one cannot establish what the upper limit to the first law efficiency is in a given application. Thus, in the example just given, one cannot tell, solely from a first law analysis, if the performance of a particular heat pump is good enough or if there is still room for significant improvements. This all-important question can be answered only through a second law analysis using either the entropy and reversibility ideas of Chapters 7 and 8 directly or the availability analysis methods of the present chapter.

The objective in the remainder of this chapter is to derive the appropriate expressions for exergy in processes not involving chemical reactions and to demonstrate their use in the complete thermodynamic analysis of the performance of typical systems and devices commonly used in engineering. Processes involving chemical reactions are excluded because only the thermodynamics of *pure substances* have been considered thus far. *Chemical substances* are those whose chemical composition can change with time and position in the system. The thermodynamics of this category of substances are not considered until Chapters 14 and 15.

Note, however, that the dead state for a substance in general corresponds to the state of *complete* (thermal, mechanical, and chemical) *thermodynamic equilibrium* with the reference environment. Thus, when the system is at a dead state, thermal equilibrium precludes heat interaction (via a heat engine) with the reference environment that might lead to production of shaft work. Mechanical equilibrium likewise excludes the possibility of work via system boundary displacement against the pressure exerted by the reference environment. Chemical equilibrium rules out the possibility of chemical reactions with the stuff of the reference environment that might lead to work production in the surroundings. This chapter is limited to cases of *restricted equilibrium*, which require only that a thermal and mechanical equilibrium condition should exist between the system and the reference environment when the system is at a dead state condition. When chemical equilibrium is also a requirement for the dead state, the *unrestricted equilibrium* situation occurs.

9.2   Exergy in Nonflow Processes

Expressions are derived in this section first for the exergy of a system that is not at the dead state. Expressions for the flow of exergy that corresponds to energy flow as heat or work are derived later. The notation introduced earlier in this chapter for exergy analysis is used consistently in the remainder of this book. Thus, the pressure and temperature of the reference environment are designated p_0 and T_0, respectively. System properties carrying the subscript $_0$ are the properties (for the system) at the dead state (p_0, T_0). The designation χ (the Greek symbol *chi*) is used for exergy, while a descriptive subscript such as "system" or "Q" is used to identify the entity whose exergy is being defined. The expressions for exergy derived here are for the *Gibbsian maximum net (or useful) work potential* and not the Maxwellian maximum gross work potential.

9.2.1   Expressions for the Exergy of a Closed System

Figure 9.2 depicts a closed system initially having a volume V, internal energy U, temperature T, entropy S, and pressure p. The reference environment is at a fixed temperature T_0 and exerts a pressure p_0 at the system boundary. The corresponding system properties at the dead state are V_0, U_0, T_0, S_0, and p_0.

The exergy of the system is the maximum work output possible as the system is reduced to the dead state. From previous discussions in Chapters 7 and 8, the maximum work is produced when all the processes involved are reversible. This can be achieved using the arrangement shown in Figure 9.2 in which it is assumed that T is greater than T_0. (It is easy to show that essentially the same conclusions will be reached about the exergy of the system when T is lower than T_0.) A heat engine has been placed between the system and the reference environment to permit shaft work extraction from the heat flow occurring between the system and the reference environment.

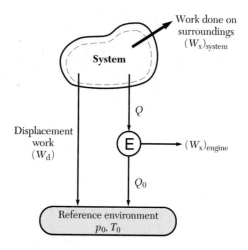

Figure 9.2 Maximizing the work output of a closed system.

Application of the First Law of Thermodynamics

The displacement work, W_d, done by the system due to boundary displacement against the opposing pressure of the reference environment can be determined using Equation 5.7, which was derived in Chapter 5. Thus,

$$W_d = p_0(V_0 - V) \tag{9.13}$$

where V represents an initial system volume and V_0 corresponds to the system volume at the dead state. The heat transferred from the system to the heat engine in Figure 9.2 is designated as Q. Applying the first law of thermodynamics to the nonflow process gives

$$(-Q) - (W_x)_{system} - W_d = U_0 - U \tag{9.14}$$

where $(W_x)_{system}$ represents the direct shaft (or net) work output of the system as it is brought to the dead state from its original state.

The first law of thermodynamics can also be applied to the heat engine cycle. Thus, if $(W_x)_{engine}$ is the net shaft work output of the engine and Q_0 is the heat flow from the engine to the reference environment, then it follows that

$$Q = Q_0 + (W_x)_{engine} \tag{9.15}$$

Application of the Second Law of Thermodynamics

By considering the system and its surroundings, the application of the second law of thermodynamics leads to the following equation, which is basically the same as Equation 8.63 of Chapter 8 or Equation 9.6 derived earlier in this chapter:

$$(\Delta S)_{universe} = (\Delta S)_{system} + (\Delta S)_{engine} + (\Delta S)_{ref\ environ}$$
$$\geq 0 \tag{9.16}$$

The following substitutions can be made for each of the entropy change terms in the inequality of Equation 9.16:

1. For the system, ΔS is given by

$$(\Delta S)_{\text{system}} = S_0 - S \tag{9.17}$$

2. Provided the engine operates in cyclic manner, the entropy change per cycle is zero.

3. For the reference environment, the process is internally reversible and the increase in entropy is given by

$$(\Delta S)_{\text{ref environ}} = \frac{Q_0}{T_0} \tag{9.18}$$

Substituting Equations 9.17 and 9.18 into the inequality of Equation 9.16 leads to the following:

$$(\Delta S)_{\text{universe}} = (S_0 - S) + \frac{Q_0}{T_0} \geq 0 \tag{9.19}$$

Eliminating Q and Q_0 between Equations 9.14, 9.15, and 9.19 leads to

$$[(W_x)_{\text{engine}} + (W_x)_{\text{system}}] + W_d \leq (U - U_0) - T_0(S - S_0) \tag{9.20}$$

Substituting the expression for W_d given by Equation 9.13 yields

$$[(W_x)_{\text{engine}} + (W_x)_{\text{system}}] \leq (U - U_0) + p_0(V - V_0) - T_0(S - S_0) \tag{9.21}$$

The left-hand side of the inequality in Equation 9.20 represents the *gross work* (W_{gross}) produced in processes that bring the closed system to the dead state. In Equation 9.21, however, the terms on the left-hand side denote the total *shaft (or net) work* (W_{net}) output produced. Maximum work is produced when all the processes are reversible. In this event, the inequalities in Equations 9.20 and 9.21 should be replaced by an equals sign. Thus, from Equation 9.20, the *maximum gross work* obtainable is given by

$$(W_g)_{\text{max}} = (U - U_0) - T_0(S - S_0) \tag{9.22}$$

Similarly, the *maximum shaft (or net) work* obtainable can be deduced from Equation 9.21 as

$$(W_x)_{\text{max}} = (U - U_0) + p_0(V - V_0) - T_0(S - S_0) \tag{9.23}$$

Strictly speaking, $(W_g)_{\text{max}}$ given by Equation 9.22 should properly be regarded as the exergy of the system, and indeed, it was so regarded by Maxwell and Stodola. However, the maximum gross work obtainable is not of much use in practice because the contribution W_d due to the system boundary displacement against the pressure of the reference environment is, in most instances, not available as useful work on the surroundings. It is thus customary instead to follow Gibbs and others and regard $(W_x)_{\text{max}}$, the maximum shaft (or net) work, as the measure of the system exergy at the specified initial state. Accordingly, this book regards the definition of exergy as the *maximum shaft (or net) work potential* of a system. In other words, from Equation 9.23, the exergy of a closed system is

$$\chi_{\text{system}} = (U - U_0) + p_0(V - V_0) - T_0(S - S_0) \qquad (9.24)$$

An equivalent expression in terms of enthalpy is

$$\chi_{\text{system}} = (H - H_0) - V(p - p_0) - T_0(S - S_0) \qquad (9.25)$$

Whenever the system pressure (p) is identical to that for the reference environment (p_0), the second term in Equation 9.25 vanishes. The use of this alternative expression is advantageous in such cases.

The following worked examples assume a dead state of $p_0 = 100$ kPa and $T_0 = 300$ K.

EXAMPLE 9.1

Determine the exergy of the following closed systems:

(a) 5 kg of water at $p = 100$ kPa and 90°C

(b) 2 kg of ice at $p = 100$ kPa and -10°C

(c) 0.1 kg of steam at $p = 4$ MPa and 500°C

(d) 0.5 kg of wet steam at $p = 10$ kPa with a quality of 0.85

Given: Systems

(a) 5 kg of liquid water at $p = 1$ bar, $T = 90$°C

(b) 2 kg ice at $p = 1$ bar, $T = -10$°C

(c) 0.1 kg steam at $p = 40$ bar, $T = 500$°C

(d) 0.5 kg wet steam at $p = 0.1$ bar, $x = 0.85$

Find: χ_{system} for (a) through (d)

Solution: Water (H_2O) is the substance in all of these cases. Either the tables of properties in Appendix A or the **STEAM** code can be used to determine the dead state properties for water as well as the properties at the other specified conditions. The dead state properties are

$$u_0 = 113.1 \text{ kJ/kg}, \quad h_0 = 113.2 \text{ kJ/kg}$$

$$v_0 = 0.001005 \text{ m}^3/\text{kg}, \quad s_0 = 0.395 \text{ kJ/kg} \cdot \text{K}$$

Equation 9.24 for the exergy of a system can be written as

$$\chi_{\text{system}} = m[(u + p_0v - T_0s) - (u_0 + p_0v_0 - T_0s_0)]$$

Now,

$$(u_0 + p_0v_0 - T_0s_0) = h_0 - T_0s_0$$
$$= 113.2 - 300(0.395) \text{ kJ/kg}$$
$$= -5.3 \text{ kJ/kg}$$

This quantity has the same value for all four cases since the same substance is involved.

(a) For water at 100 kPa and 90°C, the **STEAM** code gives the following property values:

$$u = 376.9 \text{ kJ/kg}, \quad h = 377 \text{ kJ/kg}$$

$$v = 0.001035 \text{ m}^3/\text{kg}, \quad s = 1.193 \text{ kJ/kg} \cdot \text{K}$$

Since $p = p_0$ in this case, it follows that

$$(u + p_0 v - T_0 s) = (u + p v - T_0 s) = h - T_0 s$$

$$= 377 - 300(1.193) = 19.1 \text{ kJ/kg}$$

Hence,

$$X_{\text{system}} = 5[19.1 - (-5.3)] \text{ kJ}$$

$$= \underline{122 \text{ kJ}}$$

(b) In this case, $p = p_0$ also, and the alternative Equation 9.25 can be used for the exergy of the system. Thus,

$$X_{\text{system}} = m[(h - h_0) - v(p - p_0) - T_0(s - s_0)]$$

$$= m[(h - h_0) - T_0(s - s_0)]$$

Using either the property tables in the appendix or the **STEAM** code gives the following property values for ice at 100 kPa and $-10°C$:

$$h = -354.1 \text{ kJ/kg}, \quad s = -1.298 \text{ kJ/kg} \cdot \text{K}$$

$$\therefore X_{\text{ice}} = 2[-354.1 - 113.2 - 300(-1.298 - 0.395)] \text{ kJ}$$

$$= \underline{81.2 \text{ kJ}}$$

Note in this case that, although the temperature of the system is less than T_0, the exergy of the system is greater than zero, and work could be obtained from the system.

(c) In this instance, $p \neq p_0$ and the $V(p - p_0)$ term in Equation 9.25 does not vanish. Using Equation 9.25 rather than Equation 9.24 for the exergy of the system is thus of no particular advantage. The **STEAM** code gives the following properties for steam at 40 bar and 500°C:

$$u = 3099.8 \text{ kJ/kg}, \quad h = 3445.3 \text{ kJ/kg}$$

$$v = 0.08637 \text{ m}^3/\text{kg}, \quad s = 7.090 \text{ kJ/kg} \cdot \text{K}$$

$$\therefore (u + p_0 v - T_0 s) = 3099.8 + 100(0.08637) - 300(7.09) \text{ kJ/kg}$$

$$= 981.4 \text{ kJ/kg}$$

Hence,

$$X_{\text{steam}} = 0.1[981.4 - (-5.3)] = \underline{98.7 \text{ kJ}}$$

(d) Here again, $p \neq p_0$, and there is no particular advantage in using Equation 9.25 rather than Equation 9.24 for the exergy of the system. The following property values are obtained for wet steam at 10 kPa and 0.85 quality using the **STEAM** code:

$$u = 2100.3 \text{ kJ/kg}, \quad h = 2225.2 \text{ kJ/kg}$$

$$v = 12.47 \text{ m}^3/\text{kg}, \quad s = 7.023 \text{ kJ/kg} \cdot \text{K}$$

$$\therefore (u + p_0 v - T_0 s) = 2100.3 + 100(12.47) - 300(7.023) \text{ kJ/kg}$$

$$= 1240.4 \text{ kJ/kg}$$

Hence,

$$\chi_{\text{wet steam}} = 0.5[1240.4 - (-5.3)] = \underline{623 \text{ kJ}}$$

Worked Example 9.1 should serve to demonstrate the following two points:

1. Whenever the system is a substance for which property tables (or computer codes) are available, recourse should be made to the tables (or computer code) to determine the appropriate values of system properties for substitution in the equation for exergy of a system.

2. The only occasion when you can correctly drop the $V(p - p_0)$ term in Equation 9.25 is when p is equal to p_0.

EXAMPLE 9.2

Determine the exergy of 1 m³ of complete vacuum.

Given: System is 1 m³ of complete vacuum

$$p_0 = 100 \text{ kPa}, \quad T_0 = 300 \text{ K}$$

Find: χ_{vacuum}

Solution: You can use either Equation 9.24 or 9.25 for the computation of exergy in this case. Since a vacuum has zero mass,

$$U = 0, \quad H = 0, \quad \text{and} \quad S = 0$$

Likewise, if the vacuum were reduced to the dead state, it follows that

$$U_0 = 0, \quad H_0 = 0, \quad S_0 = 0, \quad \text{and} \quad V_0 = 0$$

The prevailing pressure p for the vacuum is, by definition, zero, but the volume $V = 1$ m³ and is nonzero. Thus, using Equation 9.24,

$$\chi_{\text{vacuum}} = (0) + 100(1 - 0) - 300(0) \text{ kJ}$$

$$= \underline{100 \text{ kJ}}$$

Using Equation 9.25 instead gives

$$\chi_{vacuum} = (0) - 1(0 - 100) - 300(0) \text{ kJ}$$
$$= \underline{100 \text{ kJ}}$$

as before.

It may seem surprising at first that a vacuum should have an exergy. If you reflect on the fact that vacuum pumps need an exergy input (work) to enable them to pull a vacuum, you should not be surprised to find that the vacuum produced is an exergy product. The same conclusion is reached if you consider, as an alternative, the use of an air motor between the atmosphere and the evacuated space. The pressure difference is all that is needed to obtain a work output from the motor. Eventually, the pressure in the previously evacuated chamber becomes equal to that of the ambient atmosphere, and then work production by the air motor ceases. It is evident, therefore, that the original evacuated space is a system that has an exergy or "work capability."

Exergy for Ideal Gas Systems

When the exergy of an ideal gas system is evaluated, the simplest approach is to use the appropriate expressions for property changes given in Chapters 4 and 8. Thus, based on Equations 4.32 and 4.1, respectively, $U - U_0$ and $V - V_0$ can be written as follows for the ideal gas:

$$U - U_0 = mc_v(T - T_0)$$
$$V - V_0 = m(v - v_0)$$
$$= m\left(\frac{RT}{p} - \frac{RT_0}{p_0}\right)$$

Hence,

$$p_0(V - V_0) = mRT_0\left(\frac{T}{T_0}\frac{p_0}{p} - 1\right)$$
$$= mRT_0\left(\frac{V}{V_0} - 1\right)$$

Also, based on Equation 8.24 with c_v constant, $S - S_0$ can be written as follows:

$$S - S_0 = m\left(c_v \ln \frac{T}{T_0} + R \ln \frac{V}{V_0}\right)$$
$$= m\left[c_v \ln \frac{T}{T_0} - R \ln \left(\frac{p}{p_0}\frac{T_0}{T}\right)\right]$$

Substituting these expressions in Equation 9.24 yields the following for the exergy of an ideal gas system:

$$\chi_{\text{ideal gas}} = mc_v T_0 \left(\frac{T}{T_0} - 1 - \ln \frac{T}{T_0} \right)$$
$$+ mRT_0 \left(\frac{V}{V_0} - 1 - \ln \frac{V}{V_0} \right) \tag{9.26}$$

or

$$\chi_{\text{ideal gas}} = mc_v T_0 \left(\frac{T}{T_0} - 1 - \ln \frac{T}{T_0} \right)$$
$$+ mRT_0 \left(\frac{p_0}{p^*} - 1 - \ln \frac{p_0}{p^*} \right) \tag{9.27}$$

where

$$p^* = \frac{mRT_0}{V} = p\frac{T_0}{T} \tag{9.27a}$$

Based on Equation 9.27, the exergy for the ideal gas system can be loosely interpreted as comprising a thermal exergy component (χ_T) and a pressure exergy term (χ_{p^*}) given as

$$(\chi_T)_{\text{ideal gas}} = mc_v T_0 \left(\frac{T}{T_0} - 1 - \ln \frac{T}{T_0} \right) \tag{9.27b}$$

and

$$(\chi_{p^*})_{\text{ideal gas}} = mRT_0 \left(\frac{p_0}{p^*} - 1 - \ln \frac{p_0}{p^*} \right) \tag{9.27c}$$

The thermal exergy term vanishes whenever $T = T_0$. Notice, however, that the pressure exergy term only vanishes when $p^* = p_0$. We can obtain yet another expression for the exergy of an ideal gas based on Equation 9.25:

$$\chi_{\text{ideal gas}} = mc_p T_0 \left(\frac{T}{T_0} - 1 - \ln \frac{T}{T_0} \right)$$
$$+ mRT_0 \left[\frac{T}{T_0} \left(\frac{p_0}{p} - 1 \right) - \ln \frac{p_0}{p} \right] \tag{9.28}$$

EXAMPLE 9.3

Calculate the exergy (in kJ) for the following closed systems:

(a) 0.1 kg of air at $p = 10$ bar and 300 K

(b) 0.2 kg of air at $p = 1$ bar and 600 K

Given: Systems

(a) 0.1 kg air at 10 bar, 300 K

(b) 0.2 kg air at 1 bar, 600 K

Find: χ_{system} for (a) and (b)

Solution: Air at the stated conditions can be regarded as an ideal gas system. Thus, any of the Equations 9.26 to 9.28 can be used for the evaluation of the exergy of the system. For air, the following values can be assumed for the specific heats and R:

$$c_v = 0.718 \text{ kJ/kg} \cdot \text{K}, \quad c_p = 1.005 \text{ kJ/kg} \cdot \text{K}, \quad \text{and} \quad R = 0.2871 \text{ kJ/kg} \cdot \text{K}$$

(a) Use of Equations 9.27 to 9.27c appears appropriate in this case. From Equation 9.27b, the thermal exergy is zero since $T = T_0$. To calculate the pressure exergy, one needs to determine p^* first. Equation 9.27a gives $p^* = p = 10$ bar. Hence, using Equation 9.27c for the pressure exergy gives

$$(\chi_{p*})_{air} = (0.1)(0.2871)(300)(1/10 - 1 + \ln 10) = 12.08 \text{ kJ}$$

Thus

$$\chi_{air} = (\chi_T)_{air} + (\chi_{p*})_{air} = \underline{12.1 \text{ kJ}}$$

(b) In this case, $p^* = p(T_0/T) = 1(300/600) = 0.5$ bar. Thus, using Equations 9.27b and 9.27c, one can determine the thermal and pressure exergy components:

$$(\chi_T)_{air} = (0.2)(0.718)(300)[600/300 - 1 - \ln(600/300)] = 13.2 \text{ kJ}$$

and

$$(\chi_{p*})_{air} = (0.2)(0.2871)(300)[1/0.5 - 1 - \ln(1/0.5)] = 5.29 \text{ kJ}$$

Hence,

$$\chi_{air} = (\chi_T)_{air} + (\chi_{p*})_{air} = \underline{18.5 \text{ kJ}}$$

Alternatively, Equation 9.28 can be used to advantage since $p = p_0$ and therefore the pressure-related terms in Equation 9.28 cancel out. In other words, in this instance,

$$\chi_{air} = mc_p T_0 \left(\frac{T}{T_0} - 1 - \ln \frac{T}{T_0} \right)$$

$$= (0.2)(1.005)(300)[600/300 - 1 - \ln(600/300)] \text{ kJ}$$

$$= \underline{18.5 \text{ kJ}}$$

Exergy for Substances in the Solid and Liquid Phases

Tables of properties are often not available for substances in the solid or liquid phase. Solids and liquids can, in general, be regarded as incompressible, which means that the term in Equation 9.24 involving $V - V_0$ is negligible. If attention is restricted initially to cases in which the substance exists as a solid (or liquid) in both its present state and the dead state, the following expressions can be used for the internal energy change and the entropy change terms in Equation 9.24:

$$U - U_0 = mc(T - T_0)$$

and, based on Equation 8.21,

$$S - S_0 = mc \ln(T/T_0)$$

where c is the specific heat of the substance, which is considered to be constant. Thus, the system exergy is given by

$$\chi_{\text{system}} = (U - U_0) + p_0(V - V_0) - T_0(S - S_0)$$

$$= mc(T - T_0) + 0 - mcT_0 \ln \frac{T}{T_0} \qquad (9.29)$$

$$= mcT_0 \left(\frac{T}{T_0} - 1 - \ln \frac{T}{T_0} \right)$$

EXAMPLE 9.4

Calculate the exergy (in kJ) of 2 kg of copper ($c = 385 \text{ J/kg} \cdot \text{K}$) block at 2 bar and 400 K. (Assume $T_0 = 300$ K and $p_0 = 1$ bar.)

Given: System: 2 kg copper block ($c = 385 \text{ J/kg} \cdot \text{K}$) at 2 bar, 400 K

$$p_0 = 1 \text{ bar and } T_0 = 300 \text{ K}$$

Find: χ_{copper}

Solution: From Equation 9.29,

$$\chi_{\text{copper}} = mcT_0[T/T_0 - 1 - \ln(T/T_0)]$$

$$= (2)(0.385)(300)[400/300 - 1 - \ln(400/300)] \text{ kJ}$$

$$= \underline{10.5 \text{ kJ}}$$

There is a growing interest in the use of *phase change materials (PCMs)* in a variety of applications, including thermal energy storage. The actual phase in which a PCM exists may be different from the phase at the dead state. A similar situation is often encountered when fish, beef, and other livestock are frozen for long-term storage. In such cases, the substance may exhibit a different specific heat at ambient temperature than it has in the chilled or frozen condition. Also, a latent heat of fusion (h_{if}) is associated with freezing of the substance at a discernible freezing point. Example 9.5 provides an illustration of how the exergy of a substance can be determined when it is at a state that corresponds to a different phase from that in the dead state.

EXAMPLE 9.5

A mass of 1000 kg of fish, initially at 300 K and 1 bar, is to be cooled to $-20°C$. The freezing point of fish is $-2.2°C$, and the specific heats of the fish below and above the freezing point c_s, c_l are, respectively, 1.7 kJ/kg · K and 3.2 kJ/kg · K. The latent heat of fusion $\langle h_{if} \rangle$ for the fish can be taken as 235 kJ/kg. Calculate the exergy produced in the chilling process. Assume $T_0 = 300$ K and $p_0 = 1$ bar.

Given: System: 1000 kg fish at $T_1 = 300$ K and $p_1 = 1$ bar

Final temperature, $T_2 = -20°C$ (253 K)

For the fish, $c_s = 1.7$ kJ/kg · K, $c_l = 3.2$ kJ/kg · K, and $h_{if} = 235$ kJ/kg

The freezing point, $T_s = -2.2°C$ (271 K)

Find: $\chi_{product}$

Solution: Assuming that $p_2 = p_1 = p_0$, one can determine the exergy produced using the reduced form of Equation 9.25:

$$\chi_{product} = \chi_2 - \chi_1 = H_2 - H_1 - T_0(S_2 - S_1)$$

The significance of this example is that no property table exists for fish from which values of the enthalpy or the entropy can be read. Figure 9.3 depicts the phase characteristics for the fish. With reference to the figure, one can establish the following relationships for the changes in enthalpy and entropy:

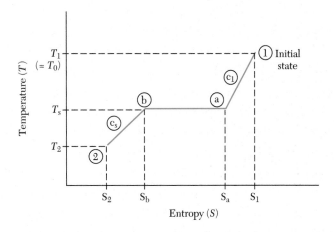

Figure 9.3 T-S diagram for the fish-freezing problem of Example 9.5.

$$H_2 - H_1 = (H_2 - H_b) + (H_b - H_a) + (H_a - H_1)$$
$$= mc_s(T_2 - T_s) - mh_{if} + mc_l(T_s - T_1)$$

and

$$S_2 - S_1 = (S_2 - S_b) + (S_b - S_a) + (S_a - S_1)$$
$$= mc_s \ln \frac{T_2}{T_s} - m \frac{h_{if}}{T_s} + mc_l \ln \frac{T_s}{T_1}$$

Equations 8.19 and 8.21 were used to derive the expression for the entropy change between state 1 and state 2. Substituting the expressions for enthalpy and entropy changes into Equation 9.25 gives

$$\chi_{product} = mc_s\left(T_2 - T_s - T_0 \ln \frac{T_2}{T_s}\right)$$
$$+ mh_{if}\left(\frac{T_0}{T_s} - 1\right) + mc_l\left(T_s - T_1 - T_0 \ln \frac{T_s}{T_1}\right)$$

Applying the equation to the present case yields the following result:

$$\chi_{product} = 4389 \text{ kJ} + 25{,}148 \text{ kJ} + 4797 \text{ kJ}$$
$$= \underline{34.3 \text{ MJ (or 9.54 kWh)}}$$

This is the estimate of the minimum exergy that must be supplied for the chilling process. The amount of cooling is $(H_1 - H_2)$, which in this case is 358 MJ (or 99.5 kWh). The minimum work required when a hypothetical device operating on the reversed Carnot cycle is employed can be estimated by assuming $T_H = T_0$ and $T_L = T_2$. Substituting these values in Equation 7.22a for the reversed Carnot cycle yields $COP_C = 5.41$ and, hence, a minimum work input of 66.2 MJ (or 18.4 kWh). This confirms the view expressed earlier that the Carnot cycle is not the optimum cycle whenever either the heat source or the heat sink is not a constant temperature system. Note that if instead of $T_L = T_2$, T_L is assumed to equal T_m, the thermodynamic mean temperature at constant pressure, with

$$T_m = \frac{H_2 - H_1}{S_2 - S_1}$$

then the Carnot cycle analysis would yield the correct result. The exergy analysis provides a correct determination of optimum energy requirements in such situations. This point is continued in Chapters 10 and 11.

Unless otherwise stated, assume $p_0 = 100$ kPa (14.5 psia) and $T_0 = 298$ K (537°R) for the reference environment.

1. Determine the exergy of 0.5 kg of H_2O at 1 atm and
   (a) $T = -50°C$
   (b) $T = 100°C$, $x = 0$
   (c) $T = 100°C$, $x = 1$
   (d) $T = 200°C$

   Answer: (a) 27 kJ, (b) 17 kJ, (c) 243 kJ, (d) 272 kJ

2. Determine the exergy of 1 lbm of H_2O at 1 atm and
   (a) $T = -60°F$
   (b) $T = 212°F$, $x = 0$
   (c) $T = 212°F$, $x = 1$
   (d) $T = 400°F$

   Answer: (a) 19,000 ft · lbf, (b) 11,200 ft · lbf, (c) 162,000 ft · lbf, (d) 182,000 ft · lbf

3. A volume of 0.009 m³ of ambient air at 100 kPa and 298 K is compressed isentropically to a volume of 1.5×10^{-3} m³. Assume $c_v = 0.718$ kJ/kg · K and $R = 0.2871$ kJ/kg · K for air. Calculate
   (a) the mass of the air
   (b) the final temperature of the air
   (c) the exergy of the air at its final state

   Answer: (a) 0.0105 kg, (b) 610 K, (c) 1.6 kJ

4. One short ton (or 2000 lbm) of water at 14.5 psia and 77°F is turned to ice at 32°F. Determine
   (a) the cooling load (in kWh)
   (b) the minimum work input (in kWh) to an electric refrigeration device with a $COP_C = 1.8$
   (c) the exergy of the ice produced

   Answer: (a) 110 kWh, (b) 61.2 kWh, (c) 9.16 kWh

9.2.2 Equivalence Between Mechanical Energy Forms and Exergy

Exergy and Work

Since the flow of mechanical energy (or work) is the "currency" in which exergy is measured, it follows that work automatically converts to exergy flow having the same magnitude. Thus, a flow of work (W) is equivalent to the same amount of flow of exergy (χ_W), or

$$\chi_W = W \qquad (9.30a)$$

Note that strictly speaking, since exergy is defined as the *net* or *useful work*, any non-useful work associated with a change in system volume against the pressure p_0 of the reference environment should be subtracted from W to obtain the correct exergy flow equivalent.

Exergy and Mechanical Energy

Both potential energy (PE) and kinetic energy (KE) are mechanical energy forms and thus can be equated to exergy of the same magnitude. Thus,

$$\chi_{PE} = PE = mg(Z - Z_0) \tag{9.30b}$$

when dealing with gravitational potential energy. Z is the height of the object relative to an arbitrary datum, while Z_0 denotes the ground level position relative to the arbitrary datum. Also,

$$\chi_{KE} = KE = \tfrac{1}{2}m(|\mathbf{V}|)^2 \tag{9.30c}$$

EXAMPLE 9.6

A pumped storage plant described by Angrist (1976) is to deliver 100 MW of peaking power for 4 h per day during the peak load period. The upper water reservoir could be placed 152 m (500 ft) above a river. Assume that the pumping efficiency is 65%, while the electrical power generating efficiency of the plant is 85%. Determine

(a) The electrical energy (in MWh) required each day to provide adequate pumped storage capacity.

(b) The volume of water (in millions of gal) that must be pumped to the upper reservoir per day for the desired storage capacity.

Given:
Pumped storage plant

    Required capacity—to deliver 100 MW for 4 h each day

    Elevation—152 m (500 ft) above river level

Process efficiencies

    First law pumping efficiency = 65%

    First law electrical power generating efficiency (from PE of reservoir water) is 85%

Find:

(a) χ_{input} to pump (in MWh/day)

(b) Volume of water (in millions of gal) required per day for the pumped storage

Solution:

(a) Let PE denote the pumped storage potential energy daily requirement in MWh/day. Consider the pumping operation first.

$$\chi_{product} = \chi_{PE} = PE \text{ MWh/day}$$

In a completely reversible operation, the required exergy input is equal to the exergy product. The process efficiency for a pumping operation is given by Equation 8.114, with $W_s = (\chi_{input})_{rev}$ and $W_a = (\chi_{input})_{actual}$. Hence, the required exergy input can be determined as follows:

$$(\chi_{input})_{actual} = \frac{(\chi_{input})_{rev}}{\eta_{\substack{adiabatic \\ pump}}}$$

$$= \frac{PE}{0.65}$$

Next let's consider the reservoir discharge operation. The exergy input for the process is PE in MWh/day. In a completely reversible operation, the exergy product is equal to PE in MWh/day. The process efficiency of the actual operation is defined by Equation 8.109, with $W_s = (\chi_{product})_{rev}$ and $W_a = (\chi_{product})_{actual}$. Therefore,

$$(\chi_{product})_{actual} = \eta_{\substack{adiabatic \\ turbine}} (\chi_{product})_{rev}$$

$$= (0.85)\ PE = (0.85)(0.65)(\chi_{input})_{actual}$$

Hence,

$$(\chi_{input})_{actual} = \frac{(\chi_{product})_{actual}}{(0.85)(0.65)}$$

Now,

$$(\chi_{product})_{actual} = (100\ MW)(4h) = 400\ MWh\ (= 0.85\ PE)$$

Therefore, PE = 470.6 MWh (or 1.694×10^{12} J) and the required exergy input to the pump, $(\chi_{input})_{actual} = \underline{724\ MWh/day}$. This is the required daily supply of electrical energy.

(b) From the definition of PE, one can determine the mass of water to be pumped each day:

$$m_{water} = \frac{PE}{g\ \Delta Z}$$

$$= \frac{1.694 \times 10^{12}\ J}{(9.81\ m/s^2)(152\ m)} = 1.136 \times 10^9\ kg$$

Taking the density of water as 10^3 kg/m^3, one can calculate the volume of water required daily:

$$V_{water} = 1.136 \times 10^6\ m^3 \left(\frac{264.2\ gal}{1\ m^3} \right) = \underline{300\ million\ gal}$$

Unless otherwise stated, assume $p_0 = 100$ kPa (14.5 psia) and $T_0 = 298$ K (537°R) for the reference environment.

1. An electric heating coil is used to raise the temperature of a fixed mass of a liquid. Define the liquid and the heating coil as the system. If 66 kJ of electricity is supplied to the heating coil, what exergy flow is this equivalent to?

2. The volume of a compressible system increases by 0.01 m³ while the pressure of the system remains constant at 300 kPa. The expansion is fully resisted by a combination of a fixed weight that is raised and the ambient pressure of 100 kPa. Determine
   (a) the work done (W) by the system
   (b) the exergy flow equivalent (χ_W) of the work transfer (W) from the system to the surroundings.

   Answer: **(a)** 3 kJ, **(b)** 2 kJ

3. A waterfall is 100 ft high. What is the exergy equivalent of the potential energy of 1 million gal of water at the top of the fall? (Assume 1 gal of water has a mass of 8.34 lbm.)

   Answer: 314 kWh

9.2.3   Flow of Exergy (χ_Q) Associated with Heat Flow (Q)

The proper measure of the traditionally termed *mechanical equivalent of heat* ought to be the exergy content of a heat flow. Let Q denote the heat flow from the surroundings to a system across a boundary surface area at a uniform temperature T_{surface}, as shown in Figure 9.4. The action is equivalent to the supply of the heat Q from a thermal

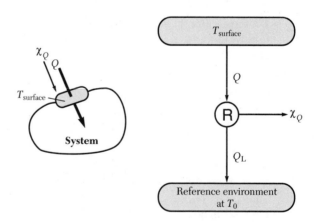

Figure 9.4 Flow of exergy associated with heat flow.

reservoir. Therefore, the maximum work possible when a reversible heat engine is interposed between the reservoir at $T_{surface}$ and the reference environment at T_0 is the exergy associated with the heat flow. The maximum work is the product of Q times the Carnot cycle efficiency. In other words,

$$\chi_Q = \left(1 - \frac{T_0}{T_{surface}}\right) Q \tag{9.31}$$

Notice that the exergy flow is in the same direction as Q when $T_{surface} > T_0$. When $T_{surface} < T_0$, however, χ_Q is in the opposite direction. In other words, when a medium already at a lower temperature than the reference environment is cooled further, the exergy of that medium is actually increasing. This inference makes correct physical sense. This is what happens, for example, in ice production when a refrigeration device is used to produce a cooling (or heat removal) action, thereby increasing the exergy of the system (H_2O).

Now that appropriate expressions have been considered for the exergy of various entities involved in nonflow processes for a closed system, both the first law and second law efficiencies can be examined for such processes. The following worked examples are typical of those often encountered in engineering.

EXAMPLE 9.7

The temperature of 1 kg of water initially at 25°C is raised to 90°C using an electric heating coil. Assuming that heat losses to the surroundings are negligible, calculate the first law and second law efficiencies of the process. Assume that the environment is at 300 K.

Given: 1 kg of water, heated using an electric coil

$T_1 = 298$ K, $T_2 = 363$ K, $T_0 = 300$ K

Negligible heat losses

Find: (a) η_1, (b) η_2

Solution:

(a) *First law efficiency*

The system is illustrated in Figure 9.5 and comprises 1 kg of water and the heating coil (assumed as having negligible thermal capacity).

$$Q_{12} = 0$$

W_d denotes the displacement work done by the system and W_X is the shear or shaft work done by the system. The electricity supplied to the heating coil counts as shaft work on the system, and hence,

$$W_X = -\text{electrical energy supplied to the coil.}$$

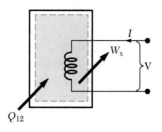

Figure 9.5 Water heating
process of Example 9.7 (using
an electric heating coil).

Assume that the water is heated at constant atmospheric pressure, in which case,

$$W_d = p_0(V_2 - V_1)$$

Thus, applying the first law yields

$$Q_{12} - (W_d + W_X) = U_2 - U_1 = m(u_2 - u_1)$$

or

$$-W_X = (U_2 + p_0V_2) - (U_1 + p_0V_1)$$
$$= (U_2 + p_2V_2) - (U_1 + p_1V_1)$$
$$= H_2 - H_1 = m(h_2 - h_1)$$

Either the steam tables (Table A.1) or the **STEAM** code gives

$$u_1 = 105 \text{ kJ/kg}, \quad h_1 = 105 \text{ kJ/kg}$$
$$u_2 = 377 \text{ kJ/kg}, \quad h_2 = 377 \text{ kJ/kg}$$

Thus,

$$-W_X = 272 \text{ kJ}$$

For the first law efficiency, one can regard the desired energy output as the increase produced in the internal energy of the system (water). Thus,

$$\eta_1 = \frac{\text{desired (energy) output}}{\text{required (energy) input}} \cdot 100\%$$

$$= \frac{U_2 - U_1}{(-W_X)} \cdot 100\% = \frac{272}{272} \cdot 100\%$$

$$= 100\%$$

Thus, the first law efficiency is 100%.

(b) *Second law efficiency*

Based on Equation 9.30a, $\chi_{input} = -W_X = 272$ kJ. Since $p_1 = p_2 = p_0$, it is appropriate to use Equation 9.25 for the system exergy. Hence,

$$\chi_{product} = (H_2 - H_1) - T_0(S_2 - S_1)$$

From Table A-1 in Appendix A,

$$s_1 = 0.367 \text{ kJ/kg} \cdot \text{K}$$

$$s_2 = 1.193 \text{ kJ/kg} \cdot \text{K}$$

Now,

$$T_0 = 300 \text{ K}$$

Hence,

$$\chi_{product} = 24.2 \text{ kJ}$$

The second law efficiency is calculated using Equation 9.8:

$$\eta_2 = \frac{\text{exergy product}}{\text{exergy input}} \cdot 100\%$$

$$= \frac{24.2}{272} \cdot 100\% = \underline{8.9\%}$$

Thus, the second law efficiency achieved is a mere 9%. In other words, over 90% of the exergy supplied is destroyed in the electrical energy to internal energy conversion process. This result highlights the colossal waste in heating water with an electric heating coil.

EXAMPLE 9.8

Consider the heating of water in Example 9.7 and assume that the heating is accomplished by supplying heat from a constant temperature system at **(a)** 1500 K and **(b)** 400 K. Determine the second law efficiency in each case.

Given: Heating of 1 kg of water from $T_1 = 298$ K to $T_2 = 363$ K

Heat source temperature is **(a)** 1500 K, **(b)** 400 K

Assume $T_0 = 300$ K

Find: η_2 for **(a)** $T_{source} = 1500$ K, **(b)** $T_{source} = 400$ K

Solution: The system is illustrated in Figure 9.6. The solution for Example 9.7 can be applied here:

1. The exergy product is the same and is equal to 24.2 kJ.
2. In place of a shaft work input (the electricity supply of Example 9.7), the present example has a heat input Q_{12} which, applying the first law of thermodynamics, can

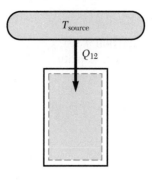

Figure 9.6 Water heating process of Example 9.8 (using thermal reservoirs).

be shown to be equal in magnitude to the shaft work input determined in Example 9.7. Thus, $Q_{12} = 272$ kJ.

The exergy input can be determined using Equation 9.31:

$$\chi_{Q_{12}} = \left(1 - \frac{T_0}{T_{source}}\right) Q_{12}$$

Thus, for $T_{source} = 1500$ K, the exergy input is $(1 - 300/1500)(272) = 217.6$ kJ. When $T_{source} = 400$ K, the exergy input is $(1 - 300/400)(272) = 68$ kJ.

The second law efficiency is computed using Equation 9.8. The results are as follows:

(a) $\eta_2 = \underline{11.1\%}$

(b) $\eta_2 = \underline{35.6\%}$

The inference that can be drawn from these estimates is clear. To minimize waste of exergy, the process should be accomplished using a heat source that is at a temperature as close as possible to the desired temperature of the water being heated. Whenever source heating is accomplished with a large temperature difference between the heat source and the system being heated, a substantial destruction of exergy occurs. This in fact is the major source of irreversibility in heat engines that are operated from heat sources produced by the combustion of fuels. Chapter 10 returns to this point and deals with heat engines in more detail.

EXAMPLE 9.9

A mass of 1000 kg of ice at 0°C is to be produced from water initially at 20°C.

(a) Determine the minimum work input to a reversible (work-absorbing) cyclic device required for the ice production.

(b) If the ice is actually produced using a vapor compression plant with a COP_C of 1.5, compare the actual compressor work requirement to the minimum work input (for the reversible cyclic device) using a ratio.

Assume that the environment is at 25°C.

Given: System comprising 1000 kg of water

$$T_1 = 293 \text{ K (water)}, \quad T_2 = 273 \text{ K (ice)}$$

$$p_1 = p_2 = \text{atmospheric pressure } (p_0) \text{ can be assumed}$$

$$COP_C \text{ for actual vapor compression plant} = 1.5$$

$(\chi_{input})_{actual}$ is for vapor compression plant with $COP_C = 1.5$

$$T_0 = 298 \text{ K}$$

Find:

(a) $W_{min} = (\chi_{input})_{min}$ for the task
(b) The ratio $(\chi_{input})_{actual}/(\chi_{input})_{min}$

Solution:

(a) *Minimum exergy required*

From Equation 9.25,

$$\chi_{product} = (H_2 - H_1) - T_0(S_2 - S_1)$$

since $p_1 = p_2 = p_0$.
Use of tables of water properties (Tables A.1 and A.4) or the **STEAM** code gives

$$H_1 = 84 \times 1000 \text{ kJ}, \quad S_1 = 0.296 \times 1000 \text{ kJ/K}$$

$$H_2 = -333.4 \times 1000 \text{ kJ}, \quad S_2 = -1.221 \times 1000 \text{ kJ/K}$$

$$T_0 = 298 \text{ K}$$

Thus, $\chi_{product} = 34,666$ kJ (or 9.63 kWh)
Now, $(\chi_{input})_{min} = \chi_{product}$, corresponding to when all processes take place reversibly.

Therefore, the minimum work input required is 34.7 MJ (or 9.63 kWh).

(b) *Ratio of actual work to minimum work required*

The cooling load is $-Q_{12}$, which by the first law is given by

$$-Q_{12} = H_1 - H_2$$

$$= 417,400 \text{ kJ}$$

Since $COP_C = 1.5$, it follows that the actual compressor work required is

$$W_{actual} = -Q_{12}/1.5 = 278,267 \text{ kJ} = 278.3 \text{ MJ (or 77.3 kWh)}$$

that is,

$$(\chi_{\text{input}})_{\text{actual}} = 278.3 \text{ MJ (or } 77.3 \text{ kWh)}$$

$$\text{Thus, } (\chi_{\text{input}})_{\text{actual}}/(\chi_{\text{input}})_{\text{min}} = \underline{8.02}.$$

Thus, in Example 9.9, the first law efficiency is 150%, while the second law efficiency is only 12.5%. Evidently, these systems have a long way to go to operate close to their optimum performance level.

EXAMPLE 9.10

A thermal energy storage system (A) composed of water receives 1000 kJ of heat and heats up at a constant volume from 27°C to 100°C. The system remains in the liquid state throughout the heating process. An alternative storage system (B) uses water at 27°C that becomes dry, saturated steam at 100°C and 1 atm while receiving 1000 kJ of heat. The reference environment is at 300 K and 1 atm pressure. Compare the exergy increase produced in the two cases.

Given:

(a) System A

$$Q_{12} = 1000 \text{ kJ}$$

$$T_1 = 300 \text{ K (liquid water)}$$

$$T_2 = 373 \text{ K (liquid water)}$$

Constant volume process

$$T_0 = 300 \text{ K}$$

(b) System B

$$Q_{12} = 1000 \text{ kJ}$$

$$T_1 = 300 \text{ K (liquid water)}$$

$$T_2 = 373 \text{ K (dry, saturated steam)}$$

Constant pressure process

$$T_0 = 300 \text{ K}$$

Find: χ_{product} for each system

Solution: In both cases, nonflow processes are involved and the exergy product, χ_{product}, can be calculated using either Equation 9.24 or 9.25. For system A, the volume stays constant, and Equation 9.24 is the appropriate choice. Since $p_1 = p_2 = p_0$ for system B, however, Equation 9.25 can be used when considering the exergy product for the system.

(a) *System A*

Equation (9.24) gives

$$\chi_{\text{product}} = \chi_2 - \chi_1$$
$$= (U_2 - U_1) - T_0(S_2 - S_1)$$
$$= m[u_2 - u_1 - T_0(s_2 - s_1)]$$

The system mass (m) can be determined using the specified value of Q_{12} and by applying the first law of thermodynamics. In other words,

$$Q_{12} = U_2 - U_1 = m(u_2 - u_1)$$

$$\therefore m = \frac{Q_{12}}{u_2 - u_1}$$

The **STEAM** code gives the following:

$$u_1 = 113 \text{ kJ/kg}, \quad s_1 = 0.395 \text{ kJ/kg} \cdot \text{K}$$
$$u_2 = 418.9 \text{ kJ/kg}, \quad s_2 = 1.307 \text{ kJ/kg} \cdot \text{K}$$

Substituting these values in the previous expressions yields the following:

$$m = (1000)/(418.9 - 113) = 3.27 \text{ kg}$$

and

$$\chi_{\text{product}} = (3.27)[418.9 - 113 - 300(1.307 - 0.395)] \text{ kJ}$$
$$= 105.6 \text{ kJ}$$

Thus, the exergy increase produced is just slightly over 105 kJ.

(b) *System B*

For this case, one can use Equation 9.25 to obtain the following expression for the increase in exergy produced:

$$\chi_{\text{product}} = \chi_2 - \chi_1$$
$$= (H_2 - H_1) - T_0(S_2 - S_1)$$
$$= m[h_2 - h_1 - T_0(s_2 - s_1)]$$

As for the system A, the system mass (m) is obtained by applying the first law of thermodynamics to the process and substituting the specified value of Q_{12}:

$$Q_{12} \underset{\substack{\text{(constant}\\\text{pressure)}}}{} = H_2 - H_1 = m(h_2 - h_1)$$

$$\therefore m = \frac{Q_{12}}{h_2 - h_1}$$

The **STEAM** code gives

$$h_1 = 113.2 \text{ kJ/kg}, \quad s_1 = 0.395 \text{ kJ/kg} \cdot \text{K}$$
$$h_2 = 2676 \text{ kJ/kg}, \quad s_2 = 7.355 \text{ kJ/kg} \cdot \text{K}$$

Substituting these values gives

$$m = (1000)/(2676 - 113.2) = 0.39 \text{ kg}$$

and

$$\chi_{product} = (0.39)[2676 - 113.2 - 300(7.355 - 0.395)] \text{ kJ}$$
$$= 185.3 \text{ kJ}$$

The exergy increase produced in this case is 185 kJ.

$$\therefore \frac{(\chi_{product})_{\text{system B}}}{(\chi_{product})_{\text{system A}}} = \underline{1.75}$$

Example 9.10 demonstrates the potential for improved thermal energy storage via the use of a latent heat storage medium. In other words, the latent heat storage system in this example stores more exergy than does the sensible heat storage medium for the same energy input as heat.

EXAMPLE 9.11

For safety reasons, the motors used in mines are often designed to run on compressed air. In a particular design, an air motor requires a minimum of 7 bar air pressure to operate while delivering 105 kJ/kg of air supplied. A compressed air bottle of volume 1.66 m³ containing air at 20 bar and 300 K is available for operating the air motor. If the ambient pressure is 1 atm and the ambient temperature is 300 K, determine

(a) the maximum work output of the air motor operated from one bottle of compressed air

(b) the second law efficiency achieved in the operation of the air motor

Given:
Compressed air supply

$V = 1.66 \text{ m}^3, \quad p = 2000 \text{ kPa}, \quad T = 300 \text{ K}$

Air motor

minimum supply pressure required = 700 kPa,   Output = 105 kJ/kg air

$p_0 = 101.3 \text{ kPa}, \quad T_0 = 300 \text{ K}$

Find:

(a) W_{max} for the air motor using one bottle of compressed air
(b) η_2 achieved

Solution: First, determine the mass of air initially in the compressed air bottle. Using Equation 4.7,

$$m_{initial} = \frac{p_{initial} V}{RT_{initial}}$$

$$= \frac{(2000 \text{ kN/m}^2)(1.66 \text{ m}^3)}{(0.2871 \text{ kJ/kg} \cdot \text{K})(300 \text{ K})} = 38.5 \text{ kg}$$

The cut-off stage in the operation of the air motor is when the supply pressure is down to 700 kPa. The mass of air left in the bottle is now given by

$$m_{final} = \frac{p_{final} V}{RT_{final}}$$

$$= \frac{(700 \text{ kN/m}^2)(1.66 \text{ m}^3)}{(0.2871 \text{ kJ/kg} \cdot \text{K})(300 \text{ K})} = 13.5 \text{ kg}$$

Actual work output of the air motor

From the previous calculations, the mass of air supplied to the motor = $38.5 - 13.5 = 25$ kg. The air motor delivers work output of 105 kJ/kg of air supplied. Therefore, the work output of the air motor when operated from one compressed air bottle is the exergy product, $\chi_{product} = 25(105) = \underline{2625 \text{ kJ}}$.

(a) *Maximum work possible*

The maximum possible work is obtained when the air initially in the compressed bottle performs work reversibly and finally arrives at the dead state. In other words, the maximum possible work is the exergy of the compressed air at the initial condition. Assuming ideal gas behavior for air, Equation 9.27 can be used to compute this exergy. Since $T_{initial}$ for the compressed air is equal to T_0, one can conclude from Equation 9.27a that $p^* = p_{initial} = 2000$ kPa. Also, the thermal exergy given by Equation 9.27b is zero. The pressure exergy is determined using Equation 9.27c:

$$(\chi_{p^*})_{air} = mRT_0 \left(\frac{p_0}{p^*} - 1 - \ln \frac{p_0}{p^*} \right)$$

$$= (38.5 \text{ kg})(0.2871 \text{ kJ/kg} \cdot \text{K})(300 \text{ K}) \left\{ \frac{(101.3 \text{ kPa})}{(2000 \text{ kPa})} - 1 - \ln \frac{(101.3 \text{ kPa})}{(2000 \text{ kPa})} \right\}$$

$$= 6740 \text{ kJ}$$

Thus, the maximum possible work output from the compressed air bottle is $\underline{6740 \text{ kJ}}$.

(b) *Second law efficiency achieved*

It was determined earlier that the work output from the air motor specified is

2625 kJ. The second law efficiency of the process is computed using the following variation on Equation 9.8:

$$\eta_2 = \frac{\text{actual work from motor } (\chi_{\text{product}})}{\text{maximum possible work } (\chi_{\text{input}})} \cdot 100\%$$

$$= \frac{2625}{6740} \cdot 100\% = 38.9\%$$

Thus, the second law efficiency achieved is 38.9%.

Note: After the cut-off stage (when $p_{\text{final}} = \overline{700\text{ kPa}}$) for the air motor, the mass of compressed air left in the bottle has an exergy of

$$m_{\text{final}}\, RT_0[p_0/p_{\text{final}} - 1 - \ln(p_0/p_{\text{final}})] = 1253 \text{ kJ}$$

The overall performance will improve if this exergy can be put to use and not lost by simply discharging the air left in the bottle to the atmosphere.

EXERCISES

for Section

9.2.3

Unless otherwise stated, assume $p_0 = 100$ kPa (14.5 psia) and $T_0 = 298$ K (537°R) for the reference environment.

1. A heat transfer of 50 kJ occurs from a mixture of ice and water at 0°C to a wet refrigerant vapor at $-20°C$. Define the ice and water mixture as system. If the system boundary is chosen such that $T_{\text{boundary}} = 0°C$, determine the magnitude and direction of the exergy flow χ_Q associated with the heat transfer Q.

   Answer: 4.58 kJ (to the system)

2. Repeat Exercise 1 assuming the system to be the wet refrigerant vapor and the system boundary to be such that $T_{\text{boundary}} = -20°C$.

   Answer: 8.89 kJ (from the system)

9.2.4 Exergy Consumption and Entropy Generation

Section 8.8 (of Chap. 8) established that for a closed system, entropy generation occurs that provides a measure of the loss of work production due to irreversibilities according to the following:

$$W_{\text{rev}} - W_{\text{actual}} = T_0 S_{\text{gen}} \tag{8.68}$$

Using the nonflow availability equation (NFAE) (Eq. 9.7) and various expressions for exergy, an equivalent relationship for exergy consumption (due to irreversibilities) can now be established.

Consider a typical nonflow process, as shown in Figure 9.7, by which a system initially at an equilibrium state 1 is taken to a final equilibrium state 2. The system can have a work interaction W_d with the reference environment in addition to a useful work output W. The incremental heat transfer, δQ_{12}, occurs such that the total heat transfer, Q_{12},

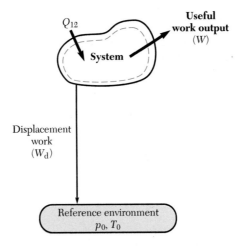

Figure 9.7 A typical nonflow process of a closed system that can have a boundary displacement work interaction with the reference environment.

is simply the integral of δQ_{12}. The inequality of Clausius can be expressed in terms of entropy change, as indicated by Equation 9.4, with the integral of $(1/T_{\text{boundary}})\,\delta Q_{12}$ denoting the entropy flux through the boundary surface. Figure 9.8 shows the exergy transfers involved in the nonflow process.

First law analysis

The nonflow energy equation (NFEE) (Eq. 6.10) applied to the process leads to the following:

$$\int_{\text{state } 1}^{\text{state } 2} \delta Q_{12} - \{W + p_0(V_2 - V_1)\} = U_2 - U_1 \qquad (9.32\text{a})$$

It is assumed that changes in the kinetic and potential energy of the system are negligible.

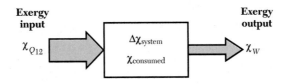

Figure 9.8 Exergy transfers in the typical nonflow process of Figure 9.7.

Second law and availability analysis

The exergy input is the exergy flow associated with the heat flow to the system. From Equation 9.31, one can write the exergy input in this instance as

$$\chi_{\text{input}} = \chi_{Q_{12}} = \int_{\text{state 1}}^{\text{state 2}} \left(1 - \frac{T_0}{T_{\text{boundary}}}\right)\delta Q_{12} \tag{9.32b}$$

The exergy product can be written as the sum of the increase in exergy of the system, $(\chi_2 - \chi_1)$, and the useful work output W. Thus, using Equations 9.24 and 9.30a, one obtains the following for the exergy product:

$$\chi_{\text{product}} = \{U_2 - U_1 + p_0(V_2 - V_1) - T_0(S_2 - S_1)\} + W \tag{9.32c}$$

Now apply the NFAE (Eq. 9.7) to obtain the following for the exergy consumed in the process:

$$\chi_{\text{consumed}} = \chi_{\text{input}} - \chi_{\text{product}}$$

$$= \left\{\int_{\text{state 1}}^{\text{state 2}} \delta Q_{12} - [W + p_0(V_2 - V_1)] - (U_2 - U_1)\right\} \tag{9.32d}$$

$$+ T_0\left(S_2 - S_1 - \int_{\text{state 1}}^{\text{state 2}} \frac{\delta Q_{12}}{T_{\text{boundary}}}\right)$$

The expression on the right-hand side of Equation 9.32d that is enclosed within the set of large braces is equal to zero from Equation 9.32a. The second expression can be modified by using the inequality of Clausius (Eq. 9.4) written as an equality:

$$S_2 - S_1 = \int_{\text{state 1}}^{\text{state 2}} \frac{\delta Q_{12}}{T_{\text{boundary}}} + S_{\text{gen,system}} \tag{9.32e}$$

where $S_{\text{gen,system}}$ is a *nonnegative* quantity that is the entropy production in the system due to *internal irreversibility*. Combining Equations 9.32a and 9.32e with Equation 9.32d thus leads to the important result for the system:

$$(\chi_{\text{consumed}})_{\text{system}} = T_0 S_{\text{gen,system}} \tag{9.32f}$$

The result is equivalent to Equation 8.68 obtained in Chapter 8. The exergy destruction indicated by Equation 9.32f is that due to internal irreversibility.

Let's also consider the exergy destruction (or consumption) associated with *external irreversibility* for the nonflow process indicated in Figure 9.7. The result is in the same form as Equation 9.32f and can be written as

$$(\chi_{\text{consumed}})_{\text{surr}} = T_0 S_{\text{gen,surr}} \tag{9.32g}$$

Finally, combining Equations 9.32f and 9.32g gives the total exergy consumption for the universe:

$$(\chi_{\text{consumed}})_{\text{universe}} = (\chi_{\text{consumed}})_{\text{system}} + (\chi_{\text{consumed}})_{\text{surr}}$$

$$= T_0\{S_{\text{gen,system}} + S_{\text{gen,surr}}\} \tag{9.32h}$$

$$= T_0\{(\Delta S_{12})_{\text{system}} + (\Delta S_{12})_{\text{surr}}\}$$

The last step in this equation follows from using the equality version of Equations 9.4 and 9.5, which include entropy generation terms. Thus, the total exergy consumption for the universe is given by

$$(\chi_{consumed})_{universe} = T_0\{(\Delta S_{12})_{system} + (\Delta S_{12})_{surr}\}$$
$$= T_0(\Delta S_{12})_{universe}$$

(9.32)

This final result is widely quoted in the literature as the *Gouy–Stodola lost work theorem*. The exergy destruction is often called by other names, such as *lost work, lost availability,* or *irreversibility*. When the latter term is used, the symbol I denotes its magnitude.

For a *power-producing* scheme, I represents the *lost potential for work* (or power) or, more simply, the difference between the reversible work output and the actual work output with both processes originating from the same initial state and terminating at the same state. The expression for irreversibility or exergy consumption applies equally when considering power-absorbing processes, or any other processes for that matter. In the case of *power-absorbing* processes, I (or $\chi_{consumed}$) represents the additional work input needed, over and above the minimum requirement in the reversible case, to produce the desired change. In any event, $\chi_{consumed}$ represents an *irretrievable loss (or dissipation) of exergy* or a loss in the *capacity to do work*.

Example 9.7 on heating water with an electric coil can be used to demonstrate the use of the lost exergy principle. Applying the NFAE (Eq. 9.7) yields

$$\chi_{consumed} = \chi_{input} - \chi_{product}$$
$$= (272 - 24.2)\ kJ = \underline{248\ kJ}$$

Alternatively, $\chi_{consumed}$ can be determined using Equation 9.32:

$$(\chi_{consumed})_{universe} = T_0[(\Delta S_{12})_{system} + (\Delta S_{12})_{surr}]$$
$$= (300\ K)(1\ kg)(1.193 - 0.367)\ kJ/kg \cdot K$$
$$= \underline{248\ kJ}$$

Zero entropy change is indicated for the surroundings since work interaction only occurs between the system and the surroundings, and as observed in Chapter 8, no entropy flow is associated with work.

Having established the fundamental principles involved in availability analysis for the nonflow processes and closed systems, let's now turn to flow processes for which, in addition to heat and work transfer between the system and its surroundings, the transfer of mass can take place. The expressions for the exergy associated with heat and work are also applicable to open systems. However, a different expression arises for the exergy of the mass flowing into or out of the control volume in the case involving flow processes.

EXERCISES

for Section

9.2.4

Unless otherwise stated, assume $p_0 = 100$ kPa (14.5 psia) and $T_0 = 298$ K (537°R) for the reference environment.

1. A volume of 450 gal of water at the ambient condition of 14.5 psia and 77°F is to be raised through 90°F each week. (1 gal of water has a mass of 8.34 lbm.) Assume

that the heating is accomplished using an electric resistance heating coil. Calculate

(a) the exergy product (in kWh)
(b) the exergy input (in kWh) using the electric heating coil
(c) the exergy destroyed (in kWh)

Answer: (a) 7.56 kWh, (b) 99 kWh, (c) 91.4 kWh

2. Compute $T_0(\Delta S)_{universe}$ for the process in Exercise 1 and compare this result with that for the exergy destroyed.

3. Assume that the heating load in Exercise 1 is met by a solar pond at 180°F. Calculate

(a) the exergy input from the solar pond for the required heating of the 450 gal of water
(b) the exergy destroyed as the difference between the exergy input and the exergy product

Also compute $T_0(\Delta S)_{universe}$ and compare this with the result in part (b).

Answer: (a) 15.9 kWh, (b) 8.37 kWh

9.3 Exergy in Steady-Flow Processes

A typical single-stream steady-flow problem is shown in Figure 9.9 with a heat flow rate $\dot{Q}$ from the surroundings to the control volume and a shaft power output $\dot{W}_x$ from the control volume.

9.3.1 Expressions for Exergy in Steady-Flow Processes

The expressions for exergy corresponding to the heat flow rate and the shaft power are of the same form as those for nonflow processes given by Equations 9.31 and 9.30a, respectively. For the mass flowing into or out of the control volume, however, a different expression can be written:

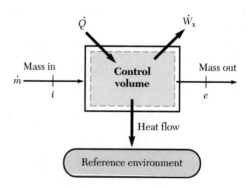

Figure 9.9 A single-stream steady-flow problem.

$$\dot{X}_{\text{mass flow}} = \dot{m}[h - h_0 - T_0(s - s_0)]$$
$$+ \dot{m}\left[\frac{V_m^2}{2} + g(Z - Z_0)\right] \qquad (9.33)$$

Included in Equation 9.33 are the kinetic energy and gravitational potential energy terms. The elevation Z at the port of entry (or exit) of the mass stream into (or out of) the control volume is measured relative to an arbitrary datum. The elevation Z_0 of the ground level is measured relative to the same datum. A comparison of Equation 9.33 with 9.25 for a closed system reveals only the noninclusion of a term in Equation 9.33 which corresponds to the $-mv(p - p_0)$ term in Equation 9.25. Indeed, for nonflow processes occurring at a constant pressure equal to p_0, the expression for exergy of the system has essentially the same form as that for a flow of mass in flow processes.

Derivation of the Expression for the Exergy of a Stream of a Substance

Formal derivation of the expression for mass stream exergy essentially follows the same procedure as outlined in the case of the exergy for a closed system in Section 9.2. Figure 9.10 shows a hypothetical scheme by which a steady stream of a substance is brought to the dead state via a set of completely reversible processes.

First Law Analysis

Application of the SFEE (Eq. 6.29) to the stream through the control volume (CV_1) yields

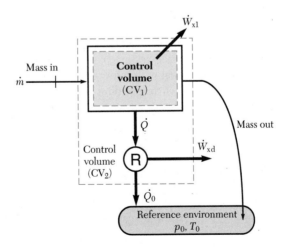

Figure 9.10 A hypothetical and completely reversible scheme for bringing a material stream to the dead state.

$$(-\dot{Q}) - (\dot{W}_{x1})_{rev} = \dot{m}\left[(h_0 - h) + \left(0 - \frac{V_m^2}{2}\right) + g(Z_0 - Z)\right] \quad (9.33a)$$

In Figure 9.10, the exit stream is shown as being at the dead state and somehow blending with the reference environment. Reversible cyclic devices must be employed for the heat exchange between the control volume and the reference environment. The overall energy balance for these reversible cyclic devices is

$$\dot{Q} - \dot{Q}_0 - (\dot{W}_{xd})_{rev} = 0 \quad (9.33b)$$

Second Law Analysis

The second law of thermodynamics can be applied using Equation 8.82 with respect to the control volume (CV_2) in the sketch of Figure 9.10:

$$\dot{m}(s_0 - s) + \frac{\dot{Q}_0}{T_0} = 0 \quad (9.33c)$$

This equation applies when steady-flow conditions exist and complete reversibility of the processes can be assumed.

Eliminating the heat transfer rates between Equations 9.33a to 9.33c yields this result:

$$(\dot{W}_{x1})_{rev} + (\dot{W}_{xd})_{rev} = \dot{m}[h - h_0 - T_0(s - s_0)]$$

$$+ \dot{m}\left[\frac{V_m^2}{2} + g(Z - Z_0)\right]$$

which is identical with Equation 9.33 stated earlier for the exergy flow rate of a material stream.

Exergy Flow Rate Associated with Work and Heat

It was pointed out earlier that the expressions for the exergy flow corresponding to work and heat in flow processes are basically the same as for nonflow processes. Flow processes frequently deal with rates of work and heat flow, and one can formally write the expressions for the corresponding exergy flow rates as follows:

$$\dot{\chi}_{\dot{W}_x} = \dot{W}_x \quad (9.34)$$

and

$$\dot{\chi}_{\dot{Q}} = \left(1 - \frac{T_0}{T_{boundary}}\right)\dot{Q} \quad (9.35)$$

EXAMPLE 9.12

A steady stream of geothermal water at 95°C and 1 atm is available for mechanical power generation. The mass flow rate of the water is 10 kg/s. If the ambient environment is at 1 atm and 25°C, determine the maximum shaft power obtainable from the geothermal water stream.

Given: 10 kg/s geothermal water at 95°C and 1 atm

$$p_0 = 1 \text{ atm}, T_0 = 25°C \text{ (298 K)}$$

Find: Maximum shaft power obtainable

Solution: Consider a typical scheme such as that illustrated in Figure 9.9. The device in the control volume is a hypothetical reversible engine that can interact reversibly with the reference environment and produce the maximum possible shaft power from the geothermal water supply. The exergy of the stream of water is obtained using Equation 9.33. Assume that the contributions due to any kinetic energy and potential energy of the water are negligible. Thus,

$$\dot{X}_{\text{water stream}} = \dot{m}[h - h_0 - T_0(s - s_0)]$$

$$= (10 \text{ kg/s}) [398 - 105 - (298)(1.25 - 0.367)] \text{ kJ/kg}$$

$$= 298.7 \text{ kW}$$

The h and s values are those for water at 95°C and 1 atm, while h_0 and s_0 are for water at the dead state, which is 25°C and 1 atm in this case.

Referring to the steady-flow availability equation (SFAE) (Eq. 9.9), one can conclude that in the reversible case, the exergy production rate is equal to the exergy input rate. Thus, the maximum possible shaft power is just over 298 kW.

The conditions specified in Example 9.12 are identical to those in Example 8.8, in which the inventor claimed that her device could produce a mechanical shaft power of 400 kW. This is in considerable excess of the theoretically maximum power obtainable in the circumstances specified.

EXAMPLE 9.13

A manufacturer of vortex tubes (shown in Fig. 9.11) gives the following performance characteristics for a particular design. When the compressed air stream is supplied at 240 kPa and 300 K, the cold stream and the hot stream exit at temperatures of 272 K and 328 K, respectively. Assume that the mass flow rates for the cold and hot streams are equal and that both streams exit at the ambient pressure of 100 kPa. The ambient temperature is 300 K. If the vortex tube is to be used only for a cooling application, determine the second law efficiency of operation for the device.

Given: A vortex tube cooling operation is illustrated in Figure 9.11.

$$p_1 = 240 \text{ kPa}, p_2 = 100 \text{ kPa} = p_3 = p_0$$

$$T_1 = 300 \text{ K} = T_0, T_2 = 272 \text{ K}, T_3 = 328 \text{ K}$$

$$\dot{m}_2 = \dot{m}_3 = \tfrac{1}{2}\dot{m}_1 (= \dot{m})$$

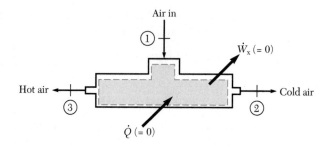

Figure 9.11 A vortex tube cooling operation.

Find: η_2 for the process

Solution: The exergy input rate is the exergy flow rate of the inlet stream of compressed air. Thus, from Equation 9.33,

$$\dot{X}_{input} = \dot{X}_1 = \dot{m}_1[h_1 - h_0 - T_0(s_1 - s_0)]$$

$$= (2\dot{m})\left[c_p(T_1 - T_0) - T_0\left(c_p \ln \frac{T_1}{T_0} - R \ln \frac{p_1}{p_0}\right)\right]$$

$$= (2\dot{m}) RT_0 \ln \frac{p_1}{p_0}, \quad \text{since } T_1 = T_0 \text{ in this case.}$$

Substituting $R = 0.2871$ kJ/kg · K and the specified values of T_0, p_1, and p_0 leads to

$$\dot{X}_{input} = 150.8 \, \dot{m} \text{ kW}$$

The exergy production rate is the exergy flow rate of the cold air stream, which again can be determined from Equation 9.33. In this case, $p_2 = p_0$, and therefore the expression for the exergy flow rate of the cold air stream simplifies to

$$\dot{X}_{product} = \dot{X}_2 = \dot{m}_2[h_2 - h_0 - T_0(s_2 - s_0)]$$

$$= (\dot{m})\left[c_p(T_2 - T_0) - T_0\left(c_p \ln \frac{T_2}{T_0} - R \ln \frac{p_2}{p_0}\right)\right]$$

$$= \dot{m}c_p T_0\left(\frac{T_2}{T_0} - 1 - \ln \frac{T_2}{T_0}\right)$$

Substituting $c_p = 1.005$ kJ/kg · K for air and the specified values of T_0 and T_2 gives

$$\dot{X}_{product} = 1.4 \, \dot{m} \text{ kW}$$

Using Equation 9.10, an estimate of 0.93% is obtained for the second law efficiency.

This is an extremely low efficiency, which probably explains why the vortex tube is not widely used except for specialized applications, such as those mentioned in Chapter 8 where the device was first discussed.

EXAMPLE 9.14

A flow of hot water at 90°C is used to heat relatively cold water at 25°C to a temperature of 50°C in the heat exchanger arrangements, shown in Figures 9.12(a) and (b). The cold water flows at the rate of 1 kg/s. Consider the operation of the heat exchanger in either the *parallel-flow* mode or the *counter-flow* mode. When the heat exchanger is operated in the parallel-flow mode, the exit temperature of the hot water stream must not be less than 60°C. In the counter-flow mode, in contrast, the exit temperature of the hot water stream can be as low as 35°C. Compare the second law efficiencies achieved in the two modes of operation. Assume the reference environment is at 300 K.

Given:

(a) Parallel-flow heat exchanger design (Fig. 9.12a).

$$T_1 = 25°C, \ T_2 = 50°C$$

$$T_3 = 90°C, \ T_4 = 60°C$$

$$\dot{m}_c = 1 \ \text{kg/s}$$

(b) Counter-flow heat exchanger design (Fig. 9.12b).

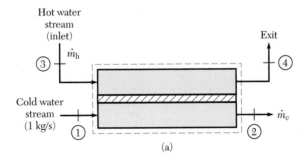

(a)

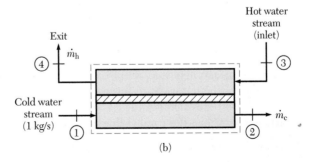

(b)

Figure 9.12 (a) Parallel-flow heat exchanger design, and (b) counter-flow heat exchanger design.

$$T_1 = 25°C, \; T_2 = 50°C$$

$$T_3 = 90°C, \; T_4 = 35°C$$

$$\dot{m}_c = 1 \text{ kg/s}$$

$$T_0 = 300 \text{ K}$$

Find: η_2 for **(a)** and **(b)**

Solution:

First law analysis

The multiple stream version of the SFEE (Eq. 6.30) can be applied to determine the necessary mass flow rate for the hot water in each case. Assuming negligible changes in kinetic and potential energy terms in the equation, the reduced energy balance equation is as follows:

$$0 = (\dot{m}_h h_4 + \dot{m}_c h_2) - (\dot{m}_h h_3 + \dot{m}_c h_1)$$

or

$$\dot{m}_h = \frac{(h_2 - h_1)}{(h_3 - h_4)} \dot{m}_c$$

The fluid is water, and hence, one can use either the steam tables or the **STEAM** code to determine the specific enthalpy values. Values of the specific entropy are needed later for the second law analysis.

For the two modes of operation, states 1 to 3 are unchanged and the property values are

$$h_1 = 104.9 \text{ kJ/kg}, \qquad s_1 = 0.367 \text{ kJ/kg} \cdot \text{K}$$

$$h_2 = 209.5 \text{ kJ/kg}, \qquad s_2 = 0.703 \text{ kJ/kg} \cdot \text{K}$$

$$h_3 = 377 \text{ kJ/kg}, \qquad s_3 = 1.193 \text{ kJ/kg} \cdot \text{K}$$

Parallel-flow mode

$T_4 = 60°C$, and hence, $h_4 = 251.3$ kJ/kg and $s_4 = 0.831$ kJ/kg · K. The mass flow rate of the hot water is thus determined from the reduced energy balance equation and is 0.832 kg/s for the parallel-flow arrangement.

Counter-flow mode

$T_4 = 35°C$, and hence, $h_4 = 146.7$ kJ/kg and $s_4 = 0.505$ kJ/kg · K. Thus, the mass flow rate of the hot water in this case is 0.454 kg/s. Note that the counter-flow arrangement uses significantly less hot water and thus promises to be the better arrangement than the parallel-flow scheme.

Second law analysis

This analysis assumes that the hot water stream on exit from the heat exchanger is simply dumped into the drain. The exergy flow rate of the hot water stream at entry is therefore considered as the proper measure of exergy input rate to the process. This exergy input rate is determined using Equation 9.33 and assuming negligible contributions from kinetic and potential energy of the water. The dead state properties of water (at 27°C) are $h_0 = 113.2$ kJ/kg and $s_0 = 0.395$ kJ/kg · K. Thus, the exergy input rate is

$$\dot{X}_{input} = \dot{X}_3 = \dot{m}_h[h_3 - h_0 - T_0(s_3 - s_0)]$$

$$= 24.4 \, \dot{m}_h \text{ kW}$$

For the parallel-flow case, the exergy input rate is thus $(24.4)(0.832) = 20.3$ kW. For the counter-flow case, the exergy input rate is $(24.4)(0.454) = 11.1$ kW.

The exergy production rate is identical in the two cases and is measured by the increase in the exergy flow rate of the cold water stream. Equation 9.33 can be used again, and assuming that the contributions from kinetic and potential energy changes are negligible, the exergy production rate can be computed from

$$\dot{X}_{product} = \dot{X}_{12} = \dot{m}_c[h_2 - h_1 - T_0(s_2 - s_1)]$$

$$= 3.8 \, \dot{m}_c = 3.8 \text{ kW}$$

(a) *Parallel-flow mode*

The second law efficiency is calculated using Equation 9.10. In this case, the second law efficiency is $(3.8/20.3)100\% = 18.7\%$

(b) *Counter-flow mode*

The second law efficiency is $(3.8/11.1)100\% = 34.2\%$.

Thus, η_2 for the counter-flow arrangement is nearly twice the value for operation in the parallel-flow mode. The former clearly uses more of the exergy of the hot water stream than is possible with the parallel-flow arrangement.

EXERCISES

for Section

9.3.1

Unless otherwise stated, assume $p_0 = 100$ kPa (14.5 psia) and $T_0 = 298$ K (537°R) for the reference environment.

1. A mass flow rate of 5 kg/s of geothermal water at 100 kPa and 90°C is available for power generation. What is the maximum possible (mechanical) power output from this stream?

   Answer: 130 kW

2. The water level in a hydroelectric dam is 100 ft above the location of the hydraulic turbines. What is the maximum work (in kWh) that can be produced from each 1 million gal of water passing through the hydraulic turbines? (1 gal of water has a mass of 8.34 lbm.)

   Answer: 314 kWh

3. Steam at 50 bar and 500°C flows steadily through an adiabatic turbine. Assume that expansion through the turbine is isentropic and that the steam pressure at the exit to the turbine is 0.05 bar. Calculate

   (a) the decrease in the exergy per unit mass of the steam between the entry and exit states

(b) the specific work output of the turbine

Answer: (a) 1310 kJ/kg

9.3.2   Exergy Dissipation and Entropy Generation

Derivation of the lost work theorem for flow processes follows essentially the same procedure as used in Section 9.2.4. Referring to the typical flow process of Figure 9.9, one can write the following for the exergy input rate and the exergy production rate based on Equations 9.33, 9.34, and 9.35:

$$\dot{X}_{\text{input}} = \sum_{\substack{\text{control} \\ \text{surface}}} \left(1 - \frac{T_0}{T_{\text{boundary}}}\right) \dot{Q} \tag{9.36a}$$

and

$$\dot{X}_{\text{product}} = \dot{m}[h_2 - h_1 - T_0(s_2 - s_1)]$$
$$+ \dot{m}\{\tfrac{1}{2}((V_m)_2^2 - (V_m)_1^2) + g(Z_2 - Z_1)\} + \dot{W}_x \tag{9.36b}$$

The exergy consumption rate is the difference between the exergy input rate and the exergy production rate, as given by Equation 9.9. Thus,

$$\dot{X}_{\text{consumed}} = \dot{X}_{\text{input}} - \dot{X}_{\text{product}}$$

$$= T_0 \left[\dot{m}(s_2 - s_1) - \sum_{\substack{\text{control} \\ \text{surface}}} \frac{\dot{Q}}{T_{\text{boundary}}} \right]$$

$$+ \left(\sum_{\substack{\text{control} \\ \text{surface}}} \dot{Q} - \dot{W}_x - \dot{m}\left\{ h_2 - h_1 + \left[\frac{(V_m)_2^2}{2} - \frac{(V_m)_1^2}{2} \right] \right. \tag{9.36c}$$

$$\left. \left. + g(Z_2 - Z_1) \right\} \right)$$

The expression within the large parentheses on the right-hand side of Equation 9.36c is equal to zero from the SFEE (Eq. 6.29) applied to this case. Furthermore, from Equation 8.80, the expression within the set of square brackets can be replaced with the entropy generation rate in the control volume. Hence, Equation 9.36c is finally reduced to

$$(\dot{X}_{\text{consumed}})_{\text{CV}} = T_0 \dot{S}_{\text{gen,CV}} \tag{9.36}$$

An expression similar to Equation 9.36 is easily obtained for the corresponding exergy consumption in the surroundings:

$$(\dot{X}_{\text{consumed}})_{\text{surr}} = T_0 \dot{S}_{\text{gen,surr}} \tag{9.37a}$$

Thus, for the entire universe,

$$(\dot{X}_{\text{consumed}})_{\text{universe}} = (\dot{X}_{\text{consumed}})_{\text{CV}} + (\dot{X}_{\text{consumed}})_{\text{surr}}$$

$$= T_0(\dot{S}_{\text{gen,CV}} + \dot{S}_{\text{gen,surr}}) \tag{9.37}$$

$$= T_0\dot{S}_{\text{gen,universe}}$$

Equation 9.37 is again the Gouy–Stodola lost work theorem for steady-flow processes.

EXERCISES

for Section
9.3.2

Unless otherwise stated, assume $p_0 = 100$ kPa (14.5 psia) and $T_0 = 298$ K (537°R) for the reference environment.

1. A steady stream of dry, saturated refrigerant-22 at -20°C is compressed to a pressure of 20 bar. The compression process is adiabatic, but with a process efficiency of 0.85. Determine, per unit mass of the refrigerant,
   (a) the work input required to the compressor
   (b) the increase in the exergy of the refrigerant resulting from the process
   (c) the exergy destroyed in the process
   Compare your result in part (c) with $T_0(s_e - s_i)$ for the adiabatic compression process.
   Answer: (a) 63.8 kJ/kg, (b) 55.9 kJ/kg, (c) 7.84 kJ/kg

2. The superheated refrigerant-22 vapor at the end of the compression process in Exercise 1 is condensed by passing it through a condenser that rejects heat to room air at $T_0 = 25$°C. The refrigerant leaves the condenser as a liquid at 20 bar and 40°C. Determine, per unit mass of the refrigerant,
   (a) the heat transfer from the condenser to the room air
   (b) the decrease in the exergy of the refrigerant resulting from the condensation process
   (c) the exergy destroyed in the condensation process
   Answer: (a) 211 kJ/kg, (b) 19.9 kJ/kg, (c) 19.9 kJ/kg

9.3.3 Alternative Expressions for the Exergy Flow Rate Associated with a Flow of Mass

Starting with Equation 9.33 for the exergy flow rate associated with a flow of mass, one can show that for a single component, single-phase, chemically nonreactive substance, the exergy flow rate can be calculated as follows:

$$\dot{X}_{\text{mass flow}} = \dot{X}_T + \dot{X}_p + \dot{X}_{\text{KE}} + \dot{X}_{\text{PE}} \tag{9.38}$$

The exergy terms on the right-hand side of Equation 9.38 are defined as follows:

Thermal exergy flux

$$\dot{X}_T = \dot{m} \int_{T_0}^{T} c_p \left(1 - \frac{T_0}{T}\right) dT \tag{9.38a}$$

Pressure exergy flux

$$\dot{X}_p = \dot{m} \int_{p_0}^{p} (v)_{T_0} \, dp \tag{9.38b}$$

Kinetic energy (exergy flux)

$$\dot{\chi}_{KE} = \tfrac{1}{2}\dot{m}V_m^2 \tag{9.38c}$$

Potential energy (exergy flux)

$$\dot{\chi}_{PE} = \dot{m}g(Z - Z_0) \tag{9.38d}$$

The last two terms in Equation 9.38 are the kinetic and potential energy terms in Equation 9.33. The thermal exergy and pressure exergy terms can be shown to be identical to the expression involving $[h - h_0 - T_0(s - s_0)]$ in Equation 9.33. The interested reader can refer to Rodriguez (1980) for example, for details of the steps needed to arrive at Equations 9.38a and 9.38b for the quantities that have been labeled *thermal* and *pressure exergy* flux. For certain multicomponent substances, Equation 9.38 can be used if the thermal and pressure exergy contributions are determined for each component and then added up for the mixture.

For a chemical substance, an additional term, $\dot{\chi}_c$, denoting *chemical exergy* flow rate should be added to Equation 9.38 such that

$$\dot{\chi}_{\text{mass flow}} = \dot{\chi}_T + \dot{\chi}_p + \dot{\chi}_c + \dot{\chi}_{KE} + \dot{\chi}_{PE} \tag{9.39}$$

Chapter 14 further discusses chemical exergy and covers combustion and other chemical reactions.

Exergy Flow Rate for an Incompressible Fluid Stream

Assume that the fluid is single component, single phase, and chemically nonreactive. The term *single phase* is used here to mean the same phase in the current and dead states for the substance. If c_p is reasonably constant over the temperature range of interest, then Equation 9.38a gives

$$\dot{\chi}_T = \dot{m}c_pT_0\left(\frac{T}{T_0} - 1 - \ln\frac{T}{T_0}\right) \tag{9.40a}$$

Furthermore, if the fluid is incompressible, one can assume that v does not change significantly with pressure. Hence, from Equation 9.38b,

$$\dot{\chi}_p = \dot{m}v(p - p_0) \tag{9.40b}$$

The total exergy is obtained by substituting these expressions for thermal and pressure exergy in Equation 9.38.

Exergy for an Ideal Gas Stream

For ideal gases, one can use Equation 9.40a for the thermal exergy if the specific heat, c_p, does not vary significantly over the temperature range of interest. For the pressure exergy, however, one can use the ideal gas pvT rule (which is $(v)_{T_0} = RT_0/p$) in Equation 9.38b to obtain (for an ideal gas)

$$\dot{\chi}_p = \dot{m}\int_{p_0}^{p}(v)_{T_0}\,dp = \dot{m}RT_0\int_{p_0}^{p}\frac{dp}{p} = \dot{m}RT_0\ln\frac{p}{p_0} \tag{9.40c}$$

It is easy to show that Equation 9.40c reduces to Equation 9.40b when $(p - p_0)$ is much smaller in magnitude than p_0.

EXAMPLE 9.15

The exhaust stream of a 5-MW gas turbine flows at the rate 119,000 kg/h and is at 1 atm and 790 K. Determine the power output of a reversible heat engine that uses the exhaust stream as a heat source and the ambient environment at 300 K and 1 atm as a heat sink. (The specific heat, c_p, for the exhaust gases can be taken as 1.15 kJ/kg · K.)

Given: Heat source is the exhaust (gas) stream at $p = 1$ atm and $T = 790$ K.

$$\dot{m} = 119{,}000 \text{ kg/h, } c_p = 1.15 \text{ kJ/kg} \cdot \text{K}$$

$$T_0 = 300 \text{ K, } p_0 = 1 \text{ atm}$$

Find: Maximum power output possible.

Solution: The maximum power output is obtained when all processes are reversible and the exhaust stream is brought to the dead state, as illustrated in Figure 9.10. Referring to the SFAE (Eq. 9.9), the maximum power output corresponds to the exergy flow rate of the exhaust gas stream from the turbine. Neglecting the contributions due to kinetic energy and potential energy in Equation 9.38, the exergy flow rate of the gas stream is made up of the thermal and pressure exergy flow rate terms only. Since $p = p_0$, the pressure exergy flow rate is zero. Thus, from Equations 9.38, 9.40a, and 9.40b,

$$\dot{X}_{\text{exhaust gas}} = \dot{X}_T = \dot{m} c_p T_0 \left(\frac{T}{T_0} - 1 - \ln \frac{T}{T_0} \right)$$

Substituting the specified values of mass flow rate, specific heat, and temperature, the calculated exergy flow rate for the exhaust gas stream is 7.58 MW. Thus, the maximum power obtainable from the reversible heat engine is 7.58 MW.

EXERCISES

for Section

9.3.3

Unless otherwise stated, assume $p_0 = 100$ kPa (14.5 psia) and $T_0 = 298$ K (537°R) for the reference environment.

1. A steady stream of air at 1100 K and 6 bar enters a turbine and expands reversibly and adiabatically through the turbine. The exit pressure is 1 bar. Assume $c_p = 1.117$ kJ/kg · K and $R = 0.2871$ kJ/kg · K for the air. Determine, per unit mass of the air and prior to entry into the turbine,
   (a) the thermal exergy
   (b) the pressure exergy
   (c) the total exergy

   Answer: (a) 461 kJ/kg, (b) 153 kJ/kg

2. For the air expansion through the turbine in Exercise 1, determine
   (a) the exit temperature of the air
   (b) the thermal exergy per unit mass of the air stream leaving the turbine
   (c) the total exergy per unit mass of the air stream leaving the turbine
   (d) the specific work output of the gas turbine

   Answer: **(a)** 694 K, **(b)** 161 kJ/kg, **(d)** 453 kJ/kg

3. Suppose the exhaust gas stream in Exercise 1 was simply discharged into the atmosphere. What percentage of the initial exergy of the gas does this unused or wasted exergy represent? Suggest possible uses for the exergy of the exhaust gas in situations where the gas turbine is employed as a power unit in an industrial or factory plant setting.

   Answer: 26.2%

9.4   *Exergy Flow and Optimum Thermodynamic Cycles*

Thus far, this chapter has considered availability in a sufficiently broad manner intended to facilitate the application of the second law of thermodynamics to the myriad of processes encountered in engineering. The remainder of the chapter is devoted to the use of the availability method in performance evaluation of heat engines and refrigeration devices in particular. Chapters 10 and 11 deal with heat engines and refrigeration systems, which between them probably account for most of the world's energy (or, more accurately, exergy) expenditure. Significant improvement in the efficiency of these devices would undoubtedly enhance the prospects for making the world's dwindling energy resources stretch out for a longer time.

Chapters 7 and 8 considered the efficiency of heat engines and refrigeration devices when they are operating between a pair of *thermal reservoirs*. For devices operating between thermal reservoirs (or constant temperature systems), the optimum thermodynamic cycle is the Carnot cycle or any other completely reversible cycle. Complete reversibility implies both internal reversibility for each component as well as external reversibility outside each system involved. By far the vast majority of heat sources and sinks available in nature are *finite* and cannot be assumed to be infinite, however. A *finite heat source* (or *sink*) is one whose temperature does not stay constant as heat is extracted from it (or added to it). Geothermal water is an example of a finite heat source. As heat is extracted from a geothermal water stream, a steady fall occurs in the temperature of the heat source. The heat source in a conventional steam power plant is another example of a finite heat source. The heating provided in the steam power cycle is at the expense of cooling the stream of hot products of fuel combustion in air. The cooling of a stream of air in air conditioning applications provides yet another example of a finite heat source. The stream of air is the heat source from which heat is extracted, resulting in a fall in temperature of the (open) system. In contrast to these examples, a constant temperature heat source or sink is often referred to as *unlimited*.

When dealing with heat engines or refrigeration systems for which either the heat source or the heat sink is finite, it is important to realize that the Carnot cycle is no longer the optimum thermodynamic cycle. Indeed, the only valid principle that can be inferred from the second law of thermodynamics is that for maximum performance, complete reversibility in work and heat interactions must occur between the device and the heat source and between the device and the heat sink. In most applications, the ambient environment is likely to be either the heat sink or the heat source. In other words, the ambient environment (which is a thermal reservoir) can also be regarded as a reference environment against which to measure the optimum delivery possible for a device operating between a finite heat source or sink and the ambient environment. Thus, if the exergy of the finite heat source or sink is known, one can conclude from the appropriate version of the exergy equation that the exergy product must be equal to the exergy input when complete reversibility occurs in all the processes involved. This approach has already been demonstrated in Examples 9.9 (on ice production), 9.12 (on power generation from geothermal water), and 9.15 (on power generation using the exhaust from a gas turbine). Just as Chapter 8 considered the conceptual operation of the Carnot cycle, it may be useful to imagine what the optimum cycle would be like if either the heat source or the heat sink were finite. Wilson and Radwan (1977) give an excellent consideration of this and also introduce terms such as the *trilateral cycle* and the *quadlateral cycle* for the appropriate heat engine models in the type of situation currently being discussed.

Optimum Engine Cycle When the Heat Source is Finite

Figure 9.13 illustrates a frequently encountered situation in which the heat source for a heat engine cycle is a stream of high temperature fluid. As heat is delivered from the fluid to the engine cycle, the temperature of the heat source falls. When the temperature of the heat source fluid finally falls to that of the ambient environment, heat can no

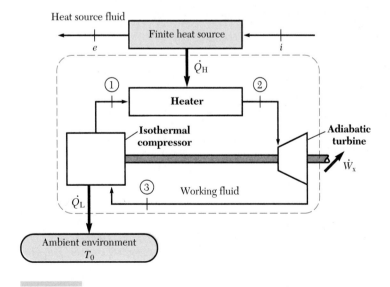

Figure 9.13 A heat engine operating between a finite heat source and an unlimited heat sink.

longer be extracted for the purpose of producing mechanical power from the cyclic device. For simplicity, assume that the heat source fluid is at ambient pressure so that its exergy at the inlet to the control volume (the heat exchanger) is made up entirely of the thermal exergy flow rate at that state. The hypothetical engine shown in Figure 9.13 has three major parts: the *heater* in which the working fluid is heated by the heat source fluid, a reversible and adiabatic *turbine*, and a reversible and isothermal *compressor*. The *working fluid* is the fluid used by the engine to transport energy among the various components of the device. As the working fluid passes through the heater, heat flows from the finite heat source to the fluid. For this heating process to be reversible, a *counter-flow* heat exchanger like the one described in Example 9.14 is needed. Also, the heat source fluid temperature should not exceed the working fluid temperature by more than an infinitesimal amount. When the working fluid passes through the compressor, heat rejection to the ambient environment (the sink) is necessary, and for external reversibility, this must take place at a temperature that does not exceed the ambient temperature T_0 by more than an infinitesimal amount. Since T_0 is constant, the compressor must be isothermal (as opposed to an adiabatic compressor that would be used in the Carnot cycle).

Conceptually, one can view reversible heat flow between the heat source fluid and the engine working fluid in much the same way as done in Section 8.9, which considered the shaft work done in quasi-equilibrium processes. Thus, as indicated in Figures 9.13 and 9.14, one can view each steady mass flow as occurring through a succession of miniature control volumes each of which instantaneously holds a fixed mass at a near-equilibrium condition. In addition, as indicated in Figure 9.14, the corresponding miniature control volumes for the heat source fluid and the working fluid must have fluids at temperatures that do not differ by more than an infinitesimal amount. Thus, if δQ is the heat flow from the heat source fluid to the engine working fluid in the time interval t to $(t + \Delta t)$, application of Equation 8.82 yields the following:

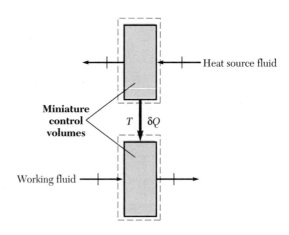

Figure 9.14 A conceptual representation of reversible heat flow from a heat source fluid stream to a working fluid stream.

$$\frac{(-\delta Q)}{T} = (\dot{m}_{source} \, \Delta t)(ds)_{source} \tag{9.41a}$$

and

$$\frac{\delta Q}{T} = (\dot{m}_{working \atop fluid} \, \Delta t)(ds)_{working \atop fluid} \tag{9.41b}$$

Let δQ_H denote the total heat flow in the same time interval between the heat source fluid and the engine working fluid. The term δQ_H can be obtained by integration of either Equation 9.41a or 9.41b:

$$\frac{\delta Q_H}{\Delta t} = -\int_{s_i}^{s_e} T\dot{m}_{source}(ds)_{source} = \int_{s_1}^{s_2} T\dot{m}_{working \atop fluid}(ds)_{working \atop fluid} \tag{9.41c}$$

For convenience, let's define for each flow of mass an entropy flow rate S^* that is a product of the mass flow rate and the corresponding specific entropy of the fluid:

$$S^* = \dot{m}s \tag{9.41d}$$

Equation 9.41c for the heat input rate to the engine cycle can thus be written in terms of the integral of T with respect to dS^* for each fluid stream as follows:

$$\dot{Q}_H = \frac{\delta Q_H}{\Delta t} = -\int_{S_i^*}^{S_e^*} T(dS^*)_{source}$$

$$= \int_{S_1^*}^{S_2^*} T(dS^*)_{working \atop fluid} \tag{9.41}$$

The form of Equation 9.41 suggests a representation of the thermodynamic cycle and the reversible process path for the heat source fluid on a T-S^* diagram as shown in Figure 9.15. The process path for the heat source fluid is the line $i{\rightarrow}e$ in the T-S^*

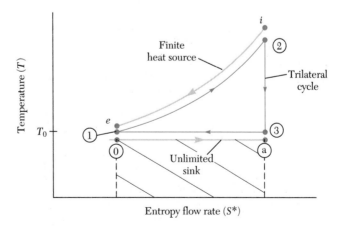

Figure 9.15 T-S* (entropy flow rate) diagram for reversible flow processes in a trilateral cycle.

diagram. At the state marked i for the heat source fluid, S^* is calculated using the specific entropy of the fluid at the inlet condition, while for the state marked e, the specific entropy (s_0) for the heat source fluid at the dead state is used. The heat transfer rate to the engine cycle is the area under the curve ie on the T-S^* diagram in accordance with Equation 9.41. Similarly, the process path for the working fluid must be along line $1{\rightarrow}2$ as shown in Figure 9.15 since the heat source fluid temperature can only differ by an infinitesimal amount from that for the working fluid. By a judicious choice of each reference state or datum for the specific entropy of each of the fluids, one can ensure that S_1^* for the working fluid is the same as S_e^* for the heat source fluid and thus that S_2^* for the working fluid is equal to S_i^* for the heat source fluid. The latter inference is based on Equation 9.41 which implies that $S_i^* - S_e^*$ must be equal to $S_2^* - S_1^*$. Thus, the two paths $i{\rightarrow}e$ and $1{\rightarrow}2$ are shown approximately on top of each other.

The flow of the working fluid through the reversible and adiabatic turbine is isentropic and therefore S_3^* must be equal to S_2^*. The process is thus represented by the vertical line $2{\rightarrow}3$ in Figure 9.15. For the isothermal fluid compression process $3{\rightarrow}1$, one can apply Equation 8.82 to obtain the following expression for the heat rejection rate from the compressor:

$$\dot{Q}_{\mathrm{L}} = T_0(S_3^* - S_1^*)_{\substack{\text{working}\\\text{fluid}}} \tag{9.42}$$

The process can thus be represented in Figure 9.15 by the horizontal line $3{\rightarrow}1$ corresponding to a constant temperature process. The heat transfer rate is the area under the line on the T-S^* diagram. The corresponding increase in entropy of the heat sink, which is regarded as unlimited, can be obtained using Equation 8.54. Thus,

$$(\Delta S^*)_{\substack{\text{heat}\\\text{sink}}} = \frac{\dot{Q}_{\mathrm{L}}}{T_0} \tag{9.43}$$

where ΔS^* for a closed system can be interpreted as the rate of increase in entropy of the system. From Equations 9.42 and 9.43, one can again infer that the process path line $0{\rightarrow}a$ for the sink and $3{\rightarrow}1$ for the working fluid must lie on top of each other for complete reversibility of the process. The three-sided cycle $1{\rightarrow}2{\rightarrow}3{\rightarrow}1$ is what Wilson and Radwan (1977) have termed the *trilateral cycle*, which is the optimum cycle for a heat engine operating between a finite source and an unlimited sink.

The work output rate from the trilateral cycle is obtained by applying the first law of thermodynamics to the engine cycle:

$$\dot{W}_{\mathrm{x}} = \dot{Q}_{\mathrm{H}} - \dot{Q}_{\mathrm{L}} \tag{9.44}$$

In other words, the maximum possible work output corresponds to the area enclosed by the trilateral cycle. This work is equal in magnitude to the area enclosed by the process paths for the heat source, the heat sink, and the connecting isentropic line in Figure 9.15. This interpretation can be linked to Equation 9.33, which was derived earlier for the exergy of a flow of mass. If the kinetic energy and potential energy terms in Equation 9.33 are assumed negligible, the following expression is obtained for the exergy flow rate of the heat source fluid:

$$\begin{aligned}\dot{\chi}_{\text{source}} &= \dot{m}_{\text{source}}[h_i - h_e - T_0(s_i - s_e)] \\ &= \dot{Q}_{\mathrm{H}} - T_0(S_i^* - S_e^*)\end{aligned} \tag{9.45}$$

This equation is based on the assumption that the heat source fluid leaves at the dead state. Also, the first term on the right-hand side (the second line) of the equation was obtained by applying the SFEE (Eq. 6.29) to the heat source fluid flow. The second term (on the second line) is equal to $T_0(S_2^* - S_1^*)$, which in turn is equal to $T_0(S_3^* - S_1^*)$ or the rate of heat rejection from the engine cycle. Thus, the first term (on the second line) in Equation 9.45 can be identified as the area under the heat input rate curve on the T-S^* diagram, while the second term is the area under the heat rejection rate curve. The difference is the net power output, which corresponds to the area enclosed by the trilateral cycle in Figure 9.15. The necessary heat rejection rate is referred to by some authors somewhat loosely as *unavailable energy* in contrast to the *available energy* represented by the area enclosed by the trilateral cycle. Further discussion of the optimum engine cycle in relation to the nature of the heat source and sink is deferred to Chapter 10.

Optimum Refrigeration Cycle When the Heat Source is Finite

A similar approach to that indicated for the heat engine can be used for a refrigeration system operating between a finite heat source and an unlimited sink. The minimum work input is required when all processes take place reversibly. This minimum can be determined once the exergy product has been calculated. In applications involving flow processes, the use of the T-S^* diagram is encouraged. Areas under a T-S^* curve correspond to heat transfer rates. The area enclosed by the path for the refrigeration cycle gives the work input required for the cycle.

9.5 Summary

This chapter presented the concept of availability as the proper measure of the potential (or capacity) to do work (or produce change). Whereas the first law of thermodynamics implies that the total energy of a system and its surroundings remains constant, regardless of the processes involved, this chapter has shown, as a major consequence of the second law, that the total exergy must decrease in all real processes. Indeed, the total exergy remains constant only when all the processes are reversible.

The quantitative measure of exergy for a system has been given as the maximum net (or shaft) work that can be produced by the system if it were to be reduced to the dead state via reversible processes. Equation 9.24 accordingly defines the exergy, χ_{system}, for a closed system. Similarly, for the steady flow of a substance, the exergy flow rate is defined as the maximum shaft power that can be obtained if the mass flowing were to be brought to the dead state condition via reversible processes. The expression for the exergy flow rate of a steady-state steady-flow mass is given by Equation 9.33.

Although defined in terms of capacity for doing work, exergy is the proper index for performance evaluation even when nonwork-producing processes are involved. Accordingly, the concept of second law efficiency has been introduced, which enables a comparison of the exergy conserved in real processes with the initial quantity of exergy that would have been conserved if all the processes had been reversible. The second law efficiency cannot exceed 100%. A high value of second law efficiency implies that processes are operating in a near-perfect manner. A low second law efficiency, however, indicates waste and irreversibilities in the processes involved.

The chapter concludes by considering the optimum cycles for heat engines and refrigeration systems when either the heat source or the sink is finite and not unlimited. Quantitative determination of the optimum work output (for heat engines) or work input (for refrigeration systems) was shown to be based on the calculation of the exergy input (for heat engines) or product (for refrigeration systems) associated with the finite heat source or sink. Furthermore, when dealing with open systems, the use of the T-S^* diagram was encouraged, where S^* represents the entropy flow rate in reversible processes. On the T-S^* diagram, the area under a curve can be shown to correspond to a heat flow rate, while the area enclosed by a cyclic path gives the net work output (or input) rate involved.

References

Angrist, S. W. 1976. *Direct Energy Conversion.* 3rd ed. Boston: Allyn and Bacon.

Bosnjakovic, F. 1938. Arch Warmewirt. Vol. 19. No. 1.1–2.

Denbigh, K. 1966. *The Principles of Chemical Equilibrium.* 2nd ed. Cambridge, U.K: Cambridge University Press.

Gaggioli, R. A. 1980. "Principles of Thermodynamics." *ACS Symposium Series 122.* Washington, D.C.: American Chemical Society. 3–13.

Gibbs, J. W. 1873. "A method of geometrical representation of the thermodynamic properties of substances by means of surfaces." *Transactions Connecticut Academy of Arts and Science.* Vol. II, p. 382 in *The Collected Works of J. Willard Gibbs* 1931 (London: Longmans Green.)

Gouy, M. 1889. "Sur les transformations et l'équilibre en thermodynamique." *Comptes Rendus*, Paris. Vol. 108, p. 509.

Hatsopoulos, G. N. and Keenan, J. H. 1965. *Principles of General Thermodynamics.* New York: John Wiley & Sons.

Haywood, R. W. 1974a. "A Critical Review of the Theorems of Thermodynamic Availability, with Concise Formulations. Part 1. "Availability." *Journal of Mechanical Engineering Science.*

Haywood, R. W. 1974b. "A Critical Review of the Theorems of Thermodynamic Availability, with Concise Formulations. Part 2. Irreversibility." *Journal Mechanical Engineering Science.* Vol. 16, No. 4. 258–267.

Haywood, R. W. 1980. *Equilibrium Thermodynamics for Engineers and Scientists.* New York: John Wiley & Sons.

Keenan, J. H. 1932. "A steam chart for second law analysis." *Mechanical Engineering.* Vol. 54. 195–204.

Maxwell, J. C. 1871. *Theory of Heat.* 1st edition. London: Longmans Green.

Rant, Z. 1956. "Exergie, ein neues Wort fur 'technische Arbeitsfähigkeit": (Exergy, a new word for 'technical work capacity'). *Forsch.-Ing.-Wes.* Vol. 22.

Rodrigues, L. 1980. "Calculation of Available-Energy Quantities." *ACS Symposium Series 122.* Washington, D.C.: American Chemical Society. 39–59.

Stodola, A. 1898. "Die Kreisprozesse der Gasmachine" (Gas engine cycles). *Z. Ver. dt. Ing.* Vol. 42. p. 1088.

Tsatsaronis, G. In press. "Thermoeconomic Analysis and Optimization of Energy Systems." *Progress in Energy and Combustion Science.*

Wilson, S. S. and Radwan, M. S. 1977. "Appropriate Thermodynamics for Heat Engine Analysis and Design. *International Journal of Mechanical Engineering Education*. Vol. 5, No. 1. 68–80.

Questions

1. Briefly explain each of the following concepts:
   (a) exergy or availability
   (b) datum or reference environment
   (c) the dead state
   (d) complete thermodynamic equilibrium
   (e) second law efficiency
   (f) first law efficiency

2. Suggest an appropriate reference environment for each of the following situations:
   (a) operation of a domestic refrigerator in the kitchen area of a house
   (b) a 3-ton heat pump operating in a family house
   (c) a gas-fired water heater in a large catering facility

3. Write the general expression for the exergy of a closed system that can interact with a reference environment at a pressure p_0 and a temperature T_0.

4. A rigid container of capacity V is completely evacuated. What is the exergy of the vacuum?

5. Compressed air at a pressure p is held in a receiver of volume V. The ambient air pressure is p_0 and the temperature T_0 is the same as that for the compressed air. Write an expression for the exergy of the compressed air in terms of p, V, p_0, and T_0 and any relevant characteristics of air.

6. Blocks of ice are mixed with hot water in a well-insulated container. Is the ensuing mixing process reversible or irreversible? Do you expect the exergy of the mixture to be less or greater than the total exergy of the ice and the hot water prior to the mixing process? Explain your answer.

7. An inventor offers you the cyclic device illustrated in Figure Q9.7, which he claims draws heat (Q) from the ambient air and produces an equivalent amount of work (W). What is the exergy input to the device? What is the exergy output? Is such a device possible? Explain your answer.

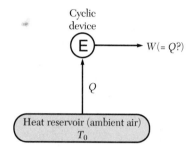

Figure Q9.7 Cyclic heat engine that produces work while drawing heat from ambient air.

8. A refrigeration device is needed for maintaining a system at a temperature T_L by removing heat at a rate $\dot{Q}_L$. The ambient temperature is T_0 and is higher than T_L. What is the minimum exergy input rate needed to the refrigeration device?

9. A heat source comprises a steady stream of a hot gas at a temperature T_P and a pressure equal to the ambient pressure. The mass flow rate of the gas is $\dot{m}_g$, and the specific heat at constant pressure is c_{pg}. The temperature of the ambient air is T_0. Write an expression for the exergy flow rate of the hot gas.

10. What is the maximum mechanical power obtainable from a heat engine operating between the ambient air as the sink and the heat source described in Question 9? Illustrate the cycle for the optimum heat engine using the T-S^* (entropy flow rate) coordinates.

11. Explain how you can use the exergy method to determine whether or not a hypothetical process or device is feasible.

Problems

Unless otherwise stated, assume $p_0 = 100$ kPa (14.5 psia) and $T_0 = 298$ K (537°R) for the reference environment.

9.1. Determine the exergy of the following systems:
 (a) 50 kg of water at $p = 100$ kPa and 80°C
 (b) 5 kg of ice at $p = 100$ kPa and -20°C
 (c) 0.2 kg of H_2O at $p = 4$ MPa and 500°C
 (d) 0.6 kg of H_2O at $p = 10$ kPa with a quality of 0.8

9.2. Determine the exergy of 0.2 m^3 of complete vacuum.

9.3. The specific heat of copper at room temperature is 385 J/kg · K. A block of copper ($m = 4$ kg) at a temperature of 400 K is dropped into a large expanse of water at 300 K. The large expanse of water can be taken as the reference environment. Compute
 (a) the exergy of the block of copper before it is dropped into the water pool
 (b) the exergy destroyed in the process

9.4. Determine the increase in the exergy of the system (in italics) in the following processes:
 (a) 20 kJ heat to *ice water* mixture that remains at 0°C
 (b) heating of *5 kg of water* from 25°C to 45°C at a constant pressure of 100 kPa
 (c) cooling of *5 kg of water* from 25°C to 5°C at a constant pressure of 100 kPa
 (d) 50 kJ heat from freezing *water* at 0°C

9.5. Calculate the exergy (in kJ) of the following systems:
 (a) 0.2 kg of air at $p = 50$ kPa and $T = 298$ K
 (b) 0.2 kg of air at $p = 200$ kPa and $T = 298$ K
 (c) 0.2 kg of air at $p = 100$ kPa and $T = 596$ K

9.6. A heat engine is operating between two constant temperature reservoirs at 1000 K and 300 K, respectively. If 1000 kJ of heat is supplied to the heat engine from the high temperature reservoir, what is the maximum work that can be done by the engine? (Take $T_0 = 300$K in this case.)

9.7. A heat engine is operating between two constant temperature reservoirs at 70°F and -30°F, respectively. If 1000 Btu of heat is rejected to the low temperature reservoir, what is the maximum work that can be done by the engine? Assume the surroundings are at 70°F.

9.8. Suppose that in a certain process, the exergy input was 500 kJ, the exergy product was 300 kJ, and the exergy consumed during the process was 130 kJ. What was the unused exergy for the process?

* 9.9. Determine the exergy of each of the following closed systems:
 (a) 1 kg of steam at 150 bar and 500°C
 (b) 2 kg of wet steam at 20 bar with a quality of 0.6
 (c) 3 kg of ice at 1 bar and -20°C
 (d) 4 kg of water at 1 bar and 60°C

* 9.10. Determine the exergy of each of the following closed systems:
 (a) 1 lbm of steam at 100 psia and 500°F
 (b) 1 lbm of wet steam at 300 psia with a quality of 0.5
 (c) 1 lbm of water at 20 psia and 150°F
 (d) 1 lbm of ice at 40 psia and 0°F

9.11. Determine the exergy of 2 m^3 of air at 10 kPa and 25°C.

* 9.12. A mass of 2 kg of refrigerant-22 is stored in a tank as a saturated vapor at -10°C. What is the exergy of the refrigerant? Ignore the chemical exergy of the refrigerant.

* 9.13. Calculate the exergy (in kJ) for
 (a) 2 kg of air at 1 MPa and 400 K
 (b) 1 kg of air at 100 kPa and 1000 K

* 9.14. Calculate the exergy for 10 lbm of copper ($c_p = 0.096$ Btu/lbm°R) at 100 psia and 675°F.

* 9.15. Calculate the exergy of 2 kg of lemonade stored in a pitcher at 4°C. Assume lemonade has the same thermodynamic properties as water.

* 9.16. Hot water is stored at atmospheric pressure in a copper tank. The temperature of the water and the tank is 50°C. The mass of the hot water is 110 kg, and the mass of the copper tank is 10 kg. What is the exergy of the system, which includes both the tank and the water?

9.17. A rigid, well-insulated vessel contains 2 kg of a perfect gas having $c_v = 1.1$ kJ/kg · K and initially at 25°C. A stirring paddle is inserted into the vessel, and the work done on the paddle by the stirrer motor is 66 kJ. Determine
 (a) the exergy input to the gas via the paddle

(b) the exergy increase of the gas

(c) the second law efficiency of the process

9.18. An identical vessel as in Problem 9.17 is heated through the same temperature difference by immersing it in a hot bath at a constant temperature of 90°C. What is the second law efficiency of the process?

9.19. One ton of refrigeration is defined as the rate of cooling involved in producing 1 (short) ton or 2000 lbm of ice at 32°F from water at 32°F in a 24-h period. This cooling rate is approximately equal to 12,000 Btu/h.

(a) Show that 1 ton of refrigeration is equivalent to 3.52 kW.

(b) Determine the exergy production rate equivalence (in Btu/h) of 1 ton of refrigeration.

*⁎ 9.20.** A mass of 80 kg of beef at atmospheric pressure is to be cooled from 300 K to −30°C. The freezing point of the beef is −3°C, and c_s, c_l, and h_{if} for the beef are 1.8 kJ/(kg · K), 3.5 kJ/(kg · K), and 500 kJ/kg, respectively. Calculate the exergy of the beef after it is cooled.

*⁎ 9.21.** A pitcher of iced tea is sitting on a countertop. The mass of the ice in the pitcher is 4 lbm, while the mass of the liquid is 5 lbm. Neglecting the pitcher, what is the exergy of the system?

9.22. A mass of 5 kg of water is heated from 25°C to 50°C at a constant pressure of 100 kPa using an electric heating coil. Determine the second law efficiency of the process.

9.23. A mass of 5 kg of hot water at 50°C is obtained by mixing 2.5 kg water at 25°C with 2.5 kg water at 75°C. Determine the second law efficiency of the process.

9.24. Suppose the water heating in Problem 9.22 is carried out using a large volume of solar-heated water that stays at an approximately constant temperature of 65°C. What is the second law efficiency achieved?

*⁎ 9.25.** A mass of 200 lbm of water at atmospheric pressure is heated from 70°F to 130°F with an electric heating coil. The surroundings are at 70°F. During the process, 1000 Btu of heat is lost to the surroundings. Calculate the first law and second law efficiencies for the process.

*⁎ 9.26.** Iced tea is produced by adding 0.2 kg of ice at −5°C to 0.8 kg of hot tea at 95°C. The thermodynamic properties of the hot tea can be taken

as approximately the same as those for water. If the vessel holding the iced tea is well-insulated, determine:

(a) the final temperature of the brew

(b) the second law efficiency of the process

*⁎ 9.27.** 1000 kg of water at 100 kPa and 25°C is turned completely to ice at −5°C using a refrigeration device with a sink at 298 K. Calculate

(a) the cooling load (in kJ)

(b) the exergy increase produced (in kJ)

(c) the minimum exergy input required (in kJ) for the task

(d) the second law efficiency achieved if the refrigeration device operated on the reversed Carnot cycle with a source at −15°C and a sink at 298 K

(e) the second law efficiency achieved when a refrigeration device with a COP_C of 1.5 is used

*⁎ 9.28.** Ice at −10°C is produced from water initially at the temperature of the surroundings, which is 25°C.

(a) Determine the minimum work input required (per kg of ice) for a reversible cyclic device.

(b) If the ice is actually produced using a vapor compression plant with a coefficient of performance for cooling of 2, compare the actual compressor work requirement to the minimum work input required for the reversible device.

*⁎ 9.29.** A pumped storage plant (Fig. P9.29 on page 402) is to deliver 1000 MWh of electrical work output each day during the peak demand period. The upper water reservoir is 100 m above a river. The pumping efficiency is 70%, while the electrical power generating efficiency of the plant is 88%.

(a) What is the electrical energy input required each day to provide adequate pumped storage capacity?

(b) How much of the electrical energy input is wasted during the process?

(c) What is the volume of water that must be pumped to the upper reservoir each day to provide the desired capacity?

*⁎ 9.30. (a)** Determine the exergy of compressed air in a receiver of volume 0.5 m³ if the pressure and temperature of the air are 10 atm and 25°C, respectively.

(b) The receiver in **(a)** is connected via a valve to an evacuated receiver, also of volume 0.5 m³.

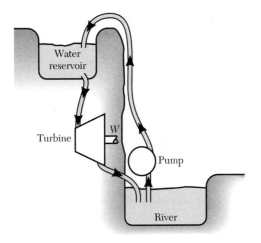

Figure P9.29 Pumped-storage plant for electrical power generation.

Initially, the valve is closed. Both receivers are well-insulated. If the valve is open, determine:

(i) the final pressure in each receiver

(ii) the final temperature of the air in each receiver

(iii) the exergy destruction that occurs

(iv) the second law efficiency of the process

* **9.31.** A rigid vessel of volume 0.1 m³ contains air at the ambient pressure of 100 kPa and temperature of 298 K.

(a) Determine the minimum work input required to produce a 40-kPa vacuum in the vessel. Assume that the temperature of the air left in the vessel remains constant at 298 K.

(b) What is the minimum work input needed to completely evacuate the vessel?

* **9.32.** An evacuated, rigid bottle of volume 1 m³ has its valve opened until the bottle is full at the ambient condition of 100 kPa and 298 K. Assume that the bottle fills slowly and that the bottle is uninsulated. Calculate

(a) the mass of air finally in the bottle

(b) the work done by the atmosphere

(c) the exergy destroyed in the process

* **9.33.** An air motor is operated for 30 min during which time it draws air at the rate of 0.5 kg/min from a large receiver holding air at 750 kPa and 298 K. The power output of the motor averages 0.5 kW

while the air pressure in the receiver remains approximately constant at 750 kPa. Determine, for the period of operation of the motor,

(a) the mass of air drawn by the motor from the receiver

(b) the exergy of the air taken from the receiver

(c) the work output of the motor

(d) the second law efficiency of the operation

* **9.34.** A Carnot heat engine operates between two heat reservoirs at 1000 K and 500 K, respectively.

(a) What is the first law efficiency for this engine?

(b) What is its second law efficiency?

* **9.35.** A reversed Carnot engine operates to provide refrigeration for a meat storage facility. The engine maintains the meat at −10°C while rejecting the waste heat at 40°C. The surroundings are at 25°C. What is the second law efficiency for this reversed Carnot engine?

* **9.36.** A reversed Carnot heat engine operates to provide heat for a building. The building is maintained at 25°C while the surroundings are at 5°C. Assume source and sink temperatures of 0°C and 30°C, respectively, for the reversed Carnot engine. What is the second law efficiency for this reversed Carnot engine?

* **9.37.** Two systems are under consideration for energy storage for a brick kiln. In one system, 10^5 Btu is added as heat to 2000 lbm of bricks initially at 80°F [c_p = 0.2 Btu/(lbm · °F) for the brick]. In the second system, the same quantity of heat is added at constant pressure to 100 lbm of water initially at 80°F.

(a) Compare the exergy increase produced in each case.

(b) Explain which system is preferable from exergy considerations only. Explain the advantages and disadvantages of each system.

* **9.38.** A cyclic heat engine operates between a finite heat source and an infinite sink at T_0. Initially, the finite heat source has a total energy E, a total volume V, and a total entropy S. Assume an operation of the heat engine such that the source is eventually brought to the dead state with a corresponding total energy E_0, a total volume V_0 and a total entropy S_0. The infinite sink can be taken as the reference environment. Also assume that the prevailing pressure is the same for both the heat source system and the heat sink environment. Determine

(a) the exergy of the finite heat source at its initial state
(b) the maximum useful work output that can be obtained in the operation of the cyclic heat engine
(c) the first law efficiency for the optimum heat engine operating between the heat source and the heat sink

Note: The optimum heat engine is one that delivers the maximum possible useful work output while the finite heat source is taken from its initial state to the dead state.

* **9.39.** Figure P9.39 gives an idealized thermodynamic model for a complete absorption refrigeration system. T_L is the evaporator temperature, T_H, the absorber/condenser temperature, and T_G, the generator temperature. The heat sources and sinks can be regarded as thermal reservoirs. The infinite sink at T_H serves as the reference environment.

(a) Determine
  (i) the exergy input (from the generator unit) to the cyclic device in terms of Q_G, T_G, and T_H
  (ii) the exergy product defined in terms of the cooling effect produced in the reservoir at T_L
(b) Use your results in part (a) to show that the maximum possible COP_C, defined as the ratio Q_C/Q_G, is

$$(COP_C)_{max} = \frac{T_L\,(T_G - T_H)}{T_G\,(T_H - T_L)}$$

(c) A typical design of a solar-heated absorption refrigeration system effectively has $T_L = 270$ K, $T_H = 300$ K, and $T_G = 340$ K. Determine the maximum possible value of COP_C.

* **9.40.** Compute the exergy input and the exergy product in each of the following, and hence, determine whether or not the process indicated could actually occur.

(a) A supply of 100 kJ of heat is input from a thermal reservoir at 420 K to a cyclic work-producing device that rejects heat to the ambient environment at 298 K and produces 40 kJ of useful work.
(b) Spontaneous cooling of 10 kg of water at 50°C occurs in an adiabatic enclosure to a temperature of 25°C resulting in a positive work output. The water pressure is the same as the ambient pressure of 100 kPa. The reference environment is at a temperature of 298 K.
(c) A fixed mass of air at 298 K and 200 kPa (abs) pressure is confined by a thin diaphragm to a region of volume 0.1 m³ in a rigid vessel that is well insulated thermally (Fig. P9.40). The total volume of the vessel is 0.2 m³. The thin diaphragm is ruptured so that the air fills the entire volume of the vessel. Assume that the process is isentropic and thus follows the law $pV^{1.4} = $ constant. Is this possible from a second law perspective?
(d) Assume that the process in part (c) is isothermal. Is this possible?

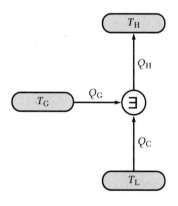

Figure P9.39 An idealized thermodynamic model for a complete absorption refrigeration system.

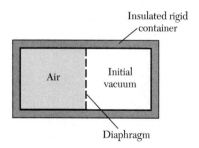

Figure P9.40 Expansion of air initially contained by a diaphragm in one-half of a rigid vessel.

(e) A heat engine has a source at 64°C, a sink at 27°C, and a thermal efficiency of 12%.

(f) A refrigerator operates between a source at −5°C and a sink at 35°C and achieves a COP_C of 3.

9.41. Determine the exergy per unit mass for the steady flow of each of the following:

(a) steam at 1.5 MPa and 500°C

(b) air at 1.5 MPa and 500°C

(c) water at 4 MPa and 300 K

(d) air at 4 MPa and 300 K

(e) air at 1.5 MPa and 300 K

9.42. (a) Determine the exergy (in kW) of a 25-kg/s steady stream of geothermal water at 100 kPa and 90°C.

(b) An inventor claims to be able to produce 1100 kW of mechanical power from the steady stream of geothermal water in part (a). Do you believe her? Give reasons for your answer.

9.43. An inventor proposes to produce mechanical power from 50 kg/s of geothermal water ($c_p = 4.2$ kJ/kg · K) at 365 K. The water stream is at the ambient pressure of 100 kPa. Assume $T_0 = 300$ K.

(a) What is the maximum possible mechanical power output (in kW)?

(b) An actual heat engine design is tested and found to produce 450 kW from the geothermal water stream. What is the second law efficiency achieved?

9.44. Determine (i) the exergy input rate, (ii) the (useful) exergy production rate, and (iii) the second law efficiency achieved in the following steady-flow processes:

(a) A rate of 2 kg/s water at 100 kPa is heated from 300 K to 350 K by using an electric heating coil that is supplied with electricity from a main supply (Fig. P9.44).

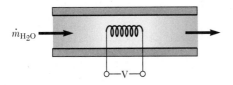

Figure P9.44 Heating of water in steady-flow process.

(b) A rate of 2 kg/s air at 1 MPa and 298 K flows steadily through an adiabatic expansion valve to a pressure of 100 kPa.

(c) Adiabatic throttling of 0.5 kg/s of steam occurs between 1.5 MPa and 100 kPa with an exit temperature of 150°C.

(d) A rate of 0.5 kg/s steam with negligible entry velocity expands through an adiabatic nozzle from 3 MPa and 400°C to 400 kPa and 270°C at the exit to the nozzle.

(e) A rate of 11 kg/s hot water is produced by adiabatic mixing of 10 kg/s of 298 K water mixed with 1 kg/s dry, saturated steam at 100°C.

9.45. A rate of 2 kg/s air at 400 kPa and 300 K flows into a *Hilsch–Ranque vortex tube* and emerges as two streams each at a rate of 1 kg/s and a pressure of 100 kPa but with differing temperatures of 260 K and 340 K, respectively. The ambient temperature is 300 K. Compute the *exergy in* and the *exergy out*, and hence, determine whether or not the process indicated could actually have taken place.

9.46. Steam enters an adiabatic nozzle with a negligible velocity and leaves with a velocity of 729 m/s. Assume the following:

	Entry	**Exit**	**Dead State**
h(in kJ/kg)	3231	2965	113.1
s(in kJ/kg · K)	6.921	7.379	0.395

T_0 may be taken as 300 K. Determine, per kg of steam,

(a) the exergy of steam at entry to the nozzle

(b) the exergy of steam at exit from the nozzle

(c) the exergy destroyed in the process

What is the second law efficiency achieved?

9.47. A liquid (with $c_p = 6$ kJ/kg · K) is heated at an approximately constant pressure from 298 K to 90°C by passing it through tubes immersed in a furnace (Fig. P9.47). The mass flow rate is 0.2 kg/s. Determine

(a) the heating load (in kW)

(b) the exergy production rate (in kW) corresponding to the temperature rise of the liquid

9.48. Assume that the furnace in Problem 9.47 behaves as a thermal reservoir at 1400°C. What is the second law efficiency of the process?

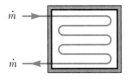

Figure P9.47 Heating of a liquid in a furnace.

9.49. Suppose the heating in Problem 9.47 is provided by an electrical resistance heating element. Determine the second law efficiency of the process.

*** 9.50.** An air motor produces work output when high pressure air is supplied to the motor. In one particular design shown in Figure P9.50, an air motor produces output work when air with a pressure of at least 60 psia is supplied to the motor. The work output from the motor is 2×10^4 ft · lbf per lbm of air supplied to the motor. A compressed air bottle with a volume of 50 ft^3 contains air at 120 psia and 70°F. This bottle is used to operate the air motor. If the ambient pressure is 1 atm and the ambient temperature is 70°F, determine
(a) the maximum work output of the air motor operated from one bottle of compressed air
(b) the second law efficiency achieved in the operation of the air motor

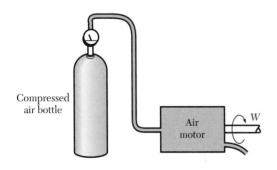

Figure P9.50 Work from compressed air motor.

*** 9.51.** A flow of hot water at 80°C is used to heat cold water from 20°C to 45°C in a heat exchanger arrangement, as shown in either Figure 9.12(a) or (b). The cold water flows at the rate of 2 kg/s. When the heat exchanger is operated in the parallel-flow mode, the exit temperature of the hot water stream cannot be less than 55°C, while in the counter-flow mode, the exit temperature of the hot water stream can be as low as 30°C. Assuming the surroundings are at 300 K, compare the second law efficiencies for the two modes of operation.

*** 9.52.** A flow of hot water at 180°F is used to heat cold water from 60°F to 120°F in a heat exchanger arrangement, as shown in either Figure 9.12(a) or (b). The cold water flows at the rate of 1 lbm/s. When the heat exchanger is operated in the parallel-flow mode, the exit temperature of the hot water stream cannot be less than 130°F, while in the counter-flow mode, the exit temperature of the hot water stream can be as low as 70°F. Assuming the surroundings are at 70°F, compare the second law efficiencies for the two modes of operation.

9.53. The rate of exergy consumed for the universe for a certain process is 200 kJ/s. Assuming the surroundings are at 300 K, what is the rate of entropy generation for the universe for the process?

9.54. For a certain process, the rate of entropy generation for a control volume is 600 kW/K, while the rate of entropy generation for the surroundings is 300 kW/K. If the surroundings are at 300 K, what is the rate of exergy being consumed in the universe?

*** 9.55.** Water at 90°C is flowing in a pipe. The pressure of the water is 3 bar, the mass flow rate is 10 kg/s, the velocity is 0.5 m/s, and the elevation of the pipe is 200 m above the exit plane of the pipeline (to be regarded as ground level). For this steady-state steady-flow stream, compute
(a) thermal exergy flux
(b) pressure exergy flux
(c) exergy flux from kinetic energy
(d) exergy flux from potential energy
(e) total exergy flux of the material stream

*** 9.56.** Air at 150°F is flowing in a pipe. The pressure of the air is 50 psia, the mass flow rate is 5 lbm/s, the velocity is 3 ft/s, and the elevation of the pipe is 100 ft above the exit plane of the pipeline (to be regarded as ground level). For this steady-state steady-flow stream, compute
(a) thermal exergy flux
(b) pressure exergy flux
(c) exergy flux from kinetic energy

(d) exergy flux from potential energy

(e) total exergy flux of the material stream

* 9.57. The exhaust stream of a gas turbine is at 1 atm and 600 K with a flow rate of 30 kg/s. Determine the power output of a heat engine, operating reversibly, with the exhaust stream as the heat source and the surroundings at 1 atm and 300 K as the heat sink. Assume c_p for the exhaust gases to be 1.1 kJ/kg · K.

* 9.58. The exhaust stream of a gas turbine is at 1 atm and 1200°R with a flow rate of 20 lbm/s. Determine the power output of a heat engine, operating reversibly, with the exhaust stream as the heat source and the surroundings at 1 atm and 530°R as heat sink. Assume c_p for the exhaust gases to be 0.25 Btu/lbm · °R.

* 9.59. Use the appropriate concepts on exergy to show that, for all reversible heat engines operating with constant temperature thermal reservoirs at a sink temperature T_L and a source temperature T_H, the thermal efficiency is the Carnot cycle efficiency η_C given by

$$\eta_C = \left(1 - \frac{T_L}{T_H}\right)100\%$$

* 9.60. From a second law perspective, evaluate

(a) a typical heat engine operating between two thermal reservoirs

(b) a typical reversed heat engine operating between two thermal reservoirs

Thermodynamics of Heat Engine Cycles

10.1 Introduction

Prime movers are machines or devices for producing mechanical power, usually from other sources of energy. Such devices are fundamentally energy converters (or energy transformers). Prime movers can be classified in several ways. In one classification, a prime mover is said to be of the *first kind* if the input to it is a *primary energy resource*, such as fossil fuel (coal, oil, or gas), nuclear fuel, wind energy, or water power. Gas engines, steam engines, steam power plants, windmills, and hydraulic turbines are all prime movers of the first kind. If, however, the input to a prime mover is *secondary energy*, such as electricity or compressed air, then the prime mover is said to be of the *second kind*. An electric motor and an air motor are examples of prime movers of the second kind. Such classifications are arbitrary and should by no means be regarded as universal. The main concern in this chapter is with *heat engines*, which constitute an important category of prime movers of the first kind and were defined earlier as devices that convert heat from a source to work while rejecting some heat to a sink.

Historical Perspective

History has shown that the development of prime movers has closely paralleled the stages of development of human society (see, for example, Loftness, 1984, and Angrist and Hepler, 1967). Thus, in the earliest period of human history, people apparently relied wholly on *muscle power* to get work done. Searching for a way to get work done without having to expend as much personal energy, humans then moved on to the use of *animal power*. Indeed, both human and animal power were widely exploited by the early Egyptians and Romans. Again, one finds that even in the early stages of the industrial revolution era, much reliance was initially placed on animal power before the transition to the prime movers in use today. For example, the early forms of transportation at the beginning of the industrial revolution were invariably horse-driven carriages and barges.

The principal limitation on the use of both human and animal power has to do with the low level of work output that is obtainable. It is estimated, for instance, that the maximum power output of a human, an ox, and a horse are only about 75 W, 0.2 kW, and 0.75 kW, respectively. Thus, the search for power engines did not stop with the discovery of animal power. Devices were later developed to harness *wind* and *water power*. Neither the windmill nor the water wheel needed wages (required by people) or fodder (required by animals) to keep it going. Sails on boats fostered the use of wind power in transportation by sea. A number of industrial processes could also be powered using windmills. In the industrial revolution era, many industries, such as textiles, were sited close to waterfalls where water power was available. Levels of power output that can be obtained have been estimated at about 30 kW to 1000 kW for windmills, 80 kW to 500 kW for water wheels, and 500 kW to 200 MW for water turbines (Loftness, 1984). Larger windmills have been tested and put to use in recent times that are each capable of generating power of several megawatts (MW). Obvious limitations exist, however, to the widespread use of prime movers operating on either wind or water power. Wind power is available only intermittently and in geographically favorable zones. Similarly, water power is available only at waterfalls and can therefore be tapped only in regions with favorable topography and with an abundance of water flow.

The next most significant phase in the history of prime movers was the development of the *steam engine*. Although a form of steam engine had been contrived and built by Hero about a half a century before the birth of Christ, no significant practical development in the use of steam for power generation was made until the turn of the 17th century. As recently as the early part of this century, the use of steam for power production remained very extensive in both transportation and power generation. The extensive use of steam, especially in power generation, has persisted until today. Steam engines have been built with power output capacity in the range of 1 kW to 3000 kW. The *steam turbines* in use in steam power plants have a power output capacity in the range of 2 MW to 2000 MW (Loftness, 1984).

In the wake of the initial successes recorded in the use of steam for the generation of power, most attention was focused on achieving better efficiencies and on diversifying the applications of the steam engine. It was not until the second half of the 19th century that a determined venture into the development of the *internal combustion* (gasoline and diesel) *engine* began. In 1860, a French engineer named Lenoir came up with the first practical internal combustion machine. The common feature of all internal combustion machines is that the fuel is burned within the engine itself. Steam plants, in contrast, use steam obtained typically from a boiler that is fired by the combustion of fuel external to the engine itself. Such machines are *external combustion engines*.

Lenoir's gas engine of 1860 did not compete well with the steam engine, but it did provide a beginning for subsequent improvements that eventually established internal combustion machines as a superior alternative to steam engines for a variety of tasks requiring power. Today, the use of internal combustion machines is dominant in such important sectors as transportation and agriculture. In power generation, gas turbines are widely used, often as stand-by provisions in power stations, but also for direct power generation, as in certain designs of nuclear plants.

Most of the machines just mentioned belong to an important category of prime movers known as *heat engines*, which were introduced in Chapter 7. A heat engine is a device that converts heat to useful mechanical work. In thermodynamics, a heat engine is modeled as a device that receives heat from a *high temperature source*, produces mechanical work, and rejects heat to a *low temperature sink*. Figure 10.1 is a chart showing how the various categories of heat engines can be classified. A variety of criteria for classification are in use. In the classification in Figure 10.1, the primary division is between internal combustion engines and external combustion engines. While the concern here is with the thermodynamic analysis of heat engine cycles, generally attention is focused on vapor power cycles and gas power cycles (including gas turbines and the reciprocating internal combustion engines) which, respectively, represent important types of external combustion and internal combustion engines in use today.

By way of conclusion to this brief survey of power-producing devices, it is important to realize that, historically, the development of prime movers has played a decisive role in ushering in the mechanical age. In several instances, the pattern of development was characterized by a conceptual stage followed by the construction of a prototype that functioned, although generally with a low efficiency. Many of the pioneers in the development effort were not necessarily academic people; rather, most were simply highly skilled individuals who demonstrated remarkable ingenuity. As new industries were de-

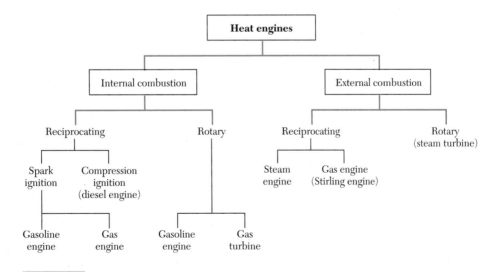

Figure 10.1 Classification scheme for heat engines.

veloped, the demand for power increased rapidly, and the need grew to improve on the designs of the prime movers that had hitherto been developed. The initial response was often directed at achieving small practical improvements that led only to marginally greater efficiencies. Eventually, it also became necessary to review some of the fundamental ideas relating to the generation of mechanical power from heat sources.

It was soon recognized that the efficiency of power plants depended not only on practical details of construction but also on certain particulars of a fundamental nature. The science of thermodynamics was born. The work of such people as Carnot, Clausius, Joule, Kelvin, Gibbs, Rankine, Poincare, Planck, and a host of others contributed in no small measure to the laying of the foundations of thermodynamics which, today, is an indispensable tool for the analysis and evaluation of the processes and performance of energy-based devices.

10.2   Thermodynamic Modeling of Heat Engines

Thermodynamic modeling of a process or system generally involves abstraction and approximation. *Abstraction* basically entails the identification of the essential features of whatever process or phenomenon you may wish to study. *Approximation*, however, has to do with replacing real processes with idealized processes that are easier to analyze. Indeed, in classical (or equilibrium) thermodynamics, complete analysis is possible only when processes occur in a quasi-equilibrium manner. Thus, a real process occurring in real life often must be modeled by identifying idealized, reversible processes that are reasonably close to the actual process when considering the essential details of the phenomenon. As explained earlier, the results obtained in such analyses can, at best, only provide a rough indication of how the actual device will perform. Beyond this, empirically determined process efficiencies must be used to relate indicated performance for the ideal case to that for the corresponding actual process or device. The following section shows how a typical external combustion engine can be thermodynamically modeled.

10.2.1   Ideal Heat Engine Cycles

All the heat engines to be considered here use a *working fluid* to which heat from a heat source is transferred. The working fluid in turn produces work as it expands through an *expander* or a *turbine*.

External Combustion Heat Engine

The steam (turbine) power plant is typical of the external combustion engine class. Figure 10.2 is a schematic diagram showing the essential features of a vapor power plant. The principal components are the *vapor generating unit*, the *turbine*, the *condenser*, and the *feed pump*. External heating is provided typically via the combustion of fuel in air. As a consequence, the working fluid is changed from the liquid state, at which it enters the boiler, to the vapor state. The vapor generating unit usually includes a *boiler* and a *superheater*. The saturated (or superheated) vapor at high temperature (and high

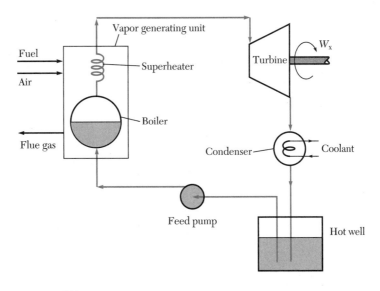

Figure 10.2 Schematic diagram of a vapor power plant.

pressure, usually) leaving the vapor generating unit expands through the turbine, thereby producing a positive work output. Complete condensation of the low pressure (and slightly wet) vapor leaving the turbine is accomplished in the condenser unit. The cycle is completed as the feed pump raises the pressure of the working fluid (now liquid) and feeds the liquid into the boiler.

The vapor power cycle described is often referred to as a *pressure limited cycle*, which means that the engine is operated between two pressure limits. The heating process takes place when the working fluid is at the higher pressure, also known as the *boiler pressure*, while the vapor condensation occurs at the lower pressure, or the *condenser pressure*.

Actual heat engines do not operate reversibly and therefore their behavior cannot be represented in a diagram designed to indicate the path of the processes involved. As a first approximation, however, real heat engine cycles are modeled using idealized quasi-static processes as an approximation to real processes. The resulting cycle is an *ideal heat engine cycle*. In the present case, the ideal cycle just described is termed a *Rankine cycle*.

Detailed analysis of the Rankine cycle and other ideal vapor power cycles is reserved for Section 10.3. To conclude this illustration on how to model an external combustion engine, the process of abstraction needs to be taken one step further. The model needs to be reduced to the classic one for a heat engine, as depicted in Figure 10.3. Figure 10.3 represents the steam power plant as a cyclic device (E) that receives heat from a finite heat source of the type described in Section 9.4 (Chap. 9) and rejects some of the energy as heat to the sink while producing work. When a finite heat source is employed, the source temperature continues to fall as heat is extracted from it. The heat source in this example is an open system comprising a mixture of gases that are products of the combustion process taking place in a furnace. As indicated in Figure 10.3, the flow

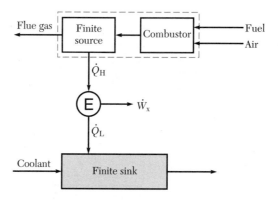

Figure 10.3 Thermodynamic model of a steam power plant.

process (which includes the combustion process) can be represented in two stages: (1) an adiabatic combustion in flow through a *combustor* followed by (2) a heat flow from the stream of hot products of combustion to the working fluid in the engine. It should be noted that the heat source in the present case is *finite* (or limited) since the temperature of the combustion products will fall as heat is extracted for vapor generation in the engine cycle.

The condenser unit effectively serves as the heat sink for the plant. A *condenser* is basically a heat exchanger that uses a cold water stream (the *coolant*) for the extraction of heat from the vapor stream leaving the turbine of the engine. The temperature of the coolant rises as it extracts heat from the condenser. Thus, the heat sink must also be considered finite rather than unlimited. In theory, when a steam power plant is located close to a large water supply such as a lake or a river, the temperature rise of the coolant can be made small by increasing the mass flow rate. This way, the sink could become nearly *unlimited*. In practice, though, increasing the coolant flow rate results in higher pumping costs and can be uneconomical. In addition, to protect the heat exchanger surfaces, treated water must be used whenever the water supply is seawater or brackish water. In such cases, it is generally more economical to recycle the treated water than to treat freshly drawn water from the supply. Thus, realistically, the heat sink in this case is also limited rather than unlimited.

Proper categorization of the heat source and heat sink is important to avoid the common pitfall of holding out the Carnot engine as the optimum design in every conceivable situation. The Carnot cycle is frequently presented as the most efficient heat engine cycle possible. This is correct only when both the heat source and the heat sink are thermal reservoirs in the strict sense defined in previous chapters. As indicated in Section 9.4, the optimum engine cycle based on the second law of thermodynamics depends on the nature of the heat source and the heat sink, whether finite or unlimited, in addition to other factors. The cycle with the optimum thermodynamic efficiency must involve processes that are completely reversible. Section 10.6 returns to this point later.

Internal Combustion Heat Engine

Modeling of internal combustion engines is often more complicated than external combustion engines because most internal combustion engines operate on an open cycle rather than a closed cycle. An *open cycle* is one in which the working fluid is not recycled, and thus, from a strict thermodynamics standpoint, an open system cannot be described as executing a thermodynamic cycle. The process of abstraction applied to such situations often takes the form of replacing the operation of such engines with an equivalent closed-cycle operation using air as the working fluid. The resulting ideal cycles are termed *air standard cycles*. In some cases, such as reciprocating internal combustion engines, the air standard cycle comprises a set of nonflow processes involving a fixed mass. Examples of such cycles (considered in Section 10.5) include the Otto cycle (named after Nikolaus Otto) for gasoline engines and the Diesel cycle (named after Rudolf Diesel) for diesel engines. In other instances, the air standard cycle comprises a series of flow processes through the various components of the engine. The Brayton (or Joule) cycle is such an air standard cycle and is the ideal cycle generally assumed for the operation of the gas turbine. Section 10.4 discusses this cycle further. Discussion of optimum cycles for the internal combustion engines is deferred to Section 10.6.

10.2.2 Performance Criteria for Heat Engines

Relatively few innovations have occurred in recent times in the available range of power-producing devices. Significant improvements in *performance*, however, have been made through detailed studies of existing designs based on the principles of thermodynamics. Most of these studies have been based almost exclusively on first law considerations. Indeed, most of the currently available textbooks on thermodynamics have analyzed heat engine cycles mainly from the first law perspective; second law considerations are often only briefly mentioned. Comparatively recent studies (for example, Gaggioli and Petit, 1977) have shown, however, that the most valuable insights on how to improve performance can be gained only if a second law or exergy analysis is included. Therefore, the analysis of thermodynamic cycles here covers both first law and second law perspectives.

Thermal Efficiency (η_1) and Efficiency Ratio (ε)

A variety of indices of performance for a heat engine have been established over the years, most of which are based on first law considerations. The most common of these is the *thermal efficiency* (η_1), which is a first-law efficiency. Equation 7.2 (in Chap. 7) defined η_1 as

$$\eta_1 = \frac{W}{Q_H} \cdot 100\%$$

The first law efficiency for an actual heat engine is of rather limited utility as an index of performance by itself. If, however, the thermal efficiency (η_{ideal}) of the corresponding ideal heat engine cycle (that is, one with all processes internal to the engine occurring reversibly) is also known, one can define an *efficiency ratio* (ε) that compares the actual efficiency achieved with the best that could be achieved based on the ideal cycle:

$$\varepsilon = \frac{\eta_1}{\eta_{\text{ideal}}} \tag{10.1}$$

It is important to note that η_{ideal} is not necessarily the optimum efficiency of power generation attainable from a heat engine operating between a given heat source and heat sink. While heat engines operating on an ideal cycle do so with all the processes internal to the heat engine occurring reversibly, quite often external irreversibility cannot be ruled out. The *optimum thermodynamic efficiency* is achieved when there is *complete* (internal and external) *reversibility* for all the processes involved.

The Work Ratio (r_w)

Another index of performance is the *work ratio*, which is a very important measure of how close to the performance of the ideal engine that of the actual engine is likely to be. Surprisingly, many books on thermodynamics omit reference to this ratio altogether. The work ratio (r_w) for a heat engine is defined as the ratio of the net work output to the positive work output:

$$r_w = \frac{\text{net work output } (W)}{\text{positive work } (W_{+\text{ve}})} \tag{10.2}$$

The net work output is the positive work output minus the input work in each cycle. Note that the subscript "$+\text{ve}$" is used for positive quantities, while "$-\text{ve}$" is used for negative quantities.

In general, the work ratio depends on the cycle as well as the working fluid. A high work ratio is indicative of a high efficiency ratio, whereas a low work ratio suggests that a low efficiency ratio will be measured in practice. This direct correlation between the work ratio and the efficiency ratio is demonstrated in the following example.

Consider two ideal devices, A_I and B_I, both producing the same net work of 40 kJ each, while the positive work output for each device is, respectively, 100 kJ and 50 kJ. If the same heat input is required in the two cases, the thermal efficiencies for A_I and B_I will be the same, but the work ratios will be 0.4 and 0.8, respectively.

Let A_R and B_R denote real devices with ideal cycles corresponding, respectively, to those on which A_I and B_I operate. Suppose that for the two real devices, the irreversibilities result in the actual positive work output being only 80% of the ideal positive work, while the actual negative work is 20% higher (in magnitude) than the ideal negative work.

For device A_R:

$$\text{Actual positive work} = (80/100)100 = 80 \text{ kJ}$$

$$\text{Actual negative work} = -(120/100)60 = -72 \text{ kJ}$$

$$\therefore \text{Actual net work} = 8 \text{ kJ}$$

Hence, the efficiency ratio $= (8/40) = 0.2$.

For device B_R:

$$\text{Actual positive work} = (80/100)50 = 40 \text{ kJ}$$

$$\text{Actual negative work} = -(120/100)10 = -12 \text{ kJ}$$

$$\therefore \text{Actual net work} = 28 \text{ kJ}$$

Hence, the efficiency ratio $= (28/40) = 0.7$.

This example clearly demonstrates that the nearer the work ratio is to unity, the less adversely the actual performance of a heat engine (relative to that of the ideal engine) is affected by the irreversibilities that are present in real processes. Thus, the work ratio provides an indication of how sensitive the cycle is to irreversibilities that are encountered in the actual processes of the cycle. It should be noted further that, while the work ratio is defined for an ideal cycle, the efficiency ratio can be determined only after the thermal efficiency of the actual device has been ascertained. Note, however, that a high work ratio does not necessarily mean a high cycle efficiency. What it really means is that for any given cycle, the actual efficiency is close to the ideal cycle efficiency when the work ratio is close to unity.

The work ratio for the Carnot cycle is rather low, regardless of the working fluid employed. This is one of the reasons why heat engines are generally not designed to operate on the Carnot cycle. The following worked example demonstrates that the work ratio for the Carnot cycle is generally low, resulting in a rather low cycle efficiency for an actual heat engine based on the Carnot cycle as ideal.

EXAMPLE 10.1
Work Ratio and Efficiency Ratio

A heat engine using steam as the working fluid is designed to operate ideally on a Carnot cycle, as shown in Figure 10.4. The engine has a boiler pressure of 10 MPa (100 bar) and a condenser pressure of 5 kPa (0.05 bar). The cycle comprises flow processes, and $w_{23} = 0$ and $w_{41} = 0$ since both processes (2→3 and 4→1) occur at a constant pressure. A cycle analysis yields the following results for the ideal cycle:

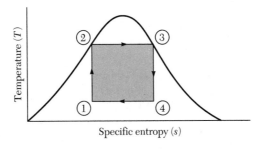

Figure 10.4 Ideal Carnot cycle for the heat engine in Example 10.1.

$$\text{Positive work output } (w_{+ve}) = w_{34} = 1015 \text{ kJ/kg}$$

$$\text{Negative work } (w_{-ve}) \qquad = w_{12} = -387 \text{ kJ/kg}$$

$$\text{Heat input to the cycle } (q_{in}) = q_{23} = 1318 \text{ kJ/kg}$$

where the subscript "in" indicates input.

(a) For the ideal Carnot cycle, determine
  (i) the net work output (w), in kJ/kg
  (ii) the work ratio (r_w)
  (iii) the thermal efficiency (η_{ideal})

(b) Assume that the actual engine operates with isentropic efficiencies of 0.8 and 0.6 for the turbine and compressor, respectively. Determine
  (i) the actual net work output (w_A), in kJ/kg
  (ii) the actual thermal efficiency (η_{actual})
  (iii) the efficiency ratio (ε)

Note: Lowercase q and w denote the heat and work, respectively, per unit mass of the working fluid employed in the engine.

Given:
(a) Ideal engine operating on the Carnot cycle with

$$w_{+ve} = 1015 \text{ kJ/kg}$$

$$w_{-ve} = -387 \text{ kJ/kg}$$

$$q_{in} = 1318 \text{ kJ/kg}$$

(b) Actual engine has isentropic efficiencies of 0.8 for the turbine and 0.6 for the compressor

Find:
(a) (i) $w(\text{kJ/kg})$, (ii) r_w, (iii) η_{ideal}
(b) (i) w_A (kJ/kg), (ii) η_{actual}, (iii) ε

Solution:
(a) *Ideal Carnot cycle*
  (i) $w = w_{+ve} + w_{-ve}$
    $= 1015 - 387 = \underline{628 \text{ kJ/kg}}$
  (ii) $r_w = w/w_{+ve}$
    $= 628/1015 = \underline{0.619}$
  (iii) $\eta_{ideal} = (w/q_{in})100\%$
    $= (628/1318)100\% = \underline{47.6\%}$

Recall that for the Carnot cycle,

$$\eta_C = (1 - T_L/T_H)100 = (1 - 305.9/584)100 = \underline{47.6\%}$$

which agrees with the result just obtained.

(b) *Actual engine*

(i) Equation 8.109 for the process efficiency of an adiabatic turbine gives

$$(w_{+ve})_{actual} = (w_{+ve})_{ideal} \cdot \eta_{adia.\ turbine}$$

$$= (1015)(0.8) = 812\ kJ/kg$$

Likewise, using Equation 8.111 for the process efficiency of an adiabatic compressor gives

$$(w_{-ve})_{actual} = (w_{-ve})_{ideal}/\eta_{adia.\ compr.}$$

$$= -387/0.6 = -645\ kJ/kg$$

Now,

$$w_A = (w_{+ve})_{actual} + (w_{-ve})_{actual}$$

$$= 812 - 645 = \underline{167\ kJ/kg}$$

(ii) $\eta_{actual} = (w_A/q_{in})100\%$
$= (167/1318)100\% = \underline{12.7\%}$

(iii) $\varepsilon = (\eta_{actual}/\eta_{ideal})$
$= 12.7/47.6 = \underline{0.266}$

EXAMPLE 10.2
Work Ratio and Efficiency Ratio

The heat engine in Example 10.1 is modified to operate on the ideal Rankine cycle, shown in Figure 10.5, again with a boiler pressure of 10 MPa (100 bar) and a condenser pressure of 5 kPa (0.05 bar). A cycle analysis yields the following for the ideal cycle:

$$\text{Positive work output } (w_{+ve}) = w_{34} = 1015\ kJ/kg$$

$$\text{Negative work } (w_{-ve}) \qquad = w_{12} = -9.8\ kJ/kg$$

$$\text{Heat input to the cycle } (q_{in}) = q_{23} = 2577.2\ kJ/kg$$

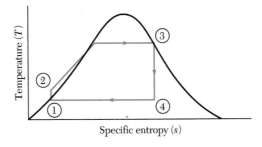

Figure 10.5 Ideal Rankine cycle for the heat engine in Example 10.2.

(a) For the ideal cycle, determine

    (i) the net work output (w), in kJ/kg

    (ii) the work ratio (r_w)

    (iii) the thermal efficiency (η_{ideal})

(b) Assume that the actual engine operates with isentropic efficiencies of 0.8 and 0.6 for the turbine and feed water pump, respectively. Determine

    (i) the actual net work output (w_A), in kJ/kg

    (ii) the actual thermal efficiency (η_{actual})

    (iii) the efficiency ratio (ε)

Given:

(a) Heat engine operating ideally on Rankine cycle with

$$w_{+ve} = 1015 \text{ kJ/kg}$$

$$w_{-ve} = -9.8 \text{ kJ/kg}$$

$$q_{in} = 2577 \text{ kJ/kg}$$

(b) Actual engine has isentropic efficiencies of 0.8 (turbine) and 0.6 (pump)

Find:

(a) (i) w (kJ/kg), (ii) r_w, (iii) η_{ideal}

(b) (i) w_A(kJ/kg), (ii) η_{actual}, (iii) ε

Solution:

(a) *Ideal (Rankine) cycle*

    (i) $w = w_{+ve} + w_{-ve}$
        $= 1015 - 9.8 = \underline{1005.2 \text{ kJ/kg}}$

    (ii) $r_w = w/w_{+ve}$
         $= 1005.2/1015 = \underline{0.99}$

    (iii) $\eta_{ideal} = (w/q_{in})100\%$
          $= (1005.2/2577)100 = \underline{39.0\%}$

(b) *Actual cycle*

    (i) $(w_{+ve})_{actual} = (w_{+ve})_{ideal} \cdot \eta_{adia.\ turbine}$ (Eq. 8.109)
         $= (1015)(0.8) = 812 \text{ kJ/kg}$

      $(w_{-ve})_{actual} = (w_{-ve})_{ideal}/\eta_{adia.\ pump}$ (Eq. 8.114)
         $= -9.8/0.6 = -16.3 \text{ kJ/kg}$

      $w_A = (w_{+ve})_{actual} + (w_{-ve})_{actual}$
        $= 812 - 16.3 = \underline{796 \text{ kJ/kg}}$

    (ii) $\eta_{actual} = (w_A/q_{in})100\%$
         $= (795.7/2577)100\% = \underline{30.9\%}$

    (iii) $\varepsilon = (\eta_{actual}/\eta_{ideal})$
         $= 30.9/39.0 = \underline{0.792}$

Examples 10.1 and 10.2 clearly confirm certain trends that were noted previously:

1. Whenever the work ratio is low, the actual efficiency is much lower than the corresponding ideal cycle efficiency. Thus, in Example 10.1, the ideal cycle efficiency (for a Carnot engine) is 47.6%, whereas the actual cycle efficiency is a mere 12.7%. The work ratio is 0.619.

2. When the work ratio is high, the actual efficiency is close to the corresponding ideal cycle efficiency. The work ratio for the Rankine cycle in Example 10.2 is 0.99, and consistent with previous observations, the ideal and actual cycle efficiencies are close, at 39% and 30.9%, respectively.

3. For cycles such as these, the actual efficiency for the Rankine-like cycle is probably going to be higher than that for the Carnot-like cycle.

The Second Law Efficiency (η_2)

The index of performance based on the second law of thermodynamics is the *second law efficiency* (η_2), which for a heat engine can be defined on the basis of Equation 9.8 or 9.10 (Chap. 9) as

$$\eta_2 = \frac{\dot{W}_x}{\dot{X}_{source}} \cdot 100\% \qquad (10.3)$$

As Chapter 9 indicated, the second law efficiency is the proper measure of thermodynamic efficiency for any device. A low value for the second law efficiency implies that most of the exergy supplied to the heat engine is destroyed as a consequence of the irreversibilities in the processes involved. The best possible performance corresponds to the case when all processes are reversible, resulting in the work produced being equal to the exergy supplied.

Whether dealing with either an internal or external combustion engine, one can unambiguously define the exergy input in terms of the chemical exergy of the fuel supplied. A fairly detailed treatment of chemical exergy in relation to chemical substances is provided in Chapter 14 on combustion. For the present, it suffices to note that the exergy equivalence of the chemical energy stored in the fuels commonly used in heat engines can be determined and utilized for the computation of the second law efficiency achieved in specified applications.

EXERCISES

for Section 10.2.2

1. An ideal cycle for the gasoline engine comprises an adiabatic compression process $(1{\rightarrow}2)$, a constant volume heating process $(2{\rightarrow}3)$, an adiabatic expansion process $(3{\rightarrow}4)$ and a constant volume cooling process $(4{\rightarrow}1)$. From a first law analysis, the work and heat transfer per unit mass of the working fluid are obtained as follows:

$$w_{12} = -279 \text{ kJ/kg}$$

$$w_{34} = 446 \text{ kJ/kg}$$

$$q_{23} = 295 \text{ kJ/kg} \ (= q_{in})$$

Determine

(a) w_{net}

(b) η_{ideal} (thermal efficiency)

(c) r_w

Answer: (a) 167 kJ/kg, (b) 56.5%, (c) 0.373

2. Assume that for the actual cycle corresponding to the conditions in Exercise 1, the process efficiency is 85% for both the compression process ($1 \rightarrow 2$) and the expansion process ($3 \rightarrow 4$). The heat input per unit mass of the working fluid is 246 kJ/kg. Determine

(a) w_{net}

(b) η_{actual}

(c) the efficiency ratio (ε)

Answer: (a) 50.9 kJ/kg, (b) 20.7%, (c) 0.366

3. Assume the heat inputs in Exercises 1 and 2 to be effectively from a thermal reservoir at a temperature of 1800 K. Assuming $T_0 = 300$ K, determine the second law efficiency for

(a) the ideal case

(b) the actual case

Answer: (a) 67.9%, (b) 24.8%

10.2.3   *Procedure for Heat Engine Cycle Analysis*

The following step-by-step procedure is suggested for a complete thermodynamic analysis and evaluation of heat engine cycles.

Step 1. First, identify whether the cycle comprises nonflow or flow processes. If nonflow processes are involved, define a suitable system and, with the aid of suitable sketches, show the system boundary and identify the heat and work interactions across the boundary in each process. If the cycle comprises flow processes, control volumes should be defined for each flow process and the heat and shaft work interactions should be identified.

Step 2. Display the cycle on a T-s diagram and other appropriate diagrams such as a p-V diagram (for nonflow processes) or the h-s diagram (generally used in steam turbine cycle analysis).

Step 3. If the ideal cycle comprises nonflow processes, the first law analysis (or energy balance) only requires the application of the nonflow energy equation (NFEE) to each process in turn and possibly the use of the pertinent process path equations along with the corresponding expressions for work and heat interactions. If, however, the cycle is made up of flow processes, application of the steady-flow energy equation (SFEE) along with the appropriate process path equations and expressions for work and heat will, in general, be all that is required for a complete energy balance.

Step 4. An exergy analysis for each process can be conducted by using the appropriate expressions for exergy (given in Chap. 9) to determine the exergy supply to each process, the exergy product that results, and hence, the exergy consumption (and/or exergy loss) in the process. For the entire operation, an overall second law efficiency can be defined in terms of the exergy product (which in this instance is the shaft work, or power, produced) and the exergy input to the heat engine. If, for example, the heat source derives from the combustion of fuel, one can simply define the exergy input to the entire operation as the chemical exergy flux of the fuel. This flux can be easily computed once the chemical exergy (x_c) per unit mass of the fuel is known.

Step 5. The analyses in steps 3 and 4 can be improved by assigning reasonable values for process efficiencies so that the actual processes are recognized as real and irreversible, not ideal.

Subsequent sections of this chapter use several worked examples to demonstrate the use of this procedure for a complete thermodynamic analysis of heat engine cycles from both the first law and second law (or exergy) perspectives.

10.3   *Vapor Power Cycles*

Vapor power plants are normally external combustion engines. Figure 10.2 showed a typical vapor power plant layout. Steam is the working fluid used in most conventional thermal power plants. Although by no means a perfect working fluid, steam nevertheless has a number of advantages over other fluid media:

1. Water is relatively inexpensive and easy to obtain.
2. It is chemically stable over the normal range of pressures and temperatures encountered in the operation of such plants.
3. It is reasonably noncorrosive, particularly when properly treated.
4. It has a large value for the enthalpy of vaporization (h_{fg}), thereby permitting large energy transfers using relatively compact units.
5. High heat transfer rates to the working fluid are easily achieved, thus minimizing problems in transfer of heat to the water.

Water has awkward critical parameters, however, thus severely limiting its effectiveness in transforming the exergy released in fuel combustion to useful work. (The critical temperature of water is 647.126 K and the critical pressure is 220.55 bar.) When water is used as the working fluid in a vapor power cycle, the maximum typical pressure allowed rarely exceeds about 150 bar (15 MPa). Supercritical pressure plants operating with boiler pressures higher than 220 bar are relatively new, and only a few such plants have been operated commercially. The saturation temperature at a steam pressure of 150 bar is about 615 K, which is considerably lower than the 2000 K temperature level of typical fuel combustion products. Thus, a large exergy consumption occurs in a power plant using water.

The *basic Rankine cycle* uses dry, saturated steam at the entry to the turbine. To reduce the thermal irreversibility in the vapor-generating phase of the cycle, a number of modifications are made to the basic Rankine cycle. These modifications include superheating and reheating, both of which are discussed in more detail later in this section. The modifications also serve another important role in ensuring an acceptable steam quality at the exit to the turbine. If water droplets are present to an appreciable extent in the steam passing through the turbine, significant erosion of the turbine blades can occur. Reheating and superheating of steam have the beneficial effect of improving the steam quality leaving the turbine.

10.3.1   Thermodynamic Analysis of the Basic Rankine Cycle

The essential features of the basic Rankine cycle are shown in Figures 10.6(a) and (b), which are the *T-s* and *h-s* diagrams. The *h-s* diagram for steam is also referred to as the *Mollier chart* and is commonly employed in vapor power cycle analyses because of the relative ease with which the pertinent property values can be determined for the various fluid states in the cycle.

The Rankine cycle is an ideal heat engine cycle, and it comprises the following flow processes:

a. isentropic feed pump work (1→2)

b. steam generation at constant boiler pressure (2→3)

c. isentropic steam expansion through the turbine (3→4)

d. condensation of steam at the condenser pressure (4→1)

The basic Rankine cycle uses saturated water at the inlet to the pump (state 1). The steam produced enters the turbine in a dry, saturated condition (state 3).

As explained in Section 10.2, a complete thermodynamic analysis of the cycle requires both first law considerations and an exergy analysis. For the former, the steady-flow energy equation (Eq. 6.29) is used for each component of the heat engine treated as a control volume through which a steady stream of the working fluid passes. Similarly, equations derived in Chapter 9 are used for the flow of exergy to arrive at a performance evaluation of the cycle based on the second law of thermodynamics.

The bulk of the computations in steam power cycle analyses generally requires finding the steam property values at the various thermodynamic states involved. This can be

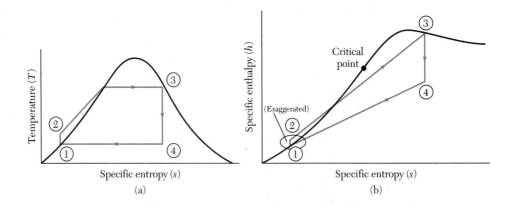

Figure 10.6 Basic Rankine cycle on (a) *T-s* coordinates and (b) *h-s* coordinates.

very cumbersome if numerous operating conditions are considered and when one has to work from tables of properties of steam. Tables of properties are provided in Appendices A (SI units) and B (USCS units) for steam and refrigerant-22. In addition, the computer codes **STEAM** for steam properties and **R22** for refrigerant-22 are provided to minimize the effort required in the determination of property values for water and steam or refrigerant-22, thereby freeing the reader to focus attention on the application of thermodynamic principles in power cycle analyses. The following worked examples demonstrate the procedures involved in a first law and second law analysis of steam power cycles.

EXAMPLE 10.3
Ideal Basic Rankine Cycle

A steam power plant receives heat from a heat source at a rate of 100 MW. The plant operates with a boiler pressure of 40 bar (4 MPa) and a condenser pressure of 0.075 bar (7.5 kPa).

   If the plant is designed to operate ideally on the basic Rankine cycle, calculate

(a) the cycle efficiency

(b) the work ratio for the cycle

(c) the power output (in MW) of the plant

(d) the required mass flow rate (in kg/h) of the working fluid

(e) the specific steam consumption[1] (abbreviated s.s.c., in kg/kWh)

Given: Engine operating on ideal basic Rankine cycle shown in Figure 10.7

$$\text{Heat input rate} = 100 \text{ MW}$$

Find: (a) η_{ideal}, (b) r_w, (c) $\dot{W}_x$ (in MW), (d) $\dot{m}_s$ (kg/h), (e) s.s.c. (kg/kWh)

---

[1] The specific steam consumption can be formally defined as the mass flow rate of steam (in kg/h) per unit kW of net power output.

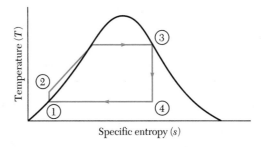

Figure 10.7 Basic Rankine cycle of Example 10.3.

Solution: The cycle comprises flow processes and therefore, the SFEE (Eq. 6.29) is applicable. Assuming that changes in kinetic and potential energy are negligible, the SFEE can be written (on a unit mass of working fluid basis) as

$$q - w = \Delta h$$

Process 1→2 (isentropic compression)

This process is reversible and adiabatic, and hence, $q_{12} = 0$. Also, $s_2 = s_1$. Applying the SFEE thus gives

$$w_{12} = h_1 - h_2 \approx v_1(p_1 - p_2)$$

The last step is based on Equation 8.42.

Process 2→3 (reversible constant pressure process)

It is normally assumed that the heating process takes place at a constant pressure. *Internal reversibility* is also assumed, which means that (from Eq. 8.99) the shaft work w_{23} is assumed to be zero. Applying the SFEE thus leads to

$$q_{23} = h_3 - h_2$$

Process 3→4 (isentropic expansion)

In this case, $q_{34} = 0$ (and $s_4 = s_3$). Applying the SFEE gives

$$w_{34} = h_3 - h_4$$

Process 4→1 (reversible constant pressure process)

Note that this part of the analysis is not needed for any of parts (a) to (e) of the problem and is only included for completeness. The process is reversible and at a constant pressure, and hence, $w_{41} = 0$. Applying the SFEE gives

$$q_{41} = h_1 - h_4$$

The values of h for H_2O at states 1 to 4 must be determined so that the q and the w can be calculated. These values were obtained using the **STEAM** code. Table 10.1 gives the summary of the two independent intensive properties that were specified in each case. The reader is encouraged to cross-check the h values in Table 10.1 by using the steam tables in Appendix A as well as the **STEAM** code.

TABLE   10.1

Summary of Properties Specified in Obtaining h Values Using the STEAM Code

State Number	Two Independent Intensive Properties Specified	h (kJ/kg)
1	$p_1 = 0.075$ bar (7.5 kPa), $x_1 = 0$	169
2	$p_2 = 40$ bar (4 MPa), $s_2 = s_1 = 0.576$ kJ/kg · K	172.8
3	$p_3 = p_2 = 40$ bar, $x_3 = 1$	2801
4	$p_4 = p_1 = 0.075$ bar, $s_4 = s_3 = 6.068$ kJ/kg · K	1890.2

Substituting the appropriate h values in the previous equations yields the following values for the w and the q:

$$w_{12} = h_1 - h_2 = -3.8 \text{ kJ/kg}$$

$$w_{34} = h_3 - h_4 = 910.8 \text{ kJ/kg}$$

$$q_{23} = h_3 - h_2 = 2628 \text{ kJ/kg}$$

Also, $q_{41} = h_1 - h_4 = -1721$ kJ/kg. These results can be cross-checked by calculating the net heat input and the net work output for each cycle:

$$\oint q = q_{23} + q_{41} = 907 \text{ kJ/kg}$$

$$\oint w = w_{12} + w_{41} = 907 \text{ kJ/kg}$$

Equation 6.1a is the cyclic statement of the first law of thermodynamics by which the net heat input must be equal to the net work output for a cycle. Such is the case as indicated by the results.

(a) The ideal cycle efficiency can be calculated from Equation 7.2:

$$\eta_{\text{ideal}} = \frac{w}{q_{\text{in}}} \cdot 100\% = \frac{(w_{12} + w_{34})}{q_{23}} \cdot 100\%$$

$$= \frac{907 \text{ kJ/kg}}{2628 \text{ kJ/kg}} \cdot 100\% = 34.5\%$$

(b) The work ratio is determined using Equation 10.2:

$$r_{\text{w}} = \frac{w}{w_{+\text{ve}}} = \frac{(w_{12} + w_{34})}{w_{34}}$$

$$= \frac{907 \text{ kJ/kg}}{910.8 \text{ kJ/kg}} = 0.996$$

(c) The mechanical power output can be calculated using Equation 7.2 and substituting the known values of heat input rate and the ideal cycle efficiency:

$$\dot{W}_{\text{x}} = \dot{Q}_{\text{H}} \frac{\eta_{\text{ideal}} (\%)}{100} = 34.5 \text{ MW}$$

(d) The mass flow rate of the working fluid can be determined from the relationship between the total heat input rate and q_{23} or the mechanical power output and w. Using the first of these gives

$$\dot{m}_{\text{s}} = \frac{\dot{Q}_{\text{H}}}{q_{23}} = \frac{10^5 \text{ kJ/s}}{2628 \text{ kJ/kg}} \left(\frac{3600 \text{ s}}{1 \text{ h}} \right) = 1.37 \times 10^5 \text{ kg/h}$$

(e) The *specific steam consumption* (s.s.c) is a design parameter that gives an indication of the size of the plant. It is a measure of how much steam (in kg) is cycled to produce 1 kWh of net work output. The larger the numerical value of the specific steam consumption, the bigger the plant has to be. From this definition, the s.s.c can be calculated by

$$\text{s.s.c} \equiv \frac{\dot{m}_s \ (\text{kg/h})}{\dot{W}_x \ (\text{kW})} \equiv \frac{\dot{m}_s \ (\text{kg/s})}{\dot{W}_x \ (\text{kW})} \left(\frac{3600 \ \text{s}}{1 \ \text{h}} \right) = \frac{3600}{w} \ \text{kg/kWh}$$

In the present case, s.s.c = 3600/907 = 3.97 kg/kWh.

Note that you must be careful when using specific steam consumption because the specific volume can vary, thus changing component sizes.

EXAMPLE 10.4
Real Rankine-Like Cycle

Repeat the calculations for parts (a), (c), (d), and (e) in Example 10.3 for an actual heat engine that operates on the basic Rankine-like cycle but has an adiabatic turbine with a process efficiency of 85%.

Given: The engine cycle is illustrated in Figure 10.8, which is based on the Rankine cycle but with adiabatic turbine process efficiency = 85%.

$$\text{Heat input rate} = 100 \ \text{MW}$$

$$\text{Boiler pressure} = 40 \ \text{bar} \ (4 \ \text{MPa})$$

$$\text{Condenser pressure} = 0.075 \ \text{bar} \ (7.5 \ \text{kPa})$$

Find: (a) η_{actual}, (b) $\dot{W}_x$ (in MW), (c) $\dot{m}_s$ (in kg/h), (d) s.s.c (in kg/kWh)

Solution: With reference to Figure 10.8, note that process 3→4 differs from what it was in Example 10.3. While the process in the present instance is adiabatic, it is irreversible, and hence, dashed lines must be used to represent it in the *T-s* diagram. The continuous line 3→4' corresponds to the isentropic expansion through a reversible turbine, and the work done (per unit mass), $w_{34'}$, is the same as the w_{34} of Example 10.3. Applying Equation 8.109 for the process efficiency of an adiabatic turbine leads to the following for the actual work done in the present example:

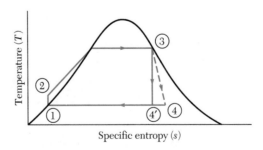

Specific entropy (s)

Figure 10.8 Actual Rankine-like cycle of Example 10.4.

$$w_{34} = w_{34'} \, \eta_{\text{adiabatic}\atop\text{turbine}}$$

$$= (910.8)(0.85) = 774.2 \text{ kJ/kg}$$

Note that the computed value of w_{34} can be used to compute h_4 since, by the SFEE (Eq. 6.29), the adiabatic expansion through the turbine gives $w_{34} = h_3 - h_4$. Substituting for w_{34} and h_3 (determined from Example 10.3) gives $h_4 = 2027$ kJ/kg. Since h_1, h_2, and h_3 have the same values as for Example 10.3, the previous solution gives

$$w_{12} = h_1 - h_2 = -3.8 \text{ kJ/kg}$$

$$q_{23} = h_3 - h_2 = 2628 \text{ kJ/kg}$$

Note that $q_{41} = h_1 - h_4$ is now equal to -1858 kJ/kg. Observe a nearly 8% increase in the heat rejection rate in this example compared with Example 10.3, which operated with complete internal reversibility.

$$w_{+\text{ve}} = w_{34} = 774.2 \text{ kJ/kg}$$

$$w_{-\text{ve}} = w_{12} = -3.8 \text{ kJ/kg}$$

$$w = w_{+\text{ve}} + w_{-\text{ve}} = 770.4 \text{ kJ/kg}$$

$$q_{\text{in}} = q_{23} = 2628 \text{ kJ/kg}$$

(a) The cycle efficiency is calculated using Equation 7.2:

$$\eta_{\text{actual}} = (770.4/2628)100\% = \underline{29.3\%}$$

This represents a 15% reduction in cycle efficiency compared with the ideal case of Example 10.3. The efficiency ratio ε from Equation 10.1 is $(29.3/34.51) = 0.849$.

(b) The power output can be determined using Equation 7.2 and substituting the specified value of the heat input rate as well as the calculated cycle efficiency. Thus,

$$\dot{W}_x = \dot{Q}_H \frac{\eta_{\text{actual}}\,(\%)}{100} = \underline{29.3 \text{ MW}}$$

which is less than the output of 34.5 MW obtained in the ideal case.

(c) Since q_{23} and the heat input rate are the same in both Examples 10.3 and 10.4, the mass flow rate has the same value of 1.37×10^5 kg/h as that obtained in Example 10.3.

(d) The specific steam consumption (s.s.c) in this case is

$$\text{s.s.c} = 3600/w = 3600/770.4 \text{ kg/kWh}$$

$$= \underline{4.67 \text{ kg/kWh}}$$

This value is nearly 18% higher than for the ideal case. In other words, the actual steam plant must be larger in size to accommodate a higher flow of steam for the same work output.

EXAMPLE 10.5
Combined Energy and Exergy Analysis for the Ideal Basic Rankine Cycle

A steam power plant is designed to operate on the ideal basic Rankine cycle, as in Example 10.3. The boiler pressure is 40 bar while the condenser pressure is 0.075 bar. The heating rate remains at 100 MW. Suppose that the heating for steam generation in the boiler unit is provided by a steady stream of hot gases initially at 2000 K and 1 atm pressure. The hot gases exhaust at a temperature of 450 K to the ambient atmosphere, which is at 300 K and 1 atm. The mean specific heat at constant pressure for the hot gases can be taken as 1.1 kJ/kg · K.

  Calculate the following:

(a) the mass flow rate (in kg/h) of the stream of hot gases required

(b) the exergy flow rate (in MW) for the hot gases at the initial condition

(c) the exergy loss rate (in MW) associated with the exhaust stream

(d) the exergy consumption (in MW) in the steam generation process

(e) the exergy consumption (in MW) in the vapor condensation process

(f) the second law efficiency achieved in the entire operation

Given: An ideal steam power plant operates on the basic Rankine cycle with a finite heat source, as illustrated in Figure 10.9(a).

$$\text{Boiler pressure} = 40 \text{ bar (4 MPa)}$$

$$\text{Condenser pressure} = 0.075 \text{ bar (7.5 kPa)}$$

Finite heat source

$$T_i = 2000 \text{ K}, \quad p_i = 1 \text{ atm}$$

$$T_e = 450 \text{ K}, \quad p_e = 1 \text{ atm}$$

$$c_{pg} = 1.1 \text{ kJ/kg} \cdot \text{K}$$

$$\text{Heat input rate to the engine cycle} = 100 \text{ MW}$$

$$\text{Ambient } p_0 = 1 \text{ atm and } T_0 = 300 \text{ K}$$

Find:

(a) $\dot{m}_{source}$ (kg/h)

(b) $(\dot{X}_i)_{source}$ (MW)

(c) $(\dot{X}_e)_{exhaust}$ (MW)

(d) $(\dot{X}_{consumed})_{\substack{steam \\ generation}}$ (MW)

(e) $(\dot{X}_{consumed})_{\substack{vapor \\ condensation}}$ (MW)

(f) $(\eta_2)_{plant}$

Solution:

(a) To determine the mass flow rate of the hot gases, define the finite heat source as the control volume and apply the SFEE (Eq. 6.29) with the kinetic energy and

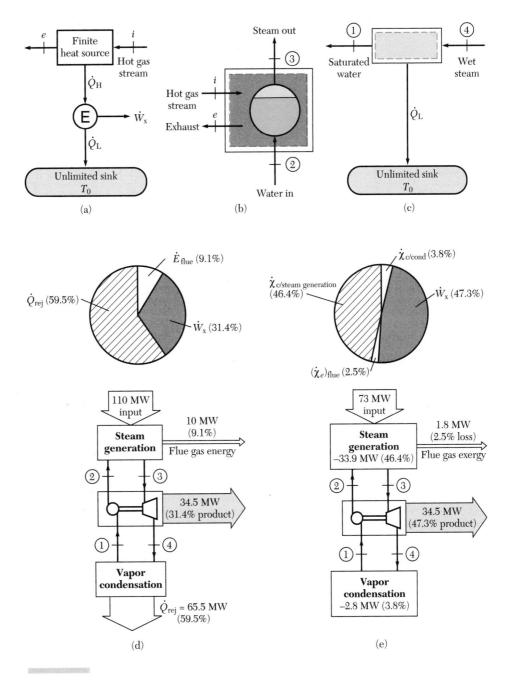

Figure 10.9 (a) Model for the ideal steam power plant of Example 10.5. (b) Control volume for the steam generation process. (c) Schematic diagram of the vapor condensation process. (d) Energy disposition chart. (e) Exergy dispositon chart for the steam plant of Example 10.5.

potential energy terms neglected. Thus, assuming that the fluid can be treated as a perfect gas,

$$(-\dot{Q}_H) - 0 = \dot{m}_{source}(h_e - h_i) = \dot{m}_{source}c_{pg}(T_e - T_i)$$

Hence,

$$\dot{m}_{source} = \frac{10^5 \text{ kJ/s}}{(1.1 \text{ kJ/kg} \cdot \text{K})(2000 - 450) \text{ K}}\left(\frac{3600 \text{ s}}{1 \text{ h}}\right) = \underline{2.11 \times 10^5 \text{ kg/h}}$$

(b) The exergy flow rate of the stream of hot gases can be calculated using Equation 9.40a. The pressure exergy term is zero since the pressure of the source fluid is the same as that of the reference environment. Thus,

$$(\dot{X}_i)_{source} = \dot{m}_{source}c_{pg}T_0\left(\frac{T_i}{T_0} - 1 - \ln\frac{T_i}{T_0}\right)$$

$$= (58.7 \text{ kg/s})(1.1 \text{ kJ/kg} \cdot \text{K})(300 \text{ K})\left(\frac{2000 \text{ K}}{300 \text{ K}} - 1 - \ln\frac{2000}{300}\right)$$

$$= \underline{73 \text{ MW}}$$

(c) The exergy loss rate for the exhaust stream is calculated using the same equation as in part (b):

$$(\dot{X}_e)_{exhaust} = \dot{m}_{source}c_{pg}T_0\left(\frac{T_e}{T_0} - 1 - \ln\frac{T_e}{T_0}\right)$$

$$= (58.7 \text{ kg/s})(1.1 \text{ kJ/kg} \cdot \text{K})(300 \text{ K})\left(\frac{450 \text{ K}}{300 \text{ K}} - 1 - \ln\frac{450}{300}\right)$$

$$= \underline{1.83 \text{ MW}}$$

The exergy loss rate is only about 2.5% of the initial exergy of the source fluid.

(d) For the exergy consumption rate in the steam generation process, the steam generation process needs to be isolated, as illustrated in Figure 10.9(b). You can then apply the steady-flow availability equation (Eq. 9.9) once the input rate and the useful production rate have been ascertained. The exergy input rate may be determined as

$$\dot{X}_{input} = (\dot{X}_i)_{source} - (\dot{X}_e)_{exhaust} = 71.2 \text{ MW}$$

from the results obtained in parts (b) and (c).

The exergy production rate is calculated for the steam generation rate using Equation 9.33 with the kinetic energy and potential energy changes assumed negligible:

$$\dot{X}_{product} = \dot{m}_s[h_3 - h_2 - T_0(s_3 - s_2)]$$

$$= \frac{1.37 \times 10^5 \text{ kg/s}}{3600}[2801 - 172.8 - 300(6.068 - 0.576)] \text{ kJ/kg}$$

$$= \underline{37.3 \text{ MW}}$$

The h and s values for the water or steam are the same as those in Example 10.3. The values indicated were obtained using the **STEAM** code.

From Equation 9.9, the exergy consumption rate in the steam generation process can be determined:

$$(\dot{\chi}_{consumed})_{\substack{steam \\ generation}} = \dot{\chi}_{input} - \dot{\chi}_{product} = \underline{33.9 \text{ MW}}$$

The exergy consumption rate in the steam generation process is 46.4% of the total input rate from the heat source.

(e) Assume that the vapor condensation is achieved by rejecting heat to an infinite sink at T_0, as illustrated in Figure 10.9(c). In this case, the exergy input rate is equal to the decrease in the exergy flow rate of the wet steam at state 4 as it condenses to saturated water at state 1. This input rate can be calculated as

$$\dot{\chi}_{input} = \dot{m}_s[h_4 - h_1 - T_0(s_4 - s_1)]$$

$$= \frac{1.37 \times 10^5 \text{ kg/s}}{3600} [1890 - 169 - 300(6.068 - 0.576)] \text{ kJ/kg}$$

$$= 2.79 \text{ MW}$$

Since the heat rejection is to the reference environment, the exergy production rate must be deemed zero. (This is consistent with Equation 9.35 for the exergy flow rate associated with a heat flow rate to the reference environment with $T_{boundary} = T_0$.) Thus, applying Equation 9.9 gives

$$(\dot{\chi}_{consumed})_{\substack{vapor \\ condensation}} = \dot{\chi}_{input} - \dot{\chi}_{product} = \underline{2.79 \text{ MW}}$$

The exergy consumption rate for the vapor condensation process is thus $(2.79/73)100\% = 3.8\%$ of the total exergy input rate from the source fluid.

(f) From Example 10.3, the useful exergy production rate is the mechanical power output, which was found to be 34.5 MW. The exergy input rate from the finite heat source is 73 MW. Hence, applying Equation 9.10 gives the overall second law efficiency of the operation as

$$\eta_2 = \frac{\dot{\chi}_{product}}{\dot{\chi}_{input}} \cdot 100\% = \underline{47.3\%}$$

While this value may appear impressive, recall that almost the same percentage of the exergy input rate is consumed or destroyed in the vapor generation process due to the irreversibilities associated with the large ΔT values.

Figures 10.9(d) and (e) are the energy and exergy disposition diagrams. The energy disposition diagram gives the impression that the major loss of energy is in the vapor condensation process (59.5% of the energy input rate from the stream of hot gases). This energy rejection, however, occurs at a temperature close to the ambient temperature and therefore corresponds to a very low exergy value. The major exergy destruction due to irreversibilities is taking place in the steam generation process, as is clearly shown in the exergy disposition diagram. Thus, if a significant improvement in performance of the steam plant is to be achieved, steps must be taken to match the finite source temperatures more closely with those for the working fluid in the vapor generation process.

<table>
<tr><td>

EXERCISES

for Section
10.3.1

</td><td>

1. A steam power plant operates on the basic Rankine cycle, as illustrated in Figure 10.6(a). The condenser pressure and the boiler pressure are 7.5 kPa and 80 bar, respectively. Determine
   (a) w_{12}, q_{23}, w_{34}, and q_{41}
   (b) w and s.s.c
   (c) η_{ideal} and r_w
   What is the steam quality at the exit to the turbine?

   Answer: (b) 962 kJ/kg, 3.74 kg/kWh; (c) 37.3%, 0.992 ($x_4 = 0.673$)

2. Assume that the plant in Exercise 1 operates on an actual Rankine-like cycle, as illustrated in Figure 10.8. The isentropic efficiency for the expansion process 3→4 is 0.9. Determine
   (a) w_{34} and q_{41}
   (b) w and s.s.c
   (c) η_{actual} and ε
   What is the steam quality at the exit to the turbine? Is this acceptable?

   Answer: (b) 865 kJ/kg, 4.16 kg/kWh; (c) 33.5%, 0.898 ($x_4 = 0.714$)

3. Assume the heat inputs in Exercises 1 and 2 to be effectively from a thermal reservoir at 1800 K and that $T_0 = 300$ K. Determine the second law efficiency for
   (a) the ideal cycle
   (b) the actual cycle

   Answer: (a) 44.7%, (b) 40.2%

</td></tr>
</table>

10.3.2 Modifications to the Basic Rankine Cycle

The basic Rankine cycle can be improved in certain important respects, which are briefly reviewed here. This section discusses only the more traditional modifications, which include the Rankine cycle with superheat, Rankine cycle with reheat, Rankine cycle with regeneration, and the binary cycles using a combination of Rankine cycles. A more recent innovation based on second law considerations was introduced by Kalina (1984) and is named the Kalina cycle after its inventor. Successful demonstration of the cycle has been claimed (see Cape Range NL 1992, for example).

Rankine Cycle with Superheat

One principal constraint on the pressure and temperature conditions allowed in power plant design is a consequence of metallurgical considerations. It is found that a combination of high temperature and high pressure imposes severe strain on the metal parts of the plant's structure. The temperature limit imposed by such metallurgical considerations is referred to as the *metallurgical limit* and is typically in the region of 800 K

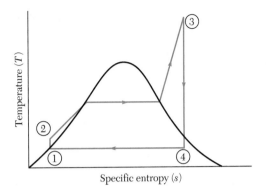

Figure 10.10 Rankine cycle with superheat.

(about 1000°F) for vapor power plants and 1000 K (about 1350°F) for gas turbines. For a steam plant, the highest steam temperature and pressure in the cycle should be under 800 K and about 150 bar (2175 psia), respectively. Therefore, a steam plant operating on the basic Rankine cycle uses steam at a maximum temperature of 615 K (647°F), corresponding to the saturation temperature of steam at 150 bar, which is significantly below the maximum allowable temperature of 800 K.

The basic Rankine cycle can be improved by adding a *superheater* to the boiler so that the steam entering the turbine is now at a temperature as high as possible without exceeding the metallurgical limit. The resulting cycle is illustrated in Figure 10.10 and is referred to as the *Rankine cycle with superheat*. This cycle expands the constant pressure heating process to the superheated state 3. An additional benefit is derived in operating on a cycle with superheating. As the boiler pressure is increased, the quality of steam at the turbine exit decreases when the cycle is the basic Rankine cycle. Wet steam causes significant erosion of the turbine blades, which are normally designed for a gas phase instead of the liquid droplets that are entrained in a stream of wet vapor. In typical steam turbine designs, the quality of steam leaving the turbine should not fall below about 0.9. Thus the operation of the cycle with superheating helps to avoid this erosion problem, which would be encountered with the basic Rankine cycle.

EXAMPLE 10.6
Rankine Cycle with Superheat

A steam power plant receives heat from a heat source at a rate of 100 MW, as in Example 10.3. The plant operates with a boiler pressure of 40 bar and a condenser pressure of 0.075 bar. In the present case, however, the plant is designed to operate ideally on the Rankine cycle with superheating to a temperature of 500°C. Calculate

(a) the cycle efficiency

(b) the work ratio for the cycle

(c) the power output (in MW) of the plant

(d) the required mass flow rate (in kg/h) of the working fluid

(e) the specific steam consumption (in kg/kWh)

Given: A steam plant operates on the ideal Rankine cycle with superheat, as illustrated in Figure 10.10.

$$\text{Boiler pressure} = 40 \text{ bar (4 MPa)}, T_3 = 773 \text{ K (500°C)}$$

$$\text{Condenser pressure} = 0.075 \text{ bar (7.5 kPa)}$$

$$\text{Heat input rate} = 100 \text{ MW}$$

Find: (a) η_{ideal}, (b) r_w, (c) $\dot{W}_x$ (in MW), (d) $\dot{m}_s$ (in kg/h), (e) s.s.c (in kg/kWh)

Solution: The cycle comprises flow processes and thus the SFEE (Eq. 6.29) can be applied for the first law analysis. Neglecting the kinetic energy and potential energy changes, the SFEE can be written as follows on a unit mass basis:

$$q - w = \Delta h$$

First, however, determine the values of the specific enthalpy h at each state using the **STEAM** code:

State Number	h (kJ/kg)	s (kJ/kg · K)
1	169	0.576
2	172.8	0.576
3	3445	7.09
4	2211	7.09

The values for q and w can now be obtained by applying the SFEE to each flow process in turn.

Process 1→2 (isentropic compression)

This process is reversible and adiabatic, and hence, $q_{12} = 0$. Also, $s_2 = s_1$. Applying the SFEE thus gives

$$w_{12} = h_1 - h_2 = -3.8 \text{ kJ/kg (as in Example 10.3)}$$

Process 2→3 (reversible constant pressure process)

Since the heating process takes place at a constant pressure and has internal reversibility, it is concluded from Equation 8.99 that the shaft work w_{23} is zero. Applying the SFEE thus leads to

$$q_{23} = h_3 - h_2 = 3272 \text{ kJ/kg}$$

Process 3→4 (isentropic expansion)

In this case, $q_{34} = 0$ (and $s_4 = s_3$). Applying the SFEE gives

$$w_{34} = h_3 - h_4 = 1234 \text{ kJ/kg}$$

The steam quality at state 4 is now approximately 0.85 compared with the value of 0.72 in Example 10.3 for the basic Rankine cycle. The improved value for the Rankine cycle with superheat is, however, slightly short of the desired design lower limit of about 0.9.

(a) The ideal cycle efficiency can be calculated from Equation 7.2:

$$\eta_{ideal} = \frac{w}{q_{in}} \cdot 100\% = \frac{(w_{12} + w_{34})}{q_{23}} \cdot 100\%$$

$$= \frac{1230 \text{ kJ/kg}}{3272 \text{ kJ/kg}} \cdot 100\% = \underline{37.6\%}$$

(b) The work ratio is determined using Equation 10.2:

$$r_w = \frac{w}{w_{+ve}} = \frac{(w_{12} + w_{34})}{w_{34}}$$

$$= \frac{1230 \text{ kJ/kg}}{1234 \text{ kJ/kg}} = \underline{0.997}$$

(c) The mechanical power output can be calculated using Equation 7.2 and substituting the known values of heat input rate and the ideal cycle efficiency:

$$\dot{W}_x = \dot{Q}_H \frac{\eta_{ideal}(\%)}{100} = \underline{37.6 \text{ MW}}$$

(d) The mass flow rate of the working fluid can be determined from the relationship between either the total heat input rate and q_{23} or the mechanical power output and w. The first of these gives

$$\dot{m}_s = \frac{\dot{Q}_H}{q_{23}} = \frac{10^5 \text{ kJ/s}}{3272 \text{ kJ/kg}} \left(\frac{3600 \text{ s}}{1 \text{ h}}\right) = \underline{1.10 \times 10^5 \text{ kg/h}}$$

(e) From the definition for the specific steam consumption given in the solution to Example 10.3, the following result is obtained:

$$s.s.c \equiv \frac{\dot{m}_s \text{ (kg/h)}}{\dot{W}_x \text{ (kW)}} \equiv \frac{\dot{m}_s \text{ (kg/s)}}{\dot{W}_x \text{ (kW)}} \left(\frac{3600 \text{ s}}{1 \text{ h}}\right)$$

$$= \frac{3600}{w} \text{ kg/kWh} = \frac{3600}{1230} \text{ kg/kWh}$$

$$= \underline{2.93 \text{ kg/kWh}}$$

The value of 2.93 kg/kWh obtained for the Rankine cycle with superheat is about 74% of that for the basic Rankine cycle of Example 10.3. The design incorporating a superheater is thus expected to lead to a more compact or smaller unit.

A comparison is given in Table 10.2 between the performance characteristics for the basic Rankine cycle and those for the Rankine cycle with superheat. The results show a definite superiority of the Rankine cycle with superheat over the basic Rankine cycle.

TABLE 10.2

Comparison of the Performance of a Basic Rankine Cycle with a Rankine Cycle with Superheat			
Cycle Parameter	Basic Rankine (Example 10.3)	Rankine with Superheat (Example 10.6)	Percentage Increase (%)
(a) η_{cycle} (%)	34.5	37.6	+9
(b) r_w	0.996	0.997	+0.1
(c) $\dot{W}_x$ (in MW)	34.5	37.6	+9
(d) $\dot{m}_s$ (kg/h)	1.37×10^5	1.10×10^5	−19.7
(e) s.s.c. (kg/kWh)	3.97	2.93	−26.2

In the next example, another comparison is made between the two cycles, this time on the basis of the second law of thermodynamics.

EXAMPLE 10.7
Comparative Second Law Evaluation of the Basic Rankine Cycle and the Rankine Cycle with Superheat

Compare the steam power plants described in Examples 10.3 (basic Rankine cycle) and 10.6 (Rankine cycle with superheat) on a second law basis and give your opinion as to which of the two options is the better choice.

Given:
Basic Rankine cycle of Example 10.3
Rankine cycle with superheat of Example 10.6

Find:
(a) Compare the cycles on a second law basis
(b) Which option is better, based on part (a)?

Solution:
(a) The analysis on the basis of the second law of thermodynamics can be carried out in terms of exergy. This was done in Example 10.5 for the basic Rankine cycle of Example 10.3, and the results were illustrated in Figure 10.9(e). If a finite heat source identical to the one in Example 10.5 is used for the Rankine cycle with superheat, you can apply the following results to the two cycles under consideration:

$$(\dot{\chi}_i)_{source} = \dot{m}_{source} c_{pg} T_0 \left(\frac{T_i}{T_0} - 1 - \ln \frac{T_i}{T_0} \right) = \underline{73 \text{ MW}}$$

$$(\dot{\chi}_e)_{exhaust} = \dot{m}_{source} c_{pg} T_0 \left(\frac{T_e}{T_0} - 1 - \ln \frac{T_e}{T_0} \right) = \underline{1.83 \text{ MW}}$$

Rankine cycle with superheat

The exergy consumption rates associated with the steam generation and the vapor condensation processes can be determined for the Rankine cycle with superheat in the same way as in Example 10.5 for the basic Rankine cycle. Thus, for the steam generation process, the exergy input rate and the useful exergy production rate are given by the following:

$$\dot{X}_{input} = (\dot{X}_i)_{source} - (\dot{X}_e)_{exhaust} = 71.2 \text{ MW}$$

$$\dot{X}_{product} = \dot{m}_s[h_3 - h_2 - T_0(s_3 - s_2)]$$

$$= \frac{1.10 \times 10^5 \text{ kg/s}}{3600} [3445 - 172.8 - 300(7.090 - 0.576)] \text{ kJ/kg}$$

$$= 40.3 \text{ MW}$$

Values of h and s are taken from the solution to Example 10.6. The exergy consumption rate in the steam generation process is thus

$$(\dot{X}_{consumed})_{\substack{steam \\ generation}} = \dot{X}_{input} - \dot{X}_{product} = \underline{30.9 \text{ MW}}$$

This consumption rate amounts to 42.3% of the 73 MW exergy input rate of the source stream of hot gases.

For the vapor condensation process, the exergy consumption rate is equal to the exergy input rate since the production rate associated with the heat rejection to the ambient environment is zero. For the Rankine cycle with superheat, the exergy input rate is

$$\dot{X}_{input} = \dot{m}_s[h_4 - h_1 - T_0(s_4 - s_1)]$$

$$= \frac{1.10 \times 10^5 \text{ kg/s}}{3600} [2211 - 169 - 300(7.090 - 0.576)] \text{ kJ/kg}$$

$$= 2.68 \text{ MW}$$

Applying Equation 9.9 gives the exergy consumption rate as

$$(\dot{X}_{consumed})_{\substack{vapor \\ condensation}} = \dot{X}_{input} - \dot{X}_{product} = \underline{2.68 \text{ MW}}$$

This is a mere 3.7% of the 73 MW exergy flow rate of the source stream.

The useful exergy production rate is the mechanical power output of 37.6 MW for the cycle that was found in the solution to Example 10.6. Thus, the overall second law efficiency achieved by applying Equation 9.10 is $(37.6/73)(100\%) = 51.5\%$.

The exergy disposition for the Rankine cycle with superheat is illustrated in Figure 10.11 and can be compared with Figure 10.9(e) for the basic Rankine cycle. In addition, Table 10.3 provides a summary of the results obtained for the second law analysis of the two cycles.

(b) The summary in Table 10.3 shows that the Rankine cycle with superheat is the better choice from a thermodynamic viewpoint. In both cycles, however, very sig-

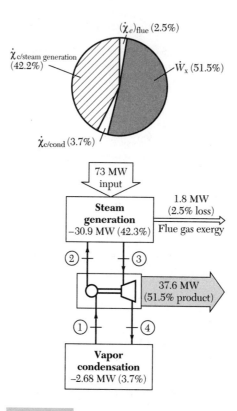

Figure 10.11 Exergy disposition for the Rankine cycle with superheat.

TABLE 10.3

TABLE 10.3

Summary of Exergies Computed for the Basic Rankine Cycle of Example 10.3 and the Rankine Cycle with Superheat of Example 10.6

Cycle Parameter	Basic Rankine Cycle (Example 10.3)	Rankine Cycle with Superheat (Example 10.6)
$(\dot{\chi}_{input})_{source}$ (MW)	73	73
Useful power output $\dot{W}_x$ (MW)	34.5	37.6
Exergy loss rate $(\dot{\chi}_e)_{flue\ gas}$ (MW)	1.83	1.83
Exergy consumption rates $(\dot{\chi}_{consumed})_{steam\ generation}$ (MW)	33.8	30.9
$(\dot{\chi}_{consumed})_{vapor\ condensation}$ (MW)	2.79	2.68
$(\eta_2)_{plant}$ (%)	47.3%	51.5

nificant irreversibility is encountered in the steam generation process. By super-heating the steam, an approximate 9% reduction occurs in the exergy consumption rate associated with the steam generation process.

Reheat Cycles

Whenever steam expansion takes place from a typical high (boiler) pressure (on the order of 80 bar to 150 bar) to the very low (condenser) pressure (on the order of 0.1 bar) in a single turbine stage, the steam at the exit to the turbine is likely to be rather wet regardless of whether the cycle is the basic Rankine cycle or the Rankine cycle with superheat. If the steam quality is not high and is not sufficiently close to unity, the water droplets present will in time cause serious physical erosion of the turbine blades. (In practice, the steam quality should not be lower than 0.9 at the exit to the turbine.) To safeguard against this, a multiple stage turbine design is used that reheats the steam between successive turbine stages. The steam pressure at the exit to each turbine stage is such that the steam quality does not fall below the lower limit of about 0.9. Figure 10.12 is a component schematic diagram of the steam plant operating on the Rankine cycle with reheat.

Figure 10.13 illustrates a typical Rankine cycle with reheat that uses two turbine stages: a high pressure (or HP) turbine stage and a low pressure (or LP) turbine stage. This modification results also in a reduced level of thermal irreversibility in the heating processes as compared with the basic Rankine cycle and, therefore, an improved performance efficiency. Example 10.8, presented later, provides an illustration of how to analyze the Rankine cycle with reheat.

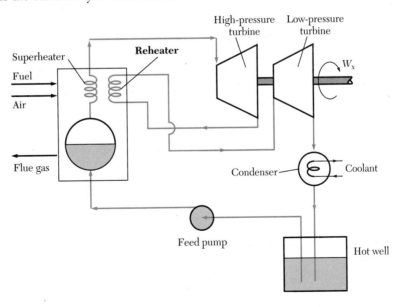

Figure 10.12 Schematic diagram of the steam plant operating on the Rankine cycle with reheat.

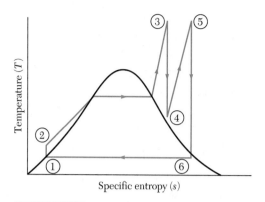

Figure 10.13 Typical Rankine cycle with reheat.

Regenerative Cycles

The purpose of a regenerative cycle is to approximate the Carnot cycle by having the heating provided by an external heat source at a constant high source temperature. Figure 10.14 shows a basic Rankine cycle modified to a regenerative cycle. Figure 10.15 is the corresponding component diagram. The regeneration is said to be complete if the water heating required from state 2 to 3 is achieved entirely from the steam that is expanding (from state 4 to 5) through the turbine. The only heating required from an external source is that for the phase change process 3→4. Thus, this is a reversible engine that only exchanges heat with two reservoirs at constant temperatures of T_3 and T_1. Therefore, the thermal efficiency must be the same as that for a Carnot cycle operating with a source temperature T_3 and a sink temperature T_1.

The Carnot cycle using steam requires the compression of a very wet vapor. It is difficult to design the cycle to handle this wet vapor efficiently. Also, the work ratio for the Carnot cycle is low because of the substantial negative work required for vapor compression. In contrast, the requirement in the regenerative cycle is for compression

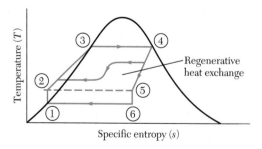

Figure 10.14 Basic Rankine cycle with regeneration.

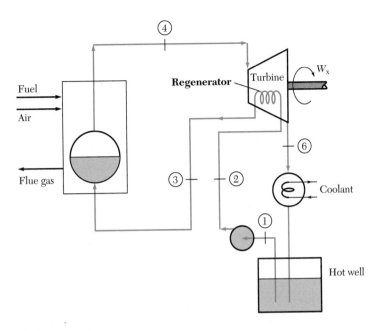

Figure 10.15 Component diagram for the steam plant operating on the basic Rankine cycle with regeneration.

of a liquid, and the negligibly small negative work involved ensures a high work ratio for the cycle, even though the positive turbine work output is reduced relative to that of the Carnot cycle. The net reduction of turbine work output in the regenerative cycle when compared to the Carnot cycle is equal to the energy needed for the regenerative heating process (2→3) in the regenerative cycle.

From a second law perspective, both the regenerative cycle and the Carnot cycle are appropriate when the heat source is at a constant temperature. They are both inappropriate, however, when the heat source is finite, for reasons which have been repeatedly discussed. Example 10.8 illustrates the analysis of the basic Rankine cycle with regeneration.

Binary Cycles

Binary cycles are used basically to achieve the best possible match between the characteristics of the available working fluids and those of the heat source and sink. Figure 10.16 illustrates how this is accomplished by having binary (or two) vapor power cycles such that the heat rejection process from one serves as part of the heat input process for the other.

Mercury is a popular choice of working fluid for the vapor power cycle operating at a very high temperature level. The saturation temperature of mercury for a typical operating boiler pressure of 75 bar is about 1025 K. For condensation at 0.1 bar, the

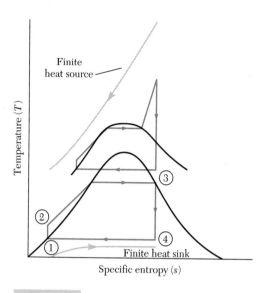

Figure 10.16 Illustrative sketch of the binary cycle.

corresponding condensation temperature is about 525 K. Thus, the vapor power plant using mercury as a working fluid could operate on a Rankine cycle with superheat while rejecting heat to a steam power cycle operating at a relatively lower temperature regime.

EXAMPLE 10.8
Comparative First and Second Law Analysis of Alternative Rankine Cycles for a Steam Power Plant

A steam power plant is to operate with a boiler pressure of 80 bar and a condenser pressure of 0.1 bar. The heat source is the type illustrated in Figure 10.9(a) and comprises a stream of exhaust gases from a gas turbine discharging at 560°C and 1 atm pressure. The minimum temperature allowed for the exhaust gas stream is 450 K (177°C), while the maximum steam temperature is 500°C. The mass flow rate of the stream of hot gases is such that the heat input rate to each steam cycle is 100 MW. The ambient temperature (T_0) and pressure (p_0) can be taken as 300 K and 1 atm, respectively. Determine η_1, r_w, and η_2 achieved with each of the following cycles:

(a) basic Rankine cycle

(b) Rankine cycle with superheat

(c) Rankine cycle with reheat (such that steam expands in the high pressure turbine until it exits as dry, saturated steam)

(d) basic Rankine cycle with complete regeneration, with the exit temperature of the exhaust gas stream taken as 320°C because the saturation temperature of steam at

80 bar is close to 300°C and the exhaust gas must exit at a temperature high enough to provide heat input for the phase change of the working fluid at 300°C.

Given:

Finite heat source (of the type in Fig. 10.9a)

$$T_i = 833 \text{ K } (500°\text{C}), p_i = 1 \text{ atm}$$

$$T_e = 450 \text{ K } (177°\text{C}), \text{ except for part (d), for which } T_e = 593 \text{ K}$$

$$p_e = 1 \text{ atm} = p_0, T_0 = 300 \text{ K}$$

Heat input rate to the cycle = 100 MW

Steam cycles

$$\text{Boiler pressure} = 80 \text{ bar}$$

$$\text{Condenser pressure} = 0.1 \text{ bar for all cycles}$$

$$\text{Maximum steam temperature allowed} = 500°\text{C}$$

Find: η_1, r_w, and η_2 for each of the following cycles:

(a) basic Rankine cycle (Fig. 10.17a)

(b) Rankine cycle with superheat (Fig. 10.17b)

(c) Rankine cycle with reheat (Fig. 10.17c)

(d) basic Rankine cycle with complete regeneration (Fig. 10.17d)

Solution: For a first law analysis of each cycle, knowledge of the h values at each of the states indicated in Figures 10.17(a) to (d) is required. The SFEE (Eq. 6.29) can then be applied to determine the work and heat transfers per unit mass of the working fluid, and hence, the cycle efficiency and the work ratio for the cycle. For the second law efficiency, the exergy input rate corresponds to the exergy of the source stream at the initial condition of 560°C and 1 atm, while the useful exergy production rate is the mechanical power output of each steam cycle.

(a) *Basic Rankine cycle*

The steam cycle is illustrated in Figure 10.17(a). Using the **STEAM** code, the following values are obtained for the h values at states 1 to 4:

State Number	1	2	3	4
h (kJ/kg)	192	200	2758	1817

Applying the reduced version of the SFEE, as done in Example 10.3, yields the following results for the w values and q_{23}:

$$w_{12} = h_1 - h_2 = -8 \text{ kJ/kg}$$

$$w_{34} = h_3 - h_4 = 941 \text{ kJ/kg}$$

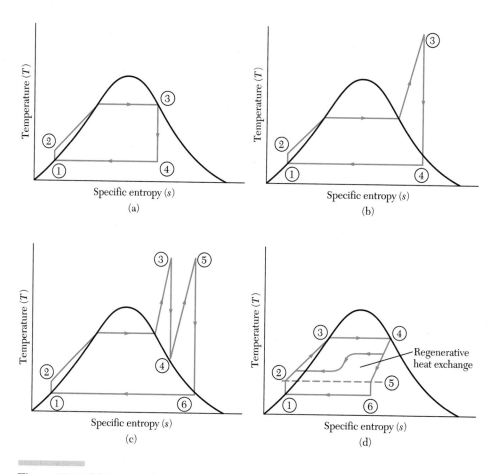

Figure 10.17 (a) Basic Rankine cycle. (b) Rankine cycle with superheat. (c) Rankine cycle with reheat. (d) Basic Rankine cycle with complete regeneration.

$$\therefore w = w_{12} + w_{34} = 933 \text{ kJ/kg}$$

$$q_{23} = h_3 - h_2 = 2558 \text{ kJ/kg}$$

Thus, Equations 7.2 and 10.2 give

$$\eta_1 = \frac{w}{q_{23}} \cdot 100\% = \underline{36.5\%}$$

and

$$r_w = \frac{w}{w_{+ve}} = \frac{w}{w_{34}} = \underline{0.991}$$

The steam quality at the exit to the turbine in this example is 0.679, and on this count alone, the cycle will be unattractive as a basis for an actual steam plant design.

For the second law efficiency achieved, one needs to determine the mechanical power output and the exergy input rate to the entire operation of the plant. Since the heat input rate to the steam cycle is given, the calculated cycle efficiency can be used with Equation 7.2 to determine the power output of the engine:

$$\dot{W}_x = \dot{Q}_H \frac{\eta_1 \, (\%)}{100} = 36.5 \text{ MW}$$

The exergy input rate of the hot gas stream is calculated using Equation 9.40a, assuming that the gas can be considered as an ideal gas and noting that the pressure exergy is zero since $p_i = p_0$. Thus, the following expression for the exergy supply rate to the plant is obtained:

$$(\dot{X}_{input})_{source} = \dot{m}_{source} c_{pg} \left(T_i - T_0 - T_0 \ln \frac{T_i}{T_0} \right)$$

$$= 227(\dot{m}_{source} c_{pg}) \text{ kW}$$

The numeric coefficient of 227 in this expression was obtained by putting $T_i = 833$ K and $T_0 = 300$ K. An expression for the product of mass flow rate and the specific heat at constant pressure for the source stream can be derived by applying the SFEE (Eq. 6.29) to the control volume through which the stream passes to produce heating of the water or steam in the vapor cycle. The result can be written as follows:

$$(\dot{m}_{source} c_{pg}) = \frac{\dot{Q}_H \, (kW)}{(T_i - T_e) \text{ K}}$$

where T_i and T_e have the same values of 833 K and 450 K, respectively, for all the cycles (a) to (c). Substituting for these in the above expressions leads to an exergy supply rate of 59.3 MW for these cases.

Hence, for the basic Rankine cycle, the second law efficiency is $(36.5/59.3)100\% = 61.6\%$.

(b) *Rankine cycle with superheat*

The steam cycle is illustrated in Figure 10.17(b). Using the **STEAM** code, the following values are obtained for the h values at states 1 to 4:

State Number	1	2	3	4
h (kJ/kg)	192	200	3398	2130

Applying the reduced version of the SFEE as done in part (a) above yields the following results for the w values and q_{23}:

$$w_{12} = h_1 - h_2 = -8 \text{ kJ/kg}$$

$$w_{34} = h_3 - h_4 = 1268 \text{ kJ/kg}$$

$$\therefore w = w_{12} + w_{34} = 1260 \text{ kJ/kg}$$

$$q_{23} = h_3 - h_2 = 3198 \text{ kJ/kg}$$

Thus, Equations 7.2 and 10.2 give

$$\eta_1 = \frac{w}{q_{23}} \cdot 100\% = \underline{39.4\%}$$

and

$$r_w = \frac{w}{w_{+ve}} = \frac{w}{w_{34}} = \underline{0.994}$$

The steam quality at the exit to the turbine in this example is 0.81, which is higher than it was for the basic Rankine cycle but still not high enough for an actual engine design.

For the second law efficiency, Equation 7.2 is used with the specified heat input rate and the calculated cycle efficiency to determine the mechanical power output, which in this case is 39.4 MW. The exergy supply rate is the same as that in part (a) and is 59.3 MW. Hence, the second law efficiency for the Rankine cycle with superheat is $(39.4/59.3)100\% = \underline{66.4\%}$.

Improvements in both the first law and second law efficiencies are achieved with superheating. The work output per unit mass of the working fluid is also increased. Therefore, conventional vapor power plants are almost always operated on the Rankine cycle with at least some superheat.

(c) *Rankine cycle with reheat*

The steam cycle is illustrated in Figure 10.17(c). In this case, the h values for six states need to be determined. The **STEAM** code gives the following values:

State Number	1	2	3	4	5	6
h (kJ/kg)	192	200	3398	2761	3482	2522

Determination of state 4 was accomplished with the code using a trial-and-error method since neither the pressure nor the temperature was known in advance. For $p_4 = 6.67$ bar and $x_4 = 1$, it is found that the specific entropy $s_4 = s_3$ for the isentropic expansion through the high pressure turbine stage. Applying the reduced version of the SFEE, as done in the earlier cases, yields the following results for the w values and q_{input}:

$$w_{12} = h_1 - h_2 = -8 \text{ kJ/kg}$$

$$w_{34} = h_3 - h_4 = 637 \text{ kJ/kg}$$

$$w_{56} = h_5 - h_6 = 960 \text{ kJ/kg}$$

$$w_{+ve} = w_{34} + w_{56} = 1597 \text{ kJ/kg}$$

$$\therefore w = w_{-ve} + w_{+ve} = w_{12} + w_{+ve} = 1589 \text{ kJ/kg}$$

$$q_{23} = h_3 - h_2 = 3198 \text{ kJ/kg}$$

$$q_{45} = h_5 - h_4 = 721 \text{ kJ/kg}$$

$$\therefore q_{input} = q_{23} + q_{45} = 3919 \text{ kJ/kg}$$

Thus, Equations 7.2 and 10.2 give

$$\eta_1 = \frac{w}{q_{\text{in}}} \cdot 100\% = \underline{40.5\%}$$

and

$$r_{\text{w}} = \frac{w}{w_{+\text{ve}}} = \underline{0.995}$$

The steam quality at the exit to the low pressure turbine in this example is 0.974, which is high and thus acceptable for an actual design.

The mechanical output is now 40.5 MW, while the exergy input rate is still 59.3 MW. Thus, the second law efficiency for the Rankine cycle with reheat is $(40.5/59.3)100\% = \underline{68.3\%}$. Compared with the basic Rankine cycle, the second law efficiency for the Rankine cycle with reheat shows an increase of about 11%. It should thus not be a surprise to learn that most of the large conventional steam power plants in use today operate on the Rankine cycle with reheat.

(d) *Basic Rankine cycle with complete regeneration*

This cycle is illustrated in Figure 10.17(d). Since regeneration is complete, the cycle is effectively a Carnot cycle with (external) source heating at $T_3 \, (= T_4 = 568 \text{ K})$ and heat rejection effectively to a sink at $T_6 \, (= T_1 = 318.8 \text{ K})$. Temperatures T_3 and T_6 correspond to the saturation temperatures at the boiler pressure (80 bar) and the condenser pressure (0.1 bar), respectively. Applying Equation 7.21 for the Carnot cycle efficiency leads to the following for the cycle efficiency in the present case:

$$\eta_1 = \left(1 - \frac{T_1}{T_3}\right) \cdot 100\% = \underline{43.9\%}$$

The external heat input per unit mass of the working fluid is now given by

$$q_{34} = h_4 - h_3 = 2758 - 1316 = 1442 \text{ kJ/kg}$$

Therefore, applying Equation 7.2 and using the known values of cycle efficiency and the heat input per unit mass yields the following for the net work output per unit mass of the working fluid:

$$w = q_{\text{in}} \frac{\eta_1 \, (\%)}{100} = 633 \text{ kJ/kg}$$

The negative work is the work input to the feed pump for the isentropic process 1→2 and is given by

$$w_{-\text{ve}} = w_{12} = h_1 - h_2 = -8 \text{ kJ/kg}$$

as before. The positive work done can thus be calculated from the following:

$$w_{+\text{ve}} = w - w_{-\text{ve}} = 641 \text{ kJ/kg}$$

Hence, the work ratio r_{w} is $(633/641) = \underline{0.988}$.

We mentioned previously that the work ratio is typically low for the Carnot cycle. The Rankine cycle with regeneration provides the high efficiency of the Carnot cycle without the disadvantage of a low work ratio.

For the second law efficiency, the mechanical power output is obtained by using the calculated cycle efficiency together with the specified heat input rate of 100 MW. The power output is thus 43.9 MW. The exergy input rate uses a different value for T_e, which is given as 593 K, in the expressions derived earlier for the basic Rankine cycle. Thus, an exergy input rate of 94.6 MW is obtained. The second law efficiency is thus $(43.9/94.6)100\% = \underline{46.4\%}$ for the basic Rankine cycle with complete regeneration. The second law efficiency is lower for regeneration because of the more substantial loss of exergy carried by the effluent or flue gas at a T_e of 593 K. This loss is approximately 39% of the exergy input rate to the plant. However, if the plant were operated using a *regeneratable heat source* (such as the coolant of a nuclear reactor), the exergy input to the plant would be 57.7 MW instead of the 94.6 MW assumed for the finite heat source. The corresponding second law efficiency is thus about 76% when the heat source is regeneratable.

Further discussion of other possible modifications to the vapor power cycles is deferred to Section 10.6.

EXERCISES

for Section 10.3.2

1. A steam plant operates on the Rankine cycle with superheat, as illustrated in Figure 10.10. The condenser and boiler pressures are 0.075 bar and 90 bar, respectively. The temperature (at state 3) of the superheated steam is 600°C. Determine
   (a) w and s.s.c.
   (b) η_1 and r_w
   What is the steam quality at the exit to the turbine?
   Answer: (a) 1455 kJ/kg, 2.47 kg/kWh; (b) 42.1%, 0.994 ($x_4 = 0.832$)

2. A steam plant operates on the Rankine cycle with reheat as illustrated in Figure 10.13. Assume $T_3 = T_5 = 600°C$, $x_6 = 0.9$, $p_1 = p_6 = 0.075$ bar, and $p_3 = p_2 = 90$ bar. Determine
   (a) w and s.s.c.
   (b) η_1
   Answer: (a) 1700 kJ/kg, 2.12 kg/kWh; (b) 44%

10.4   Gas (Turbine) Power Cycles

Gas engines are designed most frequently as *internal combustion (IC) engines*. Internal combustion engines use air as the working fluid. This air also serves as the oxidant for the hydrocarbon fuels commonly used. The *combustion* of fuel results first in the transformation of the chemical exergy of the fuel to a slightly reduced quantity of thermal exergy, which is then converted by the engine to a work output. The loss of exergy that occurs in the combustion process is due to irreversibilities associated with uncontrolled chemical reactions and is discussed in a comprehensive manner in Chapter 14. The standard operation of internal combustion engines is in either the *pressure-limited* mode (that is, between a minimum pressure and a maximum pressure limit) or the *volume-limited* mode (between a maximum and a minimum volume in a cylinder). *Gas turbines* are typical of the former, while *reciprocating internal combustion (IC) engines* (discussed in Section 10.5) are designed to operate in the volume limited mode.

The ideal cycle for a gas turbine is the Brayton (or Joule) cycle, which is considered here first. For reciprocating IC engines, the ideal or air standard cycles are the Otto cycle (for gasoline engines) and the Diesel cycle (for diesel engines). The ideal cycles use only air as the working fluid. The combustion process that occurs in the actual heat engine is represented in the ideal cycle as an equivalent heating process either at constant pressure (for the gas turbine cycle and the Diesel cycle) or at constant volume (for the gasoline engine).

Actual internal combustion engines operate on an *open cycle* in view of the fact that the exhaust stream, comprising the products of a combustion process, is unable to support fuel combustion in subsequent engine cycles. The ideal cycles, in contrast, assume a *closed-cycle operation*. The jettisoning to the atmosphere of the exhaust gases that occurs in the actual case is approximated in the corresponding air standard cycle by a heat rejection process that completes a closed-cycle sequence.

In contrast to vapor power plants, gas power cycles have low work ratios because the compression work requirement is high. In fact, the earliest efforts to develop gas turbines were frustrated largely because the power requirement for operating a poorly designed compressor easily exceeded the actual power output of the turbine used. As a result, it was impossible to obtain even a self-sustained operation for the gas turbine. As indicated previously, a low work ratio means that the actual performance of the heat engine is significantly worse than that estimated for the ideal cycle.

Unlike the case of vapor power plants (for which tables of thermodynamic properties for the working fluid are required), the thermodynamic analysis of gas power cycles can be made using the various relationships established in previous chapters for an ideal gas. Specifically, several of the equations for an ideal gas from Chapters 4 though 9 will be used.

10.4.1   Simple Gas Turbine

The simple gas turbine operates on an open cycle, as illustrated in Figure 10.18. The cycle proceeds from the *intake* of ambient air at state 1, through the *compression* stage (1→2), to the *combustion* of fuel in the compressed air stream (2→3), followed by a positive work output stage (3→4) as the hot gases *expand* through the turbine to the atmosphere. The exhaust gas stream is usually at a significantly higher temperature than that of ambient air, thus resulting in a substantial loss of thermal exergy unless a use can be found for the exhaust gas stream.

Air Standard Brayton Cycle

The *T-s* diagram for the corresponding air standard cycle is shown in Figure 10.19. This ideal cycle, known as the *Brayton* (or *Joule*) *cycle*, comprises the following four processes:

a. reversible, adiabatic compression (1→2)

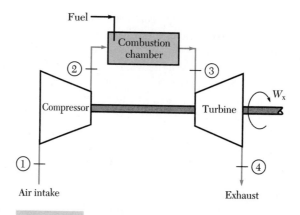

Figure 10.18 Schematic drawing of the simple gas turbine (open cycle).

b. reversible, constant pressure heating (2→3)

c. reversible, adiabatic expansion (3→4)

d. reversible, constant pressure cooling (4→1)

General equations are derived in the remainder of this section for the work done and for various performance indices both for the ideal cycle and for deviations from the ideal case. Since the cycle comprises flow processes, a first law analysis is carried out by applying the SFEE (Eq. 6.29). It is usual to assume that changes in kinetic and potential energy are negligible. Thus, for a control volume, if q denotes the heat flow per unit mass of the working fluid and w the (shaft) work done, the SFEE reduces to

$$q - w = \Delta h$$

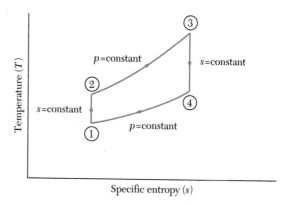

Figure 10.19 Idealized or air standard Brayton cycle for the simple gas turbine operation.

Now consider each of the processes comprising the Brayton cycle in turn.

Process 1→2 (reversible and adiabatic compression)

Since the process is adiabatic, $q_{12} = 0$. Applying the SFEE thus gives

$$w_{12} = h_1 - h_2$$
$$= c_{po}(T_1 - T_2) \tag{10.4a}$$

The second step in this equation is a result of applying Equation 4.31 on the assumption that the air may be treated as a perfect gas with constant specific heats.

For a reversible and adiabatic process involving a perfect gas, Equation 8.31 gives the relationship between the temperature ratio and the pressure ratio. The *pressure ratio* (r_p) is defined for a gas turbine cycle as the ratio of the higher pressure (p_2) to the lower pressure (p_1). Thus, T_2 can be expressed in terms of T_1 and r_p from Equation 8.31 and can be substituted in Equation 10.4a to obtain the following equation for w_{12}:

$$w_{12} = c_{po}(T_1 - T_2)$$
$$= c_{po}T_1 \left(1 - \frac{T_2}{T_1} \right) = c_{po}T_1(1 - r_p^{(k-1)/k}) \tag{10.4b}$$

It should be noted that while Equation 10.4a applies to both reversible and nonreversible adiabatic compression of a perfect gas, Equation 10.4b applies only to the case of isentropic compression.

Process 2→3 (reversible constant pressure process)

In the ideal case, the heating process occurs at a constant pressure and with internal reversibility. From Equation 8.99, one can conclude that the shaft work done is zero. Substituting $w_{23} = 0$ in the SFEE leads to the following equation for the heat input rate per unit mass flow rate of the working fluid:

$$q_{23} = h_3 - h_2$$
$$= c_{po}(T_3 - T_2) = c_{po}(T_3 - T_1 r_p^{(k-1)/k}) \tag{10.5}$$

Process 3→4 (reversible and adiabatic expansion)

The process is adiabatic and hence $q_{34} = 0$. The SFEE thus reduces to

$$w_{34} = h_3 - h_4$$
$$= c_{po}(T_3 - T_4) \tag{10.6a}$$

Equation 10.6a applies to both reversible and the nonreversible adiabatic expansion through the turbine. For isentropic expansion through the turbine, however, Equation 8.31 relates the temperature ratio to the pressure ratio. Thus, the following expression can be derived for the ideal process:

$$w_{34} = c_{po}(T_3 - T_4)$$
$$= c_{po}T_3 \left(1 - \frac{T_4}{T_3} \right) = c_{po}T_3 \left(1 - \frac{1}{r_p^{(k-1)/k}} \right) \tag{10.6b}$$

It is now possible to obtain general expressions for the net specific work output (w), the cycle efficiency (η_1), and the work ratio (r_w) for the ideal cycle.

Ideal net work output per unit mass

$$
\begin{aligned}
w &= w_{12} + w_{34} \\
&= c_{po}T_1(1 - r_p^{(k-1)/k}) + c_{po}T_3\left(1 - \frac{1}{r_p^{(k-1)/k}}\right) \\
&= c_{po}\left(1 - \frac{1}{r_p^{(k-1)/k}}\right)(T_3 - T_1 r_p^{(k-1)/k})
\end{aligned}
\tag{10.7}
$$

Ideal cycle work ratio

$$
\begin{aligned}
r_w &= \frac{w}{w_{34}} = \frac{w_{12}}{w_{34}} + 1 \\
&= 1 - \frac{T_1}{T_3} r_p^{(k-1)/k}
\end{aligned}
\tag{10.8}
$$

Cycle efficiency

$$
\begin{aligned}
\eta_1 &= \frac{w}{q_{23}} \cdot 100\% \\
&= \left(1 - \frac{1}{r_p^{(k-1)/k}}\right) \cdot 100\%
\end{aligned}
\tag{10.9}
$$

EXAMPLE 10.9
Ideal Brayton Cycle Analysis

A gas turbine uses air drawn in at 1 atm and 300 K. The air is compressed to 6 atm, and the maximum cycle temperature is limited to 1100 K by using a large air to fuel ratio. Assume that the gas turbine operates on the ideal Brayton cycle and that the heating or combustion process is equivalent to an energy input of 100 MW to the air at 6 atm pressure. Ambient condition is given by $p_0 = 1$ atm and $T_0 = 300$ K. For the ideal cycle, determine

(a) thermal efficiency

(b) work ratio

(c) power output (in MW)

(d) exergy flow rate (in MW) of the exhaust gas leaving the turbine

Given: The ideal Brayton cycle such as that shown in Figure 10.19 with

$$T_1 = 300 \text{ K } (= T_0), \ p_1 = 1 \text{ atm } (= p_4 = p_0)$$

$$T_3 = 1100 \text{ K}, \quad p_3 = 6 \text{ atm} \ (= p_2)$$

Heat input rate $= 100$ MW

Working fluid is air

Find: **(a)** η_1, **(b)** r_w, **(c)** $\dot{W}_x$ (in MW), **(d)** $\dot{\chi}_{\text{exhaust}}$ (in MW)

Solution: The problem can be solved either from first principles or by using Equations 10.4a to 10.9, which were derived for the Brayton cycle. When working from first principles, it is usually good practice to leave expressions for work and heat transfers in terms of temperatures at states 1 to 4. The temperatures can be evaluated using the appropriate equation of state and process path equations for the ideal gas. Substituting these temperatures in the expressions for work and heat transfers will thus yield the desired solution.

(a) The cycle efficiency can be determined using Equation 10.9. The pressure ratio r_p in this case is 6. For air, $k = 1.4$ can be assumed. Hence,

$$\eta_1 = \left(1 - \frac{1}{r_p^{(k-1)/k}} \right) \cdot 100\%$$

$$= \left(1 - \frac{1}{6^{0.4/1.4}} \right) \cdot 100\% = \underline{40.1\%}$$

(b) The work ratio can be calculated from Equation 10.8.

$$r_w = 1 - \frac{T_1}{T_3} r_p^{(k-1)/k}$$

$$= 1 - \frac{300 \text{ K}}{1100 \text{ K}} (6^{0.4/1.4}) = \underline{0.545}$$

Note that the temperatures T_1 and T_3 must be absolute temperatures.

(c) Once the thermal (or cycle) efficiency is known, you can use Equation 7.2 to calculate the mechanical power output for the heat input rate specified:

$$\dot{W}_x = \frac{\eta_1 \, (\%)}{100} (\dot{Q}_H) = 0.401(100 \text{ MW}) = \underline{40.1 \text{ MW}}$$

(d) The exergy flow rate of the exhaust gas stream can be determined using Equation 9.38 with Equations 9.40a and 9.40c for the thermal and pressure exergy of an ideal gas stream. In this case, $p_4 = p_0$ and therefore the pressure exergy is zero. If the kinetic energy and potential energy terms in Equation 9.38 are neglected, the exergy flow rate of the exhaust gas stream can be determined from Equation 9.40a for the thermal exergy component. First, however, both T_4 and the mass flow rate of the working fluid must be calculated, both of which are included in the expression for the exergy flow rate. T_4 can be calculated using Equation 8.31 for the isentropic process 3→4:

$$T_4 = \frac{T_3}{r_p^{(k-1)/k}} = \frac{1100 \text{ K}}{6^{0.4/1.4}} = 659 \text{ K}$$

The mass flow rate can be determined by applying the SFEE (Eq. 6.29) and making use of the positive power output of $(40.1 \text{ MW})/r_w$ determined from parts (b) and (c):

$$\dot{W}_{34} = \dot{m}c_{p0}(T_3 - T_4)$$

$$\therefore \dot{m} = \frac{\dot{W}_x/r_w}{c_{p0}(T_3 - T_4)} = \frac{\dfrac{40.1 \times 10^3}{0.545} \text{ kW}}{1.005 \text{ kJ/kg} \cdot \text{K} (1100 - 659 \text{ K})} = 166 \text{ kg/s}$$

The exergy flow rate of the exhaust gas stream can finally be computed from Equation 9.40a as follows:

$$\dot{X}_{\text{exhaust}} = \dot{m}c_{p0}T_0 \left(\frac{T_4}{T_0} - 1 - \ln \frac{T_4}{T_0} \right)$$

$$= \underline{20.5 \text{ MW}}$$

The exergy of the exhaust gas stream in Example 10.9 is a little over half the power output of the plant. This represents a substantial waste in addition to the increased thermal pollution caused when the exhaust is simply dumped into the atmosphere.

Before going on to consider deviations from the ideal cycle, it is instructive to take note of certain distinctive features of the Brayton cycle as indicated by Equations 10.7 to 10.9. Figure 10.20 plots the three dimensionless parameters $w^* = w/(c_{p0}T_3)$, η_1 and r_w against the natural logarithm of the pressure ratio r_p as the abscissa. The plots are

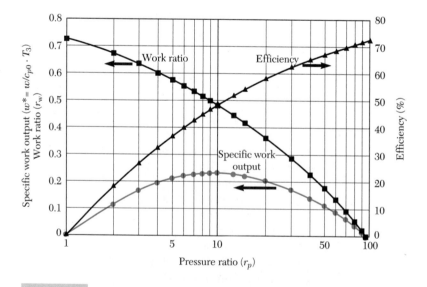

Figure 10.20 Performance characteristics of the ideal gas turbine.

for $T_1 = 300$ K and $T_3 = 1100$ K, which were the values specified in Example 10.9. The following features are noteworthy:

1. The pressure ratio is a critical design parameter for the gas turbine. As r_p increases, the work ratio and the cycle efficiency behave in a contrary manner; while r_w steadily decreases, η_1 rises steadily. The work output w^*, however, rises first, peaks at a particular pressure ratio, and thereafter declines until it reaches a zero value at a limiting pressure ratio, which is termed the *maximum allowable pressure ratio*, $(r_p)_{max}$. One can deduce from Equation 10.7 for w that w is zero when $r_p = 1$ and also when $r_p = (r_p)_{max}$, which is given by the following:

$$(r_p)_{max} = \left(\frac{T_3}{T_1}\right)^{k/(k-1)} \tag{10.10}$$

For the specified T_1 and T_3, the maximum allowable pressure ratio calculated using Equation 10.10 is 94.4.

2. For fixed T_1 and T_3, the maximum (net) work output is obtained when the following mathematical conditions are satisfied:

$$\frac{\partial w}{\partial r_p} = 0 \quad \text{and} \quad \frac{\partial^2 w}{\partial r_p^2} < 0 \tag{10.11}$$

It can be shown by obtaining partial derivatives of w from Equation 10.7 that the conditions stated in Equation 10.11 are satisfied when r_p is equal to what is termed the *optimum pressure ratio*, $(r_p)_{opt}$, given by the following equation:

$$(r_p)_{opt} = \left(\frac{T_3}{T_1}\right)^{\frac{k}{2(k-1)}} = \sqrt{(r_p)_{max}} \tag{10.12}$$

For the specified T_1 and T_3, the calculated optimum pressure ratio is 9.72.

By making the appropriate substitutions in Equations 10.7 to 10.9, the following expressions are obtained for $w^*(= w/c_{p0}T_3)$, η_1, and r_w at the optimum pressure ratio:

$$\left(\frac{\eta_1(\%)}{100}\right)_{opt} = 1 - \sqrt{\frac{T_1}{T_3}} = (r_w)_{opt} \tag{10.13}$$

$$\left(\frac{w}{c_{p0}T_3}\right)_{opt} = \left(1 - \sqrt{\frac{T_1}{T_3}}\right)^2 = \left(\frac{\eta_1(\%)}{100}\right)_{opt} (r_w)_{opt} \tag{10.14}$$

From Equations 10.13 and 10.14, the calculated optimum values for the above parameters when $T_1 = 300$ K and $T_3 = 1100$ K are

$$w^* = 0.228, \quad (\eta_1)_{opt} = 47.8\%, \quad \text{and} \quad (r_w)_{opt} = 0.478$$

Note that the optimum condition is not that corresponding to the maximum thermal efficiency but rather the one at which the net work output is maximum for a specified T_1 and T_3.

Deviations from the Ideal Brayton Cycle

For an actual gas turbine cycle, several deviations from the ideal cycle do occur, the more important of which are noted briefly here:

1. The compression process in the actual case will not be isentropic. This can be factored into the expression for the actual compression work input rate by using an appropriate process efficiency as defined in Equation 8.111. Historically, the isentropic or process efficiencies of the compressors used in gas turbines have been rather low. This was the major obstacle that thwarted the earliest efforts to develop a working gas turbine. Today, compressors with high isentropic efficiencies of between 80% and 90% are available, thus making the gas turbine quite competitive with other engines, especially in the area of air transportation.

2. The combustion process in an actual gas turbine cycle is irreversible to some degree. Combustion is an uncontrolled chemical reaction that involves some consumption or destruction of exergy. In addition, a much larger quantity of air than is required for combustion is normally used to ensure that the materials of construction of the gas turbine are not exposed to excessively high temperatures. Conceptually, the actual process is equivalent to the mixing of a high temperature gas stream with an air stream at a lower temperature. Viewed from the perspective of the second law of thermodynamics, such mixing is irreversible and additional consumption of exergy is involved. The major consequence of these irreversibilities is that more fuel chemical exergy is needed for the process than would otherwise be required if the heating process took place reversibly.

3. The gas expansion through the turbine in the actual case is not isentropic. This can be analyzed by using the process or isentropic efficiency for the turbine as defined in Equation 8.109. Turbines generally have better isentropic efficiencies than the compressors. For gas turbines in use today, values of isentropic efficiencies in excess of 90% for the expansion processes are not uncommon.

For an actual gas turbine cycle, equations can be derived that are comparable to Equations 10.7 and 10.9 for the ideal cycle by making use of the process efficiencies identified in points (1) and (3) above. Until combustion processes are considered in Chapter 14, however, it is not yet possible to evaluate the effect of the irreversibility noted in point (2) on the performance of an actual gas turbine engine. Let's denote the isentropic efficiencies of the compressor and turbine in the actual case by η_{aC} and η_{aT}, respectively. Figure 10.21 gives a comparison between the ideal Brayton cycle and an actual gas turbine cycle. For an actual engine, it should be noted that both the compression process (1→2) and the expansion process (3→4) are shown with broken lines to emphasize that they are nonreversible processes. The corresponding lines for the ideal cycle are the constant entropy lines 1→2' and 3→4' indicated in Figure 10.21. The same pressure ratio is assumed for the ideal cycle and the actual turbine cycle. Also, T_1 and T_3 are taken as being the same for both cycles.

Process 1→2 (adiabatic compression)

The SFEE is applicable, and for the compression process, Equation 10.4a obtained earlier can be assumed. However, Equation 10.4b could be used instead for the specific

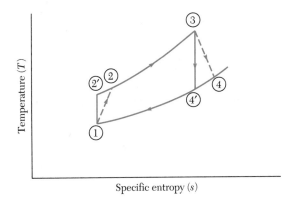

Figure 10.21 Deviations of an actual engine cycle from the ideal Brayton cycle.

work done in the isentropic compression process $1\rightarrow2'$, and then Equation 8.111 could be applied for the process efficiency of an adiabatic compressor to obtain the following for the actual work done in the adiabatic compression process:

$$w_{12} = \frac{1}{\eta_{aC}}c_{p0}T_1(1 - r_p^{(k-1)/k}) = c_{p0}(T_1 - T_2) \tag{10.15a}$$

The last step is given by Equation 10.4a. From this expression for w_{12} the following equation for temperature T_2 can be obtained:

$$\frac{T_2}{T_1} = 1 - \frac{1}{\eta_{aC}}(1 - r_p^{(k-1)/k}) \tag{10.15b}$$

Process 2→3 (constant pressure heating)

For the heat input per unit mass q_{23}, one can use Equation 10.5 with the expression for T_2 given by Equation 10.15b substituted:

$$q_{23} = c_{p0}\left(T_3 - T_1 + \frac{T_1}{\eta_{aC}}(1 - r_p^{(k-1)/k})\right) \tag{10.16}$$

Process 3→4 (adiabatic expansion through turbine)

Comparable expressions to Equations 10.15a and 10.15b can be obtained for w_{34} and T_4 by making use of Equations 10.6a, 10.6b, and 8.109:

$$w_{34} = (\eta_{aT})c_{p0}T_3\left(1 - \frac{1}{r_p^{(k-1)/k}}\right) \tag{10.17a}$$

and

$$\frac{T_4}{T_3} = 1 - \eta_{aT}\left(1 - \frac{1}{r_p^{(k-1)/k}}\right) \tag{10.17b}$$

For the actual cycle, expressions for w^* and η_1 can now be obtained using the same approach as that for deriving Equations 10.7 and 10.9 for the ideal Brayton cycle. The work ratio is particularly useful for providing an indication of how close the efficiency of the actual cycle is to that of the ideal cycle. When considering an actual cycle, however, it is not necessary to derive an expression for the work ratio. The results for the net work output per unit mass and the cycle efficiency can be written in the following final forms for the actual cycle:

Net work output per unit mass (w)

$$\frac{w}{c_{p0}T_3} = (\eta_{aT})\left(1 - \frac{1}{r_p^{(k-1)/k}}\right)\left(1 - \frac{1}{(\eta_{aT}\eta_{aC})}\frac{T_1}{T_3}r_p^{(k-1)/k}\right) \tag{10.18}$$

Actual cycle efficiency (η_1)

$$\frac{\eta_1\,(\%)}{100} = \frac{(\eta_{aT})\left(1 - \frac{1}{(\eta_{aT}\eta_{aC})}\frac{T_1}{T_3}r_p^{(k-1)/k}\right)}{1 - \frac{T_1}{T_3}\left(1 - \frac{1}{\eta_{aC}} + \frac{r_p^{(k-1)/k}}{\eta_{aC}}\right)}\left(1 - \frac{1}{r_p^{(k-1)/k}}\right) \tag{10.19}$$

Equation 10.18 gives the following for the maximum allowable pressure ratio, $(r_p)_{max}$, and the optimum pressure ratio, $(r_p)_{opt}$:

$$(r_p)_{max} = \left(\eta_{aT}\,\eta_{aC}\,\frac{T_3}{T_1}\right)^{k/(k-1)} \tag{10.20}$$

and

$$(r_p)_{opt} = \left(\eta_{aT}\,\eta_{aC}\,\frac{T_3}{T_1}\right)^{\frac{k}{2(k-1)}} = \sqrt{(r_p)_{max}} \tag{10.21}$$

As for the ideal cycle, the optimum pressure ratio is that for which the net work output is a maximum when T_1 and T_3 are specified.

EXAMPLE 10.10
Analysis of an Actual Gas Turbine Cycle

For the gas turbine described in Example 10.9, assume that the actual compression and expansion processes took place with isentropic or process efficiencies of 0.75 and 0.8, respectively. Determine

(a) actual thermal efficiency

(b) actual power output (in MW)

(c) efficiency ratio

(d) exergy flow rate (in MW) of the exhaust gas stream

Given: An actual turbine cycle, such as that shown in Figure 10.21. The ideal cycle is the Brayton cycle. The specified conditions are

$$T_1 = 300 \text{ K} (= T_0), \ p_1 = 1 \text{ atm} (= p_4 = p_0)$$

$$T_3 = 1100 \text{ K}, \qquad p_3 = 6 \text{ atm} (= p_2)$$

Heat input rate $= 100$ MW

The working fluid is air

Process efficiencies are $\eta_{aC} = 0.75$ and $\eta_{aT} = 0.8$

Find: (a) η_1, (b) $\dot{W}_x$ (in MW), (c) Efficiency ratio (ε), (d) $\dot{X}_{exhaust}$ (in MW)

Solution: The problem can be solved by making use of the various equations derived for the actual gas turbine cycle depicted in Figure 10.21.

(a) *Actual cycle efficiency*

The actual cycle efficiency is calculated by substituting in Equation 10.19. The result obtained is a mere 16% (compared with 40.1% for the ideal cycle).

(b) *Actual power output*

Since the heat input rate is 100 MW, using Equation 7.2 gives 16 MW for the net mechanical power output of the actual gas turbine.

(c) *Efficiency ratio*

The efficiency ratio is defined as the ratio of the actual cycle efficiency to the ideal cycle efficiency. In this case, $\varepsilon = (16)/(40.1)$, which is very nearly 0.4. The low efficiency ratio correlates with the low work ratio indicated earlier in the analysis of the ideal cycle.

(d) *Exergy of the exhaust gas stream*

One must determine both T_4 and the mass flow rate of air to be able to compute the exergy of the exhaust gas stream from Equation 9.40a. From Equation 10.16, the heat input per unit mass of air is found to be $q_{23} = 535$ kJ/kg. Thus, from the specified input rate of 100 MW, it is concluded that the flow rate of the working fluid is $(10^5 \text{ kW})/(535 \text{ kJ/kg}) = 187$ kg/s. T_4 can be calculated using Equation 10.17b. In this case, $T_4 = 747$ K. Thus, the exergy flow rate of the exhaust gas stream is calculated as

$$\dot{X}_{exhaust} = \dot{m} c_{p0} T_0 \left(\frac{T_4}{T_0} - 1 - \ln \frac{T_4}{T_0} \right)$$

$$= \underline{32.6 \text{ MW}}$$

The exhaust gas stream carries an increased amount of exergy in the actual case compared with the ideal case, which means that the loss rate is further exacerbated in the real engine. When allowance is made for the irreversibilities in the compression and expansion processes, the optimum pressure ratio computed from Equation 10.21 is different from that based on the ideal cycle. In this case, the optimum pressure ratio is 3.97 when the specified process efficiencies are assumed.

EXAMPLE 10.11
Actual Gas Turbine with Combustion Process Specified

A gas turbine operates on an open cycle with air entering the compressor at 300 K at a rate of 54,000 kg/h. The pressure ratio is 6:1, and the isentropic efficiencies for the compressor and turbine can be taken as 0.8 and 0.85, respectively. For the compression process, c_p and k are 1.005 kJ/kg · K and 1.4, respectively, while for the expansion process, the corresponding average values are 1.11 kJ/kg · K and 1.33. The fuel used is kerosene having a heating value of 43,300 kJ/kg and a chemical exergy of 46,080 kJ/kg. The air–fuel ratio (by mass) is 100:1, and an average value of 1.11 kJ/kg · K can be assumed for c_p during the (combustion) heating process. Determine

(a) gas turbine (net) power output

(b) thermal efficiency

(c) second law efficiency

Given: The gas turbine cycle, which is illustrated in Figure 10.22

$$\dot{m}_a = 54{,}000 \text{ kg/h}, \frac{\dot{m}_a}{\dot{m}_f} = 100$$

$$\text{The pressure ratio } r_p = 6$$

Compression process

$$T_1 = 300 \text{ K}, c_{p0} = 1.005 \text{ kJ/kg · K}, k = 1.4, \eta_{aC} = 0.8$$

Combustion (heating) process

$$\text{Heating value, HV}_{\text{fuel}} = 43{,}300 \text{ kJ/kg of fuel}$$
$$\text{Chemical exergy, } x_c = 46{,}080 \text{ kJ/kg of fuel}$$
$$\text{For the combustion products, } c_{p0} = 1.11 \text{ kJ/kg · K}$$

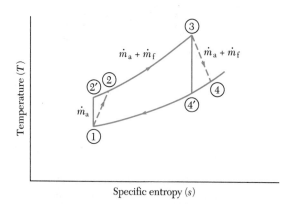

Figure 10.22 Actual gas turbine cycle with combustion (heating) process specified.

Expansion process

$$c_{po} = 1.11 \text{ kJ/kg} \cdot \text{K}, \, k = 1.33, \, \eta_{aT} = 0.85$$

$$T_0 = 300 \text{ K and } p_0 = 1 \text{ atm}$$

Find: **(a)** net power output, **(b)** η_1, **(c)** η_2

Solution: The first law analysis simply entails the application of the SFEE (Eq. 6.29) to the flow processes making up the open cycle. Let's consider each of the three processes in turn.

Compression process (1→2)

w_{12} and T_2 can be calculated using Equations 10.15a and 10.15b, respectively. Using the data given in this case gives

$$w_{12} = -251.9 \text{ kJ/kg}$$

$$T_2 = 550.7 \text{ K}$$

The mass flow rate of the air stream is 54,000 kg/h, as given. From this, the power input for the compression process is

$$-\dot{W}_{12} = -\dot{m}_a w_{12} = 3779 \text{ kW}$$

Combustion (heating process (2→3)

This problem is an appropriate place to introduce a simple procedure for the first law analysis of processes that include combustion. Figure 10.23 depicts the combustion process taking place in an adiabatic combustion chamber. $(H_R^*)_{T_2}$ is used for the total enthalpy flow rate of the *reactants*, which are made up of the mixture of the air and fuel streams prior to combustion. The subscript T_2 is the temperature of the reactants prior to entering the control volume. $(H_P^*)_{T_3}$ denotes the total enthalpy flow rate of the products of combustion leaving the control volume at a temperature T_3. If one assumes that changes in kinetic energy and potential energy are negligible and that the process occurs under adiabatic condition, application of the SFEE reduces to

$$(H_P^*)_{T_3} - (H_R^*)_{T_2} = 0$$

The left-hand side of the reduced form of SFEE can be rewritten as follows:

Figure 10.23 Combustion process in an adiabatic combustion chamber.

$$(H^*_P)_{T_3} - (H^*_R)_{T_2} = (H^*_P)_{T_3} - (H^*_P)_{T_2} + \{(H^*_P)_{T_2} - (H^*_R)_{T_2}\}$$

$$= (\dot{m}_a + \dot{m}_f)(c_{p0})_{products}(T_3 - T_2) + (\Delta H^*)_{T_2}$$

where

$$(\Delta H^*)_{T_2} = (H^*_P)_{T_2} - (H^*_R)_{T_2}$$

$(H^*_P)_{T_2}$ is the total enthalpy flow rate of the products of combustion at the lower temperature T_2. When the application of the first law of thermodynamics to combustion is considered in Chapter 14, it will be shown that the $(\Delta H^*)_T$ quantity is related to a quantity termed the *enthalpy of combustion* for the fuel, which in turn is approximately related to the heating value of the fuel, $(HV)_{fuel}$. Thus, the following expression for the $(\Delta H^*)_T$ quantity can be written:

$$(\Delta H^*)_{T_2} \approx -\dot{m}_f(HV)_{fuel}$$

Strictly speaking, the heating value of a fuel changes with temperature, although the change in the temperature range from 300 K to about 1000 K is small compared with the magnitude of $(HV)_{fuel}$ for most fuels. The application of the SFEE in the present case thus leads finally to the following equation:

$$T_3 - T_2 = \frac{\dot{m}_f(HV)_{fuel}}{(\dot{m}_a + \dot{m}_f)(c_{p0})_{products}}$$

Substituting the known quantities in this equation yields

$$T_3 = 937 \text{ K}$$

Expansion process (3→4)

The work done per unit mass of working fluid can be calculated using Equation 10.17a, while T_4 can be determined using Equation 10.17b. Thus, in the present example,

$$w_{34} = 317.3 \text{ kJ/kg}$$

and

$$T_4 = 651 \text{ K}$$

The mass flow rate of the products of combustion is $(1.01)(54,000/3600) = 15.15$ kg/s, and hence, the positive power output of the turbine is $(15.15)(317.3) = 4,807$ kW.

(a) *Net power output of the turbine*

The net (mechanical) power output is obtained by algebraically adding the positive power output of the turbine and the negative power for the compression of the air stream. Thus,

$$\dot{W}_x = \dot{W}_{12} + \dot{W}_{34} = \underline{1.03 \text{ MW}}$$

(b) *Thermal efficiency of the engine*

A thermal efficiency based on Equation 7.2 can be defined as follows:

$$\eta_1 = \frac{\dot{W}_x}{\text{fuel power} \atop \text{supplied}} \cdot 100\% = \frac{\dot{W}_x}{\dot{m}_f(HV)_{fuel}} \cdot 100\%$$

$$= \frac{(1.028 \text{ MW})}{\left(\dfrac{54,000/3600}{100} \text{ kg/s}\right)(43.3 \text{ MJ/kg})} \cdot 100\% = \underline{15.8\%}$$

(c) *The second law efficiency*

The second law efficiency can be computed using Equation 9.10, which in this case can be written as follows:

$$\eta_2 = \frac{\dot{W}_x}{\text{fuel exergy} \atop \text{supply rate}} \cdot 100\% = \frac{\dot{W}_x}{\dot{m}_f(x_c)_{fuel}} \cdot 100\%$$

$$= \underline{14.9\%}$$

EXERCISES

for Section 10.4.1

1. A steady stream of air at 300 K and 100 kPa is compressed adiabatically to a pressure of 500 kPa. Determine the exit temperature for the air assuming:
   (a) that the compression process is reversible
   (b) that the process efficiency for the compression is 0.8

   Answer: (a) 475 K, (b) 519 K

2. Compute the work done, per unit mass of air, in the compression processes of Exercise 1, parts (a) and (b).

   Answer: (a) -176 kJ/kg, (b) -220 kJ/kg

3. A steady stream of air at 1100 K and 500 kPa expands through an adiabatic turbine to the ambient pressure of 100 kPa. Suggest appropriate values for R and c_p for air in the temperature rate of 700 K to 1100 K. Determine the exit temperature for the air assuming
   (a) that the process is reversible
   (b) that the isentropic efficiency for the turbine is 0.9

   Answer: (a) 728 K, (b) 766 K

4. Compute the work done, per unit mass of air, in the processes of Exercise 3, parts (a) and (b).

   Answer: (a) 417 kJ/kg, (b) 374 kJ/kg

5. The compression process in Exercise 1 (1→2) was followed by a constant pressure heating process (2→3) to the temperature of 1100 K and finally by the adiabatic expansion process (3→4) through the turbine in Exercise 3. Suggest a suitable mean value for c_p during the constant heating process and use this to compute q_{23} for parts (a) and (b).

   Answer: (a) 686 kJ/kg, (b) 638 kJ/kg

6. **(a)** Compute for part (a) in Exercise 5, w_{net}, η_1, and r_w.
   (b) For part (b), determine w_{net}, η_1, and ε.

   Answer: **(a)** 241 kJ/kg, 35.1%, 0.578; **(b)** 154 kJ/kg, 24.1%, 0.688

10.4.2 Modifications to the Basic Gas Turbine Cycle

This section briefly reviews the modifications to the basic gas turbine cycle that have been proposed and are aimed at significant improvement in the performance of the gas engine.

Multiple Stage Compression and Expansion

As previously discussed, the main problem with the gas turbine cycle is the low work ratio. The work ratio can be significantly increased by reducing the compressor work and, possibly, by increasing the positive work output from the turbine at the same time. Figure 10.24 shows, on a *T-s* diagram, how this can be accomplished via multiple stage compression (with intercooling) and multiple stage turbine expansion processes (with reheat between the turbine stages). Figure 10.25 is a schematic diagram of the components of the modified gas engine.

   The processes involved in the two-stage compression and two-stage expansion scheme shown in Figure 10.24 are as follows:

Process 1 →1′. A reversible, adiabatic compression occurs from the air intake pressure (p_1) to an intermediate pressure $(p_{1'})$ between p_1 and p_2.

Process 1′→2′. A cooling process occurs at the intermediate pressure $(p_{1'})$. If $T_{2'}$ is the same as T_1, the intercooling is referred to as complete.

Process 2′→2 . A reversible, adiabatic compression process occurs that takes the air to the desired upper pressure limit (p_2).

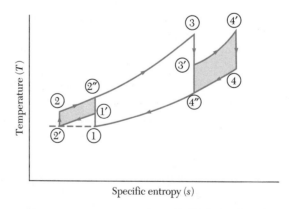

Figure 10.24 Brayton cycle with two-stage compression and two-stage turbine expansion.

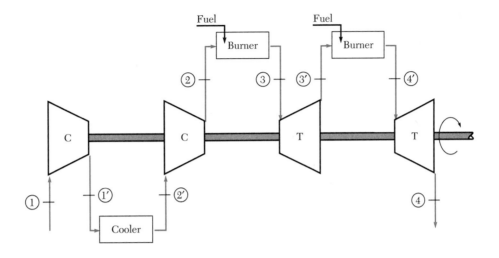

Figure 10.25 Schematic diagram of the components of a gas turbine with two-stage compression (C) and two-stage turbine (T) expansion.

The two-stage compression results in lower compression work than if the air were compressed in a single-stage device. In addition, the net work produced per cycle is increased by an amount given by the shaded area in Figure 10.24 enclosed by $1'$-$2'$-2-$2''$-$1'$.

Now let's consider the heating process, followed by the two-stage expansion through the high pressure and low pressure turbines.

Process 2→3. The heating process is at a constant pressure (p_2).

Process 3→3'. The first-stage turbine expansion occurs. The air pressure at the exit to the turbine $(p_{3'})$ lies between p_3 $(= p_2)$ and p_4 $(= p_1)$.

Process 3'→4'. A reheat process occurs at pressure $p_{3'}$. Usually the reheat is complete, and $T_{4'}$ is equal to T_3.

Process 4'→4. The second-stage turbine expansion occurs.

Process 4→1. A heat rejection process completes the cycle.

Again, the two-stage expansion through the turbine adds to the net work produced by an amount given by the shaded area enclosed by $3'$-$4'$-4-$4''$-$3'$. The positive work output from the turbines is higher than when a single-stage turbine expansion process is used between p_3 and p_4. Note, however, that with T_4 significantly higher than $T_{4''}$, a more substantial loss of thermal exergy is involved in the heat rejection process when a multiple stage turbine expansion scheme is used. Overall, the improvement to thermal efficiency resulting from these modifications is likely to be marginal. The work ratio, however, will improve significantly, thus ensuring a higher actual cycle efficiency compared with the basic cycle, which has a low work ratio. From the standpoint of a second law analysis, if the heat source is effectively a finite source, the gains in the multiple stage compression processes are slightly offset by the losses resulting from the increased thermal exergy loss in the multiple stage turbine expansion processes.

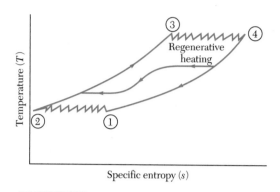

Figure 10.26 Ericsson cycle as the limiting case of isothermal compression and expansion with regeneration.

Regeneration Cycles

As with the vapor power cycles, when an approximately constant temperature heat source exists, optimum performance is obtained when heating of the working fluid from an external source occurs isothermally at a temperature close to the heat source temperature. Figure 10.26 shows a scheme for achieving this that uses multiple stage compression (with intercooling) and multiple stage expansion (with reheating) to achieve both heat rejection to the sink and heat input from the source at effectively constant temperatures. The heat rejection process (4→1) now serves, through a *regenerator*, as the heating process (2→3) for the cycle. The limiting cycle, which has isothermal (source) heating and isothermal (sink) heat rejection with regeneration between the two constant pressure processes, is known as the *Ericsson cycle*. The thermal efficiency of this cycle is identical to that for a Carnot cycle having the same sink and source temperatures.

EXERCISES

for Section 10.4.2

1. A gas turbine operates on the Ericsson cycle, which is illustrated in Figure 10.26. The cycle comprises flow processes. Assume complete regeneration and that $T_1 = 300 \text{ K} = T_2$, while $T_3 = 1100 \text{ K} = T_4$. Also, $p_1 = 100 \text{ kPa} = p_4$ and $p_2 = 500 \text{ kPa} = p_3$. Determine
   (a) w_{12} and q_{12}
   (b) w_{34} and q_{34}
   (c) w, η, and r_w

   Answer: **(a)** -139 kJ/kg; **(b)** 508 kJ/kg; **(c)** 370 kJ/kg, 72.7%, 0.727

2. A two-stage compression with complete intercooling (in other words, $T_{2'} = T_1$) is illustrated in Figure 10.24. Assume $T_1 = 300 \text{ K} = T_{2'}$, $p_1 = 100 \text{ kPa}$, $p_{1'} = 100\sqrt{5} \text{ kPa} = p_{2'}$, $p_2 = 500 \text{ kPa} = p_{2''}$, and $s_{2''} = s_{1'} = s_1$. The compression is a steady-flow process and the fluid is air. Determine, per unit mass of the fluid,
   (a) the work done in the steady-flow, two-stage compression process
      $1 \rightarrow 1' \rightarrow 2' \rightarrow 2 \rightarrow 2''$

(b) the work done in the steady-flow, single-stage compression process 1→2"

Answer: **(a)** −156 kJ/kg, **(b)** −176 kJ/kg

3. A two-stage expansion process with reheating is illustrated in Figure 10.24. Assume $T_3 = 1100$ K $= T_{4'}$, $p_3 = 500$ kPa, $p_{3'} = 100\sqrt{5}$ kPa $= p_{4'}$, $p_4 = 100$ kPa $= p_{4''}$, $s_{4''} = s_{3'} = s_3$ and $s_4 = s_{4'}$. The expansion is a steady-flow process and the fluid is air. Determine, per unit mass of the fluid,

   (a) the work done in the steady-flow, two-stage expansion process

   3→3'→4'→4→4"

   (b) the work done in the steady-flow, single-stage expansion process 3→4"

Answer: **(a)** 461 kJ/kg, **(b)** 418 kJ/kg

10.5 *Reciprocating Internal Combustion Engines*

10.5.1 *Principle of Operation*

The reciprocating internal combustion (I.C.) engine operates on a definite sequence of events, known as a *mechanical cycle*, which is repeated over and over again. The principal constructional features of a reciprocating I.C. engine are illustrated in Figure 10.27 and include an *engine cylinder* and a *piston* that moves between an *outer (bottom) dead center* position and an *inner (top) dead center*. When the piston moves from the outer dead center to the inner dead center, the distance traversed is called the *stroke* (L) while the volume displaced or "swept" is referred to as the *swept volume* (V_{sw}). When the piston is at the inner dead center, the cylinder volume above the piston is called the *clearance volume* (V_c). The *bore* (D) refers to the inside diameter of the cylinder. With reference to these, the following sequence of events comprising a mechanical cycle can be identified:

1. *induction* of charge (either air alone as in diesel engines or a mixture of fuel and air as in gasoline engines) into the cylinder

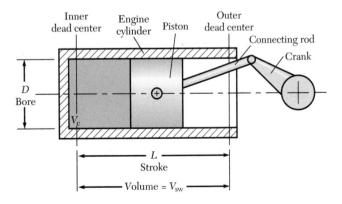

Figure 10.27 Principal features of a reciprocating internal combustion engine.

2. *compression* of the charge

3. *ignition* of the charge (with an induced air–fuel mixture), or *spontaneous burning* of injected fuel in air that is at a sufficiently high temperature after the compression process (in diesel engines)

4. *expansion* of the burning gases, which results in positive work output via the agency of the piston

5. *exhaustion* of the burnt (or spent) gases from the cylinder after which the cycle recommences

The cycle can be completed in either two or four strokes of the piston.

Four-Stroke and Two-Stroke Cycles

Most reciprocating I.C. engines in use today operate on a *four-stroke cycle*. Although Alphonse Beau de Rochas originated the idea of a four-stroke cycle in 1862, it was Nikolaus Otto who in 1876 built one that actually worked. Figure 10.28 is a schematic diagram of the operation of a four-stroke gasoline engine. The four-stroke cycle is completed in induction, compression, expansion, and exhaust strokes. The ignition of the charge occurs toward the completion of the compression stroke, or slightly before the top dead center. Four-stroke engines typically use valves that are operated by the engine via a valve mechanism (or gear). Thus, in the automobile engine, the mechanism usually comprises a cam, coupled to the engine shaft, which activates a "rocker" via a push rod. The rocker in turn opens and closes the valves at predetermined stages of the engine cycle.

While the machine cycle for a four-stroke engine occupies two shaft revolutions, the cycle for a *two-stroke engine* is completed in only one revolution of the shaft. Two-stroke engines are so named because only two strokes occur in each cycle. Such engines are in common use today in motorcycles and other applications requiring relatively small engines. In 1860 Lenoir, a French engineer, came up with the first working internal combustion machine that was essentially a two-stroke engine. The original design was

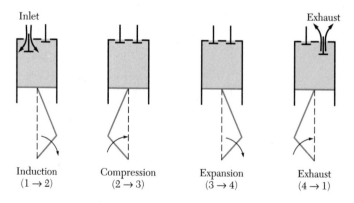

Figure 10.28 Operation of a four-stroke gasoline engine.

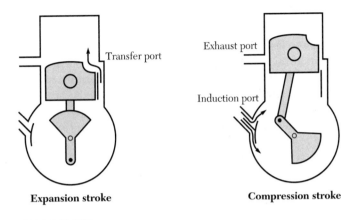

Expansion stroke   **Compression stroke**

Figure 10.29 Operation of a typical two-stroke engine.

rather bulky and inefficient. The designs in use today have much better efficiencies, and for the same cylinder size, the power output is greater than from a four-stroke machine.

Figure 10.29 illustrates the operation of a typical two-stroke engine. The two strokes of a two-stroke engine are the compression stroke and the expansion stroke. In the *compression stroke*, compression of the charge takes place while fresh charge is admitted into the crank case. The charge in the cylinder is ignited near the end of the stroke. The *expansion stroke* is the power stroke. For part of the stroke, the exhaust port opens nearly simultaneously with the transfer port so that fresh charge is admitted into the cylinder from the crank case. The incoming fresh charge helps in pushing out the spent or exhaust charge from the cylinder—a process known as *scavenging*.

Here is a brief comparison that highlights the relative merits of two-stroke and four-stroke engines:

1. The two-stroke engine does not require any mechanically operated valves. Apart from the substantial savings in parts, this feature makes it possible to run the engine in either direction without a reversing gear. This is a main attraction, particularly in marine applications.

2. For the same rotational speed, the two-stroke engine has twice the number of working strokes. Therefore, more power is obtained from a two-stroke engine than from a four-stroke engine of the same bulk and weight. Also, the greater frequency of power or working strokes in the two-stroke engine results in a more uniform torque output and, consequently, a smaller flywheel for achieving a smooth drive.

3. The two-stroke engine tends, however, to have a higher fuel consumption than the four-stroke engine. This is because of the simultaneous occurrence of the exhaust and induction processes and the shorter time available. This tends to result in either a loss of fuel or the combustion of fuel in air that already contains some of the products of combustion and is therefore less able to support combustion. Furthermore, less time is available for the fresh charge to cool the cylinder in the two-stroke engine. Overheating problems are thus more likely to occur with the two-stroke engine than with the four-stroke engine.

Spark Ignition (Gasoline) Engines and Compression Ignition (Diesel) Engines

Thus far, this chapter has not considered in any detail how the combustion of the induced charge is achieved. Table 10.4 gives a brief comparison of the two modes of ignition normally employed in the reciprocating I.C. engines. As the names suggest, a spark ignition (S.I.) engine is one that uses a "spark" to ignite a compressed air–fuel mixture, while a compression ignition (C.I.) engine relies on the high temperature attained in the compression of air to initiate the combustion of the fuel injected into the air.

The *compression ratio* is an important index that determines the performance of reciprocating I.C. engines. It is defined as the ratio of maximum volume possible to the minimum volume when the piston traverses one stroke in the cylinder. As shown later, the higher the compression ratio, the higher the thermal efficiency that can be attained.

Some of the more significant differences between the *spark-ignition (S.I.) engine* and the *compression ignition (C.I.) engine* deserve mention:

1. The induced charge in a S.I. engine is a mixture of air and fuel. Thus, if the compression ratio is very high, the mixture can attain a temperature level at which the fuel begins to burn spontaneously without ignition from an electrical spark. This is known as *pre-ignition,* and its effect on the performance of the engine is to lower its efficiency and cause "knock." To ensure that fuel combustion commences only at the correct stage for optimal performance, either the quality of the fuel must be such that pre-ignition does not occur at the high compression ratio or else the compression ratio must be reduced to a level that ensures that, for the given fuel, preignition does not occur.

2. In a C.I. engine, in contrast, the compressed charge is air only, and the higher the compression ratio, the easier it is to initiate the combustion of the atomized fuel injected into the cylinder at the appropriate stage. The problem of pre-ignition does not arise in this case. However, many C.I. engines are designed to inject the fuel into the cylinder a little before the piston reaches the top dead center (minimum volume in the cylinder). The optimum injection point is a function of crankshaft speed, and if the injection is too soon, pre-ignition also occurs with this engine. This pre-ignition causes the knock frequently heard in diesel engines

TABLE 10.4

Comparison of Modes of Ignition in I.C. Engines		
Process	**Spark Ignition (S.I.) or Gasoline Engine**	**Compression Ignition (C.I.) or Diesel Engine**
Induction	A fuel–air mixture forms the induced charge.	The induced charge consists only of air.
Compression	The fuel–air mixture is compressed and ignited at the end of the compression by an electric spark. The compression ratio varies between about 6 to 1 and 9 to 1.	The charge of air is compressed, with the compression ratio varying between about 14 to 1 and 20 to 1. At the end of the compression process, fuel is injected into the cylinder and is ignited because the temperature of the air is above the fuel ignition temperature.

operating at low speeds. The higher levels of compression ratio that can be reached with C.I. engines result in much higher temperatures for the combustion process, and hence, better thermal efficiencies than are attainable with S.I. engines. These thermal efficiencies tend to be near 35% for C.I. engines, whereas the figure is nearer 25% for S.I. engines.

3. Because of their higher compression ratios, C.I. engines tend to be of bulkier construction than S.I. engines.

4. Obvious differences exist in the physical features of S.I. versus C.I. engines. In the case of S.I. engines, a spark-producing arrangement is an essential feature, which typically includes *spark plugs*, a *contact set* (or electronically timed set), an *ignition coil*, and a *battery*. C.I. engines do not use a spark-producing scheme. Instead, a *fuel injector* introduces fuel as a fine spray into the compressed air in the cylinder. Also, some C.I. engines use *glow plugs* that add heat inside the chamber to ensure that the fuel–air mixture in the cylinder is warm at the beginning of the combustion process.

Finally, one should note that both the two-stroke and four-stroke engines can be of either the S.I. or C.I. design. Also, the C.I. two-stroke is better than the S.I. two-stroke because only air, not fuel, is lost during induction.

10.5.2 Air Standard Otto and Diesel Cycles

The approximate pattern for the cycle executed by actual reciprocating I.C. engines is obtained empirically using a device known as the *mechanical indicator*. Using this device, the variation in the air pressure in the cylinder is obtained as a function of the position of the piston. From this, a pressure (p) versus volume (V) diagram can be plotted for the charge held in the cylinder over each cycle of operation. Figures 10.30(a) and (b)

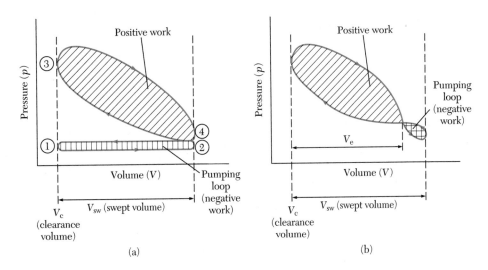

Figure 10.30 Typical *p-V* (or indicator) diagram for (a) the four-stroke gasoline engine and (b) the two-stroke engine.

illustrate the typical results obtained for the four-stroke gasoline engine and the two-stroke engine, respectively. The p-V diagram is often referred to as the *indicator diagram* for the engine. A careful study of the indicator diagrams has led to the use of certain air standard cycles as approximations to the actual cycles. The *air standard cycles* for reciprocating I.C. engines use air only in an idealized closed-cycle operation. This contrasts with the actual situation, which uses a combustible mixture of air and fuel as the working fluid in an open cycle operation. The increase in temperature and pressure produced in the working fluid as a consequence of the combustion process is modeled by defining an appropriate heat input to the air standard cycle. Likewise, the exhaust of gases to ambient air in actual engines is modeled as a heat rejection process in the air standard cycles.

Air Standard Otto Cycle

The ideal cycle approximation to the four-stroke gasoline (or spark ignition) engine is the *Otto cycle*, which is shown in Figures 10.31(a) and (b) on p-V and T-s coordinates, respectively. The following processes make up the cycle:

1. Process 1→2 is a reversible, adiabatic compression process. The volume of air at state 1 is the maximum volume of the cylinder, which is the sum of the *clearance volume* (V_c) and the *swept volume* (V_{sw}) of the cylinder. Both of these volumes are shown in Figure 10.27, which depicts the principal features of the reciprocating I.C. engine. The pressure and temperature at state 1 in Figures 10.31(a) and (b) are usually taken as those for ambient air. The volume of air at state 2 is the clearance volume of the cylinder.

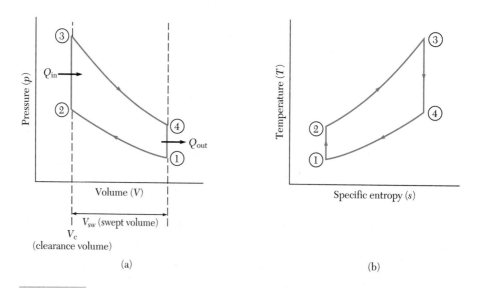

(a)

(b)

Figure 10.31 Otto cycle on (a) p-V coordinates and (b) T-s coordinates.

2. Process 2→3 is a constant volume heating process. The magnitude of the heat input determines the maximum temperature (T_3) reached in the cycle.

3. Process 3→4 is a reversible, adiabatic expansion process that terminates when $V_4 = V_1$.

4. Process 4→1 is a constant volume heat rejection process. A substantial loss of thermal and pressure exergy occurs during the heat rejection process that corresponds to the discharge of the exhaust gas into the atmosphere. Unless the exergy of the exhaust gas stream can be used in an actual situation, this loss can be a significant factor in lowering the efficiency attainable.

First Law Analysis of the Otto Cycle

Expressions for work transfer, the thermal efficiency, and the work ratio are easily derived using the various relationships for an ideal gas and following a procedure similar to that for the gas turbine cycle in Section 10.4. Nonflow processes are occurring here, so the appropriate expressions from Chapter 5 for displacement work should be used, as well as the NFEE (Eq. 6.10) in the application of the first law of thermodynamics to the processes. The latter equation takes the following reduced form if changes in kinetic energy and potential energy for the (closed) system are assumed negligible:

$$q - w = \Delta u \tag{10.22}$$

where q denotes the heat flow per unit mass of the system, w is the work done per unit mass (of the system), and u is the specific internal energy.

Process 1→2 (reversible and adiabatic compression)

Since the process is adiabatic, $q_{12} = 0$. Thus, assuming that air can be regarded as a perfect gas, application of the NFEE to process 1→2 leads to

$$-w_{12} = u_2 - u_1 = c_{v0}(T_2 - T_1) \tag{10.23a}$$

Equation 10.23a is applicable when the process is adiabatic regardless of whether it is reversible or nonreversible. For the reversible adiabatic process, however, T_2 will be different from that obtained when the process is adiabatic but not reversible. In the isentropic (reversible and adiabatic) case, Equation 8.30 can be applied, and therefore, w_{12} can be written as

$$w_{12} = c_{v0}T_1 \left(1 - \frac{T_2}{T_1}\right) = c_{v0}T_1(1 - r_v^{k-1}) \tag{10.23b}$$

where r_v is the compression ratio, which is equal to V_1/V_2 or v_1/v_2.

Process 2→3 (constant volume heating)

Since the system volume stays constant during the process, the displacement work done is zero. Hence, applying Equation 10.22 gives

$$q_{23} = u_3 - u_2 = c_{v0}(T_3 - T_2) \tag{10.24a}$$

Equation 10.24a applies whether or not the process is reversible. For the reversible cycle, however, T_2 can be replaced with $T_1 r_v^{k-1}$ based on Equation 8.30 for the isentropic process 1→2. The substitution leads to the following for q_{23}:

$$q_{23} = c_{v0} T_3 \left(1 - \frac{T_1}{T_3} r_v^{k-1} \right) \tag{10.24b}$$

Process 3→4 (reversible and adiabatic expansion)

Application of Equation 10.22 to the adiabatic expansion process for which q_{34} is zero yields the following for w_{34}:

$$w_{34} = u_3 - u_4 = c_{v0}(T_3 - T_4) \tag{10.25a}$$

If the process is also reversible, Equation 8.30 can be applied to allow substitution of T_3 / r_v^{k-1} for T_4 in Equation 10.25a. Thus, the work done per unit mass in the isentropic expansion process 3→4 is

$$w_{34} = c_{v0} T_3 \left(1 - \frac{1}{r_v^{k-1}} \right) \tag{10.25b}$$

Ideal specific (net) work output (w)

The specific net work transfer (w) is obtained from Equations 10.23b and 10.25b for the negative and positive work per unit mass of the system. Thus, w is given by

$$
\begin{aligned}
w &= w_{12} + w_{34} \\
&= c_{v0} T_3 \left(1 - \frac{1}{r_v^{k-1}} \right)\left(1 - \frac{T_1}{T_3} r_v^{k-1} \right)
\end{aligned} \tag{10.26}
$$

Ideal cycle work ratio (r_w)

The ideal cycle work ratio is obtained from Equations 10.25b and 10.26, which give the positive work output and the net work output, respectively. Thus, the work ratio as defined by Equation 10.2 reduces for the Otto cycle to the following:

$$r_w = \frac{w}{w_{34}} = 1 - \frac{T_1}{T_3} r_v^{k-1} \tag{10.27}$$

Ideal cycle thermal efficiency (η_1)

The expression for the cycle efficiency defined by Equation 7.2 is obtained, using Equations 10.24b and 10.26, for the heat input and the net work output, respectively, as

$$
\begin{aligned}
\eta_1 &= \frac{w}{q_{23}} \cdot 100\% \\
&= \left(1 - \frac{1}{r_v^{k-1}} \right) \cdot 100\%
\end{aligned} \tag{10.28}
$$

Equations 10.7, 10.8, and 10.9 for w, r_w, and η_1, respectively, for the Brayton cycle and corresponding Equations 10.26, 10.27, and 10.28 for the Otto cycle are strikingly similar. The specific heat c_{po} in the expressions for the Brayton cycle is replaced by c_{vo} in those for the Otto cycle. In addition, in place of $r_p^{(k-1)/k}$ in the equations for the gas turbine cycle, r_v^{k-1} is used in those for the Otto cycle.

Figure 10.32 shows the pattern of variation of $w^* = (w/c_{vo}T_3)$, η_1, and r_w with the compression ratio (r_v). The values of T_1 and T_3 are fixed at 300 K and 1100 K, respectively. The compression ratio can vary between unity and a maximum value, $(r_v)_{max}$, given by

$$(r_v)_{max} = \left(\frac{T_3}{T_1}\right)^{1/(k-1)} \tag{10.29}$$

When r_v is equal to 1 or $(r_v)_{max}$, the net work output is zero. The maximum thermal efficiency occurs when $r_v = (r_v)_{max}$ and is equal to the Carnot cycle efficiency, but this case is of no practical interest since work output is zero.

From the standpoint of maximizing the net work output, the optimum compression ratio is obtained by using a similar mathematical condition to that of Equation 10.11. In other words, for a fixed T_1 and T_3, the maximum net work output is obtained when

$$\frac{\partial w}{\partial r_v} = 0 \quad \text{and} \quad \frac{\partial^2 w}{\partial r_v^2} < 0 \tag{10.30}$$

By performing the partial differentiation, the optimum compression ratio, $(r_v)_{opt}$, can be obtained as

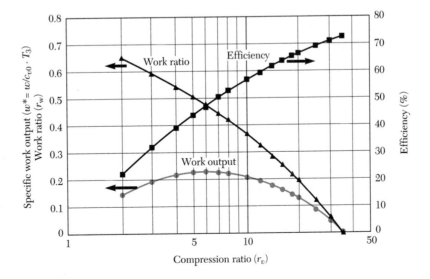

Figure 10.32 Performance characteristics for the ideal reciprocating I.C. engine operating on the Otto cycle.

$$(r_v)_{\text{opt}} = \left(\frac{T_3}{T_1}\right)^{1/[2(k-1)]} = \sqrt{(r_v)_{\text{max}}} \tag{10.31}$$

At this optimum value of r_v, the expressions for w, r_w, and η_1 are as follows:

$$(w^*)_{\text{opt}} = \left(\frac{w}{c_{v0}T_3}\right)_{\text{opt}} = \left(1 - \sqrt{\frac{T_1}{T_3}}\right)^2 \tag{10.32}$$

and

$$\left(\frac{\eta_1\ (\%)}{100}\right)_{\text{opt}} = 1 - \sqrt{\frac{T_1}{T_3}} = (r_w)_{\text{opt}} \tag{10.33}$$

For the values of T_1 and T_3 specified, these equations can be used to obtain the following:

$$(r_v)_{\text{max}} = 35.4$$

for which $k = 1.364$ is assumed based on the estimate for air at an average temperature of 700 K.

$$(r_v)_{\text{opt}} = 5.95$$

$$(\eta_1)_{\text{opt}} = 47.8\%, \quad (r_w)_{\text{opt}} = 0.478$$

$$(w^*)_{\text{opt}} = [w/(c_{v0}T_3)]_{\text{opt}} = 0.228$$

Comparing this with the gas turbine cycle performance shows that for the same T_1 and T_3, the ideal cycle efficiency and the work ratios at the optimum point are the same for both the Brayton and Otto cycles. Since c_{v0} is less than c_{p0}, however, the net specific work output of the gasoline engine is less. However, because the gasoline engine experiences the peak temperature condition T_3 only intermittently, a higher level is acceptable compared with the gas turbine cycle in which some of the components of the engine are subjected to a high temperature T_3 continuously.

Mean Effective Pressure (m.e.p.)

One other index of performance usually defined for the reciprocating I.C. engine is the *mean effective pressure (m.e.p.)*. The mean effective pressure is defined such that the work done in one engine stroke by the piston, against a uniform pressure equal to the m.e.p, is equivalent to the net work output in one cycle of the engine. Thus, with reference to Figure 10.31(a),

$$(\text{m.e.p.})V_{\text{sw}} = W_{\text{net}} = mw \tag{10.34a}$$

Now,

$$V_{\text{sw}} = V_1 - V_2 = V_1\left(1 - \frac{1}{r_v}\right) \tag{10.34b}$$

Thus, Equations 10.34a and 10.34b give

$$\text{m.e.p} = \frac{w}{v_1 \left(1 - \dfrac{1}{r_v}\right)} \qquad (10.34)$$

where v_1 is the specific volume of air at the prevailing ambient condition and is (from Eq. 4.1) given by

$$v_1 = \frac{RT_1}{p_1} \qquad (10.34c)$$

EXAMPLE 10.12
Analysis of an Air Standard Otto Cycle

In an air standard Otto cycle, the maximum and minimum temperatures are 1675 K and 300 K, respectively. The compression ratio is 6. At the beginning of the compression process, the pressure is 1 atm. The specific heats c_p and c_v can be considered constants. Determine

(a) pressure and temperature at each point in the cycle

(b) thermal efficiency

(c) work ratio

(d) mean effective pressure

Given: Air standard Otto cycle, as illustrated in Figures 10.31(a) and (b)

$$r_v = 6$$
$$p_1 = 1 \text{ atm}, \quad T_1 = 300 \text{ K}, \quad T_3 = 1675 \text{ K}$$

Find: (a) p, T at each state, (b) η_1, (c) r_w, (d) m.e.p

Solution:

(a) The appropriate process path equations can be used for the calculation of the unknown pressures and temperatures. Thus, applying Equations 8.30 and 8.32 to the isentropic compression process 1→2 gives the following results for T_2 and p_2:

$$T_2 = T_1 r_v^{k-1} = \underline{614 \text{ K}}$$

and

$$p_2 = p_1 r_v^{k} = \underline{12.3 \text{ atm}}$$

A value of $k = 1.4$ was assumed in these computations.

Process 2→3 occurs at a constant volume (isochoric), and applying Equation 4.1 for the ideal gas with $v_3 = v_2$ yields the following result for p_3:

$$p_3 = p_2 \frac{T_3}{T_2} = \underline{33.6 \text{ atm}}$$

Equations 8.30 and 8.32 can be applied again to the isentropic process 3→4, leading to the following results for T_4 and p_4:

$$T_4 = \frac{T_3}{r_v^{k-1}} = \underline{818 \text{ K}}$$

and

$$p_4 = \frac{p_3}{r_v^k} = \underline{2.73 \text{ atm}}$$

The remaining parts of the problem can be solved either from first principles or by using Equations 10.26, 10.27, 10.28, and 10.34 from the previous set that was derived for the ideal cycle.

(b) The thermal efficiency can be calculated using Equation 10.28:

$$\eta_1 = \left(1 - \frac{1}{r_v^{k-1}}\right) \cdot 100\% = \underline{51.2\%}$$

This looks impressively high. Actual cycle efficiencies are nearer the 25% to 30% range. One would expect to obtain more realistic estimates after considering the irreversibilities of the processes in the actual engine cycle.

(c) Equation 10.27 gives the following result for the work ratio:

$$r_w = 1 - \frac{T_1}{T_3} r_v^{k-1} = \underline{0.633}$$

The low value of the work ratio is a further indication that the actual cycle efficiency is significantly lower than the ideal cycle efficiency calculated in part (b).

(d) The mean effective pressure can be determined using Equation 10.34. First, however, both w and v_1 must be calculated. These are given by Equations 10.26 and 10.34c, respectively:

$$w = c_{v0}T_3 \left(1 - \frac{1}{r_v^{k-1}}\right)\left(1 - \frac{T_1}{T_3} r_v^{k-1}\right)$$

$$= c_{v0}T_3 \left(\frac{\eta_1 (\%)}{100}\right) r_w = 390 \text{ kJ/kg}$$

and

$$v_1 = \frac{RT_1}{p_1} = \frac{(0.287)(300)}{101.325} = 0.850 \text{ m}^3/\text{kg}$$

where c_{v0} was taken as 0.718 kJ/kg · K in the computation of w. Applying Equation 10.34 yields the following result for the mean effective pressure:

$$\text{m.e.p} = \frac{w}{v_1 \left(1 - \frac{1}{r_v}\right)} = \underline{551 \text{ kPa (or 5.43 atm)}}$$

This mean effective pressure is the pressure acting on the piston face through one stroke of the engine and producing the same work output as the engine does in one cycle.

Deviations from the Ideal Cycle

Similar expressions to Equations 10.15a through 10.21 for the gas turbine cycle can be derived for the gasoline engine cycle by introducing process efficiencies η_{aC} and η_{aE}, respectively, for the adiabatic compression and expansion processes. While this approach does not account for all the irreversibilities in the actual cycle, the resulting expressions provide a more realistic indication of performance of an actual engine. The final expressions that emerge are as follows:

Process 1→2 (adiabatic compression)

$$w_{12} = \frac{c_{v0}T_1}{\eta_{aC}}(1 - r_v^{k-1}) = c_{v0}(T_1 - T_2) \tag{10.35a}$$

and

$$T_2 = T_1\left(1 + \frac{(r_v^{k-1} - 1)}{\eta_{aC}}\right) \tag{10.35b}$$

Process 2→3 (constant volume heating)

$$q_{23} = c_{v0}(T_3 - T_2) = c_{v0}T_3\left[1 - \frac{T_1}{T_3}\left(1 + \frac{(r_v^{k-1} - 1)}{\eta_{aC}}\right)\right] \tag{10.36}$$

Process 3→4 (adiabatic expansion)

$$w_{34} = (\eta_{aE})c_{v0}T_3\left(1 - \frac{1}{r_v^{k-1}}\right) = c_{v0}(T_3 - T_4) \tag{10.37a}$$

and

$$T_4 = T_3\left[1 - \eta_{aE}\left(1 - \frac{1}{r_v^{k-1}}\right)\right] \tag{10.37b}$$

Actual specific net work output

$$w = w_{12} + w_{34}$$

$$= (\eta_{aE})c_{v0}T_3\left(1 - \frac{1}{r_v^{k-1}}\right)\left(1 - \frac{1}{(\eta_{aC}\eta_{aE})}\frac{T_1}{T_3}r_v^{k-1}\right) \tag{10.38}$$

Actual cycle efficiency

$$\frac{\eta_1\ (\%)}{100} = \frac{w}{q_{23}}$$

(10.39)

$$= \frac{(\eta_{aE})\left(1 - \frac{1}{(\eta_{aC}\eta_{aE})}\frac{T_1}{T_3}r_v^{k-1}\right)}{1 - \frac{T_1}{T_3}\left(1 - \frac{1}{\eta_{aC}} + \frac{r_v^{k-1}}{\eta_{aC}}\right)}\left(1 - \frac{1}{r_v^{k-1}}\right)$$

From Equation 10.38 for the specific net work output, the following expressions can be obtained for the maximum allowable compression ratio $(r_v)_{max}$ and the optimum compression ratio $(r_v)_{opt}$:

$$(r_v)_{max} = \left(\eta_{aE}\eta_{aC}\frac{T_3}{T_1}\right)^{1/(k-1)}$$

(10.40)

and

$$(r_v)_{opt} = \left(\eta_{aE}\eta_{aC}\frac{T_3}{T_1}\right)^{1/[2(k-1)]} = \sqrt{(r_v)_{max}}$$

EXAMPLE 10.13
Actual Cycle as Deviation from the Otto Cycle

Repeat parts (a), (b), and (d) of Example 10.12 for an actual cycle that has process efficiencies of 0.8 and 0.85, respectively, for compression and expansion processes.

Given: The actual cycle is illustrated in Figures 10.33(a) and (b).

$$r_v = 6 \text{ as in Example 10.12}$$

$$p_1 = 1 \text{ atm}, \ T_1 = 300 \text{ K}, \ T_3 = 1675 \text{ K}$$

$$\eta_{aC} = 0.8, \ \eta_{aE} = 0.85$$

Find: **(a)** p and T at each state, **(b)** η_1, **(c)** m.e.p.

Solution:
(a) Referring to Figure 10.33(b), use the fact that the process 1→2′ is isentropic, and therefore, from the previous analysis, $T_{2'} = 614$ K. The process efficiency for the actual compression (1→2) is 0.8. Use this with Equation 8.111 to determine the actual work done compared with the work done in the isentropic compression. By further considering the expressions for the work done resulting from the application of the NFEE (Eq. 6.10), you can determine the actual temperature T_2. Alternatively, you can make use of Equation 10.35b, which gives temperature T_2 in terms of T_1 for an adiabatic compression process:

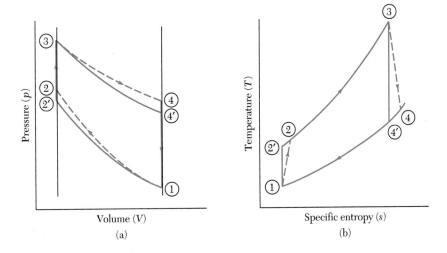

Figure 10.33 Actual cycle as deviation from the Otto cycle on (a) *p-V* coordinates and (b) *T-s* coordinates.

$$T_2 = T_1 \left(1 + \frac{r_v^{k-1} - 1}{\eta_{aC}} \right) = \underline{693 \text{ K}}$$

The ideal gas *pVT* equation can be used to calculate p_2 once T_2 is known. In other words, applying Equation 4.5 gives

$$p_2 = p_1 \left(\frac{v_1}{v_2} \right) \left(\frac{T_2}{T_1} \right) = p_1 r_v \frac{T_2}{T_1} = \underline{13.9 \text{ atm}}$$

Both a higher temperature and a higher pressure are obtained for the actual cycle than for the ideal cycle.

Since $V_3 = V_2$, the same equation can be used as for the ideal case to obtain p_3:

$$p_3 = p_2 \frac{T_3}{T_2} = \underline{33.6 \text{ atm}}$$

This result is obtained by substituting the p_2 and T_2 values that were calculated for the actual cycle.

The temperature T_4 is calculated using Equation 10.37b with $\eta_{aE} = 0.85$ and the other specified parameters substituted into the equation:

$$T_4 = T_3 \left\{ 1 - \eta_{aE} \left(1 - \frac{1}{r_v^{k-1}} \right) \right\} = \underline{947 \text{ K}}$$

This temperature is nearly 16% higher than that for the ideal cycle. The pressure p_4 is calculated using the fact that $V_4 = V_1$ and applying the ideal gas *pVT* equation. Thus,

$$p_4 = p_1 \frac{T_4}{T_1} = \underline{3.16 \text{ atm}}$$

(b) The thermal efficiency for the actual cycle can be calculated using Equation 10.39:

$$\frac{\eta_1 \, (\%)}{100} = \frac{(\eta_{aE})\left(1 - \dfrac{1}{(\eta_{aC}\eta_{aE})} \dfrac{T_1}{T_3} r_v^{k-1}\right)}{1 - \dfrac{T_1}{T_3}\left(1 - \dfrac{1}{\eta_{aC}} + \dfrac{r_v^{k-1}}{\eta_{aC}}\right)}\left(1 - \dfrac{1}{r_v^{k-1}}\right)$$

$$= \frac{(0.85)(0.461)}{(0.586)}\left(1 - \frac{1}{r_v^{k-1}}\right) = 0.342$$

The thermal efficiency for the actual cycle is thus 34.2%, which is only about two-thirds of the ideal cycle efficiency.

(c) For the mean effective pressure, calculate the actual net work output per unit mass, which is given by Equation 10.38:

$$w = (\eta_{aE})c_{v0}T_3\left(1 - \frac{1}{r_v^{k-1}}\right)\left(1 - \frac{1}{(\eta_{aC}\eta_{aE})}\frac{T_1}{T_3}r_v^{k-1}\right)$$

$$= (0.85)(c_{v0}T_3)(0.512)(0.461) = 241 \text{ kJ/kg}$$

The specific heat c_{v0} is taken as 0.718 kJ/kg · K. Substituting the calculated w and the value of $v_1 = 0.850$ m³/kg obtained for Example 10.12 into Equation 10.34 yields a mean effective pressure of 340 kPa (or 3.36 atm).

Air Standard Diesel Cycle

The ideal cycle for the four-stroke diesel (or compression ignition) engine is the Diesel cycle. Figures 10.34(a) and (b) show the air standard Diesel cycle on the p-V and T-s coordinates, respectively. The processes involved are as follows:

1. Process 1→2 is a reversible, adiabatic compression process.

2. Process 2→3 is the heating process that is assumed to occur at a constant pressure. It is usual to define a *cut-off ratio*, r_c, as the ratio of the volume of air at which the constant pressure heating ends (V_3) to the clearance volume (V_c, which equals V_2) of the cylinder. In other words,

$$r_c = \frac{V_3}{V_c} \tag{10.42}$$

3. Process 3→4 is a reversible, adiabatic expansion process.

4. Process 4→1 is a heat rejection process at a constant volume, which is the equivalent of the discharge of the exhaust gases into the atmosphere in an actual engine.

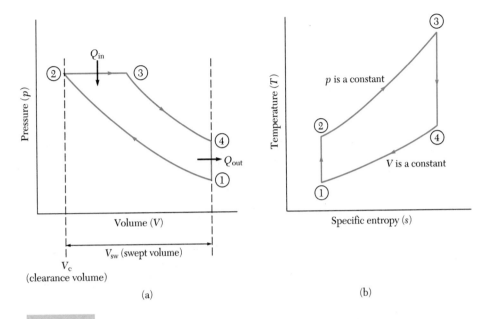

Figure 10.34 Diesel cycle on (a) p-V coordinates and (b) T-s coordinates.

First Law Analysis of the Ideal Cycle

The air standard cycle uses a fixed mass of air and therefore a first law analysis requires application of the NFEE (Eq. 6.10) in the reduced form given by Equation 10.22. Applying Equation 10.22 and using the appropriate path equations or equations of state, such as Equations 8.30, 4.1, and 4.4, gives the following results for the Diesel cycle:

Process 1→2 (reversible and adiabatic compression)

$$w_{12} = c_{v0}(T_1 - T_2) = c_{v0}T_1(1 - r_v^{k-1}) \tag{10.43}$$

Process 2→3 (constant pressure heating)

The work per unit mass w_{23} is determined using Equation 5.7 for the displacement work done in a constant pressure nonflow process. Thus,

$$\begin{aligned} w_{23} &= p_2(v_3 - v_2) = p_2 v_2(r_c - 1) \\ &= RT_2(r_c - 1) = RT_1 r_v^{k-1}(r_c - 1) \end{aligned} \tag{10.44a}$$

The expression for q_{23} is obtained by applying Equation 10.22, which is the reduced form of the NFEE. Noting from Equation 4.4 that $T_3 = T_2 r_c = T_1 r_v^{k-1} r_c$, q_{23} can be obtained as follows:

$$\begin{aligned} q_{23} &= w_{23} + (u_3 - u_2) \\ &= RT_1 r_v^{k-1}(r_c - 1) + c_{v0}(T_3 - T_2) \\ &= c_{p0}T_1 r_v^{k-1}(r_c - 1) \end{aligned} \tag{10.44b}$$

The last step in Equation 10.44b is based on the fact that for an ideal gas,

$$R = (c_{po} - c_{vo})$$

and the substitutions $T_3 = T_1 r_v^{k-1} r_c = T_2 r_c$.

Process 3→4 (reversible and adiabatic expansion)

$$w_{34} = u_3 - u_4 = c_{vo}(T_3 - T_4) \tag{10.45a}$$

The expression for w_{34} can be simplified by noting that $T_3 = T_2 r_c = T_1 r_v^{k-1} r_c$ and by further applying Equation 8.30 to the isentropic process 3→4 to obtain the following relationship between T_4 and T_3:

$$\frac{T_4}{T_3} = \left(\frac{V_3}{V_4}\right)^{k-1} = \left(\frac{V_3}{V_c}\frac{V_c}{V_4}\right)^{k-1} = \left(\frac{r_c}{r_v}\right)^{k-1} \tag{10.45b}$$

Substituting for T_3 and T_4 in Equation 10.45a thus leads to

$$w_{34} = c_{vo}T_3 \left(1 - \frac{r_c^{k-1}}{r_v^{k-1}}\right) = c_{vo}T_1 r_c(r_v^{k-1} - r_c^{k-1}) \tag{10.45}$$

For the Diesel cycle, w_{12} is the negative work (w_{-ve}), while the sum of w_{23} and w_{34} is the positive work (w_{+ve}). The following expression for the positive work can be obtained from Equations 10.44a and 10.45:

$$
\begin{aligned}
w_{+ve} &= w_{23} + w_{34} \\
&= c_{po}T_1 r_v^{k-1}(r_c - 1) - c_{vo}T_1 r_v^{k-1}(r_c - 1) + c_{vo}T_1 r_c(r_v^{k-1} - r_c^{k-1}) \\
&= q_{23} - c_{vo}T_1(r_c^k - r_v^{k-1})
\end{aligned}
\tag{10.46}
$$

From these expressions, one can determine the net specific work output (w), the work ratio (r_w), and the cycle efficiency (η_1) for the Diesel cycle.

Ideal cycle net specific work output

$$
\begin{aligned}
w &= w_{12} + w_{+ve} \\
&= q_{23} - c_{vo}T_1(r_c^k - r_v^{k-1}) + c_{vo}T_1(1 - r_v^{k-1}) \\
&= c_{po}T_1 r_v^{k-1}(r_c - 1) - c_{vo}T_1(r_c^k - 1)
\end{aligned}
\tag{10.47}
$$

Work ratio for the Diesel cycle

The work ratio is obtained by dividing the net work output by the positive work output. From Equations 10.46 and 10.47, the expression for r_w after rearranging is thus

$$r_w = \frac{k(r_c - 1)r_v^{k-1} - (r_c^k - 1)}{k(r_c - 1)r_v^{k-1} - (r_c^k - r_v^{k-1})} \tag{10.48}$$

Ideal cycle efficiency

The thermal efficiency for the ideal cycle is obtained by substituting the expressions for the net work output (w) and the heat input (q_{23}) into Equation 7.2. Thus,

$$\frac{\eta_1\,(\%)}{100} = \frac{w}{q_{23}} = 1 - \frac{(r_c^k - 1)}{k(r_c - 1)r_v^{k-1}} \qquad (10.49)$$

For the mean effective pressure, Equations 10.34 and 10.34c can be used with w now given by Equation 10.47.

EXAMPLE 10.14
Analysis of an Air Standard Diesel Cycle

An air standard Diesel cycle operates with a compression ratio of 16 and maximum and minimum cycle temperatures of 1675 K and 300 K, respectively. At the beginning of the compression process, the pressure is 1 atm. (For air, assume $k = 1.4$ and $c_p = 1.005$ kJ/kg · K.) Determine

(a) pressure and temperature at each point in the cycle

(b) specific net work output

(c) thermal efficiency

(d) work ratio

(e) mean effective pressure

Given: The air standard Diesel cycle is shown in Figures 10.34(a) and (b).

$$r_v = 16, k = 1.4, c_{p0} = 1.005 \text{ kJ/kg} \cdot \text{K}$$

$$p_1 = 1 \text{ atm}, T_1 = 300 \text{ K}, T_3 = 1675 \text{ K}$$

Find: (a) p and T for each state, (b) w, (c) η_1, (d) r_w, (e) m.e.p

Solution:

(a) The unknown pressures and temperatures can be calculated using the appropriate process path equations. Thus, applying Equations 8.30 and 8.32 to the isentropic compression process 1→2, the following results for T_2 and p_2 are obtained:

$$T_2 = T_1 r_v^{k-1} = \underline{909 \text{ K}}$$

and

$$p_2 = p_1 r_v^k = \underline{48.5 \text{ atm}}$$

The peak pressure encountered in the operation of a diesel engine is generally much higher than that in a gasoline engine because of the much higher compression ratios for which diesel engines are designed. As such, a diesel engine tends to have a more robust construction to be able to withstand the high pressures. Process 2→3 occurs at a constant pressure (isobaric) and so $p_3 = p_2 = 48.5$ atm.

Equation 4.1 for an ideal gas can be applied with $p_3 = p_2$ to obtain V_3/V_2, which is the cut-off ratio r_c:

$$r_c = \frac{V_3}{V_2} = \frac{T_3}{T_2} = 1.84$$

One can thus use Equation 10.45b to determine T_4:

$$T_4 = T_3 \left(\frac{r_c}{r_v}\right)^{k-1} = 705\ \text{K}$$

The pressure p_4 can be determined by considering the isochoric process $4 \rightarrow 1$ and applying the ideal gas pVT Equation 4.3:

$$p_4 = p_1 \frac{T_4}{T_1} = 2.35\ \text{atm}$$

(b) Since values of the temperature at each state are known, it may be easier to use the equations for w and q in terms of temperature (instead of the final expressions) for the various quantities to be determined in parts (b) through (d). Thus, Equation 10.43 gives the following for w_{12}:

$$w_{12} = c_{v0}(T_1 - T_2) = -437\ \text{kJ/kg}$$

Similarly, for w_{23}, Equation 10.44a can be rewritten as

$$w_{23} = R(T_3 - T_2) = 220\ \text{kJ/kg}$$

Also, w_{34} can be calculated from Equation 10.45a:

$$w_{34} = c_{v0}(T_3 - T_4) = 696\ \text{kJ/kg}$$

In these equations, c_{v0} and R were determined from the relationships $c_{v0} = (c_{p0} - R)$ and $k = (c_{p0}/c_{v0})$. The computed values are $c_{v0} = 0.718\ \text{kJ/kg} \cdot \text{K}$ and $R = 0.287\ \text{kJ/kg} \cdot \text{K}$. From these results for w, the positive work and the net work output can be determined as follows:

$$w_{+ve} = w_{23} + w_{34} = 916\ \text{kJ/kg}$$

$$w = w_{-ve} + w_{+ve} = w_{12} + w_{+ve} = 479\ \text{kJ/kg}$$

The specific work output is thus equal to 479 kJ/kg. This is practically the same as the value of 477 kJ/kg obtained using Equation 10.47 for the specific work output.

(c) For the thermal efficiency, one needs to compute the heat input (per unit mass), q_{23}, in addition to the specific work output already determined. Equation 10.44b for q_{23} can be rewritten as

$$q_{23} = c_{p0}(T_3 - T_2) = 770\ \text{kJ/kg}$$

Hence, Equation 7.2 yields the following for the cycle efficiency:

$$\eta_1 = \frac{w}{q_{23}} \cdot 100\% = 62.2\%$$

The same result is obtained by using Equation 10.49 for the ideal Diesel cycle efficiency.

(d) The work ratio can be calculated using the defining Equation 10.2:

$$r_{\mathrm{w}} = \frac{w}{w_{+\mathrm{ve}}} = \underline{0.523}$$

The value for the work ratio obtained using Equation 10.48 is 0.522.

While an unusually high cycle efficiency is indicated for the Diesel cycle, the low work ratio implies that the efficiency of the actual cycle is considerably lower. This fact is demonstrated in Example 10.15.

(e) The mean effective pressure is calculated using Equation 10.34. In this example, $v_1 = 0.850$ m^3/kg as previously determined, while $w = 479$ kJ/kg. Hence, the mean effective pressure is $\underline{601 \text{ kPa (or 5.93 atm)}}$.

EXAMPLE 10.15
Analysis of an Actual Diesel-like Cycle

Repeat parts (a), (b), (c), and (e) of Example 10.14 with assumed process efficiencies for the compression and power strokes of 0.8 and 0.85, respectively. Also determine the efficiency ratio.

Given: The actual cycle (based on the Diesel cycle) is shown in Figures 10.35(a) and (b).

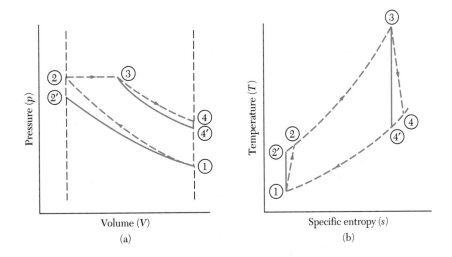

Figure 10.35 (a) *p-V* diagram for an actual cycle based on the Diesel cycle. (b) An actual cycle (based on the Diesel cycle) on *T-s* coordinates.

$$r_v = 16, \ k = 1.4, \ c_{po} = 1.005 \text{ kJ/kg} \cdot \text{K}$$

$$p_1 = 1 \text{ bar}, \ T_1 = 300 \text{ K}, \ T_3 = 1675 \text{ K}$$

$$\eta_{aC} = 0.8, \ \eta_{aE} = 0.85$$

Find: **(a)** p and T for each state **(b)** w, **(c)** η_1, **(d)** m.e.p, **(e)** ε

Solution:

(a) Equation 10.35b for the adiabatic compression of an ideal gas can be used to determine T_2 in this case:

$$T_2 = T_1 \left(1 + \frac{(r_v^{k-1} - 1)}{\eta_{aC}} \right) = 1062 \text{ K}$$

Applying the ideal gas pVT equation yields the following for p_2:

$$p_2 = p_1 \left(\frac{V_1}{V_2} \right) \left(\frac{T_2}{T_1} \right) = p_1 r_v \frac{T_2}{T_1} = 56.6 \text{ atm}$$

Process 2→3 is a constant pressure process, and hence, $p_3 = 56.6$ atm also.

By applying the ideal gas pVT equation with $p_3 = p_2$, one can determine the cut-off ratio r_c in this case as follows:

$$r_c = \frac{V_3}{V_2} = \frac{T_3}{T_2} = 1.58$$

Considering the isentropic expansion from state 3 to state 4′, one can use Equation 10.45b to determine $T_{4'}$ as follows:

$$T_{4'} = T_3 \left(\frac{r_c}{r_v} \right)^{k-1} = 663 \text{ K}$$

From Equation 10.45a, one can conclude that the work (w_s) done in the isentropic expansion from state 3 to state 4′ is

$$w_s = c_{vo}(T_3 - T_{4'})$$

From the definition of the process efficiency, as given by Equation 8.109, the actual work done during the adiabatic expansion process can be expressed as $w_a = w_s \eta_{aE}$, where η_{aE} is the specified process efficiency. Using the previous expression for w_s and applying Equation 10.45a to the actual adiabatic expansion process 3→4 yields the following for T_4:

$$T_4 = T_3 \left[1 - \eta_{aE} \left(1 - \frac{T_{4'}}{T_3} \right) \right] = 815 \text{ K}$$

Since $V_4 = V_1$, one can apply the ideal gas pVT equation to obtain the following value for the pressure p_4:

$$p_4 = p_1 \frac{T_4}{T_1} = 2.72 \text{ atm}$$

(b) Since values of the temperature at each state are known, one can make use of the equations for w and q resulting from the application of the NFEE to each of the processes. Thus, by applying the NFEE to the adiabatic compression process $1 \rightarrow 2$, the following can be obtained for w_{12}:

$$w_{12} = c_{vo}(T_1 - T_2) = -547 \text{ kJ/kg}$$

For the constant pressure expansion, the ideal work output is given by Equation 10.44a so that if one makes use of the process efficiency of 0.85 specified, the following can be obtained for w_{23}:

$$w_{23} = (\eta_{aE})R(T_3 - T_2) = 150 \text{ kJ/kg}$$

Also, w_{34} can be calculated from Equation 10.45a:

$$w_{34} = c_{vo}(T_3 - T_4) = 617 \text{ kJ/kg}$$

From these results for w, one can determine the positive work and the net work output as follows:

$$w_{+ve} = w_{23} + w_{34} = 767 \text{ kJ/kg}$$

$$w = w_{-ve} + w_{+ve} = w_{12} + w_{+ve} = 220 \text{ kJ/kg}$$

The specific work output is thus equal to 220 kJ/kg. This is less than 46% of the estimated 479 kJ/kg for the specific work output in the ideal case.

(c) For the thermal efficiency, in addition to the specific work output already determined one needs to compute the heat input (per unit mass) q_{23}, which can be determined by applying the NFEE to the actual heating process $2 \rightarrow 3$:

$$q_{23} = w_{23} + c_{vo}(T_3 - T_2) = 590 \text{ kJ/kg}$$

Hence, from Equation 7.2, the following can be obtained for the cycle efficiency:

$$\eta_1 = \frac{w}{q_{23}} \cdot 100\% = 37.3\%$$

(d) The mean effective pressure is calculated using Equation 10.34. In this case, $v_1 = 0.850 \text{ m}^3/\text{kg}$ as determined previously, while $w = 220 \text{ kJ/kg}$ with $r_v = 16$. Hence, the mean effective pressure is 276 kPa (or 2.72 atm).

(e) The efficiency ratio (ε) of the actual engine is given by Equation 10.1, which in this case is equal to $37.3/62.2 = 0.6$. This correlates with the low work ratio for the ideal cycle. The cycle efficiencies for actual diesel engines can be as high as about 35%.

For the Diesel cycle, the following relationship exists between r_c and r_v:

$$r_c = \frac{T_3}{T_1} \frac{1}{r_v^{k-1}} \qquad (10.50a)$$

The cut-off ratio must lie between 1 and r_v to have a physical meaning. In other words,

$$1 < r_c < r_v \tag{10.50b}$$

If r_c given by Equation 10.50a is substituted in the inequality of Equation 10.50b, one obtains the following for the range of possible values of r_v when T_1 and T_3 are specified:

$$(r_v)_{min} = \left(\frac{T_3}{T_1}\right)^{1/k} < r_v < \left(\frac{T_3}{T_1}\right)^{1/(k-1)} = (r_v)_{max} \tag{10.50c}$$

The expressions obtained earlier for q_{23}, w, r_w, and η_1 can be written in the following alternative forms in terms of r_c (and excluding r_v) by using Equation 10.50a for the relationship between r_v and r_c:

$$q_{23}^* = \frac{q_{23}}{c_{v0}T_3} = k\left(1 - \frac{1}{r_c}\right) \tag{10.51}$$

$$w^* = \frac{w}{c_{v0}T_3} = k\left(1 - \frac{1}{r_c}\right) - (r_c^k - 1)\frac{T_1}{T_3} \tag{10.52}$$

$$r_w = \frac{k(r_c - 1) - r_c(r_c^k - 1)\dfrac{T_1}{T_3}}{k(r_c - 1) + 1 - \dfrac{T_1}{T_3}r_c^{k+1}} \tag{10.53}$$

$$\frac{\eta_1\,(\%)}{100} = 1 - \frac{r_c(r_c^k - 1)}{k(r_c - 1)}\frac{T_1}{T_3} \tag{10.54}$$

The optimum compression ratio can be obtained by first finding $(r_c)_{opt}$ such that at $r_c = (r_c)_{opt}$,

$$\frac{\partial w^*}{\partial r_c} = 0 \quad \text{and} \quad \frac{\partial^2 w^*}{\partial r_c^2} < 0 \tag{10.55}$$

The corresponding $(r_v)_{opt}$ is then determined from Equation 10.50a. By differentiating w^* in Equation 10.52 with respect to r_c, it can be shown that the optimum condition corresponds to

$$(r_c)_{opt} = \left(\frac{T_3}{T_1}\right)^{1/(k+1)} \tag{10.56}$$

On substitution in Equation 10.50a, one obtains the following for the optimum compression ratio:

$$(r_v)_{opt} = \left(\frac{T_3}{T_1}\right)^{k/(k^2-1)} \tag{10.57}$$

Thus, by substituting in Equations 10.52, 10.53, and 10.54, the following results can be obtained for the optimum $w^*\left(= \dfrac{w}{c_{v0}T_3}\right)$, η_1 and r_w:

$$(w^*)_{\text{opt}} = k + \frac{T_1}{T_3} - (k + 1)\left(\frac{T_1}{T_3}\right)^{1/(k+1)} \tag{10.58}$$

and

$$\left(\frac{\eta_1\,(\%)}{100}\right)_{\text{opt}} = 1 - \frac{1}{k}\frac{\left(\frac{T_1}{T_3}\right)^{1/(k+1)} - \frac{T_1}{T_3}}{1 - \left(\frac{T_1}{T_3}\right)^{1/(k+1)}} = (r_{\text{w}})_{\text{opt}} \tag{10.59}$$

For a comparison with some of the earlier results, set $T_1 = 300$ K, $T_3 = 1100$ K, and $k = 1.364$. The optimum compression ratio is 7.84 and the cut-off ratio is 1.73. The corresponding values of η_1, r_{w}, and w^* are 47.2%, 0.472, and 0.272, respectively. Comparing these with the Otto cycle shows that the cycle efficiencies and the work ratios at the optimum condition are about the same for the Otto and Diesel cycles (see Figures 10.32 and 10.36). The optimum specific work output for the Diesel cycle, however, is nearly 20% higher than it is for the Otto cycle. Also the Diesel cycle operates with a higher compression ratio than does the Otto cycle. The diesel engine, designed for operation at its optimum point, is therefore larger in size than a comparable gasoline engine, designed for operation at its optimum point. As observed earlier in the discussion of the gasoline engine, actual diesel engines are designed to operate at much higher levels of peak temperature T_3 since this temperature level occurs intermittently and for a relatively short time in each machine cycle. Figure 10.36 shows the performance characteristics for the ideal reciprocating I.C. engine operating on the Diesel cycle with $T_1 = 300$ K, $T_3 = 1675$ K, and $k = 1.4$. The optimum compression ratio this time is 12.3, and the corresponding cut-off ratio is 2.05.

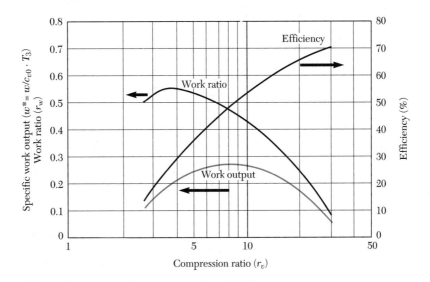

Figure 10.36 Performance characteristics for the ideal reciprocating I.C. engine operating on the Diesel cycle.

1. A four-stroke spark ignition (S.I.) engine works on an Otto-like cycle. Given $T_1 = 300$ K and $T_3 = 1800$ K, determine the optimum compression ratio assuming the following:
   (a) the cycle is the Otto cycle using air with $k = 1.4$
   (b) the cycle is the Otto-like cycle with process efficiencies for compression and expansion of 0.8 and 0.85, respectively

   Answer: (a) 9.39, (b) 5.8

2. For parts (a) and (b) in Exercise 1, determine the following at the corresponding optimum compression ratio: η_1, w, and m.e.p.

   Answer: (a) 59.2%, 453 kJ/kg, 588 kPa; (b) 34.9%, 280 kJ/kg, 393 kPa

3. For the Otto-like cycle in part (b) of Exercise 1, assume $r_v = 8$ and $p_1 = 100$ kPa. Calculate η_1, w, and m.e.p.

   Answer: 37.2%, 271 kJ/kg, 360 kPa

4. An air standard Diesel cycle operates with a compression ratio of 15. The maximum and minimum cycle temperatures are 1800 K and 300 K, respectively. Assume $p_1 = 100$ kPa and that $k = 1.4$ and $c_p = 1.005$ kJ/kg · K for air. Determine
   (a) the cut-off ratio, r_c
   (b) the specific net work output, w
   (c) the thermal efficiency, η_1
   (d) the work ratio, r_w
   (e) the mean effective pressure, m.e.p.

   Answer: (a) 2.03, (b) 553 kJ/kg, (c) 60.2%, (d) 0.568, (e) 688 kPa

5. An actual Diesel-like cycle operates with $r_v = 15$, as in Exercise 4. The maximum and minimum cycle temperatures are again 1800 K and 300 K, respectively, and $p_1 = 100$ kPa. Assume process efficiencies of 0.8 and 0.85 for the compression and power strokes, respectively. Determine
   (a) the cut-off ratio, r_c
   (b) the specific net work output, w
   (c) the thermal efficiency, η_1
   (d) the efficiency ratio, ε
   (e) the mean effective pressure, m.e.p.

   Answer: (a) 1.74, (b) 295 kJ/kg, (c) 40%, (d) 0.664, (e) 367 kPa

6. Determine the optimum r_v and r_c for the air standard Diesel cycle of Exercise 4.

   Answer: 13.6, 2.11

10.6   *Optimum Power Cycles*

The heat engines considered in Sections 10.3 through 10.5 are representative of the major categories in use today. The approach in each case was to begin with an analysis of the ideal engine cycle and then to apply the results to the actual heat engine via the use of process efficiencies that can be determined empirically. For any heat engine design, the appropriate ideal engine cycle is one that closely matches the operation of the actual engine. A key assumption made concerning ideal cycles is that all the processes are internally reversible. One cannot, however, rule out the possibility of external irreversibility, and therefore the various ideal cycles that have been considered are not necessarily the optimum power cycles. Both Sections 9.4 and 10.2 identified the *thermodynamically optimum cycle* in any given situation as one in which all processes are completely (internally and externally) reversible. Thus, a Carnot cycle between a pair of thermal reservoirs is the optimum cycle, provided the heat input and the heat rejection take place at the source and sink temperatures, respectively. If, instead, a heat engine operates on the ideal Carnot cycle between a finite (or limited) heat source and an unlimited sink, the cycle is no longer the optimum cycle since the heat input to the engine now takes place in an (externally) irreversible manner.

The main purpose in the rest of this chapter is to consider from the perspective of the second law of thermodynamics the theoretical limits of performance when one is faced with the typical constraints encountered in the operation of actual heat engines. One such constraint is that available heat sources are invariably finite and limited. In addition, the characteristics of the various working fluids are not always favorable and may in fact have a decisive effect on how close to the optimum cycle actual heat engines can ever get. Another constraint is one imposed by metallurgical considerations, as previously discussed. The *metallurgical limit* was defined as the maximum temperature allowable in the cycle so as to safeguard against engine component failure. The following discussions consider the nature of the heat sources and sinks that are available in the environment, as well as the characteristics of the most often used working fluids that are popular because they are relatively inexpensive and safe to use. A model is developed for the optimum cycle in a similar manner to what was done in Section 9.4, except that the scope of the discussion here also includes the internal combustion engine.

Nature of Available Heat Sources and Sinks in the Environment

Infinite or *unlimited* heat sources do not occur naturally in the environment. Geysers and hot water springs represent naturally occurring examples of heat sources with large thermal capacities. Unfortunately, they are only found in limited areas. Also, as heat is extracted from a stream of geothermal water, the temperature of the water stream falls. Thus, this heat source is finite in the sense defined in Sections 9.4 and 10.2. Solar ponds, using a large reservoir of water, constitute an approximation to a thermal reservoir, provided the temperature of the water in the pond remains constant when heat is extracted from it. In the case of solar ponds, however, the source temperature is rather low, on the order of 350 K, so that with a sink temperature of about 300 K, the Carnot cycle efficiency attainable is only about 14%.

Heat sources for most heat engines are not naturally occurring. Rather, they are created, usually via the combustion of hydrocarbon fuels. The combustion of such fuels effectively transforms the chemical exergy of the fuel to the thermal exergy of the combustion products whose temperature may be as high as about 2000 K. If the combustion products then serve as the heat source for a heat engine, one would expect the temperature of the heat source to fall during the heat exchange process as the engine draws more and more heat from it. This is the typical example of a *finite heat source*, which was discussed in Sections 9.4 and 10.2. For a finite heat source, the best that can be achieved is the complete conversion of the exergy of the heat source to work. The optimum cycle in such cases in not the Carnot cycle but rather the *trilateral cycle*.

One other type of heat source that was identified by Wilson and Radwan (1977) is a *regeneratable heat source*. An example of this is provided by a nuclear power plant, in which the coolant after passing through the reactor core serves as the source of heat to the working fluid used in the engine cycle. Subsequently, the coolant is regenerated by recirculating it through the reactor core. When the heat source is regeneratable, the optimum cycle is nearest to what Wilson and Radwan have termed the *quadlateral cycle*.

The real situation for heat sinks is generally more favorable. Large sources of water such as rivers and lakes are common, and these can serve effectively as constant temperature heat reservoirs. This is a major reason for siting thermal power stations close to large bodies of water. Important exceptions to this favorable situation do exist, however. Gas turbines and reciprocating I.C. engines, which are dominant in transportation, operate on an open cycle and do not make use of water reservoirs as sinks. Such constraints are important in ascertaining how closely an actual engine can match the optimum cycle performance.

Let's now briefly consider the characteristics of working fluids in relation to the optimum heat engine cycle dictated by the nature of the heat source and sink available. The working fluids commonly used in heat engines are air and water, both of which are inexpensive and abundant in the Earth's environment.

Characteristics of Working Fluids

The design of heat engines is not normally based on the Carnot cycle for reasons that have been repeatedly mentioned in this chapter. Low work ratios are typical of the Carnot cycle; therefore, the efficiency of an actual heat engine based on the Carnot cycle is likely to be much lower than the ideal cycle efficiency. Also, when the heat source is finite, the Carnot cycle is no longer the optimum cycle.

Other reasons exist for not basing heat engine designs on the Carnot cycle. For example, while the Carnot cycle requires the heat input and the heat rejection processes to occur at constant temperatures, in practice, such processes are more easily designed to take place either at constant pressures or at constant volumes. If the working fluid is water, a constant pressure process is also a constant temperature process only for fluid states in the saturation region. The critical temperature of water is near 650 K, whereas typical temperatures achieved in the combustion of hydrocarbon fuels are about 2000 K. Thus, a Carnot cycle using water as the working fluid is constrained to operate with a heat input at a much lower temperature than that of the heat source. With air as the working fluid, the temperature rises sharply with heat addition either at a constant pressure or a constant volume. In either case, it is easy to see that the Carnot cycle is generally not the optimum cycle to aim at in heat engine design.

Use of working fluids other than water and air are more appropriate in specific situations. For instance, when the heat source temperature is low, the use of a *refrigerant fluid* as the working fluid in a vapor power plant is often the only realistic option. In such circumstances, the characteristics of the refrigerant fluids are more appropriate than those of water. However, the danger to the environment posed by some refrigerants currently available excludes their use on a large scale despite their extreme suitability on other grounds. Mercury is another fluid that has excellent characteristics for applications involving both a very high source temperature and a moderately high sink temperature. However, it suffers from the fact that it is a "nonwetting" fluid, and even more importantly, it poses serious danger to life if its vapor is inhaled by humans. In addition, mercury is expensive, which together with the cost of ensuring leak-proof containment of the fluid can make designs of mercury vapor plants relatively uneconomical.

Optimum Cycles for External Combustion Engines

Most external combustion engines operate on cycles that comprise flow processes. A notable exception is the reciprocating engine operating on the *Stirling cycle*, which comprises a set of nonflow processes. The various vapor power cycles discussed in Section 10.3 operate on cycles consisting of flow processes; the same is true of the closed-cycle gas turbine mentioned in Section 10.4. The gas turbines operating on the open cycle could be considered here too, by making the usual assumption of an equivalent air standard cycle, as indicated in Section 10.4. Discussion of optimum cycles for external combustion engines is focused on the cases involving flow processes, as illustrated in Figure 10.37.

From the discussion in Section 9.4, the optimum cycle for any external combustion engine operating between a finite heat source and the ambient environment as the sink is the trilateral cycle, which was illustrated on the T-S° diagram of Figure 9.15. Recall that S° is the entropy flow rate defined by Equation 9.41d and is the product of the fluid mass flow rate and the specific entropy. The use of S° was considered appropriate when dealing with flow processes involving the flow of more than one type of substance.

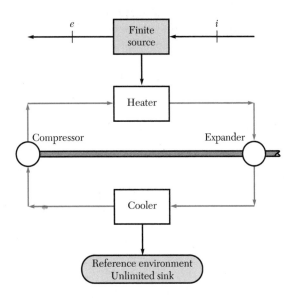

Figure 10.37 Model of a typical external combustion engine with the engine cycle comprising flow processes.

For ideal or quasi-equilibrium flow processes, the integral of T and dS° gives the heat flow rate in much the same way that the integral of T and dS represents the reversible heat flow to a closed system.

When the heat source is a stream of the gaseous products of combustion of a hydrocarbon fuel, it is unwise to allow the gases to be cooled to the point at which condensation of the moisture in the mixture occurs. If allowed, such condensation could result in the corrosion of the metallic components of the power plant. Wilson and Radwan (1977) have suggested keeping the temperature of the stream of the combustion products above 425 K to safeguard against the possibility of metal corrosion in the construction of the plant. Figure 10.38 illustrates the more realistic or constrained optimum cycle when the flue gas temperature has to be kept higher than the ambient temperature. The four-sided cycle indicated has been termed the *quadlateral cycle* by Wilson and Radwan (1977). The loss of exergy is indicated by the shaded region in Figure 10.38 and is equivalent to the residual exergy of the flue gas stream. It should be noted, however, that for a heat engine operating between a regeneratable heat source and the reference environment at T_0, the quadlateral cycle is actually the optimum cycle. The heat source fluid in such an instance reenters the regenerator (for example, the core of a nuclear reactor) at T_e and leaves at T_i to supply heat to the working fluid of the engine.

The heat flow from the heat source to the working fluid normally takes place through a device called the *heat exchanger*. This device was introduced in Example 9.14 (in Chap. 9). One can obtain an expression for the irreversibility associated with the source heating process in the case of a heat engine operating on an ideal cycle (not necessarily the optimum cycle) by applying the first and second laws of thermodynamics to miniature control volumes defined for the heat exchanger in much the same way as done in Section 9.4. Figure 10.39 illustrates for the heat exchanger the variation in temperature as a function of the surface area of the heat exchanger for both the heating fluid (the heat source fluid) and the heated fluid (the working fluid.) Figure 10.40 is a conceptual representation of the miniature control volumes for heat flow from the heat source fluid

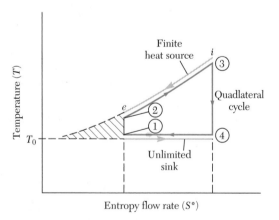

Figure 10.38 Constrained optimum cycle for a heat engine operating with a finite or regeneratable heat source and an unlimited heat sink.

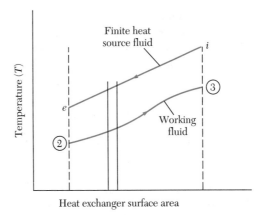

Figure 10.39 Temperature distributions for the heat source fluid and the working fluid in a counterflow heat exchanger design.

stream to the working fluid stream. With reference to each of the miniature control volumes shown in the figure, an energy balance is obtained by applying the SFEE (Eq. 6.29). Thus, if δQ is the local heat flow in the time interval t to $t + \Delta t$ from the heat source fluid to the working fluid, the following is obtained by applying the reduced form of the energy equation:

$$-\delta Q = (\dot{m}_{source}\, \Delta t)\,(dh)_{source} \tag{10.60a}$$

and

$$\delta Q = (\dot{m}_{\substack{working\\fluid}}\, \Delta t)\,(dh)_{\substack{working\\fluid}} \tag{10.60b}$$

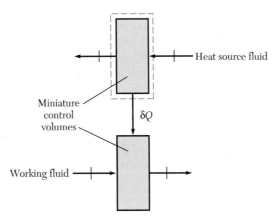

Figure 10.40 Conceptual representation of the heat flow from the heat source fluid stream to the working fluid stream.

The corresponding exergy input and exergy product can be determined by applying Equation 9.33 with the kinetic energy and the potential energy terms assumed negligible. Considering the heat source fluid stream and the working fluid flow in turn thus leads to the following:

$$\Delta \chi_{\text{input}} = (\dot{m}_{\text{source}} \, \Delta t)[-(dh)_{\text{source}} + T_0(ds)_{\text{source}}] \qquad (10.61a)$$

and

$$\Delta \chi_{\text{produced}} = (\dot{m}_{\text{working} \atop \text{fluid}} \, \Delta t)[(dh)_{\text{working} \atop \text{fluid}} - T_0(ds)_{\text{working} \atop \text{fluid}}] \qquad (10.61b)$$

Furthermore, if one assumes that the pressure is constant for both fluid streams, Equation 8.14 shows that $T \, ds = dh$ for each stream. Using this with Equations 10.60a and 10.60b enables the expressions for incremental exergy to be rewritten in the following alternative form:

$$\Delta \chi_{\text{input}} = \delta Q \left(1 - \frac{T_0}{T_{\text{source}}} \right) \qquad (10.62a)$$

and

$$\Delta \chi_{\text{produced}} = \delta Q \left(1 - \frac{T_0}{T_{\text{working} \atop \text{fluid}}} \right) \qquad (10.62b)$$

The exergy consumption (from Eq. 9.9) is the difference between the exergy input and the exergy produced. Equations 10.62a and 10.62b give the following for the exergy consumed:

$$\delta \chi_{\text{consumed}} = \delta Q \cdot T_0 \left(\frac{1}{T_{\text{working} \atop \text{fluid}}} - \frac{1}{T_{\text{source}}} \right) \qquad (10.63)$$

An independent equation for δQ in terms of the temperatures of the heat source fluid and the working fluid can be obtained by applying the rate Equation 5.31 for the heat transfer process. Thus, if U_e denotes an effective or overall heat transfer coefficient for heat transfer between the two fluid streams and ΔA_s is the heat exchanger surface area through which the δQ passes, the following additional equation can be obtained for the heat flow:

$$\delta Q = (U_e \, \Delta A_s)(T_{\text{source}} - T_{\text{working} \atop \text{fluid}})(\Delta t) \qquad (10.64)$$

Substituting this expression for δQ in Equation 10.63 for the exergy consumption leads to the following final expression for the local exergy consumption rate:

$$\delta \dot{\chi}_{\text{consumed}} = T_0 \frac{(\Delta T)^2}{T_{\text{working} \atop \text{fluid}} T_{\text{source}}} (U_e \, \Delta A_s) = T_0 \, \delta \dot{S}_{\text{gen}} \qquad (10.65)$$

In Equation 10.65, ΔT is the local temperature difference between the heat source fluid and the working fluid. The connection indicated in the equation between the entropy generation rate and the exergy consumption rate is based on Equation 9.37, which is a form of the Gouy–Stodola lost work theorem for steady-state flow processes. Equations 10.64 and 10.65 give the following for the total heat input rate to the engine cycle and the total exergy consumption rate due to the irreversibility of the source heating process:

$$\dot{Q}_H = \int_{A_s} (\Delta T) U_e \, dA_s = (U_e A_s)(\Delta T)_m \tag{10.66}$$

and

$$\dot{X}_{consumed} = T_0 \dot{S}_{gen} = T_0 \int_{A_s} \frac{(\Delta T)^2}{T_{\substack{working \\ fluid}} T_{source}} U_e \, dA_s$$

$$\approx \frac{T_0 (\Delta T)_m}{\left(T_{\substack{working \\ fluid}} T_{source} \right)_m} \dot{Q}_H \tag{10.67}$$

where the subscript "m" designates an appropriate mean for the integration with respect to dA_s and A_s denotes the heat exchanger surface. In heat transfer analysis, $(\Delta T)_m$ in Equation 10.66 is the popular *logarithmic mean temperature difference* for the two fluid media involved. The approximate result in Equation 10.67 should be good for an order of magnitude estimate of the exergy consumption rate in the source heating process as compared with the heat input rate to the cycle.

One can carry out an analysis similar to this to obtain an expression for the exergy consumption rate in the heat rejection to the unlimited sink assumed in Figure 10.37. The result is similar to Equation 10.67 and can be written as follows:

$$(\dot{X}_{consumed})_{\substack{heat \\ rej}} = T_0 \dot{S}_{gen} = T_0 \int_{A_s} \frac{(\Delta T)^2}{T_{\substack{working \\ fluid}} T_{sink}} U_e \, dA_s$$

$$\approx \frac{(\Delta T)_m}{\left(T_{\substack{working \\ fluid}} \right)_m} \dot{Q}_L \tag{10.68}$$

where $(\Delta T)_m$ is an appropriate mean temperature difference between the working fluid and the unlimited heat sink when the former is rejecting heat to the sink. Both Equations 10.67 and 10.68 confirm the widely held notion that the exergy consumption rate is directly related to the magnitude of ΔT between the fluid media involved. The equations can also serve as the basis for an approximately quantitative and qualitative comparison of ideal cycles with the constrained optimum cycle from the perspective of the second law of thermodynamics. The following example demonstrates how this can be done.

EXAMPLE 10.16
Comparison of an Ideal Cycle with an Optimum Cycle

A heat engine is to be constructed that uses as its heat source a hot stream of gaseous products of combustion at 1 atm pressure and 2000 K. The heat sink is atmospheric air at 1 atm and 300 K. Assume that the ambient air is also the reference environment. From practical considerations, the temperature of the stream of the gaseous products of combustion is set at 450 K at discharge to the atmosphere. Assuming a source heat input rate of 100 MW, determine the maximum possible mechanical power output for each of the following ideal heat engine cycles:

(a) a *quadlateral cycle*, such as that illustrated in Figure 10.38

(b) a simple *vapor power cycle* operating with a maximum temperature of 800 K and a ΔT of at least 15 K between fluid streams in a heat exchanger

Given: A finite heat source comprises a gaseous stream at 1 atm (see Fig. 10.37) with $T_i = 2000$ K and $T_e = 450$ K. The heat sink is unlimited, and is the reference environment for which $p_0 = 1$ atm and $T_0 = 300$ K.
Source heat input rate = 100 MW
Two ideal cycles are to be considered:

(a) the quadlateral cycle of Figure 10.38

(b) a simple vapor power cycle with cycle $T_{max} = 800$ K and $(\Delta T)_{min} = 15$ K (for heat exchangers)

Find: The mechanical power output in (a) and in (b).

Solution:

(a) For the quadlateral cycle illustrated in Figure 10.38, one can infer from the discussions in Section 9.4 that the area enclosed by the path 1–2–3–4–1 on the T-S^* diagram gives the mechanical power output of the engine operating on the cycle. In other words, the power output is given by

$$(\dot{W}_x)_{quad} = \int_{S_e^*}^{S_i^*} T \, dS^* - T_0(S_i^* - S_e^*)$$

$$= \dot{Q}_H - T_0(S_i^* - S_e^*)$$

By applying the SFEE (Eq. 6.29) to the control volume for the stream of the combustion products, the following can be written for the heat input rate to the engine cycle:

$$\dot{Q}_H = H_i^* - H_e^*$$

where H^* denotes the enthalpy flow rate and is defined as the product of the mass flow rate and the specific enthalpy of the fluid. Substituting this in the expression for the power output of the quadlateral cycle yields the following:

$$(\dot{W}_x)_{quad} = \dot{Q}_H\left(1 - \frac{T_0}{T^*}\right)$$

where

$$T^* = \frac{H_i^* - H_e^*}{S_i^* - S_e^*} = \frac{h_i - h_e}{s_i - s_e}$$

Assuming that the heat source fluid can be taken as an ideal gas, a simplified expression for T^* is obtained by applying Equations 4.31 and 8.25 and using the fact that $p_e = p_i = p_0$. These lead to the following simple expression for T^*:

$$T^* = \frac{T_i - T_e}{\ln \dfrac{T_i}{T_e}}$$

Substituting $T_i = 2000$ K and $T_e = 450$ K gives $T^* = 1039$ K, and hence, the following is the mechanical power output for an engine operating on the quadlateral cycle:

$$(\dot{W}_x)_{\text{quad}} = \dot{Q}_H \left(1 - \frac{T_0}{T^*}\right) = \underline{71.1 \text{ MW}}$$

The cycle efficiency in this case is 71.1%. It should also be noted that the shaft power of 71.1 MW for the constrained optimum cycle is also the exergy input rate to the simple vapor power cycle in part (b).

(b) Figure 10.41 gives the most useful representation of the engine cycle in relation to the process path for the finite heat source. Both processes are shown on the T-s coordinates, except that a scaling factor must be used for the specific entropy of either the heat source fluid or the working fluid so that the specific entropy at states i and e for the heat source fluid match those for the working fluid at states 3 and 2, respectively. The specified minimum ΔT of 15 K can be used for the heat exchangers to obtain $T_1 = 300 + 15 = 315$ K $= T_4$ and $T_2 = 450 - 15 = 435$ K. $T_3 = 800$ K is specified. An estimate of the exergy consumption rate in

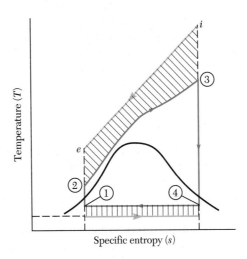

Figure 10.41 Process paths for a finite heat source and a simple vapor power cycle on T-s coordinates.

the source heating process can be obtained using the approximate form of Equation 10.67 with the following approximations:

$$(T_{\text{source}})_m \approx \frac{1}{2}(T_e + T_i) = 1225 \text{ K}$$

$$(T_{\text{working fluid}})_m \approx \frac{1}{2}(T_2 + T_3) = 618 \text{ K}$$

Hence,

$$(\Delta T)_m \approx 1225 - 618 = 607 \text{ K}$$

Substituting in the approximate form of Equation 10.67 yields an estimate of 24 MW for the exergy consumption rate in the irreversible heat flow from the source fluid to the working fluid across a finite ΔT.

For the exergy consumption rate in the heat rejection process, $(\Delta T)_m$ is 15 K, while the average temperature for the working fluid is 315 K. Applying the approximate form of Equation 10.68 thus yields

$$(\dot{X}_{\text{consumed}})_{\text{heat rej}} \approx 0.05\dot{Q}_L = 0.05[\dot{Q}_H - (\dot{W}_x)_{\text{vap cycle}}]$$

The shaft power produced is the exergy input rate of 71.1 MW minus the exergy consumption rates in the heat input and heat rejection processes. Thus,

$$(\dot{W}_x)_{\text{vap cycle}} = 71.1 - 24 - 0.05(100 - (\dot{W}_x)_{\text{vap cycle}}) \text{ MW}$$

From this, an estimate of about 44 MW is obtained for the shaft power produced.

The corresponding estimate for the exergy consumption rate in the heat rejection process is 2.8 MW. The cycle efficiency is about 44%. The peak efficiencies for steam power plants that have been built are on the order of 40%. This analysis also indicates that the more significant exergy consumption due to the irreversibility associated with heat flow across a finite ΔT occurs during the source heating process. The shaded regions in Figure 10.41 provide a qualitative indication of the exergy consumption level in the heat input and the heat rejection processes.

Optimum Cycle for Gas Engines

This discussion of optimum heat engine cycles is concluded by briefly considering gas engines. Most gas engines are internal combustion engines for which the ideal cycles frequently assumed are the air standard cycles. The closed-cycle gas turbine is a possible exception for which the Brayton cycle model would be a reasonably accurate representation. Figure 10.42 illustrates the process path for a regeneratable heat source for the closed-cycle gas turbine alongside the Brayton cycle for the engine. The diagram uses T-s coordinates in a similar fashion to Figure 10.41. Exergy consumption is likely to be significant in both the heat addition process (from the source) and the heat rejection process (to the sink), as indicated by the shaded regions in Figure 10.42. Figure 10.43 illustrates the case when the heat source is again regeneratable and the gas engine operates in a closed cycle. The irreversibility associated with the heat rejection process can be reduced by substituting an isothermal compression process for the isentropic

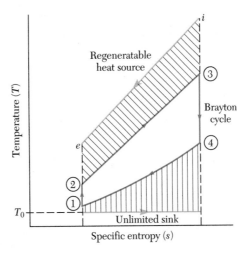

Figure 10.42 Process paths for a regeneratable heat source and a Brayton cycle on T-s coordinates.

compression process of the Brayton cycle, as indicated in Figure 10.43. A regenerator is then employed between the constant pressure cooling process 4→1 and the initial heating along the constant pressure process path 2→3. The heat rejection to the unlimited sink now occurs during the isothermal compression process 1→2, and the thermal irreversibility associated with this is considerably less than that illustrated in Figure 10.42. This modification is incorporated in the *Ericsson cycle*, which was mentioned in Section 10.4.

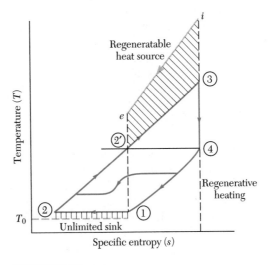

Figure 10.43 Process paths for a regeneratable heat source and a (closed) gas turbine cycle with regenerative heating.

Consideration of exergy consumption for the internal combustion gas engines cannot be very meaningful until one has considered the irreversibility associated with actual combustion processes. The thermodynamics of combustion and chemical reactions are discussed in Chapters 14 and 15. At this stage, the most useful approach to the second law analysis of the internal combustion processes is probably the type used in Example 10.11.

10.7 Summary

This chapter began with a survey of prime movers in general and then focused on heat engines, which are prime movers of the first kind. The impetus for prime mover development can be attributed to the perennial desire in every society to substitute machine power for human or animal power. Heat engines, though a relatively recent development, are extremely versatile and are the most widely used power-producing devices today.

Section 10.2 discussed modeling of heat engines using ideal heat engines for reference. Performance criteria for heat engines were introduced. Such criteria include the thermal or first-law efficiency, the efficiency ratio, the work ratio, and the second-law efficiency. The work ratio and the second-law efficiency in particular were considered key indices of performance. While the efficiency ratio compares actual heat engine efficiency with that of the ideal heat engine, the work ratio provides a priori indication of the susceptibility of the engine cycle to the adverse effects caused by irreversibilities occurring in actual processes. A low work ratio points to a low efficiency ratio, while a work ratio close to unity indicates that the efficiency of the actual engine will be close to that of the ideal heat engine. Throughout the chapter the design and performance of heat engines were discussed from the standpoint of the first and second laws of thermodynamics.

Vapor power cycles were considered in detail in Section 10.3. Steam is by far the most common choice of working fluid for vapor power plants, and the basic cycle is the Rankine cycle, not the Carnot cycle. The work ratio is generally close to unity for Rankine cycles, and as a result, the efficiency of an actual engine operating on the Rankine cycle is close to that for the ideal cycle. In contrast the work ratio is low for the Carnot cycle, resulting in much lower thermal efficiency for actual engines operating on this cycle. The list of ideal cycles considered in Section 10.3 includes the basic Rankine cycle and several modifications of this basic cycle. The modifications are designed to improve performance as well as overcome difficulties encountered with operation on the basic Rankine cycle. These other cycles include the Rankine cycle with superheat, the Rankine cycle with reheat, the Rankine cycle with regeneration, and the binary cycle. It was noted that from the second-law standpoint, vapor power plants using steam as working fluid suffer from a substantial consumption of exergy occurring in the steam generation process. This external irreversibility is most severe when the heat source is the hot gaseous stream produced in the combustion of fuel in air. It was also noted that the Carnot cycle would be even more inappropriate for the vapor power plant when the heat source is finite in view of the substantial exergy loss associated with the exit stream of the heat source fluid.

Gas turbines were discussed in detail in Section 10.4. Although gas turbines in practice operate on open cycles, the ideal engine operates on an air standard cycle called the Brayton or Joule cycle. Expressions were developed for the specific work output, the work ratio, and the thermal efficiency for the ideal cycle by applying the steady flow energy equation to each flow process and making use of isentropic relationships between properties based on the second law of thermodynamics. Gas turbine cycles have rather low work ratios and are therefore significantly affected by irreversibilities in the processes for an actual engine. Deviations of actual cycles from the ideal were discussed mainly in terms of the internal irreversibilities occurring in the compression and expansion processes. A major design parameter for the gas turbine is the pressure ratio, and an optimum pressure ratio can be determined that gives the maximum specific work output when the minimum and maximum cycle temperatures are specified. Modifications to the Brayton cycle were discussed, including multiple-stage compression (with intercooling), multiple-stage expansion (with reheat), and regeneration cycles. The Ericsson cycle was mentioned as the limiting case of isothermal compression and expansion with complete regeneration. This cycle has the same thermal efficiency as the Carnot cycle, and its use is appropriate only when the heat source is unlimited or regeneratable.

Reciprocating internal-combustion engines were considered in Section 10.5. The air standard cycle for the spark ignition (S.I.) or gasoline engine is the Otto cycle, while for the compression ignition (C.I.) or diesel engine, the ideal cycle is the Diesel cycle. Expressions for the specific work output, the work ratio, and the thermal efficiency were developed for both cycles using the nonflow energy equation (first law analysis) and the isentropic relationships between properties based on the second law of thermodynamics. The compression ratio is an important parameter for both the gasoline engine and the diesel engine. The optimum compression ratio (corresponding to maximum specific work output) was determined in terms of the minimum and maximum cycle temperatures for each cycle. Deviations of actual cycles from the ideal cycle were discussed mainly with reference to internal irreversibilities.

Optimization of power cycles was discussed for both external-combustion and internal-combustion engines in Section 10.6. The nature of heat sources and sinks and the characteristics of the working fluid were considered to be the deciding factors in the choice of the best or optimum cycle for a heat engine. Typical constraints on the operation of actual heat engines were discussed. These constraints include (1) an upper limit on cycle temperatures allowed based on metallurgical considerations, and (2) limitation of the working fluid to constant pressure or constant volume processes during the heat input and heat rejection processes. Approximate expressions were developed for the exergy consumption rate during the heat addition and heat rejection processes for external-combustion engines. Similar considerations were made in a qualitative manner for internal-combustion engines using the air standard cycles as the model for such engines.

References

Angrist, S. W. and Hepler, L. G. 1967. *Order and Chaos: Laws of Energy and Entropy.* New York: Basic Books.

Cape Range NL. 1992. "Kalina cycle successfully demonstrated," *Professional Engineering*. Vol. 5, No. 6. 40.

Gaggioli, R. A. and Petit, P. J. 1977. "Use the Second Law, First," *Chemtech*. 496–506.

Kalina, A. I. 1984. "Combined-Cycle System with Novel Bottoming Cycle," ASME Journal of Engineering for Gas Turbines and Power. Vol. 106. 737–742.

Loftness, R. L. 1984. *Energy Handbook*. 2nd ed. New York: Van Nostrand Reinhold.

Wilson, S. S. and Radwan, M.S. 1977. "Appropriate Thermodynamics for Heat Engine Analysis and Design," *International Journal of Mechanical Engineering Education*. Vol. 5, No. 1. 68–80.

Questions

1. Distinguish between prime movers of the first kind and those of the second kind. Give two examples of each.

2. Identify the form of energy input to each of these work-producing devices:
   (a) the windmill
   (b) a hydraulic turbine
   (c) the steam engine
   (d) the aircraft gas turbine
   (e) the gasoline engine
   (f) the ox

3. Suggest typical conversion efficiencies and power output ranges for the devices listed in Question 2.

4. Define the heat engine and explain how one can model an internal combustion engine as a heat engine.

5. Distinguish between *unlimited, finite,* and *regeneratable* heat sources and give a practical example of each.

6. Under what circumstances is the Carnot cycle truly the optimum engine cycle?

7. Define the work ratio for an ideal heat engine cycle and explain why it is a particularly important index of how close the actual cycle efficiency is to the efficiency of the corresponding ideal cycle.

8. What is the obvious distinguishing feature in the outward appearance of the gasoline engine (without fuel injection) and the diesel engine?

9. Indicate whether or not the proposed use of the following prime movers is feasible and give reason(s) for your answer:
   (a) diesel engine for flying an airplane
   (b) steam engine for the automobile
   (c) gas turbine for locomotives
   (d) gas turbine for stand-by power unit for an electric utility
   (e) steam engine for the tractor unit on a farm

10. Water is a popular choice of fluid for vapor power plants. List two points in favor of the choice and one point against it.

11. List the processes that comprise the mechanical cycle of a reciprocating I.C. engine. Compare the advantages and disadvantages of the four-stroke engine relative to the two-stroke engine.

12. Illustrate by means of sketches, the operation of the two-stroke and the four-stroke spark ignition reciprocating I.C. engine. Give sketches of typical indicator diagrams for these engines. Briefly discuss the relative merits of the two types of engine.

Problems

10.1. A heat engine operating on an ideal cycle receives 100 kJ of heat from its source and produces 40 kJ of work. The positive work component is 60 kJ. Determine
(a) the thermal efficiency for the ideal cycle
(b) the work ratio (r_w)

10.2. In actual operation, the heat engine in Problem 10.1 does positive work with a process efficiency of 80%, while negative work is executed with a process efficiency of 75%. The heat input per cycle is 100 kJ. Determine
(a) the actual positive work output per cycle

(b) the actual negative work done per cycle

(c) the actual net work output per cycle

(d) the actual cycle efficiency

(e) the efficiency ratio for the heat engine

10.3. Water at 0.1 bar and 300 K is compressed reversibly and adiabatically to a pressure of 50 bar in a steady-flow process using a feed pump device. What is the work input required per unit mass of the fluid?

10.4. The feed pump in Problem 10.3 operates with an adiabatic efficiency of 75%. Determine the actual specific work input required.

10.5. Dry, saturated steam at 50 bar expands through an adiabatic turbine to a pressure of 0.1 bar. Determine the specific work output of the turbine

(a) when the turbine operates reversibly

(b) if the adiabatic efficiency of the turbine is 85%

What is the steam quality upon exit from the turbine in parts (a) and (b)? Use the steam property tables for this problem.

10.6. Wet steam at 0.1 bar is compressed reversibly and adiabatically in a steady-flow process such that the fluid is in the saturated liquid state at 50 bar upon exit from the compression device. Determine the specific work input required. Use the steam tables for the fluid properties.

10.7. If an adiabatic compression device is used in Problem 10.6 with the same wet steam condition upon entry to the device and an identical exit pressure of 50 bar, determine the actual specific work input that is required assuming that the adiabatic efficiency of the device is 55%.

10.8. Air at 1 bar and 300 K is compressed reversibly and adiabatically to a pressure of 50 bar. Determine the specific work input required in

(a) a nonflow compression process

(b) a steady-flow compression process

10.9. If the compression in Problem 10.8 were accomplished using adiabatic devices with process efficiencies of 75%, determine the actual specific work input required in parts (a) and (b).

10.10. Air at 50 bar and 1050°C expands through an adiabatic turbine to a pressure of 1 bar in a steady-flow process. Determine the specific work output when

(a) the turbine operates reversibly

(b) the turbine operates with a process efficiency of 85%

10.11. Determine the maximum thermal efficiencies possible for heat engines with a sink temperature of 300 K and the following source temperatures:

(a) 2000 K (typical level of temperatures in the combustion of hydrocarbon fuels)

(b) 800 K (typical acceptable limit, based on metallurgical considerations)

(c) 90°C (typical of geothermal sources)

(d) 65°C (typical of flat plate solar collectors)

* 10.12. Given a pair of thermal reservoirs at 300 K and 1000 K, calculate the thermal efficiency and the second law efficiency for an ideal heat engine operating on the Carnot cycle with heat input temperature T_H and heat rejection temperature T_L as given below. Assume $T_0 = 300$ K.

(a) $T_L = 300$ K, $T_H = 1000$ K

(b) $T_L = 300$ K, $T_H = 900$ K

(c) $T_L = 400$ K, $T_H = 1000$ K

(d) $T_L = 400$ K, $T_H = 900$ K

* 10.13. A heat engine operates internally reversibly on the Carnot cycle with heat input to the cycle occurring at 1000 K, while the rejection of heat from the cycle is at a constant temperature of 300 K. The external heat source is at 2000 K and the sink is at $T_0 = 300$ K. In each cycle, the heat input to the cycle is 100 kJ.

(a) Calculate

   (i) the thermal efficiency of the engine

   (ii) the work output (in kJ) of the engine per cycle

   (iii) the exergy input (in kJ) from the external heat source

   (iv) the second law efficiency achieved

(b) Use a *T-S* diagram to show the process paths for

   (i) the external heat source

   (ii) the engine cycle

   (iii) the heat sink

Indicate the area on the diagram that corresponds to the exergy destruction due to external irreversibility.

10.14. A steam plant operates on the basic Rankine cycle between a condenser pressure of 8 kPa and a boiler pressure of 50 bar. Process 1→2 is the compression process, 2→3 the heating in the boiler at the boiler pressure given, 3→4 the steam expansion through the turbine, and 4→1 the condensation process. Show the cycle on the *T-s* diagram. Calculate

(a) w_{12}, q_{23}, and w_{34}

(b) the work ratio and the cycle efficiency

(c) the specific steam consumption (in kg/kWh)

What is the steam quality (x_4) upon exit from the turbine?

10.15. Repeat Problem 10.14 for a boiler pressure of 80 bar but the same condenser pressure of 8 kPa. Is the steam quality at the exit to the turbine higher or lower than what is acceptable in practice?

10.16. Repeat Problem 10.14 but assume that the plant is to operate on the Carnot cycle. Assume that the working fluid is in the saturated liquid condition at the end of the compression process (state 2), while it is in the dry, saturated vapor condition (state 3) upon entry to the turbine. Based on the results of your calculations here and in Problem 10.14, which cycle would you choose for your plant design?

10.17. An actual steam plant is designed to operate on the basic Rankine-like cycle between a condenser pressure of 8 kPa and a boiler pressure of 50 bar, as in Problem 10.14. The adiabatic efficiency for the turbine is 85%. Determine

(a) the actual specific (net) work output, w (in kJ/kg)

(b) the actual cycle efficiency

(c) the efficiency ratio achieved

10.18. Suppose the steam plant in Problem 10.17 was instead designed to operate on a Carnot-like cycle, as in Problem 10.16. The adiabatic efficiencies for the compression and expansion processes can be taken as 55% and 85%, respectively. Determine

(a) the actual specific (net) work output, w (in kJ/kg)

(b) the actual cycle efficiency

(c) the efficiency ratio achieved

Based on the results of your analysis for the actual plants in Problems 10.17 and 10.18, indicate your choice of cycle for the steam plant design.

10.19. A heat engine employs steam as the working fluid and is designed to operate ideally on a Carnot cycle, as shown in Figure 10.4. The boiler pressure is 150 bar and the condenser pressure is 0.2 bar. The H_2O entering the boiler is saturated liquid at 150 bar, while the H_2O exiting the boiler is saturated vapor at the same pressure.

(a) For the ideal Carnot cycle, determine

(i) the net work output (w), in kJ/kg

(ii) the work ratio (r_w)

(iii) the thermal efficiency

(b) For an actual Carnot-like cycle, assume states 1 and 3 are the same in part (a). Assuming that the actual engine operates with isentropic efficiencies of 0.85 and 0.65 for the turbine and the compressor, respectively, determine

(i) the actual net work output (w_A), in kJ/kg

(ii) the actual thermal efficiency

(iii) the efficiency ratio

10.20. A heat engine employs steam as the working fluid and is designed to operate ideally on a Rankine cycle, as shown in Figure 10.7. The boiler pressure is 2000 psia and the condenser pressure is 1 psia. The working fluid exits the condenser as saturated liquid and exits the boiler as saturated vapor.

(a) For the ideal Rankine cycle, determine

(i) the net work output (w), in Btu/lbm

(ii) the work ratio (r_w)

(iii) the thermal efficiency

(b) Assuming that the actual engine operates with isentropic efficiencies of 0.85 and 0.65 for the turbine and the pump, respectively, determine

(i) the actual net work output (w_A), in Btu/lbm

(ii) the actual thermal efficiency

(iii) the efficiency ratio

10.21. A steam power plant is designed to operate ideally on the basic Rankine cycle. The working fluid exits the condenser as saturated liquid and exits the boiler as saturated vapor. The pressure of the H_2O in the boiler is 1000 psia and its pressure in the condenser is 2 psia. Determine

(a) the cycle thermal efficiency

(b) the work ratio for the cycle

10.22. A steam power plant is designed to operate ideally on the basic Rankine cycle. The heat input to the boiler is at the rate of 50 MW. The H_2O exits the condenser as saturated liquid and exits the boiler as saturated vapor. The pressure of the H_2O in the boiler is 120 bar and its pressure in the condenser is 0.04 bar. Determine

(a) the cycle efficiency

(b) the work ratio for the cycle

(c) the power output (in MW) of the plant

(d) the required mass flow rate (in kg/h) of the working fluid

(e) the specific steam consumption (in kg/kWh)

10.23. Repeat parts (a), (c), (d), and (e) in Problem 10.22 for an actual power plant that operates on the basic Rankine cycle but has an adiabatic turbine with a process efficiency of 88% and an adiabatic pump with a process efficiency of 90%.

* 10.24. A steam power plant is operating with the conditions given in Problem 10.22. The heat input to the boiler is provided by a steady stream of hot gases initially at 2200 K and 1 atm. The hot gases exhaust at 600 K and 1 atm to the surroundings which are at 15°C and 1 atm. The mean specific heat at constant pressure for the hot gases can be taken as 1.1 kJ/(kg · K). Determine

(a) the mass flow rate (in kg/h) of the stream of hot gases required

(b) the exergy flux (in MW) which the hot gases possess initially

(c) the exergy loss rate (in MW) associated with the exhaust stream

(d) the exergy consumption (in MW) in the steam generation process

(e) the exergy consumption (in MW) in the vapor condensation process

(f) the second law efficiency for the entire operation

10.25. A steam plant is to be operated on the Rankine cycle with superheat such that the steam quality at exit to the turbine is at least 0.9. The boiler pressure is 50 bar while the condenser pressure is 8 kPa. Determine

(a) the minimum temperature to which the steam must be superheated

(b) the corresponding specific (net) work output, w (in kJ/kg)

(c) the cycle efficiency and the work ratio

(d) the specific steam consumption (in kg/kWh)

10.26. A steam power plant operates on the Rankine cycle with superheat. The steam exits the boiler at 800°F and 1000 psia and exits the condenser as a saturated liquid at 2 psia. Determine

(a) the cycle thermal efficiency

(b) the work ratio for the cycle

Compare your results with similar results obtained in Problem 10.21 for the basic Rankine cycle.

10.27. A steam power plant operates on the Rankine cycle with superheat. The steam exits the boiler at 800 K and 120 bar and exits the condenser as a saturated liquid at 0.04 bar. Determine

(a) the cycle thermal efficiency

(b) the work ratio for the cycle

Compare your results with similar results obtained in Problem 10.22 for the basic Rankine cycle.

10.28. A heat engine operates with H_2O on the Rankine cycle with superheat. The pressure in the condenser is 2 psia, and the working fluid exits the condenser as a saturated liquid. The state of the steam entering the turbine is 1200°F and 200 psia. Calculate the thermal efficiency for the cycle.

10.29. A heat engine operates with H_2O on the Rankine cycle with superheat. The pressure in the condenser is 1.0 bar, and the working fluid exits the condenser as a saturated liquid. The state of the steam entering the turbine is 700°C and 70 bar. Calculate the thermal efficiency for the cycle.

* 10.30. Compare the steam power plants described in Problem 10.21 (basic Rankine cycle) and Problem 10.26 (Rankine cycle with superheat) on a second law basis and give your assessment as to which option is the better choice. Assume that the heating for steam generation is provided by a steady stream of hot gases (c_p = 0.265 Btu/lbm °R) initially at 3600°R and 1 atm pressure. The hot gases exhaust at 800°R and 1 atm. The ambient atmosphere is at 77°F and 1 atm.

* 10.31. Compare the steam power plants described in Problem 10.22 (basic Rankine cycle) and Problem 10.27 (Rankine cycle with superheat) on a second law basis and give your assessment as to which option is the better choice. Assume that the heating for steam generation is provided by a steady stream of hot gases (c_p = 1.1 kJ/kg · K) initially at 2000 K and 1 atm and exhausting at 450°K and 1 atm. The ambient atmosphere is at 15°C and 1 atm.

* 10.32. A hypothetical constant temperature source at 1500 K is used for a vapor power plant operating on the Rankine cycle with superheat between a condenser pressure of 8 kPa and a boiler pressure of 80 bar. The working fluid is steam that flows at a rate of 9000 kg/h. The maximum temperature of the steam is 900 K. Process 1→2 is the compression process, 2→3 the heating in the vapor generating unit at the boiler pressure given, 3→4 the steam expansion through the turbine, and 4→1 the condensation process. The ambient environment is at a constant temperature of 300 K. Show the steam cycle on a *T-s* diagram. Also sketch the process paths for the heat source and the heat sink. Determine

(a) the specific enthalpy for each of the states 1 to 4 for the working fluid
(b) the heat input q_{23} (in kJ/kg) per unit mass of the working fluid, and hence, the heat input rate $\dot{Q}_H$ (in kW) to the steam cycle
(c) the net mechanical power output (in kW) of the cycle
(d) the exergy input rate (in kW) from the heat source
(e) the second law efficiency achieved

* **10.33.** A designer proposes a heat engine operating on the Rankine cycle with superheat for producing power from a regeneratable heat source that comprises a steady stream of 5 kg/s of solar heated water at 350 K and 1 atm. The specific heat of water can be taken as 4.2 kJ/kg · K. The ambient environment is at 300 K and 1 atm. The proposed engine is illustrated in Figure P10.33. Assume that process 1→2 is isentropic compression, 2→3 is heating at a constant pressure, 3→4, isentropic expansion through a turbine, and 4→1 is a vapor condensation process. The heat engine cycle can be assumed to be internally reversible, and the working fluid employed is such that the specific enthalpies at states 1 to 4 are as follows:

State Number	p (bar)	T (°C)	h (kJ/kg)
1	12.54	32	84.61
2	25	32.01	84.62
3	25	70	272.17
4	12.54	32	256.12

The heat transfer rate from the hot water stream to the proposed engine is 780 kW. Calculate

(a) the exergy flow rate (in kW) supplied by the hot water stream to the engine cycle

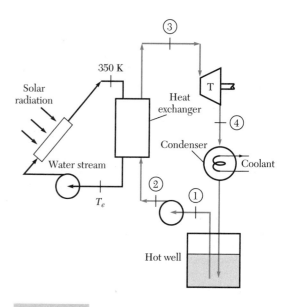

Figure P10.33 Schematic diagram of the proposed heat engine of Problem 10.33.

(b) the thermal efficiency of the proposed engine cycle
(c) the power output (in kW) of the proposed engine
(d) the second law efficiency achieved

* **10.34.** A steam plant operates on the Rankine cycle with superheat. The boiler pressure is 10 bar while the condenser pressure is 0.75 bar. The steam generated in the boiler is heated to 350°C before it enters the turbine. The heat source is a regeneratable source that comprises 8.5 × 10⁶ kg/h of hot gas at 370°C from a nuclear reactor. The gas is returned to the reactor core at 177°C. c_p for the gas can be taken as 1.06 kJ/kg · K. The ambient temperature is 300 K. Calculate

(a) the heat input rate (in MW) to the steam cycle
(b) the thermal efficiency of the plant
(c) the plant power output (in MW)
(d) the exergy input rate (in MW) to the steam cycle from the regeneratable heat source
(e) the second law efficiency of the plant

* **10.35.** Repeat parts (b), (c), and (e) of Problem 10.34 for an actual steam cycle having an adiabatic turbine efficiency of 85%.

* 10.36. A steam power plant operates on the Rankine cycle with reheat, as shown in Figure 10.13. The steam exits the boiler at 700°C and 70 bar, and exits the condenser as a saturated liquid at 0.05 bar. The steam expands isentropically through the first turbine stage and exits as a saturated vapor. From this state, the steam is reheated to 700°C. Then it expands isentropically through the second stage of the turbine and exits from this stage of the turbine at a pressure of 0.05 bar. Determine
(a) the cycle thermal efficiency
(b) the work ratio for the cycle

* 10.37. A steam power plant operates on the Rankine cycle with reheat, as shown in Figure 10.13. The steam exits the boiler at 800°F and 1000 psia and exits the condenser as a saturated liquid at 2 psia. The steam expands isentropically through the first turbine stage and exits as a saturated vapor. From this state, the steam is reheated at 800°F. Then it expands isentropically through the second stage of the turbine and exits from this stage at a pressure of 2 psia. Determine
(a) the cycle thermal efficiency
(b) the work ratio for the cycle
Compare your results with those from Problem 10.26 for the Rankine cycle with superheat.

* 10.38. A steam power plant operates on the Rankine cycle with reheat, as shown in Figure 10.13. The steam exits the boiler at 800 K and 120 bar and exits the condenser as a saturated liquid at 0.04 bar. The steam expands isentropically through the first turbine stage and exits as a saturated vapor. From this state, the steam is reheated to 800 K. It then expands isentropically through the second stage of the turbine and exits from this stage at a pressure of 0.04 bar. Determine
(a) the cycle thermal efficiency
(b) the work ratio for the cycle
Compare your results with those from Problem 10.27 for the Rankine cycle with superheat.

* 10.39. A steam power plant operates on the Rankine cycle with two stages of reheat. The steam exits the boiler at 500°C and 80 bar and exits the condenser as a saturated liquid at 0.03 bar. The steam expands isentropically through the first turbine stage and exits at a pressure of 6 bar. From this state, the steam is reheated to 500°C,

and it then expands isentropically through the second stage of the turbine to a pressure of 0.3 bar. The steam is again reheated to 500°C, and it again expands isentropically through the third stage of the turbine to an exit pressure of 0.03 bar. The steam enters the condenser from the exit of the third stage of the turbine to continue the cycle. Determine
(a) the cycle thermal efficiency
(b) the work ratio for the cycle

* 10.40. A steam power plant operates on the basic Rankine cycle with complete regeneration. The temperature for the operation of the boiler is 600°F, while the temperature for the operation of the condenser is 110°F. Determine the cycle efficiency and compare with that of a Carnot heat engine operating between the same two temperatures.

* * 10.41. A steam power plant operates with a boiler pressure of 70 bar and a condenser pressure of 0.05 bar. The heat source is a stream of exhaust gases discharging from a gas turbine at 600°C and 1 atm. The minimum temperature allowed for the exhaust gas stream is 400 K, while the maximum steam temperature is 550°C. The mass flow rate of the stream of hot gases is such that the heat input rate to the steam cycle is 50 MW. The ambient temperature and pressure are 288 K and 1 atm. Determine (i) the cycle thermal efficiency, (ii) the work ratio, and (iii) the second law efficiency achieved with each of the following cycles:
(a) the basic Rankine cycle
(b) the Rankine cycle with superheat
(c) the Rankine cycle with reheat, such that the steam exits from the first (high pressure) stage of the turbine as a saturated vapor
(d) the basic Rankine cycle with complete regeneration

* 10.42. Repeat parts (i) and (iii) for the cycles described in (a) and (b) in Problem 10.41 assuming that the pumps and turbines have an efficiency of 0.9.

10.43. An ideal gas turbine cycle comprises two isentropic processes (1→2 and 3→4) and two isobaric processes (2→3 and 4→1). Show the cycle on a T-s diagram. Given $\dot{m}$ = 2 kg/s, c_p = 1 kJ/kg · K, T_1 = 300 K, T_2 = 475 K, T_3 = 1250 K, and T_4 = 790 K, calculate the following for the Brayton cycle:

(a) w_{12}, w_{23}, w_{34}, and w_{41}
(b) q_{12}, q_{23}, q_{34}, and q_{41}
(c) $\dot{W}_{net}$ (in kW) and $\dot{Q}_{input}$ (in kW)
(d) η_1 and r_w

10.44. A gas turbine uses air drawn in at 1 atm and 300 K. The maximum cycle temperature allowed is 1000 K. Assuming that the turbine operates on the ideal Brayton cycle, calculate the following for pressure ratios of 2, 4, 6, 8, and 10:
(a) the thermal efficiency
(b) the work ratio
Comment briefly on your results.

* **10.45.** A gas turbine uses air drawn in at 70°F and 1 atm. The air is compressed to 5 atm, and the maximum cycle temperature is limited to 1500°F by use of a large air–fuel ratio. Assume that the gas turbine operates on the ideal Brayton cycle and that the heating or combustion process is equivalent to a heat input of 50×10^6 Btu/h to the air at 5 atm. The surroundings are at 70°F and 1 atm. Determine for the ideal cycle:
(a) the thermal efficiency
(b) the work ratio
(c) the power output (in Btu/h)
(d) the exergy flux (in Btu/h) of the gas leaving the turbine

* **10.46.** In an actual Brayton-like cycle with operating conditions as given in Problem 10.45, the compression and expansion processes occur with adiabatic process efficiencies of 80% and 85%, respectively. Determine for the actual cycle
(a) the actual thermal efficiency
(b) the actual power output (in Btu/h)
(c) the efficiency ratio
(d) the exergy flux (in Btu/h) of the gas stream leaving the turbine.

* **10.47.** A gas turbine operates on an open cycle with air entering the compressor at 300 K at a rate of 10,000 kg/h. The pressure ratio is 8:1 and the flow processes in the compressor and turbine are both isentropic. For the compression process, c_p is 1.0 kJ/(kg · K) and k is 1.4. For the expansion process, the corresponding average values are 1.1 kJ/(kg · K) and 1.3. The fuel used is kerosene with a heating value of 43,000 kJ/kg and a chemical exergy of 46,000 kJ/kg. The air–fuel ratio (by mass) is 80:1, and c_p has an average value of 1.1 kJ/(kg · K) during the (combustion) heating process. Determine for this cycle
(a) the gas turbine (net) power output

(b) the thermal efficiency (defined in terms of the power output and the fuel energy input rate)
(c) the second law efficiency

* **10.48.** Rework Problem 10.47 for the case in which the isentropic efficiencies for the turbine and compressor are 0.87 and 0.82, respectively. Compare your results with those obtained in Problem 10.47.

* **10.49.** The heat source for a simple gas turbine is a hot fluid stream (c_p = 1.06 kJ/kg · K) at 370°C from a nuclear reactor which is returned to the reactor core at 147°C. The mass flow rate of the reactor coolant is 8.5×10^6 kg/h. The simple gas turbine operates approximately on the Brayton cycle with minimum and maximum temperatures of 300 K and 320°C, respectively. The pressure ratio is 2.74. Assume c_p = 1.05 kJ/kg · K for the air in the gas turbine cycle; k can be taken as 1.4. The ambient temperature is 300 K. Calculate
(a) the cycle efficiency ($\eta_{thermal}$)
(b) the work ratio
(c) the power output of the gas turbine
(d) the decrease in exergy flow rate of the heat source fluid stream as it cools from 370°C to 147°C
(e) the second law efficiency of the gas turbine based on parts (c) and (d)

* **10.50.** A gas turbine operates with intake air at 1 atm and 300 K and a pressure ratio of 6. The maximum cycle temperature is 1100 K, and the isentropic efficiencies for the compressor and turbine can be taken as 0.8 and 0.85, respectively. The heating process is deemed equivalent to using a stream of hot gaseous products of combustion cooled from an initial temperature of 2000 K to 600 K at a constant pressure of 1 atm. Ambient condition is given by p_0 = 1 atm and T_0 = 300 K. For air, c_p is 1.005 kJ/kg · K and k is 1.4. Determine
(a) the specific (net) work output
(b) the thermal efficiency
(c) the second law efficiency

** **10.51.** Repeat Problem 10.50 but assume that the gas cycle is now a Carnot-like cycle with maximum and minimum temperatures of 1100 K and 300 K. The heating process is accomplished with the same hot gas stream that is cooled from 2000 K to a temperature of 1130 K. The isentropic efficiencies of the compressor and the turbine are unchanged. Assume for the Carnot-like cycle that the heat addition and the heat rejection processes are internally reversible and that $p_4 = 0.01$ atm. Which of the two cycles is better on the basis of (i) a first law analysis and (ii) a second law analysis?

** **10.52.** A simple gas turbine takes in a mixture of air and natural gas at 300 K and at an atmospheric pressure of 1 atm and compresses it through a pressure ratio of 5. The natural gas has a heating value of 50,000 kJ/kg and is burnt in the air at an approximately constant pressure during passage of the mixture through the combustion chamber. The gases leave the combustion chamber at 1000 K and are expanded through the turbine to atmospheric pressure. The isentropic efficiencies of the compressor and turbine are 0.8 and 0.85, respectively; k can be taken as 1.4. Assume $c_p = 1.03$ kJ/kg · K in the compressor and $c_p = 1.1$ kJ/kg · K in the combustion chamber and turbine. Calculate the cycle efficiency and the mass of air–fuel ratio
(a) without a heat exchanger
(b) with a regenerative heat exchanger fitted and having an efficiency of 70% (see Figure P10.52)

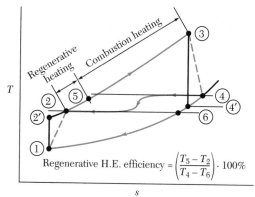

Figure P10.52 *T-s* diagram for the gas turbine with a regenerative heat exchanger as in Problem 10.H52(b).

10.53. The Otto cycle comprises two isentropic nonflow processes ($1 \rightarrow 2$ and $3 \rightarrow 4$) and two isochoric nonflow processes ($2 \rightarrow 3$ and $4 \rightarrow 1$). Sketch the cycle on *T-s* coordinates. If $\dot{m} = 2$ kg/s, $c_v = 0.715$ kJ/kg · K, $T_1 = 300$ K, $T_2 = 690$ K, $T_3 = 1800$ K, and $T_4 = 785$ K, calculate

(a) w_{12}, w_{23}, w_{34}, and w_{41}
(b) q_{12}, q_{23}, q_{34}, and q_{41}
(c) $\dot{W}_{net}$ (in kW) and $\dot{Q}_{input}$ (in kW)
(d) η_1 and r_w

10.54. In an ideal reciprocating engine, the compression ratio is 8 and all processes are internally reversible. Air initially at 100 kPa and 300 K is compressed isentropically into a clearance volume and then heated at constant volume until the pressure rises to 3000 kPa. The cycle is completed by isentropic expansion to the maximum cylinder volume, followed by constant volume cooling. For air, take $k = 1.4$ and $c_v = 0.718$ kJ/kg · K. Calculate the specific work output.

10.55. In an air standard Otto cycle, the maximum and minimum temperatures are 2400°F and 70°F, respectively. The compression ratio is 5. At the beginning of the compression process, the pressure is 1 atm. The specific heats c_p and c_v can be considered constants with values of 0.26 Btu/(lbm · °R) and 0.19 Btu/(lbm · °R), respectively. Determine
(a) the pressure and temperature at each point in the cycle
(b) the thermal efficiency
(c) the work ratio
(d) the mean effective pressure

10.56. Repeat parts (a), (b), and (d) in Problem 10.55 for an actual cycle that has process efficiencies of 0.82 and 0.88 for the compression and expansion processes.

10.57. In an air standard Otto cycle, the maximum and minimum temperatures are 1675 K and 300 K. The compression ratio is 6, and the pressure at the beginning of the compression process is 100 kPa. For air, assume $k = 1.4$ and $c_v = 0.718$ kJ/kg · K. Determine
(a) the pressure and temperature at each point in the cycle
(b) the thermal efficiency
(c) the mean effective pressure (in bar)

10.58 An air standard Diesel cycle operates with a compression ratio of 18 and maximum and minimum cycle temperatures of 2000 K and 300 K, respectively. At the beginning of the compression process, the pressure is 1 atm. The specific heats c_p and c_v can be considered constants with values of 1.16 kJ(kg · K) and 0.87 kJ/(kg · K), respectively. Determine

(a) the pressure and temperature at each point in the cycle

(b) the specific net work output

(c) the thermal efficiency

(d) the work ratio

(e) the mean effective pressure

10.59. Repeat parts (a), (b), (c), and (e) in Problem 10.58 for an actual Diesel-like cycle that has process efficiencies of 0.82 and 0.88 for the compression and power strokes. Also determine the efficiency ratio.

10.60. A diesel engine operates on the air standard Diesel cycle with a compression ratio of 16 to 1 and maximum and minimum cycle temperatures of 2000 K and 300 K. The maximum cycle pressure is 49 bar. Calculate

(a) the pressure and temperature at each state in the cycle

(b) the cut-off ratio (r_c)

(c) the thermal efficiency (η_1)

(d) the mean effective pressure (in bar)

* **10.61.** Power generating devices in use invariably use finite heat sources. For each of the power cycles illustrated in Figures P10.61(a) through (c), the heat source is taken as equivalent to a hot gas stream initially at T_i = 2000 K and 1 atm pressure. After exchanging heat with the working fluid, the heat source gas stream leaves at T_e, which has a different value for each power cycle. The mass flow rate of the hot gas is 5 kg/s, and the specific heat at constant pressure can be taken as 1.1 kJ/kg $\cdot$ K. The sink is the ambient environment, which stays at a constant temperature of 20°C and pressure of 1 atm. For the working fluid, assume c_p = 1.005 kJ/kg $\cdot$ K and

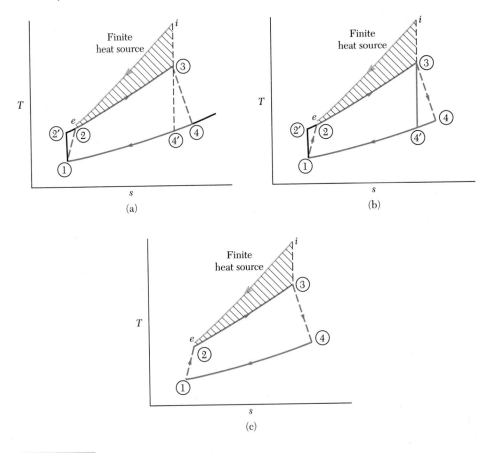

Figure P10.61 (a) Brayton cycle, (b) Otto cycle, and (c) Diesel cycle.

$k = 1.4$. The thermodynamic states for each power cycle are

Brayton-like cycle (Fig. P10.61a)

$T_e = 520$ K, $T_1 = 293$ K, $T_{2'} = 465$ K (isentropic efficiency for the compression process is 0.8), $T_3 = 1200$ K, $T_4 = 800$ K

Otto-like cycle (Fig. P10.61b)

$T_e = 770$ K, $T_1 = 293$ K, $T_2 = 770$ K, $T_3 = 1000$ K, $T_4 = 490$ K

Diesel-like cycle (Fig. P10.61c)

$T_e = 1030$ K, $T_1 = 293$ K, $T_2 = 1030$ K, $T_3 = 1500$ K, $T_4 = 745$ K

Calculate the following for each of the three cycles:
(a) the exergy supply rate (in kW)
(b) the heat input rate to the cycle (in kW)
(c) the thermal efficiency, η_1, for the cycle
(d) the power output (in kW) of the cycle
(e) the second law efficiency achieved

∗∗10.62. A heat engine is constructed that uses as a heat source a hot stream of gaseous products of combustion at 2200 K and 1 atm. The surroundings are at 300 K and 1 atm.
(a) Compute the thermal efficiency attained when the heat engine operates on the optimum cycle for this case with a finite heat source.
(b) What is the second law efficiency in this case?
(c) What is the highest second law efficiency attainable for a heat engine operating on the Carnot cycle with these conditions? Assume

the heat input to the Carnot engine is at a temperature T (between 300 K and 2200 K). Which of the two cycles is more efficient and why?

∗∗10.63. A heat engine is constructed that uses as a heat source a hot stream of gaseous products of combustion at 3000°F and 1 atm. The surroundings are at 70°F and 1 atm.
(a) Compute the thermal efficiency attained when the heat engine operates on the optimum cycle for this case with a finite heat source.
(b) What is the second law efficiency in this case?
(c) What is the highest second law efficiency attainable for a heat engine operating on the Carnot cycle with these conditions?
Which of the two cycles is more efficient and why?

∗10.64. A reversible heat engine operates on the Carnot cycle. The source and sink temperatures are 401°F and 126°F, respectively, which correspond to saturation pressures for steam of 250 psia and 2 psia, respectively. The working fluid flows steadily through each component (heater, turbine, cooler, and compressor or pump) of the heat engine in each cycle. Calculate the work ratio when
(a) the working fluid is air and the minimum and maximum pressures in the cycle are 2 psia and 250 psia.
(b) the working fluid is H_2O, which enters the turbine as dry, saturated steam at 250 psia and exits the compressor as saturated liquid at 250 psia.

Design Problems

10.1 *Steam power plant design*

You have been commissioned to produce a thermodynamic design of a 100 MW steam power plant that uses natural gas as fuel for the steam generation process. For steam condensation, you may assume that water at about 25°C is available from a nearby lake. For simplicity, you may also assume that the natural gas sample is 100% methane with a lower heating value (LHV) equal to 50.2 MJ/kg fuel and a chemical exergy per unit mass of 52.1 MJ/kg fuel.

1. Produce a sketch of the plant layout to include all the facets involved in transforming the exergy of the fuel to a mechanical power output.
2. Figure DP10.1(a) illustrates a thermodynamic model for the proposed steam power plant. By applying the first law of thermodynamics to the source heating process, which results from the combustion of fuel in air, the following expression can be obtained for the heat input rate to the engine cycle:

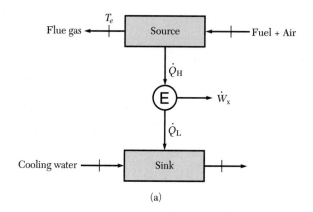

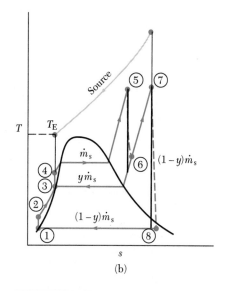

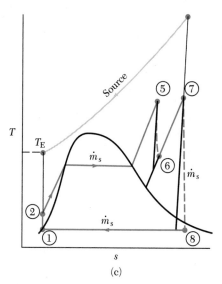

Figure DP10.1 (a) Model for a steam power plant, (b) Rankine cycle with reheat and regeneration, (c) Rankine cycle with reheat.

$$\dot{Q}_H = \dot{m}_{\text{fuel}} \, (\text{LHV})$$
$$- \dot{m}_{\text{flue gas}} \, c_{pE}(T_E - T_0) \quad (1)$$

where $\dot{m}_{\text{fuel}}$ denotes the mass flow rate of the fuel, $\dot{m}_{\text{flue gas}}$ is the mass flow rate of the flue gas (or combustion products), c_{pE} is the mean specific heat of the combustion products, T_E is the exit temperature at which the flue gas is discharged into the atmosphere, and T_0 denotes the ambient air temperature. Assume $c_{pE} = 1.11$ kJ/kg · K for the combustion products. Use Equation (1) to show that if the cycle efficiency, η_1, is known, the mass flow rate of fuel required can be determined from

$$\dot{m}_{\text{fuel}} = \frac{\dot{W}_x}{\left(\dfrac{\eta_1}{100}\right)[\text{LHV} - (\text{AF}_{\text{grav}} + 1)c_{pE}(T_E - T_0)]} \quad (2)$$

where AF_{grav} is the ratio of the mass of air to the mass of fuel and may be taken as 19 in this application.

3. Focus next on the two steam cycles illustrated in Figures DP10.1(b) and DP10.1(c). Figure DP10.1(b) shows a Rankine cycle with reheat as well as feed water heating from state 2 to 3 using a fraction (y) of the steam at state 6 exiting the high pressure turbine stage. The

second cycle is a Rankine cycle with reheat. For both cycles, fix the maximum temperature allowable (that is, the *metallurgical limit*.) Select a reasonable condenser pressure and assume a maximum pressure of 15 MPa for each cycle. Your analysis should make use of realistic values for the isentropic efficiency of each turbine. Determine for each cycle optimum values for the net specific work output and the corresponding cycle efficiency.

4. Having identified the optimum operating conditions for each of the two cycles in part 3, you should complete the project by carrying out the following additional steps:
   (a) Determine the appropriate value of T_E in each case, and use Equation 2 to calculate the fuel mass flow rate needed for producing 100 MW of mechanical power. What is the fuel energy (in Btu) needed for 1 h of operation of the plant? Find out the local cost of gas per million Btu, and hence, obtain the fuel cost for 1 h of operation of the plant. Based on this cost alone, determine the minimum amount which ought to be charged for each kWh of mechanical energy produced (that is, electricity generated).
   (b) Provide for each cycle an energy balance

for the operation of the plant over a 1 h period.
   (c) Produce for each cycle an exergy disposition chart which shows, for a 1 h operation, the fuel exergy input (in MWh), the unused exergy of the flue gas, the exergy consumption in the vapor generating process, the useful exergy produced, and the other exergy consumptions in the steam plant operation.

10.2. *Feedwater heater use in steam power plants*
An analysis of a simple steam power plant is to be conducted. A schematic diagram with suggested thermodynamic properties is given in Figure DP10.2. This is the same steam power plant as shown in Figure 1.12. The mass flow rate is 4×10^5 kg/h through the boiler. The amount of steam bled from the turbine to be used in the feedwater heater is 20%. Both turbine sections are 80% efficient and the pump is considered to be 70% efficient. The ambient temperature is taken as 290 K.

1. Sketch the T–S diagram showing this cycle.
2. Calculate the heat input to the boiler, the total work output of the turbine, and the cycle efficiency (using both first and second law analyses).

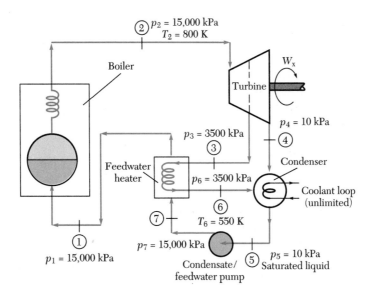

Figure DP10.2 Schematic diagram of a simple steam power plant.

3. The use of a feedwater heater is to improve the efficiency of a steam power plant. Does this design help or hurt the plant efficiency? Support this with calculations for a cycle that expands all the steam through the turbine to a pressure of 10 kPa and eliminates the feedwater heater completely.

4. Assuming that point 6 remains in the superheated region, can raising the bleed pressure (that is, the pressure at state 3), the amount of steam extracted (at point 3), and the exit temperature (point 6) of the steam from the feedwater heater make this design any more efficient than without the feedwater heater? Explain any inherent fault to this design from a thermodynamic view.

5. Redesign the feedwater heater placement to improve the cycle efficiency. You must use the closed feedwater heater shown (the bled steam and the condensate do not mix when in the heat exchanger) and reroute the piping. A small drip pump or steam trap is available for use in the new design. These devices are used to collect condensation from steam at one pressure and transfer the hot liquid to a different pressure. The drip pump requires a small amount of work input while the trap uses some of the steam energy to move the liquid. (Often the devices operate intermittently and allow the liquid to build up before

dumping. However, for the problem, assume they are continuous flow devices.) A trap can only be used when going from a high pressure to a lower pressure. Calculate only the required work for any pump used and neglect the trap losses. The design should include
(a) a diagram of the system with the thermodynamic state specified at all locations
(b) a brief description of the system operation
(c) the first and second law efficiencies
(d) work ratio, r_w

6. State some practical design considerations regarding why feedwater heaters are used in real power plant situations.

10.3 *Feedwater heaters and cooling towers in nuclear power plants*

A nuclear reactor is utilized to approximate a constant temperature heat source (reservoir) in a typical steam power plant (Figure DP10.3). The sodium loop maintains the steam generator at 600°F regardless of the steam load. An open feedwater heater is incorporated into the design. This uses some of the high pressure steam to reheat the steam generator feedwater. The open feedwater heater actually mixes the steam and condensate at a common intermediate pressure (100 psia for this case). Only enough steam is bled to heat the water and completely condense all the bleed steam so that the mixture exits as a saturated liquid at the

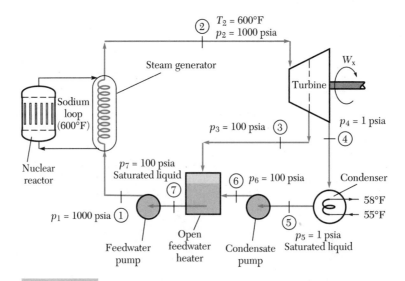

Figure DP10.3 Schematic diagram of a nuclear power plant.

feedwater heater pressure. The pump and turbine efficiencies are 75% and 85%, respectively.

1. Calculate the percent of steam needed to be bled at point 3 and the steam mass flow required (1bm/h) if the actual net turbine output is specified at 1 million horsepower.

2. Determine the first law cycle efficiency and compute it for the same system without the feedwater heater and its associated pump (i.e., all the steam passes through a single turbine to state 4, condenses to saturated liquid, and is pumped back to 1000 psia by one pump). From this first law analysis, does the feedwater heater increase the cycle efficiency? Explain the principle for this change in efficiency.

3. Does the assumption of the constant temperature heat source have an effect on the previous analysis? Briefly explain your answer.

4. From second law analysis, would you expect the feedwater heater to increase or decrease the efficiency? Calculate the second law cycle efficiency and compare it to the system without the feedwater heater (and associated pump) where the entire flow exits the turbine at 1 psia.

5. The cooling water requirement can be substantial even for the system with a feedwater heater. Strict limits on the discharge water temperature for environmental reasons make the amount of water needed even greater. For most nuclear plants, even a large river nearby (which is generally considered an unlimited heat sink) possesses some practical limits for the cooling water. Often, expensive cooling towers are added to the plant for economic reasons alone. Demonstrate this by the following calculations:

   (a) Calculate the coolant water required in gallons per hour if an environmental regulation is imposed which allows only a 3°F maximum temperature rise for the coolant water. Assume the river water comes in at 55°F, neglect any pressure change, and assume incompressible flow.

   (b) What is the coolant water pump horsepower needed if the total pipe and elevation resistance (from the river through the condenser and out again) is 20 feet of head?

   Note: 20 ft head

$$= 20 \text{ ft} \left(\frac{\text{lbf} \cdot \text{s}^2}{32.2 \text{ ft} \cdot \text{lbm}} \right) (32.2 \text{ ft/s}^2)$$

$$= 20 \text{ ft} \frac{\text{lbf}}{\text{lbm}}$$

   (c) How tall must two 200 foot diameter towers (cylinders) be to reduce the cooling water requirement by 50%? For this simple analysis assume that the towers will evaporate 3 lbm/h for every square foot of interior wall by circulating air through the center. This phenomena is termed evaporative cooling (see Chapter 13). The exit stream of evaporated water leaves as saturated vapor at 14.7 psia. Assume that the cooling water comes from the river with an enthalpy of 23 Btu/lbm and returns with a maximum of 26 Btu/lbm. Also, the entire process remains at the same atmospheric pressure of 14.7 psia.

10.4. *Second law evaluation of coal-fired steam power plant*

A coal-fired boiler supplies the heat for a simple steam power plant with a regenerative feedwater heater. The combustion gases that supply the heat enter the boiler at 2000°F and have a 1500°F drop in temperature (finite heat source). The ambient air is 75°F. The condenser coolant water is considered an unlimited heat sink as the temperature rise is kept to a minimum. The pump and turbine efficiencies are 75% and 85%, respectively. A schematic diagram of the system is shown in Figure DP10.4. The plant is identical to the one described in DP10.3 except for the source of heat.

1. Determine the plant's efficiency using first law and then second law analysis if the desired work output is 1 million horsepower.

2. For such a system, is the feedwater heater a reasonable feature to include? Compare against a system that expands all the steam through the turbine. Show the reasoning to keep or delete the heater and associated pump by first and second law efficiencies.

3. If the specific heat, c_p, of the combustion gas is assumed constant at 1.3 Btu/lbm·°F, how much will the gas mass flow rate change when the feedwater heater and pump are deleted?

4. Consider the rejection of the condenser heat, where ocean water is available at 55°F but a restriction of a 1°F temperature increase is imposed for environmental reasons. Do you

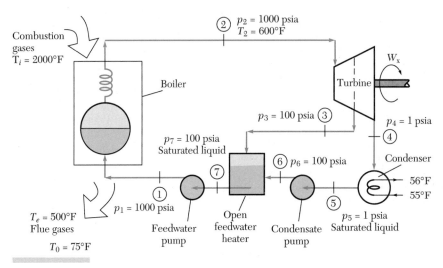

Figure DP10.4 Schematic diagram of a coal-fired steam power plant.

see practical problems with the cooling water assumption of an unlimited heat sink? Can you suggest a modification to reduce the cooling water requirements?

5. Determine the steam quality at both exit points of the turbine. Do you think they are acceptable? What could be done to the design to correct either stream if there is a problem with the quality?

6. Normally the flue gases are maintained at a high temperature when exhausted to the atmosphere (typical of the 500°F exit temperature given). If the flue gases are allowed to be reduced to the water temperature entering the boiler without using the feedwater heater, the second law efficiency will actually be greater than with the feedwater heater and flue gases at 500°F. Show that this is true by calculating the second law efficiency with $T_e = 102°F$. Explain why the flue gas from a real boiler is normally kept at a higher value despite the "waste of energy" (and exergy) involved in doing this.

7. How low could the flue gas exit temperature, T_e, be when using the feedwater system? Use this temperature to recalculate the second law efficiency. Would you recommend the installation of the feedwater heater with this gas exit temperature? Explain your reasoning.

10.5 *Open-ended design of steam power plant*
Design a steam power generation plant using De-

sign Problems 10.2 to 10.4 as guides. The objective of your design should be to maximize plant efficiency while meeting all constraints such as environmental factors, temperature limits for materials, etc. The plant's output should be 25 MW. One feedwater heater (open or closed) can be incorporated into the design. The source of heat can be coal, oil, natural gas, or nuclear fuel. Research the fuel heating value, safety factors, cost, time requirements for construction, etc., typical of the source selected. Note appropriate assumptions used in the design calculations. Select reasonable pressures, temperatures, and other operating limits as needed. Be realistic and defend each selection with an appropriate source to support each specification.

A limit is set for four separate turbines (each bleed point requires a separate turbine) and five water pumps with efficiencies of 75% and 60%, respectively. Unlimited cooling water is available at 290 K, but a limit of 10 K rise is imposed on the discharge water. The required pump work is assumed taken from the turbine output (like all the other pump work) before meeting the 25 MW requirement.

Draw a T–S diagram and a labeled system diagram. Explain the operation. Make a table of the temperature, pressure, enthalpy, entropy, mass flow rate, and steam quality for each connection pipe. Calculate the first and second law efficiencies, exergy consumed during the entire cycle, efficiency ratio, and the work ratio.

Refrigeration Cycles

11.1   Introduction

A refrigerator and a heat pump are both heat engines operating in reverse. In operation, a reversed heat engine transfers heat from a low temperature source to a higher temperature sink. By the second law of thermodynamics (Clausius statement), work (or exergy) input to the device is required to cause the transfer of heat from a cooler to a hotter body. A refrigerator or air conditioner is constructed to remove heat from a low temperature source, while a heat pump is capable of delivering heat to a high temperature sink, such as a room, to heat a substance in a space. The device is traditionally called a refrigerator or air conditioner if it provides for cooling only, while it is called a heat pump if it is arranged so that it can either heat or cool a particular space.

11.2   Principles of Refrigeration

The operation of the reversed heat engine is based on certain physical processes that are easily understood. Perhaps the most widely employed of these is the *evaporative cooling* principle. In devices based on the evaporative cooling principle, cooling is produced by causing a *refrigerant*, in the liquid state, to evaporate while absorbing heat from the system to be cooled. The unit in which this evaporation process takes place is referred to as the *evaporator*. To operate continuously or on a cycle, the refrigerant must be returned to the liquid state. This is accomplished (usually at a higher temperature and pressure) via a heat rejection process in a component known as the *condenser*.

11.2.1 Evaporative Cooling Principle

Vapor Compression Refrigeration System

The *vapor compression refrigeration system* operates on the evaporative cooling principle. It is by far the most common type of refrigeration system in use. It is widely used in the home, commerce, industry, and even transportation. Domestic refrigerators are usually of the vapor compression type. The refrigeration unit in an automobile air conditioner is also a vapor compression system.

Figure 11.1 illustrates the essential features of a vapor compression refrigeration system. Liquid refrigerant is evaporated in the evaporator. The saturated or slightly superheated vapor that leaves the evaporator is then compressed by the compressor, which raises the pressure to a level such that the corresponding saturation temperature of the refrigerant exceeds the temperature of the sink. Thus, the refrigerant vapor is able to reject heat to the sink and ultimately revert to the liquid state as it passes through the condenser. The refrigerant, now liquid, is returned to the evaporator by passing it through a throttle or expansion valve, which is often just a small diameter capillary tube. The pressure in the evaporator must be such that the corresponding saturation temperature of the refrigerant is lower than the temperature of the system being cooled.

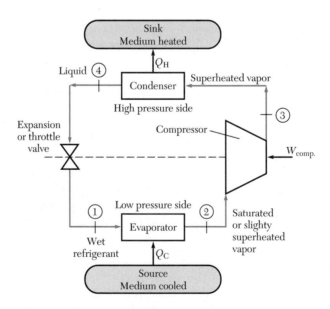

Figure 11.1 Schematic diagram of a vapor compression refrigeration system.

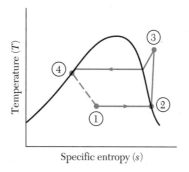

Figure 11.2 Typical vapor compression refrigeration cycle on *T-s* coordinates.

The corresponding cycle for a typical vapor compression refrigeration device is shown in Figure 11.2.

The vapor compression refrigeration cycle is basically a Rankine cycle (with super-heat) in reverse, except that in the case of a refrigeration cycle, the lowering of refrigerant pressure from the high level in the condenser to the relatively low level in the evaporator is accomplished via a throttle valve or a capillary tube. The expansion through the throttle valve is generally taken as adiabatic, in which case, application of the first law of thermodynamics indicates that the process is *isenthalpic*. This constant enthalpy process is accompanied by an increase in entropy. Since the process is also adiabatic, it follows from the principle of increase in entropy that the process is *irreversible*. Accordingly, the process is represented with a dashed line in Figure 11.2. Further discussion of the vapor compression refrigeration cycles is deferred to Section 11.4, where a procedure for the thermodynamic analysis of the cycles is also given.

Absorption Refrigeration System

Recall that a much more substantial work input is required for a specified pressure rise in the compression of a vapor (or gas) than for a liquid. The vapor compression refrigeration cycle thus requires a substantial work input because the compression of a vapor (rather than a liquid) is required. If the refrigerant vapor leaving the evaporator were absorbed into a liquid, the compression work needed would be significantly less. This principle underlies the operation of the *absorption refrigeration system*, which is illustrated in Figure 11.3.

In this system, the refrigerant is absorbed into an *absorbent–refrigerant* liquid solution in a unit called the *absorber*. A *solution pump* is then used to raise the pressure of the solution to a level corresponding to a refrigerant saturation temperature that is higher than the sink temperature. Release of the dissolved refrigerant is accomplished in the *generator* by heating the solution at the higher pressure condition. This heat is supplied at an elevated temperature and, therefore, counts as exergy input needed for heat flow from a lower source temperature to a higher sink temperature. One desirable quality of the absorbent is that it is relatively nonvolatile, so that when the solution is

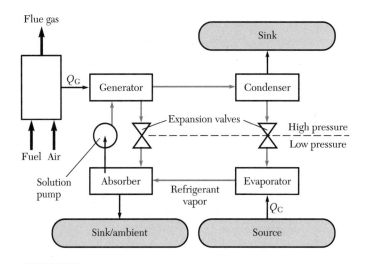

Figure 11.3 Schematic diagram of an absorption refrigeration system.

heated in the generator, only the refrigerant leaves the solution (as vapor). The refrigerant vapor thus released passes on to the *condenser* where it is cooled and condensed in precisely the same manner described for the vapor compression system. The condensed refrigerant is throttled back to the evaporator at the lower pressure, and the cycle resumes with evaporation taking place in the evaporator unit.

The prime energy input requirement for the operation of the absorption refrigeration cycle is *heat* rather than *work*. For this reason, its use is popular when heat sources such as gas, kerosene, or solar energy are available. Typically, the use of absorption refrigeration devices can be cost effective in remote areas that do not have easy access to a supply of electricity. Further consideration of the absorption refrigeration system is given in Section 11.5. However, in view of the complexity of the model and the dearth of necessary property data even for typical absorbent–refrigerant mixtures, no discussion is provided of the thermodynamic analysis of such systems. The interested reader can consult the American Society of Heating, Refrigeration, and Air-Conditioning Engineers, *Fundamentals Handbook* (ASHRAE 1989) or similar references for information on how to carry out a thermodynamic analysis of the absorption refrigeration system.

Other Evaporative Cooling Systems

Other systems based on the evaporative cooling principle include the *adsorption* (or *desiccant*) *system* and the *steam* or *vapor jet refrigeration system*. The adsorption systems use solid adsorbents in place of the liquid absorbents used in the absorption refrigeration systems. Also, the adsorption systems in general only require periodic thermal regeneration.

In the vapor jet refrigeration system, a vapor jet action in a nozzle is used to produce a low pressure (or suction) condition that results in the evaporation of the refrigerant and subsequent entrainment of the refrigerant vapor by the jet. *The Equipment Handbook* (ASHRAE 1987) is an excellent source of additional information on the various refrigeration systems based on the evaporative cooling principle.

An example of an efficient combined cycle is one that uses steam to operate both a vapor compression cycle and an absorption cycle. In such a cycle, steam is produced in a boiler, from which it goes to a turbine that drives a compressor for a vapor compression cycle. The exhaust steam from the turbine provides heat input to an absorption cycle, thereby maximizing the utilization of the exergy of the steam. An obvious disadvantage of this system is its complexity.

11.2.2 Gas Refrigeration Cycles

The operating principle for gas refrigeration systems is different from the evaporative cooling principle. *Gas refrigeration systems* operate essentially on the Brayton cycle in reverse. Figure 11.4 is a sketch showing the essential features of a typical air cooling system. The corresponding cycle is illustrated in Figure 11.5. In a typical gas refrigeration cycle, the initial compression of the gas is followed by a constant pressure cooling process during which heat is rejected (usually to the ambient environment). Adiabatic expansion of the gas through a turbine occurs to the original (or intake) gas pressure. The gas temperature at exit to the turbine is lower than that at the intake condition. Thus cooled gas is produced, which can be used to cool another substance. The work produced when the gas expands through the turbine is used to meet part of the work input required in the compression stage.

The efficiencies of actual gas refrigeration cycles are generally much lower than those for the ideal cycles due to the fact that significant irreversibilities are encountered in

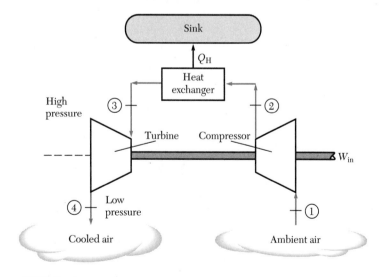

Figure 11.4 Schematic diagram of a typical air refrigeration system.

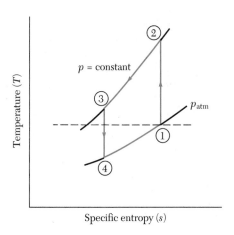

Figure 11.5 *T-s* diagram for a reversed Brayton cycle.

the compression and the turbine expansion processes, both of which involve a substantial amount of work since the fluid is a gas. Thus, the lowering of temperature that can be achieved, in practice, in a single-stage gas refrigeration cycle is usually small. If a substantial lowering of temperature is desired, a *multiple stage* or *cascade arrangement* is used that may use different gases to span the cooling range desired. Irreversible effects for the gas cycles include pressure drops due to friction and irreversible heat transfer to and from the gas. Recall also that the adverse effects due to irreversibilities in actual compression and expansion processes are more pronounced for gases than for liquids.

1. A steady stream of air at a pressure of 100 kPa and a temperature of 300 K is compressed to a pressure of 250 kPa in a reversible, adiabatic process. Determine the work done per unit mass of the fluid.
   Answer: -90.2 kJ/kg

2. Repeat Exercise 1 for saturated refrigerant-22 vapor at 1 bar which is compressed isentropically to 2.5 bar.
   Answer: -21.2 kJ/kg

3. Repeat Exercise 1 for saturated steam at 1 bar which is compressed isentropically to 2.5 bar.
   Answer: -174 kJ/kg

4. A steady stream of saturated water at 1 bar has its pressure raised to 2.5 bar in a reversible, adiabatic process. Determine the work done per unit mass of the fluid.
   Answer: -0.15 kJ/kg

5. Assume a process efficiency of 0.8 for the compression in Exercises 3 and 4 and determine the additional work input required for the compression in each case.
   Answer: 43.5 kJ/kg, 0.038 kJ/kg

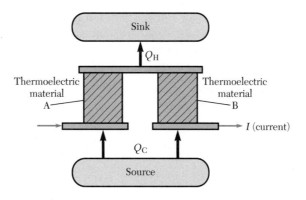

Figure 11.6 Schematic diagram of a thermoelectric cooler.

11.2.3 Thermoelectric Refrigeration

Thermoelectric refrigeration is based on the Peltier effect. When the two junctions formed by two dissimilar materials are maintained at different temperatures, an electromotive force (e.m.f.) is observed across the ends of the materials. This is known as the *Seebeck effect*, named after Seebeck who first made the observation in 1822. If instead, a current is passed through the two dissimilar thermoelectric materials, it is found that one junction heats up while the other gets colder. This inverse effect is the *Peltier effect*, which was first observed by Peltier in 1834. Figure 11.6 is a sketch showing the essential features of a thermoelectric cooler.

Thermodynamically, thermoelectric refrigeration using presently available materials is rather inefficient. There are, however, certain specialized applications for which it is ideally suited. Examples include the cooling of instruments in submarines, which must be accomplished without noise, and the cooling of electronic components that are not readily accessible. Excellent discussions of the design and analysis of thermoelectric refrigeration devices are given in *The 1977 Fundamentals Handbook* (ASHRAE 1977) and also by Angrist (1976).

11.3 Thermodynamic Evaluation of Refrigeration Cycles

The following sections discuss performance criteria for refrigeration cycles based on the first and second laws of thermodynamics. A brief discussion is included on modeling of refrigeration and heating systems with particular attention to the nature of heat sources and heat sinks.

11.3.1 Performance Criteria

The two important performance indices for refrigeration and heating systems are the coefficient of performance (COP) and the second law efficiency (η_2). The COP was first introduced in Chapter 7, and there it was noted that unlike thermal efficiency for a heat engine, the COP for refrigerators and heat pumps generally exceeds 100%. When an actual COP is compared with the optimum performance value, the corresponding index is the second law efficiency.

Coefficient of Performance (COP)

The coefficient of performance (COP) defined for a refrigeration device is a first law efficiency and is basically an energy ratio. Definitions of both the coefficient of performance for cooling (COP_C) and for heating (COP_H) were given in Chapter 7 as follows:

$$COP_C = \frac{\text{desired effect}}{\text{required input}} = \frac{Q_L \text{ (or } Q_C)}{W} \qquad (7.5)$$

and

$$COP_H = \frac{\text{desired effect}}{\text{required input}} = \frac{Q_H}{W} \qquad (7.6)$$

Figure 11.7 shows the thermodynamic model on which these definitions are based.

Second Law Efficiency (η_2)

As has been repeatedly indicated, a first law efficiency is of rather limited utility as an index for the evaluation of performance. For example, an electrically driven refrigerator might have a COP_C of about 2, while a comparable gas-fired absorption refrigeration device has a COP_C of about 0.8. A superficial comparison would lead to the conclusion that the electrically driven device is the better one. Such a conclusion, however, ignores the fact that electrical work input is a more valuable energy form than the heat input involved in the gas-fired absorption system. The second law efficiency (η_2) provides the proper measure of performance. For refrigeration systems, it can be defined as

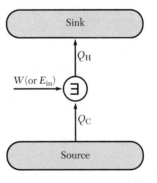

Figure 11.7 Thermodynamic model for a reversed heat engine (Ǝ).

$$\eta_2 = \frac{\text{minimum exergy supplied to refrigerated space}}{\text{actual exergy input required}} \cdot 100\% \qquad (11.1)$$

A few examples are given in Section 11.7 to demonstrate that the second law efficiency provides the proper measure of performance. Meanwhile, Sections 11.4 and 11.6 respectively, describe in greater detail the procedures for a first law analysis of the vapor compression and the gas refrigeration cycles both of which are widely employed. Note that a second law analysis of the individual processes making up a refrigeration cycle can be conducted in a similar fashion to what was done with power cycles in Chapter 10. The results of such analyses will considerably aid the designer in identifying where the prime sources of irreversibilities lie.

11.3.2 Thermodynamic Modeling of Refrigeration and Heating Systems

In an analogous manner to the modeling of heat engines, the reversed heat engine can be examined relative to its three principal components: the heat source, the heat sink, and the cyclic device used to cool the low temperature source while heating the higher temperature sink. In the case of a reversed heat engine, however, the driving force is the compressor work input (for a vapor compression refrigeration system), the generator heating provided in the case of an absorption chiller, or the electricity supplied to a thermoelectric refrigeration system. Thus, whether one is considering the first law efficiency or the second law efficiency, one must be cognizant of the actual form of energy (and exergy) input that is needed to achieve a desired cooling or heating effect.

Heat Sources and Sinks

It should be noted that for reversed heat engines, the heat source is the lower temperature medium, while the sink is the higher temperature medium. In cooling and heating applications, *sensible* cooling or heating is distinguished from *latent* cooling or heating. Sensible cooling or heating involves a change in temperature of the medium, while in the case of latent cooling or heating, the temperature of the medium is unchanged because the medium is changing phase.

Unlimited Sources and Sinks

The heat source or sink is the constant temperature (or unlimited) type whenever latent cooling or heating is involved. An example of a constant temperature heat source is encountered in the production of ice from water at 0°C which involves a phase change at the freezing point of ice. In general, the use of a phase change material at its phase change temperature either as a heat source or a heat sink provides a constant temperature source or sink. For a reversed heat engine operating between an unlimited heat source and sink, the optimum device is any reversible reversed heat engine exchanging heat with the source and sink only at the temperature T_L for the source and T_H for the sink. A device operating on the reversed Carnot cycle between the two temperatures is

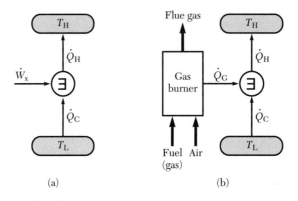

Figure 11.8 (a) Vapor compression and (b) absorption refrigeration system operating between two thermal reservoirs.

typical of such reversible reversed heat engines. The first law efficiencies for the optimum device in the present case are given by Equations 7.22a and 7.22b obtained in Chapter 7:

$$(COP_C)_{rev} = \frac{T_L}{T_H - T_L} \qquad (7.22a)$$

and

$$(COP_H)_{rev} = \frac{T_H}{T_H - T_L} \qquad (7.22b)$$

Figures 11.8(a) and (b) represent actual reversed heat engines each operating between two thermal reservoirs at T_L and T_H. The actual cooling and heating loads are designated $\dot{Q}_C$ and $\dot{Q}_H$, respectively, in each case. In the first case (Fig. 11.8a), a supply of mechanical power ($\dot{W}_X$) is used for operating the device. The second example is an absorption refrigeration system that is gas fired. The energy supply rate to the cyclic device in this case is a heat input rate ($\dot{Q}_G$) resulting from the combustion of gas.

For the vapor compression system shown in Figure 11.8(a), the COP values are given by Equations 7.5 and 7.6, which were repeated in Section 11.3.1. For the absorption system of Figure 11.8(b), however, the COP values are given by

$$COP_C = \frac{\dot{Q}_C}{\dot{Q}_G} \qquad (11.2)$$

and

$$COP_H = \frac{\dot{Q}_H}{\dot{Q}_G} \qquad (11.3)$$

The second law efficiency for the vapor compression system is easy to work out based on the defining Equation 11.1. It is necessary, however, to indicate if T_L or T_H corresponds to the temperature (T_0) of the ambient environment, which may be considered the reference environment. In the simplest case, when the system serves as a cooling device, T_H is equal to T_0, while for operation in the heating mode, the source temperature T_L is deemed equal to T_0. In either case for the vapor compression system, if, for the same desired cooling or heating load, $W_{optimum}$ represents the necessary work input for the device operating on the reversed Carnot cycle, the second law efficiency is

$$\frac{\eta_2\,(\%)}{100} = \frac{W_{optimum}}{W_{actual}}$$

$$= \left(\frac{W_{optimum}}{Q}\right)\left(\frac{Q}{W_{actual}}\right) = \frac{COP_{actual}}{COP_{optimum}} \tag{11.4}$$

where Q stands for either Q_L or Q_H depending on whether a cooling or a heating application is being considered.

EXAMPLE 11.1

A refrigerator works between two thermal reservoirs at 273 K and 300 K. The ambient temperature is 300 K.

(a) What is the maximum COP_C possible?

(b) If the refrigerator is electrically driven and has an actual COP_C of 2, determine the second law efficiency achieved.

Given: Reversed heat engine as in Figure 11.9

$$T_L = T_{source} = 273 \text{ K}$$

$$T_H = T_{sink} = 300 \text{ K} \,(= T_0)$$

$$(COP_C)_{actual} = 2$$

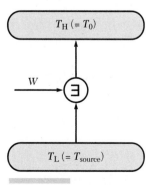

Figure 11.9 Reversed heat engine (Ⴈ) of Example 11.1.

Find:

(a) Maximum COP_C

(b) $(\eta_2)_{actual}$

Solution:

(a) Both the heat source and heat sink are unlimited, and therefore the maximum COP_C is that given by a reversed Carnot cycle, which from Equation 7.22a is

$$(COP_C)_{optimum} = \frac{T_L}{T_H - T_L} = \frac{273}{300 - 273} = \underline{10.1}$$

(b) In this example, Equation 11.4 is used to determine the second law efficiency achieved:

$$\eta_2 = \frac{COP_{actual}}{COP_{optimum}} \cdot 100\% = \frac{2}{10.1} \cdot 100\% = \underline{19.8\%}$$

In Example 11.1, note that while an actual COP_C of 2 appears flatteringly high, the second law efficiency computed for the electrically driven refrigerator indicates that more than 80% of the exergy input (in this case, electricity) is destroyed due to irreversibilities.

Returning now to the absorption system of Figure 11.8(b), the second law efficiency is no longer simply a ratio of two COP values as given by Equation 11.4. However, a simple expression for the second law efficiency can be obtained if certain parameters are used for the gas heating aspect of the system operation. Let $\dot{m}_{fuel}$ denote the mass flow rate of the gaseous fuel. The *heating value* (HV) of the fuel represents the energy released as heat when a unit mass of the fuel is completely burned. Thus, the fuel energy supply rate to the absorption system operation is given by

$$\text{Fuel energy supply rate} = \dot{m}_{fuel} \cdot HV \tag{11.5}$$

Not all of this translates to the generator heating rate ($\dot{Q}_G$), however, since the flue gas up the chimney is at a higher temperature level than the atmospheric air. One can define a burner or heater efficiency as

$$\eta_{burner} = \frac{\dot{Q}_G}{\dot{m}_{fuel} \cdot HV} \cdot 100\% \tag{11.6}$$

Finally, assume that you can look up the value of the chemical exergy per unit mass of fuel (x_c) for the gas used, in which case the exergy input rate to the entire operation is given by

$$\dot{X}_{input} = \dot{m}_{fuel} \cdot x_c \tag{11.7}$$

Thus, if a rather simple case is considered for which either T_L or T_H is the same as the ambient temperature T_0, one can define the second law efficiency of the entire operation of the absorption system as

$$\frac{\eta_2\,(\%)}{100} = \frac{\dot{W}_{\text{optimum}}}{\dot{X}_{\text{input}}}$$

$$= \left(\frac{\dot{W}_{\text{optimum}}}{\dot{Q}}\right)\left(\frac{\dot{Q}}{\dot{Q}_{\text{G}}}\right)\left(\frac{\dot{Q}_{\text{G}}}{\dot{m}_{\text{fuel}}\cdot\text{HV}}\right)\frac{\text{HV}}{x_{\text{c}}} \qquad (11.8)$$

$$= \frac{\text{HV}}{x_{\text{c}}}\left(\frac{\eta_{\text{burner}}\,(\%)}{100}\right)\frac{\text{COP}_{\text{actual}}}{\text{COP}_{\text{optimum}}}$$

where $\dot{Q}$ is used for either $\dot{Q}_{\text{C}}$ or $\dot{Q}_{\text{H}}$ and the $\text{COP}_{\text{actual}}$ is given by either Equation 11.2 or 11.3 defined earlier, depending on whether one is considering a cooling or a heating application. The $\text{COP}_{\text{optimum}}$ is that for a device operating on the reversed Carnot cycle between T_{L} (or T_{H}) and T_0.

Chapter 14 considers the thermodynamics of combustion and shows that the ratio HV/x_{c} for most gaseous hydrocarbon fuels is close to unity. A good gas burner design might operate with an efficiency on the order of 90%. Thus, Equation 11.8 for the absorption system can be simplified to the following:

$$\eta_2 = 0.9\,\frac{\text{COP}_{\text{actual}}}{\text{COP}_{\text{optimum}}}\cdot 100\% \qquad (11.9)$$

EXAMPLE 11.2

The refrigerator in Example 11.1 is actually a gas-fired absorption system with an average COP_{C} of 0.8. What is the second law efficiency achieved?

Given: Gas-fired absorption refrigeration system shown in Figure 11.10. The source and sink temperatures are 273 K and 300 K, as in Example 11.1. Thus, $\text{COP}_{\text{optimum}} = 10.1$ as before. $\text{COP}_{\text{actual}} = 0.8$.

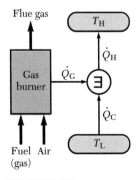

Figure 11.10 Gas-fired absorption refrigeration system.

Find: The second law efficiency

Solution: The second law efficiency can be estimated using Equation 11.9 in this case. Thus,

$$\eta_2 = 0.9 \, \frac{COP_{actual}}{COP_{optimum}} \cdot 100\% = \frac{(0.9)(0.8)}{10.1} \cdot 100\% = \underline{7.13\%}$$

Which of the two systems in Examples 11.1 and 11.2 is the better system? It is important to note that the electricity used in a vapor compression refrigerator is not a primary energy resource. Indeed, if the electricity is being produced from a gas-fired thermal power station, the fuel exergy input required is generally about three times the exergy produced in the form of electricity. Thus, relative to the primary energy resource (gas), the second law efficiency achieved with the electrically driven refrigerator is nearer 19.8/3, or 6.6%. In other words, while the COP for the vapor compression refrigerator appears to be much higher than that for the absorption system, the gas-fired absorption refrigerator may be the better choice from a second law perspective. It usually turns out that the better choice on the basis of second law efficiency is also the one with a lower operating cost.

Finite Heat Sources and Sinks

Just as for heat engines, available heat sources and sinks for reversed heat engine applications are typically finite rather than unlimited. Also, the heat source or sink is often of the regeneratable type such as when chilled water used for air conditioning is recycled through the cooling system. The definitions of first law efficiencies given by Equations 7.5, 7.6, 11.2, and 11.3 are applicable even when the heat source or sink is finite or regeneratable. In these cases, however, the reversed Carnot cycle is no longer the optimum cycle. For a finite heat source or a finite heat sink, the second law efficiency is best evaluated by determining the exergy product and the exergy input involved. Thus, the overall second law efficiency can be defined in a given operation as

$$\eta_2 = \frac{\text{desired exergy product}}{\text{required exergy input}} \cdot 100\% \qquad (11.10)$$

EXERCISES

for Section 11.3

1. A refrigeration device is operated between two thermal reservoirs at 0°C and 35°C. Determine the maximum possible COP_C (for cooling).

    Answer: 7.8

2. Assume that the actual refrigerator in Exercise 1 is electrically driven with a $COP_C = 1.8$ and that the second law efficiency for electricity production from primary energy (such as natural gas) is 30%. Compute the overall second law efficiency relative to the primary energy resource base.

    Answer: 6.92%

3. A gas-fired absorption refrigerator is used for the operation in Exercise 1. If the COP_C of the device is 0.7, determine the second law efficiency achieved.

   Answer: 8.08%

4. If electricity costs 5¢ per kWh, estimate the cost of producing 1 ton (2000 lbm) of ice at 32°F from water at 32°F using the device in Exercise 2.

   Answer: $2.35

5. Assume that 1 million Btu of natural gas costs $5. Estimate the cost of producing 1 ton (2000 lbm) of ice at 32°F from water at 32°F using the device in Exercise 3.

   Answer: $2.06

11.4 *Vapor Compression Cycles*

The principal features of construction of a typical vapor compression refrigeration device are shown in Figure 11.1, which was introduced in Section 11.2. Figure 11.2 illustrates the vapor refrigeration cycle on *T-s* coordinates. In refrigeration cycle analysis, the pressure (p) versus enthalpy (h) diagram is more commonly used since the operation of a vapor compression refrigeration system is between two pressure levels—the *evaporator pressure* (low) and the *condenser pressure* (high). Figure 11.11 shows the typical vapor compression cycle on *p-h* coordinates. The following paragraphs describe the processes that make up the cycle.

Evaporation Process (1→2)

Actual cooling in the refrigeration cycle is produced in the evaporator. To cause a liquid to evaporate, heat equivalent to the enthalpy of vaporization of the liquid must be supplied. The evaporation of the refrigerant thus produces cooling of the medium placed in contact with the evaporator unit. A variety of refrigerants have been employed, with the choice often based on the cooling temperature desired as well as on safety considerations. Ammonia, for example, is a popular choice for many refrigeration applications. However, because of its harmful qualities, its use inside buildings is prohibited by law. Another example is refrigerant-12 (dichlorodifluoromethane), which was once the most

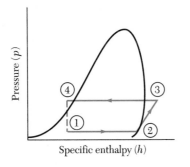
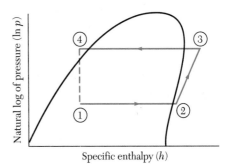

Figure 11.11 Typical vapor compression cycle on *p-h* coordinates.

commonly used refrigerant in domestic refrigeration devices and in air conditioners in automobiles. Refrigerant-12 is now thought to have a detrimental influence on the protective ozone layer in the Earth's atmosphere. Today, a less harmful refrigerant such as refrigerant-22 (monochlorodifluoromethane) is widely used in place of refrigerant-12. For this reason, Appendices A and B provide tables of properties for refrigerant-22 rather than for refrigerant-12. Also, the computer code **R22** for refrigerant-22 properties is included on the PC diskette supplied with this text. It should be noted that other refrigerants with qualities even better than those of refrigerant-22 are currently under development and may be used eventually to replace refrigerant-22.

Compression Process (2→3)

The processes following the evaporation of the refrigerant are designed to restore the vapor produced back to the original very wet vapor condition preceding the evaporation process. The compressor raises the pressure of the vapor from the low evaporator pressure at state 2 to the high condenser pressure at state 3. The compressor also causes the movement of the refrigerant through the various units making up the refrigeration device. The compression process is assumed in the ideal case to occur reversibly and adiabatically. As remarked earlier, the condenser pressure must be such that the corresponding saturation temperature of the refrigerant is higher than the sink temperature.

Condensation (3→4)

Condensation of the vapor takes place as a result of the heat rejection process (3→4), frequently to ambient air. Note that when the device is used as a heat pump, the heat rejection is to the medium maintained at the higher temperature. In such applications, the ambient air serves as the heat source and exchanges heat with the evaporator unit. The vapor temperature at the end of the compression stage is normally much higher than the ambient temperature (or that of the space to be heated). Thus, condensation to a slightly subcooled liquid condition can often be achieved using condensers suitably designed as finned heat exchangers to enhance the heat transfer rates, even when the heat transfer to the heat sink is by natural convection.

Throttling (4→1)

The throttle valve is used to restore the refrigerant to the low pressure and very wet vapor condition prior to entry into the evaporator coils. The dashed lines used on both the *T-s* and *p-h* diagrams for the throttling process are a reminder that the process is internally irreversible. (The throttling process is discussed again in some detail in Chapter 12.)

A first law analysis of the vapor compression refrigeration cycle basically involves successive application of the steady-flow energy equation (SFEE) (Eq. 6.29) to each of the processes involved. A complete analysis will require knowledge of the thermodynamic properties of the refrigerant at the various states, and for this, one can use tables of properties or charts. Charts of pressure versus specific enthalpy (*p-h*) are available for most refrigerants. As mentioned earlier, they are widely regarded as the appropriate charts to work from in refrigeration cycle calculations. The computer code that is provided for the properties of refrigerant-22 should assist in making a complete analysis of refrigeration cycles using this refrigerant. The following worked examples illustrate the procedures involved in a complete thermodynamic analysis of a vapor compression refrigeration system.

EXAMPLE 11.3

A domestic vapor compression refrigerator uses refrigerant-22 as the refrigerant. A mass of 1 kg of water at 20°C is placed in the freezer compartment of the refrigerator and cooled by evaporating the refrigerant at a pressure corresponding to a saturation temperature for the refrigerant of −10°C. The refrigerant at entry to the compressor is superheated vapor with 5°C of superheat. The condenser pressure is that corresponding to a saturation temperature of 45°C for the refrigerant. The refrigerant is condensed with 10°C of subcooling, after which it is throttled adiabatically to the evaporator pressure. The estimated isentropic efficiency of the compressor is 90%. The refrigerant mass flow rate is 6.8 kg/h. The ambient temperature can be taken as 300 K. For water, assume c_p = 4.2 kJ/kg · K and specific enthalpy of fusion = 335 kJ/kg. Calculate

(a) the COP_C for the refrigerator

(b) the power required to drive the compressor if its mechanical efficiency is 60%

(c) the time taken for the water to turn completely to ice at 0°C, stating clearly your assumptions

(d) the second law efficiency achieved in the entire ice-making process

Given: The refrigeration cycle is shown in Figures 11.12(a) and (b). Refrigerant-22 is used.

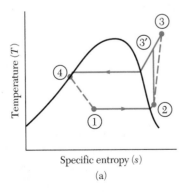

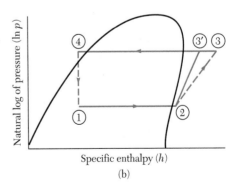

Figure 11.12 The practical refrigeration cycle of Example 11.3 on (a) *T-s* coordinates and (b) *p-h* coordinates.

$$T_1 = -10°C, \ T_2 = -5°C, \ T_4 = 35°C$$

$$T_s \ (\text{at the condenser pressure}) = 45°C$$

$$T_s \ (\text{at the evaporator pressure}) = -10°C$$

$$\dot{m}_{\text{ref}} = 6.8 \text{ kg/h}$$

and

Isentropic efficiency of the compressor = 90%

Mechanical efficiency for the compressor drive = 60%

Ambient temperature (T_0) = 300 K

The *system* is 1 kg of water, initially at 20°C, to be turned to ice at 0°C

$$c_p = 4.2 \text{ kJ/kg} \cdot \text{K and specific heat of fusion} = 335 \text{ kJ/kg}$$

Find:
(a) COP_C
(b) $(\dot{W}_x)_{\text{compressor}}$
(c) time for cooling
(d) second law efficiency achieved

Solution: The refrigeration cycle comprises flow processes that are assumed to be steady for the entire operation. The SFEE (Eq. 6.29) can thus be applied to each process. Neglecting changes in kinetic energy and potential energy, this equation can be written on a unit mass basis as follows:

$$q - w = \Delta h$$

Process 1→2 (constant pressure evaporation)

For a reversible, constant pressure flow process, the shaft work is zero. The SFEE thus reduces to the following for process 1→2:

$$q_{12} = h_2 - h_1$$

Process 2→3 (adiabatic compression)

Process 2→3 is adiabatic but nonreversible. Substituting $q_{23} = 0$ in the SFEE gives for w_{23}

$$w_{23} = h_2 - h_3$$

The isentropic efficiency η_{aC} as defined in Equation 8.111 or 8.112 gives a relationship between the actual work done (w_{23}) and the work done in the reversible adiabatic process 2→3'. Thus, knowing h_2 and $h_{3'}$, one can compute h_3 from application of Equation 8.112:

$$h_3 - h_2 = \frac{h_{3'} - h_2}{\eta_{\text{aC}}}$$

Process 4→1 (adiabatic throttling)

For the adiabatic throttling process, $q_{41} = 0$ and $w_{41} = 0$. Hence, $h_1 = h_4$ from application of the SFEE. Thus, knowing h_4 gives the value for h_1.

The values of the h_2, $h_{3'}$, and h_4 were determined using the **R22** code; h_1 and h_3 were subsequently determined using the appropriate relationships from those previously given. The summary is as follows:

$$h_1 = 88.4 \text{ kJ/kg} \quad (s_1 = 0.3429 \text{ kJ/kg} \cdot \text{K}, \ T_1 = -10°\text{C})$$

$$h_2 = 249.6 \text{ kJ/kg} \quad (s_2 = 0.9552 \text{ kJ/kg} \cdot \text{K}, \ T_2 = -5°\text{C})$$

$$h_{3'} = 290.6 \text{ kJ/kg} \quad (s_{3'} = 0.9552 \text{ kJ/kg} \cdot \text{K}, \ T_{3'} = 75.6°\text{C})$$

$$h_3 = 295.2 \text{ kJ/kg} \quad (s_3 = 0.9681 \text{ kJ/kg} \cdot \text{K}, \ T_3 = 80.8°\text{C})$$

$$h_4 = 88.4 \text{ kJ/kg} \quad (s_4 = 0.3242 \text{ kJ/kg} \cdot \text{K}, \ T_4 = 35°\text{C})$$

From these, one can calculate q_{12} and w_{23}. The values obtained are 161.2 kJ/kg and -45.6 kJ/kg, respectively.

(a) The COP_C is defined by Equation 7.5, which in the present case can be written as follows:

$$\text{COP}_\text{C} = \frac{\text{desired effect}}{\text{required input}} = \frac{\dot{m}_{\text{ref}} \cdot q_{12}}{\dot{m}_{\text{ref}} \cdot (-w_{23})}$$

Substituting the values of q_{12} and w_{23} thus gives $\text{COP}_\text{C} = 3.54$.

(b) The rate of doing work in the vapor compression process is given by the product of the refrigerant mass flow rate and w_{23}. Thus,

$$(\dot{W}_\text{x})_{23} = \left(\frac{6.8}{3600}\right)(-45.6) = -0.0861 \text{ kW}$$

The required power input is obtained by dividing this rate of doing work by the mechanical efficiency of 0.6 specified:

$$(\dot{W}_\text{x})_{\text{input}} = \frac{-(\dot{W}_\text{x})_{23}}{\eta_{\text{mech}}} = 144 \text{ W}$$

(c) Assume that the cooling rate stays constant throughout the operation. This cooling rate is obtained by multiplying q_{12} by the refrigerant mass flow rate. The result is 304 W. If cooling at this rate is carried out for a period of t seconds, the total heat flow from the system being cooled is then $(0.304t)$ kJ.

The system being cooled is 1 kg of water at 20°C, which is to be turned to ice at 0°C. Applying the NFEE (Eq. 6.10) with kinetic energy and potential energy changes assumed negligible, one obtains the following for the heat flow between the system and the evaporator coil of the refrigerator:

$$Q_{\text{system}} = (H_{\text{final}} - H_{\text{initial}})_{\text{system}}$$

$$= m_{\text{system}} [(h_{\text{ice}})_{0°\text{C}} - (h_{\text{water}})_{20°\text{C}}]$$

$$= (1 \text{ kg})[(-335) + 4.2(0 - 20) \text{ kJ/kg}] = -419 \text{ kJ}$$

Equating $(-Q_{system})$ to $0.304t$ gives $t = 1378$ s. Thus, the time taken for turning the water completely to ice at $0°C$ is 23 min.

(d) The second law efficiency is calculated using Equation 9.8. The exergy input is measured by the total work done in driving the compressor $= (0.144)(1378) = 198$ kJ. The exergy product is the increase in the exergy of the water turned to ice, which can be determined using Equation 9.25. Assuming that the water pressure is the same as that for the ambient environment, one obtains from Equation 9.25 the following for the exergy product:

$$X_{product} = m_{system} \{(h_{ice})_{0°C} - (h_{water})_{20°C} - T_0[(s_{ice})_{0°C} - (s_{water})_{20°C}]\}$$

Values of specific enthalpy and entropy for water and ice are obtained by using the property tables in Appendix A or the **STEAM** code. Substituting the appropriate property values in the previous equation gives an exergy product of 36.1 kJ. Applying Equation 9.8 yields a second law efficiency of 18.2%.

EXAMPLE 11.4

A practical vapor compression refrigeration cycle uses refrigerant-22 as the refrigerant between a 25°C (77°F) ambient environment and a freezer compartment to be maintained at $-1.11°C$ (30°F). To achieve the desired cooling, the evaporator pressure and the condenser pressure are such that the corresponding saturation temperatures for the refrigerant are $-10°C$ (14°F) for the evaporator and 35°C (95°F) for the condenser. The refrigerant vapor leaving the evaporator and the liquid leaving the condenser are both saturated. The temperature of the refrigerant after compression is 60°C (140°F). The cooling load is 3.5 kW (which is approximately 1 ton of refrigeration, where *1 ton of refrigeration* is defined as the cooling rate required to freeze 1 ton (2000 lbm) of water over a 24-h period. This quantity is given by $\dot{Q} = (2000 \text{ lbm/d})(144 \text{ Btu/lbm})(1 \text{ d/24 h})$, or $\dot{Q} = 12,000$ Btu/h. Determine the following:

(a) the exergy production rate (in kW) by the refrigeration device

(b) the exergy consumption (or destruction) rate, in kW, in each of the following processes:
   (i) cooling process resulting from the evaporation of the refrigerant in the evaporator
   (ii) compression (assumed adiabatic) of the refrigerant vapor
   (iii) condensation process in the condenser
   (iv) adiabatic throttling of the saturated (liquid) refrigerant

(c) the COP_C for the actual refrigeration cycle

(d) the COP_C for a reversed Carnot cycle between the 25°C ambient environment and the $-1.11°C$ heat source temperature

(e) the second law efficiency of the cycle

Given: The refrigeration cycle is illustrated in Figures 11.13(a) and (b). Refrigerant-22 is used, and the following conditions are specified:

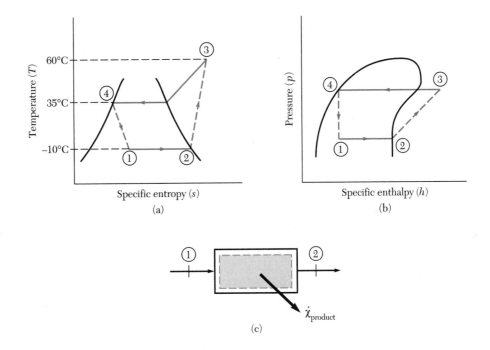

Figure 11.13 The actual refrigeration cycle of Example 11.4 on (a) *T-s* coordinates and (b) *p-h* coordinates. (c) The control volume for the evaporation process in Example 11.4.

$$T_2 = -10°C, x_2 = 1$$

$$T_3 = 60°C, p = p_s (35°C)$$

$$T_4 = 35°C, x_4 = 0$$

Freezer compartment is to be maintained at $T_L = 272$ K $(-1.11°C)$

$$\text{Cooling load} = 3.5 \text{ kW}$$

$$T_0 = 25°C (= 298 \text{ K}) = T_H$$

Find:

(a) $\dot{X}_{product}$ (in kW)

(b) $\dot{X}_{consumed}$ (in kW) for processes (i) 1→2, (ii) 2→3, (iii) 3→4, and (iv) 4→1

(c) $(COP_C)_{actual}$

(d) $(COP_C)_{reversed\ Carnot}$ with T_L and T_H specified

(e) η_2 for actual cycle

Solution:

(a) The heat source and the heat sink are both constant temperature systems and thus can be regarded as thermal reservoirs. The useful product in this case is the heat extraction at the rate 3.5 kW from the reservoir at $T_L = 272$ K $(-1.11°C)$.

Thus, if the freezing compartment is considered as the system, and maintaining it at a constant temperature is considered as internally reversible, Equations 9.7 and 9.31 give the following for the useful exergy production rate:

$$\dot{X}_{\text{product}} = (-\dot{Q}_{\text{L}})\left(1 - \frac{T_0}{T_{\text{L}}}\right) = \underline{0.335 \text{ kW}}$$

(b) For the determination of exergy consumption rates in the processes indicated, one needs to have the h and s values at states 1, 2, 3, and 4. The **R22** code gives the following values:

State Number	h (kJ/kg)	s (kJ/kg $\cdot$ K)
1	88.42	0.3429
2	246.07	0.9420
3	282.34	0.9507
4	88.42	0.3242

The refrigerant flow rate must also be calculated. This can be done simply by applying the SFEE to the evaporation process 1→2:

$$\dot{Q}_{\text{L}} = \dot{m}_{\text{ref}}(h_2 - h_1)$$

$$\therefore \dot{m}_{\text{ref}} = \frac{\dot{Q}_{\text{L}}}{h_2 - h_1} = 0.0222 \text{ kg/s}$$

(i) Figure 11.13(c) illustrates process 1→2, which results in the cooling rate of 3.5 kW and exergy production rate of 0.335 kW obtained in part (a). The exergy input rate is calculated using Equation 9.33:

$$\dot{X}_{\text{input}} = \dot{m}_{\text{ref}}[h_1 - h_2 - T_0(s_1 - s_2)] = 0.464 \text{ kW}$$

Applying Equation 9.9 yields the following for the exergy consumption rate in the cooling process:

$$\dot{X}_{\text{consumed}} = \dot{X}_{\text{input}} - \dot{X}_{\text{product}} = \underline{0.129 \text{ kW}}$$

(ii) For the exergy consumption in the vapor compression process, the exergy input rate is the power input required to the compressor for the process, while the production rate is the increase in exergy flow rate produced from state 2 to state 3. The power input required to the compressor is obtained by applying the SFEE (Eq. 6.29) to process 2→3:

$$(-\dot{W}_{\text{x}})_{23} = \dot{m}_{\text{ref}}(h_3 - h_2) = 0.805 \text{ kW}$$

(For this calculation the compression process is assumed adiabatic.) The exergy production rate is calculated by applying Equation 9.33:

$$\dot{X}_{\text{product}} = \dot{m}_{\text{ref}}[h_3 - h_2 - T_0(s_3 - s_2)] = 0.748 \text{ kW}$$

The exergy consumption rate is then calculated by applying Equation 9.9:

$$\dot{X}_{\text{consumed}} = \dot{X}_{\text{input}} - \dot{X}_{\text{product}} = \underline{0.057 \text{ kW}}$$

(iii) For the vapor condensation process, heat rejection is to the reference environment, and therefore the exergy production rate is zero. The exergy input rate can be determined by applying Equation 9.33:

$$\dot{X}_{input} = \dot{m}_{ref}[h_3 - h_4 - T_0(s_3 - s_4)] = 0.160 \text{ kW}$$

Therefore, from Equation 9.9, it is concluded that the exergy consumption rate is 0.16 kW.

(iv) The throttling process is adiabatic. Neither a heat flow nor a work flow occurs between the refrigerant flowing through the throttle valve and the surroundings. Thus, the exergy production rate is zero, and the exergy input rate is computed in terms of the decrease in the exergy flow rate of the refrigerant from state 4 to state 1:

$$\dot{X}_{input} = \dot{m}_{ref}[h_4 - h_1 - T_0(s_4 - s_1)] = 0.124 \text{ kW}$$

Thus, from Equation 9.9, it is concluded that the exergy consumption rate is 0.124 kW.

Table 11.1 gives a summary of the previous estimates of exergy disposition. The table shows that the most significant exergy consumption occurs in the condensation process. Significant irreversibilities occur also in the actual refrigeration (or cooling) process and the throttling process. The irreversibilities associated with the evaporator and the condenser processes can be reduced by design of better heat exchangers, which involve smaller ΔT values between the fluid media involved.

(c) The coefficient of performance is calculated from Equation 7.5:

$$(COP_C)_{actual} = \frac{\dot{Q}_L}{-(\dot{W}_x)_{23}} = 4.35$$

TABLE 11.1

Exergy Disposition for a Vapor Compression Cycle Between Two Thermal Reservoirs

	Exergy (kW)	Percent of Input Exergy (%)
$\dot{X}_{input}$	0.805	100
$\dot{X}_{product}$	0.335	41.6
$\dot{X}_{consumed}$		
Process 1→2	0.129	16.0
Process 2→3	0.057	7.1
Process 3→4	0.160	19.9
Process 4→1	0.124	15.4
Subtotal	0.470	58.4

(d) The COP_C for a reversed Carnot cycle operating between the 25°C ambient temperature and the T_L of 272 K can be calculated from Equation 7.22a:

$$(COP_C)_{\substack{reversed \\ Carnot}} = \frac{T_L}{T_0 - T_L} = \underline{10.4}$$

(e) The second law efficiency for the entire operation is determined by applying Equation 9.8:

$$\eta_2 = \frac{0.335 \text{ kW}}{0.805 \text{ kW}} \cdot 100\% = \underline{41.6\%}$$

Section 11.7 provides a more detailed discussion of the components of exergy destruction in typical vapor compression refrigeration applications. The treatment also includes modifications that can be made to the basic cycle that can result in improved performance.

EXERCISES

for Section 11.4

1. A practical vapor compression cycle, such as that illustrated in Figures 11.12(a) and (b), has evaporator and condenser pressures of 1.5 bar and 20 bar, respectively. Refrigerant-22 is used. The fluid leaves the evaporator as a saturated vapor and is then compressed adiabatically with the exit temperature from the compressor being 155°C. The refrigerant is condensed and subcooled to $T_4 = 40°C$ prior to the adiabatic throttling process (4→1). Use the property tables provided in Appendix A to determine h and s for each of the states 1, 2, 3', 3, and 4.

2. Cross check your property values in Exercise 1 using the **R22** code.

3. For the cycle in Exercise 1, determine, per unit mass of the refrigerant,
   (a) the refrigeration effect (q_{12})
   (b) the compressor work (w_{23})
   (c) q_{34}
   Answer: (a) 142 kJ/kg, (b) -85.4 kJ/kg, (c) -227 kJ/kg

4. For the cycle in Exercise 1, determine
   (a) COP_C
   (b) the process efficiency for the compression process (2→3)
   Answer: (a) 1.66, (b) 0.796

5. Assume that $T_0 = 300$ K and that the refrigerator in Exercise 1 was used for producing ice at 0°C from water at 0°C. Compute the exergy destruction, per unit mass of the refrigerant, in each of the following processes:
   (a) vapor compression (2→3)
   (b) throttling (4→1)
   (c) condensation (3→1), assuming the heat rejection is to the ambient air at T_0
   (d) evaporation (1→2)
   Answer: (a) 13.9 kJ/kg, (b)14.9 kJ/kg, (c) 22 kJ/kg, (d) 20.7 kJ/kg

6. Determine the second law efficiency achieved in the operation of the refrigerator in Exercise 1 for the ice production by computing the following:

(a) the exergy input to the cycle, the exergy destruction, and hence, the useful exergy produced in each cycle (per unit mass of the refrigerant)

(b) the maximum possible COP_C and using Equation 11.4

Answer: 16.3%

11.5 Absorption Refrigeration Cycles

The thermodynamic analysis of absorption refrigeration cycles is complex because at least two fluid media, an absorbent and a refrigerant, are involved as compared with the vapor compression system, which uses only one fluid (the refrigerant). Also, the absorbent–refrigerant mixture is usually not a pure substance since the composition is typically different for the vapor phase than for the liquid phase. For binary mixtures, one generally needs to specify (in addition to two independent intensive properties) the mass fraction for one of the constituents of the mixture in order to fix the state of the system.

Renewed general interest in absorption cooling systems can be attributed in part to the growing concern about ozone layer depletion, which has been blamed largely on the widespread use of the fluorocarbons in vapor compression refrigeration systems. Also, an absorption cooling system can be driven using waste or relatively low grade heat such as solar radiation. Absorption cooling systems, however, are often flawed because of low values for the coefficient of performance (COP). The COP is a first law efficiency and cannot properly be regarded as a performance evaluation index. A proper performance evaluation requires a second law analysis, and if such analysis is required for the individual processes in an absorption cycle, entropy data are needed for the refrigerant and the absorbent–refrigerant mixture.

Most absorption chillers use either the lithium bromide–water pair or the ammonia–water (aqua-ammonia) combination. The relative lack of comprehensive equilibrium data for the potential and even for the common absorbent–refrigerant media has continued to pose additional difficulty in performing a complete thermodynamic analysis of absorption cycles. The following section provides a summary of the few available sources of property data for the two most common absorbent–refrigerant mixtures.

Review of Literature on Properties of Absorbent–Refrigerant Mixtures

For the lithium bromide–water solutions, thermodynamic property tables have been published by Foote Mineral Company (Boryta 1971) based largely on the work of Uemura and Hasaba (1964). The formulation by McNeely (1979) is comprehensive, but does not include entropy in the list of thermodynamic properties offered for aqueous solutions of lithium bromide. The formulations by Gupta and Sharma (1976), Herold and Moran (1987) and Koehler et al. (1987) do include entropy alongside the other thermodynamic properties of lithium bromide–water solutions.

For the ammonia–water mixtures, the formulations that include mixture entropy are also limited. Macriss et al. (1964) provide vapor pressure and enthalpy data for ammonia–water mixtures in tabular form. Jain and Gable (1971) have produced equilibrium data equations from the data which are applicable over a restricted range of conditions.

A similar exercise was carried out by Adebiyi (1986) with applications to solar refrigeration. The reduction of equilibrium properties to data equations facilitates the use of the computer in the design and analysis of the absorption refrigeration system (for example, see Dhar et al. 1977). Jennings' (1981) publication of the thermodynamic properties of ammonia–water mixtures is comprehensive in scope, but does not include entropy in the list of properties. Formulations for the entire list of thermodynamic properties include those of Schulz (1973), Gupta and Sharma (1975), and Herold et al. (1988). These latter formulations make certain simplifying assumptions pertaining to the ideality of mixtures, but in spite of such assumptions, the resulting equations are quite complex and require the development of computer codes for their use in typical thermodynamic design calculations.

Applications to Cycle Analysis

The 1989 Fundamentals Handbook (ASHRAE 1989) was given as a good reference for the description of a design procedure for absorption cycles. The following two sources of information on design procedures can be added, one of which is based on the second law of thermodynamics. The publication by Dhar et al. (1977) describes a computer-aided design method for the absorption system, although it is based entirely on first law considerations. Anand et al. (1984) present a second law analysis of absorption cooling cycles which indicates that the most significant irreversibilities are encountered in the refrigerant absorption process (occurring in the absorber unit) and in the separation of the refrigerant from solution in the generator. These irreversibilities are more pronounced for the ammonia–water cycle than they are for the lithium bromide–water cycle. For typical conditions analyzed by Anand et al. (1984), the coefficients of performance for the ammonia–water cycle and a single-effect lithium bromide–water cycle are about 0.58 and 0.72, respectively, while the corresponding second law efficiencies are 10% and 24%. Such analyses will undoubtedly be useful in screening potential absorbent–refrigerant pairs for optimum cycle performance.

11.6   Air Standard Gas Refrigeration Cycle

Gas refrigeration is commonly used for air conditioning of aircraft cabins. As indicated earlier, the standard air refrigeration cycle is the reversed Brayton cycle. The following examples demonstrate the procedure for the thermodynamic analysis of gas refrigeration cycles.

EXAMPLE 11.5

An air refrigeration system is designed to operate on the reversed Brayton cycle. The air temperature is 20°C and the pressure is 1 atm at the compressor inlet, and the pressure ratio is 3. The air leaving the compressor is cooled to 45°C before it enters the turbine. The cooled space is to be maintained at 20°C. The isentropic efficiencies of both compressor and turbine can be taken as 85%. The ambient temperature and pressure are 300 K and 1 atm. If the mass flow rate of air is 450 kg/h, calculate

(a) the air temperature upon exit from the turbine

(b) the refrigerating or cooling effect (in kW)

(c) the exergy production rate (in kW)

(d) the net power input (in kW) required, assuming that the power output of the turbine is used to meet part of the requirement for driving the compressor

(e) the second law efficiency achieved

Given: The actual cycle (based on reversed Brayton cycle) is illustrated in Figure 11.14. The temperature and pressures given are:

$$T_1 = 293 \text{ K } (20°\text{C}), \; p_1 = 1 \text{ atm} = p_4$$

$$T_3 = 318 \text{ K } (45°\text{C}), \; p_3 = 3 \text{ atm} = p_2$$

$$\eta_{aC} = 0.85 = \eta_{aT}$$

$$\text{Mass flow rate of air} = 450 \text{ kg/h}$$

$$p_0 = 1 \text{ atm}, \; T_0 = 300 \text{ K}$$

Find:

(a) T_4

(b) $\dot{Q}_L$ (in kW)

(c) $\dot{X}_{product}$ (in kW)

(d) $(\dot{W}_{net})_{input}$

(e) η_2

Solution:

(a) $T_{4'}$ (and $T_{2'}$) for isentropic processes can be determined from Equation 8.31:

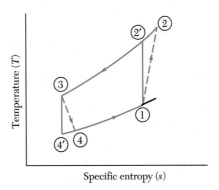

Figure 11.14 The actual air refrigeration cycle (based on the reversed Brayton cycle) of Example 11.5.

$$\frac{T_3}{T_{4'}} = r_p^{(k-1)/k} = \frac{T_{2'}}{T_1}$$

where r_p is the pressure ratio, which is 3, and k is the ratio of principal specific heats, which can be taken as 1.4 for air. Substituting these values into the equation gives $T_{4'} = 232$ K and $T_{2'} = 401$ K. T_4 can now be calculated using Equation 8.110 for the isentropic efficiency of an adiabatic turbine. Thus, using the value of 0.85 given for this efficiency, one obtains $T_4 = \underline{245 \text{ K} (-28°C)}$.

T_2 is determined by applying Equation 8.112 for the isentropic efficiency of an adiabatic compressor. The result in this case is $T_2 = 420$ K (147°C).

(b) The refrigerating effect is determined by applying the SFEE (Eq. 6.29) to the heat load removal process 4→1:

$$\dot{Q}_L = \dot{m}_{air}c_{p0}(T_1 - T_4) = \underline{6.03 \text{ kW}}$$

where $c_{p0} = 1.005$ kJ/kg · K was assumed for the air.

(c) The exergy production rate is the increase in the exergy flow rate for the air stream between the intake condition (state 1) and state 4 at the exit to the turbine. Applying Equation 9.40a thus yields

$$\dot{X}_{product} = \dot{m}_{air}c_{p0}\left(T_4 - T_1 - T_0 \ln \frac{T_4}{T_1}\right)$$

$$= \underline{0.713 \text{ kW}}$$

The pressure exergy term is not included in this expression since $p_4 = p_1$, thus making the term zero.

(d) The net power input required is obtained from the power input required for the compression process and the power output from the adiabatic turbine. Thus, applying the SFEE to the adiabatic compression process (1→2) and the adiabatic expansion process (3→4) gives the following for the net power input required:

$$(\dot{W}_{net})_{input} = \dot{m}_{air}c_{p0}[(T_2 - T_1) - (T_3 - T_4)]$$

$$= \underline{6.78 \text{ kW}}$$

The corresponding COP_C is the ratio of the cooling load and the net power input required, which works out as 0.889 in this case.

(e) The second law efficiency is determined from Equation 9.10. The exergy production rate is 0.713 kW, while the input rate is 6.78 kW. Thus, the second law efficiency is $\underline{10.5\%}$.

EXERCISES

for Section 11.6

1. A steady stream of compressed air at a pressure p (in kPa) and a temperature of 318 K is supplied to an adiabatic turbine. The air exits the turbine at 285 K and 100 kPa pressure. The isentropic efficiency of the turbine is 0.8. Determine
   (a) the exit temperature if the air had passed through an isentropic turbine from the specified inlet condition to the exit pressure of 100 kPa

(b) the entry pressure p of the compressed air

(c) the work output (per unit mass of air) of the actual turbine

Answer: **(a)** 277 K, **(b)** 163 kPa, **(c)** 33.2 kJ/kg

2. Cold air at 285 K and 100 kPa is supplied to a space to maintain it at a desired comfort level. The air leaves the room at 295 K and 100 kPa and is then compressed adiabatically to a pressure of 163 kPa. The isentropic efficiency for the compression process is 0.8. Determine

   (a) the refrigerating effect produced per unit mass of the air

   (b) the work input (per unit mass of the air) to the compressor

   Answer: **(a)** 10.1 kJ/kg, **(b)** 55.5 kJ/kg

3. An air refrigeration cycle is completed for the air in Exercises 1 and 2 by cooling the compressed air stream in Exercise 2 to the condition of 318 K and 163 kPa followed by an expansion through an adiabatic turbine to the condition of 285 K and 100 kPa in Exercise 1. The heat rejection is to ambient air at 100 kPa and 300 K. Determine

   (a) the net exergy (work) input to the cycle per unit mass of the air

   (b) the useful exergy product (per unit mass of the air) corresponding to the refrigerating effect of Exercise 2(a)

   (c) the effective COP_C (for cooling)

   (d) the second law efficiency achieved

   Answer: **(a)** 22.3 kJ/kg, **(b)** 0.348 kJ/kg, **(c)** 0.453, **(d)** 1.56%

11.7 Second Law Considerations

This chapter has shown thus far how some of the different refrigeration cycles can be analyzed to determine appropriate coefficients of performance. A proper performance evaluation of any system must be based on the second law of thermodynamics. The discussion of optimum cycles for the heat engines in Chapter 10 considered the constraints that are imposed due to such factors as the nature of heat sources and heat sinks available as well as the characteristics of the working fluid. Several of the considerations in Chapter 10 are also relevant when dealing with refrigeration systems.

Nature of Heat Sources and Heat Sinks

Whenever the ambient or atmospheric air can be considered as either the heat source or the heat sink, one can assume that the source (or sink) is unlimited. In most applications of the reversed heat engine, however, a substantial amount of sensible cooling or heating is required. The system being cooled or heated cannot be regarded in such applications as a constant temperature source or sink. Rather, the system should be recognized either as a finite or regeneratable source (or sink) in the manner defined in Section 10.6.

When a reversed heat engine operates between two thermal reservoirs, the optimum cycle is any completely reversible reversed heat engine cycle such as the reversed Carnot cycle. For complete reversibility, all the heat input to the cycle must be at the temperature T_L of the heat source, which is the thermal reservoir at the lower temperature. The heat rejection to the sink must likewise take place at the temperature T_H of the thermal reservoir which is at the higher temperature. The maximum COP for cooling and for heating can be determined from Equations 7.22a and 7.22b, respectively, while the corresponding second law efficiency for an actual cycle is given by Equation 11.4.

If, however, the reversed heat engine operates between a thermal reservoir and a finite system, the optimum cycle will be the *trilateral cycle* similar to the one discussed in Sections 9.4 and 10.6. Consider as an example the illustration given in Figure 11.15(a) in which a stream of ambient air is cooled by a completely reversible reversed heat engine. The ambient or atmospheric air at T_0 is the heat sink and can also be regarded as the reference environment. Figure 11.15(b) shows the process path $i{\rightarrow}e$ for the finite heat source (the air stream) and $a_1{\rightarrow}a_2$ for the unlimited sink. Both of these paths are indicated on T-S* coordinates, where S* for the air stream is the mass flow rate times the specific entropy. For the large mass of ambient air, the change in S* represents the increase in entropy of the system (the unlimited heat sink). For a reversible heat flow from a finite heat source stream to a working fluid of a reversed heat engine, the process path $1{\rightarrow}2$ for the working fluid must lie just below $i{\rightarrow}e$ with the temperature difference between the two fluids not exceeding an infinitesimal amount. Likewise, in the heat rejection process, path $2{\rightarrow}3$ traced by the working fluid of the reversed heat engine must not exceed T_0 by more than an infinitesimal amount. To return the reversed heat engine to state 1 without any further heat exchange with the environment, an isentropic process must connect states 3 and 1. The three-sided cycle $1{\rightarrow}2{\rightarrow}3{\rightarrow}1$ is the corresponding trilaterial cycle for a reversed heat engine operating between a finite heat source and an unlimited heat sink. The area enclosed within the cycle corresponds to the minimum power input required for producing the desired cooling of air from T_i to T_e.

Another example can be considered that involves a *regeneratable heat source* as shown in Figure 11.16(a). Such examples are encountered frequently in air conditioning applications where chilled water is required for the air conditioning. Cold water is used to cool an air stream, and then the water, which has been warmed, is chilled again by a

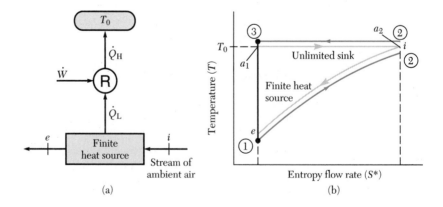

Figure 11.15 (a) Completely reversible scheme for the sensible cooling of a stream of ambient air. (b) Representation of the process paths for the reversible cooling process in part (a).

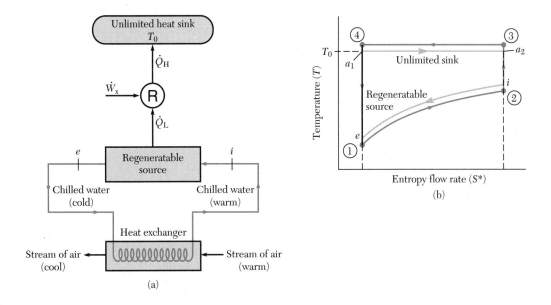

Figure 11.16 (a) Operation of a reversible reversed heat engine with a regeneratable heat source and an intermediate chilled water loop for the sensible cooling of a stream of air. (b) Process paths for the reversible reversed heat engine of (a).

refrigeration device. Figure 11.16(b) illustrates on T-S^* coordinates what the optimum cycle is in this situation. The process paths for the working fluid of the refrigeration device are from $1 \rightarrow 2 \rightarrow 3 \rightarrow 4 \rightarrow 1$. As in the previous example, the process paths for the working fluid when exchanging heat with the heat source and the heat sink must lie only infinitesimally apart from the process paths $i \rightarrow e$ and $a_1 \rightarrow a_2$ for the heat source fluid and ambient air, respectively. In this case, however, an isentropic process must connect states 2 and 3 for the working fluid since no heat is exchanged with the surroundings during the process. In addition, the path for the chilled water is from $i \rightarrow e$ when it is being regenerated prior to being used for cooling of an air stream. The resulting optimum cycle corresponds to the *quadlateral cycle* identified in Section 10.6 for the heat engine between a regeneratable heat source and an unlimited heat sink. The area enclosed by the optimum cycle again gives the minimum power necessary for the required cooling rate of the source fluid stream.

Characteristics of Typical Working Fluids

The typical refrigeration systems considered in this chapter provide a broad indication of the choices that are currently available. The majority of applications rely on the vapor compression refrigeration system that uses refrigerants as working fluids. Figure 11.17 shows a typical vapor compression cycle and the process paths for a regeneratable heat source and an unlimited heat sink at T_0 on a T-s diagram. The specific entropy for the refrigerant is represented by s, which can also be used for a scaled entropy value for the heat source fluid and the heat sink fluid, similar to what was done in Section 10.6 for heat engine cycles. The shaded regions correspond to the exergy consumption oc-

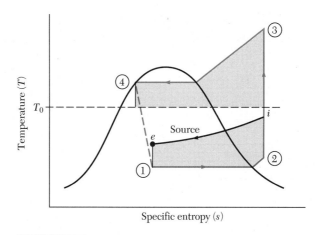

Figure 11.17 Thermal irreversibilities encountered in the operation of a typical vapor compression refrigeration system between a regeneratable heat source and an unlimited heat sink.

curring as a result of finite ΔT values between the working fluid and the heat source and sink fluids during the heat exchange processes with the source and sink. One suggestion to reduce the level of irreversibility is to use multiple stage compression in place of single-stage compression (see ASHRAE 1977). A more recent study (Adebiyi 1991) showed, however, that a more significant reduction in the thermal irreversibilities could be achieved via the use of better heat exchangers, which ensure substantially reduced levels of ΔT in the heat exchange processes involving the sink and the source.

In air refrigeration that involves direct use of cooled air, the exergy consumption in the source heating process is limited to any internal irreversibilities that occur in raising the temperature of the cooled air from T_e back to T_i. As in the limiting case of the Ericsson cycle (which was mentioned in connection with gas turbine cycles), one could imagine multiple stage compression with intercooling for achieving an almost isothermal heat rejection process to the ambient air in process 3→4 shown in Figure 11.16(b). If a small ΔT results between the working fluid and the ambient air during the process, the overall performance of the air refrigeration cycle improves. Recall, however, that the difference between positive work and negative work is comparatively small whenever air or any gas is the working fluid. Thus, internal irreversibility may ultimately determine how viable a system using air is, compared with other systems. Note that the vapor compression refrigeration cycle also involves the compression of fluid in the vapor phase, which also tends to lower the efficiency ratio defined as the actual efficiency divided by the ideal efficiency.

11.8 Summary

This chapter has reviewed the principal options for refrigeration, which include the evaporative cooling principle, work-producing adiabatic expansion of gases (gas refrigeration), and thermoelectric refrigeration. Most of the refrigeration systems in use operate on the evaporative cooling principle and are of the vapor compression design.

The traditional index of performance for refrigeration devices is the coefficient of performance (COP), which is a first law efficiency. In terms of the COP, the performance of vapor compression systems in particular appears impressive, with values typically greater than one. Since the COP is defined so that it is possible for COP > 1, the issue is not that COP > 1 but how close the magnitude of COP is to the thermodynamic limit. When evaluated from a second law perspective to compare with the thermodynamic limit, most of the currently used refrigeration systems turn out to be grossly inefficient.

Procedures for carrying out complete first law and second law analyses were explained in the chapter, and demonstration was provided by applying the methods to the analysis of vapor compression cycles and the reversed Brayton air cycle.

Section 11.7 concluded discussions on the second law evaluation of refrigeration systems by considering optimum cycles in relation to the nature of the heat source and the heat sink, and the characteristics of the working fluid. While the reversed Carnot cycle is the optimum cycle for operation between two thermal reservoirs, other cycles are more appropriate when either the heat source or the heat sink is finite or regeneratable. Two such optimum cycles are the trilateral cycle, when the heat source or heat sink is finite, and the quadlateral cycle, when the heat source or sink is regeneratable.

References

Adebiyi, G. A. 1986. "Equilibrium Property Data Equations for Aqua-Ammonia Solution in Solar Refrigeration." *The Nigerian Engineer*. Vol. 21, No. 2. 56–63.

Adebiyi, G. A. 1991. "Thermodynamic Analysis of Refrigeration Systems Using Refrigerant-22." *Mechanical Engineering Systems*. Vol. 1, No. 6. 22–36.

American Society of Heating, Refrigeration and Air-Conditioning Engineers 1977. *The 1977 Fundamentals Handbook*. New York: ASHRAE.

American Society of Heating, Refrigeration and Air-Conditioning Engineers 1987. *The 1987 Equipment Handbook*. New York: ASHRAE.

American Society of Heating, Refrigeration and Air-Conditioning Engineers 1989. *The 1989 Fundamentals Handbook*. New York: ASHRAE.

Anand, D. K., Lindler, K. W., Schweitzer, S., and Kennish, W. J. 1984. "Second Law Analysis of Solar Powered Absorption Cooling Cycles and Systems." *Journal of Solar Energy Engineering*. Vol. 106. 291–298.

Angrist, S. W. 1976. *Direct Energy Conversion*, 3rd ed. Boston: Allyn and Bacon.

Boryta, D. A. 1971. *Lithium Bromide: Technical Data (Bulletin 145)*. Exton, PA: Foote Mineral Company.

Dhar, P. L., Jain, V. K., and Gupta, R. M. 1977. "Computer Aided Design and Analysis of an Aqua-Ammonia Vapour Absorption Refrigeration System." *International Journal of Mechanical and Engineering Education*. Vol. 5, No. 4. 357–370.

Gupta, C. P. and Sharma, C. P. 1975. "Entropy and Availability Function Values of Saturated Ammonia-Water Mixtures." *ASME*. Paper 75-WA/PID-2.

Gupta, C. P. and Sharma, C. P. 1976. "Entropy Values of Lithium Bromide–Water Solutions and Their Vapors." *ASHRAE Transactions*. Vol. 82, Part 2. 35–46.

Herold, K. E. and Moran, M. J. 1987. "Thermodynamics Properties of Lithium Bromide/Water Solutions." *ASHRAE Transactions*. Vol. 93, Part 1. 35–47.

Herold, K. E., Han, K., and Moran, M. J. 1988. "AMMWAT: A Computer Program for Calculating the Thermodynamic Properties of Ammonia and Water Mixtures Using a Gibbs Free Energy Formulation." *Proceedings of the ASME Winter Annual Meeting*. AES-Vol. 4. 65–75.

Jain, P. C. and Gable, G. K. 1971. "Equilibrium Property Data for Aqua-Ammonia Mixture." *ASHRAE Transactions*. Vol. 77, Part 1. 149–151.

Jennings, B. H. 1981. "The Thermodynamic Properties of Ammonia-Water Mixtures: A Reassessment in Tabular Format." *ASHRAE Transactions*. Vol. 87, Part 2. 419–433.

Koehler, W. J., Ibele, W. E., Soltes, J., and Witner, E. R. 1987. "Entropy Calculations for Lithium Bromide Aqueous Solutions and Approximation Equation." *ASHRAE Transactions*. Vol. 93, Part 2. 2379–2388.

Macriss, R. A., Eakin, B. E., Ellington, R. T., and Huebler, J. 1964. *Physical and Thermodynamic Properties of Ammonia–Water Mixtures*. Chicago: Institute of Gas Technology, Research Bulletin No. 34.

McNeely, L. A. 1979. "Thermodynamic Properties of Aqueous Solutions of Lithium Bromide." *ASHRAE Transactions*. Vol. 85, Part 1. 413–434.

Schulz, S. C. G. 1973. "Equations of State for the System Ammonia-Water for Use with Computers." *Progress in Refrigeration Science and Technology*. Vol. 2. 431–436.

Uemura, T. and Hasaba, S. 1964. "Studies on the Lithium Bromide Water Absorption Refrigerating Machine." *Technology Report of Kansai University*. Vol. 6. 31–55.

Questions

1. What is the Clausius statement of the second law of thermodynamics? What is the major implication of the law in relation to achieving refrigeration or cooling of a system?

2. Categorize each of the following systems according to whether they are unlimited, finite, or regeneratable sources or sinks. Explain your answer in each case.
   (a) cooling of a given mass of water from room temperature to 5°C
   (b) freezing of water at 0°C (assume the ice is also at 0°C)
   (c) chilling of water for air conditioning in the summer
   (d) heating of a supply air stream for space heating in winter
   (e) domestic hot water stream production using a heat pump

3. Briefly explain each of the following refrigeration principles and give, for each case, two practical applications of the principle:

   (a) the evaporative cooling principle
   (b) cooling via adiabatic work-producing gas expansion
   (c) thermoelectric refrigeration

4. Suggest the most appropriate choice of refrigeration system for each of the following proposed applications. Give reasons for your choice.
   (a) cooling of an aircraft cabin
   (b) air conditioning of living areas in a submarine
   (c) automobile air conditioning
   (d) freezing of pork
   (e) air conditioning needs in an orbital space environment
   (f) domestic space conditioning in summer and winter
   (g) food preservation in groceries
   (h) preservation of food in transportation for several days
   (i) cooling of microprocessor chips in high speed computer units

5. Discuss promising strategies for dealing with the urgent problem of replacing CFCs (chlorofluoro-

carbons) with other refrigerants that are less harmful to the environment.

6. Why is H_2O a poor choice as a refrigerant for the vapor compression refrigeration system in typical refrigeration applications?

7. Suggest typical COP values for each of the following systems or applications:
   (a) a domestic refrigerator that is a vapor compression device
   (b) a thermoelectric cooler
   (c) an electric heat pump being operated during a mild winter season
   (d) an absorption refrigerator in a trailer unit

8. Figure Q11.8 illustrates one option for achieving refrigeration in a remote location. The power needed by the refrigeration device is produced by a heat engine whose overall efficiency η is defined as

$$\eta = \frac{\text{power } (\dot{W}_x)}{\begin{array}{c}\text{fuel energy}\\ \text{supply rate}\end{array}} \cdot 100\%$$

The refrigerator produces cooling at a rate of $\dot{Q}_C$. The combined efficiency of the unit is defined as the ratio of the cooling rate to the fuel energy supply rate. Suggest typical values for η and the COP of the refrigerator, and hence, give an order of magnitude estimate for the overall efficiency of the unit.

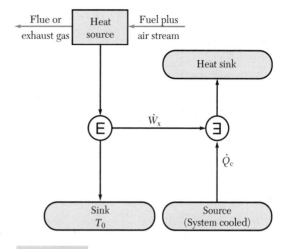

Figure Q11.8 Schematic diagram of a hybrid scheme for refrigeration in a remote location.

9. Several applications in refrigeration and space heating involve either a finite or a regeneratable heat source (or sink). Explain why the reversed Carnot cycle is not the optimum cycle for such applications.

10. Identify and discuss the promising options for achieving solar refrigeration.

Problems

11.1. A steady stream of refrigerant-22 is compressed in a reversible, adiabatic process from a slightly superheated vapor condition of $-5°C$ and 3.551 bar (corresponding $T_{sat} = -10°C$) to 75.48°C and 17.27 bar ($T_{sat} = 45°C$). The corresponding specific enthalpies for the fluid are 249.56 kJ/kg and 290.53 kJ/kg, respectively. What is the specific work input, in kJ/kg, to the compressor?

11.2. If the process efficiency for the compression in Problem 11.1 is 85%, determine the actual work input, per unit mass of refrigerant, to the compressor.

11.3. In the absorption refrigeration systems, liquid (solution) compression occurs instead of vapor compression as in a vapor compression system. Assume that the thermodynamic properties of a given aqua–ammonia liquid solution are approx-

imately the same as those for pure water and calculate the specific work input (in kJ/kg) required for the steady-flow, isentropic compression of the solution from a pressure of 3.551 bar to 17.27 bar. (The density of the mixture can be taken as 1000 kg/m³.)

11.4. A steady stream of air at 1 bar and 300 K is compressed reversibly and adiabatically to a pressure of 5 bar. What is the specific work input (in kJ/kg) required?

11.5. If the compression in Problem 11.4 were accomplished with an isentropic efficiency of 85%, determine the actual work input per unit mass of the air.

11.6. Compressed air at 5 bar and 310 K flows steadily through a reversible, adiabatic turbine to a pressure of 1 bar. Calculate
   (a) the exit temperature of the air

(b) the specific work output (in kJ/kg) of the turbine

11.7. Repeat Problem 11.6 on the assumption that the adiabatic turbine operates with a process efficiency of 85%.

11.8. Compute the first law efficiency for refrigeration when the cooling load $(\dot{Q}_C)$ and the energy input rate $(\dot{E}_{input})$ are as follows:
 (a) $\dot{Q}_C = 350$ W, $\dot{E}_{input} = 110$ W
 (b) $\dot{Q}_C = 10$ kW, $\dot{E}_{input} = 4$ kW

11.9. Determine the first law efficiency for heat pump applications when the loads are as in Problem 11.8.

11.10. A device operating on the reversed Carnot cycle is used to provide cooling for a building. The outside air temperature is 95°F, and the indoor air is to be maintained at 75°F. Compute the maximum coefficient of performance for cooling for the device.

11.11. A device operating on the reversed Carnot cycle is used to provide heating for a building. The outside air temperature is 0°C, and the indoor air is to be maintained at 25°C. Compute the maximum coefficient of performance for heating for the device.

11.12. A lithium bromide–water absorption refrigeration system is used to cool food. The cooling load is 5 kW, and the system provides this rate of cooling when a heat input of 8 kW is supplied to the generator. What is the coefficient of performance for cooling for this system?

11.13. An ammonia–water absorption system is used to heat a commercial building. If the system supplies 50 kW of heat to the building when 30 kW of heat is supplied to the generator of the system, what is the coefficient of performance for heating for the system?

* 11.14. A refrigerator works between a low temperature medium at 0°C and a high temperature environment at 300 K. The ambient temperature T_0 is 300 K.
 (a) What is the maximum possible COP_C?
 (b) If the refrigeration device is electrically driven and has an actual $\text{COP}_C = 3$, what is the second law efficiency achieved?

* 11.15. A refrigerator operates between two thermal reservoirs at 32°F and 70°F, respectively. The ambient temperature is 70°F.
 (a) What is the maximum possible COP_C?

(b) If the refrigerator is electrically driven and has an actual $\text{COP}_C = 2.4$, what is the second law efficiency achieved?

* 11.16. An absorption refrigeration system operates between two thermal reservoirs at 32°F and 70°F, respectively, while the ambient temperature is 70°F. If the system operates with an average $\text{COP}_C = 0.6$, what is the second law efficiency achieved by the system?

* 11.17. Given $T_L = 260$ K and $T_H = 293$ K, calculate the $(\text{COP}_H)_{max}$. If an actual electrically driven heat pump is used with $\text{COP}_H = 2$ and $\dot{Q}_H = 3$ kW, find
 (a) $\dot{E}_{input}$ (in kW)
 (b) $\dot{Q}_L$ (in kW)
 (c) the second law efficiency achieved
 Assume $T_0 = 260$ K.

* 11.18. A vapor compression refrigeration system uses refrigerant-22 as the working fluid. The fluid exits the evaporator as a saturated vapor at -12°C and exits the condenser as a saturated liquid at 40°C. The compression process is isentropic, and the throttling process is isenthalpic. Determine the COP_C for the system.

* 11.19. A vapor compression refrigeration system uses refrigerant-22 as the working fluid. The fluid exits the evaporator as a saturated vapor at 10°F and exits the condenser as a saturated liquid at 120°F. The compression process is isentropic, and the throttling process is isenthalpic. Determine the COP_C for the system.

* 11.20. A vapor compression refrigeration system uses refrigerant-22 as the working fluid. The pressures in the evaporator and condenser are 3.55 bar and 15.31 bar, respectively. The working fluid exits the evaporator at the evaporator pressure with 5°C of superheat and exits the condenser after being cooled 5°C below its saturated liquid state. The compression process is isentropic, and the throttling process is isenthalpic. Determine the COP_C for the system.

* 11.21. A vapor compression refrigeration system uses refrigerant-22 as the working fluid. The pressures in the evaporator and condenser are 50 psia and 240 psia, respectively. The refrigerant exits the evaporator at the evaporator pressure with 5°F of superheat and exits the condenser after being cooled 4°F below its saturated liquid state. The isentropic efficiency of the compression process is 80%, and the throttling process is constant enthalpy. Determine the COP_C for the system.

* 11.22. A heat pump system uses refrigerant-22 as the working fluid. When operating in the heating mode, the pressures in the evaporator and condenser are 60 psia and 210 psia, respectively, while corresponding pressures for the cooling mode are 80 psia and 250 psia. In both the heating and cooling modes, the refrigerant exits the evaporator with 5°F of superheat and exits the condenser after being cooled 5°F below its saturated liquid state. For all operating conditions, the isentropic efficiency of the compressor is 85%, and the throttling process is constant enthalpy. Determine the COP_C for the system for the cooling mode and the COP_H for the system for the heating mode of operation.

* 11.23. A domestic vapor compression refrigerator uses refrigerant-22 as the working fluid. A mass of 20 lbm of water at 60°F is placed in the freezer compartment of the refrigerator and cooled by evaporating the refrigerant at a pressure of 60 psia. The refrigerant enters the compressor as superheated vapor with 5°F of superheat. The condenser pressure is 225 psia, and the refrigerant exits the condenser as saturated liquid. The throttling process is constant enthalpy, and the isentropic efficiency of the compressor is 92%. The mechanical efficiency of the compressor is 70%, and the refrigerant mass flow rate is 2 lbm/min. The ambient temperature is 70°F. Determine

(a) the COP_C for the refrigerator
(b) the power required to drive the compressor
(c) the time required for the water to turn completely to ice at 32°F, stating your assumptions
(d) the second law efficiency achieved in the entire ice-making process

* 11.24. The refrigerator in Problem 11.23 is operated so as to produce ice at 32°F from water at 32°F. Determine the following for the system when it is operating at full capacity:

(a) the exergy production rate of the refrigeration device
(b) the exergy consumption (or destruction) rate for each of the following processes
  (i) the cooling process resulting from the

evaporation of the refrigerant in the evaporator
  (ii) the compression (assumed adiabatic) of the refrigerant vapor
  (iii) the condensation process in the condenser
  (iv) the adiabatic throttling of the refrigerant
(c) the second law efficiency of the cycle

* 11.25. A domestic vapor compression refrigerator uses refrigerant-22 as the refrigerant. A mass of 2 kg of water at 15°C is placed in the freezer compartment of the refrigerator and cooled by evaporating the refrigerant at 3 bar to a state with 5°C of superheat. The refrigerant is condensed at 18 bar to a state with 10°C of subcooling after which it is passed through an adiabatic expansion valve into the evaporator. The estimated isentropic efficiency of the compressor is 90%. The refrigerant mass flow rate is 7 kg/h. Assume the ambient temperature $T_0 = 298$ K. Show the refrigeration cycle on both the T-s and the p-h diagrams. Calculate

(a) the average rate of cooling, in W, produced by the refrigerator
(b) the required power supply rate, in W, to the compressor of the refrigerator
(c) the COP_C for the refrigerator
(d) the time taken for the water to turn completely to ice at 0°C, stating your assumptions
(e) the second law efficiency achieved in the entire process of turning water to ice

* 11.26. A refrigeration device is used to cool 2000 lbm of fish, initially at 77°F, to 0°F. The freezing point of fish is 28°F and the specific heats for the fish above and below the freezing point can be taken as 0.76 and 0.41 Btu/lbm · °R, respectively. The specific heat of fusion is 101 Btu/lbm. The ambient temperature, T_0, is 540°R.

(a) How much refrigeration, in Btu, is needed for the process?
(b) What is the minimum exergy supply, in kWh, needed for accomplishing the desired refrigeration?
(c) If an electrically driven refrigeration device with a $COP_C = 2$ were used, what would be the second law efficiency achieved?

* **11.27.** A refrigeration device, as shown in Figure P11.27, is used to cool 200 kg of fish from 20°C to −10°C. The freezing point of the fish is −2.2°C, and the specific heats for the fish above and below the freezing point can be taken as 3.2 and 1.7 kJ/(kg · K), respectively. The specific heat of fusion for the fish is 235 kJ/kg; the ambient temperature is 300 K.

   (a) How much refrigeration is needed for the process?

   (b) What is the minimum exergy supply needed for accomplishing the desired refrigeration?

   (c) If an electrically driven vapor compression refrigeration device with a $COP_C = 2.5$ were used, what would be the second law efficiency for the process?

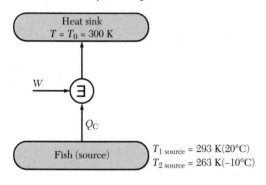

Figure P11.27 Cooling of fish with a refrigeration device.

** **11.28.** A refrigeration device is used to cool 800 kg of beef at atmospheric pressure from 25°C to −30°C. The freezing point of the beef is −3°C, and the specific heats for the beef above and below the freezing point can be taken as 3.5 and 1.8 kJ/(kg · K), respectively. The specific heat of fusion for the meat is 300 kJ/kg; the ambient temperature is 25°C.

   (a) What is the minimum exergy supply needed to accomplish the desired refrigeration?

   (b) If an electrically driven vapor compression refrigeration device with a $COP_C = 2.2$ were used, what would be the second law efficiency for the process?

* **11.29.** During the winter season, it is frequently necessary to cool the interior of a retail store while heating the perimeter area inside the building. Such interior cooling could be done by using outside air, but this is often not practical. Instead, the interior zones are cooled by an air conditioning device while the inside perimeter zone is heated. Suppose that an ammonia–water absorption device is used to provide both heating and cooling for such a building (Fig. P11.29).

   (a) Suggest an expression for the overall coefficient of performance for the absorption device operating in this manner.

   (b) If the device provides 300 kW of cooling and 700 kW of heating on a certain day and $\dot{Q}_C = 50$ kW, what is its overall coefficient of performance?

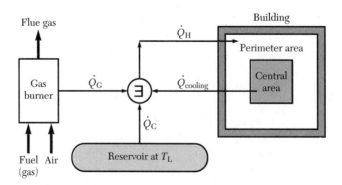

Figure P11.29 Heating and cooling of a building with an absorption device.

* **11.30.** An air refrigeration system is designed to operate on a reversed Brayton cycle. The air temperature and pressure at the compressor inlet are 25°C and 1 atm, and the pressure ratio is 4. The air leaving the compressor is cooled to 50°C before it enters the turbine. The cooled space is to be maintained at 25°C. Both the compression and expansion processes are isentropic, and the ambient temperature and pressure are 35°C and 1 atm. The mass flow rate of air in the system is 500 kg/h. Calculate
  - **(a)** the air temperature upon exit from the turbine
  - **(b)** the refrigerating effect (in kW)
  - **(c)** the exergy production rate (in kW)
  - **(d)** the net power input (in kW) required, assuming that the power output of the turbine is used to meet part of the requirement for driving the compressor
  - **(e)** the second law efficiency achieved

* **11.31.** An air refrigeration system is designed to operate on a reversed Brayton cycle. The air temperature and pressure at the compressor inlet are 70°F and 1 atm, and the pressure ratio is 3.5. The air leaving the compressor is cooled to 115°F before it enters the turbine; the cooled space is to be maintained at 70°F. The isentropic efficiencies of both compressor and turbine are 90%. The ambient temperature and pressure are 90°F and 1 atm, and the mass flow of air in the system is 1000 lbm/h. Calculate
  - **(a)** the air temperature upon exit from the turbine
  - **(b)** the refrigerating effect (in Btu/h)
  - **(c)** the exergy production rate (in Btu/h)

  - **(d)** the net power input (in Btu/h) required, assuming that the power output of the turbine is used to meet part of the requirement for driving the compressor
  - **(e)** the second law efficiency achieved

* **11.32.** A refrigeration device operates on the ideal vapor compression cycle with refrigerant-22 as the working fluid. The saturation temperature corresponding to the condenser pressure is 40°C, and the evaporator temperature is $-10°C$. For this cycle, the refrigerant exits the condenser as a saturated liquid and exits the evaporator as a saturated vapor (Fig. P11.32). The compression process is isentropic. Determine the COP_C for this cycle and compare it with the COP_C for the reversed Carnot cycle operating between the temperatures of 40°C and $-10°C$.

* **11.33.** A heat pump operates on the ideal vapor compression cycle with refrigerant-22 as the working fluid. The saturation temperature corresponding to the condenser pressure is 110°F, and the evaporator temperature is 20°F. For this cycle, the refrigerant exits the condenser as a saturated liquid and exits the evaporator as a saturated vapor. The compression process is isentropic. Determine the COP_H for this cycle and compare it with the COP_H for the reversed Carnot cycle operating between the temperatures of 110°F and 20°F.

** **11.34.** The COP_H for a heat pump declines when the inside air temperature is fixed and the outside air temperature drops. As the outside air temperature drops, it may be advisable to cease the operation of the heat pump at some point and use other heating devices (such as an electric

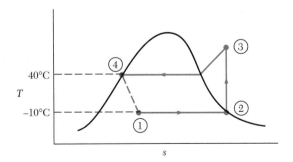

Figure P11.32 Ideal vapor compression cycle.

strip heater) because of lowered efficiency and lowered capacity. For the heat pump described in Problem 11.33, assume that the saturation temperature at the condenser pressure is fixed at 110°F. Then plot a curve of COP_H versus evaporator temperature with values from 40°F to $-20°F$. Assume an isentropic efficiency of 80% for the vapor compression process. If you were designing this system for residential application, at what evaporator temperature, if any, would you consider it advisable to switch to another heating source for the residence?

* **11.35.** An electrically driven vapor compression heat pump unit is used to heat water for a laundry. Refrigerant-22 is used. The evaporator temperature is 15°C, and the saturation temperature corresponding to the condenser pressure is 60°C. The refrigerant exits the condenser as a saturated liquid and exits the evaporator as a saturated vapor. The compression process is isentropic. What percentage of the electrical energy is saved by use of the heat pump instead of electrical resistance heating?

* **11.36.** A gas-fired absorption system with an average COP_C of 0.6 is used. The gas firing is accomplished using a steady stream of hot gases initially at 2100 K and 1 atm. The gases are cooled to 400 K as they heat the solution in the generator of the absorption system. As a result, a mass of water at 0°C is turned to ice at 0°C. Ambient conditions are 300 K and 1 atm. What is the second law efficiency achieved?

* **11.37.** A steam power plant that operates with a second law efficiency of 40% is used to provide electrical energy input to a vapor compression device that operates to produce ice at 0°C with a $COP_C = 2$.
  (a) Compute the second law efficiency of the vapor compression device.
  (b) Calculate the second law efficiency for the combined power plant and vapor compression system.
  (c) Compare your results with those obtained in Problem 11.36.

** **11.38.** A reversed Brayton cycle is used for cooling. The turbine efficiency is η_{aT} and the compressor efficiency is η_{aC}. Derive an expression for the COP_C in terms of the pressure ratio, the ratio of inlet temperature of the turbine to inlet tem-

perature of the compressor, and the turbine and compressor efficiencies.

** **11.39.** A sensible heat system of mass m and specific heat c is proposed for "cool storage" of electricity during the off-peak period. The system is initially at an ambient temperature T_0 and is subsequently cooled to a temperature T_Q. Determine the thermal exergy stored in the cool storage system. From your results show that the maximum possible COP_C for the device used to produce the cooling is

$$COP_C = \frac{T_Q^*}{T_0 - T_Q^*}$$

where

$$T_Q^* = \frac{T_0 - T_Q}{\ln\left(\dfrac{T_0}{T_Q}\right)}$$

Show further that if a reversed heat engine operating on a reversed Carnot cycle with $T_L = T_Q$ and $T_H = T_0$ were used, the second law efficiency, η_2, achieved would be given by

$$\eta_2 = \frac{T_0/T_Q^* - 1}{T_0/T_Q - 1} \cdot 100\%$$

11.40. Evaluate the thermal exergy stored, the $(COP_C)_{max}$, and η_2 for a reversed Carnot engine using the expressions given in Problem 11.39 when the cool storage system is 1000 kg of water with $T_0 = 300$ K and $T_Q = 278$ K. Assume $c = 4.2$ kJ/kg · K for water.

** **11.41.** A proposed cool storage system is to use both sensible heat and latent heat storage in a scheme to store electrical energy during the off-peak period. The system has a mass m and a (sensible) specific heat c. The specific heat of fusion is h_{if}, and the phase change (solid to liquid) temperature is T_s, which is lower than the ambient temperature T_0. Assume that the storage system is initially at a temperature T_0 and is cooled until the entire mass is frozen solid at T_s. Determine
  (a) the cooling load (Q_C) involved
  (b) the exergy stored, χ_{stored}, in the cool storage system.
  Hence, show that the maximum possible coefficient of performance (based on work input) for the refrigeration device is

$$(COP_C)_{max} = \frac{1 + Ste}{(T_0/T_s - 1) + Ste(T_0/T_s^* - 1)}$$

where Ste (Stefan number) $= c(T_0 - T_s)/h_{if}$ and

$$T_s^* = \frac{T_0 - T_s}{\ln\left(\dfrac{T_0}{T_s}\right)}$$

* **11.42** A proposed ice-storage scheme uses 1000 kg of water in the manner described in Problem 11.41. The specific heat of water and the specific heat

of fusion of ice can be taken as 4.2 kJ/kg · K and 333.4 kJ/kg, respectively. Assume $T_0 = 300$ K and $T_{fi} = 273$ K. Determine $(COP_C)_{max}$. If an electrically driven refrigeration device with a $COP_C = 2$ is used to produce the ice, evaluate the second law efficiency achieved. Also determine

(a) the cooling load involved
(b) the actual work done in producing the desired cooling
(c) the exergy stored in the cool storage system

Design Problems

11.1 *Aircraft air-conditioning*

Figure DP11.1(a) illustrates a cooling scheme for a commercial aircraft. Adequate cooling is required when the aircraft is on the ground and the ambient air temperature is as high as about 38°C (100°F). The system must also be adequate when the aircraft is in flight at an altitude of 20,000 ft with an outdoor air condition of −4°C and 47 kPa. (Ram-air heating could raise the unpressurized air temperature to about 27°C.) Whether the aircraft is on the ground or at the high altitude, assume that the cabin air should be maintained at about 24°C and 1 atm. Assume a cabin air supply re-

quirement of 105 kg/min at 1 atm and temperature between 12°C and 16°C. Two possible air cooling systems are to be considered. The first is the system illustrated in Figure DP11.1(b), which operates on the air cycle. The second system uses the vapor compression refrigeration device for producing the required cooled air.

You are to carry out a design analysis and comparative evaluation of the two systems. The scope of work should include the following:

1. A complete first law analysis, which gives the net power requirement, the cooling effect of the chilled air produced, and the effective

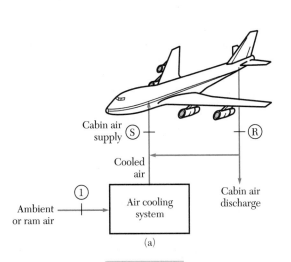

(a)

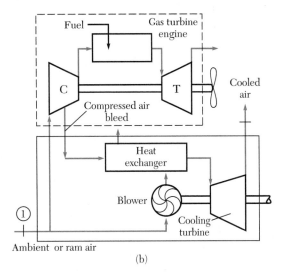

(b)

Figure DP11.1 (a) Aircraft cabin cooling scheme. (b) Air cooling system (air cycle).

COP_C for each system. Realistic values for process efficiencies should be used wherever necessary.

2. A second law analysis is also needed that pinpoints which processes contribute the most to exergy consumption.

Which of the two systems would you recommend? Give reasons for your recommendation.

11.2 *Air conditioning of residence*

You are to select a vapor compression air conditioning system for cooling a residence in New Mexico. The air conditioner uses refrigerant-22 and is to be similar to the system shown in Figure 11.1. The residence is located in a dry climate with a design-point summertime temperature of 93°F. Your system should maintain the residence at 78°F. The house has 1600 ft² of heated area and is designed so that the overall heat transfer coefficient (UA, which includes conduction, convection, and infrared radiation through the walls, doors, floors, ceilings, and windows) is 2500 Btu/h · °F, where the heat transfer is given by $q_c = (UA)(T_0 - T_i)$, with T_0 being the outside (ambient) temperature and T_i the inside temperature of the building. In addition to the above heat transfer, heat is also transferred by air infiltration and by solar insolation. The heat transfer due to air infiltration is given by $q_{infil} = 1200 \, (T_0 - T_i)$ Btu/h. The incident solar flux, which is transmitted through the glass in the windows and doors and absorbed inside the residence, is 1000 Btu/h at the summertime design-point.

Determine the capacity of the air conditioner required (in tons) and select an appropriate unit available in your locale to meet the cooling needs for the residence. Estimate the air flow required between the cooling coil and the residence, and ensure that the unit you select has the capacity to meet the need. Find out the efficiency ratings for available units and select your recommendation for the residence based on your evaluation of initial costs, operating costs, life cycle costs, and efficiency. (Note that commercial units are frequently rated by an energy efficiency ratio (EER) instead of COP_C, where EER is equivalent to COP_C except that EER has units and COP_C does not.)

Write a report summarizing your results. Give reasons for your decisions and choices.

11.3 *Heating of residence with a heat pump*

You are to select a heat pump for heating the residence described in Design Problem 11.2. For the wintertime design condition, the indoor temperature of the residence should be maintained at 68°F, while the outside air (ambient) temperature is 35°F. The wintertime design point occurs at night so that there is no absorbed solar flux, but there are air infiltration losses and overall heat transfer losses from the residence. Will the heat pump you select for heating also serve to satisfactorily cool the residence? If necessary, modify your design and select a heat pump for heating *and* cooling based on costs and efficiency as outlined in Design Problem 11.2. Write a report summarizing your results.

11.4 *Heating of water using heat recovery from an air conditioner*

Heat recovery is being used in many applications to reduce cost and maximize the use of valuable resources. One system now available for both residential and commercial use is a heat recovery unit to add to a standard vapor compression air conditioning system. The heat recovery unit includes a heat exchanger, which is installed downstream of the compressor outlet (point 3 in Figures 11.1 and 11.2), in order to heat water with part of the waste heat normally rejected from the condenser of the air conditioning system. Water is circulated by a pump or by natural convection between the heat recovery unit and a hot water storage tank. When such a system provides hot water for residential use, heat transfer in the heat exchanger is through forced convection on the refrigerant side, and either forced or free convection on the water side.

Assume that a heat recovery unit is used with the air conditioner of Design Problem 11.2 to provide hot water for a family of four living in the residence. The hot water thermostat in the residence is normally set between 125°F and 140°F, and input water to the water heater is usually at about 75°F. Should the flow in the heat exchanger be parallel flow or counterflow for the best results? Why? Would you expect the efficiency of the air conditioner to increase or decrease as a result of the installation of such a heat recovery unit? Why?

Estimate the hot water requirement for the family, and estimate the portion of this requirement that can be met during the cooling season by use of a heat recovery unit. Find out the installed cost of such a unit in your locale and determine whether or not a purchase would be economical. Prepare a report of your results.

11.5 *Performance improvement of a heat pump*
Many different methods have been used to improve the performance of air conditioners and heat pumps. An obviously important parameter is the temperature difference between the inside air and the outside air. Thus President Carter during the "energy crisis" of 1975–1977 urged all Americans to set their thermostats low in the winter and high in the summer. Of similar importance for heat pumps to the thermostat setting for inside air is the temperature level for the phase change process of the refrigerant during the heat exchange process with the outside air. Thus, it is desirable to use forced convection with an efficient heat exchanger during this heat exchange with the outside air, which involves the refrigerant condensation (outdoors) during the cooling season, and the evaporation process (outdoors) during the heating season. Also, it is desirable to provide outside air with a temperature as low as possible during the cooling season and as high as possible during the heating season. One approach to improve performance that was designed and tested by one of the authors (Russell) is described in the following paragraph.

If a residence is built with a conventional foundation so that there is a crawl space under the house, then the outside air input to the outside heat exchanger can be drawn from the crawl space. If the floor is heavily insulated so that minimal losses occur through the floor, then the air drawn through the crawl space will be coupled to the ground so that the ambient air is cooled during the cooling season and heated during the heating season. In addition, such air flow helps to eliminate moisture problems in the crawl space during the cooling season. Data collected by the author for such a system were as follows:

Outside air temperature (°F)	40	50	60	70	80	90
Air temperature at inlet to outdoor unit (°F)	48	56	64	70	77	86

Estimate the improvement in the performance of a heat pump during the heating season and during the cooling season as a result of the use of such a system. If such a unit costs $300 to install for the residence in Design Problems 11.2 and 11.3, would you recommend such an installation? Justify your recommendations. Write a report presenting your recommendations and results.

Thermodynamic Property Relationships

12.1 Introduction

Thermodynamics has been described as a science primarily concerned with heat, work, and changes in the properties of a system. The first law of thermodynamics not only establishes energy as a property of a system but also defines the relationship among heat, work, and changes in the energy of a system. The second law of thermodynamics likewise provides the basis for the evaluation of processes from the entropy changes taking place. In all of these, knowledge of the properties of substances employed is essential for a meaningful analysis to be possible. Several categories of substances are encountered in engineering, but by far the largest group is the class known as *pure substances*. Only pure substances have been considered in previous chapters, and the development to this point can be aptly described as the thermodynamics of pure substances. This chapter provides a more rigorous treatment of thermodynamic property relationships for pure substances based on the foundations that have been laid in earlier chapters. In subsequent chapters, some discussion is provided on the thermodynamic property relationships for other substances, such as moist air–liquid mixtures (not a pure substance) and chemically reactive systems (termed *chemical substances*).

Chapters 2 and 3 noted that for substances in general, a complete listing of values for all properties at a particular equilibrium state might be necessary to define that state for the system. For pure substances, however, any two independent intensive properties suffice. This phenomenological rule was referred to as the *two-property rule*. It was also noted that a similar rule known as the *state postulate* or the *state principle* applied in the case of simple systems. A *simple system* is often regarded simply as a macroscopically homogeneous and isotropic system (Haywood 1980, Bejan 1988). In either case, the significant consequence is that equations of state can, in principle, be found for relationships between every intensive property (z) and any two other independent intensive properties x and y. Some thermodynamic property data for a substance can be obtained via empirical or experimental measurements. Such data exist for most commonly used substances and generally include pressure (p), specific volume (v) or density (ρ), temperature (T), and specific heat measurements. Using such available data, values for other system properties that cannot be measured directly, such as internal energy and entropy, must be obtained from appropriate equations of state that are formulated on the basis of the laws of thermodynamics. Also, available empirically determined property data may be incomplete or may not extend to regions of interest for a particular application. In such cases, formulations based on the laws of thermodynamics can be used for the additional property values needed.

Chapter 8 introduced the *combined first and second laws* of thermodynamics or the $T\,dS$ equations, which will serve as the basis for the formulations for the thermodynamic properties of pure substances in this chapter. When written exclusively in terms of intensive properties, the equations are often referred to as *fundamental* or *characteristic equations of state* for pure substances. These are referred to as fundamental equations of state because all the other thermodynamic properties can be evaluated for the substance from the solution to any one of these equations. Recall that while the derivation of the intensive $T\,ds$ equations was made with reference to a reversible process, the resulting equations are valid for irreversible processes as well, since they merely express relationships among properties of the system. Recall, however, that the equations do not apply when irreversible chemical reactions take place. Since this chapter covers pure substances whose chemical composition stays constant throughout, this exception is not relevant.

The remainder of this chapter first introduces a number of mathematical relationships that relate to the task of establishing relationships among the thermodynamic properties of a substance from the fundamental equations of state. Applications to the formulations for pure substances are discussed in detail. Procedures often used in constructing tables of properties for substances using a comparatively limited property data base are included.

12.2 Mathematical Considerations

Independent and Dependent Variables

Chapter 3 stated that only two independent intensive thermodynamic properties are needed to specify the thermodynamic state of a simple, compressible pure substance. Several thermodynamic properties were considered, such as pressure, specific volume, temperature, internal energy, enthalpy, entropy, the Gibbs function, and the Helmholtz function. Any two of these properties that are independent of each other could be used to specify the thermodynamic state of a substance. The choice of the two properties to specify is arbitrary. Once they are specified, these two properties become the independent variables while all other thermodynamic properties become dependent variables. But note that some properties are never independent; for example, density and specific volume are reciprocal quantities and are not independent of each other. Other properties are not independent in certain regions, such as pressure and temperature in the two-phase region.

If one considers a continuous function z of two independent variables x and y so that $z = z(x, y)$, then z is a dependent variable while x and y are the independent variables. Furthermore, a partial derivative of z with respect to x means that z is differentiated with respect to x while holding y constant, that is $(\partial z/\partial x)_y$. Likewise, the partial derivative of z with respect to y means that the function is differentiated with respect to y while holding x constant, that is $(\partial z/\partial y)_x$. Thus, in either of the notations $(\partial z/\partial x)_y$ or $(\partial z/\partial y)_x$, z is the dependent variable and x and y are the independent variables.

The total differential of a function involves evaluation of the net change in the function when all independent variables are allowed to change. Thus, for z which is a continuous function of x and y, the total differential dz is given by

$$dz = \left(\frac{\partial z}{\partial x}\right)_y dx + \left(\frac{\partial z}{\partial y}\right)_x dy \tag{12.1}$$

where the partial derivatives are given by the terms on the right-hand side for the case when the two independent variables are x and y.

Exact and Inexact Differentials

As stated in Chapter 2, infinitesimal changes in the thermodynamic properties are given by exact differentials, while infinitesimal transfers of heat and work are given by inexact differentials. It is possible to determine whether or not a differential is exact by use of mathematical relationships concerning exact differentials. Consider the differential function given by Equation 12.1. The differential can be rewritten in the following form:

$$dz = M \, dx + N \, dy \tag{12.2}$$

The function dz is an exact differential if

$$\left(\frac{\partial M}{\partial y}\right)_x = \left(\frac{\partial N}{\partial x}\right)_y \tag{12.3}$$

Equation 12.3 gives a necessary and sufficient condition for dz to be an exact differential; it is known as a *test for exactness*. In addition, the following is a direct consequence of exactness:

$$\oint dz \, (x, y) = 0$$

In words, the integral of dz around every closed path is zero if z is an exact differential. Furthermore, the integral

$$\int_{z(x_1,y_1)}^{z(x_2,y_2)} dz\,(x,\,y) = z(x_2,\,y_2) - z(x_1,\,y_1)$$

is a function only of end points and is independent of the path for the exact differential. These characterizations are true for thermodynamic properties, and one should expect the differentials for properties to be exact. In other words, the statements concerning these two integrals and Equation 12.3 must be true for thermodynamic properties.

When the quantity $M\,dx + N\,dy$ is not an exact differential, it can at times be made exact by multiplying through by an integrating factor. The fact that δQ for a reversible process is inexact while dS (which is equal to $\delta Q/T$ for the reversible process) represents an exact differential is, however, a powerful consequence of the second law of thermodynamics and is not a mathematical deduction. Nevertheless, the theorems for exact differentials can be applied, and thus several useful relationships among properties of a substance involving dS can be derived.

Some Mathematical Relationships

Some additional mathematical relationships of value in dealing with thermodynamic properties are as follows. The *cyclic relationship* given as follows (without proof) expresses a relationship between partial derivatives among $(x,\,y,\,z)$, such as $x = f(y,\,z)$ or $f(x,\,y,\,z) = 0$:

$$\left(\frac{\partial z}{\partial x}\right)_y \left(\frac{\partial x}{\partial y}\right)_z \left(\frac{\partial y}{\partial z}\right)_x = -1 \tag{12.4}$$

Also, for a continuous single-valued function,

$$\left(\frac{\partial y}{\partial x}\right)_z = \frac{1}{\left(\dfrac{\partial x}{\partial y}\right)_z} \tag{12.5}$$

Another useful relationship is called the *substitution rule*. This allows change of independent variables. Suppose the function $f(x,\,y,\,z) = 0$ is desired in the form $y = y(x,\,z)$. In this situation, the following substitution rule can be employed

$$\left(\frac{\partial f}{\partial x}\right)_z = \left(\frac{\partial f}{\partial x}\right)_y + \left(\frac{\partial f}{\partial y}\right)_x \left(\frac{\partial y}{\partial x}\right)_z \tag{12.6}$$

The following equation is also valid:

$$\left(\frac{\partial z}{\partial y}\right)_x \left(\frac{\partial y}{\partial f}\right)_x \left(\frac{\partial f}{\partial z}\right)_x = 1 \tag{12.7}$$

12.3   *Maxwell Relations*

The *Maxwell relations* are equations that relate the derivatives of the properties p, v, T, and s for a simple, compressible pure substance. The Maxwell relations can be developed from the combined first and second laws as introduced in Chapter 8 as Equations 8.13 and 8.14

$$T \, ds = du + p \, dv \tag{8.13}$$

$$T \, ds = dh - v \, dp \tag{8.14}$$

along with the Helmholtz and the Gibbs functions, which were defined in Chapter 3 as Equations 3.2 and 3.3 (given here per unit mass)

$$a = u - Ts \tag{3.2}$$

$$g = h - Ts \tag{3.3}$$

Differentiating the Helmholtz function in Equation 3.2 and combining results with Equation 8.13 yields

$$da = -p \, dv - s \, dT \tag{12.8}$$

Similarly, differentiating the Gibbs function in Equation 3.3 and combining this with Equation 8.14 yields

$$dg = v \, dp - s \, dT \tag{12.9}$$

Now it is desirable to rewrite Equations 8.13 and 8.14 in the form of Equation 12.2 as

$$du = T \, ds - p \, dv \tag{12.10}$$

$$dh = T \, ds + v \, dp \tag{12.11}$$

The set of Equations 12.8 through 12.11 are the fundamental or characteristic equations of state defined previously. They are all of the form of Equation 12.2, and the differentials are exact for each of the thermodynamic properties as functions of the others. The theorem expressed by Equation 12.3 can be applied to establish relationships between the coefficients in Equations 12.8 through 12.11. Thus, applying the theorem to Equation 12.8 gives

$$\left(\frac{\partial p}{\partial T} \right)_v = \left(\frac{\partial s}{\partial v} \right)_T \tag{12.12}$$

Similarly for Equations 12.9 through 12.11, one obtains

$$\left(\frac{\partial v}{\partial T} \right)_p = -\left(\frac{\partial s}{\partial p} \right)_T \tag{12.13}$$

$$\left(\frac{\partial T}{\partial v} \right)_s = -\left(\frac{\partial p}{\partial s} \right)_v \tag{12.14}$$

$$\left(\frac{\partial T}{\partial p}\right)_s = \left(\frac{\partial v}{\partial s}\right)_p \qquad (12.15)$$

Equations 12.12 through 12.15 are the *Maxwell relations* for a pure simple compressible substance. They relate pressure, temperature, and specific volume to entropy and allow values for entropy to be computed from measurements involving pressure, temperature, and specific volume. Note that in these equations, it is tacitly assumed that each dependent variable is a function of two other independent properties. Recall that for a pure substance, pressure and temperature are not independent in the mixed phase region. However, if only homogeneous substances (or single phases of a pure substance) are considered, the pressure and temperature are independent properties.

Fundamental Equations 12.8 through 12.11 can also be written in the form of Equation 12.1:

$$da = \left(\frac{\partial a}{\partial v}\right)_T dv + \left(\frac{\partial a}{\partial T}\right)_v dT \qquad (12.16)$$

$$dg = \left(\frac{\partial g}{\partial p}\right)_T dp + \left(\frac{\partial g}{\partial T}\right)_p dT \qquad (12.17)$$

$$du = \left(\frac{\partial u}{\partial s}\right)_v ds + \left(\frac{\partial u}{\partial v}\right)_s dv \qquad (12.18)$$

$$dh = \left(\frac{\partial h}{\partial s}\right)_p ds + \left(\frac{\partial h}{\partial p}\right)_s dp \qquad (12.19)$$

Now comparing Equation 12.16 with 12.8 gives

$$\left(\frac{\partial a}{\partial v}\right)_T = -p \quad \text{and} \quad \left(\frac{\partial a}{\partial T}\right)_v = -s \qquad (12.20)$$

Similarly, comparing Equations 12.17 through 12.19 with Equations 12.9 through 12.11 gives

$$\left(\frac{\partial g}{\partial p}\right)_T = v \quad \text{and} \quad \left(\frac{\partial g}{\partial T}\right)_p = -s \qquad (12.21)$$

$$\left(\frac{\partial u}{\partial s}\right)_v = T \quad \text{and} \quad \left(\frac{\partial u}{\partial v}\right)_s = -p \qquad (12.22)$$

$$\left(\frac{\partial h}{\partial s}\right)_p = T \quad \text{and} \quad \left(\frac{\partial h}{\partial p}\right)_s = v \qquad (12.23)$$

All of these property relationships for a simple, compressible pure substance (excluding regions of phase change) are based on the combined first and second laws of thermodynamics in the form of Equations 8.13 and 8.14. Recall from Section 8.4 that Equation 8.13 was developed for the case where the only work done was $p\, dv$ work. For the cases where other forms of reversible work occur, the combined first and second law can be written in scalar differential form as

$$T \, ds = du + p \, dv + f \, dx + f \, dL + G \, d\theta + \sigma \, dA^* + \text{EI} \, dt + \cdots$$

Other Maxwell relations can be developed for cases involving other forms of work. For example, if the only form of work is surface tension work, then the combined first and second law is

$$du = T \, ds - \sigma \, dA^*$$

and the Maxwell relation for this case is

$$\left(\frac{\partial T}{\partial A^*} \right)_s = - \left(\frac{\partial \sigma}{\partial s} \right)_{A^*}$$

where A^* is the surface area per unit mass of the system. Other relationships could be written for this case as well as for those involving other forms of work. Note that the original Maxwell relations are those given by Equations 12.12 through 12.15. The others are also referred to as Maxwell relations only because their derivation is based on an identical principle.

1. The following definition is proposed for an ideal gas in terms of the specific Gibbs function:

$$g = g^0(T) + RT \ln p$$

where the pressure p is measured in atm and $g^0(T)$ is the specific Gibbs function for the gas at 1 atm and temperature T. Use Equation 12.21 to obtain the following for v and s for the ideal gas:

$$v = \frac{RT}{p} \quad \text{and} \quad s = -\frac{dg^0}{dT} - R \ln p$$

2. Use the relationship between g and h as well as the results in Exercise 1 to show that for the ideal gas

$$h = g^0(T) - T\frac{dg^0}{dT}$$

12.4   Specific Heat (at constant p, constant v)

It is desirable to obtain relationships for all thermodynamic properties in terms of those properties that can be experimentally determined. These include pressure, specific volume, and temperature. In addition, specific heat is relatively easy to measure. Therefore, it is desirable to obtain relationships between specific heat and other properties to facilitate the tabulation of these other properties based on available experimental data. Specifically, data for specific heat values over a wide range of conditions can be extremely useful for the computation of other properties such as enthalpy, internal energy, and entropy.

The principal specific heats at constant pressure (c_p) and at constant volume (c_v), defined in Chapter 3 are

$$c_p = \left(\frac{\partial h}{\partial T}\right)_p \tag{3.5}$$

and

$$c_v = \left(\frac{\partial u}{\partial T}\right)_v \tag{3.6}$$

Now consider the function $s = s(T, p)$. Following Equation 12.1, one can write

$$ds = \left(\frac{\partial s}{\partial T}\right)_p dT + \left(\frac{\partial s}{\partial p}\right)_T dp \tag{12.24}$$

Similarly, for $h = h(T, p)$, one obtains

$$dh = \left(\frac{\partial h}{\partial T}\right)_p dT + \left(\frac{\partial h}{\partial p}\right)_T dp \tag{12.25}$$

The first partial derivative on the right-hand side of Equation 12.25 is recognized as c_p. Combining Equation 8.14 or 12.11 with Equation 12.25 in such a manner as to eliminate dh gives

$$ds = \frac{c_p}{T} dT + \frac{1}{T}\left[\left(\frac{\partial h}{\partial p}\right)_T - v\right] dp \tag{12.26}$$

Equating the coefficients of the dT terms in Equations 12.24 and 12.26 yields

$$c_p = T\left(\frac{\partial s}{\partial T}\right)_p \tag{12.27}$$

Following a similar procedure for $s = s(T, v)$ and $u = u(T, v)$ gives the following expression for c_v:

$$c_v = T\left(\frac{\partial s}{\partial T}\right)_v \tag{12.28}$$

Equations 12.27 and 12.28 are useful for the determination of entropy changes from specific heat data and additional information concerning p, v, and T in any single phase region.

An advantage when dealing with gases is that considerable data are available for specific heat values of common gases as a function of temperature at low pressures. Such data are called *zero pressure* specific heats. These data can be used to obtain specific heat information for gases at higher pressures. The approach is shown as follows.

Combining Equation 12.24 with Equation 12.27 and the Maxwell relation in Equation 12.13 gives

$$ds = \frac{c_p}{T} dT - \left(\frac{\partial v}{\partial T}\right)_p dp \tag{12.29}$$

Following a similar procedure for $s = s(T, v)$ yields

$$ds = \frac{c_v}{T} dT + \left(\frac{\partial p}{\partial T}\right)_v dv \tag{12.30}$$

Applying the theorem on exactness to Equation 12.29 gives

$$\left(\frac{\partial c_p}{\partial p}\right)_T = -T\left(\frac{\partial^2 v}{\partial T^2}\right)_p \tag{12.31}$$

Similarly, for Equation 12.30,

$$\left(\frac{\partial c_v}{\partial v}\right)_T = T\left(\frac{\partial^2 p}{\partial T^2}\right)_v \tag{12.32}$$

Note that Equation 12.31 gives a relationship for how constant pressure specific heat changes with pressure in terms of pvT information, while Equation 12.32 gives information about how constant volume specific heat changes with volume in terms of pvT information. To obtain information about c_p at elevated pressures, Equation 12.31 must be integrated at constant temperature to obtain

$$c_p - c_{p0} = -T\int_0^p \left(\frac{\partial^2 v}{\partial T^2}\right)_p dp \quad \text{along a constant } T \tag{12.33}$$

This expression can be used to obtain specific heat information at elevated pressures from a knowledge of the zero pressure specific heat and a knowledge of the pvT relationship for the substance. If such pvT data are available in analytical form, then the integral on the right-hand side can be obtained directly. If not, the integral can be evaluated numerically from tabulated data if available.

It is sometimes of interest to know the magnitude of the difference between c_p and c_v. Generally, it is easier to measure c_p than c_v. Therefore, if the relationship is available for the difference between the two quantities, this relationship along with c_p data could be used to determine the magnitude of c_v. Subtracting Equation 12.30 from Equation 12.29 gives

$$\left(\frac{c_p - c_v}{T}\right) dT = \left(\frac{\partial v}{\partial T}\right)_p dp + \left(\frac{\partial p}{\partial T}\right)_v dv \tag{12.34}$$

Now temperature can be expressed as a function of pressure and specific volume as $T = T(p, v)$. From this expression, the differential dT can be written as

$$dT = \left(\frac{\partial T}{\partial p}\right)_v dp + \left(\frac{\partial T}{\partial v}\right)_p dv \tag{12.35}$$

After simplifying Equation 12.34, the coefficients of the dp terms in Equation 12.34 and Equation 12.35 can be set equal to obtain

$$c_p - c_v = T\left(\frac{\partial v}{\partial T}\right)_p \left(\frac{\partial p}{\partial T}\right)_v \tag{12.36}$$

Using the cyclic rule for pvT relationships yields

$$\left(\frac{\partial p}{\partial T}\right)_v = -\left(\frac{\partial v}{\partial T}\right)_p \left(\frac{\partial p}{\partial v}\right)_T \tag{12.37}$$

Substituting Equation 12.37 into 12.36 gives

$$c_p - c_v = -T\left(\frac{\partial v}{\partial T}\right)_p^2 \left(\frac{\partial p}{\partial v}\right)_T \tag{12.38}$$

Several important conclusions can be drawn from Equation 12.38. First, the last term on the right-hand side is known from experimental observation to be negative for all substances in all phases. Since temperature is always positive and the squared term is always positive, it is concluded that $c_p - c_v$ must always be positive or zero for all substances in all phases. Second, it can be seen from the equation that c_p approaches c_v as T approaches zero. Third, it is known from experimental data for solids and liquids that specific volume changes with temperature at constant pressure are relatively small. Therefore, when this term is squared, the result gives a relatively small difference between c_p and c_v for solids and liquids. Consequently, it is generally assumed that c_p is approximately equal to c_v for such substances. Thus, many tables give only one value for specific heat for solids and liquids (generally c_p). Finally, if the ideal gas equation of state is used with Equation 12.38, it can be shown that $c_p - c_v = R$ for an ideal gas.

EXERCISES

for Section 12.4

1. An ideal gas was defined in Chapter 4 as a gas that obeys $pv = RT$. Use this definition with Equation 12.38 to obtain $c_p - c_v = R$ for an ideal gas.

2. The pvT equation for a gas in a limited temperature and pressure range can be approximated by $v = (RT/p) + b$ where b is a constant. Show by using Equation 12.38 that $c_p - c_v = R$ for this gas also.

12.5 Enthalpy, Internal Energy, and Entropy

This section gives equations to determine changes in enthalpy, internal energy, and entropy for a pure substance in a single phase in terms of specific heat and pvT data. Many different equations could be developed; the ones here are representative of equations that can be used to determine changes for the properties in a reasonably direct manner.

Enthalpy

Combining Equation 3.5, which defines c_p, and Equation 12.25 for dh gives an expression for the change of enthalpy in a pure substance as follows:

$$dh = c_p \, dT + \left(\frac{\partial h}{\partial p}\right)_T dp \tag{12.39}$$

The coefficients of dp in Equations 12.24 and 12.26 can be equated:

$$T\left(\frac{\partial s}{\partial p}\right)_T = \left(\frac{\partial h}{\partial p}\right)_T - v \tag{12.40}$$

Substituting the Maxwell relation Equation 12.13 into Equation 12.40 and rearranging gives

$$\left(\frac{\partial h}{\partial p}\right)_T = v - T\left(\frac{\partial v}{\partial T}\right)_p \tag{12.41}$$

Combining Equations 12.39 and 12.41 leads to

$$dh = c_p \, dT + \left[v - T\left(\frac{\partial v}{\partial T}\right)_p\right] dp \tag{12.42}$$

This is a general equation for a change in enthalpy for a simple compressible substance in terms of pvT and specific heat data. The equation can be integrated to obtain the change in enthalpy for a change of state as

$$h_2 - h_1 = \int_{T_1}^{T_2} c_p \, dT + \int_{p_1}^{p_2} \left[v - T\left(\frac{\partial v}{\partial T}\right)_p\right] dp \tag{12.43}$$

This equation can be evaluated along a constant pressure line from knowledge of only c_p, while evaluation along an isotherm requires information concerning pvT for the substance.

Internal Energy

Combining Equation 12.10 with Equation 12.30 gives

$$du = c_v \, dT + \left[T\left(\frac{\partial p}{\partial T}\right)_v - p\right] dv \tag{12.44}$$

This equation can be integrated to find the change in internal energy associated with a change of state as

$$u_2 - u_1 = \int_{T_1}^{T_2} c_v \, dT + \int_{v_1}^{v_2} \left[T\left(\frac{\partial p}{\partial T}\right)_v - p\right] dv \tag{12.45}$$

Thus, the change of internal energy associated with a change of state can be found from a knowledge of the constant volume specific heat plus knowledge of the pvT relationships for the substance.

Entropy

The change of entropy associated with a change of state can be found from Equations 12.29 and 12.30, which involve only specific heat and pvT relationships. Therefore, these equations can be integrated to yield two expressions for computing the change in entropy associated with the change of state. These expressions are

$$s_2 - s_1 = \int_{T_1}^{T_2} c_p \frac{dT}{T} - \int_{p_1}^{p_2} \left(\frac{\partial v}{\partial T}\right)_p dp \tag{12.46}$$

$$s_2 - s_1 = \int_{T_1}^{T_2} c_v \frac{dT}{T} + \int_{v_1}^{v_2} \left(\frac{\partial p}{\partial T}\right)_v dv \tag{12.47}$$

Either of these equations gives the same results between the same two end states. The choice of which equation to use depends on the form of the pvT relationship that is available for the substance.

EXERCISES

for Section 12.5

1. Use Equation 12.41 to prove that for an ideal gas, defined as a system that obeys $pv = RT$, $h = h(T)$ only.

2. Suppose the pvT equation for a gas under prescribed conditions is given approximately by $v = RT/p + b$, where b is a constant. Use Equation 12.41 to show that for this gas, the specific enthalpy h can be written as $h = pv + f(T)$. From your result, show that the specific internal energy for the gas is a function only of temperature.

3. The accompanying figure illustrates on T-p coordinates a combination of isothermal and isobaric processes for moving from the dead state (p_0, T_0) to the actual state (p, T) for a single-phase pure substance. Derive the following, using Equations 12.43 and 12.46 and performing integration along $0 \rightarrow a$ followed by $a \rightarrow 1$:

$$h - h_0 = \int_{T_0}^{T} c_p(p, T)\, dT + \int_{p_0}^{p} \left[v - T\left(\frac{\partial v}{\partial T}\right)_p\right]_{T=T_0} dp$$

$$s - s_0 = \int_{T_0}^{T} \frac{c_p(p, T)}{T}\, dT - \int_{p_0}^{p} \left[\left(\frac{\partial v}{\partial T}\right)_p\right]_{T=T_0} dp$$

Use these results to show that the flow exergy per unit mass is

$$x = h - h_0 - T_0(s - s_0)$$

$$= \int_{T_0}^{T} c_p(p, T)\left(1 - \frac{T_0}{T}\right) dT + \int_{p_0}^{p} v_{T_0}\, dp$$

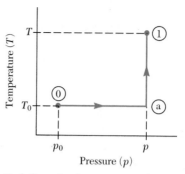

Path from dead state to actual
state for a single-phase pure
substance.

12.6 Clapeyron Equation

Section 12.5 presented relationships for determining the change in enthalpy, internal
energy, and entropy for a change of state of a single-phase, compressible pure substance.
In many cases, it is desirable to determine changes in properties associated with a change
of phase. The *Clapeyron equation* relates changes in properties for a change of phase
of a substance. Figure 12.1 shows a typical vapor pressure curve for a simple compressible pure substance such as water. As shown in Figure 12.1, the typical vapor
pressure curve is presented with pressure as the ordinate and temperature as the abscissa. As can be seen from the curve, pressure and temperature are dependent during
a change of phase, and the specification of either property automatically fixes the other.
The saturation pressure–temperature data are generally available for many different
substances, and the Clapeyron equation uses such data for certain other property relationships.

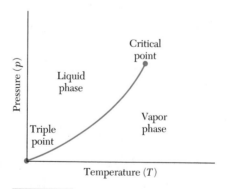

Figure 12.1 Vapor pressure curve for a
simple, compressible pure substance such
as water.

To develop the Clapeyron equation, one begins with the Maxwell relation Equation 12.12:

$$\left(\frac{\partial p}{\partial T}\right)_v = \left(\frac{\partial s}{\partial v}\right)_T \tag{12.12}$$

In the phase change region, the pressure is a function of temperature only (as illustrated in Figure 12.1), and thus the term on the left-hand side of Equation 12.12 becomes a total derivative. Furthermore, for an *isothermal phase change process*, the term on the right-hand side can be determined by making use of Equations 3.18 and 3.15 for changes in entropy and specific volume for a wet vapor with a quality x and at a specified saturation pressure or temperature. Thus, from

$$s = s_f + xs_{fg} \tag{3.18}$$

and

$$v = v_f + xv_{fg} \tag{3.15}$$

one obtains the following for the phase change region:

$$\left(\frac{\partial s}{\partial v}\right)_T = \frac{\Delta s}{\Delta v} = \frac{s_{fg}\,\Delta x}{v_{fg}\,\Delta x} = \frac{s_{fg}}{v_{fg}} \tag{12.48}$$

Therefore, Equation 12.12 can be written for a phase change as

$$\left(\frac{dp}{dT}\right)_{sat} = \frac{s_{fg}}{v_{fg}} \tag{12.49}$$

which is sometimes called the Clapeyron equation for the general case. Since the term on the left-hand side is the slope of the vapor pressure curve, Equation 12.49 can be used to compute s_{fg} if data are available concerning v_{fg} and the vapor pressure curve.

For a constant pressure and constant temperature phase change, Equation 8.14 can be used to determine that

$$h_{fg} = Ts_{fg} \tag{12.50}$$

Combining Equations 12.49 and 12.50 gives

$$\left(\frac{dp}{dT}\right)_{sat} = \frac{h_{fg}}{Tv_{fg}} \tag{12.51}$$

This is widely referred to as the *Clapeyron equation* or the *Clausius–Clapeyron equation*. Note that this equation is in terms of the specific enthalpy of vaporization and the temperature, instead of the specific entropy of vaporization used in Equation 12.49. It can be used to compute the change in enthalpy associated with the phase change when the vapor pressure curve and data concerning v_{fg} are available for the substance.

If h_{fg} and v_{fg} data are available, however, the Clapeyron equation can be integrated to develop the vapor pressure curve for the substance. In this case, the integral takes the form of

$$(p_2 - p_1)_{sat} = \int_{T_{s1}}^{T_{s2}} \frac{h_{fg}}{T_s v_{fg}} dT_s \qquad (12.52)$$

where T_s is the saturation temperature. Equation 12.49 can be used to compute entropy change for a phase change if information concerning v_{fg} and the vapor pressure curve is available. Similarly, Equation 12.51 can be used to compute h_{fg}. The change in internal energy for the phase change can be computed from

$$u_{fg} = h_{fg} - p v_{fg} \qquad (12.53)$$

Thus, from knowledge of v_{fg} and the vapor pressure curve, one can obtain s_{fg}, h_{fg}, and u_{fg}.

Although the Clapeyron equation was presented here for the phase change from a liquid to a vapor, it can be generalized to apply to any phase change.

12.7 Physical Coefficients

In many cases, particularly involving solids, one is interested in how the volume of the substance changes as a function of temperature and pressure. If v is considered a function $v(T, p)$, then, as in Equation 12.1, one can write

$$\frac{dv}{v} = \frac{1}{v} \left(\frac{\partial v}{\partial T} \right)_p dT + \frac{1}{v} \left(\frac{\partial v}{\partial p} \right)_T dp \qquad (12.54)$$

where the equation has been divided through by v. The coefficient of dT is known as the *coefficient of thermal expansion* and is written as

$$\alpha = \frac{1}{v} \left(\frac{\partial v}{\partial T} \right)_p \qquad (12.55)$$

This coefficient is sometimes also called the *volume expansivity* or the *coefficient of volume expansion*.

The negative of the coefficient of the dp term is called the *coefficient of isothermal compressibility*. It can be written as

$$\beta_T = -\frac{1}{v} \left(\frac{\partial v}{\partial p} \right)_T \qquad (12.56)$$

where the minus sign has been added because the partial derivative is always negative and the minus sign allows β_T to be positive. The reciprocal of the isothermal compressibility is called the *isothermal modulus*, and it can be written as

$$B_T = -v \left(\frac{\partial p}{\partial v} \right)_T = \frac{1}{\beta_T} \qquad (12.57)$$

Combining Equation 12.38 with Equations 12.55 and 12.56 shows that the difference in specific heats $(c_p - c_v)$ for a substance can be determined by

$$c_p - c_v = \frac{\alpha^2 v T}{\beta_T} \qquad (12.58)$$

Note that any derivatives of properties are also properties. Therefore, the coefficients expressed here are also properties of substances. Values of the physical properties discussed here are given for numerous substances in many standard handbooks.

Joule–Thomson Coefficient

The throttling process was briefly discussed in Example 6.16 of Chapter 6. In this process, a constriction is used to reduce the pressure of a flowing fluid in the absence of heat or work for the process. In such a case, Example 6.16 showed that the enthalpy at the exit of the throttling device equals the enthalpy at the inlet. It is important in the throttling process whether the temperature of the fluid increases or decreases during the process. The Joule–Thomson coefficient can be used to assess this temperature change. The *isenthalpic Joule–Thomson coefficient* is defined by

$$\mu_h = \left(\frac{\partial T}{\partial p}\right)_h \tag{12.59}$$

Equation 12.59 shows that the temperature drops when the pressure drops (at constant enthalpy) if the coefficient is positive, while the temperature increases for a pressure drop at a constant enthalpy if the coefficient is negative. If the isenthalpic Joule–Thomson coefficient is zero, no change in temperature occurs whether the pressure rises or drops while the enthalpy is kept constant.

Figure 12.2 shows a typical T-p diagram for a fluid that might undergo throttling. The constant enthalpy lines are shown with arrows to indicate the direction of the throttling process, which produces a reduction in pressure. Note that a state of maximum temperature exists for some of the constant enthalpy lines. This state of maximum temperature for the different constant enthalpy lines is joined by the inversion line. A

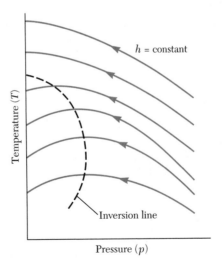

Figure 12.2 Typical T-p diagram for a fluid that might undergo throttling.

constant enthalpy process involving a reduction in pressure on the right side of the inversion line results in an increase in temperature, while a similar process on the left side of the inversion line results in a reduction in temperature. Thus, the Joule–Thomson coefficient is negative to the right of the inversion line and positive to the left side of the inversion line. Refrigerants have a positive Joule–Thomson coefficient under the typical conditions that exist for the throttling process. Note that a reduction in pressure at constant enthalpy for the enthalpy lines at the top of the diagram will never result in a cooling effect. Thus, some states exist from which a substance cannot be cooled by throttling. For example, hydrogen and helium have maximum inversion temperatures of 202 K and about 40 K, respectively, and therefore cannot be cooled by throttling at ambient temperatures, which are considerably higher than the maximum inversion temperatures.

As noted in Example 6.16, a throttling process is highly irreversible and is not necessarily a constant enthalpy process. It is only known that the enthalpy at the exit plane is equal to the enthalpy at the inlet plane. Thus, the constant enthalpy lines of Figure 12.2 do not represent processes but rather each line is a locus of points of equilibrium states with the same specific enthalpy values. In any case, the assumption of equal enthalpy at the inlet and exit planes is sufficient to determine whether the temperature of a substance rises or falls during a throttling process. The Joule–Thomson coefficient is thus a valuable coefficient for determining whether or not a fluid can be used for cooling processes by dropping the pressure under adiabatic condition.

EXERCISES

for Section 12.7

1. Use the cyclic relationship given by Equation 12.4 to obtain the following alternative expression for the Joule–Thomson coefficient:

$$\mu_h = -\frac{1}{c_p}\left(\frac{\partial h}{\partial p}\right)_T = -\frac{v(1 - \alpha T)}{c_p}$$

2. Use the expression for μ_h in Exercise 1 to prove that the Joule–Thomson coefficient for an ideal gas is zero.

12.8 Development of Property Tables from Experimental Data for Real Substances

Numerical values for thermodynamic properties are needed to make calculations for thermodynamic systems. Therefore, numerical values for thermodynamic properties are normally compiled in tables or in computer retrieval systems to provide the desired information. These tables are based on experimental measurements of certain properties for the substances over a wide range of conditions. However, it is not feasible to make experimental measurements for all property values that might be needed. Therefore, an experimental program is only designed to make an adequate number of measurements so as to develop an accurate pvT equation of state for the substance. These, together with specific heat (c_p or c_v) data and appropriate thermodynamic relationships such as Equations 12.43, 12.45, 12.47, 12.49, and 12.51 are then used to calculate properties for a wide range of the variables.

The measurements normally made in the experimental program include vapor pressure (such as that shown in Figure 12.1), specific heat (c_p or c_v), and relationships among p, v, and T. The measurements to relate p, v, and T are frequently made with the substance in a closed fixed-volume container. The temperature is varied throughout the desired temperature range, and pressure measurements are made while the specific volume remains constant. Such measurements can be made for a large number of values of the specific volume to obtain data that relate p, v, and T over a wide range of conditions. In addition to experimental measurements, appropriate mathematical relations are required to develop complete thermodynamic tables for all the properties.

The most general approach to the development of thermodynamic tables is to use the data to develop a characteristic equation of state for the substance. Such a characteristic equation of state can be used to determine all of the thermodynamic properties of the substance by appropriate mathematical manipulations. A general approach for developing the characteristic equation is discussed in the following paragraphs.

Any of these four fundamental equations

$$da = -p\ dv - s\ dT \tag{12.8}$$

$$dg = v\ dp - s\ dT \tag{12.9}$$

$$du = T\ ds - p\ dv \tag{12.10}$$

$$dh = T\ ds + v\ dp \tag{12.11}$$

can be used to develop a characteristic equation of state. Each term on the left-hand side of these four equations represents a total differential for a thermodynamic property in terms of other thermodynamic properties that are given on the right-hand side of the equations. The properties given by total differentials on the right-hand side can be chosen as the independent variables, and the properties on the left-hand side can be chosen as the dependent variables. The following functional relationships for the four equations can then be obtained:

$$u = u(s, v) \tag{12.60}$$

$$h = h(s, p) \tag{12.61}$$

$$a = a(v, T) \tag{12.62}$$

$$g = g(p, T) \tag{12.63}$$

Each of these equations is called a *characteristic equation of state* for a substance.

Now consider the Helmholtz function from Equation 12.62 to illustrate how the characteristic equation is used to obtain thermodynamic properties for a substance. The total differential for da from Equation 12.62 can be expressed as

$$da = \left(\frac{\partial a}{\partial v}\right)_T dv + \left(\frac{\partial a}{\partial T}\right)_v dT \tag{12.64}$$

Comparison of Equation 12.64 with 12.8 gives

$$\left(\frac{\partial a}{\partial v}\right)_T = -p \tag{12.65}$$

$$\left(\frac{\partial a}{\partial T}\right)_v = -s \tag{12.66}$$

Having chosen Equation 12.62 as the characteristic equation, the independent variables are v and T. Also, a is computed from the characteristic equation. Consequently, p and s can be computed from Equations 12.65 and 12.66. From the definition of the Helmholtz function, u can now be computed as

$$u = a + Ts \tag{12.67}$$

Also, the definition of enthalpy allows the computation of h as

$$h = u + pv \tag{12.68}$$

Thus, the use of characteristic Equation 12.62 allows one to evaluate a, p, s, u, and h once v and T are specified. Consequently, one is able to obtain all of the basic thermodynamic properties from this characteristic equation. Similar results can be developed using any of the other characteristic equations given by Equations 12.60, 12.61, and 12.63.

Although any of the four characteristic equations can be used mathematically to determine *all* of the basic thermodynamic properties, the two involving entropy as an independent variable [that is, $u = u(s, v)$ and $h = h(s, p)$] are not normally used because s cannot be measured directly. Instead, either $a = a(v, T)$ or $g = g(p, T)$ are normally used because the independent variables involve p, v, and T, all of which can be readily measured.

Probably the most accurate currently available property formulation for steam is that provided in the NBS/NRC Steam Tables (Haar et al. 1984). These tables were developed from an international effort involving many different individuals and are based on more than 25,000 individual experimental data points. Results are given for the widest range of values in any table currently available. A characteristic equation in terms of the Helmholtz function was developed by Haar et al. (1984) for use in the NBS/NRC tables. The Helmholtz function was specified in terms of the independent variables of density and temperature (instead of specific volume and temperature). The base function for the Helmholtz function was derived from the theoretical equation for a dilute gas (called the Ursell–Mayer equation). Two correction terms were added to the base function to correlate with the experimental data. The first correction term, called a *residual function*, was developed to correct for deviations from the base equation at high densities and at conditions near the critical point. The second correction term was developed to account for ideal gas behavior at very low densities when the other terms are approaching zero. Thus, the Helmholtz function used in the NBS/NRC tables was given by

$$a(\rho, T) = a_{\text{base}}(\rho, T) + a_{\text{residual}}(\rho, T) + a_{\text{ideal gas}}(T) \tag{12.69}$$

For an ideal gas, the correction term is a function of temperature only. Extensive efforts obviously went into the development of the appropriate correlations to ensure an accurate characteristic equation. The result was a very complex mathematical expression for the Helmholtz function. The base equation $a_{\text{base}}(\rho, T)$ is an expression that includes six terms involving logarithms, exponential terms, and several implicit expressions in terms of density. The residual function $a_{\text{residual}}(\rho, T)$ consists of 40 terms that involve density and temperature and include implicit exponential and power terms. The ideal gas function $a_{\text{ideal gas}}(T)$ includes 18 terms and also involves logarithms and powers of temperature. Thus, although the Helmholtz function is an analytical expression, manipulations involving the function are rather difficult and generally require the use of the computer.

Once the Helmholtz function was developed, all of the basic thermodynamic properties could be obtained from appropriate manipulation with the Helmholtz function. In addition, some of the other thermodynamic properties were obtained from the following expressions:

$$g = a + p/\rho \tag{12.70}$$

$$c_v = -T\frac{\partial^2 a}{\partial T^2} \tag{12.71}$$

$$c_p = c_v + \frac{T\left(\frac{\partial p}{\partial T}\right)^2}{\rho^2\left(\frac{\partial p}{\partial \rho}\right)} \tag{12.72}$$

Using the approach outlined, the NBS/NRC Steam Tables (Haar et al. 1984) were developed, and they give very accurate values over a wide range of conditions. However, the use of the associated computer program to obtain properties is sometimes difficult. The difficulties include the following:

1. Because the properties are evaluated with numerical differentiation, sometimes raised to a power as in Equation 12.72, any computation requires at least 16 digit accuracy to obtain reasonable results.

2. Calculations involving implicit relationships that include over 50 terms require a considerable amount of computer time when one has available a property such as p or T and wishes to know another property such as s or ρ. Such a calculation requires high machine precision and considerable iteration to obtain the properties.

The authors have developed computer codes that are based on an alternate approach. These codes are explained in Appendix D and basically involve integration of functions containing c_p rather than differentiation of a fundamental equation of state. The specific heat data are correlated and used in developing integral expressions for enthalpy and entropy. Furthermore, the equations are developed in a more simplified manner for specific regions of the data to minimize the complexity of the computational procedure. Since numerical integrations normally round off errors, while numerical differentiation accentuates errors, the property codes in the present effort yield results that show remarkable consistency and accuracy when compared with those in standard tables.

12.9 Generalized Charts for Real Gases

The principle of corresponding states discussed in Section 4.6 allows the correlation of pvT data for a substance in terms of a compressibility factor Z, which is

$$Z = Z(p_R, T_R) \tag{4.38}$$

where p_R and T_R indicate normalized pressure and temperature based on the critical values. Thus, the principle of corresponding states leads to the conclusion that one can correlate pvT data for any real gas in terms of the critical properties of the substance. Furthermore, one should also be able to correlate any other thermodynamic property for a real gas in terms of the critical properties. Thus, relationships and charts can be developed in generalized form which can be used to approximate properties for substances when adequate data are not available. Generalized charts of this type that are based on the principle of corresponding states are presented and discussed in the following sections.

12.9.1 Chart for Enthalpy

A generalized chart for enthalpy is given in Appendix C. It can be developed as follows. Equation 12.42 is an expression for enthalpy in terms of pvT and specific heat data for a simple, compressible pure substance:

$$dh = c_p \, dT + \left[v - T\left(\frac{\partial v}{\partial T}\right)_p \right] dp \qquad (12.42)$$

The goal is to express such a relationship for the change in enthalpy in terms of the critical properties and the compressibility factor. Therefore, Equation 4.35, $v = ZRT/p$, is combined with Equation 12.42 to obtain, after simplification,

$$dh = c_p \, dT - \frac{RT^2}{p}\left(\frac{\partial Z}{\partial T}\right)_p dp \qquad (12.73)$$

Now, Equations 4.36 and 4.37, which give the pressure and temperature in terms of the critical properties, are combined with Equation 12.73 to obtain

$$dh = c_p T_c \, dT_R - \frac{RT_c T_R^2}{p_R}\left(\frac{\partial Z}{\partial T_R}\right)_{p_R} dp_R \qquad (12.74)$$

Since Z is a constant ($Z = 1$) for an ideal gas, one can see from Equation 12.73 that $dh = c_p \, dT$ for an ideal gas. Therefore, it is desirable to integrate Equation 12.74 from an ideal gas state to some other state. Integration of Equation 12.74 at a constant temperature from zero pressure to some higher pressure yields

$$\frac{h° - h}{RT_c} = \int_{p_R=0}^{p_R} \frac{T_R^2}{p_R}\left(\frac{\partial Z}{\partial T_R}\right)_{p_R} dp_R \qquad (12.75)$$

where $h°$ is the ideal gas specific enthalpy and h is the specific enthalpy at some higher pressure p_R. Note that T_R is held constant during the integration process. The value of the integral can be obtained by numerical integration using data from the generalized compressibility chart. Results give values for the left-hand side of the equation in terms of p_R and T_R. The quantity on the left-hand side of Equation 12.75 is called the *enthalpy departure function*. It is presented as a function of T_R and p_R in the generalized enthalpy chart in Appendix C.

EXAMPLE 12.1

Butane gas (C_4H_{10}) is cooled at a pressure of 10 atm in a constant pressure process from 500 K to 400 K. Calculate the heat transfer per unit mass (a) assuming ideal gas behavior, and (b) using the generalized enthalpy chart. Compare these results based on ideal gas and real gas behavior for the butane. Properties for butane (C_4H_{10}) are T_c = 425.2 K, p_c = 3.8 MPa, and c_{p0} = 1.72 kJ/kg · K.

Given: Butane gas (with T_c = 425.2 K, p_c = 3.8 MPa, and c_{p0} = 1.72 kJ/kg · K) cooled at 10 atm from 500 K to 400 K.

Find:

(a) Compute q using ideal gas equations

(b) Compute q using generalized enthalpy chart and compare to part (a)

Solution: For a constant pressure process with no work, the first law for a closed system is

$$q = (h_2 - h_1)$$

(a) Computing q for the ideal gas gives

$$q = (h_2^\circ - h_1^\circ) = c_{p0}(T_2 - T_1)$$
$$= 1.72 \text{ kJ/kg} \cdot \text{K} (400 - 500 \text{ K}) = -172 \text{ kJ/kg}$$

(b) For the real gas using the generalized chart, one needs p_R and T_R.

For the butane at 10 atm (1013. kPa), $p_R = p/p_c$ = 1013. kPa/3800 kPa = 0.26, and $T_{R2} = T_2/T_c$ = 400/425.2 = 0.94 for T_2 = 400 K. Also, $T_{R1} = 500/425.2 = $ 1.18 for T_1 = 500 K. From the generalized enthalpy chart, one obtains

$$(h^\circ - h)/RT_c = 0.19 \text{ for state 1}$$
$$= 0.38 \text{ for state 2}$$

Solving gives

$$(h_2 - h_1) = (h_2^\circ - h_1^\circ) - 0.19 \, RT_c$$

Thus, the difference in h between the ideal gas and real gas is $0.19 \, RT_c$, and

$$q = (h_2 - h_1) = (h_2^\circ - h_1^\circ) - 0.19 \, RT_c$$

or

$$q = -172 \text{ kJ/kg} - 0.19(8.314 \text{ kJ/kmol} \cdot \text{K})(1/58 \text{ kg/kmol})(425.2 \text{ K})$$
$$q = -184 \text{ kJ/kg}$$

Thus, the actual heat transfer out of the system for the real gas is approximately 7% more than for the ideal gas calculation.

12.9.2 Chart for Entropy

A generalized chart for entropy can be developed in essentially the same manner as that previously outlined for enthalpy. Equation 4.35 can be used to replace v with ZRT/p in Equation 8.14, which can then be written as follows:

$$ds = \frac{dh}{T} - \frac{R dp}{p} + (1 - Z) \frac{R dp}{p} \qquad (12.76a)$$

Let $p = \delta$ atm denote a very small, but finite pressure at which the gas behaves as an ideal gas. Integrate Equation 12.76a with respect to dp from δ to p(atm) while keeping the temperature T constant. The result is

$$
s(p, T) - s(\delta, T) = \frac{h(p, T) - h^0(T)}{T} - R \ln \left(\frac{p}{\delta}\right)
$$
$$
+ R \int_\delta^p \frac{(1 - Z) dp}{p} \qquad (12.76b)
$$

From Equation 8.29a, the specific entropy $s(\delta, T)$ can be written as

$$s(\delta, T) = s^0(T) - R \ln (\delta)$$

where $s^0(T)$ is the standardized specific entropy at the standard pressure of 1 atm and defined by Equation 8.27 as

$$s^0(T) = \int_{T_{ref}}^T \frac{c_{p0}}{T} dT \qquad (8.27)$$

(T_{ref} tends to zero degree absolute temperature (0 K) in Equation 8.29a.) Now define a specific entropy $s*(p, T)$ as

$$
s*(p, T) = s(\delta, T) - R \ln (p/\delta)
$$
$$
= s^0(T) - R \ln (p) \qquad (12.76c)
$$

Then, Equation 12.76b can be rewritten as

$$s(p, T) - s*(p, T) = \frac{h(p, T) - h^0(T)}{T} + R \int_\delta^p \frac{(1 - Z)}{p} dp \qquad (12.76d)$$

$s*(p, T)$ is at times referred to as the specific entropy at a fictitious state attained on the assumption of ideal gas behavior up to the elevated pressure p and temperature T. Since δ can be made to approach zero infinitesimally, Equation 12.76d can be written finally as

$$\frac{s*(p, T) - s(p, T)}{R} = \frac{h^0(T) - h(p, T)}{(RT_c)T_R} - \int_0^{P_R} (1 - Z) \frac{dp_R}{p_R} \qquad (12.76)$$

The first term on the right-hand side of Equation 12.76 can be obtained from the knowledge of T_R and the generalized enthalpy chart. The second term is obtained from numerical integration of the compressibility data. Results give the term on the left-hand side, which is the entropy departure function. A generalized entropy chart involving this function is given in Appendix C.

EXAMPLE 12.2

Compute the entropy change for the constant pressure process of the butane specified in Example 12.1 (a) assuming ideal gas behavior, and (b) using the generalized entropy chart. Compare the results obtained in parts (a) and (b).

Given: Butane changing state as specified in Example 12.1

Find:
(a) compute $s_2^0 - s_1^0$ using ideal gas equations
(b) $s_2 - s_1$ using the generalized entropy chart and compare to part (a)

Solution:
(a) Computing Δs for the ideal gas gives

$$s_2^0 - s_1^0 = c_{p0} \int_{T_1}^{T_2} \frac{1}{T} \, dT = c_{p0} \ln \frac{T_2}{T_1}$$

$$= -0.384 \text{ kJ/kg} \cdot \text{K}$$

(b) Since $p_2 = p_1$, it follows that $s_2 - s_1 = -0.384$ kJ/kg · K when ideal gas behavior is assumed. For the real gas, the generalized entropy chart gives

$$(s_p^0 - s_p)_T/R = 0.11 \text{ for state 1}$$

$$= 0.31 \text{ for state 2}$$

Combining these expressions (with $s_{p2} = s_2$ and $s_{p1} = s_1$) gives

$$s_2 - s_1 = (s_2^0 - s_1^0) - 0.20R$$

Thus, the difference in s is 0.20 R and

$$s_2 - s_1 = (s_2^0 - s_1^0) - 0.20 \, R$$

$$= -0.384 \text{ kJ/kg} \cdot \text{K} - 0.20(8.314/58) \text{ kJ/kg} \cdot \text{K}$$

$$= -0.41 \text{ kJ/kg} \cdot \text{K}$$

Thus, the actual entropy decrease is approximately 7% more than that calculated for the ideal gas.

Although only two generalized charts have been discussed, many other charts of this type could be developed. A generalized chart for the property called *fugacity* is available for mixtures. Generalized charts for specific heats at high pressures have also been developed.

All of the charts discussed here are based on compressibility information, which is given in terms of the two parameters T_R and p_R. Lee and Kesler (1975) developed a procedure (which is presented in the textbook by Howell and Buckius 1987) for using the corresponding state method in terms of three parameters. The three parameters are p_R, T_R, and an acentric factor that indicates the degree of deviation of the molecular structure of a particular molecule from that of a simple spherical model. Results indicate that the three-parameter model is more accurate for most hydrocarbons than the two-parameter model. In particular, the two-parameter model generally gives results with an uncertainty of about 6%, while the three-parameter model gives results with an uncertainty of about 2% or less.

12.10   Summary

This chapter presented thermodynamic property relationships for pure substances. Relationships are needed between properties because not all properties are readily available or easily measured. Therefore, mathematical relationships among the properties allow the computation of values for properties not easily measured from values that are available from less difficult measurements.

Mathematical considerations were outlined briefly, and the $T \, ds$ equations for the combined first and second law were used along with the Gibbs function and the Helmholtz function to develop the well-known Maxwell relations. These relations, along with appropriate mathematical manipulations, were used to develop equations for c_p, c_v, h, u, and s in terms of easily measured properties. The Clapeyron equation was also presented as an illustration of how to relate the pressure and temperature of a substance that undergoes a phase change.

Certain physical coefficients such as the coefficient of thermal expansion, the isothermal compressibility, and the isenthalpic Joule–Thomson coefficient were also presented and discussed. These coefficients are derived from thermodynamic properties and thus are also thermodynamic properties. Such properties are useful in a variety of engineering applications.

Procedures were outlined for the development of thermodynamic property tables for different substances. Such tables are based on experimental data and are normally obtained by the use of appropriate mathematical manipulations and available data. In some cases, inadequate data are available to develop property tables for a substance. In such cases, the principle of corresponding states can be employed to develop generalized charts for substances. These generalized charts can be used to obtain a reasonable approximation for the properties of such substances in the gaseous phase. Generalized charts for enthalpy and entropy were presented and discussed. Several other generalized charts are available in the literature.

References

Bejan, A. 1988. *Advanced Engineering Thermodynamics*. New York: John Wiley & Sons.

Haar, L., Gallagher, J. S., and Kell, G. S. 1984. *NBS/NRC STEAM TABLES: Thermodynamic and Transport Properties and Computer Programs for Vapor and Liquid States of Water in SI Units*. New York: Hemisphere Publishing Company.

Haywood, R. W. 1980. *Equilibrium Thermodynamics for Engineers and Scientists*. Chichester, U.K.: John Wiley & Sons.

Howell, J. R. and Buckius, R. O. 1987. *Fundamentals of Engineering Thermodynamics*. New York: McGraw-Hill.

Keenan, J. R. 1941. *Thermodynamics*. New York: John Wiley & Sons.

Lee, B. I. and Kesler, M. G. 1975. "A Generalized Thermodynamic Correlation Based on Three-Parameter Corresponding States." *American Institute of Chemical Engineering Journal*. Vol. 21, No. 3. 510–527.

Questions

1. What is a pure substance? Discuss the two-property rule, including how the rule dictates the form of the equations of state for pure substances.

2. Define a simple system. What is the state principle and how does the principle help in the determination of the equations of state for a simple system?

3. Write the test for exactness for a function of two variables. Explain in your own words the usefulness of the mathematical theorem on exact differentials to the establishment of relationships among thermodynamic properties.

4. Write the version of the combined first and second law equation that involves the specific enthalpy (h). What are the restrictions that apply for the equation to be valid?

5. Use Equation 12.22 to sketch the variation of the specific internal energy (u) for a pure substance with the specific entropy (s) when the specific volume (v) is constant. Indicate what happens to each constant v line as $T \to 0$ and as T tends to infinity.

6. Consider a fluid contained in a rigid, well-insulated vessel. Suppose work is done on the fluid using a stirrer. From application of the first law of thermodynamics to the process, would you expect u for the fluid to increase or decrease? Referring to your sketch in Question 12.5, would you expect s for the fluid to increase or decrease? Give reasons for your answer.

7. Equation 12.38 for the difference between the principal specific heats is

$$c_p - c_v = -T\left(\frac{\partial v}{\partial T}\right)_p^2 \left(\frac{\partial p}{\partial v}\right)_T$$

By considering each of the terms on the right-hand side of the equation, indicate why c_p cannot be less than c_v for any substance.

8. Consider the Clapeyron equation and assume that a pure substance is in a region where $h_{fg} \approx$ constant and $v_{fg} \approx v_g \approx RT/p$. Use these approximations to obtain the following form of the equation for a vapor pressure curve:

$$\ln p = a - \frac{b}{T}$$

9. Define the following coefficients and obtain expressions for each coefficient when the substance is an ideal gas:
   (a) the coefficient of thermal expansion (α)
   (b) the coefficient of isothermal compressibility (β_T)
   (c) the isothermal modulus (β_T)

10. What is the principle of corresponding states? Briefly explain how the principle is used in the compilation of the generalized compressibility chart for real gases.

Problems

12.1. Determine whether or not each of the following functions is an exact differential:
   (a) $dz = xy\,dx + xy\,dy$
   (b) $dz = y^2\,dx + x^2\,dy$
   (c) $du = T\,ds - p\,dv$
   (d) $dh = T\,ds + v\,dp$

* **12.2.** Using Equations 12.3, 12.4, 12.10, and appropriate mathematical expressions, derive the Maxwell relation given in Equation 12.15.

* **12.3.** The only form of work done by a system is mechanical displacement work. Develop a Maxwell relation for this case.

* **12.4.** The only form of work done by a system is mechanical shaft work. Develop a Maxwell relation for this case.

* **12.5.** The only form of work done by a system is stretched wire work. Develop a Maxwell relation for this case.

** **12.6.** Show that $u = u(T)$ only and $h = h(T)$ only for an ideal gas.

* **12.7.** The equation of state for an ideal gas is $pv = RT$. Verify that dT for this equation is an exact differential.

* **12.8.** Derive the following expressions:
(a) $ds = (c_p/T)\, dT - (\partial v/\partial T)_p\, dp$
(b) $ds = (c_v/T)\, dT + (\partial p/\partial T)_v\, dv$

* **12.9.** Develop an expression for $(\partial u/\partial p)_T$ that involves only p, T, and v.

* **12.10.** Develop an expression for $(\partial u/\partial v)_T$ that involves only p, T, and v.

* **12.11.** Beginning with Equation 8.16, show that the enthalpy of a fluid increases as a result of isentropic compression of the fluid.

* **12.12.** Show that constant pressure lines in the two-phase region (under the dome) of an enthalpy–entropy diagram (Fig. P12.12) are straight but not parallel.

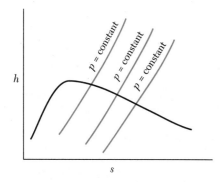

Figure P12.12 Constant pressure lines on an enthalpy–entropy diagram.

* **12.13.** Develop an expression for the slope of an isentropic line in the single-phase region on a pressure–enthalpy diagram (Fig. P12.13) in terms of p, v, and/or T only.

* **12.14.** Show that the Joule–Thomson coefficient for a gas can be expressed as

$$\mu_h = \frac{1}{c_p}\left[T\left(\frac{\partial v}{\partial T}\right)_p - v\right]$$

* **12.15.** Using results from Problem 12.14, determine the value of the Joule–Thomson coefficient for

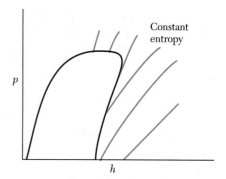

Figure P12.13 Constant entropy lines on a pressure–enthalpy diagram.

an ideal gas. Can you get either a heating or a cooling effect by throttling an ideal gas?

* **12.16.** Using results from Problem 12.14 along with the equation $pv = ZRT$, show that the Joule–Thomson coefficient can be given by

$$\mu_h = \frac{RT^2}{pc_p}\left(\frac{\partial Z}{\partial T}\right)_p$$

* **12.17.** Beginning with Equation 12.38, show that $c_p - c_v = R$ for an ideal gas.

* **12.18.** Beginning with the appropriate functional relationship, derive Equation 12.30.

** **12.19.** Beginning with $g = g(p, T)$, derive a characteristic equation of state for a perfect gas.

* **12.20.** Using Equation 12.52 and data from the steam tables, plot a vapor pressure curve for H_2O from the triple point to the critical point.

* **12.21.** Using Equation 12.52 and data from the refrigerant tables, plot a vapor pressure curve for refrigerant-22 from $-60°F$ to $160°F$.

* **12.22.** Determine the difference between c_p and c_v for copper at 300K. At this temperature,

$$\alpha = 50 \times 10^{-6}\ K^{-1}$$

$$\beta_T = 0.78 \times 10^{-11}\ m^2/N$$

$$v = 1.12 \times 10^{-4}\ m^3/kg$$

The value for c_p is approximately 2 kJ/kg · K at this temperature. Is it reasonable to assume c_p equals c_v for copper at this temperature?

* **12.23.** Estimate the final temperature that would be reached if a steady stream of nitrogen gas were

throttled from 0°C and 100 atm to 1 atm. Assume that the Joule–Thomson coefficient equals 0.22°C/atm for the process.

* **12.24.** Butane gas is cooled at a constant pressure of 15 atm from 450 K to 400 K as shown in Figure P12.24. Calculate the heat transfer per unit mass
  (a) assuming ideal gas behavior
  (b) using the generalized enthalpy chart.
  Compare your results for parts (a) and (b).

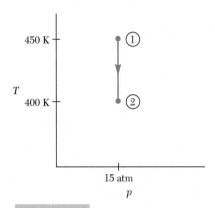

Figure P12.24 Constant pressure cooling of butane gas.

* **12.25.** Nitrogen gas is heated at a pressure of 66.9 atm in a constant pressure process from 25°C to 450 K. Assume $p_c = 3.39$ MPa and $T_c = 126.2$ K for nitrogen. Calculate the heat transfer per unit mass
  (a) assuming ideal gas behavior
  (b) using the generalized enthalpy chart.
  Compare your results.

* **12.26.** Using the generalized entropy chart, compute the specific entropy change for the constant pressure process for the butane specified in Problem 12.24. Compare your results with calculations based on the assumption of ideal gas behavior.

* **12.27.** Using the generalized entropy chart, compute the entropy change for the constant pressure process for the nitrogen specified in Problem 12.25. Compare your results with calculations based on the assumption of ideal gas behavior.

* **12.28.** Using the generalized relationships and the van der Waals pvT equation of state, show that

$$h_2 - h_1 = h°(T_2) - h°(T_1) + (p_2v_2 - p_1v_1)$$
$$+ a\left(\frac{1}{v_1} - \frac{1}{v_2}\right) - R(T_2 - T_1)$$

(*Hint*: It is suggested that you use Equation 12.44 as a starting point and express the relationship between the change in specific internal energy and the specific enthalpy using $h = u + pv$.)

* **12.29.** Using the generalized relationships and the Redlich–Kwong equation of state, develop an expression for $(h_2 - h_1)_T$.

* **12.30.** Derive an expression for the coefficient of thermal expansion for a van der Waals gas.

* **12.31.** By using data from the steam tables, determine the Joule–Thomson coefficient for steam at 600°F and 20 atm. If steam at this condition were throttled to atmospheric pressure, would the steam temperature increase or decrease during the throttling process?

* **12.32.** By using data from the tables in the appendix, determine the Joule–Thomson coefficient for refrigerant-22 at 25°C and 10 bar. If this refrigerant were used at this condition as a pressurant in a deodorant container, would the temperature of the substance increase or decrease upon throttling to room conditions?

** **12.33.** Ethylene (C_2H_4) is stored as a gas at high pressure in a storage tank. Properties for the ethylene are as follows:

$T_c = 282.4$ K

$p_c = 49.7$ atm

molecular weight = 28.054

$c_p = 43.7$ kJ/kmol · K) at 300 K

The ethylene is stored at 300 K and 500 atm.
  (a) Using results from Problem 12.16 along with the generalized compressibility chart, estimate the Joule–Thomson coefficient for the gas at this condition. If the gas is throttled, will its temperature increase or decrease?
  (b) The autoignition temperature of ethylene is 490°C. This is the temperature that combustion would occur spontaneously for the gas in the presence of air (or oxygen). If the gas were throttled adiabatically from its storage condition to a pressure of 1 atm, how

much would the temperature of the gas change? Would there be any danger of autoignition of the gas? (Note that serious problems can occur from this type of phenomenon.)

****12.34.** Repeat Problem 12.33 for the case when the gas is stored initially at 480 K and 200 atm.

****12.35.** Repeat Problem 12.33 for the case when the gas is stored initially at 400 K and 1000 atm.

****12.36.** A constant volume ideal gas thermometer is shown in Figure P12.36. An ideal gas is contained in the bulb and rigid tube, as shown. As the pressure and temperature of the gas changes, the volume of the gas is kept constant by adjusting the mercury column in the flexible tube so that the mercury level on the right side remains at the zero level. Since p/T is constant for an ideal gas at constant volume, measurements of z are proportional to the temperature T of the gas. If the pressure of the ideal gas (p_{tp}) is measured when the temperature of the gas equals the temperature of the triple point of

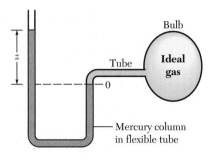

Figure P12.36 Constant-volume ideal gas thermometer.

water (273.16 K), then any other temperature (in K) can be determined from $T = 273.16\, p/p_{tp}$ by measurement of p at the new temperature. Show that measurements of temperature from this constant volume ideal gas thermometer give results that are exactly equivalent to the thermodynamic temperature. Thus, this thermometer provides results that are equivalent to the thermodynamic temperature scale.

Design Problems

12.1 *Temperature control of spacecraft cabin*

The first orbital flight by a U.S. astronaut was led by astronaut John Glenn in 1962. Water was carried along on the flight for various uses, including hygiene and drinking water, and a "water boiler" was used to provide cooling for the cabin. Waste water was dumped overboard to the vacuum of space.

Define the process for the venting of liquid water at cabin temperature to the vacuum of space. During the flight, astronaut Glenn reported "fireflies" in the vicinity of the spacecraft. Do you think these "fireflies" had any relation to the dumping of the water? Prepare a report of your results, carefully defining your assumptions. Present a design of a more appropriate alternative than the "water boiler" for maintaining the temperature of the spacecraft cabin.

12.2 *Evaluation of properties for new refrigerant*

Problems have been encountered due to the de-

pletion of ozone in the upper atmosphere of the Earth. This depletion seems to be accelerated by fluorocarbons which have been used as refrigerants. Consequently, a search is underway to develop alternate refrigerants that do not contribute to the ozone depletion. Your company has discovered a potential refrigerant that is safe and does not contribute to the ozone problem. It is your responsibility to develop and carry out a plan for determining the properties of the new refrigerant so that it can be used as a refrigerant in air conditioners and refrigeration systems. Your plan should include preliminary investigations to establish the feasibility of using the refrigerant and to assess its potential. The second phase of your plan should be to determine all properties needed for successful marketing of the refrigerant.

Write a report detailing your plan and outlining expected results. Include costs and schedules for completion of your plan.

Nonreactive Ideal Gas Mixtures

13.1 Introduction

This chapter considers *nonreactive ideal gas mixtures* in general and focuses particular attention on *air–water vapor mixtures*. Although the focus is on nonreactive mixtures, the treatment in this chapter is also intended to serve as an introduction to *reactive mixtures*, which are considered in more detail in the remaining chapters. *Moist air*, comprising the constituents of *dry air* (namely, oxygen, nitrogen, argon, and trace amounts of other gases such as carbon dioxide) plus *moisture* (or water vapor), is a pure substance since its chemical composition is uniform throughout. If, however, some of the moisture condenses and the system then comprises air, water vapor, and liquid water, the mixture can no longer be regarded as a pure substance. In this case, the chemical composition of the liquid phase is H_2O, while the gaseous phase comprises moist air (dry air constituents plus H_2O) with a different chemical composition. In general, such cases are analyzed by considering each phase separately and using the fact that each phase is a pure substance.

Multicomponent Substances and the State Postulate

The analysis of thermodynamic systems thus far has been limited to systems that are simple, compressible pure substances. A *pure substance,* as you may recall, is a system of uniform (homogeneous and invariable) chemical composition. For such a system, the *two-property rule* applies. That is, only two independent intensive thermodynamic properties are needed to define the state of the system when it is in a state of equilibrium, provided that effects due to such extraneous factors as gravity, motion, electricity, magnetism, and capillarity are negligible. The rule applies to pure substances regardless of whether they are single component or multicomponent systems. Thus, for a given sample of moist air (a multicomponent pure substance), the state is determined if any two independent intensive properties such as pressure (p) and temperature (T) are specified.

To characterize a given multicomponent system completely, however, it is necessary to specify the amount of each component present in addition to the intensive thermodynamic properties. This principle is implied in the statement of the *state postulate* as formulated by Kline and Koenig (1957, p. 31):

1. Given any macroscopic system in equilibrium, there exists a group of functional relations among its properties.

2. The number of independent properties in each of these relations for a given system is usually equal to, and in no case greater than, a positive integer n.

3. The maximum number of independent properties n for a given system depends only on the specification of the system and is found by empirical observation.

The integer n is defined as the maximum number of independent properties for the system. In the case of a mixture of k different gases, for example, the independent properties include the amounts of each gas in the mixture. The composition of the mixture can be expressed in terms of mole fractions, $y_i (= N_i/N)$, where N_i is the number of moles of the ith gaseous constituent in the mixture and N is the total number of moles for the mixture. In other words,

$$N = \sum_{i=1}^{k} N_i \tag{13.1}$$

and

$$\sum_{i=1}^{k} y_i = 1 \tag{13.2}$$

Equation 13.2 implies that if the composition is expressed in terms of mole fractions, the additional independent properties are actually $k - 1$ since only this number of mole fractions need to be fixed to characterize the mixture completely. For the mixture, therefore, the functional relationship among properties is typically of the form

$$z = z(p, T, y_1, y_2, \ldots, y_{k-1}) \tag{13.3}$$

The property z is an intensive property. As indicated in Chapters 2 and 3, any extensive property Z can be expressed as an intensive property, known as the specific property z, by dividing the property value by the total mass (or amount) of the system. The specific property then counts the same way in Equation 13.3 as any other intensive property.

Ideal Gas Behavior

Gaseous systems at low pressures and high temperatures are known to behave as ideal gases. For a pure substance in the vapor phase, a low pressure condition corresponds to a pressure that is much lower than the critical pressure for the substance. For example, the critical pressure for water is about 218 atm, while for nitrogen it is 33.5 atm. Thus, both of these substances in the gaseous or vapor condition at ambient pressure of about 1 atm behave as an ideal gas at normal atmospheric temperatures.

Chapter 4 considered at length the laws governing the behavior of ideal gases. In dealing with mixtures of ideal gases, the fundamental assumption is that *each component individually obeys the laws for an ideal gas*. One can therefore determine the behavior of the mixture as an aggregate of the behaviors of the components. This principle is enshrined in the various additive laws for mixtures that are considered in Section 13.2. It may be helpful at this juncture to review some of the important characteristics of ideal gases that were introduced previously in Chapters 4, 8, and 9.

Ideal Gas pvT Rule

The formal definition of an *ideal gas* was given in Chapter 4 as any gaseous system that obeys the ideal gas pvT equation of state:

$$pv = RT \qquad (4.1)$$

or, on a molar basis,

$$p\bar{v} = \bar{R}T \qquad (4.13)$$

This definition provides a test as to whether or not a given gaseous system can be regarded as an ideal gas. Thus, if a given gaseous system obeys the ideal gas pvT equation of state over a specified range of conditions, the system is an ideal gas for those conditions. This approach is used in Section 13.2, in conjunction with certain additive laws, to establish the fact that *a mixture of ideal gases is also an ideal gas*.

Specific Heat at Constant Pressure (c_p) and at Constant Volume (c_v) for an Ideal Gas

Chapter 12 showed that for a system obeying the ideal gas pvT equation of state, the internal energy (u) and the enthalpy (h) depend only on the temperature and are independent of pressure. This means that the principal specific heat at constant volume (c_v) and at constant pressure (c_p) for an ideal gas depend only on temperature. Definitions of c_v and c_p in terms of temperature derivatives of u and h are given in Chapter 4. Similar definitions, on a molar basis, are given here.

First, the *molar specific internal energy* and the *molar specific enthalpy* can be expressed in terms of the specific internal energy u and the specific enthalpy h, respectively, according to the following:

$$\bar{u} = uM \tag{13.4}$$

and

$$\bar{h} = hM \tag{13.5}$$

where M is the molecular weight (or molar mass) of the substance. The *molar specific heats* at constant volume $(\bar{c}_v)$ and at constant pressure $(\bar{c}_p)$ for an ideal gas are then defined in a similar manner to Equations 4.20 and 4.21, which are on a gravimetric (or mass) basis:

$$\bar{c}_v = \frac{d\bar{u}}{dT} = c_v M \tag{13.6}$$

and

$$\bar{c}_p = \frac{d\bar{h}}{dT} = c_p M \tag{13.7}$$

The principal specific heats at constant volume and at constant pressure for ideal gases are designated hereafter as c_{v0} and c_{p0} as a reminder that a near-zero pressure condition should exist if the gas is to behave as an ideal gas. *Perfect gases* constitute a special category of ideal gases that, in addition to obeying the ideal gas rule, have constant values of c_{v0} and c_{p0}. As a first-level approximation, air–water vapor mixtures are often treated as mixtures of perfect gases over the typical temperature range encountered in psychrometry. *Psychrometry* is normally defined as the study of mixtures of air and water vapor at conditions that are near atmospheric.

Enthalpy and Internal Energy Changes for an Ideal Gas

For an ideal gas, h and u can be defined relative to an arbitrary reference temperature, T_{ref}. If h_{ref} thus denotes the enthalpy of the gas when its temperature is T_{ref}, the following expressions can be used for perfect gases for the enthalpy and the internal energy at any other temperature T:

$$h = c_{p0}(T - T_{ref}) + h_{ref} \tag{13.8}$$

and

$$u = c_{v0}(T - T_{ref}) + (h_{ref} - RT_{ref}) \tag{13.9}$$

The extra term $(h_{ref} - RT_{ref})$ on the right-hand side of Equation 13.9 arises from the fact that $h = u + pv$ by definition, and therefore both u and h cannot be independently assigned a value at the reference temperature. For ideal gases generally, the expressions to use are as follows:

$$h = \int_{T_{ref}}^{T} c_{p0}(T)\, dT + h_{ref} \tag{13.10}$$

and

$$u = \int_{T_{ref}}^{T} c_{v0}(T)\, dT + (h_{ref} - RT_{ref}) \tag{13.11}$$

Equations 13.8 through 13.11 can be written in equivalent molar forms. For perfect gases,

$$\bar{h} = \bar{c}_{p0}(T - T_{\text{ref}}) + \bar{h}_{\text{ref}} \tag{13.12}$$

and

$$\bar{u} = \bar{c}_{v0}(T - T_{\text{ref}}) + (\bar{h}_{\text{ref}} - \bar{R}T_{\text{ref}}) \tag{13.13}$$

For ideal gases in general,

$$\bar{h} = \int_{T_{\text{ref}}}^{T} \bar{c}_{p0}(T)\, dT + \bar{h}_{\text{ref}} \tag{13.14}$$

and

$$\bar{u} = \int_{T_{\text{ref}}}^{T} \bar{c}_{v0}(T)\, dT + (\bar{h}_{\text{ref}} - \bar{R}T_{\text{ref}}) \tag{13.15}$$

Entropy Change for an Ideal Gas

The reference condition for entropy requires the specification of at least two properties. The two properties commonly chosen are the pressure (p_{ref}) and temperature (T_{ref}). The term s_{ref} denotes the entropy of the gas at the reference state. For a perfect gas, Equation 8.25 can be used to obtain an expression for the entropy at any other state:

$$s - s_{\text{ref}} = c_{p0} \ln \frac{T}{T_{\text{ref}}} - R \ln \frac{p}{p_{\text{ref}}} \tag{13.16}$$

For ideal gases generally, one uses Equation 8.26 instead, with c_{p0} defined as a function of the temperature (T):

$$s - s_{\text{ref}} = \int_{T_{\text{ref}}}^{T} \frac{c_{p0}(T)}{T}\, dT - R \ln \frac{p}{p_{\text{ref}}} \tag{13.17}$$

Chapter 8 defined $s^{\circ}(T)$ as follows:

$$s^{\circ}(T) = \int_{T_{\text{ref}}}^{T} \frac{c_{p0}(T)}{T}\, dT \tag{8.27}$$

Substituting this expression in Equation 13.17 gives the following alternative equation for the specific entropy of an ideal gas:

$$s - s_{\text{ref}} = s^{\circ}(T) - R \ln \frac{p}{p_{\text{ref}}} \tag{13.17a}$$

(Note that the ideal gas tables in Appendices A and B actually give values of $[s_{\text{ref}} + s^{\circ}(T)]$ for the standardized specific entropy at the standard pressure of 1 atm.)
The corresponding equations in molar terms are

$$\bar{s} - \bar{s}_{\text{ref}} = \bar{c}_{p0} \ln \frac{T}{T_{\text{ref}}} - \bar{R} \ln \frac{p}{p_{\text{ref}}} \tag{13.18}$$

for a perfect gas, and

$$\bar{s} - \bar{s}_{ref} = \int_{T_{ref}}^{T} \frac{\bar{c}_{p0}(T)}{T} dT - \bar{R} \ln \frac{p}{p_{ref}} \tag{13.19}$$

for an ideal gas that is not perfect.

Flow Exergy of an Ideal Gas

The exergy per unit mass in the steady flow of a perfect gas is designated x and is obtained by combining Equations 9.38, 9.40a, and 9.40c of Chapter 9:

$$x = \frac{\dot{X}}{\dot{m}} = c_{p0}\left((T - T_0) - T_0 \ln \frac{T}{T_0}\right) + RT_0 \ln \frac{p}{p_0} \tag{13.20}$$

where p_0 and T_0 are defined for the substance strictly at the dead state condition. For ideal gases generally, one uses Equations 9.38, 9.38a, and 9.40c:

$$x = \frac{\dot{X}}{\dot{m}} = \int_{T_0}^{T} c_{p0}(T)\left(1 - \frac{T_0}{T}\right) dT + RT_0 \ln \frac{p}{p_0} \tag{13.21}$$

In terms of molar quantities, the corresponding forms of Equations 13.20 and 13.21 are

$$\bar{x} = \frac{\dot{X}}{\dot{N}} = \bar{c}_{p0}\left((T - T_0) - T_0 \ln \frac{T}{T_0}\right) + \bar{R}T_0 \ln \frac{p}{p_0} \tag{13.22}$$

and

$$\bar{x} = \frac{\dot{X}}{\dot{N}} = \int_{T_0}^{T} \bar{c}_{p0}(T)\left(1 - \frac{T_0}{T}\right) dT + \bar{R}T_0 \ln \frac{p}{p_0} \tag{13.23}$$

13.2   Additive Laws for Ideal Gas Mixtures

The additive laws for mixtures of ideal gases are essentially empirical and can be grouped into two categories. The first set includes *Dalton's law of additive pressures* and *Amagat–Leduc's law of additive volumes*, which lead to the establishment of the pVT relationship for the mixture based on the corresponding relationships for the components of the mixture. The second set of principles deals with the estimation of the other (extensive) thermodynamic properties U, H, S, and so on, for the mixture from their values for the constituents of the mixture. The additive law in this instance is the *Gibbs–Dalton law*.

13.2.1   pVT Relationship for Ideal Gas Mixtures

Let's first define and clarify certain concepts commonly used in the analysis of gaseous mixtures. Consider a mixture of gases contained in a vessel of volume V at a uniform temperature T and pressure p.

Partial Pressure and Pressure Analysis

The *partial pressure* p_i, due to a typical component i in the mixture, is defined as the pressure that the component i would exert if it alone occupied the total volume V at temperature T. A *pressure analysis* for a gaseous mixture involves establishing the ratio of partial pressure to the total pressure p_i/p for each component of the mixture.

Partial Volume and Volumetric Analysis

The partial volume V_i for the typical component i is the volume that the component occupies if held at the total pressure p and temperature T for the entire mixture. A *volumetric analysis* defines V_i/V for all the components of the mixture.

Molar and Gravimetric Analysis

If N_i denotes the number of moles of a typical component i in the mixture and N is the total number of moles of the entire mixture, the *mole fractions* N_i/N for all the components give the *molar analysis* for the mixture. Likewise, a *gravimetric analysis* defines the *mass fractions* m_i/m for all the components of the mixture, where m_i denotes the mass of a typical component i in the mixture and m is the total mass of the mixture. With the possible exception of psychrometrics, problems that involve mixtures are usually easier to solve when based on volumetric, pressure, or molar analysis than on gravimetric analysis. Dalton's law of additive pressures and Amagat–Leduc's law of partial volumes can now be stated, from which the important theorem can be established that for a mixture of ideal gases, the partial pressure, volumetric, and molar analyses are equivalent.

Dalton's Law of Additive Pressures

Dalton's law of additive pressures states that the total pressure p of a mixture of ideal gases is the sum of the partial pressures p_i of its constituents. In other words, if the number of gas constituents in the mixture is k, the total pressure is given by

$$p = \sum_{i=1}^{k} p_i \tag{13.24}$$

Each ideal gas constituent must obey the ideal gas pVT Equation 4.14 with the gas pressure taken as the partial pressure p_i, the volume taken as the volume V for the mixture, and the temperature T as that for the mixture. In this case,

$$p_i = \frac{N_i \overline{R} T}{V} \tag{13.25}$$

This expression for the partial pressure of each constituent can be substituted in Equation 13.24 for Dalton's law of additive pressures to obtain

$$p = \sum_{i=1}^{k} p_i = \sum_{i=1}^{k} \frac{N_i \overline{R} T}{V} = \frac{\overline{R} T}{V} \sum_{i=1}^{k} N_i$$
$$= \frac{N \overline{R} T}{V} \tag{13.26}$$

The last step in Equation 13.26 follows from Equation 13.1, which indicates that the total number of moles for the mixture is the sum of the number of moles of the constituents of the mixture. It should be stressed that this is the case when the mixture is nonreacting and that for a *reactive mixture* the number of moles are not necessarily conserved if a chemical reaction takes place.

Equation 13.26 defines an ideal gas. This fact establishes that a mixture of ideal gases is also an ideal gas. Furthermore, dividing Equation 13.25 by 13.26 yields the important result that *a partial pressure analysis is identical to a molar analysis*:

$$\frac{p_i}{p} = \frac{N_i}{N} \tag{13.27}$$

Also, from Equations 13.1 and 13.24 one has

$$\sum_{i=1}^{k} \frac{N_i}{N} = \sum_{i=1}^{k} \frac{p_i}{p} = 1 \tag{13.28}$$

The ideal gas characteristics R and M for the mixture can be determined by making use of the relationship between the number of moles (N), the mass (m), and the molecular weight (M) of a substance. Thus, if the molar analysis of the mixture is known and the molecular weight for each constituent is given, the following equation can be used to determine the molecular weight of the mixture:

$$M = \frac{m}{N} = \frac{\sum_{i=1}^{k} m_i}{N} = \frac{\sum_{i=1}^{k} N_i M_i}{N}$$
$$= \sum_{i=1}^{k} \frac{N_i}{N} M_i \tag{13.29}$$

From a knowledge of the molecular weight, the specific gas constant for the mixture is easily determined from the following equation, which was given in Chapter 4 as Equation 4.11:

$$R = \frac{\overline{R}}{M} \tag{13.30}$$

If, however, the composition of the mixture is given in terms of mass fractions, the following alternative expression can be used for computing M:

$$M = \frac{m}{N} = \frac{m}{\sum_{i=1}^{k} N_i} = \frac{m}{\sum_{i=1}^{k} \frac{m_i}{M_i}}$$
$$= \frac{1}{\sum_{i=1}^{k} \frac{m_i}{m} \frac{1}{M_i}} \tag{13.31}$$

An expression for the specific gas constant R of the mixture is obtained in terms of the mass fractions and the specific gas constants of the constituents by eliminating M between Equations 13.30 and 13.31:

$$R = \overline{R} \sum_{i=1}^{k} \frac{m_i}{m} \frac{1}{M_i} = \sum_{i=1}^{k} \frac{m_i}{m} \frac{\overline{R}}{M_i} = \sum_{i=1}^{k} \frac{m_i}{m} R_i \qquad (13.32)$$

The last step in Equation 13.32 is based on the application of Equation 13.30 to each ideal gas constituent.

Amagat–Leduc's Law of Additive Volumes

Amagat–Leduc's law of additive volumes states that the volume V occupied by a mixture of ideal gases is the sum total of the partial volumes V_i of its constituents. In other words,

$$V = \sum_{i=1}^{k} V_i \qquad (13.33)$$

Again, since each constituent of the mixture is an ideal gas, the following relationship must be true for V_i in terms of N_i (for the ith component), the pressure p, and the temperature T for the mixture, where V_i is the partial volume of each component as defined earlier:

$$V_i = \frac{N_i \overline{R} T}{p} \qquad (13.34)$$

Combining Equations 13.33 and 13.34 shows again that the mixture of ideal gases is an ideal gas:

$$V = \sum_{i=1}^{k} V_i = \sum_{i=1}^{k} \frac{N_i \overline{R} T}{p} = \frac{\overline{R} T}{p} \sum_{i=1}^{k} N_i$$
$$= \frac{N \overline{R} T}{p} \qquad (13.35)$$

Combining Equations 13.34 and 13.35 gives

$$\frac{V_i}{V} = \frac{N_i}{N} \qquad (13.36)$$

Also, combining Equations 13.1 and 13.33 gives

$$\sum_{i=1}^{k} \frac{N_i}{N} = \sum_{i=1}^{k} \frac{V_i}{V} = 1 \qquad (13.37)$$

In conclusion, *a volumetric analysis is equivalent to a molar analysis, which is equivalent to a partial pressure analysis.*

The following examples demonstrate how the properties for a mixture of ideal gases can be determined from those for the components in the mixture using the previous equations.

EXAMPLE 13.1

A sample of the combustion products of methane in air comprises the following gaseous constituents: $N_{CO_2} = 1$ kmol, $N_{H_2O} = 2$ kmol, $N_{O_2} = 3$ kmol, and $N_{N_2} = 18.8$ kmol.

(a) If the mixture is at a pressure of 100 kPa, determine the partial pressure of each constituent.

(b) What is the temperature at which the water vapor will start to condense if the pressure of the mixture remains at 100 kPa?

Given: A mixture with molar composition as follows:

$$N_{CO_2} = 1 \text{ kmol}, N_{H_2O} = 2 \text{ kmol}, N_{O_2} = 3 \text{ kmol}, N_{N_2} = 18.8 \text{ kmol}$$
$$p = 100 \text{ kPa}$$

Find:

(a) each p_i (for each constituent)

(b) T for start of water vapor condensation

Solution:

(a) Equation 13.27 indicates that the partial pressure analysis p_i/p is identical to the molar analysis N_i/N for a mixture of ideal gases. The principle is used to determine the partial pressure for each constituent using the given molar composition of the mixture. It is usually most convenient to set out the solution using a tabular format, which has been done in this case. The solution is given in Table 13.1 and is summarized as follows:

$$p_{CO_2} = 4.03 \text{ kPa}, \ p_{H_2O} = 8.06 \text{ kPa}, \ p_{O_2} = 12.1 \text{ kPa}, \ p_{N_2} = 75.8 \text{ kPa}$$

(b) The temperature at which the water vapor starts to condense is the saturation temperature T_s which corresponds to the prevailing partial pressure of the water vapor in the mixture. Using the **STEAM** code, the saturation temperature at $p_s = 8.06$ kPa is 41.7°C. Hence, the temperature at which the water vapor starts to condense is 41.7°C. This temperature has a special name in psychrometry; it is called the $\overline{dew\ point}$ *temperature* and is discussed further in Section 13.3.

TABLE 13.1

Calculation of the Partial Pressures of Gases in Example 13.1			
Gas	**N_i (kmol)**	**N_i/N ($\equiv p_i/p$)** **or $N_i/\Sigma N_i$**	**p_i (kPa)** **or $(p_i/p) \times 100$ kPa**
CO_2	1	0.0403	4.03
H_2O	2	0.0806	8.06
O_2	3	0.1210	12.1
N_2	18.8	0.7581	75.81
Totals	24.8	1.000	100.00

EXAMPLE 13.2

The gravimetric analysis of air is given approximately as follows:

Constituent	Molecular Weight	Gravimetric Analysis (%)
N_2	28	75.53
O_2	32	23.14
Ar	39.9	1.28
CO_2	44	0.05

Determine R and the mean molecular weight (M) for air.

Given: Air with composition and gravimetric analysis (m_i/m) indicated in the statement of the problem.

Find: R and M for the air.

Solution: The problem can be approached in several ways. In general, the use of a tabular format for the solution is recommended. Since both the gravimetric analysis (m_i/m) and the molecular weights (M_i) are given for each constituent in the mixture, Equation 13.31 can be used to calculate M for the mixture. Once M has been computed, the specific gas constant R for the mixture can be determined using Equation 13.30. The solution for M based on this approach is set out in Table 13.2. The value of M thus computed is 29.0 kg/kmol.

The specific gas constant is calculated using Equation 13.30 and is

$$R = \frac{\bar{R}}{M} = \frac{8.3144 \text{ kJ/kmol} \cdot \text{K}}{29.0 \text{ kg/kmol}} = \underline{0.287 \text{ kJ/kg} \cdot \text{K}}$$

TABLE   13.2

Computation of M and R for the Gas Mixture of Example 13.2

Constituent	M_i (kg/kmol)	$\dfrac{m_i}{m}$	$\dfrac{1}{M_i}\left(\dfrac{m_i}{m}\right)$	
N_2	28	0.7553	0.0270	
O_2	32	0.2314	0.0072	$M = \dfrac{1}{\Sigma \dfrac{1}{M_i}\dfrac{m_i}{m}}$
Ar	39.9	0.0128	0.0003	
CO_2	44	0.0005	0.0000	
Totals		1.00	0.0345	$= 29.0$ kg/kmol

1. 1 kmol of carbon monoxide is mixed with 3 kmol of N_2. The total pressure of the mixture is 100 kPa and the temperature is 300 K. Calculate
   (a) the partial pressure for each constituent
   (b) the molecular weight (M) for the mixture
   (c) the specific gas constant (R) for the mixture
   (d) the total mass of the mixture

   Answer: (a) p_{CO} = 25 kPa; p_{N_2} = 75 kPa; (b) 28 kg/kmol; (c) 0.297 kJ/kg · K;
   (d) 112 kg

2. 1 kmol of carbon monoxide is mixed with 0.5 kmol of oxygen. The mixture is at 100 kPa and 25°C. Determine the volume of the mixture.

   Answer: 37.2 m³

3. The carbon monoxide in Exercise 2 reacts with the oxygen (O_2) forming 1 kmol of carbon dioxide. The product of reaction is cooled to 25°C, while the pressure remains at 100 kPa. What is the final volume of this product?

   Answer: 24.8 m³

13.2.2 Gibbs–Dalton Law for Ideal Gas Mixtures

The *Gibbs–Dalton law* has a particular bearing on analyses involving the application of the first and second laws of thermodynamics to gaseous mixtures. The specific internal energy u_i, the specific enthalpy h_i, and the specific entropy s_i are defined for a typical component i when it alone occupies the entire volume V of the mixture at the prevailing temperature T. Recall that a partial pressure p_i is assigned to the component under the condition described. For an ideal gas, both the specific internal energy and the specific enthalpy depend on temperature only while the specific entropy is a function of both the partial pressure and temperature. Thus, Equations 13.8 to 13.11 can be used to determine the specific enthalpy and the specific internal energy merely by substituting the appropriate specific heat for the component involved. For specific entropy, the partial pressure p_i must be substituted for p in Equations 13.16 and 13.17 along with the appropriate specific heat at constant pressure and the specific gas constant for the constituent of the mixture.

The Gibbs–Dalton law gives the internal energy U, the enthalpy H, and the entropy S of the mixture of ideal gases as the sum of the aggregates for the k constituents of the mixture. In other words, for a mixture of ideal gases, U, H, and S are given by

Internal energy additivity

$$U = mu = \sum_{i=1}^{k} m_i u_i \tag{13.38}$$

Enthalpy additivity

$$H = mh = \sum_{i=1}^{k} m_i h_i \tag{13.39}$$

Entropy additivity

$$S = ms = \sum_{i=1}^{k} m_i s_i \tag{13.40}$$

From the set of Equations 13.38 to 13.40, one can derive expressions for specific internal energy, enthalpy, and entropy for the mixture which are similar to those for a single gas, provided the reference condition (p_{ref}, T_{ref}) is the same for all the gases in the mixture. The derivation is set out here for the case when the constituents of the mixture are all perfect gases. Thus, from Equations 13.9, 13.32, and 13.38, one can establish the following:

$$
\begin{aligned}
u &= \sum_{i=1}^{k} \frac{m_i}{m} c_{v0,i}(T - T_{ref}) + \sum_{i=1}^{k} \frac{m_i}{m}(h_{i,ref} - R_i T_{ref}) \\
&= c_{v0}(T - T_{ref}) + (h_{ref} - RT_{ref})
\end{aligned}
\tag{13.41}
$$

where

$$
c_{v0} = \sum_{i=1}^{k} \frac{m_i}{m} c_{v0,i}
\tag{13.42}
$$

and

$$
h_{ref} = \sum_{i=1}^{k} \frac{m_i}{m} h_{i,ref}
\tag{13.43}
$$

Equation 13.32 was used for the specific gas constant of the mixture in arriving at the final expression for the specific internal energy of the mixture in a form that is identical to that given by Equation 13.9 for a single ideal gas. By a similar method, one can use Equations 13.8 and 13.39 to obtain the following expression for the specific enthalpy of the mixture:

$$
\begin{aligned}
h &= \sum_{i=1}^{k} \frac{m_i}{m} c_{p0,i}(T - T_{ref}) + \sum_{i=1}^{k} \frac{m_i}{m} h_{i,ref} \\
&= c_{p0}(T - T_{ref}) + h_{ref}
\end{aligned}
\tag{13.44}
$$

where

$$
c_{p0} = \sum_{i=1}^{k} \frac{m_i}{m} c_{p0,i}
\tag{13.45}
$$

and h_{ref} is given by Equation 13.43. The final expression for the specific enthalpy of the mixture is identical to that for a single ideal gas given by Equation 13.8.

The specific entropy of the mixture is obtained by using Equation 13.16 for each constituent and the Gibbs–Dalton law, Equation 13.40:

$$
\begin{aligned}
s &= \sum_{i=1}^{k} \frac{m_i}{m} c_{p0,i} \ln \frac{T}{T_{ref}} - \sum_{i=1}^{k} \frac{m_i}{m} R_i \ln \frac{p_i}{p_{ref}} + \sum_{i=1}^{k} \frac{m_i}{m} s_{i,ref} \\
&= c_{p0} \ln \frac{T}{T_{ref}} - R \ln \frac{p}{p_{ref}} + s_{ref} - \sum_{i=1}^{k} \frac{m_i}{m} R_i \ln \frac{p_i}{p}
\end{aligned}
\tag{13.46}
$$

where

$$
s_{ref} = \sum_{i=1}^{k} \frac{m_i}{m} s_{i,ref} = \sum_{i=1}^{k} \frac{m_i}{m} s_i \, (p_{ref}, T_{ref})
\tag{13.47}
$$

A comparison of Equation 13.46 with Equation 13.16 for a single ideal gas shows an additional term for the entropy of the gaseous mixture. Since the ratio of partial pressure to the total pressure of the mixture is a fraction less than unity, the last term in Equation 13.46 is positive. The term represents the *entropy of mixing*, and it corresponds to the entropy increase that occurs when the k gases, each separately occupying a volume V at a temperature T (and with a pressure equal to p_i), are mixed such that the final volume is V and the temperature is T. Such a process is irreversible.

Equations 13.38 to 13.40 expressing the Gibbs–Dalton law for a mixture of ideal gases can also be written on a molar basis as follows:

Internal energy additivity

$$U = N\bar{u} = \sum_{i=1}^{k} N_i \bar{u}_i \tag{13.48}$$

Enthalpy additivity

$$H = N\bar{h} = \sum_{i=1}^{k} N_i \bar{h}_i \tag{13.49}$$

Entropy additivity

$$S = N\bar{s} = \sum_{i=1}^{k} N_i \bar{s}_i \tag{13.50}$$

Likewise, Equations 13.41 to 13.47 can be written on a molar basis as follows:

$$\bar{u} = \sum_{i=1}^{k} \frac{N_i}{N} \bar{c}_{v0,i}(T - T_{\text{ref}}) + \sum_{i=1}^{k} \frac{N_i}{N}(\bar{h}_{i,\text{ref}} - \bar{R}T_{\text{ref}})$$
$$= \bar{c}_{v0}(T - T_{\text{ref}}) + (\bar{h}_{\text{ref}} - \bar{R}T_{\text{ref}}) \tag{13.51}$$

where

$$\bar{c}_{v0} = \sum_{i=1}^{k} \frac{N_i}{N} \bar{c}_{v0,i} \tag{13.52}$$

and

$$\bar{h}_{\text{ref}} = \sum_{i=1}^{k} \frac{N_i}{N} \bar{h}_{i,\text{ref}} \tag{13.53}$$

For the molar specific enthalpy,

$$\bar{h} = \sum_{i=1}^{k} \frac{N_i}{N} \bar{c}_{p0,i}(T - T_{\text{ref}}) + \sum_{i=1}^{k} \frac{N_i}{N} \bar{h}_{i,\text{ref}}$$
$$= \bar{c}_{p0}(T - T_{\text{ref}}) + \bar{h}_{\text{ref}} \tag{13.54}$$

where

$$\bar{c}_{p0} = \sum_{i=1}^{k} \frac{N_i}{N} \bar{c}_{p0,i} \tag{13.55}$$

The expression for the molar specific entropy of the mixture is

$$\begin{aligned}
\bar{s} &= \sum_{i=1}^{k} \frac{N_i}{N} \bar{c}_{p0,i} \ln \frac{T}{T_{\text{ref}}} - \sum_{i=1}^{k} \frac{N_i}{N} \bar{R} \ln \frac{p_i}{p_{\text{ref}}} + \sum_{i=1}^{k} \frac{N_i}{N} \bar{s}_{i,\text{ref}} \\
&= \bar{c}_{p0} \ln \frac{T}{T_{\text{ref}}} - \bar{R} \ln \frac{p}{p_{\text{ref}}} + \bar{s}_{\text{ref}} - \bar{R} \sum_{i=1}^{k} \frac{N_i}{N} \ln \frac{p_i}{p}
\end{aligned} \tag{13.56}$$

where

$$\bar{s}_{\text{ref}} = \sum_{i=1}^{k} \frac{N_i}{N} \bar{s}_{i,\text{ref}} = \sum_{i=1}^{k} \frac{N_i}{N} \bar{s}_i \, (p_{\text{ref}}, T_{\text{ref}}) \tag{13.57}$$

EXAMPLE 13.3

The approximate volumetric composition of air can be taken as 79% atmospheric nitrogen ($M = 28.15$) and 21% oxygen ($M = 32$). The molar specific heat at constant pressure is 29 kJ/kmol · K for atmospheric nitrogen and 29.4 kJ/kmol · K for oxygen. Determine

(a) the molar specific heat at constant pressure for air

(b) the specific gas constant for air

(c) c_p and c_v for the air sample

Given: Air with the following composition and properties:

Gas	(V_i/V)	M_i	$\bar{c}_{p0}$ (kJ/kmol · K)
N_2	0.79	28.15	29
O_2	0.21	32	29.4

Find: (a) molar specific heat for air, (b) R for the air, (c) c_{p0}, c_{v0} for air

Solution:
(a) The molar specific heat for the mixture can be determined using Equation 13.55 once the molar analysis is known. From Equation 13.36 one can equate the molar analysis to the volumetric analysis since they are identical for a mixture of ideal gases. Hence,

$$\begin{aligned}
\bar{c}_{p0} &= \sum_{i=1}^{k} \frac{N_i}{N} \bar{c}_{p0,i} = \sum_{i=1}^{k} \frac{V_i}{V} \bar{c}_{p0,i} \\
&= (0.79)(29) + (0.21)(29.4) \text{ kJ/kmol · K} \\
&= \underline{29.1 \text{ kJ/kmol · K}}
\end{aligned}$$

(b) After calculating M using Equation 13.29, one can evaluate R using Equation 13.30:

$$M = \sum_{i=1}^{k} \frac{N_i}{N} M_i$$

$$= (0.79)(28.15) + (0.21)(32) = 28.96 \text{ kg/kmol}$$

Therefore,

$$R = \frac{\overline{R}}{M} = 0.287 \text{ kJ/kg} \cdot \text{K}$$

(c) The specific heat values can be calculated once the molecular weight and the molar specific heat values of the mixture are known. Thus,

$$c_{p0} = \frac{\overline{c}_{p0}}{M}$$

$$= \frac{29.1}{28.96} = 1.005 \text{ kJ/kg} \cdot \text{K}$$

The specific heat at constant volume can be calculated in a similar way. Alternatively, one could use Equation 4.26, which expresses the relationship between the principal specific heats and R for an ideal gas. Equation 4.26 therefore gives

$$c_{v0} = c_{p0} - R$$

$$= 1.005 - 0.287 = 0.718 \text{ kJ/kg} \cdot \text{K}$$

Reference Condition for Enthalpy and Entropy

Perhaps the point should be reiterated that the same reference condition p_{ref}, T_{ref} must be used for all the constituents of the ideal gas mixture in the evaluation of internal energy, enthalpy, and entropy. When dealing with substances that are involved in chemical reactions, the *standard reference condition* specified is a temperature of 25°C and a pressure of 1 atm. The standardized enthalpy for an element is assigned a zero value at the standard reference condition, while for compounds, it is defined as the enthalpy of formation of the substance. This subject is considered in more detail in Chapter 14, which deals with the thermodynamics of combustion.

For nonreacting mixtures, the reference condition p_{ref}, T_{ref} can be chosen arbitrarily as long as it is the same for all the constituents of the mixture, as indicated earlier. However, the reference point enthalpy and entropy for each constituent do not have to be assigned a zero value. This can be illustrated by considering air–water vapor mixtures, for example. When dealing with air–water vapor mixtures, remember that the tabulated values of enthalpy and entropy given in steam tables normally assume a zero value for the enthalpy and entropy of liquid water at the triple point state for water. Thus, if consistency with steam table property values is desired, an appropriate choice of reference point would be T_{ref} equal to the triple point temperature of water T_t ($= 273.16$ K) and the triple point pressure of water p_t ($= 0.6117$ kPa). The corresponding h_{ref} and s_{ref} for water vapor are then given by the triple point values of h_g (= 2501 kJ/kg) and s_g (= 9.155 kJ/kg · K). For dry air, consistency may be desired with the property values indicated in the air property tables in Appendices A and B. Then the reference point enthalpy and entropy must be such that, when substituted in Equations 13.44 and 13.46, the tabulated values are obtained for the standard reference condition previously mentioned. For dry air, the enthalpy and entropy at the standard temperature of 25°C and pressure of 1 atm are 0 kJ/kg and 6.696 kJ/kg · K. Therefore,

$$h_{air,ref} = h\ (1\ atm,\ 25°C) - c_{po}(298.15 - 273.16)$$

$$= 0 - 1.005(24.99) = -25.1\ kJ/kg$$

and

$$s_{air,ref} = s\ (1\ atm,\ 25°C) - 1.005\ \ln\frac{298.15}{273.16} + 0.287\ \ln\frac{101.3}{0.6117}$$

$$= 6.696 - 0.088 + 1.466 = 8.074\ kJ/kg \cdot K$$

The thermodynamic analysis of processes involving nonreacting mixtures, however, most frequently involves the computation of changes in U, H, or S and may not require knowledge of the reference point property values.

EXAMPLE 13.4

Oxygen at 100 kPa and 300 K is contained in a rigid vessel of volume 0.5 m³. Another rigid container of capacity 2 m³ holds carbon monoxide at 200 kPa and 400 K. Both containers are well insulated thermally and are connected by a valve. If the valve is opened, determine the following for the system comprising the oxygen and the carbon monoxide:

(a) the final temperature and pressure of the system

(b) the change in entropy of the system

Given: The closed system of Figure 13.1 comprises 0.5 m³ of oxygen at 100 kPa and 300 K and 2 m³ of carbon monoxide at 200 kPa and 400 K, each contained in thermally insulated rigid vessels. The connecting valve is opened and the gases are mixed.

Find: (a) T_2, p_2, (b) $S_2 - S_1$

Solution: The initial amount of oxygen and carbon monoxide can be determined by applying the ideal gas pVT Equation 4.14:

$$N_{O_2} = \frac{p_{O_2}V_{O_2}}{\overline{R}T_{O_2}} = \frac{(100(0.5)}{(8.3144)(300)} = 0.02\ kmol$$

and

$$N_{CO} = \frac{p_{CO}V_{CO}}{\overline{R}T_{CO}} = \frac{(200)(2)}{(8.3144)(400)} = 0.1203\ kmol$$

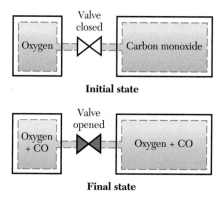

Figure 13.1 Adiabatic mixing of gases contained in rigid vessels.

Thus, for the system comprising oxygen plus carbon monoxide, the number of moles $N = 0.1403$ kmol.

Ideal gas properties for carbon monoxide and oxygen are given in Tables A.16 and A.18, respectively. The reference pressure and temperature assumed are 1 atm and 298 K, respectively. At this reference condition, the tables give

$$\bar{h}_{O_2,\text{ref}} = 0, \quad \bar{s}_{O_2,\text{ref}} = 205.06 \text{ kJ/kmol} \cdot \text{K}$$

and

$$\bar{h}_{CO,\text{ref}} = -110{,}523 \text{ kJ/kmol}, \quad \bar{s}_{CO,\text{ref}} = 197.91 \text{ kJ/kmol} \cdot \text{K}$$

Also, in the temperature range of interest, the molar specific heat values are

$$\bar{c}_{p0,O_2} = 29.5 \text{ kJ/kmol} \cdot \text{K}$$

and

$$\bar{c}_{p0,CO} = 29.2 \text{ kJ/kmol} \cdot \text{K}$$

These properties are needed in the solution to parts (a) and (b) of the problem.

(a) The NFEE (Eq. 6.10) can be applied to the nonflow adiabatic mixing process. Thus, neglecting changes in the kinetic energy and potential energy terms, one obtains

$$Q_{12} - W_{12} = U_2 - U_1$$

In this case, Q_{12} is zero since the vessels are well insulated, and $W_{12} = 0$ (both vessels are rigid and no displacement or shaft work is done). Thus, application of the NFEE results in $U_2 = U_1$.

Equation 13.13 can be used to determine the molar specific internal energy at the initial state for the oxygen and the carbon monoxide separately. First, however, let's calculate the molar specific heat at constant volume for each gas using the molar form of Equation 4.26:

$$\bar{c}_{v0} = \bar{c}_{p0} - \bar{R}$$

$$\therefore \bar{c}_{v0,O_2} = 21.2 \text{ kJ/kmol} \cdot \text{K}$$

and

$$\bar{c}_{v0,CO} = 20.9 \text{ kJ/kmol} \cdot \text{K}$$

The molar specific internal energy at the initial state for each gas is thus calculated from Equation 13.13 as follows:

$$\bar{u} = \bar{c}_{v0}(T - T_{\text{ref}}) + (\bar{h}_{\text{ref}} - \bar{R}T_{\text{ref}})$$

$$\therefore \bar{u}_{O_2} = -2435 \text{ kJ/kmol}$$

and

$$\bar{u}_{CO} = -110{,}869 \text{ kJ/kmol}$$

The total internal energy at state 1 is calculated by

$$U_1 = N_{O_2}\bar{u}_{O_2} + N_{CO}\bar{u}_{CO} = -13{,}386 \text{ kJ}$$

Since $U_2 = U_1$, it follows that the molar specific internal energy of the mixture at the final state must be $-13{,}386/0.1403 = -95{,}411$ kJ/kmol.

For the mixture, Equation 13.51 gives an independent expression for the molar internal energy in terms of the final temperature T_2. The molar specific heat at constant volume for the mixture which appears in the equation can be found by applying Equation 13.52:

$$\bar{c}_{v0} = \frac{0.02}{0.1403}(21.2) + \frac{0.1203}{0.1402}(20.9) = 20.9 \text{ kJ/kmol} \cdot \text{K}$$

Likewise, the reference value specific enthalpy for the mixture is calculated by substituting the values for the constituents in Equation 13.53:

$$\bar{h}_{\text{ref}} = \frac{0.02}{0.1403}(0) + \frac{0.1203}{0.1402}(-110{,}523) = -94{,}768 \text{ kJ/kmol}$$

Substituting these values in Equation 13.51 gives

$$\bar{u}_2 = 20.9(T_2 - 298) - 97{,}245 \text{ kJ/kmol}$$

Equating this to $-95{,}411$ kJ/kmol determined earlier gives a final temperature for the system of 386 K.

The final pressure is calculated from Equation 4.14 by substituting $T_2 = 386$ K, $V_2 = 2.5$ m³, and $N = 0.1403$ kmol. The final pressure is thus 180 kPa.

(b) S_1 is calculated based on the molar specific entropies for the oxygen and carbon monoxide at the initial condition. Applying Equation 13.18 yields the following molar specific entropies:

$$\bar{s}_{O_2} = 29.5 \ln \frac{300}{298} - 8.3144 \ln \frac{100}{101.3} + 205.06 = 205.37 \text{ kJ/kmol} \cdot \text{K}$$

and

$$\bar{s}_{CO} = 29.2 \ln \frac{400}{298} - 8.3144 \ln \frac{200}{101.3} + 197.91 = 200.85 \text{ kJ/kmol} \cdot \text{K}$$

Hence, one can determine S_1 by the following:

$$S_1 = N_{O_2}\bar{s}_{O_2} + N_{CO}\bar{s}_{CO} = 28.27 \text{ kJ/K}$$

The calculation of S_2 requires a determination of the molar specific entropy at the final state from Equation 13.56. For this, the molar specific heat at constant pressure for the mixture is calculated using Equation 13.55, which gives a value of 29.2 kJ/kmol $\cdot$ K. The reference molar specific entropy for the mixture obtained by applying Equation 13.57 is 198.93 kJ/kmol $\cdot$ K. Substituting these in Equation 13.56 gives the following for the molar specific entropy of the mixture:

$$\bar{s} = 29.2 \ln \frac{386}{298} - 8.3144 \ln \frac{180}{101.3} + 198.93$$

$$- 8.3144 \left(\frac{0.02}{0.1403} \ln \frac{0.02}{0.1403} + \frac{0.1203}{0.1403} \ln \frac{0.1203}{0.1403} \right) \text{ kJ/kmol} \cdot \text{K}$$

$$= 205.1 \text{ kJ/kmol} \cdot \text{K}$$

Thus, the total entropy at the final state S_2, which is the number of moles times the molar specific entropy of the mixture, is 28.77 kJ/K. The increase in entropy is thus $28.77 - 28.27 = \underline{0.5 \text{ kJ/K}}$.

Comments

1. The positive change in entropy confirms that the adiabatic mixing process is irreversible.

2. Only a slight increase in entropy is indicated, however, because a relatively small amount of oxygen is mixed with the carbon monoxide. If the relative proportions of oxygen and carbon monoxide had been nearly equal, a more substantial increase in entropy would have been obtained.

3. A classic paradox is encountered when one tries to apply Equation 13.56 to the mixing of an identical ideal gas species. Mathematically, the entropy of mixing term in Equation 13.56 is positive and nonzero even when only one type of gas is involved. Physically, however, there is no such entropy of mixing, and the correct equation to use is Equation 13.18 for a single component system.

Flow Exergy for Mixtures

In the treatment of exergy in Chapter 9, a distinction was made between *restricted* equilibrium (requiring only thermal and mechanical equilibrium between a system and the reference environment at the dead state) and *unrestricted* equilibrium, which also requires chemical equilibrium. When dealing with mixtures of substances, the equilibrium is generally unrestricted, and a proper identification of what Wepfer and Gaggioli (1980) term a *stable reference environment* becomes most crucial. A stable reference environment is one that is internally at complete stable equilibrium and is also very large compared to all of the other systems together (Wepfer and Gaggioli, 1980). Further consideration of the dead state for chemically reactive materials is deferred to Section 14.7 (Chap. 14). The present chapter is concerned specifically with the appropriate expressions for the flow exergy of nonreactive mixtures, typical of which is moist air.

For moist air, a realistic choice for a stable reference environment is ambient (atmospheric) air, which is assumed to be in a stable equilibrium condition with total pressure p_0, temperature T_0, and chemical potential μ_{i0} (defined later in Chap. 15). For mixtures of perfect gases, in particular, the partial pressure p_{i0} for each component i in the stable ambient environment can be used in place of the chemical potential. The dead state for the moist air sample thus corresponds to a total pressure p_0 and temperature T_0 for the sample. In addition, the partial pressure for each component i in the sample must be equal to p_{i0}. The resulting expressions for a mixture of perfect gases, based on Equations 13.20 and 13.22, are

$$x = \frac{\dot{X}}{\dot{m}} = c_{p0}\left(T - T_0 - T_0 \ln \frac{T}{T_0}\right) + T_0 \sum_{i=1}^{k} \frac{m_i}{m} R_i \ln \frac{p_i}{p_{i0}} \qquad (13.58)$$

and

$$\bar{x} = \frac{\dot{X}}{\dot{N}} = \bar{c}_{p0}\left(T - T_0 - T_0 \ln \frac{T}{T_0}\right)$$
$$+ \bar{R}T_0 \ln \frac{p}{p_0} + \bar{R}T_0 \sum_{i=1}^{k} \frac{N_i}{N} \ln \frac{p_i/p}{p_{i0}/p_0} \qquad (13.59)$$

The derivation tacitly assumes that the mixture is nonreactive and that it is made up of ideal gases that remain in the gaseous form both in the given mixture and in the reference environment.

Flow Exergy of Separation for Ideal Gas Mixtures

Physical processes are frequently employed for separating gaseous mixtures into their constituents. In cryogenics, for example, air separation into oxygen and nitrogen is carried out by first liquifying the air and subsequently separating the components by evaporation at their different boiling points corresponding to a given pressure. From Equation 13.59 for the flow exergy of an ideal gas mixture, an expression can be derived for the minimum exergy needed for separating a mixture of ideal gases into its constituents. Consider the flow of N_0 kmol of a mixture of ideal gases at a pressure p_0 and temperature T_0. The flow includes N_{i0} kmol of the ith constituent of the mixture. Assume that the separation process produces pure gases, each of which is at the same pressure and temperature as the original mixture. The mixture can be chosen as reference environment, in which case its total pressure and temperature together with the partial pressures of the constituents of the mixture may be assumed in the application of Equation 13.59.

Since the mixture is the reference environment, its flow exergy must be zero. The minimum exergy ($\chi_{separate}$) needed for separating N_0 kmol of the mixture into pure species of its constituents is thus given by the sum of the flow exergy for each of the constituents after the separation process. Putting $T = T_0$ and $p = p_0$ in Equation 13.59 (and noting that $p_i/p = 1$ while $p_{i0}/p_0 = N_{i0}/N_0$ for each separated gas) thus leads to the following expression for $\chi_{separate}$:

$$\chi_{separate} = \overline{R}T_0 \sum_{i=1}^{k} N_{i0} \ln (N_0/N_{i0}) \tag{13.59a}$$

When only one gas (A) in a mixture is to be extracted pure, leaving the others mixed, and if both gas A and the remaining mixture of gases are each at pressure p_0 and temperature T_0, application of Equation 13.59 yields the following for the minimum exergy input needed for the extraction of N_A kmol of gas A present in the original mixture flow:

$$\chi_{extract} = \overline{R}T_0 \left(N_A \ln \frac{N_0}{N_A} + (N_0 - N_A) \ln \frac{N_0}{N_0 - N_A} \right) \tag{13.59b}$$

EXAMPLE 13.5

Estimate the minimum exergy supply required per unit mass of air in separating a steady stream of air at ambient pressure p_0 and temperature T_0 into its principal constituents (oxygen and nitrogen), each being at the ambient pressure and temperature condition.

Given: The hypothetical reversible-flow process of Figure 13.2 involves the separation of air at p_0 and T_0 into oxygen and nitrogen, each at p_0 and T_0 also.

Find: The minimum exergy supply per unit mass of air (x) required

Solution: The volumetric composition of air is approximately 21% oxygen and 79% nitrogen. Consider a steady-flow process as illustrated in Figure 13.2, which involves 100 kmol of air over a period of time. Thus, 21 kmol of oxygen and 79 kmol of nitrogen are produced in the separation process. Equation 13.59 can be used to determine the molar exergy flow per unit number of moles for the air, which is a

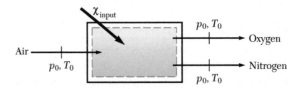

Figure 13.2 A hypothetical reversible-flow process for the separation of air into oxygen and nitrogen.

mixture of ideal gases. Since $T = T_0$, the thermal exergy component in Equation 13.59 is zero. The remaining component can be determined as follows:

$$\bar{x} = \frac{\dot{X}}{\dot{N}} = \bar{R}T_0 \sum_{i=1}^{k} \frac{N_i}{N} \ln \frac{p_i}{p_{i0}}$$

$$= \bar{R}T_0 \left(\frac{21}{100} \ln \frac{21}{21} + \frac{79}{100} \ln \frac{79}{79} \right) = 0$$

It should not be surprising that $\bar{x}$ for the air stream is zero. The air stream is ambient air, which is at the dead state and therefore possesses zero exergy. From Equation 13.59, by substituting the same values for p_i in the actual air stream sample as p_{i0} for the same constituent in the reference environment, a zero exergy is again obtained for the air stream.

For the single-component gases, one should still employ Equation 13.59 for the exergy flow since oxygen only exists in combination with nitrogen in the stable ambient environment. Thus, for the oxygen and the nitrogen one has

$$\chi_{O_2} = 21\bar{R}T_0 \ln \frac{p_0}{0.21p_0} = 32.8\bar{R}T_0 \text{ kJ}$$

and

$$\chi_{N_2} = 79\bar{R}T_0 \ln \frac{p_0}{0.79p_0} = 18.6\bar{R}T_0 \text{ kJ}$$

The exergy product is the sum of the exergy of the oxygen and the nitrogen. Taking the universal gas constant as 8.3144 kJ/kmol · K and assuming that T_0 is 300 K, the exergy product becomes 1.28×10^5 kJ per 100 kmol of air. The mass of 100 kmol of air is approximately (100)(29) kg. Thus, the exergy product per kg of air is 44.1 kJ. The minimum exergy input requirement corresponds to the case when all processes are reversible. This input is equal to the exergy product. Hence, the minimum exergy input required for the separation process is a little over 44.1 kJ/kg of air.

Note that an identical answer would be obtained by applying Equation 13.59a for the flow exergy of separation of a mixture of ideal gases. The reader is encouraged to verify this as an exercise.

EXERCISES

for Section 13.2.2

1. An ideal gas mixture comprises 1 kmol of carbon dioxide, 1 kmol of oxygen, and 7.52 kmol of nitrogen. Assume a reference condition of 1 atm pressure and 25°C. Use the ideal gas tables in Appendix A to determine the following for the mixture:
   (a) $\bar{h}_{\text{ref}}$
   (b) $\bar{s}_{\text{ref}}$
   (c) $\bar{c}_{p0}$ at 1000 K
   Answer: (a) -41.3 MJ/kmol, (b) 195.24 kJ/kmol · K, (c) 35.15 kJ/kmol · K

2. The ideal gas mixture in Exercise 1 is at a temperature of 1700 K. Use Equation 13.54 to determine the molar specific enthalpy for the mixture. Use the molar specific heat value obtained at 1000 K for this estimate.
   Answer: 7940 kJ/kmol

3. For each component of the ideal gas mixture given in Exercise 1, determine the molar specific enthalpy at 1700 K from the ideal gas tables in Appendix A. Apply the Gibbs–Dalton enthalpy additivity principle to determine the molar specific enthalpy for the mixture at 1700 K.

   Answer: 7300 kJ/kmol

4. Use Equation 13.56 to determine the molar specific entropy of the ideal gas mixture given in Exercise 1 at a temperature of 1700 K and a pressure of 1 atm.

   Answer: 261.93 kJ/kmol · K

5. Verify your answer in Exercise 4 by reading the values for the molar specific entropy of the individual components in the mixture and then applying the Gibbs–Dalton entropy additivity law.

   Answer: 252.27 kJ/kmol · K

6. For the ideal gas mixture of Exercise 1 at a temperature of 25°C and a pressure of 1 atm, determine
   (a) the molar specific enthalpy
   (b) the molar specific entropy

   Answer: (a) −41.3 MJ/kmol, (b) 200.73 kJ/kmol · K

7. Use your results from Exercises 3, 5, and 6 to determine (a) ΔH and (b) ΔS for the gas mixture in Exercise 1 when it is cooled from 1 atm and 1700 K to 1 atm and 298 K.

   Answer: (a) −463 MJ, (b) −491 kJ/K

13.3   Air–Vapor Mixtures

This section focuses on air–vapor mixtures. In *psychrometry*, the air–water vapor mixture is treated as a two-component system with air (a mixture of ideal gases) regarded as one component and water vapor as the other. The subscript "a" refers to air and "v" refers to the water vapor in the air–vapor mixture. The abbreviation "da" refers to dry air (that is, moisture-free air).

13.3.1   Humidity Parameters

Humidity, in the colloquial sense, connotes the presence of a significant amount of moisture or water vapor in the air. A comprehensive list of humidity parameters includes the following:

1. The *humidity ratio* (ω or hr) is a measure of the amount of moisture in the air.

2. The *degree of saturation* (μ) is the ratio of actual moisture held to the maximum amount that the air would hold at the saturated condition.

3. The *relative humidity* (ϕ or rh) provides another measure of how close a given sample of moist air is to the saturated condition.

4. The *dew point temperature* (T_d) is the temperature at which moisture begins to condense out of a moist air sample cooled at a constant pressure.

5. The *dry bulb temperature* (T_{db}) is the actual temperature of the air sample. The term derives from the fact that this is the temperature measured by the dry bulb thermometer of a hygrometer traditionally used for the measurement of the humidity of moist air. Moist air temperature can be designated as dry bulb by appending the abbreviation "db" after the temperature, for example, 20°C db.

6. The *wet bulb temperature* (T_{wb}) is the temperature indicated on the thermometer of a hygrometer whose bulb is kept wet by a piece of wet cloth. Formal definition of the *thermodynamic wet bulb temperature*, also referred to as the *adiabatic saturation temperature* (T^*), is given later. It is found in practice that the thermodynamic wet bulb temperature is the same as that measured using the wet bulb thermometer of the hygrometer. The abbreviation "wb" is often used to indicate a wet bulb temperature.

Relationships among some of these parameters can be established using the ideal gas rule and the empirical laws given in Section 13.2 for ideal gas mixtures. These humidity parameters are now formally defined and discussed.

Humidity Ratio

The *humidity ratio* (ω or hr) of moist air is defined as the ratio of the mass of water vapor (m_v) in air to the mass of dry air (m_a). Thus,

$$\omega = \frac{\text{mass of water vapor}}{\text{mass of dry air}} = \frac{m_v}{m_a} \tag{13.60}$$

Let N_a and N_v denote the corresponding number of moles of dry air and water vapor in the mixture, respectively. The humidity ratio can be expressed in terms of mole fractions, and hence, in terms of the ratio of partial pressures of air and water vapor, as follows:

$$\omega = \frac{m_v}{m_a} = \frac{N_v M_v}{N_a M_a} = \frac{M_v}{M_a} \frac{N_v/N}{N_a/N} \tag{13.61}$$

where N ($= N_v + N_a$) is the total number of moles of the air–water vapor mixture. Equation 13.27 expresses the identity between a molar analysis and a partial pressure analysis. Thus, if p_v and p_a denote the partial pressures of the water vapor and dry air in the mixture, respectively, Equation 13.61 can also be written as

$$\omega = \frac{M_v}{M_a} \frac{(N_v/N)}{(N_a/N)} = \frac{M_v}{M_a} \frac{(p_v/p)}{(p_a/p)} = \frac{M_v}{M_a} \frac{p_v}{p_a} \tag{13.62}$$

Equation 13.62 in its final form is often written in terms of the total pressure (p) of moist air and the partial pressure of the water vapor contained. From Equation 13.24, the partial pressure of dry air p_a is equal to $p - p_v$. Also, for steam, $M_v = 18.015$ kg/kmol, while for dry air, $M_a = 28.96$ kg/kmol. The resulting final expression for humidity ratio is thus

$$\omega = 0.622 \frac{p_v}{p - p_v} \tag{13.63}$$

EXAMPLE 13.6

A sample of moist air at 100 kPa fills a space of volume 100 m^3 at a temperature of 300 K. If the humidity ratio of the air is 0.01, determine

(a) the partial pressure of water vapor in the air

(b) the mass of water vapor contained in the air sample

Given: Moist air with $p = 100$ kPa, $V = 100$ m^3, $T = 300$ K, and $\omega = 0.01$

Find: (a) p_v, (b) m_v

Solution:

(a) Equation 13.63 can be transposed to give the partial pressure of the water vapor p_v in terms of p and ω:

$$p_v = \left(\frac{1.608\omega}{1 + 1.608\omega} \right) p = \underline{1.58 \text{ kPa}}$$

(b) The number of moles (N) for the moist air can be determined using Equation 4.14:

$$N = \frac{pV}{\overline{R}T} = 4.01 \text{ kmol}$$

From Equation 13.27, one obtains the following for the number of moles of water vapor in the sample:

$$N_v = N \frac{p_v}{p} = 0.0634 \text{ kmol}$$

Applying Equation 4.9 for the relationship between mass and the number of moles, the following is obtained for the mass of water vapor in the sample:

$$m_v = N_v M_v = \underline{1.14 \text{ kg}}$$

EXAMPLE 13.7

A steady stream of moist air having a humidity ratio of 0.005 is humidified by adding steam at the rate 0.02 kg/s. If the humidity ratio of the exiting moist air stream is 0.01, determine the mass flow rate of the moist air in kg/s da (dry air).

Given: A mass flow rate of 0.02 kg/s of steam is added to an unspecified mass flow rate of moist air, as illustrated in Figure 13.3. For the moist air stream, $\omega_i = 0.005$, $\omega_e = 0.01$

Find: $\dot{m}_a$ (in kg/s da)

Solution: Let $\dot{m}_a$ (in kg/s da) denote the mass flow rate of dry air in the moist air. This mass flow rate is also the mass flow rate of the moist air expressed in

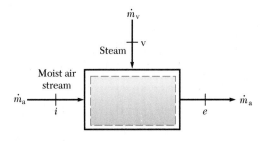

Figure 13.3 Humidification of a moist air stream by adding steam.

kg/s da (dry air). The mass flow rate of dry air is the same for both the inlet moist air stream and the exit stream since the addition of steam can only affect the moisture content. The mass balance for the moisture can be written as follows:

$$\underset{\substack{\text{moisture} \\ \text{in entering} \\ \text{moist air} \\ (\dot{m}_a\omega_i)}}{} + \underset{\substack{\text{steam} \\ \text{added} \\ (\dot{m}_v)}}{} = \underset{\substack{\text{moisture} \\ \text{in exiting} \\ \text{moist air} \\ (\dot{m}_a\omega_e)}}{}$$

$$\therefore \dot{m}_a = \frac{\dot{m}_v}{(\omega_e - \omega_i)} = 4 \text{ kg/s da}$$

Thus, the mass flow rate of the moist air is 4 kg/s da.

Saturation Humidity Ratio

A *saturated moist air* condition refers to the unique state at which air at a given pressure (p) and temperature (T) is holding as much water vapor as it can. Any attempt to introduce more water vapor results in the excess water being retained in the liquid state, usually as water droplets. The *saturation humidity ratio* (ω_s) is the value of humidity ratio for moist air at the saturated condition. It provides a yardstick by which to measure how close a given sample of moist air is to being saturated.

The saturation humidity ratio is a function of the total pressure of moist air and the dry bulb temperature. Figure 13.4 is a *T-v* diagram for steam on which the intensive state of the water vapor in air is marked as state 1. The point must lie on both the pressure $= p_v$ line and the temperature $= T_{db}$ line. The other constant pressure line on Figure 13.4 corresponds to the saturation pressure p_s when the steam temperature is T_{db}. If one therefore imagines the addition of moisture to the original moist air sample such that the total pressure and the dry bulb temperature stay constant, the intensive state of the water vapor in moist air will eventually move from state 1 to state 2, which is on the saturated vapor line. When state 2 is reached, the moist air is *saturated*, and any moisture added will now condense. The partial pressure of water vapor at state 2 is

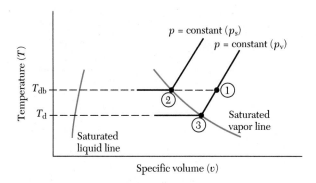

Figure 13.4 T-v diagram for steam illustrating relationships among some of the humidity parameters.

p_s, a function of T_{db}, the dry bulb temperature. From Equation 13.63, the expression for ω_s is

$$\omega_s = 0.622 \, \frac{p_s \, (T_{db})}{p \, - \, p_s \, (T_{db})} \tag{13.64}$$

Note that p_s has been written as $p_s \, (T_{db})$ in this equation to serve as a reminder that the saturation pressure is that at the dry bulb temperature.

EXAMPLE 13.8

Compute the saturation humidity ratio for moist air at 1 atm pressure for dry bulb temperatures in the range of 0°C to 60°C. Plot the results obtained on a graph of ω_s versus the dry bulb temperature.

Given: Saturated moist air at 1 atm and dry bulb temperatures from 0°C to 60°C

Find:
(a) ω_s in the specified range of dry bulb temperatures
(b) plot ω_s versus T_{db}

Solution:
(a) ω_s is calculated from Equation 13.64:

$$\omega_s = 0.622 \, \frac{p_s \, (T_{db})}{p \, - \, p_s \, (T_{db})}$$

The values of p_s at different dry bulb temperatures are determined from Table A.1 in Appendix A. Table 13.3 gives the results obtained for the saturation humidity ratio as a function of the dry bulb temperature in the specified range of temperatures.

(b) The results in Table 13.3 are plotted on Figure 13.5.

TABLE 13.3

Results for $\omega_s = \omega_s(T_{db})$ in Example 13.8		
T_{db} (°C)	p_s (kPa)	ω_s
0	0.612	0.0038
5	0.870	0.0054
10	1.23	0.0076
15	1.70	0.0106
20	2.33	0.0146
25	3.17	0.0201
30	4.24	0.0272
35	5.63	0.0366
40	7.38	0.0489
45	9.59	0.0650
50	12.35	0.0863
55	15.76	0.1146
60	19.94	0.1524

Figure 13.5 demonstrates certain important characteristics of ω_s. Water at 1 atm pressure has a saturation temperature of 100°C. At temperatures much lower than 100°C, p_s for water is much smaller than 1 atm, and the partial pressure of dry air in the denominator of Equation 13.64 (that is, the denominator $p - p_s$) can be replaced with the total pressure of 1 atm without a significant loss of accuracy. However, at higher

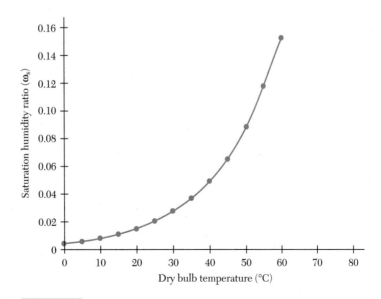

Figure 13.5 ω_s versus T_{db} for moist air at 1 atm.

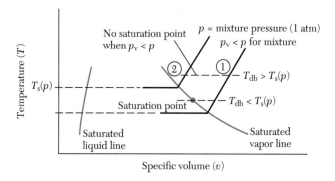

Figure 13.6 T-v diagram illustrating undefined value for ω_s when T_{db} is equal to or higher than T_s (p).

temperatures, the p_s component dominates, thus making ω_s rise sharply. As the dry bulb temperature approaches 100°C, p_s gets closer to the total mixture pressure of 1 atm and ω_s tends toward infinity.

What happens when the dry bulb temperature is higher than 100°C? Figure 13.6 shows two constant pressure lines on the T-v diagram for steam. One line corresponds to a pressure (p) of 1 atm. The second line is for a pressure equal to the partial pressure of water vapor in the moist air sample. For moist air at a pressure of 1 atm, p_v for the moisture must be less than the total pressure for the mixture and therefore the second line must lie below the 1 atm constant pressure line. The horizontal line from ① to ② shown in Figure 13.6 corresponds to a dry bulb temperature T_{db}, which is higher than 100°C. It is easy to see that a saturated moist air condition is impossible when T_{db} is higher than 100°C and the partial pressure of the water vapor cannot exceed 1 atm. Thus, no defined value exists for ω_s at dry bulb temperatures equal to or higher than the saturation temperature T_s for water vapor at a pressure p equal to the total pressure of the air–water vapor mixture.

Degree of Saturation

The *degree of saturation* (μ) measures how close to saturation a given sample of moist air is. It is defined as

$$\mu = \frac{\omega}{\omega_s} \tag{13.65}$$

where ω is the humidity ratio of the moist air at pressure p and temperature T_{db} and ω_s is the saturation humidity ratio at the same total pressure p and dry bulb temperature T_{db}. As with ω_s, note that the degree of saturation has meaning only for air temperatures below the saturation temperature T_s for steam at pressure p equal to the total pressure of the mixture. When μ is defined, the maximum value it can have is 1, corresponding to the saturated moist air condition. The degree of saturation (μ) is zero when the air is completely dry or moisture free.

Relative Humidity

Relative humidity (ϕ or rh) provides a measure of the degree of saturation of moist air, this time with reference to the partial pressure of the water vapor in moist air. Thus, relative humidity is defined as

$$\phi = \frac{p_{\mathrm{v}}}{p_{\mathrm{s}}} \cdot 100\% \tag{13.66}$$

where p_{v} is the partial pressure of water vapor in the moist air sample and $p_{\mathrm{s}} = p_{\mathrm{s}}\,(T_{\mathrm{db}})$ is the saturation pressure for steam at the same temperature as the dry bulb temperature (T_{db}) of the moist air. At the saturated moist air condition, p_{v} and p_{s} are the same and the relative humidity is 100%.

Some similarities exist between μ and ϕ. Both parameters are zero for dry (moisture-free) air; at the saturated condition, μ is 1 while the relative humidity is 100%. Nevertheless, the two parameters are defined differently and, in general, produce different results at conditions other than those for saturated air and for moisture-free air. The following relationship can be established between the degree of saturation and the relative humidity:

$$\mu\left(1 - \frac{p_{\mathrm{v}}}{p}\right) = \frac{\phi\,(\%)}{100} - \frac{p_{\mathrm{v}}}{p} \tag{13.67}$$

Thus, if the partial pressure of the water vapor is much lower than the total pressure of the mixture, $\mu \approx (\phi/100)$. The derivation uses the ideal gas rule, and the formal definitions given for ϕ, ω, and μ and is left as an exercise for the reader.

Dew Point Temperature

Consider cooling while the total pressure remains constant of a sample of moist air, initially unsaturated at a pressure p and dry bulb temperature T_{db}. The partial pressure p_{v} of the water vapor remains constant while the temperature of the mixture falls. A lower temperature level is eventually reached such that the saturation pressure for steam at that temperature is equal to the partial pressure of the water vapor. This lower temperature is termed the *dew point temperature* (T_{d}). At this temperature, the moist air is saturated with moisture, and any attempt at further cooling only results in the condensation of water vapor as droplets. This is what happens, for example, when moisture condenses on the windshields of cars whenever the ambient air temperature drops to a sufficiently low level. Thus, the temperature at which the sample of moist air becomes moisture saturated, at the prevailing pressure of the mixture, is the dew point temperature (T_{d}). Knowing the dew point temperature enables one to figure out the partial pressure (p_{v}) of water vapor in the mixture at the original condition of pressure and temperature. The partial pressure p_{v} is the saturation pressure of the water vapor at the dew point temperature T_{d} measured.

The T-v diagram for steam in Figure 13.4 illustrates the various parameters involved in the definitions of ϕ and the dew point. The initial steam condition is shown as state 1 corresponding to a partial pressure p_{v} and a mixture temperature T_{db}. One moves horizontally from state 1 to 2 to establish the dry, saturated condition for water vapor at temperature T_{db}. The corresponding saturation pressure at state 2 is the $p_s = p_s (T_{db})$. If one moves along a constant pressure (p_v) line from state 1 to 3, the dry, saturated water vapor condition at state 3 gives the dew point temperature T_d. Conversely, knowing the dew point temperature enables one to determine p_v, and hence, the relative humidity of the original moist air sample can be computed.

Dry Bulb and Wet Bulb Temperatures

The *dry bulb temperature* (T_{db}) is defined simply as the actual temperature of moist air. The *thermodynamic wet bulb temperature* (T_{wb}), however, is the same as the *adiabatic saturation temperature* (T^*), which is formally defined later in Section 13.3.3 and involves the application of the first law of thermodynamics to a hypothetical adiabatic process. Basically, it is the temperature at which the moist air sample becomes saturated in an adiabatic channel flow in which liquid water, also at T^* (or T_{wb}), is added. In practice, the temperature reading on a thermometer whose bulb is covered with a thoroughly wetted wick gives a reliable estimate of the wet bulb temperature.

In general, once the dry bulb and wet bulb temperatures are specified for moist air, the other humidity parameters can also be determined. Later, the use of the adiabatic saturation temperature (T^*) is demonstrated for the determination of relative humidity and the other humidity parameters.

EXAMPLE 13.9

Calculate the relative humidity and the humidity ratio of an air–vapor mixture at 1 bar and 30°C if the dew point temperature is 25°C.

Given: Moist air at $p = 100$ kPa, $T_{db} = 30°C$, and $T_d = 25°C$

Find: **(a)** ϕ, **(b)** ω

Solution:

(a) The relative humidity ϕ is defined by Equation 13.66. Table A.1 of Appendix A gives

$$p_v = p_s (T_d) = 3.17 \text{ kPa}$$

and

$$p_s (T_{db}) = 4.24 \text{ kPa}$$

Substituting these in Equation 13.66 gives the following for the relative humidity:

$$\phi = \frac{p_v}{p_s (T_{db})} 100\% = \underline{74.8\%}$$

(b) Substituting the values of p_v and p in Equation 13.63 yields the following for the humidity ratio:

$$\omega = 0.622 \frac{p_v}{p - p_v} = \underline{0.0204}$$

EXAMPLE 13.10

Calculate the dew point temperature and the humidity ratio of an air–vapor mixture of 60% relative humidity at a total pressure of 100 kPa and a temperature of 20°C.

Given: Moist air at $p = 100$ kPa, $T_{db} = 20°C$, and $\phi = 60\%$

Find: (a) T_d (b) ω

Solution:

(a) The partial pressure of the water vapor can be calculated from Equation 13.66 once the saturation pressure at T_{db} is found. From the calculated value of p_v, the dew point temperature, which is the saturation temperature of water vapor at p_v, can be determined. Table A.1 in Appendix A shows that $p_s (T_{db})$ is 2.33 kPa. Thus, from Equation 13.66, one obtains the following for p_v:

$$p_v = \left(\frac{\phi \, (\%)}{100}\right) p_s = 1.40 \text{ kPa}$$

T_d can be determined by linear interpolation from Table A.2: $T_d = \underline{11.2°C}$.

(b) ω is calculated from Equation 13.63:

$$\omega = 0.622 \frac{p_v}{p - p_v} = \underline{0.0088}$$

EXAMPLE 13.11

A house having internal dimensions of 1200 ft^2 floor area and 8 ft high ceilings contains moist air at 78°F and 14.7 psia with a relative humidity of 80%. Find

(a) the humidity ratio

(b) the degree of saturation

(c) the dew point temperature

(d) the total mass of water vapor in the house

Given: Moist air at $p = 14.7$ psia, $T_{db} = 78°F$, and $\phi = 80\%$

                   Volume of moist air $= (1200)(8)$ ft^3

Find: (a) ω, (b) μ, (c) T_d, (d) m_v

Solution:

(a) The **STEAM** code is used this time. Thus,

$$p_s (T_{db}) = 0.474 \text{ psia}$$

Applying Equation 13.66 gives

$$p_v = (0.8)(0.474) = 0.379 \text{ psia}$$

Therefore, from Equation 13.63, the following is obtained for the humidity ratio:

$$\omega = 0.622 \frac{p_v}{p - p_v} = \underline{0.0165}$$

(b) The saturation humidity ratio is calculated by applying Equation 13.64:

$$\omega_s = 0.622 \, \frac{p_s \, (T_{db})}{p - p_s \, (T_{db})} = 0.0207$$

The degree of saturation μ is defined by Equation 13.65. In this case, it is

$$\mu = \frac{\omega}{\omega_s} = 0.797$$

(c) The **STEAM** code is used to determine T_d, which by definition is the saturation temperature of steam at p_v. Thus, the dew point temperature is 71.2°F.

(d) The mass of moist air can be determined once N_v is calculated using Equation 13.25:

$$N_v = \frac{p_v V}{\overline{R} T_{db}}$$

$$= \frac{\left(0.379 \, \dfrac{lbf}{in.^2}\right)(9600 \, ft^3)\left(\dfrac{144 \, in.^2}{1 \, ft^2}\right)}{\left(1545 \, \dfrac{ft \cdot lbf}{lbmol \cdot °R}\right)(460 + 78°R)}$$

$$= 0.630 \, lbmol$$

From Equation 4.9, one can then determine the mass of the water vapor m_v:

$$m_v = N_v M_v = 11.3 \, lbm$$

---

1. Ambient air is at a pressure of 100 kPa and a temperature of 35°C. Determine the humidity ratio when the relative humidity is
   (a) 20%
   (b) 50%
   (c) 90%

   Answer: **(a)** 0.0071, **(b)** 0.0180, **(c)** 0.0331

2. Moist air at 77°F and 1 atm fills a space of volume 2000 ft³. What is the mass of moisture in the air when the relative humidity level is 90%?

   Answer: 2.59 lbm

3. For the moist air in Exercise 2, determine the relative humidity when the humidity ratio is 0.004.

   Answer: 20.4%

4. The air in an automobile was at 20°C and a relative humidity of 50% when the car was parked outdoors overnight in the cold. If the ambient pressure was 1 atm, determine the temperature at which moisture would begin to condense on the wind-

shield of the car. Suppose the air temperature actually falls to 4°C during the night. To what temperature must the air in the automobile (and also the windshield surface) be raised to clear the moisture off the windshield?

Answer: 9.2°C

13.3.2 Psychrometric Chart

The *psychrometric chart* graphically depicts the values of the humidity parameters of a mixture of air and water vapor at a specified barometric pressure. The chart is generally available for an atmospheric pressure of 1 atm although supplemental charts can be obtained for other barometric pressures for high altitudes or below sea level. A typical chart indicates values for the following parameters:

1. the dry bulb temperature (T_{db})
2. the wet bulb temperature (T_{wb})
3. the dew point temperature (T_d), which is coincident with the wet bulb temperature value along the saturation line (having 100% rh) on the psychrometric chart
4. the relative humidity (ϕ)
5. the humidity ratio (ω) or moisture content
6. the enthalpy of moist air (per unit mass of dry air) (h_{ma})
7. specific volume of moist air (per unit mass of dry air) (v_{ma})

Charts are available for different ranges of dry bulb temperatures, thus giving the designer a choice depending on the application. Figures 13.7(a) and (b) (and also Figures C4 and C5 in Appendix C) are typical psychrometric charts in SI and USCS units that can be used for the examples in this text. The psychrometric charts have the dry bulb temperature as the abscissa and the humidity ratio as the ordinate. Note that the humidity ratio in Figure 13(a) is given in units of grams moisture per kg da and must be divided by 1000 to obtain ω in kg moisture per kg da. If T_{db} and ω are known, the intersection of the vertical line through T_{db} and the horizontal line through ω gives the state of the moist air sample. This is illustrated in Figure 13.8 where point I marks the intersection of the two lines.

The *saturation line* or *envelope* on the left side of Figure 13.8 corresponds to the saturated moist air condition. Along the line, the relative humidity is 100%, the humidity ratio is the saturation humidity ratio (ω_s), and the wet bulb temperature, the dry bulb temperature, and the dew point temperature are all coincident and have the same value. Thus, if you need to find the dew point temperature, say, at a given humidity ratio level ω, you simply move along a horizontal line, through the specified ω, to the point of intersection A with the saturation line, as illustrated in Figure 13.8. The corresponding dry bulb temperature at state A is also the dew point temperature.

Figure 13.8 also shows a set of sloping lines of constant wet bulb temperatures. Line IB is shown passing through point I and intersecting the saturation line at point B. The vertical line through point B intersects the temperature axis at the wet bulb temperature T_{wb} for the moist air at state I. Conversely, if T_{wb} and T_{db} are specified, you would simply identify the point of intersection of the appropriate constant wet bulb temperature line

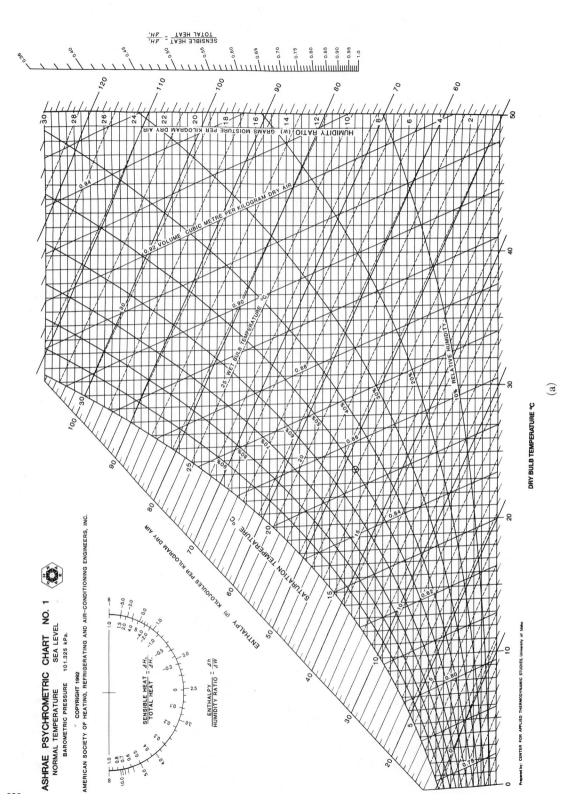

ASHRAE PSYCHROMETRIC CHART NO. 1
NORMAL TEMPERATURE   SEA LEVEL
BAROMETRIC PRESSURE   101.325 kPa.
COPYRIGHT 1992
AMERICAN SOCIETY OF HEATING, REFRIGERATING AND AIR-CONDITIONING ENGINEERS, INC.

SENSIBLE HEAT $= \dfrac{\Delta H_s}{\Delta H_t}$
TOTAL HEAT

ENTHALPY $= \dfrac{\Delta h}{\Delta W}$
HUMIDITY RATIO

$\dfrac{\text{SENSIBLE HEAT}}{\text{TOTAL HEAT}} = \dfrac{\Delta H_s}{\Delta H_t}$

HUMIDITY RATIO (W) GRAMS MOISTURE PER KILOGRAM DRY AIR

VOLUME CUBIC METRE PER KILOGRAM DRY AIR

WET BULB TEMPERATURE °C

RELATIVE HUMIDITY

SATURATION TEMPERATURE °C

ENTHALPY (h) KILOJOULES PER KILOGRAM DRY AIR

DRY BULB TEMPERATURE °C

(a)

Prepared by: CENTER FOR APPLIED THERMODYNAMIC STUDIES, University of Idaho

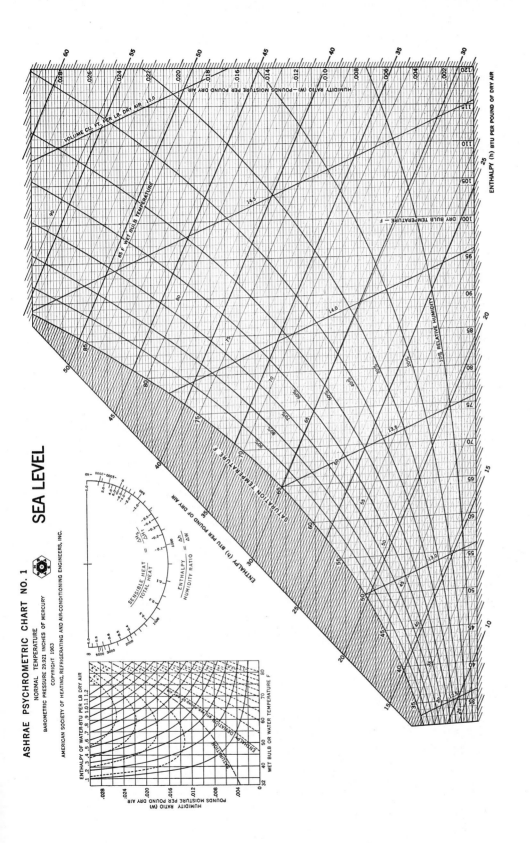

Figure 13.7 Psychrometric chart for moist air at 1 atm and normal temperatures (a) in SI units and (b) in USCS units. (Parts a and b reprinted with permission of the American Society of Heating, Refrigerating and Air Conditioning Engineers, Inc.)

(b)

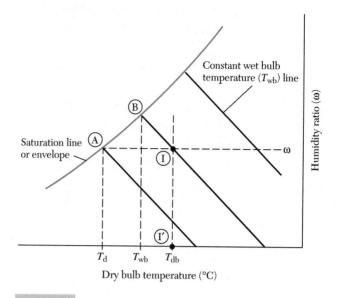

Figure 13.8 Illustration of how to read the psychrometric chart.

BI with vertical line I′I for the dry bulb temperature to determine state I for the moist air sample.

The constant wet bulb temperature lines and the constant moist air enthalpy (h_{ma}) are typically so nearly coincident that at times only one set of these lines is shown on the chart. A scale for the determination of enthalpy is annexed to the left of the saturation line so that numerical determination of enthalpies for moist air samples can be made even when only the constant wet bulb temperature lines are displayed. The datum for enthalpy of moist air shown on Figure 13.7 is chosen to ensure consistency with the datum for the steam property tables given in Appendix A, in which the enthalpy of saturated liquid is set equal to zero at the triple point for water. Figures 13.7(a) and (b) also give the lines of constant specific volume for moist air (on a unit mass of dry air basis) as well as the constant relative humidity lines. A systematic procedure for reading the chart is to aim at locating the intersection of any two constant property lines corresponding to whichever two properties are specified. Other properties can then be determined once the state point has been located.

EXAMPLE 13.12

Use the psychrometric chart to find the unknown parameters in the following:

(a) ω, when $T_{db} = 35°C$ and $T_{wb} = 30°C$

(b) T_{wb}, when $T_{db} = 35°C$ and $\phi = 20\%$

(c) T_{db}, when $\omega = 0.0142$ and $\phi = 40\%$

(d) h_{ma}, when $T_{wb} = 25°C$ and $T_{db} = 30°C$

(e) ω, when dew point temperature is 14°C

Given:

(a) $T_{db} = 35°C$ and $T_{wb} = 30°C$

(b) $T_{db} = 35°C$ and $\phi = 20\%$

(c) $\omega = 0.0142$ and $\phi = 40\%$

(d) $T_{wb} = 25°C$ and $T_{db} = 30°C$

(e) $T_d = 14°C$

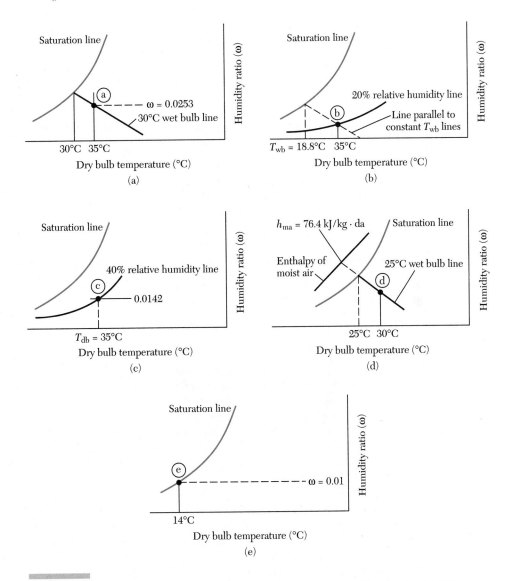

Figure 13.9 Illustrations of the solutions for (a) Example 13.12(a), (b) Example 13.12(b), (c) Example 13.12(c), (d) Example 13.12(d), and (e) Example 13.12(e).

Find: **(a)** ω, **(b)** T_{wb}, **(c)** T_{db}, **(d)** h_{ma}, **(e)** ω (Using the psychrometric chart)

Solution: Read the chart first and then compare your answers to those provided below. Three intensive properties are normally needed for the state of a two-component pure substance. One of these is the pressure p for the mixture, which is assumed to be 1 atm. Figures 13.9(a) through (e) are sketches that illustrate how the psychrometric chart [Fig. 13.7(a)] has been used for the following solutions:

(a) $\omega = \underline{0.0253}$

(b) $T_{wb} = \underline{18.8°C}$

(c) $T_{db} = \underline{35°C}$

(d) $h_{ma} = \underline{76.4 \text{ kJ/kg da}}$

(e) $\omega = \underline{0.01}$

EXERCISES

for Section

13.3.2

1. Verify your answers to the exercises for Section 13.3.1 using the psychrometric charts in Figures 13.7(a) and (b).

2. Use the psychrometric chart to find the unknown parameters in the following:
   (a) ω, when $T_{db} = 45°C$ and $T_{wb} = 34°C$
   (b) T_{wb}, when $T_{db} = 33°C$ and $\phi = 35\%$
   (c) T_{db}, when $\omega = 0.0142$ and $\phi = 42\%$
   (d) h_{ma}, when $T_{wb} = 27°C$ and $T_{db} = 32°C$
   (e) ω, when dew point temperature is $12°C$

   Answer: **(a)** 0.0298, **(b)** 21.3°C, **(c)** 34.3°C, **(d)** 85.3 kJ/kg da, **(e)** 0.0087

13.3.3 Thermodynamics of Psychrometric Processes

In air conditioning, the following psychrometric processes can be involved:

1. *Mixing* of moist air streams, which frequently occurs under adiabatic condition.

2. *Cooling* of a moist air stream. Sensible cooling is often achieved by a refrigeration device. If an increased level of humidity is allowed for the air stream, sensible cooling can also be produced by spraying the air stream with water. When a water spray is applied to a moist air stream, the process is termed *humidification*. Cooling of a moist air stream can result also in moisture removal. The process of moisture removal is referred to as *dehumidification*.

3. *Heating* of a moist air stream.

Whenever heat flow occurs to or from an air stream without being accompanied by a change in temperature of the air stream, the load is termed *latent heat load*. Conversely, if the humidity of the air stream does not change while heat flow occurs, the load is referred to as *sensible heat load* and the temperature of the air stream changes. Both heating and cooling processes can occur with dehumidification or humidification. In very hot, dry climates, for example, it is common to combine sensible cooling via a refrigeration device with some evaporative cooling and humidification produced by applying a water spray to the air stream. In this way, a desired lower temperature and higher humidity level can be achieved. In hot, humid climates, however, sensible cooling with dehumidification is needed to achieve a desired comfort level.

The thermodynamics of psychrometric processes should include both first law and second law considerations. The rest of this chapter focuses on steady-flow processes. Therefore, only the expressions for enthalpy and flow exergy of moist air and the other fluids (namely, steam and water) that are used in psychrometric processes are considered here. These expressions are given in the following.

Enthalpy and Exergy of Moist Air Stream

This book uses p_0 and T_0 to denote the prevailing ambient pressure and temperature. The ambient or atmospheric air is taken as the reference environment. The terms $\dot{m}_a$ and $\dot{m}_v$ denote the mass flow rate of dry air and of the moisture (water vapor) in the stream of moist air. The ratio $\dot{m}_v/\dot{m}_a$ is the humidity ratio ω, which was defined by Equation 13.60. The moist air stream is at a pressure p and a dry bulb temperature T_{db}. Let H^* denote the enthalpy flow rate of the moist air stream. Based on Equation 13.39, H^* can be written as

$$H^* = \dot{m}_a h_a + \dot{m}_v h_v \qquad (13.68a)$$

where h_a and h_v are the specific enthalpy for dry air and the water vapor in the mixture, respectively. The specific enthalpies can be written on the basis of Equation 13.8 as

$$h_a = c_{pa} (T_{db} - T_{ref}) + h_{a,ref} \qquad (13.68b)$$

and

$$h_v = c_{pv} (T_{db} - T_{ref}) + h_{v,ref} \qquad (13.68c)$$

where c_{pa} and c_{pv} are the specific heat at constant pressure for the dry air and the water vapor in the mixture, respectively. The enthalpy of moist air is often expressed on a unit mass of dry air basis. Thus, if h_{ma} denotes the enthalpy of moist air per unit mass of dry air, Equations 13.68a to 13.68c give the following equation for h_{ma}:

$$h_{ma} = \frac{H^*}{\dot{m}_a} = h_a + \omega h_v$$

$$\qquad (13.68)$$

$$= c_{pa} \left(1 + \frac{c_{pv}}{c_{pa}} \omega \right) (T_{db} - T_{ref}) + (h_{a,ref} + \omega h_{v,ref})$$

Section 13.2 suggested a possible choice of $T_{ref} = T_t$ (the triple point temperature of water) for an air–water vapor system if consistency is to be maintained with the steam and air tables provided in Appendix A (SI units). Such a choice is made here, except that while $h_{v,ref}$ is determined to achieve consistency with the steam table property values, $h_{a,ref}$ is set to zero at the triple point temperature of water. This automatically rules out consistency with the air tables given in the appendix. These choices are summarized as follows:

$$T_{ref} = 273.16 \text{ K } (492°\text{R})$$

$$h_{a,ref} = 0$$

$$h_{v,ref} = 2501 \text{ kJ/kg } (1075 \text{ Btu/lbm})$$

For dry air and steam at near ambient temperatures, the following specific heat values are assumed:

$$c_{pa} = 1.005 \text{ kJ/kg} \cdot \text{K} \ (0.240 \text{ Btu/lbm} \cdot {}^\circ\text{R})$$

$$c_{pv} = 1.86 \text{ kJ/kg} \cdot \text{K} \ (0.444 \text{ Btu/lbm} \cdot {}^\circ\text{R})$$

Substituting these in Equation 13.68 yields the following equations for the enthalpy of moist air in SI and USCS units, respectively:

$$h_{ma} \ (\text{kJ/kg da}) = 1.005(1 + 1.851\omega)T_{db} \ ({}^\circ\text{C}) + 2501\omega - 0.01 \quad \text{(13.69a)}$$

and

$$h_{ma} \ (\text{Btu/lbm da}) = 0.24(1 + 1.851\omega)T_{db} \ ({}^\circ\text{F}) + 1061\omega - 7.7 \quad \text{(13.69b)}$$

For the exergy flow rate of the moist air stream, Equations 13.45 and 13.58 give the following expression for the exergy per unit mass of dry air x_{ma}:

$$x_{ma} = \frac{\dot{X}}{\dot{m}}$$

$$= c_{pa}T_0 \left[\left(1 + \frac{c_{pv}}{c_{pa}} \omega \right) \left(\frac{T_{db}}{T_0} - 1 - \ln \frac{T_{db}}{T_0} \right) \right. \quad \text{(13.70a)}$$

$$\left. + \frac{R_a}{c_{pa}} \left(\ln \frac{p_a}{p_{a0}} + \omega \frac{R_v}{R_a} \ln \frac{p_v}{p_{v0}} \right) \right]$$

where p_{v0} is the partial pressure of the water vapor and p_{a0} is that for the dry air in the stable reference environment, which is the ambient or atmospheric air. Let ω_0 denote the humidity ratio for atmospheric air. Equation 13.63 yields

$$\omega_0 = 0.622 \frac{p_{v0}}{(p_0 - p_{v0})} \quad \text{(13.70b)}$$

By suitably manipulating Equation 13.63 for the actual moist air, and Equation 13.70b for atmospheric air, one obtains

$$\frac{p_v}{p_{v0}} = \left(\frac{\omega}{\omega_0} \right) \left(\frac{1 + 1.608\omega_0}{1 + 1.608\omega} \right) \frac{p}{p_0} \quad \text{(13.70c)}$$

Both Equations 13.63 and 13.70b can also be expressed in terms of p_a and p_{a0} by making the substitutions $p_v = (p - p_a)$ and $p_{v0} = (p_0 - p_{a0})$. After the substitution and the necessary algebraic manipulation, the following is obtained for p_a/p_{a0}:

$$\frac{p_a}{p_{a0}} = \left(\frac{1 + 1.608\omega_0}{1 + 1.608\omega} \right) \frac{p}{p_0} \quad \text{(13.70d)}$$

By substituting for p_v/p_{v0} and (p_a/p_{a0}) as given by Equations 13.70c and 13.70d, respectively, in Equation 13.70a and rearranging, the following is obtained for the exergy of the moist air (per unit mass of dry air):

$$\frac{x_{ma}}{c_{pa}T_0} = (1 + 1.851\omega)\left(\frac{T_{db}}{T_0} - 1 - \ln\frac{T_{db}}{T_0}\right)$$
$$+ 0.2857\left[1.608\omega \ln\frac{\omega}{\omega_0} + (1 + 1.608\omega) \ln\left(\frac{1 + 1.608\omega_0}{1 + 1.608\omega}\frac{p}{p_0}\right)\right]$$

(13.70)

The numerical constants in this equation were obtained by substituting $(c_{pv}/c_{pa}) = 1.851$ and $(R_a/c_{pa}) = 0.2857$ at near ambient temperatures. Also, $R_v/R_a = M_a/M_v = 1.608$ from Equation 4.11. Equation 13.70 is applicable in both SI and USCS units. T_0 and T_{db} must be expressed as absolute temperatures. Equation 13.70 for the exergy of a moist air stream is essentially the same as that obtained by Wepfer et al. (1979).

EXAMPLE 13.13

Outside air at 95°F (35°C) with a humidity ratio of 0.0142 is mixed with return air at 80°F (26.67°C) with a humidity ratio of 0.0112. The return air is mixed with outside air in the ratio of 1:2 on a dry air mass basis. If the mixing occurs adiabatically, determine

(a) the temperature and humidity ratio of the resulting mixed air stream

(b) the second law efficiency of the mixing process

Given: Adiabatic mixing of outside air with return air, as shown in Figure 13.10

Outside air condition

$$T_{db} = 95°F\ (35°C) = T_0, \omega_1 = 0.0142\ (= \omega_0)$$

Return air condition

$$T_{db} = 80°F\ (26.67°C), \omega_2 = 0.0112$$

$$(\dot{m}_a)_{\text{outside air}} = 2\ (\dot{m}_a)_{\text{return air}}$$

Find:

(a) T_{db} and ω for the resulting mixed air stream

(b) η_2 for the process

Solution:

(a) A mass balance and an energy balance are needed for the determination of ω and T_{db} for the mixed air stream. The mass balance condition is that the mass flow

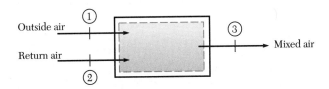

Figure 13.10 Adiabatic mixing of moist air streams.

rate of each species in the mixture (namely, the dry air and the moisture) must be conserved. For the energy balance, the multiple stream version of the SFEE (Eq. 6.30) is applied in the reduced form that is obtained by neglecting changes in the kinetic energy and the potential energy of the moist air streams.

Mass balances

Referring to Figure 13.10, use $\dot{m}_a$ for the mass flow rate of the return air in kg/s da. Therefore, the mass flow rate for the outside air is $2\dot{m}_a$. The mass balance for the dry air is

$$\dot{m}_{a3} = \dot{m}_{a1} + \dot{m}_{a2} = 3\dot{m}_a$$

and for the moisture is

$$\dot{m}_{v3} = \dot{m}_{v1} + \dot{m}_{v2}$$

$$\therefore \ \omega_3 \dot{m}_{a3} = \omega_1(2\dot{m}_a) + \omega_2\dot{m}_a$$

or

$$\omega_3 = \frac{(2\omega_1 + \omega_2)\dot{m}_a}{3\dot{m}_a} = \underline{0.0132}$$

Energy balance

When applying Equation 6.30 in the reduced form, use is made of the fact that the heat flow rate term is zero for an adiabatic process and that the shaft power term must be zero since no shaft or shear work is involved. Equation 6.30 therefore reduces to the following:

$$\dot{m}_{a3}(h_{ma})_3 = \dot{m}_{a1}(h_{ma})_1 + \dot{m}_{a2}(h_{ma})_2$$

$$\therefore \ (h_{ma})_3 = \frac{[2(h_{ma})_1 + (h_{ma})_2]\dot{m}_a}{3\dot{m}_a}$$

Equation 13.69b gives $(h_{ma})_1 = 30.8$ Btu/lbm da and $(h_{ma})_2 = 23.8$ Btu/lbm da. Substituting in the reduced energy equation yields $(h_{ma})_3 = 28.5$ Btu/lbm da. Applying Equation 13.69b again gives $(T_{db})_3 = \underline{90.3°F \ (32.4°C)}$.

(b) For the second law efficiency of the process, calculate the exergy input rate and the exergy production rate and then apply Equation 9.10. In this case, the exergy flow rates are as follows:

$$\dot{X}_{input} = \dot{m}_{a1}(x_{ma})_1 + \dot{m}_{a2}(x_{ma})_2$$

and

$$\dot{X}_{product} = \dot{m}_{a3}(x_{ma})_3$$

The exergy per unit mass of dry air in these equations is calculated for each stream using Equation 13.70. For the outside air drawn from the reference environment, $(x_{ma})_1$ is zero. In the case of the return air stream and the mixed stream, assume that $p = p_0$. Substituting the appropriate values of T_{db}, T_0 (in °R), ω, and

ω_0 in Equation 13.70 gives $(x_{ma})_2 = 7.11 \times 10^{-2}$ Btu/lbm da (0.165 kJ/kg da) and $(x_{ma})_3 = 7.08 \times 10^{-3}$ Btu/lbm da (0.0165 kJ/kg da). The second law efficiency is thus obtained as follows:

$$\eta_2 = \frac{\dot{X}_{product}}{\dot{X}_{input}} 100\%$$

$$= \frac{3\dot{m}_a(x_{ma})_3}{\dot{m}_a(x_{ma})_2} 100\% = \underline{29.9\%}$$

This mixing process is highly irreversible. Approximately 70% of the exergy of the return air is destroyed as a result of diluting it with ambient air.

Enthalpy and Flow Exergy of Water Vapor

Let's now consider a steady flow of pure steam at a low pressure p_v and a near-ambient temperature T_v. The steam can be treated as a perfect gas so that Equation 13.8 gives the following for the enthalpy per unit mass steam h_v:

$$h_v = c_{pv}(T_v - T_{ref}) + h_{v,ref} \tag{13.71}$$

The expression to use when working with SI units is

$$h_v \text{ (kJ/kg steam)} = 1.86T_v \text{ (°C)} + 2501 \tag{13.71a}$$

and for USCS units is

$$h_v \text{ (Btu/lbm steam)} = 0.444T_v \text{ (°F)} + 1061 \tag{13.71b}$$

For the exergy flow rate, use Equation 13.20 with p_v substituted for p and p_{v0} for p_0 since for pure steam at the dead state, the pressure is that of the moisture contained in the reference environment. Thus, the flow exergy per unit mass of steam (x_v) is given by

$$\frac{x_v}{c_{pv}T_0} = \frac{T_v}{T_0} - 1 - \ln\frac{T_v}{T_0} + \frac{R_v}{c_{pv}} \ln\frac{p_v}{p_{v0}} \tag{13.72a}$$

From Equation 13.70b, one can express p_{v0} in terms of p_0 as follows:

$$p_{v0} = \left(\frac{1.608\omega_0}{1 + 1.608\omega_0}\right)p_0 \tag{13.72b}$$

Substituting for p_{v0} in Equation (13.72a), one obtains the following for (x_v):

$$\frac{x_v}{c_{pv}T_0} = \frac{T_v}{T_0} - 1 - \ln\frac{T_v}{T_0} + 0.2481 \ln\left(\frac{1 + 1.608\omega_0}{1.608\omega_0}\frac{p_v}{p_0}\right) \tag{13.72}$$

Equation 13.72 is the same as the expression given by Wepfer et al. (1979).

Enthalpy and Flow Exergy for a Liquid Water Stream

To determine the enthalpy and flow exergy for a liquid water stream, the first step is to establish a relationship between the enthalpy and entropy of water and the corresponding values for dry, saturated steam at the same temperature T_w as for the water. Thus, if $h_w(T_w)$ and $s_w(T_w)$ denote the specific enthalpy and the specific entropy of water, respectively, then

$$h_w(T_w) = h_g(T_w) - h_{fg}(T_w) = c_{pw}(T_w - T_t) \tag{13.73a}$$

and

$$s_w(T_w) = s_g(T_w) - s_{fg}(T_w) \tag{13.73b}$$

where $h_g(T_w)$ and $s_g(T_w)$ are the specific enthalpy and the specific entropy of dry, saturated steam at the temperature T_w. Also, c_{pw} is the average specific heat for water, while T_t is the triple point temperature of water. Equation 13.71 for h_v can be used for h_w with T_w substituted for T_v in the equation. An expression for $s_g(T_w)$ is obtained from Equation 13.16 with the triple point temperature T_t and pressure p_t for water substituted for T_{ref} and p_{ref}, respectively:

$$s_g(T_w) = c_{pv} \ln \frac{T_w}{T_t} - R_v \ln \frac{p_s(T_w)}{p_t} + s_{v,ref} \tag{13.73c}$$

where $p_s(T_w)$ denotes the saturation pressure of the water vapor at temperature T_w. For water temperatures near typical ambient conditions, the following expression for h_{fg} can be inferred from Equation 13.73a:

$$\frac{h_{fg}(T_w)}{c_{pv}T_t} = \left(\frac{h_{v,ref}}{c_{pv}T_t} + \frac{c_{pw}}{c_{pv}} - 1\right) - \left(\frac{c_{pw}}{c_{pv}} - 1\right)\frac{T_w}{T_t} \tag{13.73d}$$

For $s_{fg}(T_w)$, one can use the following relationship between s_{fg} and h_{fg}, as indicated in Chapter 12:

$$s_{fg}(T_w) = \frac{h_{fg}(T_w)}{T_w} \tag{13.73e}$$

The exergy per unit mass of water x_w can be determined using Equation 9.33 with the kinetic energy and potential energy terms assumed negligible. In the present case, the reduced expression is as follows:

$$x_w = h_w(T_w) - h_{v0} - T_0[s_w(T_w) - s_{v0}] \tag{13.73f}$$

The dead state properties for the exergy of the water are h_{v0} and s_{v0} which are the specific enthalpy and the specific entropy, respectively, for the water vapor contained in the reference environment. After making the necessary substitutions, the following expressions are obtained for specific enthalpy h_w and the exergy per unit mass x_w for the water stream at T_w:

$$\frac{h_w(T_w)}{c_{pv}T_0} = \frac{c_{pw}}{c_{pv}}\frac{(T_w - T_t)}{T_0} \tag{13.74}$$

and

$$
\frac{x_{\mathrm{w}}}{c_{p\mathrm{v}}T_0} = \frac{T_{\mathrm{w}}}{T_0} - 1 - \ln\frac{T_{\mathrm{w}}}{T_0}
$$

$$
+\ 0.2481 \ln\left(\frac{1 + 1.608\omega_0}{1.608\omega_0}\frac{p_s\left(T_{\mathrm{w}}\right)}{p_0}\right) \tag{13.75}
$$

$$
+\ \frac{h_{\mathrm{fg}}\left(T_{\mathrm{w}}\right)}{c_{p\mathrm{v}}T_{\mathrm{t}}}\left(\frac{T_{\mathrm{t}}}{T_0}\right)\left(\frac{T_0}{T_{\mathrm{w}}} - 1\right)
$$

The temperatures in Equation 13.75 must be absolute temperatures.

Adiabatic Saturation Temperature (T*)

The adiabatic saturation temperature (T^*) provides an indication of the humidity level of a moist air stream. Figure 13.11 illustrates a hypothetical situation in which moist air at dry bulb temperature T_{db} with a humidity ratio ω enters an adiabatic conduit and then flows over the surface of (liquid) water maintained at temperature T^*. This is called an *adiabatic saturation process*, with the pressure assumed to remain constant for the entire process. The air stream exits the conduit at a saturated condition with temperature T^* and a humidity ratio equal to the saturation humidity ratio (ω_s^*) at T^*. A relationship can be established between ω and T^* via mass balance and energy balance for the flow through the conduit. The term $\dot{m}_{\mathrm{a}}$ denotes the mass flow rate of dry air and $\dot{m}_{\mathrm{w}}$ the mass flow rate of liquid water entrained. The mass balance for the water vapor in the air stream between the entry to the conduit and the exit is given by

$$
\dot{m}_{\mathrm{w}} = \dot{m}_{\mathrm{a}}[\omega_s\left(T^*\right) - \omega] \tag{13.76a}
$$

Likewise, the energy balance for the process is

$$
\dot{m}_{\mathrm{w}}h_{\mathrm{w}}\left(T^*\right) + \dot{m}_{\mathrm{a}}h_{\mathrm{ma}}\left(T_{\mathrm{db}},\omega\right) = \dot{m}_{\mathrm{a}}h_{\mathrm{ma}}\left(T^*, \omega_s^*\right) \tag{13.76b}
$$

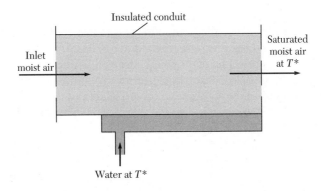

Figure 13.11 A hypothetical scheme for the measurement of the adiabatic saturation temperature (T^*).

This equation was obtained by applying the SFEE (Eq. 6.30) with the terms for changes in kinetic and potential energy assumed negligible. Also, since the process is adiabatic, the heat flow rate is zero. The term $h_w(T^*)$ is determined from Equation 13.74, while the specific enthalpies for moist air at T_{db} and T^* are determined using Equation 13.68. Eliminating $\dot{m}_a / \dot{m}_w$ between Equations 13.76a and 13.76b leads to the following equation for the humidity ratio ω of the original moist air sample:

$$\omega = \omega_s(T^*) + \left(\frac{h_{ma}(T_{db}, \omega) - h_{ma}(T^*, \omega_s^*)}{h_w(T^*)} \right) \tag{13.76c}$$

The specific enthalpies for moist air are given in terms of the dry bulb temperature and the humidity ratio by Equation 13.68. If these substitutions are made in Equation 13.76c and it is rearranged, the following final expression for ω is obtained:

$$\omega = \frac{h_{fg}(T^*) \cdot \omega_s(T^*) - c_{pa}(T_{db} - T^*)}{h_g(T_{db}) - h_w(T^*)} \tag{13.76}$$

Thus, if T_{db} and T^* are known, one can determine all the quantities on the right-hand side of Equation 13.76. Thus, the humidity ratio of the original moist air can be determined, that is, the moisture content of the entering air stream can be determined from the measurements of temperature (and pressure for a nonisobaric process) of the air stream entering and leaving the conduit (which are relatively easy to make). The adiabatic saturation temperature T^* was identified earlier as the thermodynamic wet bulb temperature. It turns out to be roughly the same as the temperature reading on the wet bulb thermometer of practical hygrometers.

EXAMPLE 13.14

A hygrometer gives a dry bulb temperature of 30°C and a wet bulb temperature of 25°C. Calculate the relative humidity of air assuming a total pressure of 1 atm.

Given: Moist air with $T_{db} = 30°C$, $T_{wb} = 25°C$, and $p = 1$ atm

Find: ϕ

Solution: The basic assumption is that the wet bulb temperature and the adiabatic saturation temperature are the same. Thus, from Equation 13.76, the humidity ratio for the air can be determined once you have calculated $\omega_s(T^*)$ and the enthalpy values involved. The partial pressure p_v of the moisture in the air sample is then determined by substituting the calculated value of ω in Equation 13.63. Once p_v is known, you can calculate the relative humidity using Equation 13.66.

Table A.1 in Appendix A gives the following:

$$p_s(T^*) = p_s(25°C) = 3.17 \text{ kPa}$$

Applying Equation 13.64 for ω_s yields the following for the saturation humidity ratio at T^*:

$$\omega_s = 0.622 \left(\frac{3.17}{101.325 - 3.17} \right) = 0.0201$$

$h_w(T^*)$ can be determined either from Table A.1 or from Equation 13.74 with c_{pw} taken as 4.2 kJ/kg · K. The value for h_w at 25°C is 105 kJ/kg. Table A.1 also gives $h_{fg}(T^*) = 2442$ kJ/kg and $h_g(T_{db}) = 2555$ kJ/kg. Substituting these in Equation 13.76 yields $\omega = 0.0180$.

Equation 13.63 gives $p_v = p\omega/(0.622 + \omega) = 2.85$ kPa. Also, from Table A.1, $p_s(T_{db}) = 4.24$ kPa. Substituting these values in Equation 13.66 yields a relative humidity for the moist air of 67.2%. Note that the estimate obtained from the psychrometric chart [Fig. 13.7(a)] is 67.5%.

EXERCISES

for Section 13.3.3

1. A mass flow rate of 0.017 kg/s of superheated steam at 1 atm and 200°C is added to a cold air stream at 10°C with a relative humidity of 1%. The mass flow rate of the cold air stream is 2 kg/s da. The mixing process is adiabatic, and the prevailing pressure can be taken as 1 atm. Determine the temperature and relative humidity of the exit mixed stream.

   Answer: $T_e = 12.9°C$, $\phi_e = 92.5\%$

2. What is the second law efficiency for the adiabatic mixing process of Exercise 1? Assume the dead state is identical to the state of the cold air stream.

   Answer: 39.2%

3. Moist air at 100 kPa and 300 K has a relative humidity of 70%. Determine R and c_p for the moist air. If a stream of the moist air is compressed isentropically to a pressure of 200 kPa, determine the temperature and the relative humidity of the exit air.

   Answer: $R = 0.290$ kJ/kg · K, $c_p = 1.018$ kJ/kg · K
   $T_e = 92.4°C$, $\phi = 6.50\%$

4. What is the dew point temperature for the exit air stream in Exercise 3? Suppose the exit air stream is cooled at a pressure of 200 kPa until the temperature falls to 315 K. Determine the rate of heat transfer per unit mass of dry air.

   Answer: $T_d = 32.8°C$, $q = -52.1$ kJ/kg da

13.3.4 Computer Code **PSY** for the Analysis of Psychrometric Processes

The computer programs supplied with this text include a package code named **PSY** that is intended to aid the thermodynamic analysis of psychrometric processes. The package includes a set of subroutines for computation of the other humidity parameters when any two are specified. Subroutines are also included for the computation of specific enthalpy and the flow exergy per unit mass for moist air, water vapor, and (liquid) water streams, which are variously encountered in psychrometrics.

The computer program for psychrometric processes uses the equations that were developed in this chapter. These equations generally represent a first-order approximation considering the fact that air and water vapor are treated as perfect gases. Better approximations are given in the literature (see, for example, Hyland and Wexler 1983 and Stewart et al. 1983), but these often involve a much more complex solution scheme.

In any case, the solutions based on the first-order approximation are judged to be good enough for most engineering situations. The code is written in BASIC for use on any IBM-compatible PC. Further information on the computer code is given in Appendix D. The procedure for operating the program is also detailed in Appendix D. The following three examples are drawn from available energy analyses of selected psychrometric processes by Wepfer et al. (1979). They are used to demonstrate the versatility of the computer code in the thermodynamic analysis and evaluation of psychrometric processes. Note that, in general, a complete analysis includes the following:

1. a mass balance (from which one can usually determine unknown humidity parameters)
2. an energy balance (or first law analysis)
3. an exergy (or second law) analysis

EXAMPLE 13.15
Use of PSY Code for Analysis of an Adiabatic Mixing Process

A stream of 1.512 kg/s da (12,000 lb/h da) at 26.67°C (80°F) db and $\omega = 0.01116$ is adiabatically mixed with 1.512 kg/s da (12,000 lb/h da) at 48.9°C (120°F) db and $\omega = 0.01799$. The dead state can be taken as air at 35°C (95°F), 101.325 kPa (1 atm), and $\omega = 0.01406$. Determine

(a) the dry bulb temperature and humidity ratio of the resulting mixture

(b) the second law efficiency of the process

Given: The adiabatic mixing process is illustrated in Figure 13.12.

$$\dot{m}_{a1} = 1.512 \text{ kg/s da}, \ T_{db1} = 26.67°C, \ \omega_1 = 0.01116$$

$$\dot{m}_{a2} = 1.512 \text{ kg/s da}, \ T_{db2} = 48.9°C, \ \omega_2 = 0.01799$$

The dead state condition is given as follows:

$$p_0 = 101.325 \text{ kPa}, \ T_{db0} = 35°C, \ \omega_0 = 0.01406$$

Note that these values for the dead state should be entered in place of the default values when executing the computer program **PSY**.

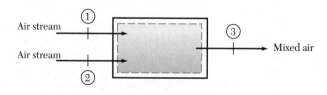

Figure 13.12 Adiabatic mixing process of Example 13.15.

Find: (a) T_{db3} and ω_3, (b) η_2

Solution:

(a) To determine the condition of the resulting mixture, carry out a mass balance and an energy balance.

Mass balance

For the dry air: $\dot{m}_{a3} = \dot{m}_{a1} + \dot{m}_{a2} = 3.024$ kg/s da

For the moisture: $\omega_3 = \dfrac{\dot{m}_{a1}\omega_1 + \dot{m}_{a2}\omega_2}{\dot{m}_{a3}} = \underline{0.0146}$

Energy balance

Use the computer program to obtain the following:

$$(h_{ma})_1 = f(T_{db1}, \omega_1) = 55.24 \text{ kJ/kg da}$$

and

$$(h_{ma})_2 = f(T_{db2}, \omega_2) = 95.72 \text{ kJ/kg da}$$

Application of the SFEE (Eq. 6.29) to the adiabatic mixing process yields the following for the enthalpy of the mixed air stream:

$$(h_{ma})_3 = \frac{\dot{m}_{a1}\,(h_{ma})_1 + \dot{m}_{a2}\,(h_{ma})_2}{\dot{m}_{a3}} = 75.48 \text{ kJ/kg da}$$

Further use of the computer code **PSY** yields

$$T_{db3} = f[(h_{ma})_3, \omega_3] = \underline{37.9°C \ (100°F)}$$

(b) *Availability analysis*

For each of the states 1, 2, and 3, the exergy per unit mass of dry air (x_{ma}) is on the list of properties output by the computer program. Thus,

$$(x_{ma})_1 = 0.162 \text{ kJ/kg da}$$
$$(x_{ma})_2 = 0.386 \text{ kJ/kg da}$$
$$(x_{ma})_3 = 0.015 \text{ kJ/kg da}$$

The second law efficiency η_2 for the process can be defined as follows:

$$\eta_2 = \frac{\dot{X}_{product}}{\dot{X}_{input}} 100\%$$

$$= \frac{\dot{m}_{a3}\,(x_{ma})_3}{(\dot{m}_{a1}\,(x_{ma})_1 + \dot{m}_{a2}\,(x_{ma})_2)} 100\% = \underline{5.47\%}$$

Note that this is about the same as the value of 5.4% obtained by Wepfer et al. 1979.

EXAMPLE 13.16
Use of PSY Code for Steam Spray Humidification Analysis

A stream of 3.024 kg/s da (24,000 lb/h da) of outdoor air at 35°C (95°F) db and ω = 0.01406 is sprayed with 0.0106 kg/s (84.1 lb/h) of dry saturated steam at 110°C. p_0 is the same as that for Example 13.15. Determine

(a) the dry bulb temperature and humidity ratio of the humidified air stream

(b) the second law efficiency of the process

Given: The steam spray humidification process is illustrated in Figure 13.3.

$$\dot{m}_a = 3.024 \text{ kg/s da}, \ T_{dbi} = 35°C \ (= T_{dbo}), \ \omega_1 = 0.01406 \ (= \omega_0)$$

$$\dot{m}_v = 0.0106 \text{ kg/s}, \ T_v = 110°C \text{ (dry sat. vapor)}$$

Find: (a) T_{dbe} and ω_e, (b) η_2

Solution:

(a) *Mass balance*

The mass of dry air in the inlet moist air stream is the same as that in the exit moist air stream. The moisture content differs, however. As in Example 13.7, an equation expressing mass balance for moisture can be written from which the humidity ratio for the exit air stream can be determined:

$$\omega_e = \omega_i + \frac{\dot{m}_v}{\dot{m}_a} = \underline{0.0176}$$

Energy balance

The pressure p_v of the dry, saturated steam at 110°C is needed to use the computer code for a determination of the specific enthalpy of the steam. From Table A.1, $p_v = p_s (110°C) = 143.3$ kPa. Thus, the **PSY** code gives the following:

$$(h_{ma})_i = f(T_{dbi}, \omega_i) = 71.22 \text{ kJ/kg da}$$

$$h_v = f(T_v, p_v) = 2691 \text{ kJ/kg steam}$$

Application of the SFEE (Eq. 6.30) to the process yields

$$(h_{ma})_e = (h_{ma})_i + \frac{\dot{m}_v}{\dot{m}_a} h_v = 80.65 \text{ kJ/kg da}$$

The computer code is further used to determine $T_{dbe} = f[(h_{ma})_e, \omega_e] = \underline{35.4°C \ (95.7°F)}$.

(b) *Availability analysis*

The **PSY** code gives the following for the exergy per unit mass for each of the streams:

$$(x_{ma})_i = 0$$

$$(x_{ma})_e = 0.057 \text{ kJ/kg da}$$

$$x_v = 606 \text{ kJ/kg steam}$$

The second law efficiency is thus determined by

$$\eta_2 = \frac{\dot{X}_{product}}{\dot{X}_{input}} 100\%$$

$$= \frac{\dot{m}_a (x_{ma})_e}{\dot{m}_a (x_{ma})_i + \dot{m}_v x_v} 100\% = \underline{2.68\%}$$

Note that Wepfer et al. (1979) obtained 2.67% for this case.

EXAMPLE 13.17
Use of the PSY Code for Analysis of an Adiabatic Water Injection Humidification Process

Adiabatic humidification of 3.024 kg/s da (24,000 lb/h da) of outdoor air at 35°C db (95°F) and $\omega = 0.01406$ is produced by injecting 0.0106 kg/s (84.1 lb/h) of water at 23.91°C (75°F). Determine the second law efficiency of the process.

Given: The adiabatic water injection humidification process is illustrated in Figure 13.13.

$$\dot{m}_a = 3.024 \text{ kg/s da, } T_{dbi} = 35°C \ (= T_{dbo}), \ \omega_1 = 0.01406 \ (= \omega_0)$$

$$\dot{m}_w = 0.0106 \text{ kg/s, } T_w = 23.9°C \text{ (water)}$$

Find: η_2

Solution: First, determine the state of the humidified air stream. As in Examples 13.15 and 13.16, a mass balance yields $\omega_e = 0.0176$ for the exiting air stream. Again, applying the SFEE coupled with the computer code **PSY** gives a dry bulb temperature of 26.7°C (80°F) for the humidified air stream.

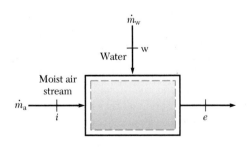

Figure 13.13 Adiabatic water injection humidification process of Example 13.17.

Availability analysis

The output of the computer program **PSY** yields the following values of the exergy per unit mass for each stream involved in the humidification process:

$$(x_{ma})_i = 0$$

$$(x_{ma})_e = 0.176 \text{ kJ/kg da}$$

$$x_w = 132 \text{ kJ/kg water}$$

The second law efficiency can be obtained as follows:

$$\eta_2 = \frac{\dot{X}_{product}}{\dot{X}_{input}} 100\%$$

$$= \frac{\dot{m}_a\, (x_{ma})_e}{(\dot{m}_a\, (x_{ma})_i + \dot{m}_w x_w)} 100\% = \underline{38.2\%}$$

Note that Wepfer et al. (1979) obtained 37.97% for this case.

Note that although the same level of humidity is achieved in both Example 13.16 and Example 13.17, the latter is more efficient and much less destructive of exergy.

EXERCISE 1. Repeat the exercises for Section 13.3.3 using the computer code **PSY**.

for Section
13.3.4

13.4 *Summary*

Mixtures of gases in general are pure substances. If both gaseous and liquid phases are present in a system, however, it is generally unlikely that the two phases will have the same chemical composition. In addition to the two independent intensive properties that must be specified for a pure substance, the relative amounts of the components of a gas mixture must also be given for a unique definition of the state of the system. The behavior of ideal gas mixtures was considered in this chapter. Such mixtures can be modeled by treating each constituent as an ideal gas and applying the additive laws, which include Dalton's law of additive pressures, Amagat–Leduc's law of additive volumes, and the Gibbs–Dalton laws (additive internal energy, enthalpy, entropy, the Gibbs function, and the Helmholtz function). One important corollary of the laws of additive pressures and volumes is that the molar, pressure, and volumetric analyses for a mixture of ideal gases are all equivalent. The Gibbs–Dalton additive laws gave rise to several equations for determination of the internal energy, enthalpy, entropy, and exergy for ideal gas mixtures.

Air–water vapor mixtures, in particular, were treated as two-component mixtures of ideal gases for which several humidity parameters were defined. The thermodynamic analysis of psychrometric processes, involving air–water vapor mixtures, is based on the derivation of appropriate expressions for the enthalpy and exergy flow rate for moist air, water vapor, and (liquid) water streams. These were derived from the equations based on the Gibbs–Dalton additive laws for ideal gas mixtures in general.

Psychrometric charts are normally available to help with the first law analysis of psychrometric processes. The various expressions required for a complete analysis of psychrometric processes are incorporated into the computer code **PSY**, which comes with the text. This should facilitate both a first law and a second law thermodynamic analysis of typical psychrometric processes.

References

ASHRAE. 1983. *Equipment Handbook*. Atlanta, GA: American Society of Heating, Refrigerating, and Air-Conditioning Engineers.

Adebiyi, G. A. 1989. "Computer-Aided First- and Second-Law Thermodynamic Evaluation and Design of Air Conditioning Systems." *Proceedings of the Third National Conference on Microcomputer Applications in Energy*. Tucson, AZ, Nov. 1–3, 1988. 195–202.

Bejan, A. 1988. Advanced Engineering Thermodynamics. New York: John Wiley & Sons.

Hyland, R. W. and Wexler, A. 1983. "Formulations for the Thermodynamic Properties of Dry Air from 173.15 K to 473.15 K and of Saturated Moist Air from 173.15 K to 372.15 K, at Pressures to 5 MPa." *ASHRAE Transactions*. Vol. 89, Part 2A. 520–535.

Kline, S. J. and Koenig, F. O. 1957. "The State Principle: Some General Aspects of the Relationships Among the Properties of Systems." *Journal of Applied Mechanics*. Vol. 24. 29–34.

McCallum, T. O. and Russell, L. D. 1983. *Device to Reduce Air Conditioning Loads by Use of Waste Heat for Dehumidification*. Final Report on DOE Grant Number DE-FG44-81R-410486 to the U.S. Department of Energy.

Stewart, R. B., Jacobsen, R. T., and Becker, J. H. 1983. "Formulations for Thermodynamic Properties of Moist Air at Low Pressures as Used for Construction of New ASHRAE SI Unit Psychrometric Charts." *ASHRAE Transactions*. Vol. 89, Part 2A. 536–548.

Wepfer, W. J., Gaggioli, R. A., and Obert, F. E. 1979. "Proper Evaluation of Available Energy for HVAC." *ASHRAE Transactions*. Vol. 85, Part 1. 214–230.

Wepfer, W. J. and Gaggioli, R. A. 1980. "Reference Datums for Available Energy." In Gaggioli, R. A., ed. 1980. *ACS Symposium Series 122 on Thermodynamics: Second Law Analysis*. Washington, DC: American Chemical Society. 77–92.

Questions

1. Which of the systems listed below are pure substances and which are not?

   (a) a nonreacting mixture of CO_2, N_2, and O_2
   (b) a reacting mixture of H_2 and O_2
   (c) a mixture of H_2 and O_2 not yet reacting
   (d) a mixture of moist air and liquid water
   (e) dry air and moisture (water vapor)

2. State the following for a mixture of ideal gases:
   (a) the pvT equation of state
   (b) the relationship among c_p, c_v, and R

(c) the change in internal energy ΔU in terms of the appropriate specific heat, the mass of the system, and the change in temperature ΔT

3. State Dalton's law of additive pressures and prove by application of the law that a mixture of ideal gases is itself an ideal gas.

4. Explain the terms *partial pressure* analysis, *volumetric* analysis, and *molar* analysis for a mixture of ideal gases. What is the relationship among the three analyses listed for an ideal gas mixture?

5. What is the approximate volumetric analysis for atmospheric air? What is the number of kmols of N_2 in atmospheric air for each kmol of O_2 in the air?

6. A sample of moist air has a relative humidity of 100% at ambient pressure p_0 and temperature T_0. Describe what happens as a result of each of the following processes:
   (a) the temperature of the moist air is raised
   (b) the temperature of the moist air is lowered
   (c) the pressure of the moist air is raised

(d) the pressure of the moist air is lowered
(e) moisture is added to the moist air sample

7. Moist air at 30°C db and 25°C wb is to be used for keeping chickens cool in a poultry farm. The air stream passes through a wet mat before reaching the birds. What is the lowest possible temperature for the air stream reaching the birds?

8. The products of combustion when natural gas is burned in air normally contain a substantial amount of moisture. If the moisture should condense on a metal, corrosion is likely to set in. Suggest steps that could be taken to safeguard against this in the design of gas burners.

9. Discuss the promising options for solar-assisted air conditioning in a hot, humid climate.

10. Describe a typical scheme for home air conditioning in the summer using a vapor compression refrigeration system. Discuss the major irreversibilities in the entire process.

Problems

13.1. A mixture of gases comprises 30% CO, 15% CO_2, and 55% H_2 by volume. Find for the mixture
(a) the gravimetric analysis
(b) the specific gas constant
(c) the molecular weight

13.2. An approximate volumetric analysis of dry air is given as 21% O_2 and 79% N_2. Determine for the air sample
(a) the gravimetric analysis
(b) the mass of O_2 contained in 1 m^3 of air at 100 kPa and 25°C

13.3. A sample of air has a mass of 1 kg and comprises 23.2% of O_2 by mass with N_2 forming the balance. The pressure of the air is 100 kPa and the temperature is 350 K. Determine
(a) the partial pressure of oxygen in the air
(b) the volumetric analysis of the sample
(c) the total volume of the mixture

13.4. A gaseous mixture consists of the following constituents (percentage by mass): 18% H_2, 4% H_2O, 25% O_2, 21% N_2, 3% CO, 13% CO_2, and 16% CH_4. Determine the specific gas constant for the mixture.

13.5. The partial pressures for the constituents of an ideal gas mixture are as follows: O_2, 5 psia; N_2, 7 psia; H_2, 0.2 psia; H_2O, 0.1 psia; CO, 1 psia; and CO_2, 2 psia. Determine the total pressure of the mixture.

13.6. What is the specific gas constant R for benzene (C_6H_6)?

13.7. An ideal gas mixture comprises CO, CO_2, and N_2. A volume of 10 liter (l) of the mixture at 100 kPa and 25°C is passed through reagents that successively remove the CO_2 and CO from the sample. After removal of the CO_2, the volume of the remaining mixture is 7.95 l when reduced to a pressure of 100 kPa at a temperature of 25°C. When the carbon monoxide is removed, the volume left is 3.85 l at 100 kPa and 25°C. Determine
(a) the volumetric analysis of the original gas mixture
(b) the specific gas constant of the gas mixture
(c) the molecular weight of the gas mixture
(d) the mass of the original sample

13.8. The volumetric analysis of air is given as follows: 78.09% N_2, 20.95% O_2, 0.93% Ar, and the rest

CO_2. The corresponding molar specific heats at constant pressure are 29.13, 29.38, 20.78, and 37.23 kJ/kmol · K. Determine

(a) the molar specific heat at constant pressure of the air sample
(b) the molecular weight of air
(c) the gravimetric analysis of the air sample
(d) c_p, c_v, and R for air

13.9. The gravimetric analysis of a sample of moist air at 1 atm and 25°C is 98.6% dry air and 1.4% moisture. Determine

(a) the partial pressure exerted by each of the two components of the moist air sample
(b) the specific gas constant for the moist air
(c) the humidity ratio of the moist air
(d) the mass of moisture in 30 m³ of the air sample

13.10. Using the ideal gas equation, show that 1 kmol of any ideal gas occupies 24.45 m³ at the standard conditions of 25°C and 1 atm. Hence, calculate the density (in kg/m³) for the moist air sample in Problem 13.9.

* 13.11. An expression for the specific heat at constant pressure for ethylene (C_2H_4) is as follows (with T in K):

$$\bar{c}_{p0} = (c_1 + c_2T + c_3T^2 + c_4T^3) \text{ kJ/kmol} \cdot \text{K}$$

where

$$c_1 = 3.4, c_2 = 0.1536,$$
$$c_3 = -7.898 \times 10^{-5}, \text{ and}$$
$$c_4 = 1.59 \times 10^{-8}$$

Using a reference temperature of $T_{\text{ref}} = 298$ K, and assuming $\bar{h}_{\text{ref}} = 52.4$ MJ/kmol for ethylene, tabulate values for $\bar{h}$ and $\bar{u}$ (in MJ/kmol) for ethylene at temperatures of 400 K, 500 K, 600 K, 700 K, 800 K, 900 K, and 1000 K.

* 13.12. Using data given in Problem 13.11 and assuming ideal gas behavior, tabulate $\bar{s}$ (in kJ/kmol · K) for ethylene at 1 atm and temperatures given in the problem. (Assume that the reference pressure is 1 atm and that $\bar{s}_{\text{ref}} = 219.60$ kJ/kmol · K for ethylene.)

* 13.13. Using data given in Problem 13.11 and assuming

ideal gas behavior, tabulate values for the flow exergy per kmol of ethylene at 2 atm and temperatures given in the problem. (Assume that the surroundings are at 1 atm and 298 K.)

13.14. Determine the mean molecular weight (M) for the gaseous mixture given in Problem 13.4.

* 13.15. Using data from the ideal gas tables or the **GAS** code, determine $\bar{h}$, $\bar{u}$, and $\bar{s}$ for the ideal gas mixture of Problem 13.5 if the temperature of the mixture is 400 K.

* 13.16. Using data from the ideal gas tables or the **GAS** code, determine the mean molar specific heat at constant pressure and the mean molar specific heat at constant volume for the mixture in Problem 13.5 over the temperature interval 25°C to 500 K.

* 13.17. Using results from Problem 13.16, compute the thermomechanical[†] flow exergy per kmol of the mixture given in Problem 13.5 when the temperature of the mixture is 500 K. (Assume that the surroundings are at 1 atm and 298 K.)

* 13.18. A steady stream of ethylene and nitrogen is at $p = 1$ atm and $T = 300$ K. The volumetric composition of the mixture is 50% ethylene and 50% nitrogen. Determine the minimum exergy supply required per unit mass of the mixture to separate the stream into its constituents with each constituent existing at $p = 1$ atm and $T = 300$ K.

* 13.19. A gaseous mixture is flowing in a steady stream at $p = 1$ atm and $T = 300$ K. The volumetric composition of the mixture is 25% N_2, 15% CO, 40% CO_2, 3% H_2, and 17% O_2. Determine the minimum exergy supply required per unit mass of the mixture to separate the stream into its constituents. Assume that each separated gas exists at $p = 1$ atm and $T = 300$ K.

* 13.20. The mole fraction of argon in ambient air is 0.0093. Determine the minimum exergy supply required to obtain 1 kmol of argon at 1 atm and 300 K from a steady stream of atmospheric air at 1 atm and 300 K. Assume that the remaining gas mixture, after removal of argon, is also at 1 atm and 300 K.

13.21. A fixed mass of moist air at 1 atm and 35°C is

[†]Thermomechanical exergy is frequently used for the restricted equilibrium case when the chemical exergy of a substance is not taken into account (see Bejan 1988, for example).

cooled at a constant pressure until condensation of the moisture just begins when the temperature has reached 20°C. Find

(a) the partial pressure of water vapor in the initial moist air sample

(b) the gravimetric analysis of the air sample

* **13.22.** The products of complete combustion of methane in air are scrubbed by passage through a water bath. The number of moles of CO_2, O_2, and N_2 in the mixture are 1 kmol, 3 kmol, and 18.8 kmol, respectively. If the total pressure of the scrubbed products is 100 kPa and the temperature is 40°C, determine the number of kmols of water vapor in the scrubbed products.

13.23. Calculate the relative humidity and the humidity ratio of an air–water vapor mixture at 1 bar and 35°C if the dew point temperature is 20°C.

13.24. Calculate the relative humidity and the humidity ratio of an air–water vapor mixture at 1 atm and 90°F if the dew point temperature is 78°F.

13.25. Calculate the dew point temperature and the humidity ratio of an air–water vapor mixture with a total pressure of 1 atm, a temperature of 80°F, and a relative humidity of 75%.

13.26. Calculate the humidity ratio in the following cases:

(a) dew point temperature $T_d = 10°C$, dry bulb temperature $T_{db} = 15°C$

(b) dry bulb temperature $T_{db} = 30°C$, relative humidity $\phi = 90\%$

Assume a total pressure of 1 atm in both cases.

13.27. Calculate the unknown parameters for moist air in the following:

(a) T_{db} when $\omega = 0.0142$ and $\phi = 40\%$

(b) T_d when the humidity ratio is 0.008

(c) ϕ and μ when temperature is 25°C db and the humidity ratio is 0.01

(d) ω_s when the temperature is 30°C db

(e) the humidity ratio when the dew point temperature is 14°C

(f) T_{db} when $\omega = 0.01$ and $\phi = 20\%$

Assume that the pressure of moist air in each case is 100 kPa.

13.28. Cross-check your answers in Problems 13.26 and 13.27 using

(a) the psychrometric chart

(b) the computer code **PSY**

* **13.29.** A house having internal dimensions of 1800 ft² floor space and ceilings 8 ft high contains moist air at 1 atm and 70°F with a relative humidity of 60%. Calculate

(a) the humidity ratio

(b) the degree of saturation

(c) the dew point temperature

(d) the total mass of water vapor in the house

* **13.30.** Use the psychrometric chart to solve Problem 13.29.

* **13.31.** Outside air at 100°F with a humidity ratio of 0.013 is mixed with return air at 78°F with a humidity ratio of 0.010. The return air is mixed with outside air at the ratio 1:1 on a dry air mass basis. If the mixing occurs adiabatically, determine

(a) the temperature and humidity ratio of the resulting mixed air stream

(b) the second law efficiency of the mixing process

* **13.32.** A steady stream of dry, saturated steam at 120°C is sprayed into an ambient air stream at 25°C with a humidity ratio of 0.01. The mass flow rate of the steam is 0.1 kg/s, while the ambient air stream flow rate is 40 kg/s da. The ambient pressure is 1 atm. Determine the second law efficiency of the process, assumed to be adiabatic.

* **13.33.** A steady stream of ambient air at 95°F with a humidity ratio of 0.012 is cooled and humidified adiabatically by spraying the stream with liquid water at 60°F. The flow rate of air is 5 lbm/s da, while the water spray is at the rate of 0.02 lbm/s. The ambient air pressure is 1 atm. Determine

(a) the exit temperature of the moist air stream

(b) the second law efficiency of the process

* **13.34.** Ambient air at 1 atm and 95°F with a humidity ratio of 0.013 is passed over the cold evaporator coil of an air conditioner. The air leaves the coil at 50°F in a saturated condition. The mass flow rate of the air is 1 lbm/s da. Determine

(a) the rate that water is being condensed from the wet air stream

(b) the cooling load

(c) the minimum exergy supply rate required for the air conditioning process

(d) the second law efficiency achieved for the process if the cooling coil is part of a refrigeration device operating with a $COP_C = 2.5$.

Sketch the process on a psychrometric chart.

* **13.35.** A hygrometer gives a dry bulb temperature of 90°F and a wet bulb temperature of 75°F. Calculate the relative humidity of the air.

* **13.36.** A hygrometer gives a dry bulb temperature of 35°C and a wet bulb temperature of 15°C. Calculate the relative humidity of the air.

* **13.37.** A stream of 12,000 lbm/h da of outdoor air at 90°F db with a humidity ratio of 0.013 is sprayed with 40 lbm/h of dry, saturated steam at 220°F. The surroundings are at 90°F and 1 atm. Determine
  (a) the dry bulb temperature and humidity ratio of the humidified air stream
  (b) the second law efficiency of the process

* **13.38.** Adiabatic humidification of 12,000 lbm/h da of outdoor air at 90°F db with a humidity ratio of 0.013 is produced by injecting 40 lbm/h of water at a temperature of 65°F. The surroundings are at 90°F and 1 atm. Determine the second law efficiency of the process.

13.39. Steam is generated at the rate of 4 kg/s in a kitchen that is to be ventilated by an air stream having a humidity ratio of 0.008. Upon exit, the humidity ratio of the air stream is not to exceed 0.0101. Determine the mass flow rate (in kg/s da) of the ventilating air stream required.

13.40. Moisture is removed from a steady stream of moist air at 35°C db and a relative humidity of 40%. The exiting air stream is saturated and is at a temperature of 10°C. If the mass flow rate of the inlet air stream is 2 kg/s da, determine the moisture removal rate (in kg/h). Assume an ambient pressure of 100 kPa.

13.41. A steady stream of ambient air at 35°C db with a humidity ratio of 0.0142 mixes adiabatically with a stream of room air at 27°C with a humidity ratio of 0.011. The mass flow rate on a dry air basis is the same for the two streams. The pressure can be taken as 1 atm for the air streams. Find
  (a) the humidity ratio of the mixture
  (b) the dry bulb temperature of the mixed air stream
  (c) the second law efficiency of the process

13.42. Air at a volume flow rate of 0.2 m³/s and at 10°C and 50% relative humidity flows through a device in which it is both heated and sprayed with (liquid) water at 25°C. The air stream exits at 35°C and 90% relative humidity. The condition of ambient air can be taken as $p_0 = 1$ atm, $T_0 = 35$°C db, and $\omega_0 = 0.0142$. Determine
  (a) the heat supplied (in kW)
  (b) the water supplied (in kg/h)

* **13.43.** It is estimated that the average person under normal conditions adds about 0.5 kg/h of moisture to his or her immediate environment. In addition, a sensible heating loss rate of about 90 W occurs from each person by convection and radiation. Consider a classroom that measures 5 m × 6 m × 3 m and has 35 people in it. The classroom is tightly sealed and is maintained at an average temperature of 25°C with a humidity ratio of 0.01 by a central air conditioning unit. The air conditioning unit fails and is inoperable for a period of 30 min. Estimate the average temperature and humidity ratio of the air in the room after this period. Clearly state your assumptions.

13.44. An ambient moist air stream, saturated at 2°C, enters a heating device at the rate of 9.5 m³/s. The air leaves the coil at 38°C. Determine
  (a) the heating rate (in kW) required
  (b) the second law efficiency of the process, assuming that the heating is produced using an electric resistance heater

* **13.45.** A stream of ambient air at 20°C db and 6.6°C wb is processed by adiabatic injection of superheated steam at 110°C and 1 atm. The exit stream has a dew point temperature of 12.5°C. The flow rate of the air is 1.5 kg/s da. Determine
  (a) the final dry bulb temperature of the moist air
  (b) the required steam supply rate (in kg/h)
  (c) the second law efficiency of the process

* **13.46.** An ambient air stream at 35°C db with a humidity ratio of 0.0142 is cooled by passage through a wetted mat before being used for ventilation on a poultry farm. Assuming the humidification process is adiabatic, determine the temperature of the water in the wetted mat that will ensure that the processed air stream is a saturated moist air stream at the same temperature as the water.

* **13.47.** Liquid mercury is stored in a classroom laboratory that has no ventilation. A mercury spill occurs, and the mercury is not totally removed from the floor. In such a case, the mercury evaporates and reaches an equilibrium condition

with the mole fraction (N_i/N) for mercury given by p_{vap}/p, where p_{vap} is the vapor pressure of the mercury and p is the total system pressure.

(a) If such a spill occurs in a laboratory at a temperature of 20°C, what will the mole fraction of mercury be at equilibrium conditions? Assume that the vapor pressure of the mercury is 1.7×10^{-6} bar at 20°C and that the room air pressure is 1 atm.

(b) Using the ideal gas law, determine the molar specific volume for air at this condition. Then determine the mass of mercury per cubic meter of air (that is, the concentration of the mercury) in the room ($M = 200.61$ for mercury).

(c) The present U.S. Federal Standard (OSHA) for exposure to mercury in air is 0.1 mg/m³ as a maximum value. Is the concentration calculated in part (b) a hazard according to the standard? Is it safe to store an open container of mercury in an unventilated area?

* **13.48.** A student leaves 5 lbm of carbon disulfide (CS_2) in an open container in an unventilated laboratory room for three days. The volume of the room is 1200 ft³, and the temperature in the room is 85°F. The vapor pressure of CS_2 at 85°F is about 0.55 atm, and its normal boiling point is about 115°F. Therefore, it is likely that the CS_2 will evaporate rapidly in such conditions.

(a) If the CS_2 evaporates as expected in the unventilated room, what will the mole fraction of the CS_2 be after evaporation?

(b) If the mole fraction of CS_2 in air exceeds 1.3%, then CS_2 is an explosive hazard. If the mole fraction exceeds this limit (that is, the lower flammable limit, or LFL), then an explosion may occur if a source of ignition is present. Would the conditions determined in part (a) constitute a potential hazard?

* **13.49.** A mixture of ideal gases has the following volumetric analysis: 10% O_2, 20% N_2, 40% CO_2, and 30% CH_4.

(a) Determine the gravimetric analysis of the mixture and its apparent molecular weight.

(b) A tank with a volume of 5 m³ contains 15 kg of this gas at 10°C. What is the pressure in the tank?

13.50. A mixture of ideal gases has the following mass analysis: 10% O_2, 20% He, 40% N_2, and 30%

CO_2. The total pressure is 30 psia and the temperature is 70°F. Determine

(a) the partial pressure of the CO_2

(b) the apparent molecular weight and the gas constant for the mixture

* **13.51.** A steady stream of ambient air at 35°C with a humidity ratio of 0.0142 is cooled and humidified adiabatically by applying a (liquid) water spray at 25°C. The flow rate of air is 3 kg/s da, while the water spray is at the rate 0.01 kg/s. The ambient air pressure is 1 atm. Determine

(a) the exit temperature of the moist air stream

(b) the second law efficiency of the process

* **13.52.** Ambient air at 1 atm and 35°C with a humidity ratio of 0.0142 is passed over the cold evaporator coil of an air conditioner. The air leaves the coil at 13.3°C and in a saturated condition. The mass flow rate of the air is 0.5 kg/s da.

(a) How much water is condensed from the wet air stream?

(b) Estimate the cooling load.

(c) What is the minimum exergy supply rate required for the air conditioning process?

(d) If the cooling coil is part of a refrigeration device operating with a COP_C of 3, estimate the second law efficiency achieved.

Sketch the process on a psychrometric chart.

* **13.53.** Typical ambient air in Starkville, MS, during the summer period can be taken as 93°F (33.89°C) db with a humidity ratio of 0.0163. In a poultry farm, the ambient air is drawn through a wet mat to lower the temperature of the air reaching the birds. If the air reaching the birds is saturated and at the wet bulb temperature, determine the reduction in temperature achieved

(a) by using a psychrometric chart for the wet bulb temperature

(b) by calculating the adiabatic saturation temperature (assumed equal to the wet bulb temperature)

Problems 13.54 to 13.58 involve a desiccant dehumidifier designed by McCallum and Russell (1983). A desiccant dehumidifier is a device for removing moisture from a moist air stream via the adsorption process (see ASHRAE 1983, Chap. 7). Figure P13.54 depicts the operation of McCallum and Russell's desiccant dehumidifier. Table 13.4 provides a typical set of performance data that was obtained. Both the process air

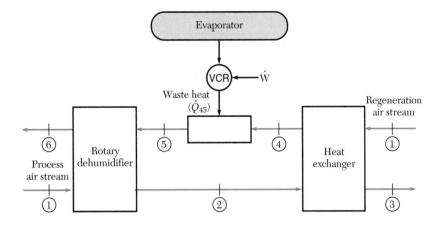

Figure P13.54 Schematic diagram of the desiccant dehumidifier designed by McCallum and Russell (1983) using the vapor-compression refrigeration (VCR) device.

TABLE   13.4

Typical Set of Performance Data Obtained for the Desiccant Dehumidifier by McCallum and Russell

State Number	Dry Bulb Temperature, T_{db} (°C)	Humidity Ratio (ω)
1	25.9	0.0153
2	39.7	0.0106
3	28.3	0.0106
4	30.3	0.0153
5	50.8	0.0153
6	36.7	0.0173

stream and the regeneration air stream are initially at the ambient condition of 25.9°C db with a humidity ratio of 0.0153. The volume flow rates of the process air and the regeneration air are 40 scfm and 110 scfm, respectively. (scfm is standard cubic feet per minute, or volumetric air flow per minute at 1 atm and 25°C.) The waste heat transfer rate ($\dot{Q}_{45}$) to the regeneration air stream is 4450 Btu/h. The COP_C for the vapor compression refrigeration (VCR) device can be taken as 2.5.

13.54. Show processes 1→2→3 and 1→4→5→6 on the psychrometric chart. Determine
  (a) the mass flow rates, on a dry air basis, of the process air and the regeneration air
  (b) the moisture removal rate (in kg/h) resulting from the dehumidification process

* **13.55.** The actual moisture removal occurs in process 1→2. Suggest a suitable definition of a first law efficiency in which the desired outcome is moisture removal. Hence, compute a first law efficiency for the dehumidification process using the waste heat transfer rate to the dehumidifier as the energy input rate.

* **13.56.** Determine the second law efficiencies achieved in
  (a) process 1→2
  (b) process 1→2→3

13.57. The vapor compression refrigeration (VCR) device could have been used to produce further cooling of the process air stream. Determine the lowest temperature that could theoretically be achieved if the process air stream, leaving the dehumidifier at state 3, were to be passed over the evaporator coils of the VCR.

* **13.58.** What is the overall second law efficiency that could be achieved in the mode of operation suggested in Problem 13.57?

Design Problems

13.1 *Aircraft air conditioning*

Repeat Design Problem 11.1 from Chapter 11. This time, however, assume that the ambient air at the ground level has a relative humidity of 90% and that the cabin air humidity should be kept below 50%.

13.2 *Residence air conditioning*

Repeat Design Problem 11.2 from Chapter 11 for the case in which the residence is located in Houston, Texas. In this case the summertime design-point condition is 93°F and 85% relative humidity for the outside air. You are required to maintain the air inside the house at 78°F with a maximum relative humidity of 50%. What is the rate of moisture condensation required for this case.

Determine a reasonable evaporator temperature for an operation in New Mexico where no moisture removal is required. Also determine the evaporator temperature required for operation in Houston. Estimate the first and second law efficiencies for the units operating in the two locations, and determine the impact on the efficiencies due to the requirement for moisture removal in Houston. Prepare a report of your results.

13.3 *Design of evaporative cooling air conditioning system*

Design an evaporative cooling air conditioning system for the residence described in Design Problem 11.2. Assuming that water is used for the evaporative cooling process, what is the rate of water use at the summertime design point? Prepare a report of your results. Show schematic diagrams and thermodynamic process diagrams for your design. State your assumptions and include an estimated cost for your unit. Compare your estimated cost with that for a vapor compression unit using refrigerant-22.

CHAPTER 14

Combustion

14.1 Introduction

Previous chapters have considered only nonreacting systems. Chemical reactions play a significant role in several applications in engineering, such as automobile engines, thermal power plants, and aircraft engines. Accordingly, this chapter is concerned with the application of the basic principles of thermodynamics to reacting systems.

The term *combustion* here means a chemical reaction involving a fuel and oxygen—the reactants—to form products. The reaction generally occurs rapidly and in an uncontrolled manner and is usually accompanied by a flame with heat released to the surroundings. Most combustion processes involve a regulated supply of fuel and the oxidizer; in contrast, fires are more likely to occur with an unregulated supply of air.

The molecules of a fuel have energy stored in the bonds between their constituent atoms. This energy is a form of internal energy called *chemical energy*. When combustion occurs, stored chemical energy of the fuel is transformed, in essence, to another form of internal energy of the reacting mixture. If the reacting system is thermally well insulated, the internal energy is unchanged, although an increase in temperature of the system is observed. However, if the system is to remain at its original temperature, the energy released as a result of the reaction must be transferred to the surroundings, usually as heat. Such a reaction is termed an *exothermic reaction*. Conversely, reactions that require absorption of heat so that the system temperature remains constant are referred to as *endothermic*. The break-up or dissociation of carbon dioxide (CO_2) into carbon monoxide (CO) and oxygen (O_2) is a typical case of an endothermic reaction.

Fuels and Combustion Products

Most of the fuels commonly used for combustion processes are hydrocarbons comprising mainly hydrogen (H_2) and carbon (C), with the most chemically stable products of combustion being water or water vapor (H_2O) and CO_2. In addition, CO is often found in the products of combustion, although it is an undesirable product for safety and efficiency reasons. Other products formed due to ionization and high temperature effects may also be present. Frequently, such products are hazards to materials, plants, animals, and humans, that is, they are environmental pollutants and may produce such effects as acid rain.

The common fuels that are familiar from every day experience include coal, gasoline, heating oil, and natural gas. All of these fuels are hydrocarbons. The combustible elements in coal are chiefly C, H_2, and a small quantity of sulfur (S). Coal also generally includes noncombustible substances such as nitrogen (N_2), H_2O, and small quantities of minerals that form ash in the combustion process.

Petroleum oils are complex mixtures of many different hydrocarbons. Such oils include gasoline, kerosene, diesel fuel, light fuel oil, and heavy fuel oil. All of these substances include both carbon and hydrogen, which can react with oxygen to form CO_2 and H_2O as products of combustion. For simplification, this text normally considers that gasoline is *octane* with a chemical formula C_8H_{18}, diesel fuel is *dodecane* with a chemical formula of $C_{12}H_{26}$, and natural gas is *methane* with a chemical formula of CH_4. Other fuels include alcohols, wood products, and similar substances that react with oxygen to form products of combustion while releasing energy as heat transfer to the surroundings. As indicated already, most fuels are really very complex mixtures of a variety of elements. To be able to write down the chemical equation for the combustion process, a chemical analysis of the fuel is generally needed. For solid and liquid fuels, an analysis called an *ultimate analysis* is frequently performed that indicates the relative amount of each element in the fuel as a percent by mass. Such an ultimate analysis for coal could be given on an as-received basis, which includes H_2O, or it could be given on a dry basis, which excludes H_2O from the analysis.

Another analysis for coal is the *proximate analysis*, which gives the percent by mass of fixed carbon, volatile matter, ash, and moisture. Obviously this analysis is not adequate for detailed chemical evaluations of the combustion process, but it does give a rough indication of the relative value of different coals. The analysis of a gaseous fuel is normally given in percent by volume of each element or compound present in the gas. Such an analysis is referred to as a *volumetric analysis*.

As already mentioned, although the primary constituents of hydrocarbon fuels are carbon and hydrogen, trace elements of sulfur and nitrogen may be present which lead to combustion products that become air pollutants or are damaging to the construction materials that the products come into contact with. To avoid such dangers, these trace elements should be removed from the fuel prior to burning or prior to discharging the exhaust gases to the atmosphere. Many different processes have been developed and are currently being studied for controlling or minimizing such hazards. Although the analysis in this chapter is limited to pure fuels, the hazards posed by potential pollutants and impurities cannot be ignored in practical applications.

14.2   Conservation of Mass and Atomic Species

In any chemical reaction, the elements and components that exist prior to the reaction are called *reactants* and those that exist after the reaction is completed are called *products*. Thus, an equation for the reaction is written in the following form:

$$\text{Reactants} \rightarrow \text{Products} \tag{14.1}$$

For such a reaction, the conservation of mass states that the total mass of the products must be the same as that for the reactants. Furthermore, as you recall from elementary chemistry, the total mass of each chemical element is conserved for the process.

Now consider the simple reaction of carbon with oxygen to form carbon dioxide as given by

$$C + O_2 \text{ (reactants)} \rightarrow CO_2 \text{ (product)} \tag{14.2}$$

This equation shows that one atom of carbon reacts with one molecule of oxygen (two atoms of O) to produce one molecule of carbon dioxide (which consists of one atom of carbon plus two atoms of oxygen). The number of atoms of each element is the same on each side of the equation, thus, it is called a *balanced equation*. The arrow represents the direction of the reaction.

One can also consider Equation 14.2 on a molar basis, as follows:

$$1 \text{ mol } C + 1 \text{ mol } O_2 \rightarrow 1 \text{ mol } CO_2 \tag{14.3}$$

Since the mass of each element is given by its molecular mass, Equation 14.3 can also be expressed (in SI units) as

$$12 \text{ kg } C + 32 \text{ kg } O_2 \rightarrow 44 \text{ kg } CO_2 \tag{14.4}$$

Recalling the fact that all ideal gases occupy equal volumes per mole at the same pressure and temperature and recognizing that solids and liquids have negligible volume in comparison to gases, Equation 14.3 can also be written in terms of volume (vol) as

$$0 \text{ vol } C \text{ (s)} + 1 \text{ vol } O_2 \rightarrow 1 \text{ vol } CO_2 \tag{14.5}$$

where it is assumed that the gases are ideal gases and the carbon reactant is a solid. If the carbon had existed as a reactant in a gaseous (g) form, the equation would have been written as

$$1 \text{ vol } C \text{ (g)} + 1 \text{ vol } O_2 \rightarrow 1 \text{ vol } CO_2 \tag{14.6}$$

Although not all reactants and products can be treated as ideal gases, most can for typical applications. In general, ideal gas behavior is assumed here for both gas reactants and products in the analyses that follow (unless stated otherwise).

Now consider the reaction of hydrogen gas with oxygen to form H_2O:

$$H_2 + \tfrac{1}{2} O_2 \rightarrow H_2O \tag{14.7}$$

In this case, one has

$$1 \text{ mol } H_2 + \tfrac{1}{2} \text{ mol } O_2 \rightarrow 1 \text{ mol } H_2O \tag{14.8}$$

or

$$2 \text{ kg } H_2 + 16 \text{ kg } O_2 \rightarrow 18 \text{ kg } H_2O \tag{14.9}$$

If the H_2O exists as a vapor in the product, one can write

$$1 \text{ vol } H_2 + \tfrac{1}{2} \text{ vol } O_2 \rightarrow 1 \text{ vol } H_2O \text{ (g)} \tag{14.10}$$

Since liquids have negligible volume in comparison to gases, the volume equation would be written as follows when the H_2O in the products is in the liquid form:

$$1 \text{ vol } H_2 + \tfrac{1}{2} \text{ vol } O_2 \rightarrow 0 \text{ vol } H_2O \text{ (l)} \tag{14.11}$$

These equations show that mass is conserved while the number of moles (or the volume) is not. If the chemical equation is appropriately balanced, one must be able to show that the total mass of the products is the same as that for the reactants and also that the mass of each element is conserved.

EXAMPLE 14.1

Gasoline (C_8H_{18}) reacts with pure oxygen gas to form products that include only CO_2 and H_2O. Determine the mass of each product per kg of fuel supplied for the reaction.

Given: Reaction of C_8H_{18} and O_2 to form CO_2 and H_2O

Find: Mass $CO_2/(\text{kg } C_8H_{18})$ and mass $H_2O/(\text{kg } C_8H_{18})$

Solution: The balanced chemical equation is

$$C_8H_{18} + 12.5\ O_2 \rightarrow 8\ CO_2 + 9\ H_2O$$

which gives

$$1 \text{ kmol } C_8H_{18} + 12.5 \text{ kmol } O_2 \rightarrow 8 \text{ kmol } CO_2 + 9 \text{ kmol } H_2O$$

Recall that gravimetric mass (m) is equal to the number of moles (N) times the molecular weight (M). Thus the mass balance is

$$114 \text{ kg } C_8H_{18} + 400 \text{ kg } O_2 \rightarrow 352 \text{ kg } CO_2 + 162 \text{ kg } H_2O;$$

the result is $352/114 = 3.09$ kg CO_2/kg fuel and $162/114 = 1.42$ kg H_2O/kg fuel. Note that 514 kg of reactants combine to form 514 kg of products.

1. Balance each of the following chemical equations:
   (a) $CO + ? O_2 \rightarrow CO_2$
   (b) $CH_4 + 4 O_2 \rightarrow ? CO_2 + ? H_2O + ? O_2$
   (c) $C_{12}H_{26} + ? O_2 \rightarrow ? CO_2 + ? H_2O$
   (d) $C + ? O_2 \rightarrow 0.9 CO_2 + ? CO$

2. Write down the molecular weights for each of the following substances:
   (a) carbon
   (b) oxygen (O_2)
   (c) hydrogen (H_2)
   (d) nitrogen (N_2)
   (e) sulfur (S)
   (f) dodecane $(C_{12}H_{26})$
   (g) ammonia (NH_3)

3. Verify the mass balances for each of the equations (as balanced) in Exercise 1.

14.3 Stoichiometry of Reactions

Stoichiometric Combustion

A *stoichiometric mixture* of a fuel and oxygen is a mixture that contains the precise amount of oxygen to burn all the combustible elements in the fuel completely. Such a reaction involves complete conversion of all carbon to CO_2 and all hydrogen to H_2O. Neither an excess nor a deficiency of oxygen is present in the reactants. Consequently, no free oxygen exists in the products of the reaction, while all the combustibles in the fuel are completely oxidized. For example, a stoichiometric reaction of propane (C_3H_8) with oxygen is given by

$$C_3H_8 + 5 O_2 \rightarrow 3 CO_2 + 4 H_2O \qquad (14.12)$$

The equation is balanced with the precise amount of oxygen present in the reactants to provide for the complete combustion of the fuel. Such a reaction is called the *theoretical* or *stoichiometric reaction*.

Complete Combustion

A reaction is termed *complete combustion* if all the combustibles in the fuel are completely oxidized. In other words, all the carbon and hydrogen in the reacting mixture now exists only in the stable forms of CO_2 and H_2O in the products. There is also the possibility that some oxygen is left over in a complete combustion process. In the previous example involving the combustion of propane in pure oxygen, if the reacting mixture had more than 5 kmol of O_2 to each kmol of C_3H_8, the excess amount of oxygen would remain in the combustion products. Thus, there could be complete combustion in this case, but the reaction would not be a theoretical or stoichiometric reaction. A theoretical reaction produces complete combustion, but complete combustion does not necessarily imply a theoretical reaction.

Incomplete Combustion

The term *incomplete combustion* is used if some unburned fuel remains in the products. This may be the case as a result of a deficiency of oxygen in the reactants leading to C,

CO, H_2, or OH in the products, or there might simply have been an inadequate exposure of the fuel to oxygen during the period of the combustion.

If an insufficient amount of oxygen is available in the reactants to completely oxidize all the carbon and hydrogen in the fuel, the hydrogen in the fuel will most likely be the element that is completely oxidized to form H_2O first because oxygen has a greater affinity for hydrogen than it does for carbon. In such a situation, the equation for H_2 going to H_2O is balanced and then CO and perhaps C are assumed to be present in the products.

Thus, in a reaction of propane in which only 4 kmol of oxygen are available per kmol of fuel in the reactants, the reaction might proceed according to the following:

$$C_3H_8 + 4\ O_2 \rightarrow 4\ H_2O + CO_2 + 2\ CO \tag{14.13}$$

Note that in Equation 14.13, all the hydrogen in the fuel is shown to be reacting with oxygen to form H_2O, while the relative proportions of CO_2 and CO in the products are assumed to give a correct balance for oxygen on both sides of the equation.

Composition of Atmospheric Air

For most applications involving combustion, air is used to supply the oxygen needed for the combustion process. For purposes of the calculations here, all elements in air are neglected except O_2 and N_2, and dry air is assumed to be 21% O_2 and 79% N_2 by volume. Therefore,

$$0.21\text{ mol }O_2 + 0.79\text{ mol }N_2 = 1\text{ mol air} \tag{14.14}$$

or, by simple proportion (with $0.79/0.21 = 3.76$),

$$1\text{ mol }O_2 + 3.76\text{ mol }N_2 = 4.76\text{ mol air} \tag{14.15}$$

Thus, every mole of oxygen is accompanied by 3.76 mol of N_2 when air is the oxidizer in a combustion process. This nitrogen exists as N_2 and is inert at atmospheric temperatures and pressures. It may react, however, at elevated temperatures to form nitrous oxides (NO_x), which are hazardous pollutants. To keep the treatment of the subject simple, N_2 is considered here to be an inert gas unless otherwise stated.

Air–Fuel Ratio (AF)

Whenever air provides the oxygen in a combustion process, the term *air–fuel ratio* (AF) is frequently used. It is defined as follows:

$$AF = \frac{\text{amount of air}}{\text{amount of fuel}}, \quad \text{by mass, number of moles, or volume} \tag{14.16a}$$

The stoichiometric reaction uses precisely the theoretical amount of air necessary to balance the equation with all fuel burned and no oxygen in the products. The air–fuel ratio in this instance is termed the *stoichiometric air–fuel ratio* (AF_s), while the amount of air required for the combustion of a unit mass (or volume) of the fuel is the *theoretical air*.

Excess Air

An excess of air provides more oxygen in the reactants than is needed to balance the equation, while a deficiency provides less. For example, the term *30% excess air* means that 1.3 times the stoichiometric quantity of air is supplied in the reactants, while the term *30% deficiency* means that 0.7 times the theoretical quantity is supplied. Likewise, 100% excess air means that 2 times the stoichiometric quantity of air is supplied in the reactants. The excess air can be computed from the following:

$$\text{Excess air} = \left(\frac{\text{AF}}{\text{AF}_s} - 1 \right) 100\% \qquad (14.16b)$$

In most applications, excess air is supplied to ensure complete combustion of the fuel, thus minimizing waste. It is extremely difficult to react the theoretical amount of oxygen with the fuel in most processes because of difficulties posed by inadequate mixing of air and fuel and also by the short time available for the reaction.

Fuel–Air Ratio (FA)

The term *fuel–air ratio* (FA) is also used to represent the ratio of fuel to air in the reactants. Thus, FA is the reciprocal of AF:

$$\text{FA} = 1/\text{AF} \qquad (14.16c)$$

Mixture Strength

Mixture strength for a fuel–air mixture is defined as the ratio of the amount of fuel in the actual mixture to that which is completely burned in a stoichiometric reaction using the same amount of air as in the actual mixture. The mixture strength (m.s.) can thus be determined using

$$\text{Mixture strength (m.s.)} = \frac{\text{actual fuel}}{\substack{\text{stoichiometric fuel} \\ \text{(with same air)}}} = \frac{\text{AF}_s}{\text{AF}} = \frac{\text{stoichiometric air}}{\text{actual air}} \qquad (14.16d)$$

Gravimetric, Molar, and Volumetric Analyses

As indicated in Equation 14.16a and discussed earlier with Equations 14.2 to 14.5, reactants and products in a combustion process can be specified on either a mass, volume, or molar basis. The term *gravimetric analysis* is used to indicate that the relative amounts of the components in a mixture are specified as a percent by mass (or weight). If, however, the mixture is characterized according to the relative amount (in kmol/kmol) present, the analysis is called *molar*. A *volumetric analysis* refers to characterization on a relative volume basis for each constituent in the mixture. Section 13.2 (Chap. 13) indicated that for ideal gas mixtures, the molar and volumetric analyses are identical.

Wet and Dry Products Analysis

Liquid water, H_2O (l), or water vapor, H_2O (g), frequently exists in the products of combustion. If the water exists as a vapor in the products, then the analysis of products is called a *wet analysis*. Whenever the water vapor is condensed and removed, the analysis of the products is referred to as a *dry analysis*. In balancing a reaction equation in which water vapor is supplied in the supply air, the H_2O (g) in the air is treated as an inert quantity and is carried through to the products in a manner similar to that discussed above for nitrogen.

The following examples provide several illustrations of how to conduct analyses of combustion reactions based on stoichiometric principles.

EXAMPLE 14.2

Methane (CH_4) reacts with dry air in a stoichiometric (or exactly correct) air–fuel ratio to form products of combustion. Determine the stoichiometric equation and calculate the air–fuel and the fuel–air ratios by volume.

Given: Reaction of CH_4 with stoichiometric amount of (dry) air to form CO_2 and H_2O

Find: Stoichiometric equation and AF and FA by volume

Solution: The stoichiometric equation can be obtained as follows:

$$CH_4 + \text{dry air} \rightarrow b\ CO_2 + c\ H_2O + d\ N_2$$

or

$$CH_4 + a(O_2 + 3.76\ N_2) \rightarrow b\ CO_2 + c\ H_2O + d\ N_2$$

Balance of atomic species

From carbon balance: $b = 1$

From hydrogen balance: $c = 2$

From oxygen balance: $2a = 2b + c = 4$ and $a = 2$

From nitrogen balance: $2(3.76)a = 2d$, or $d = 7.52$

Therefore, the stoichiometric equation is

$$CH_4 + 2(O_2 + 3.76\ N_2) \rightarrow CO_2 + 2\ H_2O + 7.52\ N_2$$

Now

$$AF = \frac{\text{volume of air}}{\text{volume of fuel}} = \frac{2(4.76)\ \text{mol}}{1\ \text{mol}} = \underline{9.52}$$

and

$$FA = \frac{1}{AF} = \frac{1}{9.52} = \underline{0.105}$$

EXAMPLE 14.3

Diesel fuel ($C_{12}H_{26}$) reacts with 80% theoretical dry air. Determine the (wet) products of combustion on a volume basis.

Given: $C_{12}H_{26}$ and 80% theoretical dry air

Find: Products (wet) of combustion on volume basis

Solution: The stoichiometric equation is

$$C_{12}H_{26} + 18.5(O_2 + 3.76\ N_2) \rightarrow 12\ CO_2 + 13\ H_2O + (18.5)(3.76)N_2$$

Therefore, the equation for 80% air is

$$C_{12}H_{26} + 0.8(18.5)(O_2 + 3.76\ N_2) \rightarrow$$
$$13\ H_2O + a\ CO_2 + b\ CO + 0.8(18.5)(3.76)N_2$$

Note that with incomplete combustion, all hydrogen is assumed to go to H_2O, but there must be some CO in the products to balance the equation.

Balance of atomic species

From carbon balance: $a + b = 12$

From oxygen balance: $0.8(18.5)2 = 13 + 2a + b$

Solving gives $a = 4.6$ and $b = 7.4$

Therefore, the wet products of combustion on a volume basis are as follows:

Product	Number of Moles	% by Volume
H_2O	13.0	16.1
CO_2	4.6	5.7
CO	7.4	9.2
N_2	55.6	69.0
	80.6	100.0

EXAMPLE 14.4

Propane (C_3H_8) reacts with 120% theoretical dry air to form products of combustion. Determine

(a) the products of combustion on a mass basis

(b) the dew point of the products if the pressure is 1 atm

Given: Reaction of C_3H_8 with 120% theoretical air to form products

Find:

(a) % by mass for products

(b) dew point of products

Solution:

(a) For complete combustion with excess air, only CO_2, H_2O, N_2, and O_2 are in products. The stoichiometric (theoretical) equation is

$$C_3H_8 + 5(O_2 + 3.76\ N_2) \rightarrow 3\ CO_2 + 4\ H_2O + 18.8\ N_2$$

With 120% theoretical air, the equation is

$$C_3H_8 + 1.2(5)(O_2 + 3.76\ N_2) \rightarrow 3\ CO_2 + 4\ H_2O + O_2 + 1.2(18.8)N_2$$

Therefore, the products on a mass basis are as follows:

Product	Number of Moles		Molecular Mass		Mass (kg) per Mole of Fuel	% by Mass
CO_2	3.0	×	44	=	132	15.2
H_2O	4.0	×	18	=	72	8.3
O_2	1.0	×	32	=	32	3.7
N_2	22.56	×	28	=	632	72.8
	30.56				868	100.0

(b) The partial pressure of H_2O (g) in the products is the number of moles of H_2O (g) divided by the number of moles of products times the total pressure:

$$p_{H_2O} = \frac{4}{30.56}\ (101.3\ kPa) = 13.26\ kPa$$

The saturation temperature corresponding to this pressure is 51.7°C, which is also the dew point temperature for the products. Note that water condensed from the products of combustion frequently contains dissolved gases that are corrosive. Therefore, the products of combustion in many real applications are kept above the dew point temperature until they are discharged to the atmosphere.

EXAMPLE 14.5

Gasoline (C_8H_{18}) is burned with 30% excess dry air, and the products are cooled to 25°C at 1 atm. Give the volumetric analysis of the products at this state. Also give a volumetric analysis of the products on a dry basis (that is, with no water vapor in the products).

Given: C_8H_{18} reacts with 130% theoretical air
Products cooled to 25°C and 1 atm

Find:

(a) volumetric analysis of products at 25°C and 1 atm
(b) volumetric analysis of products on dry basis

Solution:

(a) The theoretical equation is

$$C_8H_{18} + 12.5(O_2 + 3.76\ N_2) \rightarrow 8\ CO_2 + 9\ H_2O + 47.0\ N_2$$

With 30% excess air the reaction equation is

$$C_8H_{18} + 1.3(12.5)(O_2 + 3.76\ N_2) \rightarrow 8\ CO_2 + 9\ H_2O + 1.3(47)\ N_2 + 3.75\ O_2$$

To obtain the volumetric analysis of the products in the final state, one should first determine if the H_2O produced can all be in the vapor state at the pressure of 1 atm and temperature of 25°C for the products. Should there be more H_2O than the mixture can hold as water vapor, the next step would be to determine how much H_2O has condensed out at the specified state. This allows determination of the volumetric analysis of the products at that state. If all the H_2O produced is present as water vapor in the products, the volumetric analysis is as follows:

Product	Number of Moles	% by Volume
CO_2	8.0	9.8
H_2O	9.0	11.0
N_2	61.1	74.6
O_2	3.75	4.6
	81.85	100.0

The partial pressure of saturated water vapor at 25°C is 3.169 kPa. The mole fraction of the water vapor in the mixture (with a saturated mixture) is p_i/p = $3.169/101.3$ = 0.0313. Thus, the partial pressure of water vapor in the combustion products cannot exceed 3.13% of the total pressure (p). The earlier assumption that all the H_2O produced was present as water vapor led to a higher partial pressure for the water vapor at 11% of p, which is not possible at the specified mixture state of 1 atm and 25°C. Thus, some of the H_2O must have condensed out of the combustion products.

Now let a be the number of moles of water vapor remaining in products after cooling. Then the total number of moles of products per mole of fuel is

$$N_{prod} = 8\ CO_2 + a\ H_2O + 61.1\ N_2 + 3.75\ O_2$$

or

$$N_{prod} = 8 + a + 61.1 + 3.75$$

$$= 72.85 + a$$

Now,

$$\frac{a}{72.85 + a} = 0.0313$$

and

$$\therefore a = 2.35$$

Thus, the volumetric analysis in the final state is as follows:

Product	Number of Moles	% by Volume
CO_2	8.0	10.6
H_2O	2.35	3.1
N_2	61.10	81.3
O_2	3.75	5.0
	75.20	100.0

Note that the amount of water vapor condensed out of the gaseous products is $9 - 2.35 = 6.65$ mol H_2O/mol fuel (C_8H_{18}).

(b) As just demonstrated, cooling the products does not remove all of the water vapor. If all of the water vapor is removed by some drying agent, such as silica gel, then the products would exist in a dry condition. The volumetric analysis of the products on a dry basis is as follows:

Product	Number of Moles	% by Volume
CO_2	8.0	11.0
H_2O	0.0	0.0
N_2	61.10	83.9
O_2	3.75	5.1
	72.85	100.0

EXAMPLE 14.6

Coal is burned with 25% excess air. The ultimate analysis of the coal yields 66.0% carbon (C), 2.0% sulfur (S), 6.0% hydrogen (H_2), 1.5% nitrogen (N_2), 9.5% oxygen (O_2), 5.0% ash (noncombustible), and 10.0% moisture (H_2O).

Calculate the air–fuel ratio for the combustion (a) on a molar basis, and (b) on a mass basis.

Given: Coal + 125% theoretical air → products

                  Ultimate analysis of coal (% by mass)

Find: AF on (a) molar basis, and (b) mass basis

Solution: First, analyze the coal on a molar basis. This analysis is obtained as follows. Consider, for example, 100 kg of coal.

Note that ash is nonvolatile and does not enter into the combustion process. Much of it is exhausted from the process as a pollutant unless appropriate cleaning techniques are used to remove it either from the coal prior to combustion or from the exhaust products prior to venting these gases to the atmosphere. Also, when burning coal, carbon may be present in unreacted solid form as refuse in the products. However, for purposes of this discussion, all carbon in the fuel is assumed to be volatile and none remains as refuse in the products.

Component	m (Mass in kg)	M (Mass per kmol)	$N = m/M$ (kmol per 100 kg fuel)	% by Mole
C	66.0	12	5.50	58.1
S	2.0	32	0.06	0.6
H_2	6.0	2	3.00	31.7
N_2	1.5	28	0.05	0.5
O_2	9.5	32	0.30	3.2
Ash	5.0		(nonvolatile)	
H_2O	10.0	18	0.56	5.9
	100.0		9.47	100.0

Now write the reaction equations per kmol of volatile elements:

$$0.581 \ C + 0.581 \ O_2 \rightarrow 0.581 \ CO_2$$

$$0.006 \ S + 0.006 \ O_2 \rightarrow 0.006 \ SO_2$$

$$0.317 \ H_2 + 0.158 \ O_2 \rightarrow 0.317 \ H_2O$$

Thus,

(0.745 kmol O_2 required per kmol coal) − (0.032 kmol O_2 in coal/kmol coal)

= 0.713 kmol O_2 required from air/kmol coal

Now the complete combustion equation for 1 kmol of coal with theoretical air is

$$\text{fuel} + \text{air} \rightarrow \text{products}$$

or

$(0.581 \ C + 0.006 \ S + 0.317 \ H_2 + 0.005 \ N_2 + 0.032 \ O_2 + 0.059 \ H_2O)$

$+ \ [0.713(O_2 + 3.76 \ N_2)] \rightarrow 0.581 \ CO_2 + 0.006 \ SO_2 + 0.376 \ H_2O + 2.686 \ N_2$

For 25% excess air, the equation is

$(0.581 \ C + 0.006 \ S + 0.317 \ H_2 + .005 \ N_2 + 0.032 \ O_2 + 0.059 \ H_2O)$

$+ \ [1.25(0.713)(O_2 + 3.76 \ N_2)] \rightarrow 0.581 \ CO_2 + 0.006 \ SO_2$

$+ \ 0.376 \ H_2O + 3.356 \ N_2 + 0.178 \ O_2$

(a) Thus, the molar air–fuel ratio is

$$N_{air}/N_{fuel} = 1.25(0.713)(4.76)/1$$

$$= 4.242$$

$$AF_{molar} = \underline{4.24 \ (molar)}$$

(b) On a mass basis without considering the ash, the air–fuel ratio is

$$AF_{mass} = m_{air}/m_{fuel}$$
$$= [4.242(29)]/[0.581(12) + 0.006(32) + 0.317(2)$$
$$+ 0.005(28) + 0.032(32) + 0.0059(18)]$$
$$= 12.272 \text{ kg air/kg fuel}$$

Thus, $AF_{mass} = \underline{12.272 \text{ kg air/kg fuel (no ash)}}$

Since this ratio is the ratio of the air to 95% of the fuel by mass, the AF ratio when considering the ash is

$$AF = 12.272 \; \frac{\text{kg air}}{\text{kg fuel (no ash)}} \cdot \frac{0.95 \text{ kg (no ash)}}{1.0 \text{ kg (with ash)}}$$

$$\underline{AF = 11.7 \text{ kg air/kg fuel (mass ratio)}}$$

EXERCISES

for Section
14.3

1. Determine the stoichiometric AF ratio (**a**) by volume and (**b**) by mass for the combustion of H_2 in air. Give the volumetric analysis of the wet products on the assumption that all the H_2O is in the vapor state.
   Answer: (**a**) 2.38; (**b**) 34.3, 34.7% H_2O, 65.3% N_2

2. Assume that the total pressure of the wet products in Exercise 1 is 14.7 psia. What is the partial pressure (in psia) of the moisture in the wet products? Determine the dew point temperature (in °F) for the products.
   Answer: 5.10 psia, 163°F

3. Methane gas (CH_4) is burned with 50% excess air. Determine
   (**a**) the volumetric fuel–air (FA) ratio
   (**b**) the mixture strength
   (**c**) the molar analysis of the wet products of combustion
   (**d**) the volumetric analysis of the dry products of combustion
   Answer: (**a**) 0.07; (**b**) 0.667; (**c**) 7% CO_2, 14% H_2O, 79% N_2; (**d**) 8.14% CO_2

4. Assume that the products in Exercise 3 are cooled to 25°C and 100 kPa. How many kmols of H_2O are condensed out of the products per kmol of the fuel?
   Answer: 1.60 kmol water/kmol of fuel

14.4 Actual Combustion Processes

The previous discussion assumed that complete information was known about the reactants for the process. In many actual combustion processes, such information is not available. For example, sometimes information about the composition of the fuel is incomplete, and in other applications, information about the air–fuel ratio is lacking. It was previously assumed that all carbon in a fuel is converted to CO_2 if excess air is available. In actual cases, CO often exists in the product gases, even with excess air supplied to the reaction. This occurs because of the difficulty of ensuring adequate mixing of the air and fuel during the combustion process. In all of these cases, the gaseous products generally must be analyzed to determine information about the reaction.

One approach to analyze the gaseous products involves the use of an *Orsat apparatus*, in which various gases are absorbed selectively (on a volumetric basis) by passing them through a series of preselected solvents. The typical Orsat apparatus, illustrated in Figure 14.1, has three reagent bottles for successive removal of CO_2, O_2, and CO. With the valves to the reagent bottles closed, the products sample is admitted until it fills the gas bottle to the level marked "B" while the water level on both sides of the manometer is also at "B." Thus, an initial volume (V) of the products sample occupies the region between "A" and "B" in the gas bottle at a pressure equal to the ambient pressure (p_0) and at a temperature (T_0) equal to a constant level maintained by the water jacket around the gas bottle. Next, the valve for the admission of the products sample is closed while each reagent bottle is opened in turn. Following the removal of each constituent (i), the water manometer is manipulated until the pressure in the gas bottle is again equal to the ambient pressure. The extra volume of water admitted into the gas bottle above the "B" mark gives the partial volume (V_i) of the constituent removed from the original mixture. Note that exposure of the initial products sample to water in the gas bottle ensures that the gas mixture remains saturated with water vapor throughout. This apparatus thus provides a volumetric analysis of CO_2, O_2, and CO on a dry basis. The normal assumption when using such an apparatus is that the remaining gas in the products is N_2.

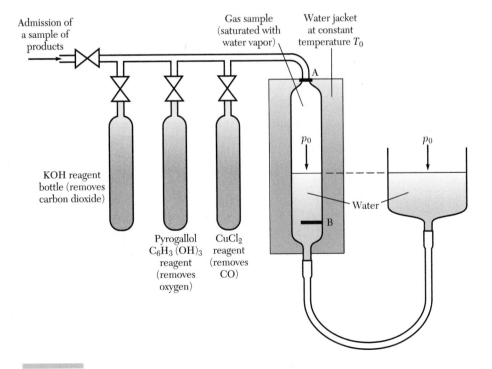

Figure 14.1 Typical Orsat apparatus.

A *gas chromatograph* can also be used to measure all the constituents of either the reactants or the products, as long as one knows in advance which chemical substances may be present.

If results from an Orsat apparatus or a gas chromatograph are available on the constituents of the reactants or products, conservation of atomic species is used to balance equations to determine the desired information concerning the reaction. The following examples demonstrate this.

EXAMPLE 14.7

Natural gas (considered here to be CH_4) is burned with dry air from the atmosphere. An Orsat analysis of the dry products gives volume measurements of 10.00% CO_2, 1.71% O_2, and 1.00% CO. What is the percent theoretical air used during the process?

Given:

$$CH_4 + \text{dry air} \rightarrow \text{products}$$

Volumetric analysis of dry products

Find: % theoretical air

Solution: First write the reaction with unknown coefficients for the reactants that will produce 100 mol of dry product gases. Assume that all dry products not measured by the Orsat apparatus are N_2.

Therefore, for 100 mol of dry product gases, the amount of N_2 is

$$100 - 10 - 1.71 - 1 = 87.29 \text{ mol}$$

The reaction equation is

$$x \text{ } CH_4 + y(O_2 + 3.76 \text{ } N_2) \rightarrow$$
$$10.00 \text{ } CO_2 + 1.00 \text{ } CO + 1.71 \text{ } O_2 + 87.29 \text{ } N_2 + z \text{ } H_2O$$

From balances for the atomic species, one obtains

$$N_2: \quad y = 23.22$$
$$C: \quad x = 11.00$$
$$H_2: \quad z = 22.00$$
$$O_2: \quad y = 23.21 \quad \text{(provides check)}$$

Therefore, the combustion equation per kmol of methane is

$$CH_4 + 2.11(O_2 + 3.76 \text{ } N_2) \rightarrow$$
$$0.91 \text{ } CO_2 + 0.09 \text{ } CO + 0.16 \text{ } O_2 + 7.94 \text{ } N_2 + 2 \text{ } H_2O$$

Thus, the actual air supplied per kmol of methane is

$$\text{Actual air} = 2.11(4.76) = 10.04 \text{ kmol air/kmol fuel}$$

The theoretical air required is found from the stoichiometric equation:

$$CH_4 + 2(O_2 + 3.76\ N_2) \rightarrow CO_2 + 2\ H_2O + 7.52\ N_2$$

Therefore, the theoretical air requirement per kmol of fuel is

$$\text{Theoretical air} = 2(4.76) = 9.52\ \text{kmol air/kmol fuel}$$

Thus, the percent of theoretical air $= (10.04/9.52)100\% = \underline{105\%}$

EXAMPLE 14.8

An unknown hydrocarbon fuel C_xH_y reacts with dry air. A volumetric analysis of the dry products gives the following results: 9.57% CO_2, 6.39% O_2, and 84.04% N_2. Determine

(a) the composition of the fuel

(b) the reaction equation for the process

(c) the percent theoretical air used in the process

Given:
$$C_xH_y + \text{dry air} \rightarrow \text{products}$$
$$\text{Volumetric analysis of products}$$

Find:

(a) x and y for C_xH_y

(b) reaction equation

(c) % theoretical air

Solution:

(a) Begin by writing the reaction equation for 100 mol of dry product gases as in the previous example:

$$C_xH_y + a(O_2 + 3.76\ N_2) \rightarrow 9.57\ CO_2 + 6.39\ O_2 + b\ H_2O + 84.04\ N_2$$

Now make elemental balances to obtain values for x, y, a, and b. From elemental balances,

$$\begin{aligned}
\text{C:} \qquad & x = 9.57 \\
\text{N}_2\text{:} \quad 3.76\ a &= 84.04, \quad a = 22.35 \\
\text{H}_2\text{:} \qquad & y = 2b \\
\text{O}_2\text{:} \quad 2a &= 9.57(2) + 6.39(2) + b
\end{aligned}$$

Solving gives $a = 22.35$, $b = 12.78$, $x = 9.57$, and $y = 25.56$. Thus, it appears that the fuel is $C_{9.57}H_{25.56}$. However, note that the ratio x/y is approximately $3/8$. Therefore, it appears that the fuel is propane (C_3H_8), but this is uncertain without determining the molecular mass M of the fuel. However, since propane is a common fuel, assume that it is propane and proceed as follows.

(b) Assuming that the fuel is propane, the reaction equation is

$$3.19 \ C_3H_8 \ + \ 22.35(O_2 \ + \ 3.76 \ N_2) \rightarrow$$

$$9.57 \ CO_2 \ + \ 6.39 \ O_2 \ + \ 12.78 \ H_2O \ + \ 84.04 \ N_2$$

or on the basis of 1 mol of fuel, the reaction equation is

$$C_3H_8 \ + \ 7(O_2 \ + \ 3.76 \ N_2) \rightarrow 3 \ CO_2 \ + \ 2 \ O_2 \ + \ 4 \ H_2O \ + \ 26.32 \ N_2$$

(c) The theoretical equation for propane is

$$C_3H_8 \ + \ 5(O_2 \ + \ 3.76 \ N_2) \rightarrow 3 \ CO_2 \ + \ 4 \ H_2O \ + \ 18.8 \ N_2$$

Comparing the theoretical equation with the actual reaction equation shows that the theoretical requirement for air–fuel is

$$AF_s \ = \ 5(4.76) \ \text{kmol air/kmol fuel}$$

while the actual air supplied was

$$AF_a \ = \ 7(4.76) \ \text{kmol air/kmol fuel}$$

Therefore,

$$\% \text{ theoretical air} \ = \ [7(4.76)/5(4.76)]100\%$$

$$= \ \underline{140\%}$$

EXERCISES

for Section 14.4

1. The volumetric analysis of the products of combustion when octane (C_8H_{18}) is burnt in air is 12.5% CO_2, 14.06% H_2O, and 73.44% N_2. Estimate the molar air–fuel ratio of the reactive mixture.

   Answer: 59.5

2. A fuel with the formula C_8H_x (where x is unknown) is burnt in air and the Orsat analysis of the products of combustion is found to be 84.3% N_2, 3.74% O_2, and the rest CO_2.
   (a) Find x.
   (b) What is the gravimetric air–fuel ratio?

   Answer: **(a)** 18, **(b)** 18.1

14.5   *Thermodynamic Analysis of Combustion Processes*

Limitations of Stoichiometric Analyses

The theoretical stoichiometric analyses of the previous sections do not always give an accurate picture of the composition of the products of a combustion reaction. For example, the following reactions are known to occur, depending on the temperatures involved:

$$CO \ + \ 0.5 \ O_2 \rightarrow CO_2$$

$$CO_2 \rightarrow CO \ + \ 0.5 \ O_2$$

The first of these reactions is an exothermic reaction (combustion), while the second, known as *dissociation*, is an endothermic reaction. In fact, both reactions occur at all temperatures, although the dissociation taking place at near-ambient temperature is practically negligible. For a given sample of products of combustion, however, the equilibrium composition cannot be determined solely from a stoichiometric analysis. One must resort to analysis based on the laws of thermodynamics, coupled with the conditions of constancy of mass and chemical atoms (based on stoichiometric principles), to determine the equilibrium composition of the products of a chemical reaction.

Application of the pertinent laws of thermodynamics is also needed for energy balance and for the determination of the efficiency achieved in given chemical reactions. The energy balance is based on the first law principle, while the performance evaluation requires the application of the second law as well.

Systems that Include Chemical Substances

Thus far, this book has restricted the various applications of the laws of thermodynamics to *pure substances*, which have been defined as systems having a uniform chemical composition throughout. Chapters 3, 12, and 13 alluded to the concept of *simple systems* whose definition is regarded, at least by some (for example, Kline and Koenig 1957, Haywood 1980) to include pure substances and also chemical substances.

In certain respects, chemical substances are similar to pure substances. A *chemical substance* can be defined formally as one that is homogeneous in composition. Chemical substances in a system may react with each other to form other compounds. Such reactions lead to changes in chemical aggregation or combination of elements. However, there is one major difference between pure substances and chemical substances. Any two independent intensive thermodynamic properties are sufficient to define the state of the system uniquely for a pure substance, whereas in the case of chemical substances, it is necessary to specify the *state of chemical aggregation* in addition to the two independent properties. This requirement might be easier to grasp with the help of the following illustration.

Consider a reactive system comprising 1 m^3 of hydrogen gas and 0.5 m^3 of oxygen at a pressure of 100 kPa and a temperature of 500 K for example. A sufficiently high temperature of 500 K has been chosen to ensure that the H_2O formed in a combustion process will be in the vapor phase at 100 kPa and 500 K. A tiny spark, the energy of which is very small and can therefore be neglected, ignites the mixture. The chemical equation for the reaction is

$$H_2 + 0.5\ O_2 \rightarrow H_2O$$

Let N_R and N_P, respectively, denote the number of moles of the reactants and the product of the reaction. The chemical equation yields a molar ratio (N_P/N_R) of $1/1.5$. Assuming that the product is brought to the same pressure and temperature as the reactants, the ideal gas pVT Equation 4.14 shows that

$$V_P/V_R = N_P/N_R$$

$$= 1/1.5$$

Thus, under conditions of 100 kPa and 500 K, the volume of H_2O produced is 1 m^3.

It thus follows that in going from the original mixture to the product of combustion, a change of state must have occurred since the volume has changed while p and T were both kept the same before and after the chemical reaction. It can be shown that if the pressure and volume are kept constant instead, the temperature for the product is not the same as for the reactants. In this example, the ideal gas pVT equation shows that T_P/T_R is equal to N_R/N_P, which is equal to 1.5 when the pressure and volume for the product of combustion are the same as for the reactants. These examples lead to the conclusion that two independent intensive properties are not adequate for the unique identification of the thermodynamic state of a chemical substance. At first glance, this appears to contradict an earlier statement that a chemical system is a simple system and that, for such systems, the state postulate holds. The *state postulate* can be formally stated as follows:

> *Any two independent intensive thermodynamic properties can be used to specify the equilibrium thermodynamic state of the substance in a simple system.*

The way to resolve this apparent contradiction is to realize that the original mixture of H_2 and O_2 in the previous examples was not really in what Hatsopoulos and Keenan (1965) and others refer to as a *stable equilibrium state*. In other words, the mixture was only in a temporary state of equilibrium, requiring just a tiny spark or catalyst to cause a chemical reaction to take place. The reaction ultimately leads to an equilibrium product comprising H_2, O_2, and H_2O in proportions that are consistent with the laws of thermodynamics. For such a condition of complete equilibrium (thermal, mechanical, and chemical), the thermodynamic state is uniquely defined by any two independent, intensive properties such as pressure and temperature.

In summary, at least two alternative approaches exist for dealing with chemical substances. In one of these, one can consider the specification of the state of chemical aggregation (or combination of elements) in addition to two independent properties as necessary for the unique definition of state of a chemical substance. In the alternative, one can regard the condition of a chemically reactive mixture as a *metastable state*, having a specified or fixed chemical aggregation in addition to any two independent intensive properties. If the chemical reaction is allowed to proceed only to a limited extent, a new metastable state exists that again requires any two independent properties and a specification of the chemical aggregation. When the reaction has progressed to the full extent possible, the stable equilibrium state reached is identified by the specification of any two independent intensive properties only. For further discussion of these principles, refer to Hatsopoulos and Keenan (1965) and Haywood (1980), among others.

Applicability of the Laws of Thermodynamics to Chemical Substances

It is necessary to underscore the fact that all systems and processes obey the laws of thermodynamics. In dealing with chemical substances, however, one should recognize that changes of state are liable to occur when chemical reactions take place with accompanying alteration of the state of chemical aggregation. In particular, the first and second laws of thermodynamics are general laws in the sense that they apply regardless of the type or nature of substances involved. Thus, while previous applications in this book were of processes involving pure substances or simple systems other than chemical substances, it should be noted that these laws are just as applicable to processes involving chemical reactions. The next two sections are devoted, respectively, to the first and second law analyses of processes involving chemical reactions.

14.6   First Law Analysis of Combustion Processes

In general, a first law analysis for nonflow processes requires knowledge of internal energy, while that for flow processes typically uses the enthalpy of the material substances involved. Most combustion reactions of interest in engineering occur in processes taking place at a constant pressure condition. In both the flow process and the constant pressure nonflow process, the property of significant interest is the enthalpy of chemical substances. Much of the remainder of this section is thus devoted to the development of the appropriate expressions for the enthalpy of chemical substances. Similar procedures can be used to determine the expressions for internal energy of chemical substances.

14.6.1   Enthalpy of Formation

Previous considerations of energy balances were restricted to processes that did not involve chemical reactions. The goal now is to extend the application of the basic laws of thermodynamics to processes that include chemical reactions. Previously, changes in enthalpy were required only for the same chemical substance. Now it is necessary to compute changes in enthalpy resulting from chemical reactions that produce chemical substances different from those at the start of the process. In other words, the problem faced is not simply one of dealing with the thermodynamics of a mixture of substances as in Chapter 13, but one of having to determine property changes when the chemical composition of the mixture actually changes.

Consider, for example, the combustion process illustrated in Figure 14.2 in which 1 kmol H_2 at 1 atm and 25°C is mixed and then burned with 0.5 kmol O_2 also at 1 atm and 25°C. The consideration of mixtures of ideal gases in Chapter 13 reveals that the enthalpy of the mixture of hydrogen and oxygen prior to the combustion process is zero.

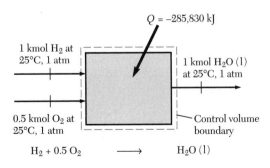

Figure 14.2 Formation of H_2O from hydrogen and oxygen.

This conclusion is based on Gibbs–Dalton's enthalpy additivity law coupled with the assignment of a zero value for the enthapy of elements at the standard reference condition of 1 atm and 25°C. Since combustion is an exothermic reaction, heat must be extracted from the combustion device to have the product (H_2O) leave at the standard condition of 1 atm and 25°C. Application of the first law to the process leads to the conclusion that the enthalpy of the combustion product at the same pressure and temperature as the entering reactants must be nonzero and different from that for the reactants.

The pattern indicated in this illustration is typical of chemical reactions in general. For a chemical substance, therefore, the *enthalpy of formation* is defined as the heat that must be added when a unit amount of the substance is formed from its constituent elements with the same temperature and pressure existing for the reactants (the elements) as for the product (the chemical substance). If the reaction leading to the formation of the substance from its constituent elements is exothermic, the enthalpy of formation is negative since heat must be extracted rather than added so as to have the same temperature for the product as for the reactants. However, the enthalpy of formation is positive if the reaction needed to form the substance is endothermic. Once the enthalpy of formation for a substance is known at a specified reference temperature and pressure, the enthalpy at any other state can be determined based on the approaches that were established for systems whose chemical composition does not change. Note from this definition that the enthalpy of formation for elements in their normal or stable state must be zero.

Datum or Reference State for Enthalpy

Some discussion of reference state for enthalpy and entropy was given in Section 13.2 (Chap. 13). To carry out an energy balance for processes involving different elements and compounds, the energy-related properties such as enthalpy must refer to the same reference state. For calculations involving changes in value of a property (rather than absolute values) when the systems have a fixed composition, an arbitrary reference state is acceptable. When dealing with different chemical species in the same energy balance, however, one must use the same reference state for all the substances for the analysis to be valid.

A logical choice of reference state might be a temperature of absolute zero. However, this reference state is not the conventional choice in the case of enthalpy. As seen later, entropy values for gases are normally referenced to 0 K based on the *third law of thermodynamics*, which postulates that the entropies of all substances tend to zero as the absolute zero temperature is approached. Enthalpy values, in contrast, are referenced to different states for different substances since no such counterpart of the third law principle exists for energy. For example, the enthalpy of H_2O is normally referenced to the triple point of water, while the enthalpy of ammonia is typically referenced to a saturated liquid at $-40°C$ (or $-40°F$). The reference state is also different for different refrigerants.

For the purpose of having a consistent reference state, the condition corresponding to a temperature of 25°C or 298.15 K (77°F or 537°R) and pressure of 1 atm is widely accepted as the *standard reference state* for the enthalpy of chemical substances. In other words, for all elements at $T = 25°C$ (77°F) and 1 atm, the enthalpy is zero, while for compounds at the standard reference state, the enthalpy is given by the enthalpy of formation of the compound at 25°C and 1 atm. The *enthalpy of formation* of a substance at the *standard reference state* is frequently designated h_f° (or sometimes Δh_f°), where the superscript $^\circ$ identifies the value as being at the standard reference state. When expressed in terms of unit amount of the substance in kmol or lbmol, the SI and USCS units for the enthalpy of formation ($\overline{h}_f^\circ$) are kJ/kmol and Btu/lbmol, respectively. The corresponding units on a unit mass basis are kJ/kg and Btu/lbm.

Determination of the Enthalpy of Formation

The process illustrated in Figure 14.2 can be considered as one involving a steady flow $\dot{N}_R$ kmol/s of reactants entering the burner, while $\dot{N}_P$ kmol/s of the product are produced. For nonflow combustion processes at constant pressure, one might instead construe N_R kmol of reactants and N_P kmol of the product with the final pressure p_{final} for the product equal to the initial pressure $p_{initial}$ for the reactants. In either case, one can use H for the total enthalpy in the nonflow formulation and the same notation for the enthalpy flux (that is, the product of the flow rate and the molar specific enthalpy) when a steady stream is involved. Application of the first law of thermodynamics to the process of forming a chemical substance from its constituent elements gives the following:

$$\dot{Q} = H_P - H_R \tag{14.17}$$

where the subscripts R and P are used for the reactants and product(s), respectively. More generally, if the reactants or the product(s) are a mixture of substances each with a flow rate $\dot{N}_i$ and a corresponding molar specific enthalpy $\overline{h}_i$ (in kJ/kmol), and additivity of enthalpy for each mixture of substances can be assumed, the first law equation can also be written as follows:

$$\dot{Q} = \sum_P \dot{N}_i \overline{h}_i - \sum_R \dot{N}_i \overline{h}_i \tag{14.18}$$

As indicated by Equations 14.17 and 14.18, the heat transfer rate during the process is the difference between the enthalpy fluxes for the products and for the reactants. If one considers a chemical reaction that produces a designated chemical compound from the constituent elements, the enthalpy for the reactants can be set equal to zero at the standard reference state. The product of the flow rate of the compound and its specific enthalpy then equals the heat transfer for the process. This quantity also equals the enthalpy of formation (for the formation of the compound from its elements) times the flow rate of the compound for the case of a reaction occurring at the standard condition of 25°C and 1 atm.

For the formation of H_2O (l) from its elements as shown in Figure 14.2, the heat transfer to the surroundings is determined from experiments (using a constant pressure calorimeter) to be 285,830 kJ/kmol H_2O (l). The reaction is exothermic, and the enthalpy of formation of H_2O (l) is

$$\overline{h}_f^\circ = -285,830 \text{ kJ/kmol } H_2O$$

Values for enthalpy of formation for compounds are generally determined experimentally, although in some cases they have been determined by methods of statistical thermodynamics. A tabulation of values of enthalpy of formation for several compounds is given in Appendices A and B.

14.6.2 Enthalpy of Chemical Substances

The enthalpy of any chemical substance, including compounds, at any state can be computed as the sum of the enthalpy of formation from the constituent elements and an enthalpy change, relative to the standard reference state, for the compound. Thus, if $h\,(T, p)$ denotes what is termed the *standardized enthalpy* of the chemical substance, the following equation can be used for computing this specific enthalpy of the substance:

$$h\,(T, p) = h_f^{\circ} + \int_{25^{\circ}C, 1\text{atm}}^{T, p} dh \tag{14.19}$$

Enthalpy for Ideal Gas Mixtures

In general, sufficient accuracy is obtained by assuming ideal gas behavior for gaseous products of combustion. Therefore, the integral in Equation 14.19 can be evaluated from specific heat data for each of the constituents of the gaseous products. Over relatively small temperature ranges, constant values can be assumed for the specific heat of each chemical species involved. To achieve a reasonable level of accuracy when the temperature range is large, however, one should use data that indicate the variation of specific heat values with temperature in performing the integration in Equation 14.19. Thus, for cases where detailed specific heat data are available for a substance, the more comprehensive data should be used to compute the enthalpy more accurately. This has been done in compiling the ideal gas property tables in Appendices A and B for a number of gases, including air, methane, carbon dioxide, carbon monoxide, nitrogen, oxygen, steam, and hydrogen. In all these cases, with the exception of air and methane, the tables span the temperature range from 298 K (25°C) to 6000 K. The computer code **GAS** has also been developed that can be used for obtaining the enthalpy and specific heat at constant pressure for these gases.

Note in particular the differences in the enthalpy for H_2O between the values given in the steam tables and those in the ideal gas tables in Appendices A and B. In the steam tables, the reference point for enthalpy is the triple point of water, while in the present case, the enthalpy of formation notion has been used together with the standard reference state of 25°C (77°F) for the computation of enthalpy.

One important point pertaining to enthalpy of formation is that it is zero for elements at the standard reference state only for the stable form of the elements at that state. Thus hydrogen, oxygen, and nitrogen that exist in the stable form as H_2, O_2, and N_2, respectively, have a zero value at the reference state for their enthalpy of formation. On the other hand, each element in the ionized state, such as O^- or H^+, has a nonzero value for the enthalpy of formation. The values given for enthalpy in the gas tables in the appendices (and in the computer code **GAS** supplied with this text) are referenced to the stable forms of elements at the standard reference state. All the property values are given as a function of temperature for the standard pressure of 1 atm, although in the case of the enthalpy of ideal gases (whose specific heats vary as a function of temperature only), the actual pressure is immaterial as long as the gas approximates the ideal gas behavior. Thus, the values in the tables are a result of substituting the correlations obtained for the ideal gas specific heat at constant pressure as a function of temperature in Equation 14.19 and performing the integration to obtain an algebraic expression for the specific enthalpy. The equivalent equation for the molar specific enthalpy of ideal gases is

$$\bar{h}\,(T) = \bar{h}\,(T, p = 1 \text{ atm}) = \bar{h}_f^\circ + \int_{25°C,1atm}^{T,1atm} \bar{c}_{p0}\, dT \qquad (14.19a)$$

EXAMPLE 14.9

Find the standardized enthalpy of hydrogen, liquid water, and water vapor, each at 50°C. Assume hydrogen behaves as an ideal gas.

Given: H_2 (g), H_2O (l), and H_2O (g) at 50°C

Find: Standardized enthalpies for

(a) H_2 (g)

(b) H_2O (l)

(c) H_2O (g)

Solution:

(a) For H_2 (g),

$$\bar{h}\,(50°C) = \bar{h}_f^\circ + \int_{25°C,1atm}^{50°C,1atm} d\bar{h}$$

The value for $\bar{h}_f^\circ$ is zero for H_2 (g), and the value for the integral can be found using ideal gas Table A.20 for hydrogen. By interpolating between the values at $T = 298.15$ K and $T = 350$ K, one obtains

$$\int_{25°C,1atm}^{50°C,1atm} d\bar{h} = \bar{h}\,(50°C) - \bar{h}\,(25°C) = 724 - 0 \text{ kJ/kmol}$$

$$= (h - h_f^\circ)M_{H_2O}$$

Thus, h (50°C, 1 atm) $= 0 + 362 = \underline{362 \text{ kJ/kg}}$, which is the standardized enthalpy for H_2 (g) at 50°C and 1 atm. Note that this value could also have been found from $\Delta h = c_p \, \Delta T = c_p(50 - 25)$ for this case with the small change in temperature. An average value of 14.3 kJ/kg · K can be assumed for c_p in the temperature range specified. Using this average value, the standardized enthalpy for hydrogen at 50°C is 358 kJ/kg.

(b) The value of h_f° for H_2O was given earlier (see Figure 14.2) and is $-285,830/18 = -15,880$ kJ/kg. The integral in Equation 14.19 can be evaluated for liquid from the steam tables as

$$\int_{25°C,1atm}^{50°C,1atm} dh = h\,(50°C, 1 \text{ atm}) - h\,(25°C, 1 \text{ atm})$$

$$= 209 - 105 \text{ kJ/kg}$$

Thus, the standardized enthalpy of liquid water is obtained from Equation 14.19:

$$h\,(50°C, 1 \text{ atm}) = -15,880 + 104 = \underline{-15,776 \text{ kJ/kg}}$$

Note that the standardized specific enthalpy for water corresponds to the value when the standard reference state of 25°C and 1 atm is enforced.

(c) For saturated water vapor at 50°C, the saturation pressure is 0.1235 bar. Clearly, H_2O cannot exist as a vapor at 50°C and 1 atm pressure. For ideal gases, however, enthalpy is independent of pressure. Therefore, the standardized enthalpy of water vapor at 50°C is calculated as in part (a) using the specific enthalpy values in ideal gas Table A.19 for water vapor since ideal gas behavior can be assumed. The value of h_f° for H_2O (g) is $-241,814/18 = -13,434$ kJ/kg. The integral in Equation 14.19 can be evaluated using the interpolated values from the ideal gas table:

$$\int_{25°C}^{50°C} dh = h\,(50°C) - h\,(25°C)$$

$$= -13377.2 - (-13424) = 46.8 \text{ kJ/kg}$$

Thus, for the water vapor at 50°C, the standardized specific enthalpy is

$$h\,(50°C) = -13,434 + 46.8 = \underline{-13,387 \text{ kJ/kg}}$$

Note that a comparable result could have been obtained for part (c) by adding h_{fg} ($= 2382$ kJ/kg at 50°C) for H_2O from the steam tables to the value obtained in part (b) for liquid water at 50°C. Thus,

$$h_{H_2O(g)} = h_{H_2O(l)} + h_{fg}$$

$$= -15,776 + 2,382 = \underline{-13,394 \text{ kJ/kg}}$$

This is the same result (to 3 significant figures) as that obtained by using the ideal gas tables.

EXERCISES

for Section 14.6.2

1. From the appropriate ideal gas table in Appendix B, determine a mean value for the specific heat at constant pressure (c_p) of CO_2 in the temperature range 77°F to 1500°F. Use this together with the enthalpy of formation for CO_2 to determine the specific enthalpy for CO_2 at 1500°F. Compare this estimate with that obtained directly from the ideal gas table.

   Answer: $(c_p)_{mean} = 0.268$ Btu/lbm · °R, $h = -3460$ Btu/lbm

2. One kmol of methane at 25°C is mixed with the stoichiometric amount of air also at 25°C. Determine the total enthalpy of the reactive mixture.

   Answer: $-74,800$ kJ

14.6.3   Enthalpy of Reaction and Heating Values

Enthalpy of Reaction

The enthalpy of formation previously discussed was defined for the restricted case of the formation of a compound from its elements. A more general chemical reaction can occur to form products from any number of reactants. In such reactions, chemical energy (that is, the potential energy stored in the bond between the constituent atoms) is either

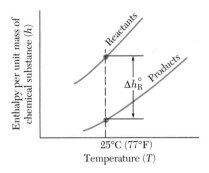

Figure 14.3 Enthalpy of reaction as a function of temperature for a typical exothermic reaction.

released or absorbed during the process. The total enthalpy change associated with a general chemical reaction when the temperature of the products equals the temperature of the reactants is called the *enthalpy of reaction* at the given temperature and is designated by Δh_R (or $\Delta \bar{h}_R$ on a molar basis). Such a change is illustrated in Figure 14.3 for an exothermic reaction, that is, a reaction in which heat transfer occurs to the surroundings to maintain a constant temperature reaction.

It appears from the discussion of properties and the thermodynamic state in Chapter 3 that another property in addition to temperature must be specified at any state to evaluate Δh_R or $\Delta \bar{h}_R$. However, it has been found from experimentation that the enthalpy of reaction generally depends on temperature only and is less sensitive to the pressure level for most substances used in combustion processes. Therefore, Δh_R or $\Delta \bar{h}_R$ is considered in this book to be a function of temperature only.

Since $Q = \Delta H$ for a constant pressure, nonflow heating process, the enthalpy of reaction is also called the *constant pressure heat of reaction*, or simply the *heat of reaction*. It is a measure of Q for the system during the reaction process. The enthalpy of reaction for any reaction can be determined by

$$\Delta \bar{h}_R = \sum_P N_i \bar{h}_i - \sum_R N_i \bar{h}_i \qquad (14.20)$$

For reaction at the standard reference state of 25°C and 1 atm,

$$\Delta \bar{h}_R^\circ = \sum_P N_i (\bar{h}_f^\circ)_i - \sum_R N_i (\bar{h}_f^\circ)_i \qquad (14.21)$$

Thus, the enthalpy of reaction at the standard reference state depends only on the enthalpies of formation and the number of moles of the reactants and products. Note that the enthalpy of any product or reactant that is an element is by definition zero at the standard reference state.

EXAMPLE 14.10

Determine the enthalpy of reaction for the reaction of methane with 20% excess dry air at 25°C and 1 atm. Assume that all water exists as vapor in the products.

Given:
$$CH_4 + 1.2 \text{ theoretical air} \rightarrow \text{products}$$

Reaction at standard reference state

Product H_2O (g)

Find: $\Delta \bar{h}_R$ for this reaction

Solution: The reaction equation is

$$CH_4 + 1.2(2)(O_2 + 3.76 \ N_2) \rightarrow CO_2 + 2 \ H_2O + 0.4 \ O_2 + 9.02 \ N_2$$

The only compound in the reactants is CH_4, while the only compounds in the products are CO_2 and H_2O. All the elements present are in their stable form with a zero value of enthalpy at the standard reference state. Therefore,

$$\Delta \bar{h}_R^\circ = (\bar{h}_f^\circ)_{CO_2} + 2 \ (\bar{h}_f^\circ)_{H_2O(g)} - (\bar{h}_f^\circ)_{CH_4}$$

$$= (1)(-393{,}536 \text{ kJ/kmol}) + 2(-241{,}826 \text{ kJ/kmol}) - 1(-74{,}848 \text{ kJ/kmol})$$

$$= \underline{-802{,}000 \text{ kJ/kmol methane}}$$

Enthalpy of Combustion

The *enthalpy of combustion*, also called the *heat of combustion*, is the term generally used for the burning of fuels in air. The standard enthalpy of combustion is the difference between the enthalpy of products at the standard reference state and the enthalpy of reactants at the same reference state when 1 kmol of fuel reacts with 100% theoretical air at constant pressure at the given reference state. More generally, the enthalpy of combustion at a given reference state is the same as the enthalpy of reaction for the combustion of the fuel in air at the specified reference condition. The enthalpy of combustion is designated by Δh_c or $\Delta \bar{h}_c$, and it provides a standard for comparing different fuels. From an energy viewpoint, a higher numerical value of the heat of combustion indicates a "better" or more energy-yielding fuel.

Heating Values for Fuels (LHV and HHV)

Another term often used for comparing fuels is the *heating value*. The heating value of a fuel is the heat transfer (per unit mole or per unit mass of fuel) that occurs from the complete combustion of the fuel in air (in a steady-flow process) at the standard reference state of 25°C and 1 atm. Note that the heating value is the negative of the heat of combustion for a combustion process at the standard reference state, that is,

$$\text{Heating value (HV)} = -\Delta \bar{h}_c^\circ \tag{14.22}$$

When the H_2O in the products is liquid, H_2O (l), the heating value is called the *higher heating value* (HHV). The *lower heating value* (LHV) is the heating value when the H_2O in the products is vapor, H_2O (g). For some fuels, considerable difference exists between the higher and lower heating values. For most applications, however, the lower heating value is used because liquid water in the products is undesirable since its presence usually results in corrosion and contamination problems. The difference between the higher and lower heating values is given by

$$\text{HHV} = \text{LHV} + \left(\frac{N_{H_2O}}{N_{fuel}}\right)(\overline{h}_{fg})_{H_2O} \tag{14.23}$$

where N_{fuel} is the number of moles of fuel in the reactants and N_{H_2O} is the number of moles of H_2O in the products.

EXAMPLE 14.11

Methane reacts with 100% theoretical air at 25°C and 1 atm. Determine the enthalpy of reaction, the enthalpy of combustion, the heat of combustion, the lower heating value, and the higher heating value.

Given: $\qquad\qquad\qquad$ CH$_4$ + theoretical air → products

$\qquad\qquad\qquad\qquad$ Reaction at standard reference state

Find:

(a) $\Delta\overline{h}_R$

(b) $\Delta\overline{h}_c$

(c) heat of combustion

(d) LHV

(e) HHV

Solution: The reaction equation is

$$CH_4 + 2(O_2 + 3.76 \ N_2) \rightarrow CO_2 + 2 \ H_2O + 7.52 \ N_2$$

(a) The enthalpy of reaction in this example is the same as for Example 14.10 if one assumes H_2O (g) in the products. Thus, assuming that the H_2O in the products is vapor,

$$\Delta\overline{h}_R = -802,000 \text{ kJ/kmol}$$

Note that the different amounts of the air in the two examples do not change the enthalpy of reaction because the air has a zero enthalpy of formation.

(b) For a reaction involving stoichiometric proportions of a fuel and air at the standard reference state, the enthalpy of combustion equals the enthalpy of reaction. Therefore,

$$\Delta\overline{h}_c = \Delta\overline{h}_c^\circ = \Delta\overline{h}_R^\circ = -802,000 \text{ kJ/kmol}$$

(c) The heat of combustion is identical to the enthalpy of combustion. Therefore, the heat of combustion = $-802,000$ kJ/kmol.

(d) The lower heating value is the negative of the enthalpy of combustion with H_2O (g) in the products. Therefore, the LHV = 802,000 kJ/kmol fuel.

(e) The higher heating value can be computed from Equation 14.23 as

$$\text{HHV} = \text{LHV} + \left(\frac{N_{H_2O}}{N_{fuel}}\right)(\bar{h}_{fg})_{H_2O}$$

$$= 802{,}000 \text{ kJ/kmol fuel} + \frac{2 \text{ kmol } H_2O}{1 \text{ kmol fuel}}(2441)(18 \text{ kJ/kmol } H_2O)$$

Therefore, the HHV = 890,000 kJ/kmol fuel. Note that the difference between the higher and lower heating values is about 11% of the lower heating value.

EXERCISES

for Section 14.6.3

1. One kmol of carbon is burned in a stoichiometric amount of air. By determining the total enthalpy for the reactants and products at the reference state of 25°C and 1 atm, obtain the enthalpy of combustion for the fuel.

   Answer: −394 MJ/kmol fuel

2. What is the heating value of the fuel in Exercise 1?

   Answer: 32.8 MJ/kg fuel

3. Calculate the lower and higher heating values for aviation fuel that has an approximate molar composition of 48.8% C and 51.2% H_2.

   Answer: LHV = 45.9 MJ/kg fuel, HHV = 49.2 MJ/kg fuel

4. Assume that the fuel in Exercise 1 is burned with 200% theoretical air. Determine
   (a) the enthalpy of combustion
   (b) the heating value of the fuel

   Answer: (a) −394 MJ/kmol fuel, (b) 32.8 MJ/kg fuel

14.6.4 Application of the First Law to Flow Processes

Application of the first law of thermodynamics to an open, steady-flow process, such as that shown in Figure 14.4, with a chemical reaction occurring within the control volume and with negligible changes in kinetic and potential energy between the inlets and exits gives

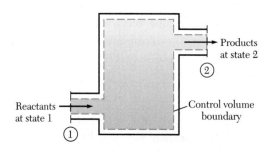

Figure 14.4 A typical steady-flow reacting process.

$$\dot{Q} - \dot{W} = \sum_P \dot{N}_i \bar{h}_i - \sum_R \dot{N}_i \bar{h}_i$$
$$= H_P - H_R \tag{14.24}$$

where H_R and H_P represent the total enthalpy fluxes for the reactants and the products, respectively. The change in enthalpy between the entering reactants and the exiting products must correspond to the appropriate values for the change in state and the change in chemical composition for what entered as reactants and exited as products.

For the typical steady-flow reacting system illustrated in Figure 14.4, the reactants enter at state 1 while the products exit at state 2. States 1 and 2 for the reactants and products are marked on Figure 14.5, which is an enthalpy flux versus temperature diagram for a steady-flow reacting system. If specific enthalpy values are available for all reactants at state 1 and for all products at state 2, one can apply Equation 14.24 to analyze the process. However, such data are generally not available for all substances. Instead, one is most likely to have knowledge of the specific heat for the reactants and products and the enthalpy of reaction (or enthalpy of combustion) for the reacting system at the standard reference state. Therefore, it is convenient to evaluate the enthalpy change from state 1 to 2 in Figure 14.5 as the series of changes from state 1→a→b→2 with states a and b being the standard reference condition for the reactants and products, respectively. The net difference is the same for $H_1 - H_2$ no matter what path is followed because enthalpy is a property.

If both reactants and products are assumed to behave as ideal gases, the enthalpy changes from state 1→a and from state b→2 are solely a function of temperature and can be computed from specific heat data. Such changes involving a single phase and the same chemical composition are called *sensible enthalpy* changes. The change from state a→b, which involves a change in chemical composition, can be computed from the

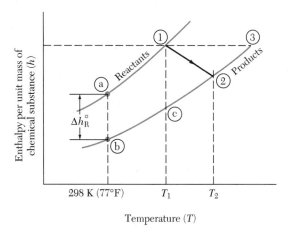

Figure 14.5 Enthalpy flux versus temperature diagram for a steady-flow reacting system.

enthalpy of reaction data for the chemical reaction at the standard reference state. Thus, the sensible enthalpy change between state a and 1 is as follows:

$$H_R - H_{R0} = \sum_R \dot{N}_i[\bar{h}_i\,(T_1) - \bar{h}_i^\circ]$$
$$= \sum_R \dot{N}_i \bar{c}_{pi}\,(T_1 - T_0) \tag{14.25a}$$

where $\bar{h}_i^\circ$ is the molar specific enthalpy of the ith reactant constituent at the standard reference temperature T_0 of 298 K (77°F), $\bar{h}_i\,(T_1)$ is the molar specific enthalpy of the ith constituent at temperature T_1, and $\bar{c}_{pi}$ is the average molar specific heat at constant pressure for the ith constituent in the temperature range T_0 to T_1. Similarly, the sensible enthalpy of the products at T_2 relative to the standard reference state (change from state b→2) is given by

$$H_P - H_{P0} = \sum_P \dot{N}_i[\bar{h}_i\,(T_2) - \bar{h}_i^\circ]$$
$$= \sum_P \dot{N}_i \bar{c}_{pi}\,(T_2 - T_0) \tag{14.25b}$$

The enthalpy flux change for the change in chemical composition at the standard reference state corresponds to that in the process from state a→b in Figure 14.5 and is equal to $H_{P0} - H_{R0}$. This quantity is related to the enthalpy of reaction at the standard reference state, which is given by Equation 14.21. Thus, if $\dot{N}_{chem}$ denotes the kmol/s of the chemically reacting substance (that is, of the reacting fuel) and $\Delta\bar{h}_R^\circ$ the enthalpy of reaction at the standard reference state, the enthalpy flux change $H_{P0} - H_{R0}$ can be written as follows:

$$H_{P0} - H_{R0} = \dot{N}_{chem}\,\Delta\bar{h}_R^\circ \tag{14.25c}$$

Now the total enthalpy change for the process from state 1→a→b→2 can be written as

$$H_P - H_R = H_P - H_{P0} + H_{P0} - H_{R0} + H_{R0} - H_R$$

where, $H_P - H_{P0}$ is given by Equation 14.25b, $H_{P0} - H_{R0}$ by Equation 14.25c, and $H_R - H_{R0}$ by Equation 14.25a. Thus, one can write $H_P - H_R$ as follows:

$$H_P - H_R = \sum_P \dot{N}_i[\bar{h}_i\,(T_2) - \bar{h}_i^\circ]$$
$$- \sum_R \dot{N}_i[\bar{h}_i\,(T_1) - \bar{h}_i^\circ] + \dot{N}_{chem}\,\Delta\bar{h}_R^\circ \tag{14.25}$$

This expression can be substituted in Equation 14.24 for the first law applied to a steady-flow process involving a chemically reacting system. An alternative expression for $H_P - H_R$ is obtained by substituting in Equation 14.25 the enthalpy of reaction as defined by Equation 14.21:

$$H_P - H_R = \sum_P \dot{N}_i\{[\bar{h}_i\,(T_2) - \bar{h}_i^\circ] + (\bar{h}_f^\circ)_i\}$$
$$- \sum_R \dot{N}_i\{[\bar{h}_i\,(T_1) - \bar{h}_i^\circ] + (\bar{h}_f^\circ)_i\} \tag{14.26}$$

Combining this equation with Equation 14.24 gives the following for the first law for a reacting system:

$$\dot{Q} - \dot{W} = \sum_P \dot{N}_i\{(\bar{h}_f^\circ)_i + [\bar{h}_i\,(T_P) - \bar{h}_i^\circ]\}$$
$$- \sum_R \dot{N}_i\{(\bar{h}_f^\circ)_i + [\bar{h}_i\,(T_R) - \bar{h}_i^\circ]\} \tag{14.27}$$

where T_R and T_P are used in place of T_1 and T_2 for the temperatures of the reactants and products, respectively, and $\bar{h}_i\,(T_P)$ and $\bar{h}_i\,(T_R)$ denote molar specific enthalpies at temperatures T_P and T_R, respectively. Note that each constituent of the reactants can enter and each constituent of the products can exit at a different temperature with this nomenclature.

Comparison of the terms in braces in Equation 14.27 with the expression for standardized enthalpy of Equation 14.19 shows equivalence except for pressure variations. Therefore, the first law can also be written in terms of the standardized enthalpy as

$$\dot{Q} - \dot{W} = \sum_P \dot{N}_i \bar{h}_i\,(T_P, p) - \sum_R \dot{N}_i \bar{h}_i\,(T_R, p) \tag{14.28}$$

when pressure variations are also considered. As indicated earlier, negligible differences exist in enthalpies due to pressure variations around atmospheric pressure for most substances.

Equations 14.25 and 14.26 show that the enthalpy change for a chemically reacting system can be computed from either enthalpy of reaction data or enthalpy of formation data. Identical results should be obtained from calculations using either approach.

Consider Figure 14.5 again. The change in enthalpy for the process could also be computed by evaluating the enthalpy of reaction from state 1→c and then the sensible enthalpy of products from state c→2. However, enthalpy of reaction data are generally available only at the standard reference state, and one usually cannot use this simpler approach.

EXAMPLE 14.12

A steady-flow calorimeter such as that shown in Figure 14.4 is used to determine the enthalpy of reaction for gaseous ethylene (C_2H_4) at the standard reference state. The fuel enters the system at 25°C and 1 atm and reacts with 150% theoretical air, which is also at 25°C and 1 atm. The products exit at 400 K and 1 atm. The heat transfer to the surroundings during the reaction is measured and found to be 1.25×10^6 kJ/kmol of fuel. Determine the enthalpy of reaction. Check your results using enthalpy of formation data.

Given: $\underset{\text{(at 25°C, 1 atm)}}{C_2H_4\,(g)}$ $+$ $\underset{\text{(at 25°C, 1 atm)}}{\text{1.5 theoretical air}} \rightarrow$ $\underset{\text{(at 400 K, 1 atm)}}{\text{products}}$

$$\frac{\dot{Q}}{\dot{N}_{\text{fuel}}} = -1.25 \times 10^6 \text{ kJ/kmol fuel}$$

Find: (a) $\Delta \bar{h}_R^\circ$, (b) Check results with $\bar{h}_f^\circ$

Solution: The reaction equation is

$$C_2H_4\,(g) + 1.5(3)(O_2 + 3.76\ N_2) \rightarrow 2\ CO_2 + 2\ H_2O + 1.5\ O_2 + 16.92\ N_2$$

(a) The enthalpy of reaction can be calculated from Equation 14.25 combined with Equation 14.24, which in this instance is written as

$$\frac{\dot{Q}}{\dot{N}_{fuel}} = \sum_P \frac{\dot{N}_i}{\dot{N}_{fuel}} [\bar{h}_i \ (400 \ \text{K}) - \bar{h}_i^\circ]$$

$$- \sum_R \frac{\dot{N}_i}{\dot{N}_{fuel}} [\bar{h}_i \ (298.15 \ \text{K}) - \bar{h}_i^\circ] + \Delta \bar{h}_R^\circ$$

since the shaft power is zero. Note further that the summation term for the reactants in this equation must be zero since, by definition, the standardized enthalpy at 25°C, $\bar{h}_i$ (298.15 K) = $\bar{h}_i^\circ$. Thus, the enthalpy of reaction is given by

$$\Delta \bar{h}_R^\circ = \frac{\dot{Q}}{\dot{N}_{fuel}} - \sum_P \frac{\dot{N}_i}{\dot{N}_{fuel}} [\bar{h}_i \ (400 \ \text{K}) - \bar{h}_i^\circ]$$

Each of the terms under the summation for the products in this equation can be determined as follows using values in the gas tables in Appendix A:

$$\frac{\dot{N}_{CO_2}}{\dot{N}_{fuel}} [\bar{h}_{CO_2} \ (400 \ \text{K}) - \bar{h}_{CO_2}^\circ] = 2[-389,531 - (-393,536)] \ \text{kJ/kmol fuel}$$

$$\frac{\dot{N}_{H_2O}}{\dot{N}_{fuel}} [\bar{h}_{H_2O} \ (400 \ \text{K}) - \bar{h}_{H_2O}^\circ] = 2[-238,376 - (-241,826)] \ \text{kJ/kmol fuel}$$

$$\frac{\dot{N}_{O_2}}{\dot{N}_{fuel}} [\bar{h}_{O_2} \ (400 \ \text{K}) - \bar{h}_{O_2}^\circ] = 1.5[3027 - 0] \ \text{kJ/kmol fuel}$$

$$\frac{\dot{N}_{N_2}}{\dot{N}_{fuel}} [\bar{h}_{N_2} \ (400 \ \text{K}) - \bar{h}_{N_2}^\circ] = 16.92[2962 - 0] \ \text{kJ/kmol fuel}$$

Thus, the enthalpy of reaction is

$$\Delta \bar{h}_R^\circ = \frac{\dot{Q}}{\dot{N}_{fuel}} - \sum_P \frac{\dot{N}_i}{\dot{N}_{fuel}} [\bar{h}_i \ (400 \ \text{K}) - \bar{h}_i^\circ]$$

$$= -1,250,000 - (8010 + 6900 + 4540.5 + 50,117) \ \text{kJ/kmol}$$

$$= -1320 \ \text{MJ/kmol fuel}$$

(b) To check against enthalpy of formation data, use Equation 14.21 with the tabulated enthalpy of formation data given in Appendix A:

$$\Delta \bar{h}_R^\circ = \sum_P \frac{\dot{N}_i}{\dot{N}_{fuel}} (\bar{h}_f^\circ)_i - \sum_R \frac{\dot{N}_i}{\dot{N}_{fuel}} (\bar{h}_f^\circ)_i$$

$$= [2(\bar{h}_f^\circ)_{CO_2} + 2(\bar{h}_f^\circ)_{H_2O} + 1.5(\bar{h}_f^\circ)_{O_2} + 16.92(\bar{h}_f^\circ)_{N_2}]$$

$$- [1(\bar{h}_f^\circ)_{C_2H_4(g)} + 4.5(\bar{h}_f^\circ)_{O_2} + 16.92(\bar{h}_f^\circ)_{N_2}]$$

From the tables in Appendix A, the enthalpies of formation for CO_2, H_2O, and C_2H_4 (g) are $-393,536$ kJ/kmol CO_2, $-241,826$ kJ/kmol H_2O, and 52,260 kJ/kmol C_2H_4 (g). For the elements O_2 and N_2, the enthalpy of formation is by definition zero. Substituting these values of enthalpies of formation in the equation for the enthalpy of reaction gives $\Delta \bar{h}_R^\circ = -1{,}323$ MJ/kmol fuel. Thus, about the same result is obtained by each approach (with both results rounded-off as indicated).

While combustion processes are more often flow processes, occasionally one has to consider a first law analysis for the nonflow processes as well. For a closed system containing ideal gases, the first law for a process involving a chemical reaction can be expressed as

$$Q - W = U_P - U_R \tag{14.29}$$

Using the definition of enthalpy as $H = U + pV$, the first law for the closed system can also be written as

$$Q - W = H_P - H_R - [(pV)_P - (pV)_R] \tag{14.30}$$

Combining this equation with the ideal gas equation gives

$$Q - W = H_P - H_R - \left(\sum_P N_i \bar{R} T_P - \sum_R N_i \bar{R} T_R \right) \tag{14.31}$$

One can further combine Equations 14.31 and 14.26 to obtain the following form for the first law applied to nonflow chemical reaction processes involving ideal gases:

$$
\begin{aligned}
Q - W = & \sum_P N_i \{ (\bar{h}_f^\circ)_i + [\bar{h}_i\,(T_P) - \bar{h}_i^\circ] - \bar{R} T_P \} \\
& - \sum_R N_i \{ (\bar{h}_f^\circ)_i + [\bar{h}_i\,(T_R) - \bar{h}_i^\circ] - \bar{R} T_R \}
\end{aligned}
\tag{14.32}
$$

A typical device to measure heating values of fuels is called a *bomb calorimeter*. The measurement involves the reaction of a fuel with oxygen (or air) in a closed, gas-tight vessel immersed in a water calorimeter. Normally the bomb is charged with the fuel sample and with oxygen at high pressure to ensure complete combustion. The temperature rise in the water calorimeter is measured and used to compute the heat transfer from the reacting system during the combustion. The fuel mass and the quantity of water employed in the water calorimeter are normally chosen so that the water temperature rises only a few degrees (about 3–4°C), and the temperature and pressure of products return almost to that of the reactants. The measurement thus essentially provides a measure of the *internal energy of combustion,*

$$\Delta \bar{u}_c^\circ = \frac{U_{P0} - U_{R0}}{N_{fuel}}$$

at the standard reference state when the test is conducted at 25°C. For most fuels, the difference between the enthalpy of combustion and the internal energy of combustion is small; thus, the bomb calorimeter also provides an approximate measure of the enthalpy of combustion.

EXAMPLE 14.13

Ethylene [C_2H_4 (g)] reacts in a bomb calorimeter with pure oxygen at 25°C. Calculate the heat transfer per kmol of fuel from the "bomb" during the process. Also compute the internal energy of combustion for the reaction and compare this value with the enthalpy of combustion for the fuel. Assume all products are gaseous.

Given:
$$C_2H_4 \text{ (g)} + 3\ O_2 \rightarrow \text{products (g)}$$

Reactants and products approximately at 25°C

Find: (a) Q/N_{fuel}, (b) $\Delta\bar{u}_c^\circ$, (c) Compare $\Delta\bar{u}_c^\circ$ with $\Delta\bar{h}_c^\circ$ for fuel

Solution: The reaction equation is

$$C_2H_4 \text{ (g)} + 3\ O_2 \rightarrow 2\ CO_2 \text{ (g)} + 2\ H_2O \text{ (g)}$$

(a) Equation 14.32 can be used to compute q ($= Q/N_{\text{fuel}}$). Assuming $W = 0$ and $T_P = T_R = 298$ K for the process, Equation 14.32 simplifies to

$$q = \sum_P N_i[(\bar{h}_f^\circ)_i - \bar{R}T_P] - \sum_R N_i[(\bar{h}_f^\circ)_i - \bar{R}T_R]$$

$$= [2(\bar{h}_f^\circ)_{CO_2(g)} + 2(\bar{h}_f^\circ)_{H_2O(g)} - 4\bar{R}T_P]$$

$$\quad - [1(\bar{h}_f^\circ)_{C_2H_4(g)} + 3(\bar{h}_f^\circ)_{O_2(g)} - 4\bar{R}T_R] \text{ kJ/kmol fuel}$$

Thus, the tables for the enthalpies of formation in Appendix A give

$$q = 2(-393{,}536) + 2(-241{,}826) - 1(52{,}226) \text{ kJ/kmol fuel}$$

$$= \underline{-1323 \text{ MJ/kmol fuel}}$$

(b) For this process at constant volume and constant temperature (at 25°C), one obtains the following from the first law:

$$q = \frac{U_{P0} - U_{R0}}{N_{\text{fuel}}} = \Delta\bar{u}_c^\circ$$

$$\therefore \Delta\bar{u}_c^\circ = \underline{-1323 \text{ MJ/kmol fuel}}$$

(c) With the process assumed to occur at constant temperature (that is, with $T_P = T_R$), Equation 14.31 shows that the change in enthalpy for the process is also equal to the change in internal energy (since in this example the number of moles of products equals the number of moles of reactants). Therefore,

$$\Delta\bar{h}_c^\circ = \Delta\bar{u}_c^\circ = \underline{-1323 \text{ MJ/kmol fuel}}$$

EXERCISES

for Section 14.6.4

1. A flow rate of 1 lbmol/h of methane at 77°F and 1 atm is burned in 23.8 lbmol/h of air also at 77°F and 1 atm. The products of combustion leave the burner at 1500°R. Assuming that the combustion is complete, determine the heat transfer rate to the burner.

Answer: $-166{,}000$ Btu/h

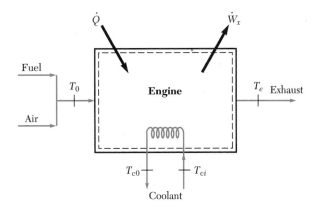

Thermodynamic model for the gasoline engine.

2. Repeat Exercise 1 but this time assume that 150% theoretical air is supplied for the combustion of methane. The steady stream of methane still has a flow rate of 1 lbmol/h, and the exit temperature of the products of combustion is 1500°R.

   Answer: $-232,000$ Btu/h

3. The figure above depicts a thermodynamic model for the gasoline engine. By applying the SFEE, show that the energy balance equation for the engine performance can be written as follows:

$$\dot{m}_{\text{fuel}}\text{LHV}/M_{\text{fuel}} = \dot{W}_x + (-\dot{Q})$$
$$+ \dot{m}_{\text{exhaust}}(c_p)_{\text{exhaust}}(T_e - T_0)$$
$$+ \dot{m}_{\text{coolant}}(c_p)_{\text{coolant}}(T_{c0} - T_{ci})$$

   Give a physical interpretation to each of the terms in the equation.

14.6.5 *Adiabatic Flame Temperature*

The maximum temperature that can occur for a combustion process is often important when designing a system because of materials and other considerations. The maximum temperature that the combustion products can reach is achieved when the system is perfectly insulated so that no heat transfer occurs to the surroundings and all the energy released during the process is used to raise the temperature of the products. Such a process is adiabatic, and the maximum temperature for the products (based on first law considerations alone) in the adiabatic combustion process is called the *adiabatic flame temperature*. The adiabatic flame temperature is normally defined with reference to a steady-flow process.

For an ideal reaction, the maximum temperature is reached for a stoichiometric reaction of the fuel, but this maximum temperature can be reduced by use of excess air in the system. Also, the maximum temperature is higher for a stoichiometric reaction using pure oxygen than for one using air (because the other components of air absorb energy and thereby reduce the maximum temperature).

For a steady-flow, adiabatic combustion process with zero work and negligible changes in potential and kinetic energy, the first law of thermodynamics reduces to $H_P = H_R$, where the H's represent enthalpy fluxes for the inlet reactants and the exit products. As can be seen from Figure 14.5 for a process with reactants initially at state 1, the final state for this adiabatic process is state 3 (since $H_P = H_R$). Note from the figure that the higher the initial temperature of the reactants, the higher the temperature of the products will be. The first law as given in Equation 14.27 reduces in this case to

$$\sum_P \dot{N}_i \{(\bar{h}_f^\circ)_i + [\bar{h}_i (T_P) - \bar{h}_i^\circ]\}$$
$$= \sum_R \dot{N}_i \{(\bar{h}_f^\circ)_i + [\bar{h}_i (T_R) - \bar{h}_i^\circ]\} \tag{14.33}$$

The right-hand side of this equation can be evaluated from specified information about the reactants, while the left-hand side can be used to determine the temperature of the products (which is the adiabatic flame temperature). Since several components normally exist in the products, an iterative procedure must be used to compute the adiabatic flame temperature. This procedure is demonstrated in Example 14.14.

If the mean specific heats at constant pressure are known for the products and the reactants, however, Equation 14.33 can be written in an alternative form that enables the adiabatic flame temperature to be determined in a noniterative manner. Equation 14.33 can be recast as follows:

$$\sum_P \dot{N}_i [\bar{h}_i (T_P) - \bar{h}_i^\circ] = \sum_R \dot{N}_i [\bar{h}_i (T_R) - \bar{h}_i^\circ]$$

$$- \left[\sum_P \dot{N}_i (\bar{h}_f^\circ)_i - \sum_R \dot{N}_i (\bar{h}_f^\circ)_i \right]$$

In other words,

$$\dot{N}_P (\bar{c}_p)_P (T_p - T_0) = \dot{N}_R (\bar{c}_p)_R (T_R - T_0) - \dot{N}_{fuel} \Delta \bar{h}_c^\circ \tag{14.33a}$$

Thus, from a knowledge of the mean specific heats $(\bar{c}_p)_P$ and $(\bar{c}_p)_R$ for the products and the reactants, respectively, and the enthalpy of combustion for the fuel, Equation 14.33a enables T_P to be determined in an explicit manner.

EXAMPLE 14.14

Gaseous ethane [C_2H_6 (g)] at 25°C and 1 atm is burned with air at 25°C and 1 atm in a steady-flow process. Assuming complete combustion and ideal gas behavior of the reactants and products, determine the adiabatic flame temperature for 100% theoretical air.

Given:

$$\begin{array}{ccc} C_2H_6 \text{ (g)} & + & \text{theoretical air} \rightarrow \text{products} \\ \text{(at 25°C, 1 atm)} & & \text{(at 25°C, 1 atm)} \end{array}$$

Steady-flow adiabatic process

Find: Adiabatic flame temperature for 100% theoretical air

Solution: The stoichiometric reaction equation is

$$C_2H_6 \text{ (g)} + 3.5(O_2 + 3.76 \, N_2) \rightarrow 2 \, CO_2 + 3 \, H_2O \text{ (g)} + 13.16 \, N_2$$

The adiabatic flame temperature can be determined using Equation 14.33a. First, obtain reasonable estimates for the mean specific heat for the products in the temperature range from 298.15 K to the yet unknown T_P. Suppose you guess that T_P is on the order of 2400 K. The mean specific heat for the products in the temperature range indicated can be determined in the following manner:

$$\dot{N}_P(\bar{c}_p)_P(T_2 - T_0) = \sum_P \dot{N}_i[\bar{h}_i\,(T_2) - \bar{h}_i\,(T_0)]$$

where T_2 is the upper temperature limit for the determination of the mean specific heat. Using the values of specific enthalpy in the gas tables in Appendix A, you obtain an estimate of 37.77 kJ/kmol $\cdot$ K for the products mean specific heat at constant pressure in the temperature range 298.15 K to 2400 K. The enthalpy of combustion for ethane can be determined either from the table for enthalpies of combustion (provided in Appendix A) or by calculation from Equation 14.21 based on the tabulated enthalpies of formation given also in Appendix A. Using the latter approach, an estimate of $-1,427,830$ kJ/kmol gaseous ethane is obtained. Applying Equation 14.33a and noting that $T_R = T_{R0}$ thus gives

$$T_P - T_0 = -\frac{\dot{N}_{\text{fuel}}\,\Delta\bar{h}_c^\circ}{\dot{N}_P(\bar{c}_p)_P} = \frac{1,427,830}{18.16(37.77)}\,\text{K} = 2082\,\text{K}$$

Hence, the adiabatic flame temperature T_P is 2380 K.

An alternative approach involves the use of Equation 14.33 and a trial-and-error procedure to obtain T_P. The gas tables in Appendix A as well as the **GAS** code provide values of enthalpy directly for each of the gaseous products, and therefore the following version of Equation 14.33 may be more appropriate:

$$\sum_P \frac{\dot{N}_i}{\dot{N}_{\text{fuel}}}\,\bar{h}_i\,(T_P) = \sum_R \frac{\dot{N}_i}{\dot{N}_{\text{fuel}}}\,(\bar{h}_f^\circ)_i$$

The sensible enthalpy change for the reactants is zero since $T_R = T_{R0}$. The expression on the right-hand side of this equation reduces to $1(-84,720)$ kJ/kmol C_2H_6 (g) since the enthalpy of formation is zero for the oxygen and nitrogen in air. The problem thus reduces to finding T_P such that H_P given by the left-hand side of this equation is equal to $-84,720$ kJ/kmol C_2H_6 (g). H_P is obtained by reading from the tables in the appendix the standardized enthalpy of each gas constituent in the products of combustion at the specified T_P:

$$H_P = 2\bar{h}_{CO_2}\,(T_P) + 3\bar{h}_{H_2O(g)}\,(T_P) + 13.16\bar{h}_{N_2}\,(T_P)$$

An iteration on T_P leading to a solution is illustrated in the following table:

T_P (K)	$\bar{h}_{CO_2}$ (kJ/kmol)	$\bar{h}_{H_2O(g)}$ (kJ/kmol)	$\bar{h}_{N_2}$ (kJ/kmol)	H_P (kJ)
2200	$-289,990$	$-158,807$	63,354	$-222,662$
2400	$-277,781$	$-148,237$	70,634	$-70,730$
(2381.6)	(Estimate obtained by linear interpolation)			$(-84,720)$
2381.6	$-278,907$	$-149,218$	69,963	$-84,755$

Values of H_P obtained for temperatures of 2200 K and 2400 K clearly indicate that the adiabatic flame temperature must lie between the two temperatures. An estimate of 2381.6 K is then obtained by linear interpolation. The entry for H_P on the last row in the table is obtained using the **GAS** code, which verifies that the difference between H_P and H_R is less than 0.05% of the magnitude of H_R when T_P is 2381.6 K. The adiabatic flame temperature can thus be taken as 2380 K (to three significant figures).

As indicated earlier, the adiabatic flame temperature is reduced if excess air is used. Heat transfer from the system also reduces the temperature of the combustion products. *Dissociation* and *ionization* generally occur for most combustion products at high temperatures; energy is absorbed during these phenomena, thereby reducing the temperature of the products. Furthermore, real systems at high temperatures involve other reactions in addition to those defined here for complete combustion. These additional reactions are usually endothermic and lead to lower temperatures for the products. As a consequence of all these effects, the actual temperature for a real combustion process is likely to be several hundred degrees lower than the adiabatic flame temperature.

When considering an actual combustion process involving air, it is sometimes possible to increase the temperature of the products by carrying out the reaction with a slight deficiency of air (instead of with a stoichiometric proportion). This is possible because less products are available to absorb the energy released in the combustion process, thereby allowing higher temperatures for the products from the reaction (see Denbigh 1964, 1966).

EXERCISES

for Section 14.6.5

1. Determine the adiabatic flame temperature for the combustion of hydrogen gas with 100% theoretical air.

   Answer: 2526 K (2253°C)

2. What is the adiabatic flame temperature for the combustion of hydrogen gas in 200% theoretical air?

   Answer: 1647 K (1374°C)

14.7   Second Law Analysis

A thermodynamic analysis of combustion processes based on just the first law and stoichiometric principles is incomplete without the application of the second law of thermodynamics. As indicated previously, one cannot decide on the basis of stoichiometry and the first law of thermodynamics alone what the composition of the products of a chemical reaction involving a reactive mixture will be. Chapter 15 shows that the establishment of criteria for chemical equilibrium and the determination of the equilibrium composition of the products of a chemical reaction are based on the second law of thermodynamics. One also relies on the second law of thermodynamics to establish the direction of chemical reactions and whether or not a specified reaction can take place. Chapter 15 also covers these aspects of the second law application.

Entropy is a property derived from the second law in much the same way that internal energy is a property derived from the first law. Before dealing with the various aspects of second law application to chemical reactions, one must therefore determine how to evaluate entropy for chemical substances. Knowing how to evaluate entropy will facilitate the second law analysis of chemically reactive systems. Several aspects of the following treatment were discussed previously in Chapters 8 and 13.

14.7.1 Entropy Change for Reacting Systems

Values for the entropy of the products and the reactants are frequently needed in order to apply the second law of thermodynamics to reacting systems. Although most calculations involve changes from one state to another, a common and consistent reference state is needed for all substances so as to make meaningful calculations in a situation where the system has different constituents and its composition is changing during the process. This problem is in certain respects similar to that discussed earlier for enthalpy. For example, the entropy of H_2O is normally referenced to the liquid at the triple point in the steam tables, while entropies of refrigerants are referenced to other temperatures.

The most reasonable reference point for entropy is the absolute zero temperature because the third law of thermodynamics states that the entropy of a pure crystalline substance approaches zero as the temperature of the substance approaches zero. This is the reference point that is normally adopted for the gas tables, which cover the range of conditions corresponding to ideal gas behavior for the substance. Although the entropy of a pure crystalline substance does not vary with pressure at absolute zero temperature, it does vary with pressure at higher temperatures. Therefore, the entropy values are tabulated as a function of temperature at a reference pressure set at 1 atm. Such is the case with the properties in the ideal gas tables in Appendices A and B.

The values for the absolute entropy $s°(T)$ at a pressure of 1 atm given in the ideal gas tables in Appendices A and B are based on a reference point of absolute zero temperature and 1 atm, which is

$$s°(T) = \int_0^T \frac{c_{p0}}{T} \, dT$$

The same reference condition was assumed in the development of the ideal gas property code **GAS** supplied with this text. It can be used for the enthalpy and entropy values for a number of gases, including air, oxygen, nitrogen, methane, hydrogen, steam, carbon dioxide, and carbon monoxide. The absolute entropy values for a larger number of substances covering a wide range of temperatures at the standard pressure of 1 atm are given in the JANAF tables (Thermal Research Laboratory 1965) and in other sources in the literature.

Procedures for the calculation of entropy changes for pure substances were discussed in Chapter 8 and for ideal gas mixtures in Chapter 13. Since it is generally acceptable in typical combustion processes to assume ideal gas behavior for both reactant and product gases, the procedures described in Chapter 13 for computing the entropy of ideal gas mixtures are only briefly outlined here while noting that such computations must be relative to a consistent reference point since the combustion process produces a change in composition of the original reactive mixture.

Absolute Entropy at Standard Condition

It is usual to define the absolute entropy s°_{ref} for chemical substances at the standard reference condition of 298.15 K and 1 atm. Thus, the absolute entropy at the standard reference condition can be written as

$$s^\circ_{\text{ref}} = \int_0^{T_{\text{ref}}(=298.15\text{K})} \frac{c_{p0}}{T}\, dT$$

Values for s°_{ref} are published in the literature for several substances, and once the value is known for a particular substance, one can compute the absolute entropy at other pressure and temperature conditions by integration of the appropriate version of the $T\,ds$ equations given in Chapter 8.

Since both s° and s°_{ref} are defined for the standard pressure of 1 atm, a relationship can be established between them using Equation 8.28:

$$s^\circ(T) = s^\circ_{\text{ref}} + \int_{T_{\text{ref}}}^T \frac{c_{p0}}{T}\, dT \tag{14.34a}$$

where c_{p0} is a function of temperature only since the pressure remains at 1 atm.

Absolute Entropy for Reactive Ideal Gas Mixtures

The entropy change for a pure substance was given earlier as

$$ds = \frac{c_{p0}}{T}\, dT - \frac{R}{p}\, dp \tag{8.23}$$

which can be integrated between any two states subject only to the requirement that the substance remains in the ideal gas condition throughout. Since each component of an ideal gas mixture behaves as if it existed alone at the temperature (T) and volume (V) of the mixture, the entropy of an ideal gas in a mixture (of fixed composition) can be obtained by integrating this equation for ds from the standard reference state of 298.15 K and 1 atm to the actual temperature (T) and partial pressure (p_i) of the gas in the mixture.

Integration of the equation on a molar basis gives

$$\bar{s}_i(T, p_i) = \bar{s}^\circ_i(T) - \bar{R} \ln \frac{p_i}{p_0} \tag{14.34}$$

where p_0 is the reference pressure of 1 atm and $\bar{s}^\circ_i(T)$ is

$$\bar{s}^\circ_i(T) = \bar{s}^\circ_{i,\text{ref}} + \int_{T_{\text{ref}}}^T \frac{\bar{c}_{pi0}}{T}\, dT$$

which is based on Equation 14.34a and given in the ideal gas tables for a number of substances. Since $p_0 = 1$ atm, if p_i is also expressed in atm, Equation 14.34 can be written as

$$\bar{s}_i(T, p_i) = \bar{s}^\circ_i(T) - \bar{R} \ln p_i \tag{14.35}$$

For a mixture of ideal gases, one may invoke Gibbs–Dalton's law of additive entropies given by Equation 13.40 (Chap. 13) to obtain

$$S_{\text{mix}} (T, p) = \sum_i N_i \bar{s}_i (T, p_i) \tag{14.36}$$

The difference between the entropy of the products and of the reactants for a chemically reacting system is given by

$$\Delta S = S_P - S_R = \sum_P N_i \bar{s}_i (T_P, p_i) - \sum_R N_i \bar{s}_i (T_R, p_i) \tag{14.37}$$

Combining Equations 14.37 and 14.35 gives

$$\Delta S = S_P - S_R = \sum_P N_i [\bar{s}_i^{\circ} (T_P) - \bar{R} \ln p_i] - \sum_R N_i [\bar{s}_i^{\circ} (T_R) - \bar{R} \ln p_i] \tag{14.38}$$

Since the mole fraction $y_i = p_i/p$, Equation 14.38 can also be written as

$$\begin{aligned}
\Delta S = S_P - S_R = \sum_P N_i [\bar{s}_i^{\circ} (T_P) - \bar{R} \ln y_i p] \\
- \sum_R N_i [\bar{s}_i^{\circ} (T_R) - \bar{R} \ln y_i p]
\end{aligned} \tag{14.39}$$

where p is the total pressure (in atm) when the reaction takes place. Calculation of the entropy change for a combustion process is illustrated in the following example.

EXAMPLE 14.15

Methane (CH_4) reacts with oxygen in a stoichiometric ratio with both products and reactants at 25°C and 1 atm. Complete combustion occurs. Assume all H_2O in the products is liquid. Calculate

(a) the entropy change for the reacting system in kJ/K per kmol of fuel

(b) the entropy change per kmol fuel for the surroundings

(c) the total entropy change per kmol fuel for the universe

Given: $CH_4 + O_2 \rightarrow$ products [with H_2O (l)]

All products and reactants at 25°C and 1 atm

Find: (a) ΔS_{system}, (b) ΔS_{surr}, (c) $\Delta S_{\text{universe}}$

Solution: The balanced reaction equation is

$$CH_4 + 2\,O_2 \rightarrow CO_2 + 2\,H_2O$$

The mole fractions are $y_{CH_4} = \frac{1}{3}$ and $y_{O_2} = \frac{2}{3}$ for the reactants and $y_{CO_2} = 1$ and $y_{H_2O} = 0$ for gaseous products (which do not include the liquid water).

(a) Using Equation 14.39 yields

$$\begin{aligned}
\Delta S_{\text{sys}} &= S_P - S_R \\
&= 1[\bar{s}_{CO_2}^{\circ} (T_P) - \bar{R} \ln y_{CO_2} p] + 2[\bar{s}_{H_2O(l)}^{\circ} (T_P)] \\
&\quad - 1[\bar{s}_{CH_4}^{\circ} (T_R) - \bar{R} \ln y_{CH_4} p] - 2[\bar{s}_{O_2}^{\circ} (T_R) - \bar{R} \ln y_{O_2} p]
\end{aligned}$$

The standardized specific entropy for H_2O (l) can be obtained by first computing the specific entropy for dry saturated water vapor at 25°C using Equation 14.35

$$\bar{s}_g \, (25°C) = \bar{s}^°_{H_2O(g)} - \bar{R} \ln p_s = 217.5 \text{ kJ/kmol} \cdot \text{K}$$

where $\bar{s}^°_{H_2O(g)}$ is taken from gas tables and p_s is saturation pressure in atm for water vapor at 25°C. Assuming negligible difference in $\bar{s}_{H_2O(l)}$ (25°C) due to pressure (between p_s and 1 atm), the standardized specific entropy for H_2O (l) is then calculated from

$$\bar{s}^°_{H_2O(l)} = \bar{s}_f \, (25°C) = \bar{s}_g \, (25°C) - \bar{s}_{fg} \, (25°C)$$

$$= 217.5 - 8.189(18.015) = 70.0 \text{ kJ/kmol} \cdot \text{K}$$

When this and other values taken from the ideal gas tables for the standardized specific entropy of methane, carbon dioxide, and oxygen are substituted together with the mole fractions as previously computed, $p = 1$ atm, and $T_R = T_0 = 25°C$, one obtains $\Delta S_{sys} = -258.5 \text{ kJ/K}$ per kmol of fuel.

The negative change for entropy of the system is not a violation of the second law because a heat transfer occurred from the system to the surroundings during the process, thereby increasing the entropy of the surroundings. A violation of the second law occurs only if the total entropy change for both system and surroundings is negative.

(b) The entropy change of the surroundings is

$$\Delta S_{surr} = Q_{surr}/T_0 = -(Q_{sys}/T_0)$$

Assume that the surroundings are at a temperature of 25°C. The heat transfer to the system Q_{sys} can be calculated by applying Equation 14.27, which reduces to the following simpler form since $T_P = T_R = 298.15$ K and since no shaft work is done:

$$\frac{\dot{Q}_{sys}}{\dot{N}_{fuel}} = \sum_P \frac{\dot{N}_i}{\dot{N}_{fuel}} (\bar{h}^°_f)_i - \sum_R \frac{\dot{N}_i}{\dot{N}_{fuel}} (\bar{h}^°_f)_i$$

$$= 1(\bar{h}^°_f)_{CO_2} + 2(\bar{h}^°_f)_{H_2O(l)} - 1(\bar{h}^°_f)_{CH_4}$$

The enthalpy of formation for oxygen is not included in this equation because its value is zero by definition. The enthalpy of formation for H_2O (l) is that for H_2O (g) minus the molar specific heat of vaporization at 25°C and is equal to $-285,800$ kJ/kmol H_2O. The gas tables in Appendix A give the values $-393,536$ kJ/kmol CO_2 and $-74,848$ kJ/kmol CH_4 for the other enthalpies of formation. Substituting these in the above expression for the heat transfer per kmol fuel, one obtains the result $\dot{Q}_{sys}/\dot{N}_{fuel} = 888,688$ kJ/kmol fuel. Thus, one can determine the increase in entropy for the surroundings from $\Delta S_{surr} = (Q_{surr}/T_0) = (888,688/298.15) = 2980 \text{ kJ/K}$ per kmol of fuel.

(c) $\Delta S_{total} = \Delta S_{sys} + \Delta S_{surr} = 2720 \text{ kJ/K}$ per kmol of fuel.

EXERCISES

for Section 14.7.1

1. The accompanying figure illustrates a hypothetical reaction in which CO_2 at p_0 and T dissociates into CO and O_2 as a result of the heat transfer rate $\dot{Q}$ from a thermal reservoir at T. Assume that $p_0 = 1$ atm and $T = 25°C$. The CO_2 is supplied at a rate of 1 kmol/h, while the mixture comprising 1 kmol/h of CO and 0.5 kmol/h of O_2 leaves at a pressure p_0 and temperature T. Determine $\dot{Q}$ and the entropy generation rate for the control volume. Is this hypothetical reaction possible or impossible?

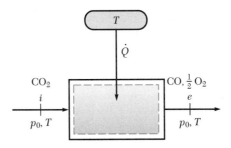

Hypothetical reaction involving dissociation of CO_2 into CO and O_2 as a result of heat transfer from a thermal reservoir.

Answer: $\dot{Q} = 283{,}000$ kJ/h, $\dot{S}_{gen,CV} = -863$ kJ/h $\cdot$ K

2. Repeat Exercise 1 assuming $T = 4000$ K instead of 25°C.
   Answer: $\dot{Q} = 268{,}000$ kJ/h, $\dot{S}_{gen,CV} = 13.7$ kJ/h $\cdot$ K

14.7.2 Exergy Analysis for Reacting Systems

As discussed earlier in Chapter 9, the important measure for the useful output from any thermodynamic process is the exergy product for the process. Exergy effectively defines the maximum work potential of a system in a particular state. This maximum useful work obtainable for a system is measured relative to the *dead state*, previously defined as the state at which the system is in complete (thermal, mechanical, and chemical) thermodynamic equilibrium with the stable reference (or datum) environment.

Dead State for Chemically Reactive Systems

The concept of a stable reference environment was briefly discussed in Section 13.2.2. In essence, it refers to a system that is internally at complete stable equilibrium and that is also very large compared to all of the other systems together (Wepfer and Gaggioli 1980). For combustion processes, the ambient or atmospheric air represents a convenient choice for such a stable reference environment.

At the dead state, a chemically reactive system must be at the same temperature T_0 as the reference environment (for thermal equilibrium) and also at the same pressure p_0 as the reference environment (for mechanical equilibrium). In addition, the system must be in chemical equilibrium with the reference environment which (as Section 15.3 shows) requires, for each chemical species in the system, a chemical potential $\bar{\mu}$ identical to that $(\bar{\mu}_{i0})$ for the corresponding stable form of the chemical species present in the reference environment. Examples are given in Rodriguez (1980), from the work by Gaggioli and Petit (1977) and Petit and Gaggioli (1980), on the choice of a stable reference environment and fixing of the dead state for processes involving the combustion of hydrocarbons. Thus, the stable configurations in the reference environment corresponding to carbon, oxygen, and nitrogen, in a reactive mixture are taken as CO_2, O_2, and N_2, respectively, in the proportions in which they exist in air saturated with water vapor at the prevailing ambient temperature and pressure. Likewise, the corresponding stable configuration for hydrogen in a reactive mixture is H_2O (l) in equilibrium with saturated air at T_0 and p_0.

The choices of T_0 and p_0 for a stable reference environment for processes involving combustion generally are not necessarily determined to be 298.15 K and 1 atm (the reference values normally used for the enthalpy of chemical substances). In practice, however, the pressure and temperature for typical atmospheric conditions are not very different from these values; therefore, the choices of $p_0 = 1$ atm and $T_0 = 298.15$ K are often made for the reference environment.

Calculation of Exergy for Chemical Substances

Disregarding magnetic and other forms of stored energy, such as gravitational potential energy, the total flow of exergy associated with a steady-flow of mass is given by

$$\dot{X}_{\text{mass flow}} = \dot{X}_T + \dot{X}_p + \dot{X}_c \qquad (9.39)$$

where $\dot{X}_T$, $\dot{X}_p$, and $\dot{X}_c$ are the thermal exergy flux, pressure exergy flux, and chemical exergy flux, respectively. If one is considering a fuel–air mixture (reactants) at the temperature and pressure of the reference environment, both the thermal exergy flux and the pressure exergy flux are zero while the chemical exergy flux for the mixture is that for the fuel in the mixture.

The more fundamental expression for the exergy flux for a flow of mass is that given by Equation 9.33, which involves the computation of the steady-flow exergy flux as the difference between the function $H - T_0S$ for the substance at its present state and $H_0 - T_0S_0$ for the substance when it is reduced to the dead state condition. (For a mass flow, H and S actually represent enthalpy and entropy fluxes.) When dealing with chemical substances, one can determine both H and S for the substance in its present state using the procedures already outlined in Sections 14.6.2 and 14.7.1. The corresponding computations for the dead state condition must, however, be made on the basis that the necessary mass exchange with the stable reference environment takes place so as to ensure that the chemical reactions are carried through to completion, thus leading to chemical equilibrium with the reference environment. Thus, in terms of the Gibbs function $G (= H - TS)$, the chemical exergy flux for a flow of reactants (neglecting kinetic and potential energy changes) can be written as

$$\dot{X}_{\text{reactants}} = G_{R0} - G_0$$

where G_{R0} is the Gibbs function flux for the reactants, while G_0 is the Gibbs function for the original mixture when reduced to the dead state. The reactants in this case are at the reference conditions p_0 and T_0, which are taken to be the same as the pressure and temperature corresponding to the dead state condition.

If the products of the reaction are also reduced to the reference pressure and temperature, the exergy flow rate of the products are likewise given by

$$\dot{\chi}_{\text{products}} = G_{\text{P0}} - G_0$$

Note that G_0 is the same in the expressions for both the exergy of the reactants and the exergy of the products of reaction.

The maximum possible exergy flow rate delivered in the previous reaction can be denoted by $(\dot{W}_x)_{\text{max}}$ and can be written

$$\begin{aligned}(\dot{W}_x)_{\text{max}} &= \dot{\chi}_{\text{reactants}} - \dot{\chi}_{\text{products}} \\ &= G_{\text{R0}} - G_{\text{P0}} \\ &= (H_{\text{R0}} - H_{\text{P0}}) - T_0(S_{\text{R0}} - S_{\text{P0}})\end{aligned} \qquad (14.40)$$

The term $H_{\text{R0}} - H_{\text{P0}}$ is related to what was defined in Section 14.6.3 and in Figure 14.3 as the enthalpy of reaction. The term $S_{\text{P0}} - S_{\text{R0}}$ is the entropy change for a combustion process at the reference condition and can be calculated using Equation 14.39, which was derived for ideal gas mixtures in Section 14.7.1.

By way of illustration, recall Example 14.15 which involved the combustion of methane gas in a stoichiometric mixture of the fuel and air at 25°C and 1 atm. If one considers a mixture with just 1 kmol of methane, $H_{\text{P0}} - H_{\text{R0}}$ is simply the enthalpy of reaction (combustion), which can be easily calculated or obtained from a table of properties. In this case, the value is $-802,000$ kJ or $-890,000$ kJ (see Example 14.11) according to whether the H_2O in the products is assumed to be in the gaseous or liquid phase. For the change in entropy $S_{\text{P0}} - S_{\text{R0}}$, one can use Equation 14.39 for the gaseous products and assume that the entropy for the liquid component of the combustion products can simply be added to that for the gaseous components. Thus, assuming that the H_2O in the products at 1 atm and 25°C is completely in the liquid phase, Equation 14.40 can be used to compute the maximum exergy that can be extracted in the specified reaction at the reference condition:

$$(\dot{W}_x)_{\text{max}}/\dot{N}_{\text{fuel}} = 890,000 - 298.15(240.5) \text{ kJ}$$

$$= 818,300 \text{ kJ/kmol fuel}$$

If, however, the H_2O remains in the gaseous form in the products, then

$$(\dot{W}_x)_{\text{max}}/\dot{N}_{\text{fuel}} = 802,000 - 298.15(-39.58) \text{ kJ}$$

$$= 813,800 \text{ kJ/kmol fuel}$$

Note that the expression given by Equation 14.40 actually differs from the exergy of the original reactive mixture by an amount equal to the exergy of the products of the reaction. If the composition of the various chemical species in the products is different from what it is in the stable reference environment, the possibility exists, at least theoretically, for further extraction of exergy equal to $G_{\text{P0}} - G_0$. Typically, however, this residual exergy is small compared to the exergy of the reactants, and therefore the estimate given by Equation 14.40 will be close to the correct value for the exergy of the original reactive mixture.

Referring back to Equation 9.39 for the exergy flux of a flow of mass, note that when the reactants are at pressure p_0 and temperature T_0 for the stable reference environment, the thermal exergy and pressure exergy components are zero. In other words, the exergy for the reactants as computed from $G_{R0} - G_0$ corresponds to the chemical exergy of the chemical substances in the reactive mixture.

Expressions in the Literature for Chemical Exergy

Rodriguez (1980) has compiled a comprehensive set of equations that can be used to calculate chemical exergies for a variety of chemical substances, including many of the fuels commonly used in combustion applications. Sets of these equations are given in Tables 14.1 and 14.2. (Additional results for coals, tars, and fuels containing sulfur are also given by Rodriguez 1980.) Note from Table 14.1 that the chemical exergy (or availability) of a substance in a mixture is based on its mole fraction and does not depend on T or p. The data for hydrocarbon fuels in Table 14.2 are based on atomic ratios and knowledge of the lower heating value of the dry fuel, LHV_d. If H_2O is present in the fuel, the lower heating value of the wet fuel, LHV_w, is related to LHV_d as follows:

$$LHV_d = LHV_w + 2{,}442\ \omega,\ kJ/kg$$

where ω is the humidity ratio of the fuel (kg water/kg moist fuel).

The "best" fuel from a thermodynamic viewpoint is that with the highest value of chemical exergy. Obviously, many other factors are involved in selection of a fuel for a specific application. Some of these factors are economics, how much of the fuel resource is available for long term use (see Chap. 1), potential health and safety problems, cor-

TABLE   14.1

Chemical Exergies of Some Ideal Gas Components	
Component	$\bar{x}_c$, **Chemical Exergy (kJ/kmol)***
Ammonia	$2478.907 \ln y_{NH_3} + 337{,}861$
Benzene	$2478.907 \ln y_{C_6H_6} + 3{,}253{,}338$
Carbon (graphite)	$410{,}535$
Carbon dioxide	$2478.907 \ln y_{CO_2} + 20{,}108$
Carbon monoxide	$2478.907 \ln y_{CO} + 275{,}224$
Ethane	$2478.907 \ln y_{C_2H_6} + 1{,}484{,}952$
Hydrogen	$2478.907 \ln y_{H_2} + 235{,}153$
Methane	$2478.907 \ln y_{CH_4} + 830{,}212$
Nitrogen	$2478.907 \ln y_{N_2} + 693$
Oxygen	$2478.907 \ln y_{O_2} + 3948$
Water vapor	$2478.907 \ln y_{H_2O} + 8595$

*y represents the component's mole fraction; for pure substances,
   $\ln y = 0$. Note that $\bar{x}_c$ does not depend on T or p.
Source: Extracted from Rodriguez (1980).

TABLE 14.2

*Chemical Exergy of Hydrocarbon Fuels**

Fuels Containing Only C, O, H, and N

Solids
$$\frac{\bar{x}_c}{(LHV)_d} = 1.0438 + 0.0158\frac{H}{C} + 0.0813\frac{O}{C} + 0.0471\frac{N}{C} \quad \text{if } \frac{O}{C} \le 0.5$$

$$\frac{\bar{x}_c}{(LHV)_d} = \frac{1.0438 + 0.0158\frac{H}{C} - 0.3343\frac{O}{C}\left(1 + 0.0609\frac{H}{C}\right) + 0.0447\frac{N}{C}}{1 - 0.4043\frac{O}{C}} \quad \text{if } \frac{O}{C} > 0.5$$

Liquids
$$\frac{\bar{x}_c}{(LHV)_d} = 1.0374 + 0.0159\frac{H}{C} + 0.0567\frac{O}{C} + 0.5985\frac{S}{C}\left(1 - 0.1737\frac{H}{C}\right)$$

Gases
$$\frac{\bar{x}_c}{(LHV)_d} = 1.0334 + 0.0183\frac{H}{C} - 0.0694\frac{1}{C}$$

*Notation: $\frac{H}{C}$, $\frac{O}{C}$, and $\frac{N}{C}$ are atomic ratios (of dry fuel).

$\bar{x}_c$ is chemical exergy, with units the same as for $(LHV)_d$.

$(LHV)_d$ is lower heating value of dry fuel (or of dried fuel per unit moist fuel from which it originates).

rosion effects produced by products of combustion, and temperature levels and associated problems with materials. But if all other factors are equal, the best fuel is the one with the highest value of chemical exergy. Sample calculations involving chemical exergy are given in the following examples.

EXAMPLE 14.16

Compute the chemical exergy for pure gaseous methane at T_0 and p_0 using the equations found in both Tables 14.1 and 14.2 and compare the results. The LHV_d for methane is 802 MJ/kmol from Example 14.11.

Given: Equations for chemical exergy $\bar{x}_c$

Find: $\bar{x}_c$ for gaseous methane

Solution: From Table 14.1, the chemical exergy for methane is

$$\bar{x}_c = 2478.907 \ln y_{methane} + 830{,}212 \text{ kJ/kmol CH}_4$$

For pure methane, $\ln y_{methane} = 0$, and therefore,

$$\bar{x}_c = \underline{830 \text{ MJ/kmol}}$$

From Table 14.2 for gases,

$$(\bar{x}_c/LHV_d) = 1.0334 + 0.0183(H/C) - 0.0694(1/C)$$

For CH_4, $H/C = 4$ and $(1/C) = 1$.
Therefore, $\bar{x}_c = (1.0372)802 \text{ MJ/kmol} = \underline{832 \text{ MJ/kmol}}$.
The results agree within about 0.2%.

EXAMPLE 14.17

Gaseous methane at T_0 and p_0 enters a reaction vessel and reacts with stoichiometric air at T_0 and p_0 as shown in Figure 14.6. The chemical reaction occurs at constant pressure in the reaction vessel, and the products are exhausted to the surroundings, which are at T_0 and p_0.

Part I. Calculate the maximum work possible per kmol fuel from the combustion process if the exhaust products leave the reaction vessel at the reaction temperature of

  (a) T_0, with all products in stable configurations in equilibrium with the surroundings

  (b) 1000 K

Part II. If the heat of reaction is supplied at the reaction temperature to a Carnot heat engine operating as shown in Figure 14.6, compute the work done by the engine per kmol of fuel if the exhaust products leave the reaction vessel at the reaction temperature of

  (c) T_0

  (d) 1000 K

Given: Methane reacting with stoichiometric air

Find: Part I: Maximum work possible per kmol fuel if products exhaust at

(a) T_0 in stable equilibrium with the surroundings

(b) 1000 K

Part II: Work by Carnot heat engine per kmol fuel if the heat is supplied to the heat engine at the reaction temperature of

(c) T_0

(d) 1000 K

Solution: The reaction equation is

$$CH_4 + 2(O_2 + 3.76\ N_2) \rightarrow CO_2 + 2\ H_2O + 7.52\ N_2$$

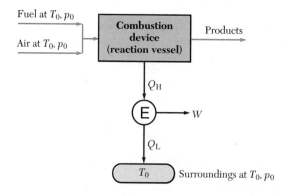

Figure 14.6 Heat engine driven by heat from products of combustion.

(a) The maximum work possible (per kmol fuel) is given by the decrease of exergy, which is $\Delta\bar{x} = \bar{x}_{in} - \bar{x}_{exit}$. The total exergy per kmol flowing is the sum of the thermal exergy $(\bar{x}_T)$, the pressure exergy $(\bar{x}_p)$, and the chemical exergy $(\bar{x}_c)$. For the flow of fuel into the vessel, both the thermal exergy and pressure exergy components are zero since $T_{in} = T_0$ and $p_{in} = p_0$. Therefore, $\bar{x}_{in} = (\bar{x}_c)_{in}$.

For this case, the exit temperature and pressure are the same as those for the reference environment and thus the thermal exergy and pressure exergy components at exit are also zero. Since the reaction produces CO_2, H_2O, and N_2, which are all stable configurations in the dead state, the chemical exergy component at exit is also zero. Thus, the exergy of the exit stream is zero. The decrease in exergy is given in this case by

$$\Delta\bar{x} = (\bar{x}_c)_{in}$$

where $(\bar{x}_c)_{in}$ for the methane can be computed from the equation given in Table 14.1:

$$\bar{x}_c = 2478.907 \ln y_{methane} + 830,212 \text{ kJ/kmol } CH_4$$

The mole fraction in this case is calculated as

$$y_{methane} = \frac{\text{mol methane}}{\text{total mol reactants}}$$

$$= 1/(1 + 2 + 2(3.76)) = 0.095$$

Hence, $\bar{x}_c = 1478.907 \ln 0.095 + 830,212 = 826,731 \text{ kJ/kmol } CH_4$. Therefore, the maximum work possible, which is the decrease in exergy for the process, is 827 MJ/kmol CH_4.

(b) For this case, the thermal exergy component of the exit stream is not zero since the products are exhausted from the reaction vessel at 1000 K, although the stream is not used to produce additional work after leaving the reaction vessel. In this case, the thermal exergy component (per kmol of the exit stream) is calculated using Equation 9.40a:

$$\bar{x}_T = \bar{c}_p \left(T - T_0 - T_0 \ln \frac{T}{T_0} \right)$$

The specific heat for the mixture is determined from Equation 13.55:

$$\bar{c}_{p0} = \sum_{i=1}^{k} \frac{N_i}{N} (\bar{c}_{p0})i \tag{13.55}$$

The mean molar specific heats at constant pressure for CO_2, H_2O, and N_2 in the temperature range 25°C to 1000 K are estimated as 47.6 kJ/kmol · K for CO_2, 37.0 kJ/kmol · K for H_2O, and 30.6 kJ/kmol · K for N_2. Applying Equation 13.55 thus gives the molar specific heat at constant pressure for the exit stream as 33.4 kJ/kmol of the exit stream, or 351.4 kJ/kmol · K for CH_4. Equation 9.40a gives a thermal exergy for the exit stream of 119,948 kJ/kmol CH_4. The pressure exergy of the exit stream is zero, and therefore, the maximum work possible in this case is $826,731 - 119,948 = 706,780 \text{ kJ/kmol } CH_4$ or 707 MJ/kmol CH_4.

Note that the rejection of the exhaust products at 1000 K directly to the atmosphere reduced the maximum work by approximately 15%.

(c) The first law efficiency of a Carnot heat engine is given by $\eta = (T_H - T_L)/T_H$ and the maximum work $W = Q_H \eta$. Since the reaction temperature for this case is at T_0, which equals T_H, and T_L also equals T_0, the maximum work for the Carnot heat engine for this case is zero.

(d) For this case, $T_H = 1000$ K and $T_L = 298.15$ K. Therefore, the maximum work is equal to $(1 - 298.15/1000)Q_H = 0.702\ Q_H$. Q for the reaction at 1000 K can be computed from Equation 14.28:

$$\frac{\dot{Q}}{\dot{N}_{fuel}} = \sum_P \frac{\dot{N}_i}{\dot{N}_{fuel}} \overline{h}_i\ (T_P, p) - \sum_R \frac{\dot{N}_i}{\dot{N}_{fuel}} \overline{h}_i\ (T_R, p)$$

$$= [(1)\overline{h}_{CO_2}\ (1000\ \text{K}) + 2\overline{h}_{H_2O}\ (1000\ \text{K})$$

$$+ 7.52\overline{h}_{N_2}\ (1000\ \text{K})] - (1)\overline{h}_{CH_4}^\circ$$

Terms for the molar specific enthalpy of the components of air at the standard reference state are not included in this expression since they are zero by definition. Using the standardized molar specific enthalpy values from the gas tables in Appendix A yields a heat transfer to the Carnot engine of 555,560 kJ/kmol fuel. Thus, the maximum possible work from the Carnot engine is 390 MJ/kmol CH_4.

Comments

1. The maximum work computed in part (a) can only be obtained for a reversible process (such as direct conversion of chemical energy to electrical energy) that converts the chemical exergy to work. Any irreversibilities (entropy generation) in the actual reaction process reduces the maximum work.

2. In part (b), the maximum work is reduced because the products at 1000 K are rejected as exhaust directly to the surroundings without utilizing the thermal exergy of the products at 1000 K.

3. Work could be obtained by some sort of direct conversion process in part (c), but not from a heat engine that uses heat from a source at an elevated temperature to produce work.

4. In case (d), Q from the combustion process is approximately equal to $\Delta \overline{x}$ in case (b). The differences are due to changes in entropy for the combustion process and also to the use of constant specific heats in the calculations for case (b) but variable specific heats in the calculations for case (d). However, the use of the heat engine degrades the value of the original chemical energy so that only about 70% of it can be used to produce work.

5. Comparing results for parts (a) and (d) shows that less than 60% of the chemical exergy of fuel can be used with the system of part (d) to produce work. Irreversibilities (entropy generation) in the combustion process and in a real heat engine would reduce the actual work output of a real engine by considerably more (perhaps a factor of two).

EXERCISES

**for Section
14.7.2**

1. Consider a stoichiometric mixture of hydrogen and air and compute the chemical exergy per kmol H_2 using the equation in Table 14.1.

   Answer: 233 MJ/kmol

2. Repeat Exercise 1 assuming a mixture with 200% theoretical air.

   Answer: 231 MJ/kmol

14.8   Summary

This chapter considered several essential principles of stoichiometry and the thermodynamics of chemical reactions, although the focus was on combustion processes. Combustion was defined as an uncontrolled chemical reaction of a fuel with oxygen to form products. Combustion processes are typically exothermic, that is, they release energy as heat to other systems for a variety of applications, such as work-production in heat engines. Actual combustion processes are accompanied by endothermic reactions as well, and therefore, complete analysis of combustion processes requires stoichiometric analysis as well as the application of the first and second laws of thermodynamics.

Sections 14.2 through 14.4 provided a comprehensive discussion of the stoichiometry of combustion processes. The essential underlying principles are those of conservation of mass and the conservation of atomic species. Various types of analysis were discussed, including gravimetric and ultimate analysis, molar analysis, volumetric analysis, and partial pressure analysis. For mixtures of ideal gases, it was recalled from Chapter 13 that the molar, volumetric, and partial pressure analyses are all identical. Other terminology introduced included reactants and products, and wet and dry products analysis. Combustion processes were classified as complete, stoichiometric (or perfect), and incomplete. Various ratios for the amount of fuel to air in the reactants were defined; these include the air-fuel ratio and the fuel-air ratio, excess air, and mixture strength. These quantities generally provide an indication of the type of reaction (whether complete or incomplete) likely to occur. It was noted, however, that actual combustion processes cannot be completely determined solely from stoichiometric principles. In some cases, the actual composition of the products can be determined experimentally using the Orsat apparatus or the gas chromatograph in order to accurately describe the chemical reaction that took place. In general, however, application of the laws of thermodynamics is necessary for the complete analysis of combustion processes.

Section 14.5 affirmed the principle that the laws of thermodynamics apply to chemical substances and processes. The definition of a chemical substance was given, and the extension of the two-property rule (introduced in Chapter 3) to chemical substances was discussed. For chemical substances, the state of chemical aggregation must be specified in addition to two independent, intensive properties if the thermodynamic state is to be uniquely defined.

The application of the first law of thermodynamics to combustion processes was discussed in Sections 14.6 and 14.7. Several concepts were introduced for the first law analysis of chemically reacting systems. These include the enthalpy of formation, the enthalpy of reaction, the enthalpy of combustion, and lower (and higher) heating values for fuels. Knowledge of these properties enables the determination of the enthalpy of chemical substances. The concept of the adiabatic flame temperature (which is the maximum possible temperature of a reacting system) was introduced, and methods for calculating it were explained and illustrated. Computation of the adiabatic flame temperature does not take account of dissociation or ionization, which occur at high temperatures of reaction. It was thus noted that application of the first law alone does not suffice for complete analysis of an actual combustion process.

Discussions of the second law application to combustion processes were provided in Sections 14.7 and 14.8. Section 14.7 described procedures for computation of entropy change for chemically reacting systems and discussed the application of the principle of increase in entropy of the universe to determine whether or not a given reaction could occur. Section 14.8 discussed chemical exergy and the computation of chemical exergy for hydrocarbon fuels which are by far the most widely employed in combustion processes. Exergy analysis for chemically reacting systems was considered. Application of the second law to the determination of criteria for maximum work production from a reacting system was discussed.

References

Denbigh, K. 1964. *The Principles of Chemical Equilibrium.* Cambridge, U.K.: Cambridge University Press.

Denbigh, K. 1966. *The Principles of Chemical Equilibrium,* 2nd ed. Cambridge, U.K.: Cambridge University Press.

Gaggioli, R. A. and Petit, P. J. 1977. "Use the Second Law, First." *Chemtech.* 496–506.

Hatsopoulos, G. N. and Keenan, J. H. 1965. *Principles of General Thermodynamics,* New York: John Wiley & Sons.

Haywood, R. W. 1980. *Equilibrium Thermodynamics for Engineers and Scientists.* New York: John Wiley & Sons.

Kline, S. J. and Koenig, F. O. 1957. "The State Principle: Some General Aspects of the Relationships Among the Properties of Systems." *Journal of Applied Mechanics.* Vol. 24. 29–34.

Petit, P. J. and Gaggioli, R. A. 1980. "Second Law Procedures for Evaluating Processes." In *Thermodynamics: Second Law Analysis,* Gaggioli, R. A., ed. *ACS Symposium Series 122.* Washington, D.C.: American Chemical Society. 15–37.

Rodriguez, L. 1980. "Calculation of Available-Energy Quantities." In *Thermodynamics: Second Law Analsyis,* Gaggioli, R. A., ed. *ACS Symposium Series 122.* Washington, D.C.: American Chemical Society. 39–59.

Thermal Research Laboratory 1965. *JANAF Thermochemical Tables.* Midland, MI: Dow Chemical Company.

Wepfer, W. J. and Gaggioli, R. A., ed. 1980. "Reference Datums for Available Energy." In *Thermodynamics: Second Law Analysis, ACS Symposium Series 122.* Washington, D.C.: American Chemical Society. 77–92.

Questions

1. Distinguish between endothermic and exothermic reactions and give one example of each type.

2. What are the constituents of the typical combustion products from (a) fuels used in thermal power stations and (b) automobile fuels? Discuss possible harmful effects on the environment from some of the constituents of these typical combustion products.

3. Explain the terms *stoichiometric, complete*, and *incomplete combustion*. Indicate which type of combustion is the most likely to occur in each of the following applications:
   (a) coal burning in a steam power plant
   (b) combustion of gasoline in a car engine being started cold in very cold weather
   (c) burning of a gas in a gas turbine

4. Is it possible to have incomplete combustion even when there is excess air? Explain your answer.

5. When oxygen is insufficient for the combustion of a hydrocarbon fuel, which of the elements C or H_2 is likely to be completely oxidized first? Why?

6. Give approximate values for
   (a) the gravimetric composition of atmospheric air
   (b) the volumetric composition of atmospheric air

7. Define the terms *excess air, air–fuel ratio, fuel–air ratio*, and *mixture strength*. What is meant when a fuel–air mixture is described as "lean" or as "rich."

8. Explain each of the following terms: wet products analysis, dry products analysis, and the Orsat analysis.

9. Describe the Orsat apparatus and explain how it can be used to determine the composition of the combustion products of a hydrocarbon fuel.

10. Enumerate and state the conservation laws that must apply to chemically reacting systems.

11. Discuss the importance and relevance of the laws of thermodynamics to the determination of
    (a) the direction in which a chemical reaction can proceed
    (b) the equilibrium composition of a chemically reacting system

12. What is a chemical substance? In what ways do chemical substances differ from pure substances?

13. Define the *enthalpy of formation* of a chemical substance. What is the enthalpy of formation for an element at the standard reference state of 25°C and 1 atm? Write down an expression for the enthalpy of a chemical substance that includes the enthalpy of formation at the standard reference state.

14. Define the *enthalpy of combustion* and the *heating value* for a fuel. Give the approximate relationship between these two concepts.

15. Distinguish between the lower heating value and the higher heating value for a fuel. Give the relationship between the lower and the higher heating values.

16. What is the adiabatic flame temperature for a combustion process? Would you expect the actual temperature attained in an adiabatic combustion process to be lower than or higher than the adiabatic flame temperature? Give reasons for your answer.

17. What is the reference temperature for the measurement of the entropy of a substance? On which law of thermodynamics is this choice based?

18. Briefly discuss the concept of a stable reference environment and the dead state for chemically reactive systems.

Problems

For the exergy analysis problems, assume $p_0 = 1$ atm and $T_0 = 298$ K for the reference environment unless otherwise stated.

14.1. Diesel fuel $(C_{12}H_{26})$ reacts with pure O_2 gas to form products of only H_2O and CO_2. Determine the mass of each product per kg of fuel supplied for the reaction.

14.2. Ethylene (C_2H_4) reacts with pure O_2 gas to form

products of only H_2O and CO_2. Determine the mass of each product per kg of fuel supplied for the reaction.

14.3. Gasoline [C_8H_{18} (g)] reacts with dry air in a stoichiometric air–fuel ratio to form products of combustion. Determine the stoichiometric equation and calculate the air–fuel and the fuel–air ratios by volume.

14.4. Determine the mass of air required for the stoichiometric combustion of 1 kg of the following fuels (as shown in Fig. P14.4):
(a) methane gas (CH_4)
(b) butane gas (C_4H_{10})
(c) kerosene with the gravimetric composition of 86.3% C, 13.6% H_2, and 0.1% S.
(d) diesel fuel (taken as dodecane, $C_{12}H_{26}$)
(e) gasoline (C_8H_{18})
Also determine in each case the gravimetric analysis of the wet products of combustion.

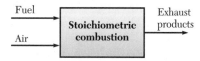

Figure P14.4 Stoichiometric combustion of fuel in air.

14.5. For each of the fuels in Problem 14.4, determine the volumetric analysis of the dry products of combustion.

* **14.6.** A sample of coal has the following ultimate analysis: 80% C, 5% H_2, 10% H_2O (moisture), and 5% ash. The coal is burned to completion with 50% excess air. Determine
(a) the theoretical air–fuel ratio (by mass)
(b) the actual air–fuel ratio (by mass)
(c) the Orsat analysis of the products

* **14.7.** A sample of pure carbon is burned in dry air in a closed vessel, as shown in Figure P14.7. At the completion of the reaction, there is no residual carbon and a volumetric analysis of the products of combustion gives 16.2% CO_2 with some CO and N_2. What is the gravimetric air–fuel ratio?

Figure P14.7 Reaction of carbon with dry air in closed container.

* **14.8.** The dry exhaust gas from a diesel engine is found to contain 8% CO_2 by weight. The fuel contains 88% C and 12% H_2 by weight, and the engine consumes 10 kg of fuel per hour. Assuming that the combustion is complete, calculate the total mass (in kg) of dry exhaust leaving the engine per hour.

14.9. The volumetric analysis of the dry products of combustion of a hydrocarbon fuel was found to be 12.5% CO_2, 3.7% CO, and 83.8% N_2. Find the air–fuel ratio (by mass) and the gravimetric analysis of the fuel.

* **14.10.** Find the gravimetric air–fuel ratio required for the stoichiometric combustion of a type of wood with the chemical formula $C_{10}H_{24}O_{11}$. Obtain the volumetric analysis of the wet products of combustion.

* **14.11.** Ethane (C_2H_6) reacts with 125% dry air to form products of combustion. Determine
(a) the products of combustion on a mass basis
(b) the dew point of the products if the pressure is 1 atm

* **14.12.** Butane [C_4H_{10} (g)] is burned with 25% excess dry air, and the products are cooled to 25°C at 1 atm. Give the volumetric analysis of the products at this state. Also give a volumetric analysis of the products on a dry basis (that is, with no water vapor in the products).

* **14.13.** Acetylene [C_2H_2 (g)] is burned with 40% excess dry air, and the products are cooled to 25°C at 1 atm. Give the volumetric analysis of the products at this state. Also give a volumetric analysis of the products on a dry basis (that is, with no water vapor in the products).

* **14.14.** Coal is burned with 30% excess air. An ultimate analysis of the coal yields the following:

Element	Percentage (by mass)
Carbon (C)	70.0
Sulfur (S)	4.0
Hydrogen (H_2)	5.0
Nitrogen (N_2)	2.0
Oxygen (O_2)	8.0
Ash (noncombustible)	4.0
Moisture (H_2O)	7.0
Total	100.0

Calculate the air–fuel ratio for the combustion on a molar basis and on a mass basis.

* **14.15.** Coal is burned with 40% excess air. An ultimate analysis of the coal yields the following:

Element	Percentage (by mass)
Carbon (C)	65
Sulfur (S)	2
Hydrogen (H_2)	6
Nitrogen (N_2)	6
Oxygen (O_2)	10
Ash (noncombustible)	3
Moisture (H_2O)	8
Total	100

Determine
(a) the theoretical air–fuel ratio on a molar basis
(b) the actual air–fuel ratio on a molar basis
(c) the Orsat analysis of the products

* **14.16.** Acetylene $[C_2H_2 (g)]$ is burned with dry air from the atmosphere. An Orsat analysis of the dry products gives volume measurements of 14.5% CO_2 and 3.64% O_2. What is the percent theoretical air used during the process?

* **14.17.** An unknown hydrocarbon fuel C_xH_y reacts with dry air. A volumetric analysis of the dry products gives the following results: 7.53% CO_2, 7.53% O_2, and 84.94% N_2. Determine
(a) the reaction equation for the process
(b) the composition of the fuel
(c) the percent theoretical air used in the process.

* **14.18.** An unknown hydrocarbon fuel C_xH_y reacts with dry air. A volumetric analysis of the dry products gives the following results: 3.3% CO_2, 16.08% O_2, and 80.62% N_2. Determine
(a) the reaction equation for the process
(b) the composition of the fuel
(c) the percent theoretical air used in the process.

* **14.19.** A combustion chamber is supplied with butane gas (C_4H_{10}) and n times the theoretical air (by volume). Combustion is complete. Determine, in terms of n, the molar analysis of the wet products of combustion. Also determine the dew point temperature of the flue gases if their pressure is 100 kPa and (i) $n = 1$, (ii) $n = 1.5$.

* **14.20.** Methane produced in a methane digester is used for operating the gas burner of a domestic water heater, as shown in Figure P14.20. A stoichiometric air–fuel mixture is supplied to the burner, and an Orsat analysis of the flue gases gives 10.4% CO_2, some CO, O_2, and N_2. Determine
(a) the air–fuel ratio (by volume)
(b) the volumetric analysis of the wet products of combustion
(c) the dew point temperature of the flue gases if their pressure is 100 kPa

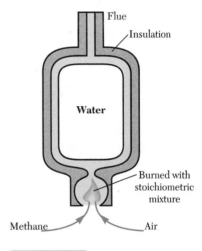

Figure P14.20 Gas water heater using methane fuel.

* **14.21.** A sample of coal has an ultimate analysis as follows: 82% C, 5% H_2, 6% O_2, 2% N_2, and 5% ash. When the coal is burned, an Orsat analysis of the flue gases gives 14% CO_2, 0.2% CO, and some O_2. Determine
(a) the gravimetric air–fuel ratio
(b) the theoretical air–fuel ratio (by mass)
(c) the percentage of excess air supplied
(d) the volumetric analysis of the dry products of combustion

* **14.22.** Methane (CH_4) reacts with 90% theoretical dry air. Determine the products of combustion on a volume basis. Clearly state your assumption(s).

* **14.23.** Ethylene (C_2H_4) reacts with 85% theoretical dry air. Determine the products of combustion on a volume basis. State your assumption(s) clearly.

14.24. Using the ideal gas equation, show that 1 kmol of any ideal gas occupies 24.45 m³ at the standard conditions of $p_0 = 1$ atm and $T_0 = 298$ K. The gross (or higher) heating values at standard conditions of pressure and temperature for the following gaseous fuels are as follows:
  (i)   methane (CH_4), 36,410 kJ/m³
  (ii) ethane (C_2H_6), 63,800 kJ/m³
  (iii) ethylene (C_2H_4), 57,720 kJ/m³
  Convert these to the equivalent values in
  (a) kJ/kmol
  (b) kJ/kg

* **14.25.** Use the ideal gas tables to calculate the standardized specific enthalpy of O_2 and CO_2, each at 60°C and 1 atm. Assume ideal gas behavior for each. Verify your results using the **GAS** code.

* **14.26.** Use the ideal gas tables to calculate the standardized specific enthalpy of CH_4, CO, and N_2, each at 140°F and 1 atm. Assume ideal gas behavior for each. Verify your results using the **GAS** code.

* **14.27.** Determine the enthalpy of reaction for the reaction of gasoline $[C_8H_{18}$ (g)] with 130% theoretical air at 77°F and 1 atm. Assume that all H_2O exists as vapor in the products.

14.28. A sample of sewage gas has a volumetric composition of 75% methane (CH_4) and 25% CO_2. The higher heating value for methane at the standard conditions of pressure and temperature can be taken as 36,410 kJ/m³. Determine the higher heating value of the sewage gas
  (a) in kJ/m³ at standard conditions of pressure and temperature
  (b) in kJ/kg

14.29. Find the lower (or net) heating values for the fuels listed in Problem 14.24. The molar specific heat of vaporization of water at 25°C can be taken as 43,990 kJ/kmol of H_2O.

* **14.30.** A sample of natural gas has a volumetric analysis of 80% CH_4, 18% C_2H_6, and 2% N_2. The enthalpies of formation at 1 atm and 25°C for the gases are $-74,850$ kJ/kmol for CH_4 and $-84,670$ kJ/kmol for C_2H_6. Determine the enthalpy of combustion at 1 atm and 25°C for the sample of natural gas. Also determine the higher and lower heating values of the fuel.

* **14.31.** Determine the enthalpy of combustion for the reaction of gasoline $[C_8H_{18}$ (g)] with air at 25°C and 1 atm. Also determine the higher heating value and the lower heating value for the fuel.

* **14.32.** Diesel fuel $[C_{12}H_{26}$ (l)] reacts with 100% theoretical air at 25°C and 1 atm. Determine
  (a) enthalpy of reaction
  (b) enthalpy of combustion
  (c) heat of combustion
  (d) lower heating value
  (e) higher heating value

* **14.33.** Diesel fuel $[C_{12}H_{26}$ (l)] reacts with 130% theoretical air at 77°F and 1 atm. Determine
  (a) enthalpy of reaction
  (b) enthalpy of combustion
  (c) heat of combustion
  (d) lower heating value
  (e) higher heating value

* **14.34.** Methane $[CH_4$ (g)] enters a steady-flow calorimeter such as that shown in Figure 14.4 at 25°C and 1 atm and reacts with 130% theoretical air, which enters at 25°C and 1 atm. The products exit at 500 K and 1 atm.
  (a) Determine the enthalpy of reaction (at 25°C and 1 atm) by use of enthalpy of formation data.
  (b) Determine the heat transferred to the surroundings during the reaction in kJ/kmol of fuel.

* **14.35.** Ethane $[C_2H_6$ (g)] enters a steady-flow calorimeter such as that shown in Figure 14.4 at 25°C and 1 atm and reacts with 125% theoretical air, which enters at 400 K and 1 atm. The products exit at 400 K and 1 atm.
  (a) Determine the enthalpy of reaction (at 25°C and 1 atm) by use of enthalpy of formation data.
  (b) Determine the heat transferred to the surroundings during the reaction in kJ/kmol of fuel.

* **14.36.** Methane $[CH_4$ (g)] reacts in a bomb calorimeter with the stoichiometric quantity of pure O_2 at 25°C and 1 atm, as shown in Figure P14.36. Calculate the heat transferred (per kmol of fuel) from the bomb during the process assuming that the products are at a final temperature of 25°C. Also compute the internal energy of combustion for the reaction and compare this value with the enthalpy of combustion for the fuel. Assume all the constituents of the products are gaseous.

* **14.37.** Gasoline $[C_8H_{18}$ (g)] reacts in a bomb calorimeter with 100% theoretical air at 25°C and 1 atm. Calculate the heat transferred (per kmol of fuel) from the bomb during the process if the prod-

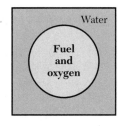

Figure P14.36 Reaction of stoichiometric mixture of fuel and oxygen in a bomb calorimeter.

ucts are at a final temperature of 25°C. Also compute the internal energy of combustion for the reaction and compare this value with the enthalpy of combustion for the fuel. Assume all the constituents of the products are gaseous.

∗ **14.38.** Hydrogen [H_2 (g)] at 25°C and 1 atm reacts with the stoichiometric amount of pure O_2 (g) at 25°C and 1 atm in a steady-flow process. Assuming complete combustion and ideal gas behavior for the reactants and products, determine the adiabatic flame temperature for the reaction. (Neglect dissociation effects.)

∗ **14.39.** Hydrogen [H_2 (g)] enters a rocket engine at 100 K and reacts with the stoichiometric amount of O_2 (g) at 100 K. The reaction occurs in a steady-flow process, and complete combustion occurs. The standardized specific enthalpy for the entering H_2 (g) and O_2 (g) are -5300 kJ/kmol and -5780 kJ/kmol, respectively. Assuming ideal gas behavior for the reactants and products, determine the adiabatic flame temperature for the reaction.

∗ **14.40.** Diesel fuel [$C_{12}H_{26}$ (l)] at 25°C and 1 atm is burned with air at 25°C and 1 atm in a steady-flow process. Assuming complete combustion and ideal gas behavior of the reactants and products, determine the adiabatic flame temperature for
   (a) 120% theoretical air
   (b) theroeretical air
   (c) 90% theoretical air

∗ **14.41.** A stoichiometric mixture of propane (C_3H_8) and air burns completely at constant pressure in a well-insulated piston-and-cylinder mechanism, as shown in Figure P14.41. The initial temper-

ature of the mixture is 25°C. Assuming that the combustion is complete, find the final temperature of the products if propane at 25°C has an enthalpy of combustion (with the H_2O in the vapor phase) of $-46,300$ kJ/kg. The c_p for N_2, CO_2, and H_2O can be taken as 1.22, 1.3, and 2.5 kJ/kg · K, respectively.

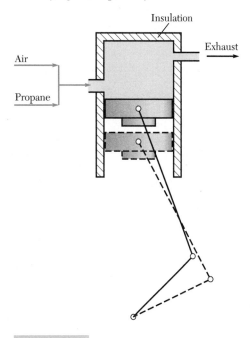

Figure P14.41 Constant pressure combustion of propane and air in a piston-and-cylinder mechanism.

∗ **14.42.** A mixture of methane (CH_4) and air is supplied at standard conditions of 1 atm and 25°C to the burner of a water heater. The volumetric air–fuel ratio is 14.29 to 1. Methane has a lower heating value of 802,000 kJ/kmol CH_4. The flow rate of methane in the mixture is equivalent to 0.5 kmol/h. The flue gases have an average c_p of 30 kJ/kmol · K and leave the heater at 400 K. Assume that complete combustion of the fuel occurs.
   (a) Write down the chemical equation for the combustion process, and hence determine the volumetric (wet) products analysis.
   (b) What is the flow rate (in kmol/h) of the flue gases?
   (c) Determine the heat transfer rate (in kW) to the water due to the combustion process.

* **14.43.** Propane gas (C_3H_8), with enthalpy of formation of $-103{,}800$ kJ/kmol, is burned in stoichiometric air in a constant pressure burner. The air–fuel mixture enters the burner at 1 atm and 25°C, while the products of combustion leave at 400 K.

  (a) Write down the chemical equation for the combustion process and obtain the volumetric analysis of the wet products of combustion.

  (b) Determine the enthalpy of combustion in kJ/kmol of the fuel. (Assume that the water in the products of combustion remains in the vapor phase.)

  (c) Determine the heat transfer (in kJ/kmol of propane) away from the burner.

  (d) Find the adiabatic flame temperature of the combustion.

* **14.44.** The following data were obtained during a trial on a single-cylinder, four-stroke oil engine of 208 mm bore and 320 mm stroke:

Item	Value
Engine speed	400 rev/min
Torque reading on dynamometer	312 N · m
Fuel consumption rate	4.13 kg/h
Heating value of fuel	42,400 kJ/kg
Exhaust gases flow rate	118.13 kg/h
Mean specific heat capacity of exhaust gases	1.09 kJ/kg · K
Engine exhaust temperature	445°C
Laboratory temperature	27°C
Engine cooling water flow rate	4.54 kg/min
Engine cooling water temperature rise	36.5°C

Draw up an energy balance using the laboratory temperature as datum. Use the result of Exercise 3 in Section 14.6.4 as the form of the energy balance required. What is the internal-combustion (I.C.) cycle efficiency of the engine?

* **14.45.** (a) Give an account of the disposal of the energy supplied in the fuel to a combustion engine.

  (b) During the trial of a single-cylinder, four-stroke oil engine of 304 mm bore and 458 mm stroke, the following observations were made:

Item	Value
Duration of trial	1 hr
Total fuel used	7.66 kg
Heating value of fuel	44,200 kJ/kg
Total number of revolutions	12,600
Brake torque	1400 N · m
Total mass of water used for cooling engine	544 kg
Inlet water temp.	25°C
Outlet water temperature	69.4°C
Total mass of air consumed by engine	368 kg
Air (ambient) temperature	27°C
Exhaust temperature	300°C
Specific heat of exhaust gas	1 kJ/kg · K

Draw up an energy balance using the format of Exercise 3 in Section 14.6.4.

* **14.46.** A mixture of methane and 300% excess air at 25°C and 1 atm pressure is fed to a constant pressure burner. Assuming that dissociation is negligible, determine the adiabatic flame temperature attained. The burner heats a 1 kg/min water stream at 25°C to 60°C. The first law efficiency for the gas burner water heater is 95%. If the flue gases exit at 370 K, determine the fuel supply rate (in kg/h) needed.

Assume $p_0 = 1$ atm and $T_0 = 298$ K in the following problems.

---

* **14.47.** Diesel fuel [$C_{12}H_{26}$ (l)] reacts with oxygen in a stoichiometric ratio with both products and reactants at 25°C and 1 atm, as shown in Figure P14.47. Complete combustion occurs. Assume all H_2O in the products is liquid. Calculate

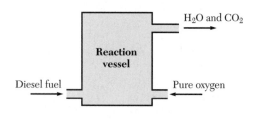

Figure P14.47 Reaction of diesel fuel and pure oxygen.

(a) the entropy change for the reacting system, in kJ/K per kmol of fuel

(b) the entropy change of the surroundings, per kmol of fuel

(c) the total entropy change of the universe, per kmol of fuel

∗ **14.48.** For the reaction of diesel fuel given in Problem 14.47, determine the maximum possible exergy delivery per kmol of fuel.

∗ **14.49.** Compute the chemical exergy for pure gaseous ethane at T_0 and p_0 using the equations found in Tables 14.1 and 14.2 and compare the results.

∗ **14.50.** Compute the chemical exergy for pure gaseous benzene at T_0 and p_0 using the equations found in Tables 14.1 and 14.2 and compare the results.

∗ **14.51.** For the sample of natural gas in Problem 14.30, determine the chemical flow exergy per unit mass of fuel.

∗ **14.52.** The higher heating value of different species of wood on a moisture-free and resin-free basis has been estimated at about 19,300 kJ/kg. For the wood sample in Problem 14.10, determine the chemical exergy per unit mass of fuel.

∗ **14.53.** Determine the second law efficiency achieved in the water-heating operation specified in Problem 14.46. The chemical exergy of methane can be taken as 51,320 kJ/kg CH_4.

∗ **14.54.** A mixture of methane and 300% excess air at 25°C and 1 atm burns in a steady-flow process under adiabatic conditions. Assuming that dissociation is negligible, determine the second law efficiency of the process.

∗ **14.55.** Repeat Problem 14.54, but this time assume that a stoichiometric air–methane mixture is fed to the burner and that the volumetric analysis of the wet products of combustion is 8.41% CO_2, 1.04% CO, 0.52% O_2, 71.12% N_2, and 18.91% H_2O.

∗∗ **14.56.** Gaseous ethane at T_0 and p_0 enters a reaction vessel to react with stoichiometric air at T_0 and p_0, as shown in Figure 14.6. The chemical reaction occurs at constant pressure in the reaction vessel, and the products are exhausted to the surroundings, which are at T_0 and p_0.

Part I. Calculate the maximum work possible per kmol of fuel from the combustion process if the exhaust products leave the reaction vessel at the reaction temperature of

(a) T_0, with all products in stable configurations in equilibrium with the surroundings

(b) 1200 K

Part II. If the heat of reaction is supplied at the reaction temperature to a Carnot heat engine operating as shown in Figure 14.6, compute the work done by the engine per kmol of fuel if the exhaust products leave the reaction vessel at the reaction temperature of

(c) T_0

(d) 1200 K

∗ **14.57.** Gasoline [C_8H_{18} (g)] at T_0 and p_0 enters a reaction vessel to react with stoichiometric air at T_0 and p_0, as shown in Figure 14.6. The chemical reaction occurs at constant pressure in the reaction vessel, and the products are exhausted to the surroundings, which are at T_0 and p_0.

Part I. Calculate the maximum work possible per kmol of fuel from the combustion process if the exhaust products leave the reaction vessel at the reaction temperature of

(a) T_0, with all products in stable configurations in equilibrium with the surroundings

(b) 2000 K

Part II. If the heat of reaction is supplied at the reaction temperature to a Carnot heat engine operating as shown in Figure 14.6, compute the work done by the engine per kmol of fuel if the exhaust products leave the reaction vessel at the reaction temperature of

(c) T_0

(d) 2000 K

Design Problems

14.1. *Water heater design*

A client needs advice on the most economical water heating system for a standard American family residence. The options to be considered are
(a) wood burning stove
(b) gas water heater
(c) kerosene water heater
(d) electric water heater

The design for options (a) through (c) can be illustrated as in Figure DP14.1.

1. Identify and reference a source of information on the weekly hot water requirement of a standard American family residence.
2. For each fuel option proposed, obtain an estimate of the amount of fuel needed to meet the anticipated weekly load.
3. Determine the weekly fuel cost for each of the options (a) through (c). Estimate also the overall second-law efficiency for each of these options and comment on your results.
4. For the option (d), determine the cost of electricity for the anticipated weekly load. Estimate also the overall second-law efficiency referenced to the supply of primary energy (such as coal or natural gas) for the production of electricity. Compare your results with those in part 3.

14.2. *Gas turbine design optimization*

Figure DP14.2 illustrates the simple gas turbine for stand-by power generation in an industrial plant. The design output of the turbine is 100 kW. Other design constraints include
(a) a metallurgical limit of 900 K
(b) a compressor with isentropic efficiency of 80%
(c) a turbine with an isentropic efficiency of 90%

Natural gas is available at the plant. Assume the gas is 100% methane (CH_4). The design objective is to minimize the amount of fuel used for the desired output.

1. Consider operation of the turbine on a Brayton-like cycle and determine an optimum pressure ratio for the actual cycle using appropriate equations from Chapter 10. Obtain estimates of temperatures at the various states shown in Figure DP14.2.
2. Carry out a first-law analysis of the combustion process to determine the necessary air–fuel ratio.
3. Use your results in parts 1 and 2 to obtain estimates of the air and fuel flow rates needed for the desired net power output of 100 kW.
4. Produce both energy and exergy disposition charts for 100 h of operation of the turbine. What are the overall first-law and second-law efficiencies for the turbine operation?
5. Find out the local cost of gas to an industrial establishment and use this to determine the

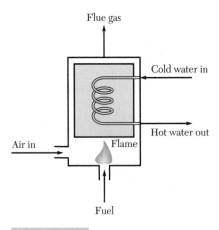

Figure DP14.1 Typical water heater design.

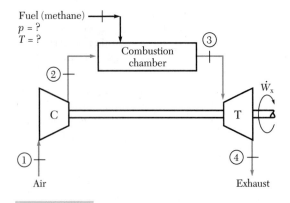

Figure DP14.2 Simple gas turbine design.

fuel cost of operating the turbine for 100 h. What then is the nominal cost of the electricity produced (in cents per kWh), based on fuel cost alone?

Suggest improvements to the simple gas turbine design in the light of your results.

14.3. *Design of unit to convert a truck engine from diesel to methane fuel*

According to data presented in Chapter 1, the United States may soon be facing shortages of gasoline and diesel fuel from domestic sources. However, methane appears to be more plentiful. Consequently, one approach that is under consideration to alleviate this potential problem is to convert truck engines so that they can burn methane. For such systems the methane could be stored as either liquid or gas.

Develop a design for a conversion unit to convert a large truck from diesel fuel to methane. Include a storage system and a preliminary injector design for the engine. Prescribe the air to fuel ratio for your design and estimate the performance for an engine using your design. Would the engine performance increase or decrease as a result of the installation of your conversion unit? What about pollution problems? What are the major advantages and disadvantages of your design? Present your results in a technical report.

Chemical Equilibrium

15.1   Introduction

The first law analyses and exergy analyses presented in Chapter 14 for combustion processes did not deal with the question of whether or not a chemical reaction will proceed to completion. Such a question is important because the quantity of energy released from an exothermic reaction and the magnitude of the exergy product depend on the degree of completion of the reaction. It is therefore necessary to establish procedures for determining the composition of the products of a chemical reaction. It is also important to know the direction of chemical reactions and to know whether or not a specific reaction can occur. Such information can be determined from considerations of chemical equilibrium and the second law of thermodynamics.

A system is in *chemical equilibrium* if there is no tendency for a net chemical reaction to occur within the system. Such a system is chemically homogeneous throughout, and no change occurs with time in its chemical composition. Equilibrium criteria can be developed from the second law of thermodynamics to determine when a reacting system is in chemical equilibrium and also to ascertain how far a chemical reaction can proceed before reaching a condition of chemical equilibrium.

15.2   Equilibrium Criteria

Chapter 8 discussed the *principle of increase of entropy*, which states that the total change in entropy for a system and its surroundings for any process is always greater than or equal to zero. No process can occur that would result in a decrease in total entropy for the system and the surroundings. Thus, a fundamental criterion deriving from the second law of thermodynamics for determining whether a chemical process will occur is that furnished by Equation 8.63:

$$dS_{universe} \ (= dS_{system} + dS_{surr}) \geq 0 \tag{8.63}$$

A process for which $dS_{universe} < 0$ is impossible and cannot occur.

Consider a chemical reaction in a closed adiabatic container having no work interaction with the surroundings. Such a system is referred to as an *isolated system*. In this case, there is neither heat nor work interaction with the surroundings, and the equilibrium criterion simplifies to what was designated as Equation 8.64:

$$dS_{isolated \atop system} \geq 0 \tag{8.64}$$

Figure 15.1 illustrates such a system with a starting point that is either 100% reactants and 0% products (on the left side of the figure) or 0% reactants and 100% products (on the right side of the figure). Starting from the extreme left position, the reaction can only occur with the entropy of the system increasing as more products are formed until the maximum value of entropy S_{max} for the isolated system is reached, as shown in the figure. Since the entropy of the system cannot decrease, the reaction stops when the peak value S_{max} is reached. If, however, one started with the mixture composition to the right of the peak shown in Figure 15.1, the mixture would continue to react in a direction to the left until the maximum entropy point was reached. In mathematical terms, this peak is specified by

$$\frac{\partial S_{system}}{\partial N_P} = 0 \quad \text{and} \quad \frac{\partial^2 S_{system}}{\partial N_P^2} < 0 \tag{15.1}$$

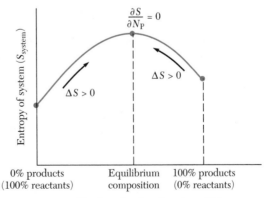

Figure 15.1 Equilibrium criterion for chemical reaction in closed adiabatic container with no work interaction.

which is a definitive criterion for equilibrium for a chemical reaction in an isolated system.

A general criterion for any closed system can be developed from consideration of the entropy generation in the system. Equation 8.65 can be expressed in the following form for a small element defined as a system with an instantaneous temperature T and a heat transfer δQ to it:

$$T \, dS - \delta Q = T \, \delta S_{gen} \tag{8.65}$$

From the second law, δS_{gen} must be positive or zero. In other words,

$$(T \, dS - \delta Q) = T \, \delta S_{gen} \geq 0 \tag{15.2}$$

Similar expressions can be written for flow processes that occur when dealing with an open system. Thus, for the interactions in an adiabatic, steady-flow process, one can write the following for the entropy generation rate in a control volume based on Equation 8.79:

$$\dot{S}_{gen,CV} = \sum_P \dot{N}_P \bar{s}_P - \sum_R \dot{N}_R \bar{s}_R \tag{15.3}$$

where Equation 15.3 is written in molar terms while Equation 8.79 was given in mass terms. Again, the entropy generation rate must be positive or zero for the reaction to occur.

If heat interaction occurs between the control volume and the surroundings, the entropy generation can be expressed as

$$\dot{S}_{gen,CV} = \sum_P \dot{N}_P \bar{s}_P - \sum_R \dot{N}_R \bar{s}_R - \frac{\dot{Q}}{T_0} \geq 0 \tag{15.4}$$

where T_0, the absolute temperature of the ambient surroundings, is also taken as the temperature at the control surface through which the heat transfer rate $\dot{Q}$ occurs.

Constant Volume Constant Temperature Reaction

Consider a closed system that is kept at a constant volume and constant temperature while a chemical reaction takes place. No work interaction occurs between the system and its surroundings. A criterion for the process can be developed as follows, based on the first and second laws of thermodynamics. With $\delta W = 0$, the first law reduces to $\delta Q = dU$, and Equation 15.2 can be written as

$$T \, dS - dU = T \, \delta S_{gen} \geq 0 \tag{15.5}$$

Now consider the Helmholtz function $A = U - TS$. A differential change in A is given by

$$dA = dU - T \, dS - S \, dT \tag{15.6}$$

Combining Equations 15.5 and 15.6 and setting $dT = 0$ for a constant temperature process gives

$$dA \, (= -T \, \delta S_{gen}) \leq 0 \tag{15.7}$$

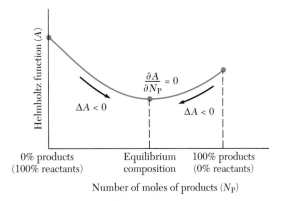

Figure 15.2 Equilibrium criterion for chemical reaction with constant volume and constant temperature and no work interaction.

Thus, for a chemical reaction involving a closed system maintained at constant volume and temperature (with no work interaction with the surroundings), the Helmholtz function for the system cannot increase. The equilibrium composition is reached when $dA = 0$. This criterion is shown graphically in Figure 15.2. With reference to this figure, the criterion is expressed mathematically as

$$\frac{\partial A}{\partial N_P} = 0 \quad \text{and} \quad \frac{\partial^2 A}{\partial N_P^2} > 0 \tag{15.8}$$

In other words, the Helmholtz function reaches a minimum value corresponding to $dA = 0$ for equilibrium.

Constant Pressure Constant Temperature Reaction

Next consider a closed system for the case of a constant pressure constant temperature process involving a chemical reaction with no work interactions other than fully resisted volume expansion work. Applying the first law of thermodynamics for the nonflow process gives $\delta Q = dH$, which combined with Equation 15.2 can be written as

$$T\,dS - dH = T\,\delta S_{\text{gen}} \geq 0 \tag{15.9}$$

In this case, the Gibbs function, $G = H - TS$, can be used to develop an equilibrium criterion. A differential change in G is given by

$$dG = dH - T\,dS - S\,dT \tag{15.10}$$

Combining Equations 15.9 and 15.10, with $dT = 0$ gives

$$dG\,(= -T\,\delta S_{\text{gen}}) \leq 0 \tag{15.11}$$

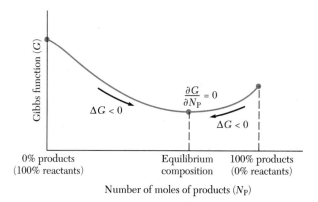

Figure 15.3 Equilibrium criterion for chemical reaction at constant pressure and constant temperature with no work interaction.

In this case, the reaction proceeds only while the Gibbs function is decreasing, and the system reaches equilibrium when $dG = 0$, as depicted in Figure 15.3. Thus, the criterion for equilibrium expressed mathematically is

$$\frac{\partial G}{\partial N_P} = 0 \quad \text{and} \quad \frac{\partial^2 G}{\partial N_P^2} > 0 \tag{15.12}$$

A General Criterion for Equilibrium

A more general equilibrium criterion for any closed system can be developed from the Gibbs function as follows. Chapter 13 indicated that for a homogeneous, single-phase mixture any extensive property can be expressed as a function of the two intensive properties T and p in addition to the amount N_i of each constituent i present in the mixture. For the extensive property G, this statement can be expressed mathematically as

$$G = G(p, T, N_1, N_2, \ldots, N_i, \ldots, N_m) \tag{15.13}$$

where N_i represents the number of moles of each chemical species present within the system at any given time. Now, from differential calculus, one can write the following for the exact differential dG:

$$dG = \left(\frac{\partial G}{\partial p}\right)_{T,N_i} dp + \left(\frac{\partial G}{\partial T}\right)_{p,N_i} dT + \sum_{i=1}^{m} \left(\frac{\partial G}{\partial N_i}\right)_{p,T,N_j} dN_i \tag{15.14}$$

The subscript N_i on each of the first two partial derivatives indicates that the number of moles for each species is held fixed during any change in p or T. N_j represents the respective number of moles of every component (except i) that is held fixed while summing on i in the last partial derivative on the right-hand side of the equation.

Equation 15.14 also applies to a mixture whose chemical composition does not change, in other words, to a pure substance. For a pure substance, the combined first and second law equation can be written as

$$dH - T \, dS = V \, dp \qquad (8.12)$$

Thus, eliminating $dH - T \, dS$ between Equations 15.10 and 8.12 yields the following for a mixture when the N_i values are kept constant:

$$dG = V \, dp - S \, dT$$

Comparing this with the expression for dG given by Equation 15.14 leads to the following deduction:

$$\left(\frac{\partial G}{\partial p} \right)_{T,N_i} = V \quad \text{and} \quad \left(\frac{\partial G}{\partial T} \right)_{p,N_i} = -S$$

The remaining terms under the summation symbol in Equation 15.14 arise only when changes in chemical composition occur. Thus, Equation 15.14 can be written for the general case as follows:

$$dG = V \, dp - S \, dT + \sum_{i=1}^{m} \left(\frac{\partial G}{\partial N_i} \right)_{p,T,N_j} dN_i \qquad (15.15)$$

Recall that for a pure substance in a state of equilibrium, both the pressure and temperature must be uniform throughout. Uniformity of pressure is associated with mechanical equilibrium, while a uniform temperature condition means that the system is also in a state of thermal equilibrium. For a system comprising chemical substance(s), the additional requirement for complete equilibrium is that the amount of each species must remain constant. Thus, in the general case at the equilibrium state, $dp = 0$, $dT = 0$, and $dN_i = 0$. Substituting these into Equation 15.15 gives

$$dG = 0 \qquad (15.16)$$

which is therefore regarded as the general criterion for thermodynamic equilibrium for closed systems.

In summary, complete thermodynamic equilibrium requires thermal, mechanical, and chemical equilibrium. For any system, the requirement for thermal equilibrium is $dT = 0$, while the requirement for mechanical equilibrium is $dp = 0$. For chemical equilibrium to exist as well, no change can occur in the amount of each species, and the last term in Equation 15.15 must be equal to zero since $dN_i = 0$.

15.3 Equilibrium and the Chemical Potential

The partial derivatives with respect to N_i (the number of moles) of the type in Equation 15.14 are generally referred to as *partial molal properties*. Thus, any extensive property Z that for a homogeneous phase of a multicomponent system can be expressed as $Z(p, T, N_1, N_2, \ldots, N_i, \ldots, N_m)$ has a corresponding partial molal property

$$\bar{Z} = \left(\frac{\partial Z}{\partial N_i} \right)_{p,T,N_j}$$

A partial molal property as thus defined gives the increase of that particular extensive property (such as G) for the system due to the addition of dN_i moles of the component i while keeping the amounts for the other components as well as the system pressure and temperature constant. The *partial molal Gibbs function* $\overline{G}$ inside the summation in Equation 15.14 has a special significance, however, and is called the *chemical potential*. The chemical potential of any chemical species is usually symbolized by $\overline{\mu}_i$ and is the partial molal Gibbs function, as defined below:

$$\overline{\mu}_i = \overline{G} = \left(\frac{\partial G}{\partial N_i}\right)_{p,T,N_j} \tag{15.17}$$

where $\overline{\mu}_i$ is the intensive property or potential function associated with changes in the amount N_i of the chemical species i.

The chemical potential (also called the *electrochemical potential*) is the driving potential for molar transfer of components. For a single-component, two-phase closed system, the mass tends to move from the phase of higher chemical potential to one of lower potential. Likewise, for a multicomponent system, a component tends to move from a region of high chemical potential to one of lower chemical potential.

Using this definition of chemical potential, Equation 15.15 can be written as follows:

$$dG = V\,dp - S\,dT + \sum_i \overline{\mu}_i\,dN_i \tag{15.18}$$

This equation can also be derived by consideration of the first and second laws of thermodynamics in relation to a homogeneous multicomponent system to which incremental amounts ΔN_i of the pure species are added in a reversible manner. The system in this case is referred to as an *open phase*, and the addition of the pure species is made through special membranes known as *semipermeable membranes*. The resulting set of equations from such application of the first and second laws of thermodynamics turn out to be extensions of the fundamental or characteristic equations of state considered in Chapter 12 to the general case of systems whose chemical composition can change. The other equations in the set are as follows:

$$dU = T\,dS - p\,dV + \sum_i \overline{\mu}_i\,dN_i \tag{15.19}$$

$$dH = T\,dS + V\,dp + \sum_i \overline{\mu}_i\,dN_i \tag{15.20}$$

$$dA = -S\,dT - p\,dV + \sum_i \overline{\mu}_i\,dN_i \tag{15.21}$$

where $\overline{\mu}_i$ is the same in all of these equations and is the chemical potential that was identified earlier. From the kind of derivation just shown, the following expression emerges for the chemical potential:

$$\overline{\mu}_i = \overline{h}_i - T\overline{s}_i = \overline{g}_i(p_i, T) \tag{15.22}$$

The molar specific enthalpy and entropy are for the pure species at pressure p_i, the partial pressure of the species in a mixture (at total pressure p), and the temperature T of the mixture. In other words, the chemical potential is equal to the molar specific Gibbs function for the pure species at pressure p_i and temperature T.

With respect to the partial molal properties, an important result is obtained when the equation for the exact differential dZ is integrated while p and T are kept constant. It is

$$Z = \sum_i N_i \bar{Z}_i \qquad (15.23)$$

Applying this to Equation 15.17, which is based on the functional relationship $G = G(p, T, N_1, N_2, \ldots, N_i, \ldots, N_m)$ yields the following result:

$$G = \sum_i N_i \bar{G}_i = \sum_i N_i \bar{\mu}_i \qquad (15.24)$$

Differentiating both sides of Equation 15.24 yields

$$dG = \sum_i N_i \, d\bar{\mu}_i + \sum_i \bar{\mu}_i \, dN_i \qquad (15.25)$$

Equating dG from Equations 15.18 and 15.25 gives

$$-S \, dT + V \, dp - \sum_i N_i \, d\bar{\mu}_i = 0 \qquad (15.26)$$

Equation 15.26 is the famous *Gibbs–Duhem equation*. One significant feature of this equation is that all the differentials involve intensive properties only. The following deductions can be made from the equation. For systems in general, pressure is the driving potential for volume change, temperature is the driving potential for the flow of entropy, and the chemical potential is the driving potential for a change in the number of moles of a chemical species due to chemical reaction or mass diffusion. Thus, a variation of $\bar{\mu}_i$ within a system causes a diffusion of mass just as a variation of T causes a diffusion of entropy. For a system to be in chemical equilibrium, therefore, the chemical potential $\bar{\mu}_i$ for each component must be the same and constant throughout the system (just as T must be uniform for thermal equilibrium). As discussed previously in Section 14.7.2, a system is in chemical equilibrium at the dead state if (1) each chemical species has a uniform chemical potential, $\bar{\mu}_i$, that is also identical with $\bar{\mu}_{i0}$ for the species in the reference environment, and (2) the only chemical species remaining are those that are stable in the reference environment. (Refer back to Section 14.7.2 for additional discussion.)

15.4   Reaction Equilibrium

Chemical reactions involve the conversion of reactants into products. Some reactions do not proceed to completion. Whether a reaction reaches completion or not, however, the laws of conservation of mass and the atomic species must hold, as discussed in Section 14.2.

Consider the reaction

$$CO + \tfrac{1}{2} O_2 \rightarrow CO_2$$

By our nomenclature, 1 kmol of CO reacts with $\tfrac{1}{2}$ kmol of O_2 to form 1 kmol of CO_2.

This reaction is accompanied by a release of energy as heat, that is, an exothermic reaction, when it proceeds in the direction of the arrow. The reaction can also proceed in the reverse direction if sufficient energy is supplied to the molecules of CO_2. Some of the molecules of CO_2 in a typical combustion process receive sufficient energy in collisions for this to occur, so that the following reaction also occurs:

$$CO_2 \rightarrow CO + \tfrac{1}{2} O_2$$

The fact that both of these reactions occur simultaneously is specified by

$$CO + \tfrac{1}{2} O_2 \leftrightarrows CO_2$$

The primary reaction is exothermic, while the reversed reaction is endothermic. At any given temperature and pressure, the proportions of CO, O_2, and CO_2 adjust until the two reactions proceed at the same rate so that the number of molecules of CO_2 being formed equals the number dissociating. If the number of moles of CO and O_2 are supplied to the reacting system in a stoichiometric ratio (that is, 2 mol CO/mol O_2), one can write a balanced chemical equation for the equilibrium state as

$$CO + \tfrac{1}{2} O_2 \rightarrow x\, CO + y\, O_2 + z\, CO_2$$

where x, y, and z give the amount (in mol) of CO, O_2, and CO_2, respectively, at the equilibrium state. In the more general case, the coefficients on the left-hand side of the equation are set by the mole fractions for the various chemical species supplied to the reacting system. The chemical equilibrium state reached is dynamic, not static, because both the combustion of CO and the dissociation of CO_2 proceed continuously and simultaneously.

Any general chemical reaction can proceed in a manner similar to that just described for CO, O_2, and CO_2 and can be represented as follows:

$$a\,A + b\,B + \cdots \leftrightarrows e\,E + f\,F + \cdots \qquad (15.27)$$

The lowercase italic letters represent the stoichiometric coefficients for a balanced chemical equation, and the uppercase letters represent the chemical species involved in the reaction. Note that the values for a, b, e, f, and so on are not the same as the N_i values (the number of moles for each constituent in the system). Neither are a and A related to the Helmholtz function, which is expressed by the same symbols. After the reaction is specified, the values for $a, b, \ldots, e, f, \ldots$ can always be obtained by balancing the chemical equation.

If the reaction specified by Equation 15.27 occurs within a system at constant temperature and pressure, then the change in the Gibbs function for an infinitesimally small change of composition is given by

$$
\begin{aligned}
dG_{T,p} &= \sum_i \bar{\mu}_i\, dN_i \\
&= (\bar{\mu}_A\, dN_A + \bar{\mu}_B\, dN_B + \cdots) + (\bar{\mu}_E\, dN_E + \bar{\mu}_F\, dN_F + \cdots)
\end{aligned}
\qquad (15.28)
$$

The dN_i terms are not independent of each other because the conservation of mass applies to the reacting system. Furthermore, the dN_i terms can be either positive or negative depending on the direction of the reaction. Thus, for the forward direction of the reaction,

$$dN_A = -ka \qquad dN_E = ke$$
$$dN_B = -kb \qquad dN_F = kf$$
$$\vdots \qquad\qquad \vdots$$

where k is a proportionality constant required for a mass balance. The negative signs are used because species A and B are disappearing while E and F are being formed.

Combining these expressions with Equation 15.28 and setting $dG_{T,p} = 0$ for equilibrium yields

$$(e\bar{\mu}_E + f\bar{\mu}_F + \cdots) - (a\bar{\mu}_A + b\bar{\mu}_B + \cdots) = 0 \qquad (15.29)$$

This equation is valid for any chemical reaction when the state of (dynamic) equilibrium has been attained. To use the equation, however, one must evaluate the chemical potential for every component. It is difficult to determine the chemical potential for substances in general. The notable exception is an ideal gas mixture, to which the remainder of this chapter is devoted. Thus, ideal gas behavior is frequently assumed for the gases involved in combustion processes. The remaining sections show how the chemical potential can be determined for such systems.

Application to Ideal Gas

On the basis of various additive laws, Chapter 13 established that an ideal gas component of an ideal gas mixture behaves as if the component alone occupied the volume of the system at the temperature of the system. Thus, each component of the mixture can be considered to be a pure component that exists within the system at temperature T of the mixture but with a partial pressure p_i such that the total pressure p for the mixture is given by $\Sigma_i p_i$. Now the question is, what is the chemical potential of a single pure gaseous component in the mixture?

Consider Equation 15.28 for a single, pure component. The chemical potential $\bar{\mu}_i$ is an intensive property that changes only when the relative amount of a species changes. Therefore, $\bar{\mu}_i$ is constant since only one species is present, and Equation 15.28 can be integrated to yield

$$G_{T,p} = \bar{\mu}_i N_i$$

or

$$\bar{\mu}_i = G_{T,p}/N_i = \bar{g}_i\,(p_i,\,T) \qquad (15.30)$$

Thus, as noted earlier in Equation 15.22, the chemical potential of each single, pure gaseous component in a mixture is equal to the specific molar Gibbs function for the component.

Since an ideal gas component exists at its partial pressure p_i in the mixture, only an expression for $\bar{g}_i\,(p_i,\,T)$ is needed for the component at p_i and T to determine the chemical potential of each component in the mixture. Equation 12.9 for a substance of fixed composition can be written in molar form as follows:

$$d\bar{g} = \bar{v}\,dp - \bar{s}\,dT \qquad (15.31)$$

where p corresponds in this case to the partial pressure for the chemical species i. This equation can be integrated to obtain an expression for $\bar{g}_i\,(T, p_i)$ for each component of the mixture. Note that the analysis is for a constant temperature and constant pressure reaction process for the mixture. Therefore, $dT = 0$ for the process, but the partial pressure p_i is changing for each component during the process even though the total pressure remains constant. Substituting $\bar{v} = \bar{R}T/p$ for an ideal gas and integrating Equation 15.31 from $p_0\,(= 1\ \text{atm})$ to p_i with $dT = 0$ for each ideal gas component gives

$$\bar{g}_i\,(T, p_i) = \bar{g}_i\,(T, p_0) + \bar{R}T \ln p_i \qquad (15.32)$$

where p_i is expressed in atm and $\bar{g}_i\,(T, p_0)$ is normally written as $\bar{g}_i^\circ\,(T)$ when $p_0 = 1$ atm. Since the chemical potential is equal to the molar specific Gibbs function for this case, the expression given by Equation 15.32 can also be written as

$$\bar{\mu}_i\,(T, p_i) = \bar{g}_i\,(T, p_i) = \bar{g}_i^\circ\,(T) + \bar{R}T \ln p_i \qquad (15.33)$$

Thus, this expression can be used to compute the chemical potential of a single, pure ideal gas component of a mixture.

15.5 Equilibrium Constant

It is now possible to determine the equilibrium composition of an ideal gas reacting mixture. This can be done by using Equations 15.29 and 15.33. Combining these two equations and rearranging gives

$$\{[e\bar{g}_E^\circ\,(T) + f\bar{g}_F^\circ\,(T) + \cdots] - [a\bar{g}_A^\circ\,(T) + b\bar{g}_B^\circ\,(T) + \cdots]\}$$
$$+ \bar{R}T[(e \ln p_E + f \ln p_F + \cdots) - (a \ln p_A + b \ln p_B + \cdots)] = 0 \qquad (15.34)$$

The quantity within the braces is referred to as the *standard state Gibbs function change* and is symbolized by $\Delta G^\circ\,(T)$. This is the change in the Gibbs function that occurs for the reaction that goes to completion at the given constant temperature of the reaction.

Using $\Delta G^\circ\,(T)$ for the term in the braces and simplifying Equation 15.34 gives

$$\Delta G^\circ\,(T) = -\bar{R}T \ln \frac{p_E^e p_F^f \cdots}{p_A^a p_B^b \cdots} \qquad (15.35)$$

or

$$\Delta G^\circ\,(T) = -\bar{R}T \ln K_p \qquad (15.36)$$

where

$$K_p = \frac{p_E^e p_F^f \cdots}{p_A^a p_B^b \cdots} \qquad (15.37)$$

Equation 15.36 can also be expressed as

$$K_p = \exp[-\Delta G^\circ\,(T)/(\bar{R}T)] \qquad (15.38)$$

The term K_p is called the *equilibrium constant*, and it is very important in the thermodynamics of chemically reacting mixtures of ideal gases. As can be seen from Equation 15.38, K_p is a function of temperature only. Although $\Delta G^\circ(T)$ is written to show that it is a function of temperature, values of the function for any particular temperature are based on the assumption that the reaction takes place at the given constant temperature T. K_p can be calculated from a knowledge of T and $\Delta G°\ (T)$ for the possible reactions. From examination of the terms in the braces of Equation 15.34, it is apparent that $\Delta G°\ (T)$ can be evaluated for any chemical reaction if the temperature and the stoichiometric equation are known for the reaction. Values of K_p are available for most reactions encountered in the combustion of gaseous fuels. Sample values tabulated in terms of $\ln K_p$ are given in Appendix A.

A few observations concerning K_p should be noted.

1. Although K_p is defined in Equation 15.37 in terms of partial pressures, the partial pressures are related in such a way that K_p does not depend on the total pressure of the reacting system. This can be seen from Equation 15.38, which shows a temperature dependence for K_p, but no dependence on total pressure.

2. Although K_p should be dimensionless, many tabulations of K_p are based on pressure in atmospheres and the remaining terms in SI units. In such tabulations, the value of K_p (or $\ln K_p$) depends on the units used in the equation to compute K_p.

3. When equilibrium is reached for a chemical reaction such as that given by Equation 15.27, a reaction is going to completion in the forward direction to yield the products E and F from the reactants A and B, while another reaction may be going to completion in the negative direction to produce A and B from E and F. In the equilibrium condition, no net change occurs in any component A, B, . . . , E, F, . . . of the system.

4. The value for K_p depends on the particular reaction occurring in the system. Furthermore, the value depends on the particular way of writing the balanced reaction equation. Consider the following forms of a balanced equation:

$$H_2 + \tfrac{1}{2} O_2 \rightarrow H_2O \tag{1}$$

$$2\,H_2 + O_2 \rightarrow 2\,H_2O \tag{2}$$

Since $\Delta G°\ (T)$ is a change in an extensive property, its value for reaction 2 is double that for reaction 1. Mathematical manipulations with Equation 15.36 for the two cases show that $\ln K_p$ in reaction 2 is equal to $2\,(\ln K_p)$ in reaction 1. Also, if the reaction is written as

$$H_2O \rightarrow H_2 + \tfrac{1}{2} O_2 \tag{3}$$

the value for $\Delta G°\ (T)$ in this case is the negative of $\Delta G°\ (T)$ in reaction 1, and K_p in reaction 3 is equal to $1/K_p$ in reaction 1. In view of these differences, the particular chemical reaction equation being used must be specified when values of K_p (or $\ln K_p$) are tabulated.

5. Some authors define K_p as the reciprocal of that given in Equation 15.37. Therefore, one must be careful when using values for K_p from other sources.

6. The magnitude of K_p at a particular temperature can be used to estimate whether or not the reaction is likely to occur at that temperature. When increased amounts of products are formed, the partial pressures of the products in the mixture increase, while smaller amounts of products mean lower partial pressures for the

products (and higher for the reactants). Examination of Equation 15.37 shows that higher partial pressures for the products lead to a higher value for K_p, while lower partial pressures for the products correspond to a lower value for K_p. As a rule of thumb, it can be stated that if $K_p > 10^3$, the system will consist primarily of products and the reaction would have gone to completion. In turn, if $K_p < 10^{-3}$, the system will consist primarily of reactants, and no reaction is likely to occur.

7. The standard state Gibbs function, $\Delta G° (T)$, if known, can be substituted in Equation 15.36 to calculate K_p. For a constant temperature process, this function is given by

$$\Delta G° (T) = \Delta H° - T \Delta S°$$

Thus, $\Delta G° (T)$ can be evaluated from the enthalpy of reaction and the entropy change for a reaction occurring at a pressure of 1 atm and temperature T. For these calculations, the enthalpy is referenced to the stable state of the elements at standard conditions of 1 atm and 25°C (77°F), while the data for entropy are normally referenced to 1 atm and 0 K (0°R).

EXAMPLE 15.1

Calculate the equilibrium constant at 1000 K for the reaction

$$H_2 + \tfrac{1}{2} O_2 \rightleftarrows H_2O$$

and compare the result with the tabulated value.

Given: Chemical reaction at 1000 K

Find: K_p

Solution: The reaction is in the form of Equation 15.27 with

$$a = 1, \quad A = H_2$$
$$b = \tfrac{1}{2}, \quad B = O_2$$
$$e = 1, \quad E = H_2O$$

From Equation 15.36,

$$\ln K_p = -\Delta G° (T)/(\overline{R}T)$$

The definition of $\Delta G° (T)$ given within the braces of Equation 15.34 gives

$$\ln K_p = \frac{-1}{\overline{R}T} \left[\overline{g}°_{H_2O} (1000\ \text{K}) - \overline{g}°_{H_2} (1000\ \text{K}) - \frac{1}{2}\overline{g}°_{O_2} (1000\ \text{K}) \right]$$

Values for $\overline{g}°_i (T)$ can be calculated by using the gas tables and the definition of $\overline{g}°_i (T) = [\overline{h}_i (T) - T\overline{s}°_i (T)]$ as follows:

$$\overline{g}°_{H_2O} (1000\ \text{K}) = \overline{h}_{H_2O} (1000\ \text{K}) - (1000\ \text{K}) \cdot \overline{s}°_{H_2O} (1000\ \text{K})$$

$$= -215{,}848\ \text{kJ/kmol} - (1000)(232.59)\ \text{kJ/kmol}$$

$$= -448{,}438\ \text{kJ/kmol}$$

$$\bar{g}^\circ_{H_2} (1000 \text{ K}) = 20,686 - (1000)(166.13) = -145,444 \text{ kJ/kmol}$$

$$\bar{g}^\circ_{O_2} (1000 \text{ K}) = 22,709 - (1000)(243.50) = -220,791 \text{ kJ/kmol}$$

Now

$$\ln K_p = \frac{-1}{(8.314 \text{ kJ/kmol} \cdot \text{K})(1000 \text{ K})} \left[-448,438 + 145,444 + \frac{1}{2}(220,791) \right] \text{ kJ/kmol}$$

or

$$\ln K_p = 23.17 \quad \text{and} \quad K_p = 1.14 \times 10^{10}$$

The tabulated value given in Appendix A for this case is $\ln K_p = 23.169$. Thus, the calculated value agrees with the tabulated value within 0.1%.

The close agreement between the calculated and tabulated value in Example 15.1 is fortuitous. The values of K_p in Table A-24 are based on data from the JANAF thermochemical tables (Thermal Research Laboratory 1965) whereas a different data base (see Appendix D for details) was used for the ideal gas properties in Appendix A. While the use of different data bases produces minimal discrepancies in the estimate of $\ln (K_p)$, the resulting discrepancy in K_p can be much larger, depending on the value of $\ln (K_p)$. The following differential form of Equation 15.36 can be used to estimate the likely discrepancy in K_p:

$$\frac{dK_p}{K_p} = \frac{d(\Delta G^\circ (T))}{\Delta G^\circ (T)} \cdot \ln (K_p)$$

Thus, a discrepancy of $a\%$ in the determined values of $\Delta G^\circ (T)$ will result in a percentage difference of $a \cdot \ln (K_p)$ in the estimates of K_p.

EXERCISES

for Section 15.5

1. Determine the equilibrium constant at 1000 K for the reaction

$$CO + \tfrac{1}{2} O_2 \leftrightharpoons CO_2.$$

   Compare the result with the tabulated value.
   Answer: 1.59×10^{10}

2. Repeat Exercise 1 for a temperature of 2000 K.
   Answer: 729

3. Repeat Example 15.1 for a temperature of 2000 K.
   Answer: 3460

15.6 *Equilibrium Compositions*

The equilibrium composition is uniquely specified by Equation 15.37. This composition can be determined from the stoichiometric coefficients and the value of K_p for any reaction at any temperature. The value of K_p can be computed from Equation 15.38 as outlined in the previous section. Thus, a procedure has been developed for computing the equilibrium composition for any reaction of ideal gases occurring at a constant temperature.

It is usually easier to deal with N_i than p_i for a reaction. Equation 15.37 is thus modified with the mole fractions substituted in place of the ratios of partial pressures to the total pressure of the system, p. Since, for an ideal gas,

$$p_i = (N_i/N)p \tag{15.39}$$

Equation 15.37 can be expressed as

$$K_p = \frac{(N_E)^e (N_F)^f \cdots}{(N_A)^a (N_B)^b \cdots} \left(\frac{p}{N}\right)^{(e+f+\cdots)-(a+b+\cdots)} \tag{15.40}$$

where p is the total pressure of the system and N is the total number of moles of the mixture at the equilibrium condition. N is given by

$$N = (N_A + N_B + \cdots) + (N_E + N_F + \cdots) + N_{\text{inerts}} \tag{15.41}$$

Thus, Equations 15.40 and 15.41 can be used with equations for conservation of mass of the atomic species to determine the equilibrium composition of a reacting mixture of ideal gases. Before demonstrating the calculation procedures, let's briefly examine how the total pressure and presence of inerts might affect the reaction taking place.

Effect of Total System Pressure on Equilibrium Composition of Reacting Mixture (Le Chatelier's Principle)

The equilibrium constant K_p is constant for any given reaction at a particular temperature since K_p is a function only of temperature for a given reaction. The exponent of p in Equation 15.40 is $[(e + f + \cdots) - (a + b + \cdots)]$. Therefore, if $(e + f + \cdots) = (a + b + \cdots)$, the pressure will not affect the equilibrium composition. However, if

$$(e + f + \cdots) > (a + b + \cdots)$$

the amount of products must decrease relative to the amount of reactants so that K_p remains constant as the total pressure is increased. Conversely, if

$$(e + f + \cdots) < (a + b + \cdots)$$

then the amount of products must increase relative to the reactants as p is increased. Thus, increasing the total system pressure for a reaction increases the amount of constituents that are on the side of the reaction having the smaller sum of stoichiometric coefficients. This conclusion is called *Le Chatelier's principle*.

EXAMPLE 15.2

In the reaction given in Example 15.1, determine the effect of pressure on the reaction. Should a high or low pressure be used to increase the amount of products at equilibrium?

Given: $H_2 + \frac{1}{2} O_2 \rightleftarrows H_2O$

Find: Impact of p on reaction

Solution: The sum of stoichiometric coefficients of the reactants is $1\frac{1}{2}$, while the sum for the products is 1. Since

$$e < (a + b)$$

the amount of products increases with increased pressure. Therefore, a <u>high pressure</u> should be used to increase the amount of products at equilibrium.

Effect of Inert Gases on Equilibrium Composition of Reacting Mixture

Following the same line of reasoning as outlined for pressure, observe from Equation 15.40 that increasing the inerts (and thereby N) produces the following effects:

When	Effect
$(e + f + \cdots) = (a + b + \cdots)$	None
$(e + f + \cdots) < (a + b + \cdots)$	Products decrease
$(e + f + \cdots) > (a + b + \cdots)$	Products increase

Thus, increasing the inerts in a reacting vessel increases the amount of products when the sum of the stoichiometric coefficients of the products is greater than the sum of the stoichiometric coefficients of the reactants.

Calculation of Equilibrium Composition for a Single Reaction

The method previously outlined for calculating the equilibrium composition of a reacting mixture is illustrated in the following examples.

EXAMPLE 15.3

Calculate the equilibrium composition at 1 atm and 1000 K for the reaction given in Example 15.1. Assume that the reactants enter the reaction vessel in a stoichiometric ratio.

Given:
$$H_2 + \tfrac{1}{2} O_2 \rightleftarrows H_2O$$

$$T = 1000 \text{ K and } p = 1 \text{ atm}$$

$$2 \text{ mol } H_2/\text{mol } O_2 \text{ enter the reaction vessel}$$

Find: Equilibrium composition

Solution: Equations 15.40 and 15.41 are used along with conservation of mass for the atomic species. For this reaction,

$$K_p = \frac{N_{H_2O}}{N_{H_2}(N_{O_2})^{1/2}} \left(\frac{1}{N}\right)^{1-(1+1/2)}$$

and

$$N = N_{H_2} + N_{O_2} + N_{H_2O}$$

If the actual chemical reaction does not go to completion, an equation for the reaction can be written in the form

$$H_2 + \tfrac{1}{2} O_2 \rightarrow x\, H_2 + y\, O_2 + z\, H_2O$$

where x, y, and z represent the amount (in mol) of H_2, O_2, and H_2O, respectively, present in the mixture at equilibrium. The value for K_p for this case was given in Example 15.1 as $K_p = 1.14 \times 10^{10}$. Now $N = x + y + z$, and the expression for K_p can be written as

$$1.14 \times 10^{10} = \frac{z}{xy^{1/2}}\, (x + y + z)^{1/2} \tag{1}$$

Mass balances on the atomic species give

$$H_2 \text{ balance:}\quad 1 = x + z \tag{2}$$

$$O_2 \text{ balance:}\quad 1 = 2y + z \tag{3}$$

This gives three equations (1, 2, and 3) and three unknowns (x, y, and z), which can be solved to obtain

$$x \approx 0, \quad y \approx 0, \quad z = 1$$

The equilibrium composition is essentially <u>100% H_2O</u>. The previous rule of thumb gives

$$K_p = 1.14 \times 10^{10} \gg 10^3$$

Therefore, the reaction is expected to go to completion.

EXAMPLE 15.4

A reaction of 1 mol gaseous C with 1 mol H_2 at 5500 K forms CH_2. The equilibrium constant $K_p = 0.003$ for the reaction. Determine the equilibrium composition if the reaction occurs at

(a) 1 atm

(b) 20 atm

(c) 1 atm, but with 5 mol inert gas in the reaction vessel

Explain the physical significance of the results

Given: 1 mol C and 1 mol H_2 in reaction $C + H_2 \rightleftarrows CH_2$ at 5500 K

$$K_p = 0.003$$

$$p = 1 \text{ atm or } p = 20 \text{ atm}$$

Additional 5 mol inerts are present for reaction at 1 atm

Find:

(a) Equilibrium composition at 1 atm

(b) Equilibrium composition at 20 atm

(c) Equilibrium composition at 1 atm with 5 mol inerts present

Explain results

Solution:

(a) Equations 15.40 and 15.41 yield

$$K_p = \frac{N_{CH_2}}{N_C N_{H_2}} \left(\frac{1}{N}\right)^{-1}$$

and

$$N = N_{CH_2} + N_C + N_{H_2}$$

If the reaction does not go to completion, a mass balance gives

$$C + H_2 \rightarrow x\,C + y\,H_2 + z\,CH_2$$

where x, y, and z are the number of moles of C, H_2, and CH_2, respectively.
   Now the equation for K_p can be written as

$$0.003 = \frac{z}{xy}\left(\frac{1}{x + y + z}\right)^{-1} \tag{1}$$

Mass balances of the atomic species give

$$\text{C balance:} \quad 1 = x + z \tag{2}$$

$$\text{H}_2 \text{ balance:} \quad 1 = y + z \tag{3}$$

Solving Equations 1, 2, and 3 simultaneously gives

$$x = 0.9987, \quad y = 0.9987, \quad z = 0.0015$$

which is the equilibrium composition at 1 atm.

(b) This is the same as Part (a) except that $p = 20$ atm. Therefore, Equations 1, 2, and 3 become

$$0.003 = \frac{z}{xy}\left(\frac{20}{x + y + z}\right)^{-1} \tag{1}$$

$$1 = x + z \tag{2}$$

$$1 = y + z \tag{3}$$

Solving gives

$$x = 0.9713, \quad y = 0.9713, \quad z = 0.0287$$

(c) For this case, the mass balance equation becomes

$$C + H_2 + N_{inerts} \rightarrow x\,C + y\,H_2 + z\,CH_2 + N_{inerts}$$

where $N_{inerts} = 5$, and

$$N = x + y + z + 5$$

Therefore, Equations 1, 2, and 3 are

$$0.003 = \frac{z}{xy}\left(\frac{1}{x + y + z + 5}\right)^{-1} \tag{1}$$

$$1 = x + z \tag{2}$$

$$1 = y + z \tag{3}$$

Solving gives

$$x = 0.9996, \quad y = 0.9996, \quad z = 0.0004$$

for the equilibrium composition.

(d) The values of z obtained demonstrate that small quantities of CH_2 were formed in every case. This was to be expected because the value of K_p at the reaction temperature was 0.003, which is only slightly above 10^{-3}, the rule-of-thumb value stated earlier for essentially no reaction to occur.

Comparison of Part (b) of this example with Part (a) shows that more products of CH_2 were formed at the higher pressure of Part (b). This is to be expected because $e < (a + b)$ or $1 < (1 + 1)$ for the reaction.

Comparison of Part (c) with Part (a) shows that the presence of inert gases caused less products of CH_2 to be formed. This could also have been predicted because $e < (a + b)$.

Calculation of Equilibrium Composition for Simultaneous Reactions

The discussion thus far has been limited to the equilibrium composition for a single reaction. In the more general case involving combustion applications, several reactions occur simultaneously in the reacting mixture. In such cases, equilibrium is reached when the total Gibbs function for all the reacting species is minimized. A reaction equation must be used for every *independent* reaction that occurs, and the conservation of mass must be satisfied for all reacting species.

A different independent equation can be written for each independent reaction. If the procedures outlined for a single reaction are followed, it can be shown that a relationship in the form of Equation 15.40 must be met independently for each independent reaction in order for the Gibbs function to be minimized for the overall system. Therefore, for a process involving r reactions and s species (in the system), one can write r equations of the same form as Equation 15.40 as well as s conservation of mass equations. These equations are then solved simultaneously to determine the equilibrium composition of the system.

EXAMPLE 15.5

A system consists of a mixture of 1 mol H_2, 1 mol CO_2, and $\frac{1}{2}$ mol O_2. The components react in a constant temperature and pressure process at 2500 K and 1 atm. Determine the equilibrium composition if only H_2, CO_2, O_2, H_2O, and CO are present at equilibrium. Assume no O_2 formed due to dissociation of CO_2.

Given: System with 1 mol H_2, 1 mol CO_2, and $\frac{1}{2}$ mol O_2

                Reactions at 2500 K and 1 atm

                Only H_2, CO_2, O_2, H_2O, and CO are present at equilibrium.

Find: Equilibrium composition

Solution: For the reactants and products given, two reaction equations are possible:

$$H_2 + \tfrac{1}{2} O_2 \rightleftarrows H_2O \tag{1}$$

$$CO_2 + H_2 \rightleftarrows CO + H_2O \tag{2}$$

Also, a mass balance can be written for the process as

$$H_2 + CO_2 + \tfrac{1}{2} O_2 \rightarrow v\, H_2 + w\, CO_2 + x\, O_2 + y\, H_2O + z\, CO$$

Mass balances on atomic species give

$$\text{H balance:} \quad 1 = v + y$$

$$\text{C balance:} \quad 1 = w + z$$

$$\text{O balance:} \quad 3 = 2w + 2x + y + z$$

Also,

$$N = v + w + x + y + z$$

From Equation 15.40, with $\ln K_p = 5.121$,

$$K_p = 167.50 = \frac{y}{vx^{1/2}} \left(\frac{1}{N}\right)^{-1/2} \qquad \text{for reaction 1}$$

and $\ln K_p = 1.805$, or

$$K_p = 6.08 = \frac{yz}{vw} \left(\frac{1}{N}\right)^{0} \qquad \text{for reaction 2}$$

where values for K_p were taken from tabulated data. Therefore, this gives six equations with six unknowns, which can be solved to obtain the equilibrium composition. Solving gives

Species	Moles at equilibrium
H_2	0.028 (v)
CO_2	0.851 (w)
O_2	0.088 (x)
H_2O	0.972 (y)
CO	0.149 (z)

In general, the solution for systems such as the one in Example 15.5 involves solving a set of simultaneous nonlinear equations. Therefore, it may be necessary to resort to the use of generalized computer programs that are available for such problems. These programs can usually account for whatever reactions might occur at the temperature and pressure of the combustion process.

Dissociation and Ionization

In a combustion process, both the fuel and oxygen (or air) are usually supplied to the system in their stable forms. As exothermic reactions occur and the temperature of the system rises, some of the compounds formed tend to *dissociate*, this is, to break down into their elements or to other compounds, as discussed earlier in this chapter. When equilibrium is reached at a certain temperature, product compounds are being formed at a certain rate, while dissociation is occurring to form other compounds and/or elements, with no net chemical reaction occurring. In general, the dissociation of a compound or of an element from a stable state to a less stable state is an endothermic reaction that tends to lower the overall energy released from the reaction. Therefore, it is important to know when to expect such dissociation reactions. This section gives a brief perspective on this phenomenon for combustion systems (taken from Glassman 1977). The expectation that certain reactions will take place in the combustion process is based on consideration of the K_p values, as discussed earlier.

Most combustion processes of interest in engineering applications involve carbon, hydrogen, oxygen, and nitrogen, and thus only the C–H–O–N system is considered here. For such a reacting system with the combustion process occurring at 1 atm and at $T_p \leq 1250$ K, the products are the normal stable species CO_2, H_2O, O_2, and N_2. Most combustion systems reach temperatures higher than 1250 K, however. In such cases, dissociations of the stable species occur. The dissociation reactions are highly endothermic, thereby lowering the total energy released from the system (a few percent dissociation will lower the adiabatic flame temperature substantially).

Some of the dissociation reactions that can occur are as follows:

$$CO_2 \rightleftarrows CO + \tfrac{1}{2} O_2 \tag{15.42}$$

$$CO_2 + H_2 \rightleftarrows CO + H_2O \tag{15.43}$$

$$CO \rightleftarrows C + O \tag{15.44}$$

$$H_2O \rightleftarrows H_2 + \tfrac{1}{2} O_2 \tag{15.45}$$

$$H_2O \rightleftarrows H + OH \tag{15.46}$$

$$H_2O \rightleftarrows \tfrac{1}{2} H_2 + OH \tag{15.47}$$

$$H_2 \rightleftarrows 2 H \tag{15.48}$$

$$O_2 \rightleftarrows 2 O \tag{15.49}$$

$$N_2 \rightleftarrows 2 N \tag{15.50}$$

$$\tfrac{1}{2} N_2 + H_2O \rightleftarrows NO + H_2 \tag{15.51}$$

$$\tfrac{1}{2} N_2 + \tfrac{1}{2} O_2 \rightleftarrows NO \tag{15.52}$$

Thus, the products of a combustion system can include CO_2, H_2O, CO, O_2, C, O, H_2, H, OH, and N. In general, all of the dissociation reactions (except for the formation of ozone and acetylene) cause a reduction in the flame temperature and a reduction in the energy released by the process.

For the C–H–O–N system in which an excess of oxygen is supplied, the primary products are CO_2, H_2O, O_2, and N_2 for any reaction at 1 atm up to about 1800 K. At temperatures above 1800 K, the reaction given by Equation 15.51 occurs. The product NO is a pollutant. Although only minute quantities of it are formed up to temperatures of about 3000 K, it is considered a serious concern for T exceeding 1800 K. It is not a significant factor in the energy balance, however, until $T >$ about 3000 K.

At temperatures above 2200 K at 1 atm (or 2500 K at 20 atm), the dissociation of CO_2 and H_2O occur in the reactions given by Equations 15.42, 15.45, and 15.47. At temperatures above 2400 K at 1 atm (or 2800 K at 20 atm), the dissociations given by Equations 15.48 and 15.49 become important. Above 3000 K, NO also forms from the reaction of Equation 15.52. Above 3500 K at 1 atm (or 3600 K at 20 atm), N_2 dissociates by Equation 15.50. These reactions adversely affect the system thermodynamically.

Therefore, a judgment can be made concerning which equations are important in a reacting system if one knows the temperature range expected for the reaction. The generalized computer programs that are available usually take account of all the possible reactions unless the user specifies that certain ones are to be ignored. In any case, adequate information is generally available for the prediction of the equilibrium composition for several reacting mixtures.

At temperatures above 3000 K at 1 atm, ionization of NO, or the loss of an electron from NO, can occur in accordance with the reaction

$$NO \rightleftarrows NO^+ + e^- \tag{15.53}$$

In this case, NO loses an electron, e^-, and becomes a positively charged ion, NO^+. Such a process is called *ionization*, and the resulting gases are called *ionized gases* or *plasmas*.

Although NO ionizes at combustion flame temperatures, most other compounds and elements require considerably higher temperatures. Nitrogen can ionize in accordance with the reaction

$$N \rightleftarrows N^+ + e^- \tag{15.54}$$

and other elements can ionize in accordance with the same type of reaction. At higher temperatures, N^+ can lose another electron according to the reaction

$$N^+ \rightleftarrows N^{2+} + e^- \tag{15.55}$$

At higher and higher temperatures, the process continues until all of the electrons are lost from the atom. Ionization increases with increasing temperature and/or decreasing pressure, thus it can be appreciable at low temperatures for conditions such as those found in the upper atmosphere and outer space. Figure 15.4 gives the equilibrium composition of air at low density. Note the presence of the ions at high temperatures.

Although thermodynamic equilibrium does not occur for highly ionized gases in the presence of an electrical field (such as that found in many applications involving plasmas), a preliminary evaluation can be made based on the assumption of thermodynamic equilibrium for a moderate field. In this case, one follows the same procedures as previously outlined for a chemical reaction. The plasma is assumed to consist of an ideal gas mixture of neutral atoms, positive ions, and electron gas.

If a general ionization reaction is considered for any atomic species A as

$$A \rightleftarrows A^+ + e^-$$
(15.56)

then Equation 15.40 can be written for this reaction as

$$K_p = \frac{(N_{A^+})(N_{e^-})}{N_A}\left(\frac{p}{N}\right)$$
(15.57)

With values available for K_p, the equilibrium composition can be calculated in the same manner as previously given for chemical reactions.

For any of the reactions just described, whether involving dissociation and ionization or not, the state of thermodynamic equilibrium can be determined from considerations of the second law. The first law can then be used to calculate the energy released from the reacting system in attaining the equilibrium condition in accordance with earlier discussions.

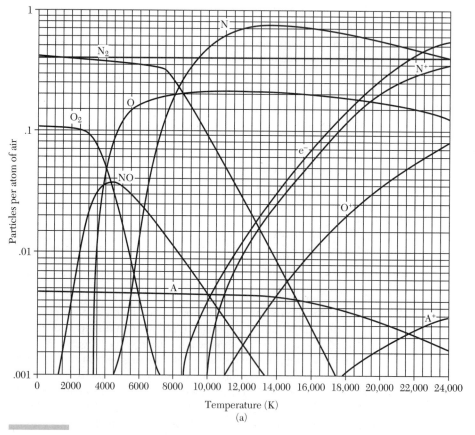

Figure 15.4 Equilibrium composition of air for (a) pressure = 1 atm and (b) pressure = 1×10^{-5} atm. (To obtain particles per initial molecule of air, multiply ordinate by 1.991.) Extracted from NACA TN 4265, pp. 6, 11.

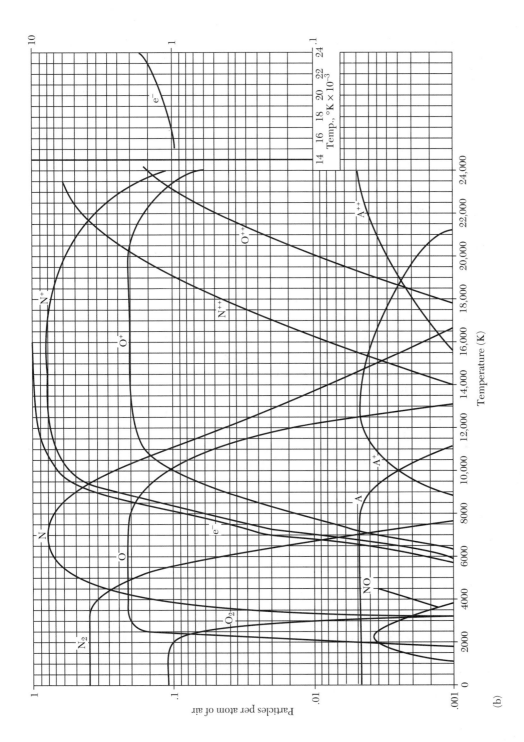

(b)

EXAMPLE 15.6

Nitrogen ionizes according to the reaction of Equation 15.54 at 10,000 K and 0.1 atm. The value of K_p for this reaction at this temperature is $K_p = 6.26 \times 10^{-4}$. Calculate the percentage of N that ionizes at this condition.

Given:

$$N \rightleftharpoons N^+ + e^-$$

$$T = 10,000 \text{ K and } p = 0.1 \text{ atm}$$

$$K_p = 6.26 \times 10^{-4}$$

Find: % N that ionizes

Solution: From Equation 15.57

$$K_p = \frac{N_{N^+} N_{e^-}}{N_N}\left(\frac{p}{N}\right)$$

Mass balance gives

$$N \rightarrow x \, N^+ + y \, e^- + z \, N$$

where x = number of moles of nitrogen ion (N^+), y = number of moles of electron gas (e^-), and z = number of moles of N remaining in the mixture at equilibrium. Since one free electron must be released for every ion formed, $x = y$. Also, for every nitrogen ion formed, one nitrogen atom disappears. Therefore, $z = (1 - x)$. Thus,

$$N = (N_{N^+} + N_{e^-} + N_N) = (1 + x)$$

Now the expression for K_p can be written as

$$K_p = \frac{x \cdot x}{1 - x}\left(\frac{p}{1 + x}\right)$$

Substituting values for K_p and p and solving gives

$$x = 0.079, \quad y = 0.079, \quad z = 0.921$$

Therefore, 7.9% of the nitrogen atoms ionize at this condition.

EXAMPLE 15.7

Gaseous carbon and oxygen are supplied in an equimolar ratio at 25°C and 1 atm to a combustion vessel in a steady-state steady-flow process. A heat exchanger is used to maintain the combustion process at 2300 K. Determine the heat transferred from the combustion gases to the heat exchanger per mole of carbon supplied if the combustion products exhaust from the system at 2300 K and 1 atm. No work is done during the process.

Given: 1 mol C (g) and 1 mol O_2 gas supplied (equimolar ratio) to combustion vessel

Reactants enter at 25°C and 1 atm
Products exhaust at 2300 K and 1 atm
Steady-state, steady-flow process with $W_x = 0$

Find: $\dot{Q}/\dot{N}_c$

Solution: The process that occurs is illustrated in Figure 15.5. First, determine what the products of combustion are. Then apply the first law for the steady-state, steady-flow (SSSF) process to determine $\dot{Q}$. With C (g) and O_2 entering at 25°C and 1 atm, and the products exiting at 2300 K and 1 atm, two possible reactions could occur:

$$C\ (g)\ +\ O_2 \rightarrow CO_2 \tag{1}$$

$$CO_2 \rightleftarrows CO\ +\ \tfrac{1}{2}\,O_2 \tag{2}$$

The first would be expected if the reaction went to completion. However, from the earlier discussion of dissociation, one can also expect reaction 2 to occur for $T > 2200$ K. Therefore, let's take both reactions into account. Values for K_p for the reactions at 2300 K are $K_p = 10^9$ for reaction 1 and $K_p = 0.011$ for reaction 2. From consideration of the value of K_p for reaction 1 (i.e., $K_p = 10^9 \gg 10^3$), expect that all C (g) will react with O_2 during the reaction, with no C (g) remaining in the products. Therefore, assume that the products consist only of CO_2, CO, and O_2.

Applying Equation 15.40 to the second reaction gives

$$0.011 = \frac{N_{CO}(N_{O_2})^{1/2}}{N_{CO_2}}\left(\frac{1}{N}\right)^{1/2} \tag{a}$$

A mass balance for the system can be written as

$$C\ (g)\ +\ O_2 \rightarrow x\ CO_2\ +\ y\ CO\ +\ z\ O_2 \tag{b}$$

where x = number of moles of CO_2, y = number of moles of CO, and z = number of moles of O_2 in the product exhaust gases. Also,

$$N = x + y + z \tag{c}$$

An element balance on reaction b gives

$$C\ \text{balance:}\quad 1 = x + y \tag{d}$$

$$O\ \text{balance:}\quad 1 = x + y/2 + z \tag{e}$$

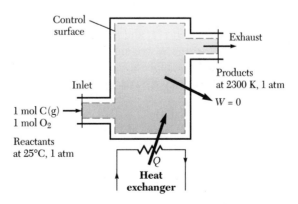

Figure 15.5 Steady-state, steady-flow combustion process.

Combining Equations a, c, d, and e and solving gives

$$x = 0.938, \quad y = 0.062, \quad z = 0.031$$

Thus, 6.2% of the CO_2 dissociates into CO at this temperature.

Now an energy balance for the SSSF system with $\dot{W} = 0$ gives, from Equation 14.28,

$$\dot{Q} = H_P - H_R$$
$$= \sum_P \dot{N}_i \bar{h}_i (T_P, p) - \sum_R \dot{N}_i \bar{h}_i (T_R, p)$$

From the gas tables in Appendix A

$$\bar{h}_{CO_2} (2300 \text{ K, 1 atm}) = -283{,}886 \text{ kJ/kmol}$$

$$\bar{h}_{CO} (2300 \text{ K, 1 atm}) = -42{,}844 \text{ kJ/kmol}$$

$$\bar{h}_{O_2} (2300 \text{ K, 1 atm}) = 70{,}658 \text{ kJ/kmol}$$

$$\bar{h}_{O_2} (25°C, 1 \text{ atm}) = 0 \text{ kJ/kmol}$$

$$\bar{h}_{C(g)} (25°C, 1 \text{ atm}) = 716{,}000 \text{ kJ/kmol} \qquad (\text{See Table A-23a})$$

Also, from the previous calculations for equilibrium composition of the products,

$$N_{CO_2} = 0.938, \quad N_{CO} = 0.062, \quad N_{O_2} = 0.031$$

Solving gives $\dot{Q}/\dot{N}_c = -983$ MJ/kmol C (g). Thus, 983 MJ is transferred *from* the system to the heat exchanger per kmol of C (g) supplied to the system.

If dissociation had been ignored and reaction 1 alone was assumed, application of Equation 14.28 would yield $\dot{Q}/\dot{N}_c = -1000$ MJ/kmol C (g). Thus, the dissociation of CO_2 to CO reduces the energy released by about 2%. If the carbon had entered in solid form, the heat released would have been significantly less. In this case, the effect of dissociation would have been about 6%.

EXERCISES

for Section 15.6

1. A steady flow of a stoichiometric mixture of carbon monoxide and oxygen is supplied at 1 atm and 25°C to a combustion chamber. The exit stream consists of only CO, CO_2, and O_2 and is at 1 atm and 2500 K. Determine the molar analysis of the exit stream. Also compute the heat loss from the combustion chamber per kmol of CO supplied.

   Answer: 12.1% CO, 81.85% CO_2, 6.05% O_2; $Q = -126$ MJ/kmol CO

2. What would the heat loss from the combustion chamber per kmol of CO have been in Exercise 1 if the exit stream did not contain any CO?

   Answer: $Q = -161$ MJ/kmol CO

15.7 *Maximizing Exergy Delivery from Chemical Reactions*

Now that both first and second law analyses for reacting systems have been considered, some additional facts should be observed when using fuels and chemical reactions for producing work. If heat is removed from the gaseous products of combustion for use in

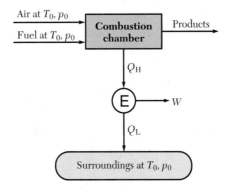

Figure 15.6 Heat engine driven by heat from the combustion chamber.

driving a heat engine, as shown in Figure 15.6, it is evident that conflicting conditions are involved in maximizing the output of the heat engine. On one hand, the maximum Q to be obtained from the combustion process is when the combustion process occurs with both reactants and products at T_0 and p_0 so that no energy is used for raising the temperature of the reactants or products above the dead state. On the other hand, the heat engine requires the supply of Q_H at a temperature much higher than T_0 if power is to be produced efficiently.

Results for calculations of the heat transfer from the combustion chamber as a function of the products' temperature are shown in Figure 15.7 for typical fuels such as methane, carbon, and hydrogen. In this case, as in Figure 15.6, the air and fuel enter the combustion vessel at T_0 and p_0. As can be seen from Figure 15.7, the heat from the combustion process is maximum at T_0 and zero at the adiabatic flame temperature T_{adia}, with an approximately linear relationship between the two temperatures (if the specific heat values of the product gases are assumed constant).

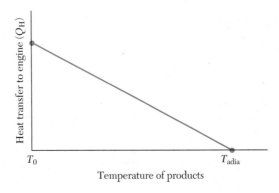

Figure 15.7 Heat rejected from combustion process as a function of temperature of products.

If such a combustion process were used as the heat source to operate a Carnot heat engine, the output of the heat engine would be zero at T_{adia} (because the input Q_H is zero) and zero also at T_0 (because $T_H = T_L = T_0$ for the heat engine). Between these extremes lies an optimum temperature at which heat from the combustion process can be delivered to the heat engine so as to produce the maximum work output from the Carnot heat engine. Based on the assumptions just outlined, it can be shown that this optimum operating temperature is $\sqrt{(T_0 T_{\text{adia}})}$. Proof of this statement is left as an exercise for the reader. Recall from Chapter 10 that a Carnot heat engine is not the optimum heat engine cycle to use when the heat supplied to the engine is from a finite heat source such as the products of combustion.

Another important factor in the use of fuels to produce work involves the irreversibility of the combustion process. The maximum work output from a heat engine can only be obtained when all the processes are reversible. Such is not the case with combustion engines because of the inherent irreversibility of the combustion process. A reversible chemical reaction, however, is conceivable, as can be demonstrated with the aid of the *van't Hoff equilibrium box*, as outlined here. Consider the reaction

$$C\ (s)\ +\ O_2 \rightarrow CO_2$$

which is occurring in the van't Hoff equilibrium box shown in Figure 15.8. The solid carbon, C (s), is contained in a reaction vessel that is equipped with two openings covered by semipermeable membranes. A semipermeable membrane is one that allows passage of only a specified chemical species and not any other. The membrane to the left allows only O_2 to pass through, while the membrane to the right is permeable only to CO_2.

A piston moving in a cylinder is used to feed O_2 slowly into the system from the left, while another piston moves in the cylinder to the right of the box to remove CO_2 from the system. The C (s) reacts inside the system with O_2 to form CO_2. Thus, only O_2 and CO_2 exist as gases within the vessel. The ratio p_{CO_2}/p_{O_2} is maintained at a value infinitesimally below the equilibrium ratio so that the C (s) burns slowly under reversible conditions. Since the reaction takes place under reversible conditions, the maximum work (which is measured by the decrease in exergy, χ, for the system) can be obtained for the reaction.

Many difficulties exist in practice, however, in the way to control such a reaction to achieve complete reversibility. First, it is nearly impossible to find membranes to meet

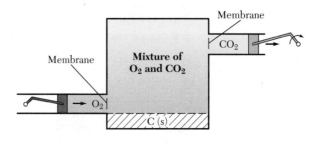

Figure 15.8 The van't Hoff equilibrium box.

the requirements. Second, the pressure differences that must be maintained cannot realistically be achieved. For example, the equilibrium ratio p_{CO_2}/p_{O_2} has a value of approximately 10^{15} at 900°C (Denbigh 1966). Therefore, if the pressure in the CO_2 cylinder were maintained at 1 atm, the pressure in the O_2 cylinder must be maintained at 10^{-15} atm to operate the reaction under reversible conditions.

Combustion processes occur irreversibly since the reaction is uncontrolled, and as indicated earlier, imponderable obstacles exist to the achievement of reversible chemical reactions in practice. Thus, in typical burners, the fuel is normally allowed to burn freely (and irreversibly), usually in a constant pressure process. The heat transfer from the reacting system is then measured by the change in enthalpy for the system. For a reversible reaction at T_0 and p_0, however, the maximum work output that can theoretically be released is equal to the decrease in exergy, which is also the decrease in the Gibbs function ΔG for the system as given by Equation 14.40. For many combustion processes, the heat from the system ($= \Delta H$) is nearly equal to ΔG for the system because the entropy change occurring as a result of the process is relatively small. Even so, from a second law perspective, the heat from the system cannot be converted completely to work via a heat engine. Therefore, the preferable reaction from a thermodynamic viewpoint is a reversible chemical reaction in which chemical exergy is converted directly to work rather than indirectly using a heat engine.

A battery or fuel cell is a system in which chemical energy is converted directly to electrical energy (and thereby to work) with minimal losses. In such systems, the electrodes act in a manner similar to the membranes in the van't Hoff equilibrium box, and the reactants and products are brought to equilibrium by maintaining a potential difference of only a few volts between the electrodes. Also, in contrast to the membranes, the processes at the electrodes take place with considerable speed. Therefore, the reaction can be maintained fairly close to equilibrium with operating conversion efficiencies greater than 60% in practice. Thus, great potential exists for electrical power generation via the use of fuel cells if economic and technical problems associated with the development of such devices can be solved.

Although combustion processes as a rule are highly irreversible, Richter and Knoche (1983) have shown that the irreversible entropy production of combustion can be decreased if immediate contact of fuel with oxygen is prevented and intermediate chemical reactions are supported. Metal oxides can be used as reactants for the intermediate reactions. These intermediate reactions reduce the irreversibility and thereby increase the potential for work output from such processes. While systems of this type have not yet been developed, the results obtained in preliminary research on such systems are encouraging.

Temperature is an important consideration in combustion processes. As discussed earlier, chemical reactions are least irreversible when they occur with the system at or near thermodynamic equilibrium. Unfortunately, combustion processes should typically take place at a high temperature to achieve a reasonable speed of reaction. This often poses the serious danger of exceeding the temperature limits for the materials of construction. The intermediate reactions discussed above might be used to achieve near-equilibrium reactions at lower temperatures (thus reducing irreversibilities) while minimizing the risk of exceeding the metallurgical limit for the materials used in the construction of the device. Typical applications that could benefit from this approach include jet engines and power plants, which should be operated at moderate temperature levels so as not to exceed safety limits for the materials used in their construction.

A further rationale exists for trying to keep reaction temperature levels moderate. For typical combustion reactions, the effects of pressure and temperature on the heat of reaction, ΔH_R, are small, typically less than 5% over a temperature range of 3000 K to 4000 K. Thus, about as much heat is released from a constant pressure, high temperature reaction that goes to completion as would be released at T_0 and p_0. However, higher temperatures tend to increase the degree of dissociation and thereby cause the equilibrium composition of a reacting mixture to change in the endothermic direction (see the discussion of Le Chatelier's principle) (Denbigh 1966). Thus, higher temperature reactions tend not to go to completion and therefore release less heat than would be released if the reaction occurred at a lower temperature. Thus, moderate temperature levels that are higher then T_0 but not extremely high are preferred for most combustion systems.

15.8 Summary

Chemical equilibrium in relation to chemically reacting systems has been considered in detail in this chapter. Equilibrium considerations in general require the application of the second law of thermodynamics. From application of the second law of thermodynamics, it is possible to determine whether or not a specific reaction can occur and if the reaction will proceed to completion. If the reaction will not proceed to completion, one can use the principles established to determine the degree of completion and the equilibrium composition.

An equilibrium constant K_p was developed for the case of chemically reacting systems that comprise only ideal gases. This constant is extremely valuable in the determination of the equilibrium composition of a reacting mixture of ideal gases. The constant K_p is a function of temperature only, and it can be used to determine the equilibrium composition of ideal gas reactions occurring at a specified temperature.

Dissociation and ionization were discussed, and it was pointed out that such phenomena generally are endothermic and tend to reduce the net quantity of energy released as heat from a reaction. These phenomena typically occur at high temperatures, resulting in the reactions not going to completion and a consequent reduction in the heat released compared to the same reactions at a lower temperature. Some general discussion was presented of the high temperature reaction phenomena to be expected at different temperature levels for different reactions.

This chapter also covered the maximization of exergy delivery from chemical reactions used for producing work. While the maximum heat transfer rate from a reacting system is obtained at T_0 and p_0, the maximum heat engine thermal efficiency corresponds to heat supply to the engine cycle at the adiabatic flame temperature. At the adiabatic flame temperature, however, the heat extracted from the reacting system is zero. Thus, to maximize the work output of a heat engine with Q supplied from chemical reaction(s), reversible engine cycles that match the heat source characteristics (as discussed in Chap. 10) must be considered. The design of alternative devices for producing work efficiently is one of the many challenges involved in the use of thermodynamics as a tool in efforts to maximize the benefit derived from the available resources for the good of society.

References

Denbigh, K. 1966. *The Principles of Chemical Equilibrium*, 2nd ed. Cambridge, U.K.: Cambridge University Press.

Moeckel, W. E. and Weston, K. C. 1958. *Composition and Thermodynamic Properties of Air in Chemical Equilibrium*. NACA Technical Note 4265. Cleveland, OH: Lewis Flight Propulsion Laboratory.

Glassman, I. 1977. *Combustion*. New York: Academic Press.

Richter, H. J. and Knoche, K. F. 1983. "Reversibility of Combustion Processes," in Gaggiolli, R. A. ed. *Efficiency and Costing*. ACS Symposium Series 235. Washington, D.C.: American Chemical Society. 71–85.

Thermal Research Laboratory 1965. *JANAF Thermochemical Tables*. Midland, MI: Dow Chemical Company.

Questions

1. Explain the conditions necessary for a system to remain in a state of chemical equilibrium.

2. Discuss the principles involved in developing equilibrium criteria for a chemically reacting system.

3. Define the chemical potential and explain how it is related to the equilibrium state for a system.

4. What is the equilibrium constant K_p and how is it derived? It is a function of what thermodynamic property? How would you calculate values for K_p for a particular reaction?

5. Explain Le Chatelier's principle.

6. What are *dissociation* and *ionization*? In general what effect do these phenomena have on the heat released from an exothermic chemically reacting system?

7. Explain the assumptions inherent in making thermodynamic calculations for a plasma.

8. Explain some of the difficulties involved in maximizing the exergy delivery from typical chemical reactions.

9. What is the van't Hoff equilibrium box and what is its conceptual value? Compare this equilibrium box to a fuel cell. What are the advantages and limitations of fuel cells?

10. The maximization of work output from a heat engine that receives heat from a chemical reaction is an optimization process. Explain why.

11. Explain how you would develop a computer program to compute the equilibrium composition(s) for combustion processes involving several different fuels.

Problems

15.1. A constant temperature reservoir at 400 K exchanges heat with the surroundings at 300 K, as shown in Figure P15.1. If 1000 kJ of heat is transferred from the reservoir to the surroundings, what is the entropy change
  (a) for the reservoir
  (b) for the surroundings
  (c) for the universe

15.2. An inventor claims that he has developed a device that takes heat from ambient air at $T_0 = 25°C$ and produces pure H_2 and O_2 from a steady flow of water. The water enters the device

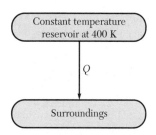

Figure P15.1 Exchange of heat from reservoir to surroundings.

at 1 atm and 25°C while the separate streams of gaseous H_2 and O_2 are each at 1 atm and 25°C.
   (a) Calculate the heat transfer to the device per kmol of water processed.
   (b) What is the entropy change per kmol of water processed, (i) for the surroundings, and (ii) for the flow through the device?
   (c) From your results in Part (b), determine the entropy change of the universe for the process.
   (d) Is the process possible as specified?

15.3. Determine the chemical exergy per kmol for H_2O (l), H_2 (g), and O_2 (g) using the formulas in Table 14.1 (Chap. 14). Use your results to determine the minimum exergy input needed to produce 1 kmol gaseous H_2 at 1 atm and 25°C, and 0.5 kmol gaseous O_2 at 1 atm and 25°C from 1 kmol of water at 1 atm and 25°C in a steady-flow process.

∗ 15.4. In a steady-flow process, 2 kmol H_2O are produced per second from a reaction of 2 kmol H_2 with 1 kmol O_2 as shown in Figure P15.4. The reaction occurs in a control volume at 1000 K. Assume that the H_2 and O_2 enter the control volume separately at 1 atm and 1000 K and that the exiting H_2O is also at 1 atm and 1000 K. What is the entropy generation rate in the control volume?

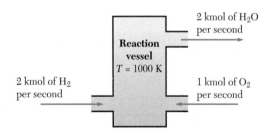

Figure P15.4 Steady-flow process.

∗ 15.5. If the heat transfer from the reaction of Problem 15.4 occurs to a heat engine at 1000 K, what is the maximum power that can be produced assuming that the ambient environment at 300 K is the heat sink for the engine. Compare this power with the lost power due to irreversibilities of the reaction in Problem 15.4.

∗ 15.6. In a steady-flow process, 2 lbmol H_2O and 1 lbmol CO_2 are produced per second from a reaction of 1 lbmol CH_4 with 2 lbmol O_2. The reaction occurs in a control volume at 1500°R. Assume that the CH_4 and the O_2 each enter the control volume at 1 atm and 1500°R and that the products leave at 1 atm and 1500°R. What is the entropy generation rate in the control volume?

∗ 15.7. In the reaction of Problem 15.6 assume that the reactants enter with each gas at 1 atm and 77°F. Determine
   (a) the heat transfer rate from the control volume
   (b) the entropy generation rate in the control volume

∗ 15.8. Calculate the equilibrium constant at 2400 K for the reaction

$$H_2 + \tfrac{1}{2} O_2 \rightleftarrows H_2O$$

and compare your result with the tabulated value in the appendix.

∗ 15.9. Calculate the equilibrium constant at 3000 K for the reaction

$$CO_2 \rightleftarrows CO + \tfrac{1}{2} O_2$$

and compare your result with the tabulated value in the appendix.

∗ 15.10. Calculate the equilibrium constant at 4500°R for the reaction

$$CO_2 + H_2 \rightleftarrows CO + H_2O$$

and compare your result with the tabulated value in the appendix.

15.11. In the reactions given in Problems 15.8 through 15.10, determine the effect of pressure on each reaction. Should a high or low pressure be used in each case to increase the amount of products at equilibrium?

∗ 15.12. Calculate the equilibrium composition at 1 atm and 2400 K for the reaction given in Problem 15.8. Assume that the reactants enter the reaction vessel in a stoichiometric ratio.

∗ 15.13. Calculate the equilibrium composition at 1 atm and 3000 K for the reaction given in Problem 15.9. Assume that the reactants enter the reaction vessel in a stoichiometric ratio.

* **15.14.** Calculate the equilibrium composition at 1 atm and 4500°R for the reaction given in Problem 15.10. Assume that the reactants enter the reaction vessel in a stoichiometric ratio.

* **15.15.** A reaction of 1 mol gaseous C with 1 mol H_2 at 3000 K forms CH_2, as shown in Figure P15.15. Determine the equilibrium composition if the reaction occurs at
  (a) 1 atm
  (b) 20 atm
  (c) 1 atm, but with 10 mol inert gas in the reaction vessel
  (d) Compare your results with those from Example 15.4 and explain the physical significance of your results.

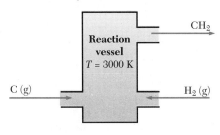

Figure P15.15 Constant temperature reaction of gaseous carbon and hydrogen.

* **15.16.** A reaction of 1 mol gaseous H_2 with 0.5 mol O_2 at 4800 K forms H_2O. Determine the equilibrium composition if the reaction occurs at
  (a) 1 atm
  (b) 40 atm
  (c) 1 atm, but with 6 mol inert gas in the reaction vessel
  (d) Explain the physical significance of your results.

** **15.17.** A system consists of a mixture of 1 mol gaseous C, 1 mol gaseous CO_2, and 3 mol H_2. The components react in a constant temperature and constant pressure process at 3200 K and 1 atm. Determine the equilibrium composition if only C, CO_2, H_2, CO, H_2O, O_2, and CH_2 are present at equilibrium. For the reactants and products specified, assume that three possible reaction equations exist.

** **15.18.** Solve Problem 15.17 if no O_2 is present at equilibrium. Compare your result with Problem

15.17. For the reactions possible at 3200 K with the reactants given, would you expect O_2 to be present at equilibrium? Explain.

* **15.19.** Nitrogen ionizes according to the reaction of Equation 15.54 at 6000 K and 0.01 atm. Calculate the percentage of N that ionizes at this condition.

* **15.20.** Nitrogen ionizes according to the reaction of Equation 15.54 at 9000°R and 0.001 atm. Calculate the percentage of N that ionizes at this condition.

* **15.21.** Nitrogen dissociates according to the reaction $N_2 \rightleftarrows 2\,N$ at 4000 K and 1 atm. Assuming no ionization, calculate the percentage of N_2 that dissociates at this condition.

* **15.22.** Assuming no ionization, calculate the percentage of O_2 that dissociates to O at 4000 K and 1 atm.

* **15.23.** Assuming no ionization, calculate the percentage of H_2 that dissociates to H at 4000 K and 1 atm.

* **15.24.** Gaseous carbon and oxygen are supplied in an equimolar ratio at 25°C and 1 atm to a combustion vessel in a steady-state, steady-flow process, as shown in Figure P15.24. A heat exchanger is used to maintain the process at 1800 K. Determine the heat transferred from the reaction gases to the heat exchanger per mole of carbon supplied if the products exhaust from the system at 1800 K and 1 atm. No work is done during the process.

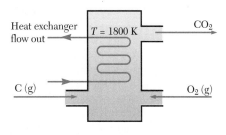

Figure P15.24 Reaction of gaseous carbon and oxygen at constant temperature in steady-state, steady-flow process.

15.25. Gaseous hydrogen and oxygen are supplied at 25°C and 1 atm in a ratio of 2 mol H_2 to 1 mol O_2 to a combustion vessel in a steady-state, steady-flow process. A heat exchanger is used to maintain the process at 1800 K. Determine the heat transferred from the reaction gases to the heat exchanger per mole of hydrogen supplied if the products exhaust from the system at 1800 K and 1 atm. No work is done during the process.

** **15.26.** Assume that (1) the heat rejected from a typical combustion process is a linear function of temperature as shown in Figure 15.7 and that (2) this heat is supplied to a Carnot heat engine. Show that the maximum work output from the Carnot heat engine is obtained when the heat is supplied to the Carnot engine at the optimum temperature of $(T_0 T_{adia})^{1/2}$.

Design Problems

15.1 *Development of computer program to calculate equilibrium compositions for reaction of methane with air*

Develop a computer program to calculate the equilibrium composition of the products of combustion for a reaction of methane with air at a pressure of 1 atm. Consider constant temperature reactions with all possible reactions that you can identify for temperatures from 500 K to 3000 K. Consider air–fuel ratios from 0.7 to 3 times theoretical air. Plot your results on appropriate charts. Present your approach and your results in a technical report.

15.2 *Development of computer program to calculate equilibrium compositions for reaction of octane with air*

Repeat Problem 15.1 for the reaction of octane (C_8H_{18}) with air at 1 atm.

15.3 *Hydrogen economy*

At the present time there is great concern about the limited reserves of petroleum fuels available, especially in the United States, and about the pollution problems caused by the use of such fuels. One idea that has been proposed is to develop a hydrogen economy. In support of this argument, it is often stated that hydrogen is plentiful and is clean burning. Hydrogen is also flexible and can be used both in a fuel cell to produce electricity and in a chemical reaction with oxygen to produce heat.

Your assignment is to investigate the potential for a hydrogen economy. Specifically, what significant applications could use hydrogen as a substitute for petroleum fuel, and how could you obtain enough hydrogen to meet the demand? Also, how would you transport and store the hydrogen? Investigate the sources of hydrogen fuel and the processes and/or reactions required.

Examples of the production of hydrogen include the use of the electrolysis process and the steam-reforming process. The electrolysis process is used on nuclear submarines to produce both hydrogen and oxygen from the excess capacity of the nuclear generating unit. From this process the oxygen is available for life-support systems, while the hydrogen is available for other uses. Such a concept has been proposed for large-scale application by proponents of nuclear power.

The steam-reforming process is the process used for most of the hydrogen produced today. This process uses methane and water in the reactions

$$CH_4 + H_2O \rightarrow 3\ H_2 + CO$$

and

$$CO + H_2O \rightarrow H_2 + CO_2$$

Determine appropriate temperature levels for the successful use of these reactions to produce hydrogen.

Investigate the first and second law efficiencies of the electrolysis process, the steam-reforming process, and any other process you think might be appropriate. Also, discuss the supply of methane necessary to sustain a hydrogen economy. Specify applications that are appropriate for the use of hydrogen. Assess the potential for a hydrogen economy primarily from a thermodynamics perspective. Present your results in a technical report.

15.4. *Air pollution*

Air pollution from steam power plants and internal combustion engines causes serious problems. One of these problems is a result of the nitrous oxides formed in the combustion process with air. One alternative that might be considered to reduce the NO_x problem is to remove the nitrogen gas from the air prior to combustion. This would appear to reduce air pollution problems while at the same time increasing the heat output from the combustion process by removing the inert nitrogen gas from the process as well as eliminating the endothermic NO_x reactions from the process. From a thermodynamic perspective, determine

1. the exergy required to separate the nitrogen from the air prior to combustion

2. the exergy gain to be derived in the combustion process as a result

State clearly your assumptions. Does this approach appear to be a practical solution to this type of air pollution problem? Compare this approach to other alternatives currently being used. Write a report of your results.

APPENDIX A

Property Tables and Constants in SI Units

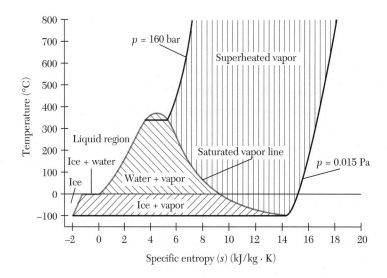

Figure A.1 Regions of applicability for values given in steam tables (SI units).

TABLE A.1 *Saturated Water and Steam Properties (Temperature Table)*

T	p_s	$v_f \times 10^3$	v_g	u_f	u_g	h_f	h_{fg}	h_g	s_f	s_{fg}	s_g
(°C)	(bar)	(m³/kg)	(m³/kg)	(kJ/kg)		(kJ/kg)			(kJ/kg · K)		
0.01	0.00612	1	206	0	2375	0	2,501	2501	0	9.154	9.154
5	0.00871	1	147.3	21	2381	21	2,489	2510	0.076	8.947	9.023
10	0.01226	1	106.5	42	2388	42	2,477	2519	0.151	8.748	8.899
15	0.0170	1.001	78.05	63	2395	63	2,465	2528	0.224	8.555	8.779
20	0.0233	1.002	57.86	84	2402	84	2,453	2537	0.296	8.369	8.665
25	0.0317	1.003	43.39	105	2409	105	2,441	2546	0.367	8.189	8.556
30	0.0424	1.004	32.91	126	2416	126	2,429	2555	0.437	8.015	8.452
35	0.0562	1.006	25.22	147	2423	147	2,417	2564	0.505	7.846	8.351
40	0.0738	1.008	19.53	168	2429	168	2,405	2573	0.572	7.683	8.255
45	0.0959	1.010	15.26	188	2436	188	2,394	2582	0.638	7.524	8.162
50	0.1235	1.012	12.03	209	2443	209	2,382	2591	0.703	7.371	8.074
55	0.1576	1.015	9.565	230	2449	230	2,370	2600	0.768	7.222	7.990
60	0.1994	1.017	7.669	251	2456	251	2,358	2609	0.831	7.077	7.908
65	0.2503	1.020	6.196	272	2462	272	2,346	2618	0.893	6.936	7.829
70	0.3118	1.023	5.042	293	2469	293	2,333	2626	0.955	6.799	7.754
75	0.3855	1.026	4.133	314	2475	314	2,321	2635	1.015	6.666	7.681
80	0.4735	1.029	3.409	335	2482	335	2,308	2643	1.075	6.536	7.611
85	0.5777	1.032	2.830	356	2488	356	2,296	2652	1.134	6.409	7.543
90	0.7004	1.036	2.363	377	2494	377	2,283	2660	1.193	6.286	7.479
95	0.8442	1.040	1.985	398	2500	398	2,270	2668	1.250	6.165	7.415
100	1.013	1.043	1.673	419	2506	419	2,257	2676	1.307	6.047	7.354
105	1.208	1.047	1.419	440	2512	440	2,244	2684	1.363	5.933	7.296
110	1.433	1.052	1.210	461	2518	461	2,230	2691	1.419	5.820	7.239
115	1.691	1.056	1.036	482	2524	483	2,216	2699	1.474	5.710	7.184
120	1.990	1.060	0.8913	504	2529	504	2,202	2706	1.528	5.602	7.130
125	2.320	1.065	0.7703	525	2535	525	2,188	2713	1.581	5.496	7.077

TABLE A.1(cont.) **Saturated Water and Steam Properties (Temperature Table)**

T	p_s	$v_f \times 10^3$	v_g	u_f	u_g	h_f	h_{fg}	h_g	s_f	s_{fg}	s_g
(°C)	(bar)	(m³/kg)	(m³/kg)	(kJ/kg)		(kJ/kg)			(kJ/kg · K)		
130	2.700	1.070	0.6680	546	2540	546	2,174	2720	1.635	5.393	7.028
135	3.130	1.075	0.5816	567	2545	568	2,159	2727	1.687	5.291	6.978
140	3.620	1.080	0.5084	589	2550	589	2,145	2734	1.739	5.191	6.930
145	4.160	1.085	0.4458	610	2555	611	2,129	2740	1.791	5.093	6.884
150	4.760	1.091	0.3923	632	2560	632	2,114	2746	1.842	4.996	6.838
155	5.440	1.096	0.3463	653	2564	654	2,098	2752	1.893	4.901	6.794
160	6.180	1.102	0.3067	675	2568	676	2,082	2758	1.943	4.807	6.750
165	7.010	1.108	0.2723	697	2572	697	2,066	2763	1.993	4.715	6.708
170	7.920	1.114	0.2425	718	2576	719	2,049	2768	2.042	4.624	6.666
175	8.930	1.121	0.2165	740	2580	741	2,032	2773	2.091	4.534	6.625
180	10.03	1.127	0.1938	762	2583	763	2,015	2778	2.140	4.446	6.586
185	11.23	1.134	0.1740	784	2587	785	1,997	2782	2.188	4.358	6.546
190	12.55	1.141	0.1564	806	2590	808	1,978	2786	2.236	4.271	6.507
195	13.98	1.149	0.1410	828	2592	830	1,959	2789	2.284	4.185	6.469
200	15.53	1.157	0.1274	851	2595	852	1,941	2793	2.331	4.100	6.431
205	17.22	1.164	0.1152	873	2597	875	1,920	2795	2.378	4.016	6.394
210	19.05	1.173	0.1044	895	2599	898	1,900	2798	2.425	3.933	6.358
215	21.03	1.181	0.0948	918	2600	921	1,879	2800	2.472	3.849	6.321
220	23.17	1.190	0.0862	941	2602	944	1,857	2801	2.518	3.767	6.285
225	25.47	1.199	0.0785	964	2603	967	1,836	2803	2.564	3.685	6.249
230	27.94	1.209	0.0716	987	2603	990	1,813	2803	2.611	·3.604	6.215
235	30.60	1.219	0.0653	1010	2603	1014	1,789	2803	2.657	3.522	6.179
240	33.44	1.229	0.0597	1033	2603	1037	1,766	2803	2.703	3.441	6.144
245	36.49	1.240	0.0547	1057	2603	1061	1,741	2802	2.748	3.360	6.108
250	39.74	1.251	0.0501	1080	2602	1085	1,716	2801	2.794	3.279	6.073
255	43.21	1.263	0.0459	1104	2600	1110	1,689	2799	2.838	3.198	6.036

TABLE A.1(cont.)   **Saturated Water and Steam Properties (Temperature Table)**

T	p_s	$v_f \times 10^3$	v_g	u_f	u_g	h_f	h_{fg}	h_g	s_f	s_{fg}	s_g
(°C)	(bar)	(m³/kg)	(m³/kg)	(kJ/kg)		(kJ/kg)			(kJ/kg · K)		
260	46.90	1.276	0.0422	1128	2598	1134	1,662	2796	2.883	3.117	6.000
265	50.83	1.289	0.0388	1153	2596	1159	1,634	2793	2.929	3.036	5.965
270	55.01	1.302	0.0356	1177	2593	1185	1,604	2789	2.975	2.954	5.929
275	59.45	1.317	0.0328	1202	2590	1210	1,575	2785	3.021	2.872	5.893
280	64.15	1.332	0.0302	1227	2586	1236	1,543	2779	3.067	2.790	5.857
285	69.13	1.348	0.0278	1253	2581	1262	1,511	2773	3.113	2.706	5.819
290	74.40	1.366	0.0256	1279	2575	1289	1,477	2766	3.160	2.622	5.782
295	79.97	1.384	0.0236	1305	2569	1316	1,442	2758	3.207	2.537	5.744
300	85.84	1.404	0.0217	1332	2563	1344	1,405	2749	3.253	2.451	5.704
305	92.04	1.425	0.0199	1359	2555	1372	1,366	2738	3.301	2.363	5.664
310	98.57	1.447	0.0183	1387	2546	1401	1,326	2727	3.349	2.274	5.623
315	105.4	1.473	0.01686	1415	2537	1431	1,283	2714	3.398	2.182	5.580
320	112.7	1.500	0.01548	1444	2525	1461	1,239	2700	3.448	2.088	5.536
325	120.3	1.528	0.01419	1474	2513	1493	1,190	2683	3.498	1.991	5.489
330	128.4	1.560	0.01299	1505	2498	1525	1,140	2665	3.550	1.891	5.441
335	136.9	1.595	0.01185	1537	2482	1558	1,087	2645	3.603	1.786	5.389
340	145.8	1.636	0.01079	1570	2464	1594	1,027	2621	3.659	1.676	5.335
345	155.2	1.685	0.00978	1605	2443	1631	964	2595	3.717	1.559	5.276
350	165.1	1.742	0.00881	1642	2418	1671	893	2564	3.777	1.433	5.210
355	175.5	1.808	0.00788	1681	2388	1713	814	2527	3.843	1.295	5.138
360	186.5	1.893	0.00696	1725	2352	1761	721	2482	3.916	1.139	5.055
365	198.1	2.011	0.00602	1777	2306	1817	608	2425	3.998	0.952	4.950
370	210.4	2.204	0.00499	1843	2235	1890	450	2340	4.109	0.700	4.809
374	220.5	3.106	0.00311	2018	2017	2086	0	2086	4.407	0.000	4.407

TABLE A.2 **Saturated Water and Steam Properties (Pressure Table)**

p	T_s	v_f	v_g	u_f	u_g	h_f	h_{fg}	h_g	s_f	s_{fg}	s_g
(bar)	°C	(m³/kg)		(kJ/kg)		(kJ/kg)			(kJ/kg · K)		
0.00612	0	0.001000	205.9	0	2375	0	2,501	2501	0	9.154	9.154
0.01	7	0.001000	129.2	29	2384	29	2,484	2513	0.106	8.869	8.975
0.02	17.5	0.001001	66.99	73	2399	73	2,460	2533	0.260	8.462	8.722
0.03	24.1	0.001003	45.65	101	2408	101	2,444	2545	0.354	8.222	8.576
0.04	29.0	0.001004	34.79	121	2414	121	2,432	2553	0.422	8.050	8.472
0.05	32.9	0.001005	28.19	138	2420	138	2,423	2561	0.476	7.917	8.393
0.06	36.2	0.001006	23.73	151	2424	151	2,415	2566	0.521	7.808	8.329
0.07	39.0	0.001008	20.52	163	2428	163	2,409	2572	0.559	7.715	8.274
0.08	41.5	0.001008	18.10	174	2431	174	2,402	2576	0.592	7.635	8.227
0.09	43.8	0.001009	16.20	183	2434	183	2,397	2580	0.622	7.563	8.185
0.1	45.8	0.001010	14.67	192	2437	192	2,392	2584	0.649	7.499	8.148
0.2	60.1	0.001017	7.647	251	2456	251	2,358	2609	0.832	7.075	7.907
0.3	69.1	0.001022	5.228	289	2468	289	2,336	2625	0.944	6.823	7.767
0.4	75.9	0.001026	3.993	318	2476	318	2,318	2636	1.026	6.643	7.669
0.5	81.3	0.001030	3.240	341	2483	341	2,304	2645	1.091	6.502	7.593
0.6	85.9	0.001033	2.731	360	2489	360	2,293	2653	1.145	6.386	7.531
0.7	89.9	0.001036	2.364	377	2494	377	2,283	2660	1.192	6.287	7.479
0.8	93.5	0.001039	2.087	392	2498	392	2,273	2665	1.233	6.201	7.434
0.9	96.7	0.001041	1.869	405	2502	405	2,266	2671	1.270	6.125	7.395
1	99.6	0.001043	1.694	417	2506	418	2,257	2675	1.303	6.056	7.359
2	120.2	0.001061	0.8855	505	2529	505	2,201	2706	1.530	5.597	7.127
3	133.6	0.001073	0.6056	561	2544	562	2,163	2725	1.672	5.320	6.992
4	143.6	0.001084	0.4623	604	2554	605	2,134	2739	1.777	5.119	6.896
5	151.9	0.001093	0.3747	640	2561	640	2,109	2749	1.861	4.961	6.822
6	158.9	0.001101	0.3155	670	2567	671	2,086	2757	1.931	4.829	6.760
7	165.0	0.001108	0.2727	696	2572	697	2,066	2763	1.992	4.716	6.708

TABLE A.2(cont.)   *Saturated Water and Steam Properties (Pressure Table)*

p	T_s	v_f	v_g	u_f	u_g	h_f	h_{fg}	h_g	s_f	s_{fg}	s_g
(bar)	°C	(m³/kg)		(kJ/kg)		(kJ/kg)			(kJ/kg · K)		
8	170.4	0.001115	0.2403	720	2577	721	2,048	2769	2.046	4.616	6.662
9	175.4	0.001121	0.2149	742	2580	743	2,031	2774	2.095	4.528	6.623
10	179.9	0.001127	0.1943	762	2583	763	2,015	2778	2.139	4.447	6.586
15	198.3	0.001154	0.1317	843	2594	845	1,946	2791	2.315	4.129	6.444
20	212.4	0.001177	0.09958	906	2600	909	1,890	2799	2.447	3.892	6.339
25	224.0	0.001197	0.07994	959	2602	962	1,840	2802	2.555	3.702	6.257
30	233.9	0.001216	0.06666	1005	2603	1008	1,795	2803	2.646	3.540	6.186
35	242.6	0.001235	0.05705	1045	2603	1050	1,753	2803	2.726	3.399	6.125
40	250.4	0.001252	0.04977	1082	2602	1087	1,714	2801	2.795	3.273	6.068
45	257.5	0.001269	0.04405	1116	2599	1122	1,676	2798	2.860	3.158	6.018
50	264.0	0.001286	0.03944	1148	2597	1154	1,640	2794	2.920	3.053	5.973
55	270.0	0.001302	0.03564	1177	2593	1184	1,605	2789	2.975	2.954	5.929
60	275.6	0.001319	0.03245	1205	2589	1213	1,571	2784	3.026	2.862	5.888
65	280.9	0.001335	0.02974	1232	2585	1241	1,537	2778	3.075	2.775	5.850
70	285.8	0.001351	0.02740	1257	2580	1267	1,505	2772	3.121	2.692	5.813
75	290.6	0.001367	0.02537	1282	2575	1292	1,473	2765	3.165	2.613	5.778
80	295.0	0.001384	0.02353	1305	2569	1316	1,442	2758	3.207	2.537	5.744
85	299.3	0.001401	0.02193	1328	2564	1340	1,410	2750	3.248	2.463	5.711
90	303.4	0.001418	0.02049	1350	2557	1363	1,379	2742	3.285	2.392	5.677
95	307.3	0.001435	0.01920	1372	2551	1385	1,348	2733	3.323	2.323	5.646
100	311.0	0.001451	0.01803	1393	2544	1407	1,318	2725	3.359	2.255	5.614
105	314.6	0.001471	0.01697	1413	2537	1429	1,286	2715	3.394	2.189	5.583
110	318.1	0.001489	0.01599	1434	2530	1450	1,255	2705	3.429	2.123	5.552
115	321.5	0.001508	0.01508	1453	2522	1471	1,224	2695	3.462	2.059	5.521
120	324.7	0.001527	0.01425	1473	2513	1491	1,193	2684	3.496	1.996	5.492
125	327.9	0.001546	0.01348	1492	2505	1511	1,162	2673	3.528	1.933	5.461

TABLE A.2(cont.)   **Saturated Water and Steam Properties (Pressure Table)**

p	T_s	v_f	v_g	u_f	u_g	h_f	h_{fg}	h_g	s_f	s_{fg}	s_g
(bar)	°C	(m³/kg)		(kJ/kg)		(kJ/kg)			(kJ/kg · K)		
130	330.9	0.001566	0.01277	1511	2496	1531	1,131	2662	3.560	1.872	5.432
135	333.9	0.001587	0.01210	1529	2486	1551	1,098	2649	3.591	1.810	5.401
140	336.7	0.001609	0.01148	1548	2476	1571	1,066	2637	3.622	1.749	5.371
145	339.5	0.001632	0.01089	1567	2466	1590	1,034	2624	3.653	1.687	5.340
150	342.2	0.001657	0.01034	1585	2455	1610	1,000	2610	3.684	1.625	5.309
155	344.8	0.001683	0.009815	1604	2443	1630	966	2596	3.715	1.563	5.278
160	347.4	0.001711	0.009317	1622	2431	1650	931	2581	3.745	1.500	5.245
165	349.9	0.001740	0.008841	1641	2419	1669	896	2565	3.775	1.437	5.212
170	352.3	0.001771	0.008384	1659	2405	1690	858	2548	3.806	1.372	5.178
175	354.7	0.001803	0.007943	1679	2390	1710	819	2529	3.839	1.305	5.144
180	357.0	0.001839	0.007515	1698	2375	1731	779	2510	3.871	1.236	5.107
185	359.2	0.001879	0.007098	1718	2358	1753	736	2489	3.904	1.164	5.068
190	361.4	0.001923	0.006689	1739	2340	1776	691	2467	3.940	1.089	5.029
195	363.6	0.001974	0.006285	1762	2320	1800	642	2442	3.975	1.365	5.340
200	365.7	0.002033	0.005884	1786	2297	1827	588	2415	4.014	0.921	4.935
205	367.8	0.002106	0.005465	1812	2271	1855	528	2383	4.056	0.824	4.880
210	369.9	0.002198	0.005019	1841	2238	1887	456	2343	4.106	0.709	4.815
215	371.9	0.002334	0.004636	1878	2190	1929	360	2289	4.169	0.559	4.728
220	373.8	0.002696	0.003659	1953	2097	2012	165	2177	4.295	0.256	4.551
220.5	374.0	0.003106	0.003107	2017	2017	2086	0	2086	4.407	0.000	4.407

TABLE A.3   **Compressed Water and Superheated Steam Properties (Saturation Pressure Table)**

p = 0.00612 bar (T_s = 0.0°C)					p = 0.05 bar (T_s = 32.9°C)				
T	*v*	*u*	*h*	*s*	*T*	*v*	*u*	*h*	*s*
(°C)	(m³/kg)	(kJ/kg)	(kJ/kg)	(kJ/kg · K)	(°C)	(m³/kg)	(kJ/kg)	(kJ/kg)	(kJ/kg · K)
0	0.001	0	0	0	0	0.001000	0	0	0
0	205.9	2375	2501	9.154	25	0.001004	105	105	0.367
25	224.8	2400	2538	9.287	32.9	0.001005	138	138	0.476
50	243.6	2442	2591	9.457	32.9	28.19	2420	2561	8.393
100	281.4	2515	2688	9.736	50	29.78	2441	2590	8.486
150	319.1	2588	2783	9.976	100	34.42	2515	2687	8.766
200	356.8	2661	2880	10.191	150	39.04	2587	2783	9.007
250	394.5	2736	2977	10.388	200	43.66	2661	2879	9.222
300	432.2	2812	3077	10.569	250	48.28	2736	2977	9.419
400	507.6	2969	3280	10.895	300	52.90	2812	3076	9.600
500	583.0	3133	3489	11.186	400	62.13	2969	3280	9.927
600	658.5	3303	3706	11.449	500	71.36	3132	3489	10.217
700	733.9	3480	3930	11.692	600	80.59	3303	3706	10.480
800	809.3	3665	4160	11.917	700	89.82	3480	3929	10.723
					800	99.06	3665	4160	10.948
p = 0.1 bar (T_s = 45.8°C)					p = 0.5 bar (T_s = 81.3°C)				
T	*v*	*u*	*h*	*s*	*T*	*v*	*u*	*h*	*s*
(°C)	(m³/kg)	(kJ/kg)	(kJ/kg)	(kJ/kg · K)	(°C)	(m³/kg)	(kJ/kg)	(kJ/kg)	(kJ/kg · K)
0	0.001000	0	0	0	0	0.001000	0	0	0
25	0.001004	105	105	0.367	25	0.001004	105	105	0.367
45.8	0.001010	192	192	0.649	50	0.001013	209	209	0.703
45.8	14.67	2437	2584	8.148	81.3	0.00103	341	341	1.091
50	14.87	2440	2589	8.164	81.3	3.240	2483	2645	7.593

TABLE A.3(cont.)   **Compressed Water and Superheated Steam Properties (Saturation Pressure Table)**

		$p = 0.1$ bar $(T_s = 45.8°C)$					$p = 0.5$ bar $(T_s = 81.3°C)$		
T	v	u	h	s	T	v	u	h	s
(°C)	(m³/kg)	(kJ/kg)	(kJ/kg)	(kJ/kg · K)	(°C)	(m³/kg)	(kJ/kg)	(kJ/kg)	(kJ/kg · K)
100	17.20	2515	2687	8.446	100	3.418	2511	2682	7.693
150	19.51	2587	2782	8.686	150	3.890	2585	2780	7.939
200	21.83	2661	2879	8.902	200	4.356	2659	2877	8.157
250	24.14	2736	2977	9.099	250	4.821	2735	2976	8.355
300	26.45	2812	3076	9.280	300	5.284	2811	3075	8.537
400	31.06	2969	3279	9.607	400	6.209	2968	3279	8.864
500	35.68	3132	3489	9.897	500	7.134	3132	3489	9.155
600	40.30	3303	3706	10.161	600	8.058	3303	3706	9.418
700	44.91	3480	3929	10.403	700	8.981	3480	3929	9.661
800	49.53	3665	4160	10.629	800	9.905	3665	4160	9.886

		$p = 1$ bar $(T_s = 99.6°C)$					$p = 1.01325$ bar $(T_s = 100°C)$		
T	v	u	h	s	T	v	u	h	s
(°C)	(m³/kg)	(kJ/kg)	(kJ/kg)	(kJ/kg · K)	(°C)	(m³/kg)	(kJ/kg)	(kJ/kg)	(kJ/kg · K)
0	0.001000	0	0	0	0	0.001000	0	0	0
25	0.001004	105	105	0.367	25	0.001004	105	105	0.367
50	0.001013	209	209	0.703	50	0.001013	209	209	0.703
99.6	0.001043	417	418	1.303	100	0.001043	419	419	1.307
99.6	1.694	2506	2675	7.359	100	1.673	2506	2676	7.355
100	1.696	2506	2676	7.361	150	1.911	2582	2776	7.607
150	1.937	2582	2776	7.613	200	2.144	2658	2875	7.827
200	2.173	2658	2875	7.833	250	2.375	2733	2974	8.026
250	2.406	2733	2974	8.033	300	2.604	2810	3074	8.209
300	2.639	2810	3074	8.215	400	3.062	2968	3278	8.537

TABLE A.3(cont.)   **Compressed Water and Superheated Steam Properties (Saturation Pressure Table)**

	p = 1 bar (T_s = 99.6°C)					p = 1.01325 bar (T_s = 100°C)			
T	v	u	h	s	T	v	u	h	s
(°C)	(m³/kg)	(kJ/kg)	(kJ/kg)	(kJ/kg · K)	(°C)	(m³/kg)	(kJ/kg)	(kJ/kg)	(kJ/kg · K)
400	3.103	2968	3278	8.543	500	3.519	3132	3488	8.828
500	3.565	3132	3488	8.834	600	3.975	3302	3705	9.092
600	4.028	3302	3705	9.098	700	4.431	3480	3929	9.334
700	4.490	3480	3929	9.341	800	4.887	3665	4160	9.560
800	4.952	3665	4160	9.566					

	p = 5 bar (T_s = 151.9°C)					p = 10 bar (T_s = 179.9°C)			
T	v	u	h	s	T	v	u	h	s
(°C)	(m³/kg)	(kJ/kg)	(kJ/kg)	(kJ/kg · K)	(°C)	(m³/kg)	(kJ/kg)	(kJ/kg)	(kJ/kg · K)
0	0.001000	0	1	0	0	0.001000	0	1	0
25	0.001004	105	105	0.367	25	0.001004	105	106	0.367
50	0.001013	209	210	0.703	50	0.001013	209	210	0.703
100	0.001042	419	419	1.307	100	0.001042	419	420	1.306
150	0.001091	632	632	1.842	150	0.001090	631	633	1.841
151.9	0.001093	640	640	1.861	179.9	0.001127	762	763	2.139
151.9	0.3747	2561	2749	6.822	179.9	0.1943	2583	2778	6.586
200	0.4251	2642	2855	7.059	200	0.2060	2621	2827	6.693
250	0.4745	2723	2960	7.270	250	0.2328	2709	2942	6.923
300	0.5227	2802	3064	7.459	300	0.2581	2792	3051	7.122
400	0.6173	2963	3272	7.793	400	0.3066	2957	3264	7.465
500	0.7109	3129	3484	8.087	500	0.3540	3124	3478	7.762
600	0.8040	3300	3702	8.352	600	0.4010	3297	3698	8.029
700	0.8968	3478	3927	8.596	700	0.4477	3476	3924	8.274
800	0.9896	3663	4158	8.822	800	0.4943	3661	4156	8.501

TABLE A.3(cont.) **Compressed Water and Superheated Steam Properties (Saturation Pressure Table)**

| \multicolumn{5}{c}{$p = 15$ bar ($T_s = 198.3°C$)} | | | | | \multicolumn{5}{c}{$p = 20$ bar ($T_s = 212.4°C$)} | | | | |

T	v	u	h	s	T	v	u	h	s
(°C)	(m³/kg)	(kJ/kg)	(kJ/kg)	(kJ/kg · K)	(°C)	(m³/kg)	(kJ/kg)	(kJ/kg)	(kJ/kg · K)
0	0.000999	0	2	0	0	0.000999	0	2	0
25	0.001004	105	106	0.367	25	0.001004	105	107	0.367
50	0.001012	209	211	0.703	50	0.001012	209	211	0.703
100	0.001042	419	420	1.306	100	0.001041	418	421	1.306
150	0.001090	631	633	1.841	150	0.001089	631	633	1.840
198.3	0.001154	843	845	2.315	200	0.001158	850	853	2.330
198.3	0.1317	2594	2791	6.444	212.4	0.001177	906	909	2.447
200	0.1324	2598	2796	6.454	212.4	0.099580	2600	2799	6.339
250	0.1521	2694	2923	6.708	250	0.1115	2679	2902	6.544
300	0.1698	2782	3037	6.917	300	0.1255	2772	3023	6.765
400	0.2030	2951	3256	7.269	400	0.1512	2945	3247	7.127
500	0.2351	3121	3473	7.570	500	0.1756	3116	3468	7.432
600	0.2667	3294	3694	7.838	600	0.1995	3291	3690	7.703
700	0.2980	3474	3921	8.084	700	0.2231	3471	3918	7.949
800	0.3292	3659	4153	8.311	800	0.2466	3658	4151	8.177

| \multicolumn{5}{c}{$p = 30$ bar ($T_s = 233.9°C$)} | | | | | \multicolumn{5}{c}{$p = 40$ bar ($T_s = 250.4°C$)} | | | | |

T	v	u	h	s	T	v	u	h	s
(°C)	(m³/kg)	(kJ/kg)	(kJ/kg)	(kJ/kg · K)	(°C)	(m³/kg)	(kJ/kg)	(kJ/kg)	(kJ/kg · K)
0	0.000998	0	3	0	0	0.000998	0	4	0
25	0.001003	105	108	0.366	25	0.001003	104	108	0.366
50	0.001012	209	212	0.702	50	0.001012	209	213	0.702
100	0.001041	418	421	1.305	100	0.001041	418	422	1.304
150	0.001088	631	634	1.839	150	0.001087	630	634	1.838

TABLE A.3(cont.)   *Compressed Water and Superheated Steam Properties (Saturation Pressure Table)*

\multicolumn{5}{c}{$p = 30$ bar ($T_s = 233.9°C$)}					\multicolumn{5}{c}{$p = 40$ bar ($T_s = 250.4°C$)}				
T	**v**	**u**	**h**	**s**	**T**	**v**	**u**	**h**	**s**
(°C)	(m³/kg)	(kJ/kg)	(kJ/kg)	(kJ/kg · K)	(°C)	(m³/kg)	(kJ/kg)	(kJ/kg)	(kJ/kg · K)
200	0.001157	850	853	2.329	200	0.001156	849	853	2.327
233.9	0.001216	1005	1008	2.646	250	0.001249	1080	1085	2.794
233.9	0.06666	2603	2803	6.186	250.4	0.001252	1082	1087	2.795
250	0.07060	2643	2855	6.286	250.4	0.04977	2602	2801	6.068
300	0.08120	2749	2992	6.538	300	0.05887	2724	2960	6.360
400	0.09936	2933	3231	6.921	400	0.07341	2920	3213	6.768
500	0.11610	3108	3457	7.234	500	0.08637	3100	3445	7.090
600	0.13230	3285	3682	7.509	600	0.09875	3279	3674	7.368
700	0.14830	3467	3912	7.757	700	0.11090	3462	3906	7.619
800	0.16410	3654	4146	7.987	800	0.12280	3650	4142	7.850

\multicolumn{5}{c}{$p = 50$ bar ($T_s = 264°C$)}					\multicolumn{5}{c}{$p = 60$ bar ($T_s = 275.6°C$)}				
T	**v**	**u**	**h**	**s**	**T**	**v**	**u**	**h**	**s**
(°C)	(m³/kg)	(kJ/kg)	(kJ/kg)	(kJ/kg · K)	(°C)	(m³/kg)	(kJ/kg)	(kJ/kg)	(kJ/kg · K)
0	0.000997	0	5	0	0	0.000997	0	6	0
25	0.001003	104	109	0.366	25	0.001002	104	110	0.366
50	0.001011	209	214	0.701	50	0.001011	208	215	0.701
100	0.001040	418	423	1.303	100	0.001040	417	424	1.302
150	0.001087	630	635	1.837	150	0.001086	629	636	1.836
200	0.001155	848	854	2.326	200	0.001153	847	854	2.324
250	0.001248	1079	1085	2.792	250	0.001246	1078	1085	2.789
264	0.001286	1148	1154	2.920	275.6	0.001319	1205	1213	3.026
264	0.03944	2597	2794	5.973	275.6	0.03245	2589	2784	5.888
300	0.04533	2697	2924	6.206	300	0.03616	2666	2883	6.065

TABLE A.3(cont.)   **Compressed Water and Superheated Steam Properties (Saturation Pressure Table)**

| \multicolumn{5}{c}{$p = 50$ bar ($T_s = 264°C$)} | | | | | \multicolumn{5}{c}{$p = 60$ bar ($T_s = 275.6°C$)} | | | | |

T	v	u	h	s	T	v	u	h	s
(°C)	(m³/kg)	(kJ/kg)	(kJ/kg)	(kJ/kg · K)	(°C)	(m³/kg)	(kJ/kg)	(kJ/kg)	(kJ/kg · K)
400	0.05782	2906	3196	6.645	400	0.04740	2892	3177	6.539
500	0.06851	3091	3434	6.975	500	0.05660	3082	3422	6.879
600	0.07860	3273	3666	7.258	600	0.06516	3267	3658	7.166
700	0.08840	3458	3900	7.511	700	0.07343	3453	3894	7.422
800	0.09805	3647	4137	7.743	800	0.08154	3643	4132	7.655

| \multicolumn{5}{c}{$p = 70$ bar ($T_s = 285.8°C$)} | | | | | \multicolumn{5}{c}{$p = 80$ bar ($T_s = 295°C$)} | | | | |

T	v	u	h	s	T	v	u	h	s
(°C)	(m³/kg)	(kJ/kg)	(kJ/kg)	(kJ/kg · K)	(°C)	(m³/kg)	(kJ/kg)	(kJ/kg)	(kJ/kg · K)
0	0.000996	0	7	0	0	0.000996	0	8	0
25	0.001002	104	111	0.365	25	0.001002	104	112	0.365
50	0.001011	208	215	0.700	50	0.001010	208	216	0.700
100	0.001039	417	424	1.302	100	0.001039	417	425	1.301
150	0.001085	629	636	1.835	150	0.001084	628	637	1.833
200	0.001152	847	855	2.322	200	0.001151	846	855	2.321
250	0.001245	1077	1085	2.787	250	0.001244	1076	1085	2.785
285.8	0.001351	1257	1267	3.121	295	0.001384	1305	1316	3.207
285.8	0.02740	2580	2772	5.813	295	0.02353	2569	2758	5.744
300	0.02947	2631	2837	5.928	300	0.02429	2999	3193	6.545
400	0.03994	2878	3157	6.447	400	0.03433	3272	3547	7.118
500	0.04809	3074	3410	6.797	500	0.04170	3065	3398	6.724
600	0.05556	3261	3649	7.088	600	0.04836	3254	3641	7.020
700	0.06273	3448	3887	7.346	700	0.05471	3444	3881	7.280
800	0.06975	3639	4127	7.581	800	0.06090	3636	4123	7.516

TABLE A.3(cont.)   *Compressed Water and Superheated Steam Properties (Saturation Pressure Table)*

| \multicolumn{5}{}{$p = 90$ bar ($T_s = 303.4°C$)} | | | | | \multicolumn{5}{}{$p = 100$ bar ($T_s = 311°C$)} | | | | |

T	v	u	h	s	T	v	u	h	s
(°C)	(m³/kg)	(kJ/kg)	(kJ/kg)	(kJ/kg · K)	(°C)	(m³/kg)	(kJ/kg)	(kJ/kg)	(kJ/kg · K)
0	0.000995	0	9	0	0	0.000995	0	10	0
25	0.001001	104	113	0.365	25	0.001001	104	114	0.365
50	0.001010	208	217	0.699	50	0.001010	208	218	0.699
100	0.001038	416	426	1.300	100	0.001038	416	427	1.299
150	0.001083	628	638	1.832	150	0.001083	627	638	1.831
200	0.001150	845	856	2.319	200	0.001149	844	856	2.318
250	0.001242	1074	1086	2.783	250	0.001241	1073	1086	2.780
300	0.001400	1331	1344	3.252	300	0.001396	1329	1343	3.248
303.4	0.001418	1350	1363	3.285	311	0.001451	1393	1407	3.359
303.4	0.02049	2557	2742	5.677	311	0.01803	2544	2725	5.614
400	0.02995	2869	3139	6.322	400	0.02643	2912	3176	6.349
500	0.03673	3056	3386	6.658	500	0.03275	3046	3374	6.597
600	0.04276	3248	3633	6.958	600	0.03827	3242	3625	6.903
700	0.04847	3439	3875	7.221	700	0.04348	3434	3869	7.168
800	0.05402	3632	4118	7.459	800	0.04852	3628	4114	7.407

| \multicolumn{5}{}{$p = 110$ bar ($T_s = 318.1°C$)} | | | | | \multicolumn{5}{}{$p = 120$ bar ($T_s = 324.7°C$)} | | | | |

T	v	u	h	s	T	v	u	h	s
(°C)	(m³/kg)	(kJ/kg)	(kJ/kg)	(kJ/kg · K)	(°C)	(m³/kg)	(kJ/kg)	(kJ/kg)	(kJ/kg · K)
0	0.000994	0	11	0	0	0.000994	0	12	0
25	0.001000	104	115	0.364	25	0.001000	104	116	0.364
50	0.001009	208	219	0.698	50	0.001009	208	220	0.698
100	0.001037	416	427	1.298	100	0.001037	416	428	1.298
150	0.001082	627	639	1.830	150	0.001081	626	639	1.829

TABLE A.3(cont.) *Compressed Water and Superheated Steam Properties (Saturation Pressure Table)*

p = 110 bar (T_s = 318.1°C)					p = 120 bar (T_s = 324.7°C)				
T	v	u	h	s	T	v	u	h	s
(°C)	(m³/kg)	(kJ/kg)	(kJ/kg)	(kJ/kg · K)	(°C)	(m³/kg)	(kJ/kg)	(kJ/kg)	(kJ/kg · K)
200	0.001148	844	856	2.316	200	0.001147	843	857	2.315
250	0.001239	1072	1086	2.778	250	0.001238	1071	1086	2.776
300	0.001392	1327	1342	3.243	300	0.001388	1324	1341	3.239
318.1	0.001489	1434	1450	3.429	324.7	0.001527	1473	1491	3.496
318.1	0.015990	2530	2705	5.552	324.7	0.01425	2513	2684	5.492
400	0.023530	2989	3248	6.441	400	0.02110	3084	3337	6.566
500	0.029490	3037	3361	6.541	500	0.02677	3027	3348	6.488
600	0.034610	3235	3616	6.851	600	0.03155	3229	3607	6.803
700	0.039400	3429	3863	7.119	700	0.03599	3425	3856	7.073
800	0.044020	3625	4109	7.359	800	0.04026	3621	4104	7.315

p = 130 bar (T_s = 330.9°C)					p = 140 bar (T_s = 336.7°C)				
T	v	u	h	s	T	v	u	h	s
(°C)	(m³/kg)	(kJ/kg)	(kJ/kg)	(kJ/kg · K)	(°C)	(m³/kg)	(kJ/kg)	(kJ/kg)	(kJ/kg · K)
0	0.000993	0	13	0	0	0.000993	0	14	0
25	0.000999	104	117	0.364	25	0.000999	104	118	0.364
50	0.001008	208	221	0.698	50	0.001008	207	221	0.697
100	0.001036	415	429	1.297	100	0.001036	415	430	1.296
150	0.001081	626	640	1.828	150	0.001080	626	641	1.827
200	0.001146	842	857	2.313	200	0.001145	842	858	2.312
250	0.001236	1070	1086	2.773	250	0.001235	1068	1086	2.771
300	0.001385	1322	1340	3.235	300	0.001381	1320	1339	3.231
330.9	0.001566	1511	1531	3.560	336.7	0.001609	1548	1571	3.622
330.9	0.01277	2496	2662	5.432	336.7	0.01148	2476	2637	5.371

TABLE A.3(cont.)   **Compressed Water and Superheated Steam Properties (Saturation Pressure Table)**

$p = 130$ bar ($T_s = 330.9°C$)					$p = 140$ bar ($T_s = 336.7°C$)				
T	v	u	h	s	T	v	u	h	s
(°C)	(m³/kg)	(kJ/kg)	(kJ/kg)	(kJ/kg · K)	(°C)	(m³/kg)	(kJ/kg)	(kJ/kg)	(kJ/kg · K)
400	0.01903	3335	3582	6.958	400	0.01724	3873	4114	7.845
500	0.02446	3018	3336	6.438	500	0.02249	3008	3322	6.390
600	0.02896	3222	3599	6.758	600	0.02674	3216	3590	6.716
700	0.03311	3420	3850	7.031	700	0.03064	3415	3844	6.991
800	0.03709	3617	4099	7.274	800	0.03437	3613	4094	7.236

$p = 150$ bar ($T_s = 342.2°C$)					$p = 160$ bar ($T_s = 347.4°C$)				
T	v	u	h	s	T	v	u	h	s
(°C)	(m³/kg)	(kJ/kg)	(kJ/kg)	(kJ/kg · K)	(°C)	(m³/kg)	(kJ/kg)	(kJ/kg)	(kJ/kg · K)
0	0.000992	0	15	0	0	0.000992	0	16	0.001
25	0.000998	104	119	0.363	25	0.000998	104	120	0.363
50	0.001007	207	222	0.697	50	0.001007	207	223	0.696
100	0.001035	415	430	1.295	100	0.001035	414	431	1.295
150	0.00108	625	641	1.826	150	0.001079	625	642	1.825
200	0.001144	841	858	2.31	200	0.001143	840	859	2.308
250	0.001233	1067	1086	2.769	250	0.001232	1066	1086	2.766
300	0.001378	1317	1338	3.227	300	0.001374	1315	1337	3.223
342.2	0.001657	1585	1610	3.684	347.4	0.001711	1622	1650	3.745
342.2	0.01034	2455	2610	5.309	347.4	0.009317	2431	2581	5.245
400	0.01567	2739	2974	5.879	400	0.01428	2720	2948	5.818
500	0.02077	2997	3309	6.344	500	0.01927	2987	3295	6.301
600	0.02482	3209	3581	6.676	600	0.02313	3202	3572	6.638
700	0.0285	3410	3837	6.954	700	0.02663	3405	3831	6.919
800	0.03201	3609	4090	7.2	800	0.02994	3606	4085	7.167

TABLE A.4   **Saturated Ice and Steam Properties (Temperature Table)**

T	p_s	v_i	v_g	u_i	u_g	h_i	h_{ig}	h_g	s_i	s_{ig}	s_g
(°C)	(Pa)	(m³/kg)		(kJ/kg)		(kJ/kg)			(kJ/kg · K)		
0	611.7	0.001091	206	−333	2375	−333	2834	2501	−1.221	10.376	9.155
−5	401.8	0.001090	307.9	−344	2368	−344	2836	2492	−1.259	10.574	9.315
−10	259.9	0.001089	467.2	−354	2361	−354	2836	2482	−1.298	10.779	9.481
−15	165.3	0.001088	720.6	−364	2354	−364	2837	2473	−1.336	10.991	9.655
−20	103.3	0.001087	1131	−374	2347	−374	2838	2464	−1.375	11.211	9.836
−25	63.29	0.001087	1809	−384	2340	−384	2839	2455	−1.414	11.439	10.025
−30	38.02	0.001086	2952	−393	2333	−393	2838	2445	−1.452	11.674	10.222
−35	22.35	0.001085	4917	−403	2326	−403	2839	2436	−1.491	11.920	10.429
−40	12.85	0.001084	8377	−412	2319	−412	2839	2427	−1.530	12.175	10.645
−45	7.206	0.001084	14610	−421	2312	−421	2839	2418	−1.569	12.441	10.872
−50	3.939	0.001083	26150	−429	2305	−429	2837	2408	−1.607	12.717	11.110
−55	2.095	0.001082	48060	−438	2298	−438	2837	2399	−1.646	13.005	11.359
−60	1.082	0.001082	90940	−446	2292	−446	2836	2390	−1.685	13.306	11.621
−65	0.5412	0.001081	1.775×10^5	−455	2285	−455	2836	2381	−1.724	13.621	11.897
−70	0.2618	0.001080	3.581×10^5	−463	2278	−463	2834	2371	−1.764	13.951	12.187
−75	0.1221	0.001080	7.489×10^5	−471	2271	−471	2833	2362	−1.803	14.296	12.493
−80	0.05478	0.001079	1.627×10^6	−478	2264	−478	2831	2353	−1.842	14.657	12.815
−85	0.02356	0.001078	3.685×10^6	−486	2257	−486	2830	2344	−1.881	15.037	13.156
−90	0.009684	0.001078	8.729×10^6	−493	2250	−493	2827	2334	−1.921	15.438	13.517
−95	0.003789	0.001077	2.170×10^7	−500	2243	−500	2825	2325	−1.961	15.860	13.899
−100	0.001405	0.001077	5.687×10^7	−507	2236	−507	2823	2316	−2.000	16.304	14.304

TABLE A.5   **Saturated Ice and Steam Properties (Pressure Table)**

p	T_s	v_i	v_g	u_i	u_g	h_i	h_{ig}	h_g	s_i	s_{ig}	s_g
(Pa)	(°C)	(m³/kg)		(kJ/kg)		(kJ/kg)			(kJ/kg · K)		
611.8	0	0.001091	206	−333	2375	−333	2834	2501	−1.221	10.376	9.155
500	−2.4	0.001090	249.8	−339	2371	−339	2835	2496	−1.239	10.471	9.232
250	−10.4	0.001089	484.9	−355	2360	−355	2837	2482	−1.301	10.796	9.495
100	−20.3	0.001087	1167	−375	2347	−375	2838	2463	−1.377	11.224	9.847
75	−23.3	0.001087	1538	−380	2343	−380	2838	2458	−1.400	11.358	9.958
50	−27.3	0.001086	2269	−388	2337	−388	2838	2450	−1.431	11.546	10.115
25	−33.9	0.001085	4416	−401	2328	−401	2839	2438	−1.483	11.867	10.384
10	−42.2	0.001084	10660	−416	2316	−416	2839	2423	−1.547	12.291	10.744
7.5	−44.7	0.001084	14060	−420	2313	−420	2838	2418	−1.566	12.423	10.857
5	−48.1	0.001083	20770	−426	2308	−426	2838	2412	−1.593	12.610	11.017
2.5	−53.7	0.001082	40520	−436	2300	−436	2838	2402	−1.636	12.927	11.291
1	−60.6	0.001081	98100	−447	2291	−447	2836	2389	−1.690	13.343	11.653
0.75	−62.7	0.001081	1.295×10^5	−451	2288	−451	2836	2385	−1.706	13.474	11.768
0.5	−65.6	0.001081	1.916×10^5	−456	2284	−456	2836	2380	−1.729	13.659	11.930
0.25	−72.8	0.001080	3.699×10^5	−467	2274	−467	2833	2366	−1.785	14.140	12.355
0.1	−76.3	0.001079	9.084×10^5	−473	2269	−473	2833	2360	−1.813	14.389	12.576
0.075	−77.9	0.001079	1.202×10^6	−475	2267	−475	2832	2357	−1.825	14.502	12.677
0.05	−80.3	0.001079	1.780×10^6	−479	2263	−479	2831	2352	−1.845	14.684	12.839
0.025	−84.6	0.001078	3.480×10^6	−485	2257	−485	2829	2344	−1.878	15.008	13.130
0.01	−89.9	0.001078	8.456×10^6	−493	2250	−493	2837	2344	−1.920	15.432	13.512
0.0015	−99.9	0.001077	5.33×10^7	−507	2236	−507	2823	2316	−1.999	16.291	14.292

TABLE A.6 **Compressed Ice and Superheated Steam Properties**

\multicolumn{5}{c}{$p = 611.7$ Pa ($T_s = 0.01°C$)}					\multicolumn{5}{c}{$p = 500$ Pa ($T_s = -2.4°C$)}				
T	v	u	h	s	T	v	u	h	s
(°C)	(m³/kg)	(kJ/kg)	(kJ/kg)	(kJ/kg · K)	(°C)	(m³/kg)	(kJ/kg)	(kJ/kg)	(kJ/kg · K)
−100	0.001077	−507	−507	−2	−100	0.001077	−507	−507	−2
−50	0.001083	−429	−429	−1.607	−50	0.001083	−429	−429	−1.607
−25	0.001087	−384	−384	−1.414	−25	0.001087	−384	−384	−1.414
0	0.001091	−333	−333	−1.221	−2.4	0.001090	−339	−339	−1.239
0	206.0	2375	2501	9.156	−2.4	249.8	2371	2496	9.232
25	224.9	2410	2548	9.319	0	252.0	2375	2501	9.248
50	243.8	2445	2594	9.469	25	275.1	2410	2548	9.412
75	262.6	2481	2641	9.609	50	298.2	2445	2594	9.562
100	281.5	2516	2688	9.740	75	321.3	2481	2641	9.702
125	300.4	2552	2736	9.863	100	344.4	2516	2688	9.833
150	319.2	2588	2783	9.979	125	367.5	2552	2736	9.956
200	357.0	2661	2880	10.194	150	390.6	2588	2783	10.072
250	394.7	2736	2978	10.391	200	436.7	2661	2880	10.287
300	432.4	2812	3077	10.572	250	482.9	2736	2978	10.484
					300	529.0	2812	3077	10.665

\multicolumn{5}{c}{$p = 100$ Pa ($T_s = -20.3°C$)}					\multicolumn{5}{c}{$p = 50$ Pa ($T_s = -27.3°C$)}				
T	v	u	h	s	T	v	u	h	s
(°C)	(m³/kg)	(kJ/kg)	(kJ/kg)	(kJ/kg · K)	(°C)	(m³/kg)	(kJ/kg)	(kJ/kg)	(kJ/kg · K)
−100	0.001077	−507	−507	−2	−100	0.001077	−507	−507	−2
−50	0.001083	−429	−429	−1.607	−50	0.001083	−429	−429	−1.607
−25	0.001087	−384	−384	−1.414	−27.3	0.001086	−388	−388	−1.431
−20.3	0.001087	−375	−375	−1.377	−27.3	2269	2337	2450	10.115
−20.3	1167	2347	2463	9.847	−25	2290	2340	2455	10.134

TABLE A.6(cont.)   *Compressed Ice and Superheated Steam Properties*

\multicolumn{5}{c}{p = 100 Pa (T_s = −20.3°C)}					\multicolumn{5}{c}{p = 50 Pa (T_s = −27.3°C)}				
T	v	u	h	s	T	v	u	h	s
(°C)	(m³/kg)	(kJ/kg)	(kJ/kg)	(kJ/kg · K)	(°C)	(m³/kg)	(kJ/kg)	(kJ/kg)	(kJ/kg · K)
0	1261	2375	2501	9.992	0	2521	2375	2501	10.312
25	1376	2410	2548	10.155	25	2752	2410	2548	10.475
50	1491	2445	2594	10.305	50	2983	2445	2594	10.625
75	1607	2481	2641	10.445	75	3214	2481	2641	10.765
100	1722	2516	2688	10.576	100	3444	2516	2688	10.896
125	1838	2552	2736	10.699	125	3675	2552	2736	11.019
150	1953	2588	2783	10.815	150	3906	2588	2783	11.135
200	2184	2661	2880	11.030	200	4367	2661	2880	11.350
250	2414	2736	2978	11.226	250	4829	2736	2978	11.546
300	2645	2812	3077	11.408	300	5290	2812	3077	11.727

\multicolumn{5}{c}{p = 10 Pa (T_s = −42.2°C)}					\multicolumn{5}{c}{p = 5 Pa (T_s = −48.1°C)}				
T	v	u	h	s	T	v	u	h	s
(°C)	(m³/kg)	(kJ/kg)	(kJ/kg)	(kJ/kg · K)	(°C)	(m³/kg)	(kJ/kg)	(kJ/kg)	(kJ/kg · K)
−100	0.001077	−507	−507	−2	−100	0.001077	−507	−507	−2
−50	0.001083	−429	−429	−1.607	−50	0.001083	−429	−429	−1.607
−42.2	0.001084	−416	−416	−1.547	−48.1	0.001083	−426	−426	−1.593
−42.2	10660	2316	2423	10.744	−48.1	20770	2308	2412	11.017
−25	11450	2340	2455	10.877	−25	22910	2340	2455	11.196
0	12610	2375	2501	11.055	0	25210	2375	2501	11.375
25	13760	2410	2548	11.218	25	27520	2410	2548	11.538
50	14910	2445	2594	11.368	50	29830	2445	2594	11.688
75	16070	2481	2641	11.508	75	32140	2481	2641	11.828
100	17220	2516	2688	11.639	100	34440	2516	2688	11.959

TABLE A.6(cont.) *Compressed Ice and Superheated Steam Properties*

	$p = 10$ Pa ($T_s = -42.2°C$)					$p = 5$ Pa ($T_s = -48.1°C$)			
T	v	u	h	s	T	v	u	h	s
(°C)	(m³/kg)	(kJ/kg)	(kJ/kg)	(kJ/kg · K)	(°C)	(m³/kg)	(kJ/kg)	(kJ/kg)	(kJ/kg · K)
125	18380	2552	2736	11.762	125	36750	2552	2736	12.081
150	19530	2588	2783	11.878	150	39060	2588	2783	12.198
200	21840	2661	2880	12.093	200	43670	2661	2880	12.413
250	24140	2736	2978	12.289	250	48290	2736	2978	12.609
300	26450	2812	3077	12.47	300	52900	2812	3077	12.790

	$p = 1$ Pa ($T_s = -60.6°C$)					$p = 0.5$ Pa ($T_s = -65.6°C$)			
T	v	u	h	s	T	v	u	h	s
(°C)	(m³/kg)	(kJ/kg)	(kJ/kg)	(kJ/kg · K)	(°C)	(m³/kg)	(kJ/kg)	(kJ/kg)	(kJ/kg · K)
−100	0.001077	−507	−507	−2	−100	0.001077	−507	−507	−2
−60.6	0.001081	−447	−447	−1.69	−65.6	0.001081	−456	−456	−1.729
−60.6	98100	2291	2389	11.653	−65.6	191600	2284	2380	11.930
−50	103000	2305	2408	11.742	−50	206000	2305	2408	12.062
−25	114500	2340	2455	11.939	−25	229100	2340	2455	12.259
0	126100	2375	2501	12.117	0	252100	2375	2501	12.437
25	137600	2410	2548	12.281	25	275200	2410	2548	12.600
50	149100	2445	2594	12.431	50	298300	2445	2594	12.751
75	160700	2481	2641	12.571	75	321400	2481	2641	12.891
100	172200	2516	2688	12.701	100	344400	2516	2688	13.021
125	183800	2552	2736	12.824	125	367500	2552	2736	13.144
150	195300	2588	2783	12.94	150	390600	2588	2783	13.260
200	218400	2661	2880	13.156	200	436700	2661	2880	13.475
250	241400	2736	2978	13.352	250	482900	2736	2978	13.672
300	264500	2812	3077	13.533	300	529000	2812	3077	13.853

TABLE A.6(cont.)   *Compressed Ice and Superheated Steam Properties*

\multicolumn{5}{c}{$p = 0.1$ Pa ($T_s = -76.3°C$)}					\multicolumn{5}{c}{$p = 0.05$ Pa ($T_s = -80.3°C$)}				
T	v	u	h	s	T	v	u	h	s
(°C)	(m³/kg)	(kJ/kg)	(kJ/kg)	(kJ/kg · K)	(°C)	(m³/kg)	(kJ/kg)	(kJ/kg)	(kJ/kg · K)
−100	0.001077	−507	−507	−2.000	−100	0.001077	−507	−507	−2.000
−76.3	0.001079	−473	−473	−1.813	−80.3	0.001079	−479	−479	−1.845
−76.3	908400	2269	2360	12.576	−80.3	1780000	2263	2352	12.839
−50	1030000	2305	2408	12.805	−50	2060000	2305	2408	13.125
−25	1145000	2340	2455	13.002	−25	2291000	2340	2455	13.322
0	1261000	2375	2501	13.180	0	2521000	2375	2501	13.500
25	1376000	2410	2548	13.343	25	2752000	2410	2548	13.663
50	1491000	2445	2594	13.494	50	2983000	2445	2594	13.814
75	1607000	2481	2641	13.633	75	3214000	2481	2641	13.953
100	1722000	2516	2688	13.764	100	3444000	2516	2688	14.084
125	1838000	2552	2736	13.887	125	3675000	2552	2736	14.207
150	1953000	2588	2783	14.003	150	3906000	2588	2783	14.323
200	2184000	2661	2880	14.218	200	4367000	2661	2880	14.538
250	2414000	2736	2978	14.415	250	4829000	2736	2978	14.734
300	2645000	2812	3077	14.596	300	5290000	2812	3077	14.915

\multicolumn{5}{c}{$p = 0.01$ Pa ($T_s = -89.9°C$)}					\multicolumn{5}{c}{$p = 0.0015$ Pa ($T_s = -99.9°C$)}				
T	v	u	h	s	T	v	u	h	s
(°C)	(m³/kg)	(kJ/kg)	(kJ/kg)	(kJ/kg · K)	(°C)	(m³/kg)	(kJ/kg)	(kJ/kg)	(kJ/kg · K)
−100	0.001077	−507	−507	−2.000	−100	0.001077	−507	−507	−2.000
−89.9	0.001078	−493	−493	−1.920	−99.9	0.001077	−507	−507	−1.999
−89.9	8456000	2250	2334	13.512	−99.9	53320000	2236	2316	14.292
−50	10300000	2305	2408	13.868	−50	68660000	2305	2408	14.743
−25	11450000	2340	2455	14.065	−25	76350000	2340	2455	14.940

TABLE A.6(cont.)   *Compressed Ice and Superheated Steam Properties*

		$p = 0.01$ Pa ($T_s = -89.9°C$)					$p = 0.0015$ Pa ($T_s = -99.9°C$)		
T	v	u	h	s	T	v	u	h	s
(°C)	(m³/kg)	(kJ/kg)	(kJ/kg)	(kJ/kg · K)	(°C)	(m³/kg)	(kJ/kg)	(kJ/kg)	(kJ/kg · K)
0	12610000	2375	2501	14.243	0	84040000	2375	2501	15.118
25	13760000	2410	2548	14.406	25	91730000	2410	2548	15.281
50	14910000	2445	2594	14.556	50	99430000	2445	2594	15.432
75	16070000	2481	2641	14.696	75	1.07×10^8	2481	2641	15.572
100	17220000	2516	2688	14.827	100	1.15×10^8	2516	2688	15.702
125	18380000	2552	2736	14.950	125	1.23×10^8	2552	2736	15.825
150	19530000	2588	2783	15.066	150	1.30×10^8	2588	2783	15.941
200	21840000	2661	2880	15.281	200	1.46×10^8	2661	2880	16.156
250	24140000	2736	2978	15.477	250	1.61×10^8	2736	2978	16.353
300	26450000	2812	3077	15.658	300	1.76×10^8	2812	3077	16.534

TABLE A.7 *Saturated Refrigerant-22 Properties (Temperature Table)*

T	p_s	$v_f \times 10^3$	$v_g \times 10^3$	h_f	h_{fg}	h_g	s_f	s_{fg}	s_g
(°C)	(bar)	(m³/kg)		(kJ/kg)			(kJ/kg · K)		
− 60	0.375	0.683	535.9	− 22.29	245.44	223.15	− 0.0998	1.1515	1.0517
− 55	0.496	0.689	413.43	− 16.71	242.36	225.65	− 0.074	1.1110	1.037
− 50	0.646	0.695	323.16	− 11.16	239.27	228.11	− 0.0488	1.0722	1.0234
− 45	0.83	0.702	255.68	− 5.59	236.12	230.53	− 0.0242	1.0349	1.0107
− 40	1.054	0.709	204.56	0	232.92	232.92	0	0.9990	0.999
− 35	1.323	0.716	165.36	5.61	229.65	235.26	0.0237	0.9643	0.988
− 30	1.643	0.724	134.97	11.25	226.30	237.55	0.047	0.9307	0.9777
− 25	2.019	0.732	111.15	16.89	222.89	239.78	0.0697	0.8983	0.968
− 20	2.458	0.74	92.3	22.51	219.43	241.94	0.092	0.8668	0.9588
− 15	2.967	0.749	77.24	28.22	215.82	244.04	0.1141	0.8361	0.9502
− 10	3.551	0.759	65.09	33.98	212.09	246.07	0.1361	0.8059	0.942
− 5	4.218	0.768	55.2	39.79	208.23	248.02	0.1577	0.7765	0.9342
0	4.984	0.779	46.99	45.61	204.26	249.87	0.1789	0.7478	0.9267
5	5.843	0.789	40.24	51.46	200.19	251.65	0.1999	0.7197	0.9196
10	6.810	0.800	34.63	57.43	195.91	253.34	0.2209	0.6919	0.9128
15	7.892	0.812	29.92	63.48	191.47	254.95	0.2418	0.6645	0.9063
20	9.098	0.825	25.95	69.61	186.86	256.47	0.2626	0.6374	0.9000
25	10.43	0.838	22.58	75.86	182.10	257.96	0.2834	0.6108	0.8942
30	11.91	0.851	19.70	82.09	177.07	259.16	0.3038	0.5841	0.8879
35	13.53	0.866	17.23	88.42	171.78	260.20	0.3242	0.5574	0.8816
40	15.31	0.883	15.11	94.93	166.21	261.14	0.3445	0.5308	0.8753
45	17.27	0.901	13.24	101.50	160.30	261.80	0.3647	0.5039	0.8686
50	19.39	0.921	11.63	108.34	154.00	262.34	0.3853	0.4766	0.8619
55	21.72	0.944	10.21	115.23	147.28	262.51	0.4057	0.4488	0.8545
60	24.23	0.969	8.96	121.98	139.92	261.90	0.4254	0.4200	0.8454
65	26.98	0.998	7.84	129.21	131.77	260.98	0.4462	0.3897	0.8359
70	29.94	1.031	6.85	137.32	122.99	260.31	0.4695	0.3585	0.8280
75	33.13	1.070	5.94	146.24	113.78	260.02	0.4949	0.3268	0.8217

TABLE A.8   Saturated Refrigerant-22 Properties (Pressure Table)

p	T_s	$v_f \times 10^3$	$v_g \times 10^3$	h_f	h_{fg}	h_g	s_f	s_{fg}	s_g
(bar)	(°C)	(m³/kg)			(kJ/kg)			(kJ/kg · K)	
0.6	−51.4	0.693	346.06	−12.74	240.15	227.41	−0.0560	1.0831	1.0271
0.8	−45.8	0.701	264.72	−6.44	236.61	230.17	−0.0280	1.0406	1.0126
1	−41.1	0.707	214.93	−1.27	233.65	232.38	−0.0055	1.0071	1.0016
1.054	−40.0	0.709	204.62	0	232.92	232.92	0	0.9990	0.9990
1.5	−32.1	0.721	147.00	8.85	227.73	236.58	0.0371	0.9449	0.9820
2	−25.2	0.732	112.14	16.62	223.06	239.68	0.0687	0.8997	0.9684
2.5	−19.6	0.741	90.83	23.04	219.08	242.12	0.0941	0.8639	0.9580
3	−14.7	0.750	76.43	28.58	215.59	244.17	0.1156	0.8341	0.9497
3.5	−10.4	0.758	66.00	33.51	212.39	245.90	0.1343	0.8084	0.9427
4	−6.6	0.765	58.09	37.98	209.44	247.42	0.1510	0.7856	0.9366
4.5	−3.1	0.772	51.86	42.04	206.71	248.75	0.1660	0.7654	0.9314
5	0.1	0.779	46.84	45.72	204.18	249.90	0.1793	0.7472	0.9265
6	5.9	0.791	39.22	52.48	199.47	251.95	0.2035	0.7149	0.9184
7	10.9	0.803	33.70	58.53	195.11	253.64	0.2247	0.6869	0.9116
8	15.5	0.813	29.52	64.04	191.05	255.09	0.2437	0.6620	0.9057
9	19.6	0.824	26.23	69.08	187.27	256.35	0.2608	0.6397	0.9005
10	23.4	0.834	23.58	73.86	183.62	257.48	0.2768	0.6191	0.8959
12.5	31.9	0.857	18.73	84.45	175.12	259.57	0.3114	0.5742	0.8856
15	39.1	0.880	15.44	93.79	167.18	260.97	0.3410	0.5353	0.8763
17.5	45.6	0.903	13.05	102.31	159.58	261.89	0.3672	0.5007	0.8679
20	51.3	0.927	11.23	110.16	152.28	262.44	0.3907	0.4693	0.8600
22.5	56.6	0.952	9.79	117.39	144.99	262.38	0.4120	0.4397	0.8517
25	61.4	0.977	8.63	123.89	137.75	261.64	0.4309	0.4117	0.8426
27.5	65.9	1.004	7.65	130.58	130.23	260.81	0.4502	0.3841	0.8343
30	70.1	1.032	6.83	137.41	122.86	260.27	0.4698	0.3579	0.8277

TABLE A.9 Superheated Refrigerant-22 Properties (Saturation Pressure Table)

$p_s = 0.6$ bar ($T_s = -51.43°C$)			$p_s = 1$ bar ($T_s = -41.13°C$)			$p_s = 2$ bar ($T_s = -25.23°C$)			$p_s = 3$ bar ($T_s = -14.69°C$)		
$h_f = -12.74$ kJ/kg; $s_f = -0.056$ kJ/kg·K			$h_f = -1.27$ kJ/kg; $s_f = -0.0055$ kJ/kg·K			$h_f = 16.62$ kJ/kg; $s_f = 0.0687$ kJ/kg·K			$h_f = 28.58$ kJ/kg; $s_f = 0.1156$ kJ/kg·K		
$h_g = 227.41$ kJ/kg; $s_g = 1.0271$ kJ/kg·K			$h_g = 232.38$ kJ/kg; $s_g = 1.0016$ kJ/kg·K			$h_g = 239.68$ kJ/kg; $s_g = 0.9684$ kJ/kg·K			$h_g = 244.17$ kJ/kg; $s_g = 0.9497$ kJ/kg·K		
T (°C)	h (kJ/kg)	s (kJ/kg·K)	T (°C)	h (kJ/kg)	s (kJ/kg·K)	T (°C)	h (kJ/kg)	s (kJ/kg·K)	T (°C)	h (kJ/kg)	s (kJ/kg·K)
-40	234.18	1.0569	-40	233.07	1.0045						
-20	246.30	1.1067	-20	245.42	1.0553	-20	243.09	0.9820			
0	258.77	1.1542	0	258.07	1.1034	0	256.20	1.0319	0	254.21	0.9875
20	271.63	1.1996	20	271.05	1.1492	20	269.52	1.0789	20	267.92	1.0359
40	284.90	1.2434	40	284.40	1.1933	40	283.12	1.1238	40	281.79	1.0817
60	298.60	1.2858	60	298.17	1.2359	60	297.07	1.1670	60	295.94	1.1254
80	312.72	1.3269	80	312.33	1.2772	80	311.37	1.2087	80	310.39	1.1676
100	327.24	1.3669	100	326.90	1.3173	100	326.05	1.2491	100	325.19	1.2083
120	342.16	1.4059	120	341.86	1.3564	120	341.11	1.2884	120	340.35	1.2479
140	357.49	1.4439	140	357.22	1.3945	140	356.54	1.3267	140	355.87	1.2864
160	373.20	1.4810	160	372.95	1.4317	160	372.35	1.3640	160	371.74	1.3239

TABLE A.9(cont.) Superheated Refrigerant-22 Properties (Saturation Pressure Table)

p_s = 4 bar (T_s = −6.56°C)			p_s = 5 bar (T_s = 0.1°C)			p_s = 6 bar (T_s = 5.86°C)			p_s = 8 bar (T_s = 15.47°C)		
T (°C)	h (kJ/kg)	s (kJ/kg·K)	T (°C)	h (kJ/kg)	s (kJ/kg·K)	T (°C)	h (kJ/kg)	s (kJ/kg·K)	T (°C)	h (kJ/kg)	s (kJ/kg·K)
−6.56 (f)	37.98	0.151	0.1 (f)	45.72	0.1793	5.86 (f)	52.48	0.2035	15.47 (f)	64.04	0.2437
−6.56 (g)	247.42	0.9366	0.1 (g)	249.9	0.9265	5.86 (g)	251.95	0.9184	15.47 (g)	255.09	0.9057
0	252.08	0.9539									
20	266.24	1.0040	20	264.48	0.9780	20	262.65	0.9558	20	258.75	0.9182
40	280.42	1.0507	40	279.00	1.0259	40	277.54	1.0050	40	274.49	0.9702
60	294.78	1.0952	60	293.60	1.0711	60	292.40	1.0510	60	289.91	1.0179
80	309.40	1.1378	80	308.39	1.1142	80	307.37	1.0946	80	305.29	1.0628
100	324.32	1.1789	100	323.45	1.1557	100	322.57	1.1365	100	320.77	1.1054
120	339.58	1.2187	120	338.81	1.1958	120	338.04	1.1769	120	336.47	1.1464
140	355.19	1.2575	140	354.50	1.2347	140	353.82	1.2160	140	352.43	1.1860
160	371.13	1.2951	160	370.51	1.2726	160	369.90	1.2540	160	368.66	1.2243

TABLE A.9(cont.) **Superheated Refrigerant-22 Properties (Saturation Pressure Table)**

$p_s = 10$ bar ($T_s = 23.44$°C)			$p_s = 12.5$ bar ($T_s = 31.88$°C)			$p_s = 15$ bar ($T_s = 39.14$°C)			$p_s = 17.5$ bar ($T_s = 45.59$°C)		
T	h	s	T	h	s	T	h	s	T	h	s
(°C)	(kJ/kg)	(kJ/kg · K)	(°C)	(kJ/kg)	(kJ/kg · K)	(°C)	(kJ/kg)	(kJ/kg · K)	(°C)	(kJ/kg)	(kJ/kg · K)
23.44 (f)	73.86	0.2768	31.88 (f)	84.45	0.3114	39.14 (f)	93.79	0.341	45.59 (f)	102.31	0.3672
23.44 (g)	257.48	0.8959	31.88 (g)	259.57	0.8856	39.14 (g)	260.97	0.8763	45.59 (g)	261.89	0.8679
40	271.26	0.9411	40	266.82	0.9090	40	261.80	0.8790			
60	287.32	0.9909	60	283.86	0.9618	60	280.15	0.9358	60	276.10	0.9115
80	303.15	1.0370	80	300.25	1.0096	80	297.31	0.9858	80	294.20	0.9643
100	318.95	1.0805	100	316.54	1.0544	100	314.08	1.0320	100	311.52	1.0120
120	334.88	1.1221	120	332.82	1.0969	120	330.72	1.0755	120	328.55	1.0565
140	351.03	1.1622	140	349.20	1.1376	140	347.38	1.1168	140	345.51	1.0985
160	367.42	1.2009	160	365.78	1.1768	160	364.17	1.1565	160	362.52	1.1387

TABLE A.9(cont.) *Superheated Refrigerant-22 Properties (Saturation Pressure Table)*

p_s = 20 bar (T_s = 51.33°C)

T (°C)	h (kJ/kg)	s (kJ/kg·K)
51.33 (f)	110.16	0.3907
51.33 (g)	262.44	0.8600
60	271.62	0.8879
70	281.49	0.9171
80	290.90	0.9441
90	300.00	0.9696
100	308.90	0.9937
110	317.67	1.0169
120	326.35	1.0393
130	334.99	1.0610
140	343.60	1.0821
150	352.21	1.1027
160	360.84	1.1228

p_s = 22.5 bar (T_s = 56.6°C)

T (°C)	h (kJ/kg)	s (kJ/kg·K)
56.6 (f)	117.39	0.4120
56.6 (g)	262.38	0.8517
60	266.27	0.8634
70	277.15	0.8956
80	287.26	0.9246
90	296.82	0.9514
100	306.08	0.9765
110	315.12	1.0004
120	324.03	1.0234
130	332.86	1.0456
140	341.64	1.0671
150	350.40	1.0880
160	359.14	1.1084

p_s = 25 bar (T_s = 61.45°C)

T (°C)	h (kJ/kg)	s (kJ/kg·K)
61.45 (f)	123.89	0.4309
61.45 (g)	261.64	0.8426
70	272.17	0.8737
80	283.24	0.9055
90	293.42	0.9339
100	303.10	0.9602
110	312.46	0.9850
120	321.63	1.0086
130	330.68	1.0313
140	339.63	1.0533
150	348.53	1.0745
160	357.41	1.0953

p_s = 30 bar (T_s = 70.08°C)

T (°C)	h (kJ/kg)	s (kJ/kg·K)
70.08 (f)	137.41	0.4698
70.08 (g)	260.27	0.8277
80	274.06	0.8674
90	285.92	0.9005
100	296.76	0.9299
110	306.89	0.9567
120	316.68	0.982
130	326.19	1.0058
140	335.54	1.0288
150	344.78	1.0508
160	353.92	1.0722

TABLE A.10 *Superheated Refrigerant-22 Properties (Saturation Temperature Table)*

$T_s = -60°C$ ($p_s = 0.375$ bar)			$T_s = -40°C$ ($p_s = 1.054$ bar)			$T_s = -20°C$ ($p_s = 2.458$ bar)			$T_s = 0°C$ ($p_s = 4.984$ bar)		
T	h	s	T	h	s	T	h	s	T	h	s
(°C)	(kJ/kg)	(kJ/kg·K)	(°C)	(kJ/kg)	(kJ/kg·K)	(°C)	(kJ/kg)	(kJ/kg·K)	(°C)	(kJ/kg)	(kJ/kg·K)
−60 (f)	−22.29	−0.0998	−40 (f)	0	0	−20 (f)	22.51	0.092	0 (f)	45.61	0.1789
−60 (g)	223.15	1.0517	−40 (g)	232.92	0.999	−20 (g)	241.94	0.9588	0 (g)	249.87	0.9267
−40	234.78	1.1038									
−20	246.78	1.1532	−20	245.30	1.0499						
0	259.16	1.2003	0	257.97	1.0981	0	255.31	1.0096			
20	271.96	1.2455	20	270.97	1.1440	20	268.80	1.0573	20	264.51	0.9784
40	285.18	1.2891	40	284.33	1.1881	40	282.52	1.1025	40	279.03	1.0263
60	298.85	1.3314	60	298.11	1.2307	60	296.55	1.1460	60	293.62	1.0715
80	312.93	1.3724	80	312.28	1.2720	80	310.92	1.1879	80	308.41	1.1146
100	327.42	1.4123	100	326.85	1.3122	100	325.66	1.2284	100	323.46	1.1561
120	342.33	1.4512	120	341.82	1.3512	120	340.76	1.2679	120	338.83	1.1962
140	357.64	1.4892	140	357.18	1.3893	140	356.24	1.3063	140	354.52	1.2351
160	373.33	1.5263	160	372.92	1.4265	160	372.07	1.3437	160	370.52	1.2729

A31

TABLE A.10(cont.)   Superheated Refrigerant-22 Properties (Saturation Temperature Table)

$T_s = 10°C$ ($p_s = 6.81$ bar)			$T_s = 20°C$ ($p_s = 9.098$ bar)			$T_s = 30°C$ ($p_s = 11.91$ bar)			$T_s = 40°C$ ($p_s = 15.313$ bar)		
T	h	s	T	h	s	T	h	s	T	h	s
(°C)	(kJ/kg)	(kJ/kg·K)	(°C)	(kJ/kg)	(kJ/kg·K)	(°C)	(kJ/kg)	(kJ/kg·K)	(°C)	(kJ/kg)	(kJ/kg·K)
10 (f)	57.43	0.2209	20 (f)	69.61	0.2626	30 (f)	82.09	0.3038	40 (f)	94.93	0.3445
10 (g)	253.34	0.9128	20 (g)	256.47	0.9	30 (g)	259.16	0.8879	40 (g)	261.14	0.8753
20	261.11	0.9397									
40	276.33	0.99	40	272.74	0.9537	40	267.94	0.9164			
60	291.4	1.0366	60	288.5	1.0025	60	284.71	0.9683	60	279.68	0.9327
80	306.53	1.0807	80	304.12	1.0481	80	300.95	1.0157	80	296.94	0.9831
100	321.84	1.1229	100	319.77	1.0912	100	317.12	1.0602	100	313.78	1.0294
120	337.41	1.1635	120	335.6	1.1325	120	333.31	1.1025	120	330.46	1.073
140	353.26	1.2029	140	351.67	1.1723	140	349.63	1.143	140	347.15	1.1144
160	369.4	1.241	160	367.98	1.2109	160	366.16	1.182	160	363.97	1.1542

TABLE A.10(cont.)   **Superheated Refrigerant-22 Properties
(Saturation Temperature Table)**

T_s = 50°C (p_s = 19.395 bar)			T_s = 60°C (p_s = 24.231 bar)		
T	h	s	T	h	s
(°C)	(kJ/kg)	(kJ/kg · K)	(°C)	(kJ/kg)	(kJ/kg · K)
50 (f)	108.34	0.3853	60 (f)	121.98	0.4254
50 (g)	262.34	0.8619	60 (g)	261.9	0.8454
60	272.74	0.8936			
80	291.71	0.9489	80	284.52	0.9114
100	309.53	0.998	100	304.03	0.9651
120	326.89	1.0433	120	322.38	1.013
140	344.06	1.0859	140	340.25	1.0574
160	361.25	1.1265	160	357.94	1.0992

TABLE A.11   *Ideal Gas Properties of Air (Gravimetric)*

T	c_p	h	$s°$		
(K)	(kJ/kg · K)	(kJ/kg)	(kJ/kg · K)	p_r	v_r
298	1.005	0	6.696	1	1
350	1.007	52	6.857	1.75	0.6695
400	1.013	103	6.992	2.80	0.4784
450	1.021	153	7.112	4.26	0.3546
500	1.030	205	7.220	6.20	0.2705
550	1.041	257	7.318	8.74	0.2110
600	1.052	309	7.409	12.01	0.1676
650	1.063	362	7.494	16.12	0.1352
700	1.075	415	7.573	21.24	0.1105
750	1.087	469	7.648	27.54	0.0913
800	1.098	524	7.718	35.21	0.0762
850	1.110	579	7.785	44.45	0.0641
900	1.121	635	7.849	55.50	0.0544
950	1.131	691	7.910	68.61	0.0464
1000	1.141	748	7.968	84.06	0.0399
1050	1.151	805	8.024	102.13	0.0345
1100	1.159	863	8.078	123.15	0.0300

TABLE A.12   *Ideal Gas Properties of Air (Molar)*

T	$\bar{c}_p$	$\bar{h}$	$\bar{s}^\circ$	p_r	v_r
(K)	(kJ/kmol · K)	(kJ/kmol)	(kJ/kmol · K)		
298	29.10	0	193.91	1	1
350	29.17	1510	198.58	1.75	0.6695
400	29.34	2972	202.49	2.80	0.4784
450	29.57	4445	205.95	4.26	0.3546
500	29.84	5930	209.08	6.20	0.2705
550	30.14	7429	211.94	8.74	0.2110
600	30.46	8944	214.58	12.01	0.1676
650	30.79	10475	217.03	16.12	0.1352
700	31.13	12023	219.32	21.24	0.1105
750	31.47	13587	221.48	27.54	0.0913
800	31.80	15169	223.52	35.21	0.0762
850	32.13	16768	225.46	44.45	0.0641
900	32.46	18382	227.31	55.50	0.0544
950	32.77	20013	229.07	68.61	0.0464
1000	33.06	21659	230.76	84.06	0.0399
1050	33.33	23318	232.38	102.13	0.0345
1100	33.57	24991	233.93	123.15	0.0300

TABLE A.13 **Ideal Gas Properties of Methane, CH₄ (Gravimetric)**

T	c_p	h	$s°$		
(K)	(kJ/kg · K)	(kJ/kg)	(kJ/kg · K)	p_r	v_r
298	2.223	−4666	11.606	1	1
350	2.361	−4547	11.973	2.03	0.5786
400	2.520	−4425	12.299	3.80	0.3529
450	2.695	−4295	12.605	6.87	0.2197
500	2.877	−4156	12.899	12.10	0.1386
550	3.061	−4007	13.181	20.88	0.0883
600	3.244	−3849	13.456	35.45	0.0568
650	3.423	−3683	13.722	59.31	0.0368
700	3.595	−3507	13.982	97.95	0.0240
750	3.760	−3323	14.236	159.80	0.0157
800	3.916	−3131	14.484	257.69	0.0104
850	4.064	−2932	14.726	410.95	0.0069
900	4.204	−2725	14.962	648.31	0.0047
950	4.336	−2512	15.193	1012.09	0.0031
1000	4.460	−2292	15.418	1564.05	0.0021
1050	4.579	−2066	15.639	2393.52	0.0015
1100	4.694	−1834	15.855	3628.89	0.0010

TABLE A.14   *Ideal Gas Properties of Methane, CH₄ (Molar)*

T	$\bar{c}_p$	$\bar{h}$	$\bar{s}°$	p_r	v_r
(K)	(kJ/kmol · K)	(kJ/kmol)	(kJ/mol · K)		
298	35.67	− 74848	186.19	1	1
350	37.87	− 72944	192.07	2.03	0.5786
400	40.43	− 70988	197.29	3.80	0.3529
450	43.23	− 68897	202.21	6.87	0.2197
500	46.15	− 66663	206.92	12.10	0.1386
550	49.11	− 64281	211.46	20.88	0.0883
600	52.04	− 61752	215.86	35.45	0.0568
650	54.91	− 59078	220.14	59.31	0.0368
700	57.67	− 56263	224.31	97.95	0.0240
750	60.31	− 53313	228.38	159.80	0.0157
800	62.82	− 50234	232.35	257.69	0.0104
850	65.19	− 47033	236.23	410.95	0.0069
900	67.44	− 43717	240.02	648.31	0.0047
950	69.55	− 40292	243.72	1012.09	0.0031
1000	71.55	− 36764	247.34	1564.05	0.0021
1050	73.46	− 33138	250.88	2393.52	0.0015
1100	75.30	− 29419	254.34	3628.89	0.0010

TABLE A.15   *Ideal Gas Properties of Carbon Dioxide, CO_2 (Molar/Gravimetric)*

Molar				Gravimetric			
T	$\bar{c}_p$	$\bar{h}$	$\bar{s}°$	T	c_p	h	$s°$
(K)	(kJ/kmol · K)	(kJ/kmol)	(kJ/kmol · K)	(K)	(kJ/kg · K)	(kJ/kg)	(kJ/kg · K)
298.15	37.14	−393536	213.66	298.15	0.844	−8944	4.856
350	39.40	−391550	219.80	350	0.895	−8899	4.995
400	41.33	−389531	225.19	400	0.939	−8853	5.118
450	43.06	−387420	230.16	450	0.979	−8805	5.231
500	44.62	−385228	234.78	500	1.014	−8755	5.336
550	46.03	−382961	239.10	550	1.046	−8704	5.434
600	47.31	−380627	243.16	600	1.075	−8651	5.526
650	48.47	−378232	246.99	650	1.102	−8596	5.613
700	49.53	−375782	250.62	700	1.126	−8540	5.696
750	50.50	−373281	254.07	750	1.148	−8484	5.774
800	51.39	−370733	257.36	800	1.168	−8426	5.849
850	52.20	−368143	260.50	850	1.186	−8367	5.921
900	52.95	−365515	263.51	900	1.203	−8307	5.989
950	53.63	−362850	266.39	950	1.219	−8247	6.054
1000	54.26	−360152	269.16	1000	1.233	−8185	6.117
1200	56.32	−349083	279.24	1200	1.280	−7934	6.346
1400	57.81	−337662	288.04	1400	1.314	−7674	6.546
1600	58.91	−325984	295.84	1600	1.339	−7409	6.724
1800	59.72	−314118	302.82	1800	1.357	−7139	6.882
2000	60.34	−302109	309.15	2000	1.371	−6866	7.026
2200	60.84	−289990	314.92	2200	1.383	−6591	7.157
2400	61.25	−277781	320.24	2400	1.392	−6313	7.278
2600	61.60	−265495	325.15	2600	1.400	−6034	7.390
2800	61.93	−253142	329.73	2800	1.407	−5753	7.494
3000	62.23	−240726	334.01	3000	1.414	−5471	7.591

TABLE A.15(cont.)   *Ideal Gas Properties of Carbon Dioxide, CO_2 (Molar/Gravimetric)*

Molar				Gravimetric			
T	$\bar{c}_p$	$\bar{h}$	$\bar{s}°$	T	c_p	h	$s°$
(K)	(kJ/kmol · K)	(kJ/kmol)	(kJ/kmol · K)	(K)	(kJ/kg · K)	(kJ/kg)	(kJ/kg · K)
3200	62.51	− 228252	338.04	3200	1.421	− 5188	7.683
3400	62.76	− 215725	341.83	3400	1.426	− 4903	7.769
3600	62.97	− 203152	345.43	3600	1.431	− 4617	7.851
3800	63.12	− 190542	348.84	3800	1.435	− 4330	7.928
4000	63.20	− 177908	352.08	4000	1.436	− 4043	8.002
4200	63.40	− 165244	355.17	4200	1.441	− 3756	8.072
4400	63.56	− 152548	358.12	4400	1.445	− 3467	8.139
4600	63.72	− 139820	360.95	4600	1.448	− 3178	8.203
4800	63.88	− 127060	363.66	4800	1.452	− 2888	8.265
5000	64.04	− 114268	366.27	5000	1.455	− 2597	8.324
5200	64.21	− 101444	368.79	5200	1.459	− 2306	8.382
5400	64.38	− 88585	371.22	5400	1.463	− 2013	8.437
5600	64.56	− 75692	373.56	5600	1.467	− 1720	8.490
5800	64.75	− 62761	375.83	5800	1.472	− 1426	8.542
6000	64.96	− 49790	378.03	6000	1.476	− 1132	8.592

TABLE A.16 *Ideal Gas Properties of Carbon Monoxide, CO (Molar/ Gravimetric)*

Molar				Gravimetric			
T	$\bar{c}_p$	$\bar{h}$	$\bar{s}°$	T	c_p	h	$s°$
(K)	(kJ/kmol · K)	(kJ/kmol)	(kJ/kmol · K)	(K)	(kJ/kg · K)	(kJ/kg)	(kJ/kg · K)
298.15	29.04	−110523	197.91	298.15	1.037	−3946	7.066
350	29.12	−109016	202.57	350	1.04	−3892	7.232
400	29.32	−107555	206.47	400	1.047	−3840	7.371
450	29.58	−106083	209.94	450	1.056	−3787	7.495
500	29.89	−104596	213.07	500	1.067	−3734	7.607
550	30.22	−103093	215.94	550	1.079	−3681	7.709
600	30.56	−101574	218.58	600	1.091	−3626	7.804
650	30.9	−100038	221.04	650	1.103	−3571	7.891
700	31.24	−98484	223.34	700	1.115	−3516	7.974
750	31.58	−96913	225.51	750	1.127	−3460	8.051
800	31.91	−95326	227.56	800	1.139	−3403	8.124
850	32.23	−93722	229.5	850	1.151	−3346	8.194
900	32.53	−92103	231.35	900	1.161	−3288	8.26
950	32.83	−90469	233.12	950	1.172	−3230	8.323
1000	33.11	−88821	234.81	1000	1.182	−3171	8.383
1200	34.1	−82096	240.94	1200	1.217	−2931	8.602
1400	34.88	−75195	246.26	1400	1.245	−2685	8.792
1600	35.49	−68155	250.96	1600	1.267	−2433	8.959
1800	35.94	−61010	255.16	1800	1.283	−2178	9.11
2000	36.27	−53787	258.97	2000	1.295	−1920	9.246
2200	36.52	−46506	262.44	2200	1.304	−1660	9.369
2400	36.72	−39181	265.62	2400	1.311	−1399	9.483
2600	36.89	−31820	268.57	2600	1.317	−1136	9.588
2800	37.04	−24428	271.31	2800	1.322	−872	9.686
3000	37.19	−17005	273.87	3000	1.328	−607	9.778

TABLE A.16(cont.)   *Ideal Gas Properties of Carbon Monoxide, CO (Molar/Gravimetric)*

Molar				Gravimetric			
T	$\bar{c}_p$	$\bar{h}$	$\bar{s}°$	T	c_p	h	$s°$
(K)	(kJ/kmol · K)	(kJ/kmol)	(kJ/kmol · K)	(K)	(kJ/kg · K)	(kJ/kg)	(kJ/kg · K)
3200	37.34	− 9553	276.27	3200	1.333	− 341	9.863
3400	37.48	− 2071	278.54	3400	1.338	− 74	9.944
3600	37.61	5439	280.69	3600	1.343	194	10.021
3800	37.68	12968	282.72	3800	1.345	463	10.094
4000	37.67	20506	284.66	4000	1.345	732	10.163
4200	37.78	28054	286.50	4200	1.349	1002	10.228
4400	37.86	35618	288.26	4400	1.352	1272	10.291
4600	37.93	43197	289.94	4600	1.354	1542	10.351
4800	38.00	50791	291.56	4800	1.357	1813	10.409
5000	38.07	58398	293.11	5000	1.359	2085	10.464
5200	38.14	66019	294.60	5200	1.362	2357	10.518
5400	38.20	73652	296.05	5400	1.364	2630	10.569
5600	38.26	81298	297.44	5600	1.366	2902	10.619
5800	38.32	88956	298.78	5800	1.368	3176	10.667
6000	38.38	96625	300.08	6000	1.370	3450	10.713

TABLE A.17 *Ideal Gas Properties of Nitrogen, N_2 (Molar/Gravimetric)*

Molar				Gravimetric			
T	$\bar{c}_p$	$\bar{h}$	$\bar{s}°$	T	c_p	h	$s°$
(K)	(kJ/kmol · K)	(kJ/kmol)	(kJ/kmol · K)	(K)	(kJ/kg · K)	(kJ/kg)	(kJ/kg · K)
298.15	29.09	0	191.49	298.15	1.038	0	6.835
350	29.06	1507	196.15	350	1.037	54	7.001
400	29.17	2962	200.03	400	1.041	106	7.140
450	29.36	4425	203.48	450	1.048	158	7.263
500	29.61	5899	206.59	500	1.057	211	7.374
550	29.89	7386	209.42	550	1.067	264	7.475
600	30.19	8888	212.04	600	1.078	317	7.568
650	30.51	10405	214.46	650	1.089	371	7.655
700	30.83	11939	216.74	700	1.100	426	7.736
750	31.15	13488	218.87	750	1.112	481	7.812
800	31.46	15054	220.90	800	1.123	537	7.885
850	31.77	16635	222.81	850	1.134	594	7.953
900	32.08	18231	224.64	900	1.145	651	8.018
950	32.37	19842	226.38	950	1.155	708	8.080
1000	32.65	21468	228.05	1000	1.165	766	8.140
1200	33.66	28102	234.09	1200	1.201	1003	8.356
1400	34.48	34919	239.34	1400	1.231	1246	8.543
1600	35.13	41883	243.99	1600	1.254	1495	8.709
1800	35.62	48960	248.16	1800	1.272	1748	8.858
2000	36.00	56124	251.93	2000	1.285	2003	8.992
2200	36.29	63354	255.38	2200	1.295	2261	9.115
2400	36.51	70634	258.55	2400	1.303	2521	9.229
2600	36.69	77955	261.48	2600	1.310	2783	9.333
2800	36.85	85309	264.20	2800	1.315	3045	9.430
3000	37.01	92695	266.75	3000	1.321	3309	9.521

TABLE A.17(cont.)   *Ideal Gas Properties of Nitrogen, N$_2$ (Molar/ Gravimetric)*

Molar				Gravimetric			
T	$\bar{c}_p$	$\bar{h}$	$\bar{s}°$	T	c_p	h	$s°$
(K)	(kJ/kmol · K)	(kJ/kmol)	(kJ/kmol · K)	(K)	(kJ/kg · K)	(kJ/kg)	(kJ/kg · K)
3200	37.16	100113	269.14	3200	1.327	3573	9.607
3400	37.31	107560	271.40	3400	1.332	3839	9.687
3600	37.45	115037	273.54	3600	1.337	4106	9.764
3800	37.54	122536	275.56	3800	1.340	4374	9.836
4000	37.56	130048	277.49	4000	1.341	4642	9.905
4200	37.67	137575	279.33	4200	1.345	4911	9.970
4400	37.75	145117	281.08	4400	1.347	5180	10.033
4600	37.82	152674	282.76	4600	1.350	5450	10.093
4800	37.90	160246	284.37	4800	1.353	5720	10.150
5000	37.97	167832	285.92	5000	1.355	5991	10.206
5200	38.04	175434	287.41	5200	1.358	6262	10.259
5400	38.12	183050	288.85	5400	1.361	6534	10.310
5600	38.19	190681	290.24	5600	1.363	6806	10.360
5800	38.27	198327	291.58	5800	1.366	7079	10.408
6000	38.35	205990	292.88	6000	1.369	7353	10.454

TABLE A.18 *Ideal Gas Properties of Oxygen, O_2 (Molar/Gravimetric)*

Molar				Gravimetric			
T	$\bar{c}_p$	$\bar{h}$	$\bar{s}°$	T	c_p	h	$s°$
(K)	(kJ/kmol · K)	(kJ/kmol)	(kJ/kmol · K)	(K)	(kJ/kg · K)	(kJ/kg)	(kJ/kg · K)
298.15	29.36	0	205.06	298.15	0.917	0	6.408
350	29.71	1531	209.79	350	0.929	48	6.556
400	30.14	3027	213.78	400	0.942	95	6.681
450	30.61	4546	217.36	450	0.957	142	6.793
500	31.10	6089	220.61	500	0.972	190	6.894
550	31.59	7656	223.60	550	0.987	239	6.987
600	32.07	9248	226.37	600	1.002	289	7.074
650	32.53	10863	228.95	650	1.017	339	7.155
700	32.97	12501	231.38	700	1.030	391	7.231
750	33.38	14159	233.67	750	1.043	442	7.302
800	33.75	15838	235.84	800	1.055	495	7.370
850	34.09	17534	237.89	850	1.065	548	7.434
900	34.38	19246	239.85	900	1.075	601	7.495
950	34.64	20971	241.72	950	1.083	655	7.554
1000	34.85	22709	243.50	1000	1.089	710	7.609
1200	35.65	29769	249.93	1200	1.114	930	7.810
1400	36.27	36963	255.48	1400	1.134	1155	7.984
1600	36.83	44274	260.36	1600	1.151	1384	8.136
1800	37.33	51691	264.72	1800	1.167	1615	8.273
2000	37.80	59205	268.68	2000	1.181	1850	8.396
2200	38.25	66811	272.31	2200	1.195	2088	8.510
2400	38.69	74505	275.65	2400	1.209	2328	8.614
2600	39.12	82287	278.77	2600	1.223	2571	8.711
2800	39.54	90153	281.68	2800	1.236	2817	8.803
3000	39.95	98103	284.42	3000	1.249	3066	8.888

TABLE A.18(cont.) *Ideal Gas Properties of Oxygen, O₂ (Molar/Gravimetric)*

Molar				Gravimetric			
T	$\bar{c}_p$	$\bar{h}$	$\bar{s}°$	T	c_p	h	$s°$
(K)	(kJ/kmol · K)	(kJ/kmol)	(kJ/kmol · K)	(K)	(kJ/kg · K)	(kJ/kg)	(kJ/kg · K)
3200	40.35	106133	287.02	3200	1.261	3317	8.969
3400	40.71	114240	289.47	3400	1.272	3570	9.046
3600	41.05	122417	291.81	3600	1.283	3826	9.119
3800	41.33	130656	294.04	3800	1.292	4083	9.189
4000	41.55	138946	296.16	4000	1.299	4342	9.255
4200	41.81	147284	298.20	4200	1.306	4603	9.319
4400	42.02	155666	300.15	4400	1.313	4865	9.380
4600	42.20	164088	302.02	4600	1.319	5128	9.438
4800	42.36	172544	303.82	4800	1.324	5392	9.494
5000	42.50	181030	305.55	5000	1.328	5657	9.548
5200	42.62	189543	307.22	5200	1.332	5923	9.601
5400	42.73	198078	308.83	5400	1.335	6190	9.651
5600	42.83	206634	310.39	5600	1.338	6457	9.700
5800	42.92	215210	311.89	5800	1.341	6725	9.747
6000	43.01	223803	313.35	6000	1.344	6994	9.792

TABLE A.19 *Ideal Gas Properties of Water Vapor, H_2O (Molar/Gravimetric)*

Molar				Gravimetric			
T	$\bar{c}_p$	$\bar{h}$	$\bar{s}°$	T	c_p	h	$s°$
(K)	(kJ/kmol · K)	(kJ/kmol)	(kJ/kmol · K)	(K)	(kJ/kg · K)	(kJ/kg)	(kJ/kg · K)
298.15	33.57	−241826	188.72	298.15	1.864	−13424	10.476
350	33.86	−240078	194.12	350	1.879	−13327	10.776
400	34.24	−238376	198.67	400	1.901	−13232	11.028
450	34.70	−236653	202.73	450	1.926	−13136	11.253
500	35.21	−234905	206.41	500	1.954	−13039	11.458
550	35.74	−233132	209.79	550	1.984	−12941	11.645
600	36.30	−231331	212.92	600	2.015	−12841	11.819
650	36.88	−229501	215.85	650	2.047	−12739	11.982
700	37.47	−227643	218.61	700	2.080	−12636	12.135
750	38.07	−225754	221.21	750	2.113	−12531	12.279
800	38.68	−223835	223.69	800	2.147	−12425	12.417
850	39.30	−221886	226.05	850	2.182	−12317	12.548
900	39.93	−219905	228.32	900	2.217	−12207	12.674
950	40.57	−217893	230.49	950	2.252	−12095	12.794
1000	41.22	−215848	232.59	1000	2.288	−11982	12.911
1200	43.73	−207355	240.33	1200	2.428	−11510	13.340
1400	45.99	−198377	247.24	1400	2.553	−11012	13.724
1600	47.94	−188980	253.51	1600	2.661	−10490	14.072
1800	49.62	−179220	259.26	1800	2.754	−9948	14.391
2000	51.07	−169147	264.56	2000	2.835	−9389	14.686
2200	52.31	−158807	269.49	2200	2.904	−8815	14.959
2400	53.36	−148237	274.09	2400	2.962	−8229	15.214
2600	54.26	−137472	278.39	2600	3.012	−7631	15.454
2800	55.02	−126541	282.44	2800	3.054	−7024	15.678
3000	55.67	−115471	286.26	3000	3.090	−6410	15.890

TABLE A.19(cont.)   *Ideal Gas Properties of Water Vapor, H₂0 (Molar/Gravimetric)*

	Molar				Gravimetric		
T	$\bar{c}_p$	$\bar{h}$	$\bar{s}°$	T	c_p	h	$s°$
(K)	(kJ/kmol · K)	(kJ/kmol)	(kJ/kmol · K)	(K)	(kJ/kg · K)	(kJ/kg)	(kJ/kg · K)
3200	56.22	−104280	289.87	3200	3.121	−5788	16.091
3400	56.71	−92985	293.30	3400	3.148	−5162	16.281
3600	57.15	−81599	296.55	3600	3.172	−4529	16.461
3800	57.56	−70128	299.65	3800	3.195	−3893	16.634
4000	57.97	−58575	302.62	4000	3.218	−3251	16.798
4200	58.27	−46953	305.45	4200	3.234	−2606	16.955
4400	58.56	−35269	308.17	4400	3.251	−1958	17.106
4600	58.83	−23530	310.78	4600	3.265	−1306	17.251
4800	59.08	−11740	313.29	4800	3.279	−652	17.390
5000	59.31	99	315.70	5000	3.292	5	17.524
5200	59.53	11982	318.03	5200	3.304	665	17.654
5400	59.74	23909	320.28	5400	3.316	1327	17.779
5600	59.94	35877	322.46	5600	3.327	1992	17.900
5800	60.15	47886	324.57	5800	3.339	2658	18.016
6000	60.35	59936	326.61	6000	3.350	3327	18.130

TABLE A.20 *Ideal Gas Properties of Hydrogen, H₂ (Molar/ Gravimetric)*

	Molar				Gravimetric		
T	$\bar{c}_p$	$\bar{h}$	$\bar{s}°$	T	c_p	h	$s°$
(K)	(kJ/kmol · K)	(kJ/kmol)	(kJ/kmol · K)	(K)	(kJ/kg · K)	(kJ/kg)	(kJ/kg · K)
298.15	28.83	0	130.59	298.15	14.298	0	64.777
350	29.07	1501	135.23	350	14.418	745	67.080
400	29.19	2958	139.12	400	14.480	1467	69.009
450	29.25	4420	142.57	450	14.510	2192	70.717
500	29.28	5883	145.65	500	14.524	2918	72.246
550	29.30	7347	148.44	550	14.533	3645	73.631
600	29.32	8813	150.99	600	14.546	4371	74.896
650	29.37	10280	153.34	650	14.567	5099	76.061
700	29.43	11750	155.52	700	14.600	5828	77.142
750	29.53	13224	157.55	750	14.646	6559	78.151
800	29.64	14703	159.46	800	14.703	7293	79.098
850	29.77	16188	161.26	850	14.769	8030	79.991
900	29.92	17680	162.97	900	14.839	8770	80.837
950	30.06	19180	164.59	950	14.909	9514	81.641
1000	30.18	20686	166.13	1000	14.972	10261	82.408
1200	30.95	26794	171.70	1200	15.354	13290	85.168
1400	31.83	33071	176.54	1400	15.787	16404	87.567
1600	32.69	39524	180.84	1600	16.216	19605	89.704
1800	33.52	46145	184.74	1800	16.625	22889	91.638
2000	34.28	52926	188.31	2000	17.003	26253	93.409
2200	34.97	59852	191.61	2200	17.347	29688	95.046
2400	35.59	66910	194.68	2400	17.656	33189	96.569
2600	36.15	74085	197.55	2600	17.932	36749	97.993
2800	36.65	81366	200.25	2800	18.178	40360	99.331
3000	37.09	88740	202.80	3000	18.399	44018	100.593

TABLE A.20(cont.)   *Ideal Gas Properties of Hydrogen, H_2 (Molar/ Gravimetric)*

Molar				Gravimetric			
T	$\bar{c}_p$	$\bar{h}$	$\bar{s}°$	T	c_p	h	$s°$
(K)	(kJ/kmol · K)	(kJ/kmol)	(kJ/kmol · K)	(K)	(kJ/kg · K)	(kJ/kg)	(kJ/kg · K)
3200	37.50	96200	205.20	3200	18.602	47718	101.787
3400	37.89	103740	207.49	3400	18.794	51458	102.921
3600	38.27	111356	209.66	3600	18.984	55236	104.000
3800	38.67	119050	211.74	3800	19.181	59052	105.032
4000	39.10	126826	213.74	4000	19.396	62910	106.021
4200	39.45	134680	215.65	4200	19.569	66806	106.972
4400	39.80	142605	217.50	4400	19.741	70737	107.886
4600	40.14	150599	219.27	4600	19.909	74702	108.767
4800	40.47	158659	220.99	4800	20.073	78700	109.618
5000	40.79	166785	222.65	5000	20.232	82731	110.441
5200	41.10	174974	224.25	5200	20.387	86793	111.237
5400	41.40	183224	225.81	5400	20.537	90885	112.009
5600	41.70	191535	227.32	5600	20.683	95007	112.759
5800	41.98	199903	228.79	5800	20.824	99158	113.487
6000	42.25	208326	230.22	6000	20.959	103336	114.195

TABLE A.21 **Specific Heat at Constant Pressure for Some Common Gases at 25°C and 1 atm**

Gas	$\bar{c}_p$ (kJ/kmol · K)	Molecular Weight
Air	29.105	28.96
Ammonia° (NH_3)	—	17.03
Argon (Ar)	20.785	39.95
Butane (C_4H_{10})	99.39	58.12
Carbon° (C)	—	12.011
Helium (He)	20.770	4.0028
Hydrogen (H_2)	28.855	2.0158
Mercury° (Hg)	—	200.6
Methane (CH_4)	35.67	16.04
Neon (Ne)	20.785	20.18
Nitrogen (N_2)	29.144	28.0134
Oxygen (O_2)	29.395	31.9988
Nitric oxide (NO)	29.864	30.008
Nitrogen dioxide (NO_2)	36.998	46.008
Carbon monoxide (CO)	29.161	28.011
Carbon dioxide (CO_2)	37.153	44.011
Refrigerant-22°	—	86.48
Water vapor° (H_2O)	—	18.0152

Data obtained from the current edition of the *JANAF Thermochemical Data*, Dow Chemical Company, Thermal Research Laboratory, Midland, MI.

°Indicates that substance is not gaseous at 25°C and 1 atm.

TABLE A.22 **Triple Point and Critical Properties of Some Common Substances**

Substance	Critical Properties T_c (K)	Critical Properties p_c (MPa)	Triple Point Temperature (K)	Triple Point Pressure (MPa)
Air	132.00	3.770	—	—
Ammonia (NH_3)	405.36	11.304	195.49	6.060×10^{-3}
Argon (A)	150.73	4.865	83.80	6.895×10^{-2}
Butane ($C_4 10_{10}$)	425.16	3.796	134.86	6.700×10^{-7}
Carbon Dioxide (CO_2)	304.22	7.383	216.59	5.180×10^{-1}
Carbon Monoxide[°] (CO)	133.16	3.498	—	—
Ethane (C_2H_6)	305.33	4.871	90.35	1.130×10^{-6}
Helium (He)	5.20	0.228	1.78	3.043
Hydrogen (H_2)	33.19	1.315	13.95	7.199×10^{-3}
Methane (CH_4)	190.56	4.595	90.68	1.172×10^{-2}
Neon (Ne)	44.45	2.664	24.56	4.338×10^{-2}
Nitrogen (N_2)	126.20	3.400	63.50	1.253×10^{-2}
Oxygen (O_2)	154.58	5.043	54.36	1.460×10^{-4}
Propane (C_3H_8)	369.80	4.242	85.47	3.000×10^{-8}
Refrigerant-12[°°] (CCl_2F_2)	385.15	4.119	117.20	—
Refrigerant-22[°°] ($CHClF_2$)	369.30	4.988	115.74	—
Water Vapor[°°°] (H_2O)	647.15	22.064	273.17	6.117×10^{-4}

Data extracted from: ASHRAE Thermodynamic Properties of Refrigerants, R. Stewart, R. Jacobsen, and
   S. Penoncello, Atlanta, Ga., 1986.
[°]Heat Exchanger Design Handbook, Vol. 5, Physical Properties, D. K. Edwards, etc., Hemisphere
   Publishing, 1984.
[°°]National Standard Reference Data Service of the U.S.S.R.: A Series of Property Tables, T. Selover, Jr.,
   English Editor, Hemisphere Publishing, 1987.
[°°°]Handbook of Chemistry and Physics, 72nd Edition, D. Lide, Editor, CRC Press, 1992.

TABLE A.23(a)   **Enthalpy of Formation and Specific Entropy at Standard Temperature of 25°C and Standard Pressure of 1 atm for Chemical Substances**

Chemical Substance	Chemical Formula	M (kg/kmol)	$(\bar{h}_f^\circ)_{298.15K}$ (MJ/kmol)	$(\bar{s}^\circ)_{298.15K}$ (kJ/kmol · K)
Acetylene (g)	C_2H_2	26.038	226.7	220.98
Benzene (g)	C_6H_6	78.114	82.9	269.20
Butane (g)	C_4H_{10}	58.124	−126.1	310.12
Carbon (graphite) (s) Carbon (Ideal gas) (g)	C	12.011	0 716	5.689 158.09
Carbon dioxide (g)	CO_2	44.011	−393.5	213.82
Carbon monoxide (g)	CO	28.011	−110.5	197.66
n-Dodecane (g) n-Dodecane (l)	$C_{12}H_{26}$	170.34	−291 −352	622.83 490.94
Ethane (g)	C_2H_6	30.068	−84.7	229.49
Ethylene (g)	C_2H_4	28.054	52.4	219.60
Hydrogen (g)	H_2	2.016	0	130.67
Hydroxyl (g)	OH	17.008	39.0	183.88
Methane (g)	CH_4	16.043	−74.9	186.23
Nitrogen (g)	N_2	28.016	0	191.63
n-Octane (g) n-Octane (l)	C_8H_{18}	114.232	−208.4 −250	466.73 360.79
Oxygen (g)	O_2	32.00	0	205.15
n-Pentane (g) n-Pentane (l)	C_5H_{12}	72.151	−146.4 −173	348.95 263.26
n-Propane (g)	C_3H_8	44.097	−103.8	269.91
Water vapor (g) Water (l)	H_2O	18.016	−241.8 −285.8	188.85 70.12

Sources: Data based mostly on the *JANAF Thermochemical Tables* (1965) and in a few instances on data from the *National Bureau of Standards Circular 500* (1961), also the CRC Handbook of *Chemistry and Physics* (1992) and *Lange's Handbook of Chemistry* (1979).

TABLE A.23(b)   *Enthalpy of Combustion and Chemical Exergy for Common Fuels*

Chemical Substance	Chemical Formula	M (kg/kmol)	Combustion in Air $(-\Delta h_c)_{298.15K}$ (MJ/kg) Lower	Higher	Chemical Exergy χ_c (MJ/kg)
Acetylene (g)	C_2H_2	26.038	48.2	49.9	49.0
Benzene (g)	C_6H_6	78.114	40.5	42.3	42.1
Butane (g)	C_4H_{10}	58.124	45.7	49.5	48.5
Carbon (graphite) (s)	C	12.011	32.8	32.8	34.2
Carbon monoxide (g)	CO	28.011	10.1	10.1	9.8
n-Dodecane (g) n-Dodecane (l)	$C_{12}H_{26}$	170.34	44.4 44.2	47.9 47.5	47.4 47.4
Ethane (g)	C_2H_6	30.068	47.4	51.9	49.9
Ethylene (g)	C_2H_4	28.054	47.1	50.3	48.8
Ethynyl (g)	C_2H	25.03	55.3	56.2	55.7
Gasoline (g)	—	—	48.2	—	51.4 (est)
Hydrogen (g)	H_2	2.016	119.8	141.8	116.6
Methane (g)	CH_4	16.043	50.0	55.5	51.8
n-Octane (g) n-Octane (l)	C_8H_{18}	114.232	44.7 44.4	48.3 47.9	47.6 47.6
n-Pentane (g) n-Pentane (l)	C_5H_{12}	72.151	45.3 45.0	49.0 48.6	48.2 48.4
n-Propane (g)	C_3H_8	44.097	46.3	50.3	49.0
Jet fuel (g)	—	—	43.5	—	46.4 (est)

Note: The enthalpies of combustion were computed using the enthalpies of formation in Table A.23(a), while computations of the chemical exergy per unit mass were made using the Rodriguez formulas in Tables 14.1 and 14.2 in Chapter 14 of the text.

TABLE A.23(c)   ***Enthalpy of Reaction of Common Fuels***

Common Combustion Equations	$\Delta \bar{h}_R^{\circ} = \sum_P^{\text{gas}} N_i (\Delta \bar{h}_f^{\circ})_i - \sum_R^{\text{gas}} N_i\ (\Delta \bar{h}_f^{\circ})_i$ **at 298.15 K and 1 MPa** **(MJ, amount of substances in kmol)**
$H_2 + \frac{1}{2} O_2 \rightarrow H_2O$	-241.8
$2\ OH + H_2 \rightarrow 2\ H_2O$	-561.6
$C + H_2 \rightarrow CH_2$	-325.6
$C + O_2 \rightarrow CO_2$	-1109.5
$H_2O_2 \rightarrow H_2O + \frac{1}{2} O_2$	-105.5
$CH_4 + 2\ O_2 \rightarrow CO_2 + 2\ H_2O$	-802.2
$2\ CH + 2\ O_2 \rightarrow CO + CO_2 + H_2O$	-1937.4
$N_2H_4 + O_2 \rightarrow 2\ H_2O + N_2$	-579.0
$N_2H_4 + N_2O_4 \rightarrow 2\ H_2O + 2\ N_2 + O_2$	-588.2
$F_2 + H_2 \rightarrow 2\ HF$	-546.6
$2\ F_2 + N_2H_4 \rightarrow 4\ HF + N_2$	-1188.6

Note: (1) The chemical substances are all assumed to be ideal gases.

   (2) Enthalpies of reaction were computed using the enthalpies of formation in Table A.23(a) and in the *CRC Handbook of Chemistry and Physics* (1992).

TABLE A.24 Equilibrium Constant, ln K_p, for Common Reactions

$$\ln K_p = -\frac{\Delta G^\circ (T)}{\overline{R}T}$$

where

$$\Delta G^\circ = e\overline{g}_E^\circ + f\overline{g}_F^\circ - a\overline{g}_A^\circ - b\overline{g}_B^\circ$$

for reactions of the type

$$a\,A + b\,B \rightleftarrows e\,E + f\,F$$

with the standard pressure, $p^\circ = 1$ atm.

Reaction Equations:

I. $H_2 \rightleftarrows 2\,H$
II. $O_2 \rightleftarrows 2\,O$
III. $N_2 \rightleftarrows 2\,N$
IV. $H_2 + \frac{1}{2}\,O_2 \rightleftarrows H_2O$
V. $2\,H_2O \rightleftarrows H_2 + 2\,OH$
VI. $N_2 + O_2 \rightleftarrows 2\,NO$
VII. $CO_2 \rightleftarrows CO + \frac{1}{2}\,O_2$
VIII. $CO_2 + H_2 \rightleftarrows CO + H_2O$
IX. $C + H_2 \rightleftarrows CH_2$
X. $N \rightleftarrows N^+ + e^-$

Temp (K)	I	II	III	IV	V	VI	VII	VIII	IX	X
200	-250.185	-285.465	-554.57	139.976	-323.019	-105.598	-159.708	-19.729	187.068	—
298	-164.018	-186.979	-367.737	92.214	-212.147	-69.902	-103.824	-11.564	121.912	-565.571
500	-92.844	-105.638	-213.424	52.697	-120.341	-40.45	-57.625	-4.929	67.865	-335.463
800	-53.148	-60.332	-127.789	30.6	-69.136	-24.145	-32.046	-1.446	37.691	-207.699
1000	-39.82	-45.16	-99.162	23.169	-51.965	-18.708	-23.535	-0.366	27.564	-164.945
1200	-30.889	-35.017	-80.04	18.189	-40.48	-15.081	-17.878	0.312	20.79	-136.357
1400	-24.479	-27.755	-66.36	14.618	-32.257	-12.489	-13.848	0.77	15.941	-115.878
1500	-21.91	-24.845	-60.88	13.187	-28.969	-11.453	-12.241	0.945	14.0	-107.668
1600	-19.65	-22.296	-56.084	11.932	-26.081	-10.546	-10.837	1.095	12.3	-100.473
1800	-15.88	-18.042	-48.08	9.836	-21.27	-9.034	-8.502	1.333	9.466	-88.457
2000	-12.854	-14.633	-41.67	8.157	-17.419	-7.824	-6.642	1.514	7.197	-78.817
2200	-10.369	-11.839	-36.418	6.78	-14.267	-6.835	-5.126	1.654	5.341	-70.906
2400	-8.292	-9.508	-32.038	5.632	-11.639	-6.012	-3.867	1.765	3.793	-64.295

TABLE A.24(cont.) **Equilibrium Constant, In K_p, for Common Reactions**

Temp (K)						Reaction				
	I	II	III	IV	V	VI	VII	VIII	IX	X
2500	-7.377	-8.483	-30.109	5.121	-10.482	-5.649	-3.315	1.805	3.113	-61.38
2600	-6.531	-7.533	-28.326	4.659	-9.415	-5.316	-2.805	1.853	2.485	-58.685
2800	-5.017	-5.839	-25.142	3.824	-7.51	-4.72	-1.899	1.924	1.362	-53.863
3000	-3.702	-4.369	-22.38	3.099	-5.859	-4.205	-1.117	1.983	0.39	-49.671
3200	-2.549	-3.082	-19.961	2.466	-4.415	-3.765	-0.435	2.031	-0.462	-45.993
3400	-1.531	-1.946	-17.824	1.905	-3.142	-3.36	0.165	2.07	-1.213	-42.739
3500	-1.064	-1.428	-16.846	1.649	-2.56	-3.179	0.438	2.087	-1.556	-41.248
3600	-0.624	-0.9349	-15.922	1.406	-2.01	-3.009	0.696	2.103	-1.88	-39.838
3800	0.189	-0.0299	-14.219	0.96	-0.998	-2.696	1.17	2.13	-2.479	-37.235
4000	0.921	0.785	-12.684	0.557	-0.087	-2.415	1.594	2.152	-3.017	-34.886
4200	1.584	1.524	-11.292	0.193	0.736	-2.162	1.977	2.17	-3.505	-32.755
4400	2.187	2.192	-10.03	-0.138	1.485	-1.933	2.324	2.185	-3.948	-30.812
4500	2.47	2.506	-9.437	-0.294	1.834	-1.827	2.485	2.191	-4.155	-29.904
4600	2.74	2.805	-8.87	-0.442	2.168	-1.726	2.639	2.197	-4.354	-29.034
4800	3.247	3.366	-7.81	-0.721	2.794	-1.536	2.928	2.206	-4.726	-27.4
5000	3.711	3.885	-6.829	-0.978	3.37	-1.361	3.192	2.213	-5.069	-25.892
5200	4.14	4.361	-5.922	-1.216	3.901	-1.201	3.435	2.218	-5.385	-24.497
5400	4.541	4.803	-5.084	-1.437	4.393	-1.054	3.659	2.222	-5.679	-23.201
5500	4.727	5.014	-4.686	-1.541	4.625	-0.985	3.764	2.223	-5.819	-22.588
5600	4.909	5.218	-4.301	-1.642	4.849	-0.918	3.866	2.224	-5.953	-21.996
5800	5.254	5.80	-3.574	-1.835	5.274	-0.792	4.059	2.225	-6.207	-20.871
6000	5.574	5.955	-2.89	-2.014	5.671	-0.675	4.238	2.224	-6.446	-19.818

All constants derived from the current edition of the *JANAF Thermochemical Tables*, Dow Chemical Company, Thermal Research Laboratory, Midland, MI.

Property Tables and Constants in USCS Units

TABLE B.1   *Saturated Water and Steam Properties (Temperature Table)*

T	p_s	v_f	v_g	u_f	u_g	h_f	h_{fg}	h_g	s_f	s_{fg}	s_g
(°F)	(psia)	(ft³/lbm)		(Btu/lbm)		(Btu/lbm)			(Btu/lbm · °R)		
32	0.08860	0.01602	3303	0	1021	0	1075	1075	0	2.1864	2.1864
35	0.09976	0.01602	2951	3	1022	3	1074	1077	0.0060	2.1699	2.1759
40	0.1215	0.01602	2448	8	1024	8	1071	1079	0.0162	2.1425	2.1587
45	0.1472	0.01602	2041	13	1025	13	1068	1081	0.0263	2.1157	2.1420
50	0.1778	0.01602	1706	18	1027	18	1065	1083	0.0361	2.0894	2.1255
55	0.2136	0.01604	1434	23	1028	23	1062	1085	0.0459	2.0636	2.1095
60	0.2558	0.01604	1209	28	1030	28	1059	1087	0.0554	2.0383	2.0937
65	0.3052	0.01605	1023	33	1032	33	1056	1089	0.0650	2.0134	2.0784
70	0.3626	0.01605	868.8	38	1034	38	1054	1092	0.0745	1.9891	2.0636
75	0.4295	0.01607	740.3	43	1035	43	1051	1094	0.0838	1.9655	2.0493
80	0.5066	0.01607	633.3	48	1037	48	1048	1096	0.0931	1.9419	2.0350
85	0.5958	0.01608	543.4	53	1038	53	1045	1098	0.1025	1.9189	2.0214
90	0.6982	0.01610	467.9	58	1040	58	1042	1100	0.1115	1.8962	2.0077
95	0.8156	0.01612	404.0	63	1042	63	1039	1102	0.1206	1.8740	1.9946
100	0.9498	0.01613	350.0	68	1043	68	1036	1104	0.1295	1.8522	1.9817
110	1.2760	0.01616	265.1	78	1046	78	1031	1109	0.1471	1.8098	1.9569
120	1.6940	0.01621	203.1	88	1049	88	1025	1113	0.1646	1.7686	1.9332
130	2.224	0.01624	157.2	98	1053	98	1019	1117	0.1818	1.7287	1.9105
140	2.891	0.01629	122.9	108	1056	108	1014	1122	0.1985	1.6903	1.8888
150	3.721	0.01634	96.97	118	1059	118	1008	1126	0.2150	1.6530	1.8680
160	4.741	0.01639	77.25	128	1062	128	1002	1130	0.2312	1.6168	1.8480
170	5.991	0.01645	62.05	138	1065	138	996	1134	0.2474	1.5817	1.8291
180	7.505	0.01652	50.25	148	1068	148	990	1138	0.2632	1.5475	1.8107
190	9.329	0.01656	41.00	158	1071	158	984	1142	0.2787	1.5143	1.7930
200	11.51	0.01663	33.67	168	1074	168	978	1146	0.2940	1.4821	1.7761
210	14.10	0.01671	27.86	178	1077	178	972	1150	0.3091	1.4505	1.7596

TABLE B.1(cont.) *Saturated Water and Steam Properties (Temperature Table)*

T	p_s	v_f	v_g	u_f	u_g	h_f	h_{fg}	h_g	s_f	s_{fg}	s_g
(°F)	(psia)	(ft³/lbm)		(Btu/lbm)		(Btu/lbm)			(Btu/lbm · °R)		
212	14.69	0.01671	26.80	180	1077	180	970	1150	0.3122	1.4443	1.7565
220	17.18	0.01677	23.15	188	1080	188	965	1153	0.3241	1.4200	1.7441
230	20.78	0.01685	19.38	198	1083	198	959	1157	0.3389	1.3901	1.7290
240	24.95	0.01692	16.32	209	1085	209	952	1161	0.3535	1.3609	1.7144
250	29.83	0.01700	13.82	218	1088	218	946	1164	0.3678	1.3323	1.7001
260	35.42	0.01709	11.76	229	1090	229	939	1168	0.3819	1.3046	1.6865
270	41.86	0.01717	10.06	239	1093	239	932	1171	0.3960	1.2771	1.6731
280	49.21	0.01725	8.641	249	1095	249	925	1174	0.4099	1.2503	1.6602
290	57.57	0.01735	7.457	259	1098	260	917	1177	0.4237	1.2241	1.6478
300	67.02	0.01745	6.464	270	1100	270	910	1180	0.4371	1.1983	1.6354
310	77.69	0.01756	5.625	280	1102	280	903	1183	0.4507	1.1730	1.6237
320	89.65	0.01765	4.913	290	1104	291	895	1186	0.4641	1.1481	1.6122
330	103.1	0.01777	4.306	301	1106	301	887	1188	0.4772	1.1238	1.6010
340	118.0	0.01788	3.787	311	1108	311	880	1191	0.4904	1.0996	1.5900
350	134.6	0.01799	3.342	321	1110	322	871	1193	0.5032	1.0760	1.5792
360	153.0	0.01810	2.957	332	1111	332	863	1195	0.5161	1.0526	1.5687
370	173.3	0.01823	2.626	342	1113	343	854	1197	0.5290	1.0295	1.5585
380	195.6	0.01836	2.337	353	1114	354	845	1199	0.5417	1.0065	1.5482
390	220.3	0.01850	2.084	364	1115	364	836	1200	0.5541	0.9838	1.5379
400	246.9	0.01865	1.866	374	1117	375	827	1202	0.5668	0.9613	1.5281
410	276.2	0.01879	1.672	385	1117	386	817	1203	0.5792	0.9394	1.5186
420	308.3	0.01894	1.503	396	1118	397	807	1204	0.5916	0.9172	1.5088
430	343.1	0.01910	1.353	407	1119	408	797	1205	0.6038	0.8954	1.4992
440	380.9	0.01926	1.219	417	1119	419	786	1205	0.6162	0.8737	1.4899
450	421.9	0.01943	1.101	429	1119	430	775	1205	0.6284	0.8520	1.4804
460	466.2	0.01961	0.9961	440	1119	442	763	1205	0.6406	0.8305	1.4711

TABLE B.1(cont.) *Saturated Water and Steam Properties (Temperature Table)*

T	p_s	v_f	v_g	u_f	u_g	h_f	h_{fg}	h_g	s_f	s_{fg}	s_g
(°F)	(psia)	(ft³/lbm)		(Btu/lbm)		(Btu/lbm)			(Btu/lbm · °R)		
470	514.0	0.01980	0.9021	451	1119	453	752	1205	0.6528	0.8089	1.4617
480	565.5	0.02001	0.8183	462	1119	464	740	1204	0.6649	0.7875	1.4524
490	620.7	0.02022	0.7432	474	1118	476	727	1203	0.6764	0.7660	1.4424
500	680.0	0.02044	0.6759	485	1117	488	714	1202	0.6886	0.7445	1.4331
520	811.6	0.02091	0.5604	509	1114	512	687	1199	0.7130	0.7012	1.4142
540	961.8	0.02145	0.4660	533	1111	537	656	1193	0.7373	0.6576	1.3949
560	1132	0.02208	0.3883	557	1105	562	625	1187	0.7622	0.6126	1.3748
580	1324	0.02278	0.3225	583	1099	589	589	1178	0.7872	0.5668	1.3540
600	1540	0.02365	0.2675	610	1090	617	549	1166	0.8130	0.5188	1.3318
620	1784	0.02464	0.2208	638	1078	646	505	1151	0.8398	0.4676	1.3074
640	2056	0.02590	0.1804	669	1063	678	454	1132	0.8680	0.4122	1.2802
660	2362	0.02768	0.1446	702	1042	715	390	1105	0.8988	0.3492	1.2480
680	2704	0.03033	0.1115	742	1011	757	310	1067	0.9353	0.2721	1.2074
700	3091	0.03646	0.07545	801	951	822	172	994	0.9891	0.1488	1.1379
705	3197	0.04976	0.04977	868	867	897	0	897	1.0526	0.0000	1.0526

TABLE B.2 **Saturated Water and Steam Properties (Pressure Table)**

p	T_s	v_f	v_g	u_f	u_g	h_f	h_{fg}	h_g	s_f	s_{fg}	s_g
(psia)	(°F)	(ft³/lbm)		(Btu/lbm)		(Btu/lbm)			(Btu/lbm · °R)		
0.1	32	0.01602	3295	0	1021	0	1075	1075	0	2.1862	2.1862
0.5	79.6	0.01607	641.3	48	1037	48	1048	1096	0.0924	1.9440	2.0364
1	101.7	0.01613	333.4	70	1044	70	1035	1105	0.1326	1.8448	1.9774
1.5	115.6	0.01620	227.6	84	1048	84	1027	1111	0.1572	1.7863	1.9435
2	126.0	0.01623	173.7	94	1052	94	1022	1116	0.1748	1.7446	1.9194
2.5	134.4	0.01628	140.8	102	1054	102	1017	1119	0.1892	1.7118	1.9010
3	141.4	0.01631	118.7	110	1056	110	1012	1122	0.2009	1.6848	1.8857
3.5	147.5	0.01632	102.7	116	1058	116	1009	1125	0.2109	1.6621	1.8730
4	152.9	0.01636	90.61	121	1060	121	1006	1127	0.2197	1.6423	1.8620
4.5	157.8	0.01639	81.13	126	1061	126	1003	1129	0.2276	1.6247	1.8523
5	162.2	0.01640	73.50	130	1063	130	1001	1131	0.2348	1.6089	1.8437
6	170.0	0.01645	61.97	138	1065	138	996	1134	0.2474	1.5817	1.8291
7	176.8	0.01648	53.63	145	1067	145	992	1137	0.2582	1.5582	1.8164
8	182.8	0.01653	47.32	151	1069	151	988	1139	0.2675	1.5379	1.8054
9	188.3	0.01656	42.39	156	1071	156	985	1141	0.2761	1.5200	1.7961
10	193.2	0.01660	38.42	161	1072	161	982	1143	0.2837	1.5041	1.7878
15	213.0	0.01672	26.29	181	1078	181	970	1151	0.3136	1.4412	1.7548
20	228.0	0.01684	20.09	196	1082	196	960	1156	0.3358	1.3961	1.7319
25	240.1	0.01692	16.29	209	1085	209	952	1161	0.3535	1.3607	1.7142
30	250.4	0.01701	13.74	219	1088	219	945	1164	0.3683	1.3313	1.6996
35	259.3	0.01708	11.89	228	1090	228	939	1167	0.3810	1.3062	1.6872
40	267.3	0.01714	10.50	236	1092	236	934	1170	0.3922	1.2845	1.6767
45	274.5	0.01721	9.399	243	1094	243	929	1172	0.4022	1.2652	1.6674
50	281.1	0.01727	8.515	250	1095	250	925	1175	0.4113	1.2475	1.6588
55	287.1	0.01733	7.786	256	1097	257	919	1176	0.4197	1.2317	1.6514
60	292.8	0.01738	7.174	262	1098	262	916	1178	0.4273	1.2169	1.6442

TABLE B.2(cont.)   *Saturated Water and Steam Properties (Pressure Table)*

p	T_s	v_f	v_g	u_f	u_g	h_f	h_{fg}	h_g	s_f	s_{fg}	s_g
(psia)	(°F)	(ft³/lbm)		(Btu/lbm)		(Btu/lbm)			(Btu/lbm · °R)		
65	298.0	0.01743	6.653	267	1099	268	912	1180	0.4345	1.2035	1.6380
70	303.0	0.01748	6.205	273	1101	273	908	1181	0.4414	1.1909	1.6323
75	307.6	0.01753	5.815	277	1101	278	904	1182	0.4476	1.1789	1.6265
80	312.1	0.01757	5.471	282	1103	282	902	1184	0.4536	1.1679	1.6215
85	316.3	0.01762	5.166	286	1104	287	898	1185	0.4591	1.1574	1.6165
90	320.3	0.01765	4.896	291	1104	291	895	1186	0.4646	1.1474	1.6120
95	324.2	0.01770	4.652	294	1105	295	892	1187	0.4696	1.1381	1.6077
100	327.9	0.01773	4.431	298	1106	299	889	1188	0.4743	1.1291	1.6034
105	331.4	0.01778	4.231	302	1106	302	887	1189	0.4791	1.1205	1.5996
110	334.8	0.01781	4.048	306	1107	306	884	1190	0.4837	1.1120	1.5957
120	341.3	0.01789	3.728	312	1108	313	878	1191	0.4920	1.0966	1.5886
130	347.4	0.01796	3.454	319	1109	319	873	1192	0.4999	1.0820	1.5819
140	353.1	0.01802	3.220	325	1110	325	868	1193	0.5073	1.0686	1.5759
150	358.5	0.01809	3.015	330	1111	331	864	1195	0.5142	1.0560	1.5702
160	363.6	0.01815	2.834	336	1112	336	860	1196	0.5209	1.0440	1.5649
170	368.5	0.01821	2.674	341	1113	341	855	1196	0.5271	1.0328	1.5599
180	373.1	0.01828	2.531	346	1113	346	851	1197	0.5329	1.0222	1.5551
190	377.6	0.01833	2.403	350	1114	351	847	1198	0.5386	1.0120	1.5506
200	381.9	0.01839	2.288	355	1114	356	843	1199	0.5441	1.0022	1.5463
250	401.0	0.01865	1.844	375	1117	376	826	1202	0.5680	0.9592	1.5272
300	417.4	0.01890	1.543	393	1118	394	809	1203	0.5885	0.9229	1.5114
350	431.8	0.01913	1.326	409	1119	410	795	1205	0.6062	0.8914	1.4976
400	444.7	0.01934	1.161	423	1119	424	781	1205	0.6220	0.8634	1.4854
450	456.4	0.01954	1.032	436	1119	437	768	1205	0.6363	0.8381	1.4744
500	467.1	0.01975	0.9279	448	1119	450	755	1205	0.6494	0.8152	1.4646
600	486.3	0.02014	0.7699	469	1118	472	732	1204	0.6721	0.7739	1.4460

TABLE B.2(cont.)   *Saturated Water and Steam Properties (Pressure Table)*

p	T_s	v_f	v_g	u_f	u_g	h_f	h_{fg}	h_g	s_f	s_{fg}	s_g
(psia)	(°F)	(ft³/lbm)		(Btu/lbm)		(Btu/lbm)			(Btu/lbm · °R)		
700	503.2	0.02051	0.6557	489	1117	491	711	1202	0.6927	0.7375	1.4302
800	518.3	0.02087	0.5690	506	1115	509	690	1199	0.7110	0.7049	1.4159
900	532.1	0.02123	0.5011	523	1112	527	669	1196	0.7278	0.6749	1.4027
1000	544.7	0.02159	0.4463	538	1110	543	649	1192	0.7433	0.6470	1.3903
1100	556.4	0.02195	0.4011	553	1107	557	631	1188	0.7579	0.6210	1.3789
1200	567.3	0.02232	0.3624	567	1103	572	612	1184	0.7715	0.5961	1.3676
1300	577.6	0.02268	0.3297	580	1100	585	594	1179	0.7841	0.5725	1.3566
1400	587.2	0.02307	0.3015	592	1096	598	576	1174	0.7963	0.5498	1.3461
1500	596.3	0.02347	0.2770	605	1092	611	558	1169	0.8080	0.5279	1.3359
1600	605.0	0.02389	0.2550	617	1087	624	539	1163	0.8195	0.5061	1.3256
1700	613.3	0.02429	0.2355	629	1083	636	520	1156	0.8307	0.4851	1.3158
1800	621.2	0.02472	0.2180	640	1077	648	502	1150	0.8415	0.4643	1.3058
1900	628.8	0.02515	0.2023	651	1072	660	483	1143	0.8520	0.4440	1.2960
2000	636.0	0.02562	0.1879	662	1066	672	464	1136	0.8620	0.4237	1.2857
2250	652.8	0.02698	0.1570	690	1050	701	415	1116	0.8876	0.3728	1.2604
2500	668.2	0.02861	0.1309	718	1031	730	362	1092	0.9129	0.3200	1.2329
2750	682.3	0.03076	0.1076	747	1006	763	298	1061	0.9403	0.2613	1.2016
3000	695.5	0.03425	0.08489	784	972	803	215	1018	0.9731	0.1870	1.1601
3198	705.2	0.04977	0.04976	868	867	897	0	897	1.0526	0.0000	1.0526

TABLE B.3   *Compressed Water and Superheated Steam Properties (Saturation Pressure Table)*

		$p = 1$ psia ($T_s = 101.7°F$)		
T	v	u	h	s
(°F)	(ft³/lbm)	(Btu/lbm)	(Btu/lbm)	(Btu/lbm · °R)
32	0.01602	0	0	0
50	0.01604	18	18	0.0361
100	0.01615	68	68	0.1295
101.7 (f)	0.01613	70	70	0.1326
101.7 (g)	333.4	1044	1105	1.9774
150	362.5	1060	1127	2.0137
200	392.5	1077	1150	2.0501
250	422.3	1094	1172	2.0834
300	452.2	1112	1195	2.1145
350	482.0	1129	1218	2.1439
400	511.8	1147	1242	2.1716
450	541.6	1165	1265	2.1980
500	571.4	1183	1288	2.2231
550	601.2	1201	1312	2.2472
600	631.0	1219	1336	2.2704
650	660.8	1238	1360	2.2926
700	690.6	1257	1384	2.3140
750	720.4	1276	1409	2.3347
800	750.2	1295	1434	2.3548
850	780.0	1314	1459	2.3742
900	809.8	1334	1484	2.3931
950	839.5	1354	1509	2.4114
1000	869.3	1374	1535	2.4293

TABLE B.3(cont.) *Compressed Water and Superheated Steam Properties (Saturation Pressure Table)*

T	v	u	h	s
(°F)	(ft³/lbm)	(Btu/lbm)	(Btu/lbm)	(Btu/lbm · °R)
32	0.01602	0	0	0
50	0.01604	18	18	0.0361
100	0.01615	68	68	0.1295
150	0.01634	118	118	0.2150
162.2 (f)	0.01640	130	130	0.2348
162.2 (g)	73.50	1063	1131	1.8437
200	78.14	1076	1148	1.8710
250	84.20	1094	1171	1.9049
300	90.23	1111	1195	1.9364
350	96.24	1129	1218	1.9660
400	102.2	1146	1241	1.9938
450	108.2	1164	1264	2.0203
500	114.2	1182	1288	2.0456
550	120.2	1201	1312	2.0697
600	126.1	1219	1336	2.0929
650	132.1	1238	1360	2.1151
700	138.1	1256	1384	2.1366
750	144.0	1275	1409	2.1573
800	150.0	1295	1434	2.1774
850	156.0	1314	1459	2.1968
900	161.9	1334	1484	2.2157
950	167.9	1354	1509	2.2341
1000	173.8	1374	1535	2.2519

$p = 5$ psia ($T_s = 162.2$°F)

TABLE B.3(cont.)   **Compressed Water and Superheated Steam Properties**
                   (Saturation Pressure Table)

$p = 10$ psia ($T_s = 193.2°F$)				
T	v	u	h	s
(°F)	(ft³/lbm)	(Btu/lbm)	(Btu/lbm)	(Btu/lbm · °R)
32	0.01602	0	0	0
50	0.01604	18	18	0.0360
100	0.01615	68	68	0.1295
150	0.01634	118	118	0.2150
193.2 (f)	0.01660	161	161	0.2837
193.2 (g)	38.42	1072	1143	1.7878
200	38.84	1074	1146	1.7924
250	41.93	1092	1170	1.8270
300	44.99	1110	1194	1.8589
350	48.01	1128	1217	1.8888
400	51.03	1146	1240	1.9169
450	54.04	1164	1264	1.9435
500	57.03	1182	1288	1.9688
550	60.03	1200	1311	1.9930
600	63.02	1219	1335	2.0162
650	66.01	1237	1360	2.0385
700	69.00	1256	1384	2.0600
750	71.98	1275	1409	2.0808
800	74.97	1295	1433	2.1009
850	77.95	1314	1458	2.1203
900	80.93	1334	1484	2.1392
950	83.92	1354	1509	2.1576
1000	86.90	1374	1535	2.1755

TABLE B.3(cont.)   *Compressed Water and Superheated Steam Properties (Saturation Pressure Table)*

\multicolumn{5}{c}{$p = 14.7$ psia ($T_s = 212°F$)}				
T	*v*	*u*	*h*	*s*
(°F)	(ft³/lbm)	(Btu/lbm)	(Btu/lbm)	(Btu/lbm · °R)
32	0.01602	0	0	0
50	0.01604	18	18	0.0360
100	0.01615	68	68	0.1295
150	0.01634	118	118	0.2150
200	0.01662	168	168	0.2940
212 (f)	0.01671	180	180	0.3122
212 (g)	26.79	1077	1150	1.7565
250	28.41	1091	1169	1.7831
300	30.52	1109	1193	1.8155
350	32.60	1127	1216	1.8456
400	34.66	1145	1240	1.8739
450	36.71	1163	1263	1.9006
500	38.76	1182	1287	1.9260
550	40.80	1200	1311	1.9503
600	42.84	1219	1335	1.9736
650	44.88	1237	1359	1.9959
700	46.91	1256	1384	2.0174
750	48.95	1275	1408	2.0382
800	50.98	1294	1433	2.0583
850	53.01	1314	1458	2.0778
900	55.04	1334	1483	2.0967
950	57.07	1354	1509	2.1151
1000	59.10	1374	1535	2.1330

TABLE B.3(cont.)   **Compressed Water and Superheated Steam Properties**
(Saturation Pressure Table)

\multicolumn{5}{c}{$p = 20$ psia ($T_s = 228°\text{F}$)}
T	v	u	h	s
(°F)	(ft³/lbm)	(Btu/lbm)	(Btu/lbm)	(Btu/lbm · °R)
32	0.01602	0	0	0
50	0.01604	18	18	0.0360
100	0.01615	68	68	0.1295
150	0.01634	118	118	0.2150
200	0.01662	168	168	0.2940
228 (f)	0.01684	196	196	0.3358
228 (g)	20.09	1082	1156	1.7319
250	20.79	1090	1167	1.7475
300	22.36	1109	1191	1.7804
350	23.90	1127	1215	1.8108
400	25.43	1145	1239	1.8393
450	26.94	1163	1263	1.8662
500	28.46	1181	1287	1.8917
550	29.96	1200	1311	1.9160
600	31.46	1218	1335	1.9394
650	32.96	1237	1359	1.9617
700	34.46	1256	1383	1.9833
750	35.96	1275	1408	2.0041
800	37.45	1294	1433	2.0242
850	38.95	1314	1458	2.0437
900	40.44	1334	1483	2.0626
950	41.94	1353	1509	2.0810
1000	43.43	1374	1534	2.0990

TABLE B.3(cont.) ***Compressed Water and Superheated Steam Properties (Saturation Pressure Table)***

$p = 40$ psia ($T_s = 267.3°F$)				
T	v	u	h	s
(°F)	(ft³/lbm)	(Btu/lbm)	(Btu/lbm)	(Btu/lbm · °R)
32	0.01602	0	0	0
50	0.01604	18	18	0.0360
100	0.01615	68	68	0.1295
150	0.01634	118	118	0.2150
200	0.01662	168	168	0.2940
250	0.01699	219	219	0.3677
267.3 (f)	0.01714	236	236	0.3922
267.3 (g)	10.50	1092	1170	1.6767
300	11.04	1105	1187	1.6993
350	11.84	1124	1212	1.7311
400	12.62	1143	1236	1.7604
450	13.40	1161	1261	1.7879
500	14.16	1180	1285	1.8138
550	14.93	1198	1309	1.8384
600	15.68	1217	1333	1.8620
650	16.44	1236	1358	1.8845
700	17.19	1255	1382	1.9062
750	17.95	1274	1407	1.9271
800	18.70	1294	1432	1.9473
850	19.45	1313	1457	1.9669
900	20.20	1333	1483	1.9859
950	20.95	1353	1508	2.0043
1000	21.70	1373	1534	2.0223

TABLE B.3(cont.)   *Compressed Water and Superheated Steam Properties*
(Saturation Pressure Table)

		$p = 60$ psia ($T_s = 292.8°F$)		
T	v	u	h	s
(°F)	(ft³/lbm)	(Btu/lbm)	(Btu/lbm)	(Btu/lbm · °R)
32	0.01602	0	0	0
50	0.01604	18	18	0.0360
100	0.01615	68	68	0.1295
150	0.01634	118	118	0.2150
200	0.01662	168	168	0.2940
250	0.01699	218	219	0.3677
292.8 (f)	0.01738	262	262	0.4273
292.8 (g)	7.174	1098	1178	1.6442
300	7.257	1101	1182	1.6496
350	7.816	1121	1208	1.6829
400	8.355	1141	1233	1.7132
450	8.882	1160	1258	1.7413
500	9.401	1178	1283	1.7676
550	9.915	1197	1307	1.7925
600	10.42	1216	1332	1.8163
650	10.93	1235	1357	1.8390
700	11.44	1254	1381	1.8608
750	11.94	1274	1406	1.8818
800	12.45	1293	1431	1.9021
850	12.95	1313	1456	1.9218
900	13.45	1333	1482	1.9408
950	13.95	1353	1507	1.9293
1000	14.45	1373	1533	1.9773

TABLE B.3(cont.)   ***Compressed Water and Superheated Steam Properties***
(Saturation Pressure Table)

\multicolumn{5}{c}{$p = 80$ psia ($T_s = 312.1°F$)}				
T	*v*	*u*	*h*	*s*
(°F)	(ft³/lbm)	(Btu/lbm)	(Btu/lbm)	(Btu/lbm · °R)
32	0.01602	0	0	0
50	0.01604	18	18	0.0360
100	0.01615	68	68	0.1295
150	0.01634	118	118	0.2149
200	0.01662	168	168	0.2940
250	0.01698	218	219	0.3677
300	0.01745	270	270	0.4372
312.1 (f)	0.01757	282	282	0.4536
312.1 (g)	5.471	1103	1184	1.6215
350	5.802	1118	1204	1.6475
400	6.219	1138	1230	1.6788
450	6.623	1158	1256	1.7076
500	7.018	1177	1281	1.7343
550	7.408	1196	1306	1.7596
600	7.795	1215	1330	1.7836
650	8.179	1234	1355	1.8065
700	8.560	1253	380	1.8284
750	8.941	1273	1405	1.8495
800	9.320	1292	1430	1.8699
850	9.698	1312	1456	1.8896
900	10.08	1332	1481	1.9087
950	10.45	1352	1507	1.9272
1000	10.83	1372	1533	1.9453

TABLE B.3(cont.)   ***Compressed Water and Superheated Steam Properties***
(Saturation Pressure Table)

$p = 100$ psia ($T_s = 327.9°F$)				
T	v	u	h	s
(°F)	(ft³/lbm)	(Btu/lbm)	(Btu/lbm)	(Btu/lbm · °R)
32	0.01602	0	0	0
50	0.01604	18	18	0.0360
100	0.01615	68	68	0.1295
150	0.01634	118	118	0.2149
200	0.01662	168	168	0.2939
250	0.01698	218	219	0.3677
300	0.01745	269	270	0.4371
327.9 (f)	0.01773	298	299	0.4743
327.9 (g)	4.431	1106	1188	1.6034
350	4.591	1115	1200	1.6190
400	4.936	1136	1227	1.6515
450	5.266	1156	1253	1.6809
500	5.588	1175	1279	1.7082
550	5.904	1195	1304	1.7338
600	6.217	1214	1329	1.7580
650	6.526	1233	1354	1.7810
700	6.833	1253	1379	1.8031
750	7.139	1272	1404	1.8244
800	7.444	1292	1430	1.8448
850	7.748	1312	1455	1.8646
900	8.050	1331	1480	1.8837
950	8.353	1352	1506	1.9023
1000	8.655	1372	1532	1.9204

TABLE B.3(cont.)   ***Compressed Water and Superheated Steam Properties (Saturation Pressure Table)***

$p = 150$ psia ($T_s = 358.5°F$)				
T	**v**	**u**	**h**	**s**
(°F)	(ft³/lbm)	(Btu/lbm)	(Btu/lbm)	(Btu/lbm · °R)
32	0.01601	0	0	0
50	0.01603	18	18	0.0360
100	0.01615	68	68	0.1295
150	0.01634	118	118	0.2149
200	0.01661	168	168	0.2939
250	0.01698	218	219	0.3676
300	0.01744	269	270	0.4371
350	0.01801	321	322	0.5033
358.5 (f)	0.01809	330	331	0.5142
358.5 (g)	3.015	1111	1195	1.5702
400	3.222	1130	1219	1.5994
450	3.456	1151	1247	1.6308
500	3.680	1172	1274	1.6593
550	3.898	1192	1300	1.6858
600	4.112	1211	1325	1.7106
650	4.322	1231	1351	1.7342
700	4.531	1251	1376	1.7566
750	4.737	1270	1402	1.7781
800	4.943	1290	1427	1.7988
850	5.147	1310	1453	1.8187
900	5.351	1330	1479	1.8380
950	5.554	1351	1505	1.8567
1000	5.757	1371	1531	1.8749

TABLE B.3(cont.)   *Compressed Water and Superheated Steam Properties (Saturation Pressure Table)*

p = 200 psia (T_s = 381.9°F)				
T	v	u	h	s
(°F)	(ft³/lbm)	(Btu/lbm)	(Btu/lbm)	(Btu/lbm · °R)
32	0.01601	0	1	0
50	0.01603	18	19	0.0360
100	0.01615	68	69	0.1294
150	0.01634	118	118	0.2148
200	0.01661	168	169	0.2938
250	0.01698	218	219	0.3675
300	0.01744	269	270	0.4370
350	0.01800	321	322	0.5032
381.9 (f)	0.01839	355	356	0.5441
381.9 (g)	2.288	1114	1199	1.5463
400	2.360	1123	1211	1.5597
450	2.549	1146	1240	1.5935
500	2.725	1168	1268	1.6234
550	2.894	1188	1295	1.6508
600	3.059	1209	1322	1.6764
650	3.220	1229	1348	1.7004
700	3.379	1249	1374	1.7232
750	3.536	1269	1400	1.7450
800	3.692	1289	1425	1.7658
850	3.847	1309	1451	1.7860
900	4.001	1329	1477	1.8054
950	4.155	1349	1503	1.8242
1000	4.308	1370	1529	1.8425

TABLE B.3(cont.)   ***Compressed Water and Superheated Steam Properties (Saturation Pressure Table)***

$p = 300$ psia ($T_s = 417.4°F$)				
T	v	u	h	s
(°F)	(ft³/lbm)	(Btu/lbm)	(Btu/lbm)	(Btu/lbm · °R)
32	0.01600	0	1	0
50	0.01603	18	19	0.0360
100	0.01614	68	69	0.1294
150	0.01633	118	119	0.2147
200	0.01661	168	169	0.2937
250	0.01697	218	219	0.3674
300	0.01743	269	270	0.4368
350	0.01799	321	322	0.5030
400	0.01867	374	375	0.5666
417.4 (f)	0.01890	393	394	0.5885
417.4 (g)	1.543	1118	1203	1.5114
450	1.636	1135	1226	1.5363
500	1.767	1159	1257	1.5698
550	1.889	1181	1286	1.5994
600	2.005	1203	1314	1.6264
650	2.117	1224	1341	1.6515
700	2.227	1244	1368	1.6750
750	2.335	1265	1395	1.6974
800	2.441	1285	1421	1.7187
850	2.547	1306	1447	1.7392
900	2.651	1326	1474	1.7590
950	2.755	1347	1500	1.7780
1000	2.859	1368	1526	1.7965

TABLE B.3(cont.)   **Compressed Water and Superheated Steam Properties**
(Saturation Pressure Table)

		$p = 400$ psia ($T_s = 444.7°F$)		
T	v	u	h	s
(°F)	(ft³/lbm)	(Btu/lbm)	(Btu/lbm)	(Btu/lbm · °R)
32	0.01600	0	1	0
50	0.01602	18	19	0.0360
100	0.01614	68	69	0.1293
150	0.01633	118	119	0.2146
200	0.01660	168	169	0.2936
250	0.01696	218	219	0.3672
300	0.01742	269	270	0.4366
350	0.01798	321	322	0.5027
400	0.01866	374	375	0.5663
444.7 (f)	0.01934	423	424	0.6220
444.7 (g)	1.161	1119	1205	1.4854
450	1.174	1122	1209	1.4898
500	1.285	1150	1245	1.5279
550	1.384	1174	1276	1.5602
600	1.477	1197	1306	1.5889
650	1.565	1219	1335	1.6152
700	1.650	1240	1362	1.6396
750	1.734	1261	1390	1.6626
800	1.816	1282	1417	1.6844
850	1.896	1303	1443	1.7053
900	1.976	1324	1470	1.7253
950	2.055	1345	1497	1.7446
1000	2.134	1366	1524	1.7633

TABLE B.3(cont.)   **Compressed Water and Superheated Steam Properties**
(Saturation Pressure Table)

| \multicolumn{5}{c}{$p = 500$ psia ($T_s = 467.1°F$)} |

T	v	u	h	s
(°F)	(ft³/lbm)	(Btu/lbm)	(Btu/lbm)	(Btu/lbm · °R)
32	0.01599	0	2	0
50	0.01602	18	19	0.0360
100	0.01614	68	69	0.1292
150	0.01633	118	119	0.2145
200	0.01660	168	169	0.2934
250	0.01695	218	220	0.3670
300	0.01741	269	271	0.4364
350	0.01797	321	322	0.5025
400	0.01864	374	375	0.5661
450	0.01944	428	430	0.6282
467.1 (f)	0.01975	448	450	0.6494
467.1 (g)	0.9279	1119	1205	1.4646
500	0.9924	1139	1231	1.4919
550	1.080	1166	1266	1.5274
600	1.159	1191	1298	1.5582
650	1.233	1214	1328	1.5857
700	1.304	1236	1356	1.6111
750	1.373	1257	1385	1.6347
800	1.440	1279	1412	1.6571
850	1.506	1300	1439	1.6783
900	1.571	1321	1467	1.6987
950	1.635	1342	1494	1.7182
1000	1.699	1363	1521	1.7371

TABLE B.3(cont.)   **Compressed Water and Superheated Steam Properties (Saturation Pressure Table)**

p = 750 psia (T_s = 511°F)

T	v	u	h	s
(°F)	(ft³/lbm)	(Btu/lbm)	(Btu/lbm)	(Btu/lbm · °R)
32	0.01597	0	2	0
50	0.01600	18	20	0.0359
100	0.01613	68	70	0.1291
150	0.01632	118	120	0.2143
200	0.01658	167	170	0.2931
250	0.01694	218	220	0.3667
300	0.01738	269	271	0.4360
350	0.01794	320	323	0.5020
400	0.01861	373	376	0.5654
450	0.01940	428	430	0.6273
500	0.02040	485	488	0.6884
511 (f)	0.02070	498	501	0.7020
511 (g)	0.6096	1116	1200	1.4228
550	0.6678	1144	1236	1.4590
600	0.7317	1173	1275	1.4963
650	0.7886	1200	1309	1.5279
700	0.8412	1224	1341	1.5561
750	0.8910	1247	1371	1.5815
800	0.9388	1270	1400	1.6052
850	0.9853	1292	1429	1.6275
900	1.031	1314	1457	1.6487
950	1.075	1336	1485	1.6689
1000	1.119	1358	1513	1.6883

TABLE B.3(cont.)   **Compressed Water and Superheated Steam Properties**
                   (Saturation Pressure Table)

$p = 1000$ psia ($T_s = 544.7°F$)				
T	v	u	h	s
(°F)	(ft³/lbm)	(Btu/lbm)	(Btu/lbm)	(Btu/lbm · °R)
32	0.01596	0	3	0.0001
50	0.01599	18	21	0.0359
100	0.01612	68	71	0.1290
150	0.01631	117	120	0.2141
200	0.01657	167	170	0.2928
250	0.01692	217	221	0.3663
300	0.01736	268	271	0.4355
350	0.01791	320	323	0.5014
400	0.01858	373	376	0.5648
450	0.01937	427	430	0.6265
500	0.02035	484	488	0.6873
544.7 (f)	0.02159	538	543	0.7433
544.7 (g)	0.4463	1110	1192	1.3903
550	0.4536	1114	1198	1.3962
600	0.5139	1153	1248	1.4444
650	0.5639	1184	1289	1.4816
700	0.6082	1212	1325	1.5132
750	0.6491	1237	1357	1.5409
800	0.6876	1261	1388	1.5662
850	0.7245	1284	1419	1.5897
900	0.7602	1307	1448	1.6118
950	0.7951	1330	1477	1.6328
1000	0.8292	1352	1506	1.6528

TABLE B.3(cont.)   **Compressed Water and Superheated Steam Properties (Saturation Pressure Table)**

$p = 1250$ psia ($T_s = 572.5°F$)				
T	v	u	h	s
(°F)	(ft³/lbm)	(Btu/lbm)	(Btu/lbm)	(Btu/lbm · °R)
32	0.01594	0	4	0.0001
50	0.01598	18	22	0.0359
100	0.01611	68	71	0.1288
150	0.01630	117	121	0.2138
200	0.01656	167	171	0.2925
250	0.01690	217	221	0.3659
300	0.01734	268	272	0.4351
350	0.01789	319	324	0.5009
400	0.01855	372	376	0.5641
450	0.01933	426	431	0.6257
500	0.02029	483	488	0.6863
550	0.02170	544	549	0.7486
572.5 (f)	0.02251	574	579	0.7777
572.5 (g)	0.3454	1101	1181	1.3619
600	0.3786	1164	1252	1.4299
650	0.4267	1202	1301	1.4754
700	0.4672	1233	1342	1.5112
750	0.5031	1261	1378	1.5418
800	0.5363	1252	1376	1.5341
850	0.5677	1276	1408	1.5589
900	0.5977	1300	1438	1.5820
950	0.6268	1324	1469	1.6037
1000	0.6551	1347	1498	1.6244

TABLE B.3(cont.)   **Compressed Water and Superheated Steam Properties (Saturation Pressure Table)**

\multicolumn{5}{c}{$p = 1500$ psia ($T_s = 596.3°F$)}				
T	v	u	h	s
(°F)	(ft³/lbm)	(Btu/lbm)	(Btu/lbm)	(Btu/lbm · °R)
32	0.01593	0	5	0.0001
50	0.01597	18	22	0.0358
100	0.01610	68	72	0.1287
150	0.01629	117	122	0.2136
200	0.01655	167	171	0.2922
250	0.01689	217	222	0.3655
300	0.01732	268	272	0.4346
350	0.01786	319	324	0.5003
400	0.01852	371	377	0.5634
450	0.01930	425	431	0.6249
500	0.02024	482	487	0.6852
550	0.02161	542	548	0.7472
596.3 (f)	0.02347	605	611	0.8080
596.3 (g)	0.2770	1092	1169	1.3359
600	0.2835	1212	1291	1.4543
650	0.3328	1263	1355	1.5138
700	0.3718	1299	1402	1.5555
750	0.4051	1330	1443	1.5895
800	0.4351	1242	1362	1.5059
850	0.4629	1268	1396	1.5322
900	0.4892	1293	1429	1.5564
950	0.5144	1317	1460	1.5789
1000	0.5389	1341	1490	1.6002

TABLE B.3(cont.)   **Compressed Water and Superheated Steam Properties (Saturation Pressure Table)**

	$p = 1750$ psia ($T_s = 617.3°F$)			
T	v	u	h	s
(°F)	(ft³/lbm)	(Btu/lbm)	(Btu/lbm)	(Btu/lbm · °R)
32	0.01592	0	5	0.0001
50	0.01595	18	23	0.0358
100	0.01608	67	73	0.1285
150	0.01627	117	122	0.2134
200	0.01653	167	172	0.2919
250	0.01687	217	222	0.3652
300	0.01730	267	273	0.4341
350	0.01784	319	324	0.4998
400	0.01849	371	377	0.5628
450	0.01926	425	431	0.6241
500	0.02019	481	487	0.6842
550	0.02152	541	548	0.7458
600	0.02348	608	616	0.8111
617.3 (f)	0.02449	634	642	0.8360
617.3 (g)	0.2265	1080	1153	1.3105
650	0.2631	1161	1246	1.3965
700	0.3023	1206	1304	1.4472
750	0.3344	1241	1349	1.4853
800	0.3623	1231	1348	1.4802
850	0.3877	1259	1384	1.5082
900	0.4115	1285	1418	1.5336
950	0.4341	1310	1451	1.5570
1000	0.4558	1335	1482	1.5790

TABLE B.3(cont.)   ***Compressed Water and Superheated Steam Properties (Saturation Pressure Table)***

$p = 2000$ psia ($T_s = 636°F$)				
T	*v*	*u*	*h*	*s*
(°F)	(ft³/lbm)	(Btu/lbm)	(Btu/lbm)	(Btu/lbm · °R)
32	0.01590	0	6	0.0001
50	0.01594	18	24	0.0357
100	0.01607	67	73	0.1284
150	0.01626	117	123	0.2131
200	0.01652	167	173	0.2916
250	0.01686	216	223	0.3648
300	0.01729	267	273	0.4337
350	0.01782	318	325	0.4992
400	0.01846	370	377	0.5621
450	0.01923	424	431	0.6233
500	0.02014	480	487	0.6831
550	0.02143	540	547	0.7443
600	0.02330	606	614	0.8090
636 (f)	0.02562	662	672	0.8620
636 (g)	0.1879	1066	1136	1.2857
650	0.2081	1222	1299	1.4371
700	0.2489	1280	1372	1.5012
750	0.2806	1320	1424	1.5449
800	0.3073	1220	1334	1.4561
850	0.3311	1250	1372	1.4860
900	0.3531	1277	1408	1.5127
950	0.3737	1303	1442	1.5371
1000	0.3934	1328	1474	1.5598

TABLE B.4 *Saturated Ice and Steam Properties (Temperature Table)*

T	p_s	v_i	v_g	u_i	u_g	h_i	h_{ig}	h_g	s_i	s_{ig}	s_g
(°F)	(psia)	(ft³/lbm)		(Btu/lbm)		(Btu/lbm)			(Btu/lbm · °R)		
32	0.08862	0.01748	3304	−143	1021	−143	1218	1075	−0.2915	2.4782	2.1867
30	0.08084	0.01748	3608	−144	1020	−144	1218	1074	−0.2936	2.4887	2.1951
20	0.05048	0.01746	5658	−149	1017	−149	1219	1070	−0.3038	2.5417	2.2379
10	0.03089	0.01745	9055	−154	1014	−154	1219	1065	−0.3141	2.5969	2.2828
0	0.01850	0.01743	14800	−159	1010	−159	1220	1061	−0.3243	2.6541	2.3298
−10	0.01083	0.01741	24730	−164	1007	−164	1221	1057	−0.3346	2.7137	2.3791
−20	0.006185	0.01740	42340	−168	1004	−168	1220	1052	−0.3448	2.7757	2.4309
−30	0.003442	0.01738	74350	−173	1000	−173	1221	1048	−0.3551	2.8404	2.4853
−40	0.001863	0.01737	134200	−177	997	−177	1220	1043	−0.3654	2.9080	2.5426
−50	0.000978	0.01737	249400	−181	994	−181	1220	1039	−0.3757	2.9786	2.6029
−60	0.000498	0.01735	478400	−185	990	−185	1220	1035	−0.3860	3.0525	2.6665
−70	0.000245	0.01733	949200	−190	987	−190	1220	1030	−0.3963	3.1299	2.7336
−80	0.000116	0.01732	1.953×10^6	−193	984	−193	1219	1026	−0.4067	3.2112	2.8045
−90	5.26×10^{-5}	0.01732	4.182×10^6	−197	981	−197	1218	1021	−0.4171	3.2967	2.8796
−100	2.29×10^{-5}	0.01730	9.343×10^6	−201	977	−201	1218	1017	−0.4274	3.3865	2.9591
−110	9.53×10^{-6}	0.01729	2.186×10^7	−205	974	−205	1217	1012	−0.4379	3.4813	3.0434
−120	3.76×10^{-6}	0.01729	5.38×10^7	−208	971	−208	1216	1008	−0.4483	3.5814	3.1331
−130	1.40×10^{-6}	0.01727	1.40×10^8	−212	967	−212	1216	1004	−0.4588	3.6873	3.2285
−140	4.93×10^{-7}	0.01725	3.86×10^8	−215	964	−215	1214	999	−0.4693	3.7994	3.3301

TABLE B.5   **Saturated Ice and Steam Properties (Pressure Table)**

p	T_s	v_i	v_g	u_i	u_g	h_i	h_{ig}	h_g	s_i	s_{ig}	s_g
(psia)	(°F)	(ft³/lbm)		(Btu/lbm)		(Btu/lbm)			(Btu/lbm · °R)		
0.0887	32	0.01748	3301	−143	1021	−143	1218	1075	−0.2915	2.4782	2.1867
0.05	19.9	0.01746	5710	−149	1017	−149	1219	1070	−0.3040	2.5425	2.2385
0.01	−11.4	0.01741	26700	−164	1007	−164	1220	1056	−0.3360	2.7221	2.3861
0.005	−23.6	0.01740	51940	−170	1003	−170	1221	1051	−0.3486	2.7989	2.4503
0.001	−49.7	0.01737	244200	−181	994	−181	1220	1039	−0.3754	2.9765	2.6011
5.0×10^{-4}	−60.0	0.01735	476100	−185	990	−185	1220	1035	−0.3860	3.0524	2.6664
1.0×10^{-4}	−81.9	0.01732	2.250×10^6	−194	983	−194	1219	1025	−0.4087	3.2274	2.8187
5.0×10^{-5}	−90.7	0.01730	4.395×10^6	−198	980	−198	1219	1021	−0.4178	3.3028	2.8850
1.0×10^{-5}	−109.1	0.01729	2.088×10^7	−205	974	−205	1218	1013	−0.4369	3.4725	3.0356
5.0×10^{-6}	−116.8	0.01729	4.084×10^7	−207	972	−207	1216	1009	−0.4450	3.5486	3.1036
1.0×10^{-6}	−133.6	0.01727	1.94×10^8	−213	966	−213	1215	1002	−0.4625	3.7263	3.2638
5.0×10^{-7}	−140.1	0.01725	3.81×10^8	−215	964	−215	1214	999	−0.4695	3.8011	3.3316

TABLE B.6 Compressed Ice and Superheated Steam Properties

$p = 0.0887$ psia ($T_s = 32°F$)

T (°F)	v (ft³/lbm)	u (Btu/lbm)	h (Btu/lbm)	s (Btu/lbm · °R)
−100	0.01730	−201	−201	−0.4274
−75	0.01733	−192	−192	−0.4015
−50	0.01737	−181	−181	−0.3757
−25	0.01740	−170	−170	−0.3500
0	0.01743	−159	−159	−0.3243
25	0.01746	−147	−147	−0.2987
32 (i)	0.01748	−143	−143	−0.2915
32 (g)	3300	1021	1075	2.1867
50	3421	1020	1076	2.1890
75	3589	1031	1090	2.2162
100	3757	1041	1103	2.2400
150	4093	1060	1127	2.2811
200	4429	1077	1150	2.3173
250	4765	1095	1173	2.3505
300	5101	1112	1196	2.3815
400	5772	1147	1242	2.4386
500	6444	1183	1288	2.4900

$p = 0.05$ psia ($T_s = 19.9°F$)

T (°F)	v (ft³/lbm)	u (Btu/lbm)	h (Btu/lbm)	s (Btu/lbm · °R)
−100	0.01730	−201	−201	−0.4274
−75	0.01733	−192	−192	−0.4015
−50	0.01737	−181	−181	−0.3757
−25	0.01740	−170	−170	−0.3500
0	0.01743	−159	−159	−0.3243
19.9 (i)	0.01746	−149	−149	−0.3040
19.9 (g)	5710	1017	1070	2.2385
25	5771	1019	1072	2.2436
50	6070	1027	1083	2.2660
75	6368	1035	1094	2.2873
100	6666	1044	1106	2.3077
150	7262	1061	1128	2.3460
200	7857	1078	1150	2.3814
250	8453	1095	1173	2.4145
300	9049	1112	1196	2.4455
400	10240	1147	1242	2.5024
500	11430	1183	1289	2.5538

TABLE B.6(cont.) *Compressed Ice and Superheated Steam Properties*

| | $p = 0.01$ psia ($T_s = -11.4°F$) | | | | | $p = 0.005$ psia ($T_s = -23.6°F$) | | | |
| T | v | u | h | s | T | v | u | h | s |
(°F)	(ft³/lbm)	(Btu/lbm)	(Btu/lbm)	(Btu/lbm · °R)	(°F)	(ft³/lbm)	(Btu/lbm)	(Btu/lbm)	(Btu/lbm · °R)
−100	0.01730	−201	−201	−0.4274	−100	0.01730	−201	−201	−0.4274
−75	0.01733	−192	−192	−0.4015	−75	0.01733	−192	−192	−0.4015
−50	0.01737	−181	−181	−0.3757	−50	0.01737	−181	−181	−0.3757
−25	0.01740	−170	−170	−0.3500	−25	0.01740	−170	−170	−0.3500
−11.4 (i)	0.01741	−164	−164	−0.3360	−23.6 (i)	0.01740	−170	−170	−0.3486
−11.4 (g)	26,700	1007	1056	2.3361	−23.6 (g)	51,940	1003	1051	2.4503
0	27,380	1010	1061	2.3976	0	54,750	1010	1061	2.4741
25	28,860	1019	1072	2.4211	25	57,730	1019	1072	2.4976
50	30,350	1027	1083	2.4435	50	60,710	1027	1083	2.5199
75	31,840	1035	1094	2.4648	75	63,690	1036	1094	2.5412
100	33,330	1044	1106	2.4852	100	66,670	1044	1106	2.5616
150	36,310	1061	1128	2.5234	150	72,620	1061	1128	2.5998
200	39,290	1078	1150	2.5589	200	78,580	1078	1150	2.6353
250	42,270	1095	1173	2.5919	250	84,530	1095	1173	2.6683
300	45,250	1112	1196	2.6229	300	90,490	1112	1196	2.6993
400	51,200	1147	1242	2.6798	400	102,400	1147	1242	2.7562
500	57,160	1183	1289	2.7313	500	114,300	1183	1289	2.8077

TABLE B.6(cont.) *Compressed Ice and Superheated Steam Properties*

	p = 0.001 psia (Ts = −49.7°F)					p = 0.0005 psia (Ts = −60°F)			
T	v	u	h	s	T	v	u	h	s
(°F)	(ft³/lbm)	(Btu/lbm)	(Btu/lbm)	(Btu/lbm · °R)	(°F)	(ft³/lbm)	(Btu/lbm)	(Btu/lbm)	(Btu/lbm · °R)
−100	0.01730	−201	−201	−0.4274	−100	0.01730	−201	−201	−0.4274
−75	0.01733	−192	−192	−0.4015	−75	0.01733	−192	−192	−0.4015
−50	0.01737	−181	−181	−0.3757	−60(i)	0.01735	−185	−185	−0.3860
−49.7 (i)	0.01737	−181	−181	−0.3754	−60 (g)	476,100	990	1035	2.6664
−49.7 (g)	244,200	994	1039	2.6011	−50	488,000	994	1039	2.6769
−25	258,900	1002	1050	2.6267	−25	517,800	1002	1050	2.7031
0	273,800	1010	1061	2.6515	0	547,500	1010	1061	2.7279
25	288,700	1019	1072	2.6750	25	577,300	1019	1072	2.7514
50	303,600	1027	1083	2.6973	50	607,100	1027	1083	2.7737
75	318,400	1036	1094	2.7186	75	636,900	1036	1094	2.7950
100	333,300	1044	1106	2.7390	100	666,700	1044	1106	2.8154
150	363,100	1061	1128	2.7773	150	726,200	1061	1128	2.8537
200	392,900	1078	1150	2.8127	200	785,800	1078	1150	2.8891
250	422,700	1095	1173	2.8457	250	845,300	1095	1173	2.9221
300	452,500	1112	1196	2.8767	300	904,900	1112	1196	2.9531
400	512,000	1147	1242	2.9336	400	1,024,000	1147	1242	3.0100
500	571,600	1183	1289	2.9851	500	1,143,000	1183	1289	3.0615

TABLE B.6(cont.) Compressed Ice and Superheated Steam Properties

	p = 0.0001 psia (T_s = −81.9°F)			
T	v	u	h	s
(°F)	(ft³/lbm)	(Btu/lbm)	(Btu/lbm)	(Btu/lbm · °R)
−100	0.01730	−201	−201	−0.4274
−81.9 (i)	0.01732	−194	−194	−0.4087
−81.9 (g)	2,250,000	983	1025	2.8187
−75	2,291,000	986	1028	2.8264
−50	2,440,000	994	1039	2.8543
−25	2,589,000	1002	1050	2.8805
0	2,738,000	1010	1061	2.9053
25	2,887,000	1019	1072	2.9288
50	3,036,000	1027	1083	2.9511
75	3,184,000	1036	1094	2.9724
100	3,333,000	1044	1106	2.9928
150	3,631,000	1061	1128	3.0311
200	3,929,000	1078	1150	3.0665
250	4,227,000	1095	1173	3.0995
300	4,525,000	1112	1196	3.1305
400	5,120,000	1147	1242	3.1874
500	5,716,000	1183	1289	3.2389

	p = 0.00005 psia (T_s = −90.7°F)			
T	v	u	h	s
(°F)	(ft³/lbm)	(Btu/lbm)	(Btu/lbm)	(Btu/lbm · °R)
−100	0.0173	−201	−201	−0.4274
−90.7 (i)	0.0173	−198	−198	−0.4178
−90.7 (g)	4,395,000	980	1021	2.8850
−75	4,582,000	986	1028	2.9029
−50	4,880,000	994	1039	2.9307
−25	5,178,000	1002	1050	2.9569
0	5,476,000	1010	1061	2.9817
25	5,773,000	1019	1072	3.0052
50	6,071,000	1027	1083	3.0276
75	6,369,000	1036	1094	3.0489
100	6,667,000	1044	1106	3.0692
150	7,262,000	1061	1128	3.1075
200	7,858,000	1078	1150	3.1429
250	8,453,000	1095	1173	3.1759
300	9,049,000	1112	1196	3.2069
400	10,240,000	1147	1242	3.2638
500	11,430,000	1183	1289	3.3153

TABLE B.7   *Saturated Refrigerant-22 Properties (Temperature Table)*

T	p_s	v_f	v_g	h_f	h_{fg}	h_g	s_f	s_{fg}	s_g
(°F)	(psia)	(ft³/lbm)		(Btu/lbm)			(Btu/lbm · °R)		
−75	5.61	0.0110	8.335	−9.31	105.37	96.06	−0.0231	0.2739	0.2508
−70	6.56	0.0110	7.210	−7.98	104.63	96.65	−0.0197	0.2685	0.2488
−65	7.63	0.0111	6.263	−6.65	103.90	97.25	−0.0163	0.2632	0.2469
−60	8.84	0.0111	5.462	−5.33	103.16	97.83	−0.0130	0.2581	0.2451
−55	10.20	0.0112	4.782	−4.00	102.42	98.42	−0.0097	0.2531	0.2434
−50	11.72	0.0112	4.202	−2.67	101.67	99.00	−0.0064	0.2481	0.2417
−45	13.41	0.0113	3.704	−1.34	100.91	99.57	−0.0032	0.2433	0.2401
−40	15.29	0.0114	3.277	0	100.14	100.14	0	0.2386	0.2386
−35	17.37	0.0114	2.908	1.34	99.36	100.70	0.0032	0.2339	0.2371
−30	19.66	0.0115	2.589	2.68	98.57	101.25	0.0063	0.2294	0.2357
−25	22.19	0.0116	2.311	4.03	97.77	101.80	0.0094	0.2249	0.2343
−20	24.96	0.0116	2.069	5.37	96.97	102.34	0.0124	0.2206	0.233
−15	27.99	0.0117	1.858	6.72	96.16	102.88	0.0155	0.2162	0.2317
−10	31.29	0.0118	1.672	8.07	95.33	103.40	0.0184	0.2121	0.2305
−5	34.89	0.0118	1.509	9.41	94.51	103.92	0.0214	0.2078	0.2292
0	38.79	0.0119	1.365	10.77	93.65	104.42	0.0243	0.2038	0.2281
5	43.02	0.0120	1.237	12.13	92.79	104.92	0.0273	0.1996	0.2269
10	47.58	0.0121	1.124	13.50	91.91	105.41	0.0302	0.1956	0.2258
15	52.51	0.0122	1.024	14.89	90.99	105.88	0.0331	0.1917	0.2248
20	57.80	0.0123	0.934	16.27	90.08	106.35	0.0360	0.1877	0.2237
25	63.49	0.0123	0.853	17.66	89.15	106.81	0.0388	0.1839	0.2227
30	69.58	0.0124	0.781	19.05	88.21	107.26	0.0416	0.1802	0.2218
35	76.25	0.0125	0.714	20.44	87.24	107.68	0.0444	0.1764	0.2208
40	83.27	0.0126	0.656	21.84	86.27	108.11	0.0472	0.1726	0.2198
45	90.76	0.0127	0.603	23.26	85.26	108.52	0.0500	0.1689	0.2189
50	98.75	0.0128	0.555	24.69	84.23	108.92	0.0528	0.1652	0.2180

TABLE B.7(cont.)   *Saturated Refrigerant-22 Properties (Temperature Table)*

T	p_s	v_f	v_g	h_f	h_{fg}	h_g	s_f	s_{fg}	s_g
(°F)	(psia)	(ft³/lbm)		(Btu/lbm)			(Btu/lbm · °R)		
55	107.2	0.0129	0.511	26.13	83.17	109.30	0.0555	0.1616	0.2171
60	116.3	0.0130	0.472	27.58	82.10	109.68	0.0583	0.1580	0.2163
65	125.9	0.0131	0.436	29.03	81.02	110.05	0.0610	0.1545	0.2155
70	136.1	0.0133	0.403	30.51	79.89	110.40	0.0638	0.1508	0.2146
75	146.8	0.0134	0.373	32.02	78.76	110.78	0.0666	0.1473	0.2139
80	158.2	0.0135	0.346	33.50	77.58	111.08	0.0693	0.1438	0.2131
85	170.2	0.0136	0.320	34.99	76.38	111.37	0.0720	0.1402	0.2122
90	182.9	0.0137	0.297	36.49	75.13	111.62	0.0747	0.1367	0.2114
95	196.2	0.0139	0.276	38.02	73.85	111.87	0.0774	0.1332	0.2106
100	210.3	0.0140	0.257	39.56	72.54	112.10	0.0801	0.1296	0.2097
105	225.1	0.0142	0.238	41.11	71.20	112.31	0.0828	0.1261	0.2089
110	240.6	0.0143	0.222	42.68	69.78	112.46	0.0855	0.1225	0.2080
115	257.0	0.0145	0.206	44.28	68.33	112.61	0.0882	0.1189	0.2071
120	274.2	0.0147	0.192	45.90	66.84	112.74	0.0909	0.1153	0.2062
125	292.1	0.0149	0.178	47.54	65.28	112.82	0.0936	0.1117	0.2053
130	311.0	0.0151	0.166	49.22	63.65	112.87	0.0964	0.1079	0.2043
135	330.7	0.0153	0.154	50.83	61.94	112.77	0.099	0.1042	0.2032
140	351.3	0.0155	0.144	52.44	60.16	112.60	0.1016	0.1003	0.2019
145	373.1	0.0158	0.133	54.10	58.27	112.37	0.1043	0.0963	0.2006
150	395.8	0.0160	0.124	55.92	56.24	112.16	0.1072	0.0922	0.1994
155	419.4	0.0163	0.115	57.83	54.16	111.99	0.1102	0.0881	0.1983
160	444.1	0.0166	0.106	59.86	52.01	111.87	0.1135	0.0839	0.1974

TABLE B.8 **Saturated Refrigerant-22 Properties (Pressure Table)**

p	T_s	v_f	v_g	h_f	h_{fg}	h_g	s_f	s_{fg}	s_g
(psia)	(°F)	(ft³/lbm)		(Btu/lbm)			(Btu/lbm · °R)		
7.5	−65.59	0.0111	6.364	−6.81	103.99	97.18	−0.0167	0.2638	0.2471
10	−55.68	0.0112	4.870	−4.18	102.52	98.34	−0.0101	0.2537	0.2436
12.5	−47.63	0.0113	3.955	−2.04	101.31	99.27	−0.0049	0.2459	0.2410
15	−40.74	0.0113	3.336	−0.20	100.25	100.05	−0.0005	0.2393	0.2388
15.29	−40.00	0.0114	3.276	0	100.14	100.14	0	0.2386	0.2386
17.5	−34.69	0.0114	2.888	1.42	99.31	100.73	0.0033	0.2337	0.2370
20	−29.31	0.0115	2.548	2.87	98.46	101.33	0.0067	0.2288	0.2355
22.5	−24.43	0.0116	2.281	4.18	97.68	101.86	0.0097	0.2245	0.2342
25	−19.93	0.0116	2.066	5.39	96.96	102.35	0.0125	0.2205	0.2330
27.5	−15.78	0.0117	1.889	6.51	96.28	102.79	0.0150	0.2169	0.2319
30	−11.91	0.0117	1.740	7.55	95.65	103.20	0.0173	0.2136	0.2309
35	−4.84	0.0118	1.504	9.45	94.48	103.93	0.0215	0.2077	0.2292
40	1.48	0.0119	1.326	11.18	93.39	104.57	0.0252	0.2025	0.2277
45	7.21	0.012	1.186	12.74	92.40	105.14	0.0286	0.1979	0.2265
50	12.52	0.0121	1.073	14.20	91.45	105.65	0.0317	0.1936	0.2253
60	21.98	0.0123	0.901	16.83	89.70	106.53	0.0371	0.1862	0.2233
70	30.31	0.0124	0.776	19.14	88.14	107.28	0.0418	0.1799	0.2217
80	37.70	0.0126	0.682	21.19	86.72	107.91	0.0459	0.1743	0.2202
90	44.52	0.0127	0.608	23.12	85.36	108.48	0.0497	0.1693	0.2190
100	50.77	0.0128	0.548	24.91	84.07	108.98	0.0532	0.1647	0.2179
110	56.54	0.0130	0.499	26.57	82.85	109.42	0.0564	0.1605	0.2169
120	61.95	0.0131	0.457	28.14	81.68	109.82	0.0594	0.1566	0.2160
130	67.09	0.0132	0.422	29.64	80.56	110.20	0.0622	0.1529	0.2151
140	71.87	0.0133	0.391	31.06	79.47	110.53	0.0648	0.1496	0.2144
150	76.44	0.0134	0.365	32.45	78.42	110.87	0.0674	0.1463	0.2137
160	80.80	0.0135	0.342	33.74	77.40	111.14	0.0698	0.1432	0.2130

TABLE B.8(cont.)   *Saturated Refrigerant-22 Properties (Pressure Table)*

p	T_s	v_f	v_g	h_f	h_{fg}	h_g	s_f	s_{fg}	s_g
(psia)	(°F)	(ft³/lbm)		(Btu/lbm)			(Btu/lbm · °R)		
170	84.95	0.0136	0.321	34.98	76.39	111.37	0.0720	0.1403	0.2123
180	88.88	0.0137	0.302	36.16	75.41	111.57	0.0741	0.1375	0.2116
190	92.68	0.0138	0.286	37.30	74.45	111.75	0.0762	0.1347	0.2109
200	96.41	0.0139	0.271	38.45	73.49	111.94	0.0782	0.1321	0.2103
225	105.0	0.0142	0.239	41.11	71.20	112.31	0.0828	0.1261	0.2089
250	112.9	0.0144	0.212	43.59	68.95	112.54	0.0870	0.1205	0.2075
275	120.2	0.0147	0.191	45.98	66.77	112.75	0.0910	0.1152	0.2062
300	127.1	0.0150	0.173	48.27	64.60	112.87	0.0948	0.1101	0.2049
325	133.5	0.0152	0.158	50.34	62.46	112.80	0.0982	0.1053	0.2035
350	139.6	0.0155	0.144	52.32	60.28	112.60	0.1014	0.1006	0.2020
375	145.4	0.0158	0.132	54.24	58.11	112.35	0.1045	0.0960	0.2005
400	150.9	0.0161	0.122	56.25	55.87	112.12	0.1077	0.0915	0.1992
425	156.2	0.0164	0.113	58.31	53.66	111.97	0.1110	0.0871	0.1981
450	161.2	0.0167	0.104	60.35	51.50	111.85	0.1142	0.0830	0.1972

TABLE B.9   *Superheated Refrigerant-22 Properties (Saturation Pressure Table)*

$p_s = 5$ psia ($T_s = -76°F$)			$p_s = 10$ psia ($T_s = -55.7°F$)		
T	h	s	T	h	s
(°F)	(Btu/lbm)	(Btu/lbm · °R)	(°F)	(Btu/lbm)	(Btu/lbm · °R)
$h_f = -9.58$ Btu/lbm; $s_f = -0.0238$ Btu/lbm · °R			$h_f = -4.18$ Btu/lbm; $s_f = 0.0101$ Btu/lbm · °R		
$h_g = 95.94$ Btu/lbm; $s_g = 0.2512$ Btu/lbm · °R			$h_g = 98.34$ Btu/lbm; $s_g = 0.2436$ Btu/lbm · °R		
−75	96.08	0.2515			
−50	99.53	0.2603	−50	99.15	0.2456
−25	103.07	0.2686	−25	102.74	0.2541
0	106.68	0.2767	0	106.40	0.2623
25	110.37	0.2845	25	110.13	0.2702
50	114.15	0.2921	50	113.93	0.2779
75	118.01	0.2995	75	117.82	0.2853
100	121.97	0.3067	100	121.79	0.2926
125	126.01	0.3138	125	125.86	0.2997
150	130.14	0.3207	150	130.00	0.3066
175	134.36	0.3275	175	134.24	0.3134
200	138.67	0.3342	200	138.55	0.3201
225	143.06	0.3407	225	142.95	0.3266
250	147.54	0.3471	250	147.44	0.3331
275	152.10	0.3534	275	152.00	0.3394
300	156.73	0.3597	300	156.65	0.3456

TABLE B.9(cont.)   **Superheated Refrigerant-22 Properties (Saturation Pressure Table)**

\multicolumn{3}{c}{$p = 15$ psia ($T_s = -40.7°F$)}			\multicolumn{3}{c}{$p = 20$ psia ($T_s = -29.3°F$)}		
T	h	s	T	h	s
(°F)	(Btu/lbm)	(Btu/lbm · °R)	(°F)	(Btu/lbm)	(Btu/lbm · °R)
−40.7 (f)	−0.200	−0.0005	−29.3 (f)	2.87	0.0067
−40.7 (g)	100.05	0.2388	−29.3 (g)	101.33	0.2355
−25	102.36	0.2442	−25	101.98	0.2370
0	106.08	0.2525	0	105.75	0.2454
25	109.85	0.2605	25	109.57	0.2535
50	113.70	0.2683	50	113.45	0.2613
75	117.61	0.2758	75	117.40	0.2689
100	121.61	0.2831	100	121.41	0.2762
125	125.69	0.2902	125	125.51	0.2834
150	129.85	0.2972	150	129.69	0.2904
175	134.09	0.3040	175	133.95	0.2972
200	138.42	0.3107	200	138.29	0.3039
225	142.83	0.3172	225	142.71	0.3105
250	147.33	0.3237	250	147.21	0.3170
275	151.9	0.33	275	151.8	0.3233
300	156.55	0.3363	300	156.46	0.3296
\multicolumn{3}{c}{$p = 25$ psia ($T_s = -19.9°F$)}			\multicolumn{3}{c}{$p = 30$ psia ($T_s = -11.9°F$)}		
T	h	s	T	h	s
(°F)	(Btu/lbm)	(Btu/lbm · °R)	(°F)	(Btu/lbm)	(Btu/lbm · °R)
−19.9 (f)	5.39	0.0125	−11.9 (f)	7.55	0.0173
−19.9 (g)	102.35	0.2330	−11.9 (g)	103.20	0.2309
0	105.41	0.2398	0	105.06	0.2350
25	109.28	0.2480	25	108.99	0.2433
50	113.20	0.2559	50	112.95	0.2513

TABLE B.9(cont.)   ***Superheated Refrigerant-22 Properties (Saturation Pressure Table)***

	$p = 25$ psia ($T_s = -19.9°F$)			$p = 30$ psia ($T_s = -11.9°F$)	
T	h	s	T	h	s
(°F)	(Btu/lbm)	(Btu/lbm · °R)	(°F)	(Btu/lbm)	(Btu/lbm · °R)
75	117.18	0.2635	75	116.96	0.2590
100	121.22	0.2709	100	121.02	0.2664
125	125.34	0.2781	125	125.16	0.2737
150	129.53	0.2851	150	129.38	0.2807
175	133.81	0.2920	175	133.66	0.2876
200	138.16	0.2987	200	138.03	0.2943
225	142.59	0.3053	225	142.47	0.3010
250	147.1	0.3117	250	146.99	0.3074
275	151.69	0.3181	275	151.59	0.3138
300	156.36	0.3244	300	156.27	0.3201

	$p = 40$ psia ($T_s = 1.5°F$)			$p = 50$ psia ($T_s = 12.5°F$)	
T	h	s	T	h	s
(°F)	(Btu/lbm)	(Btu/lbm · °R)	(°F)	(Btu/lbm)	(Btu/lbm · °R)
1.5 (f)	11.18	0.0252	12.5 (f)	14.20	0.0317
1.5 (g)	104.57	0.2277	12.5 (g)	105.65	0.2253
25	108.37	0.2358	25	107.73	0.2297
50	112.42	0.2439	50	111.87	0.2380
75	116.50	0.2517	75	116.03	0.2460
100	120.62	0.2593	100	120.21	0.2536
125	124.81	0.2666	125	124.44	0.2610
150	129.06	0.2737	150	128.73	0.2682
175	133.37	0.2806	175	133.08	0.2752
200	137.76	0.2874	200	137.49	0.2820

TABLE B.9(cont.)   **Superheated Refrigerant-22 Properties (Saturation Pressure Table)**

\multicolumn{3}{c}{p = 40 psia (Ts = 1.5°F)}			\multicolumn{3}{c}{p = 50 psia (Ts = 12.5°F)}		
T	*h*	*s*	*T*	*h*	*s*
(°F)	(Btu/lbm)	(Btu/lbm · °R)	(°F)	(Btu/lbm)	(Btu/lbm · °R)
225	142.23	0.2941	225	141.98	0.2887
250	146.77	0.3006	250	146.54	0.2952
275	151.39	0.3070	275	151.18	0.3016
300	156.08	0.3133	300	155.89	0.3079

\multicolumn{3}{c}{p = 60 psia (Ts = 22°F)}			\multicolumn{3}{c}{p = 70 psia (Ts = 30.3°F)}		
T	*h*	*s*	*T*	*h*	*s*
(°F)	(Btu/lbm)	(Btu/lbm · °R)	(°F)	(Btu/lbm)	(Btu/lbm · °R)
22 (f)	16.83	0.0371	30.3 (f)	19.14	0.0418
22 (g)	106.53	0.2233	30.3 (g)	107.28	0.2217
25	107.05	0.2244			
50	111.31	0.2330	50	110.72	0.2286
75	115.55	0.2411	75	115.05	0.2369
100	119.79	0.2489	100	119.37	0.2448
125	124.07	0.2563	125	123.70	0.2523
150	128.40	0.2636	150	128.07	0.2596
175	132.78	0.2706	175	132.48	0.2667
200	137.22	0.2775	200	136.95	0.2737
225	141.74	0.2842	225	141.49	0.2804
250	146.32	0.2908	250	146.09	0.2870
275	150.97	0.2972	275	150.76	0.2935
300	155.69	0.3035	300	155.50	0.2998

TABLE B.9(cont.)   **Superheated Refrigerant-22 Properties (Saturation Pressure Table)**

\multicolumn{3}{c}{$p = 80$ psia ($T_s = 37.7°F$)}			\multicolumn{3}{c}{$p = 90$ psia ($T_s = 44.5°F$)}		
T	h	s	T	h	s
(°F)	(Btu/lbm)	(Btu/lbm · °R)	(°F)	(Btu/lbm)	(Btu/lbm · °R)
37.7 (f)	21.19	0.0459	44.5 (f)	23.12	0.0497
37.7 (g)	107.91	0.2202	44.5 (g)	108.48	0.2190
50	110.11	0.2246	50	109.48	0.2210
75	114.54	0.2331	75	114.01	0.2297
100	118.93	0.2411	100	118.48	0.2378
125	123.32	0.2488	125	122.93	0.2456
150	127.73	0.2562	150	127.39	0.2531
175	132.18	0.2633	175	131.87	0.2603
200	136.68	0.2703	200	136.40	0.2673
225	141.24	0.2771	225	140.99	0.2741
250	145.86	0.2837	250	145.64	0.2808
275	150.55	0.2902	275	150.34	0.2873
300	155.31	0.2966	300	155.11	0.2937

\multicolumn{3}{c}{$p = 100$ psia ($T_s = 50.8°F$)}			\multicolumn{3}{c}{$p = 150$ psia ($T_s = 76.4°F$)}		
T	h	s	T	h	s
(°F)	(Btu/lbm)	(Btu/lbm · °R)	(°F)	(Btu/lbm)	(Btu/lbm · °R)
50.8 (f)	24.91	0.0532	76.4 (f)	32.45	0.0674
50.8 (g)	108.98	0.2179	76.4 (g)	110.87	0.2137
75	113.47	0.2265			
100	118.02	0.2348	100	115.62	0.2223
125	122.53	0.2427	125	120.49	0.2309
150	127.04	0.2502	150	125.26	0.2388
175	131.56	0.2575	175	129.99	0.2464
200	136.13	0.2646	200	134.72	0.2538

TABLE B.9(cont.)   *Superheated Refrigerant-22 Properties (Saturation Pressure Table)*

\multicolumn{3}{c}{p = 100 psia (Tₛ = 50.8°F)}			\multicolumn{3}{c}{p = 150 psia (Tₛ = 76.4°F)}		
T	h	s	T	h	s
(°F)	(Btu/lbm)	(Btu/lbm · °R)	(°F)	(Btu/lbm)	(Btu/lbm · °R)
225	140.74	0.2714	225	139.47	0.2608
250	145.41	0.2781	250	144.24	0.2677
275	150.13	0.2847	275	149.06	0.2743
300	154.92	0.2911	300	153.92	0.2808

\multicolumn{3}{c}{p = 200 psia (Tₛ = 96.4°F)}			\multicolumn{3}{c}{p = 250 psia (Tₛ = 112.9°F)}		
T	h	s	T	h	s
(°F)	(Btu/lbm)	(Btu/lbm · °R)	(°F)	(Btu/lbm)	(Btu/lbm · °R)
96.4 (f)	38.45	0.0782	112.9 (f)	43.59	0.0870
96.4 (g)	111.94	0.2103	112.9 (g)	112.54	0.2075
100	112.74	0.2118			
125	118.12	0.2212	125	115.46	0.2125
150	123.25	0.2298	150	121.09	0.2219
175	128.23	0.2378	175	126.42	0.2305
200	133.18	0.2454	200	131.59	0.2385
225	138.10	0.2528	225	136.69	0.2461
250	143.01	0.2598	250	141.75	0.2533
275	147.94	0.2666	275	146.81	0.2603
300	152.90	0.2733	300	151.87	0.2671

TABLE B.9(cont.)   **Superheated Refrigerant-22 Properties (Saturation Pressure Table)**

$p_s = 300$ psia ($T_s = 127.1°F$)			$p_s = 350$ psia ($T_s = 139.6°F$)		
T	h	s	T	h	s
(°F)	(Btu/lbm)	(Btu/lbm · °R)	(°F)	(Btu/lbm)	(Btu/lbm · °R)
$h_f = 48.27$ Btu/lbm; $s_f = 0.0948$ Btu/lbm · °R			$h_f = 52.32$ Btu/lbm; $s_f = 0.1014$ Btu/lbm · °R		
$h_g = 112.87$ Btu/lbm; $s_g = 0.2049$ Btu/lbm · °R			$h_g = 112.6$ Btu/lbm; $s_g = 0.202$ Btu/lbm · °R		
150	118.61	0.2145	150	115.59	0.2069
175	124.42	0.2239	175	122.14	0.2175
200	129.91	0.2323	200	128.05	0.2266
225	135.23	0.2403	225	133.64	0.2349
250	140.46	0.2477	250	139.07	0.2427
275	145.64	0.2549	275	144.41	0.2501
300	150.81	0.2618	300	149.7	0.2572

$p_s = 400$ psia ($T_s = 150.9°F$)			$p_s = 450$ psia ($T_s = 161.2°F$)		
T	h	s	T	h	s
(°F)	(Btu/lbm)	(Btu/lbm · °R)	(°F)	(Btu/lbm)	(Btu/lbm · °R)
$h_f = 56.25$ Btu/lbm; $s_f = 0.1077$ Btu/lbm · °R			$h_f = 60.35$ Btu/lbm; $s_f = 0.1142$ Btu/lbm · °R		
$h_g = 112.12$ Btu/lbm; $s_g = 0.1992$ Btu/lbm · °R			$h_g = 111.85$ Btu/lbm; $s_g = 0.1972$ Btu/lbm · °R		
175	119.54	0.2111	175	116.62	0.2048
200	126.03	0.2212	200	123.85	0.216
225	131.96	0.23	225	130.21	0.2254
250	137.63	0.2381	250	136.16	0.234
275	143.15	0.2458	275	141.87	0.2419
300	148.58	0.253	300	147.45	0.2493

TABLE B.10   **Superheated Refrigerant-22 Properties (Saturation Temperature Table)**

(T_s = −75°F) p_s = 5.609 psia			(T_s = −50°F) p_s = 11.72 psia		
T	h	s	T	h	s
(°F)	(Btu/lbm)	(Btu/lbm · °R)	(°F)	(Btu/lbm)	(Btu/lbm · °R)
−75 (f)	−9.31	−0.0231	−50 (f)	−2.67	−0.0064
−75 (g)	96.06	0.2508	−50 (g)	99	0.2417
−50	99.52	0.2595			
−25	103.05	0.2679	−25	102.61	0.2503
0	106.67	0.2760	0	106.29	0.2585
25	110.36	0.2838	25	110.03	0.2665
50	114.14	0.2914	50	113.85	0.2741
75	118.01	0.2988	75	117.75	0.2816
100	121.96	0.3060	100	121.73	0.2889
125	126.00	0.3131	125	125.8	0.2960
150	130.14	0.3200	150	129.95	0.3029
175	134.36	0.3268	175	134.19	0.3098
200	138.67	0.3334	200	138.51	0.3164
250	147.53	0.3464	250	147.4	0.3294
300	156.73	0.3589	300	156.62	0.3420

(T_s = −25°F) p_s = 22.19 psia			(T_s = 0°F) p_s = 38.79 psia		
T	h	s	T	h	s
(°F)	(Btu/lbm)	(Btu/lbm · °R)	(°F)	(Btu/lbm)	(Btu/lbm · °R)
−25 (f)	4.03	0.0094	0 (f)	10.77	0.0243
−25 (g)	101.80	0.2343	0 (g)	104.42	0.2281
0	105.60	0.2428			
25	109.45	0.2510	25	108.45	0.2366
50	113.34	0.2588	50	112.49	0.2447
75	117.30	0.2664	75	116.56	0.2525

TABLE B.10(cont.)   **Superheated Refrigerant-22 Properties (Saturation Temperature Table)**

$(T_s = -25°F)\ p_s = 22.19$ psia			$(T_s = 0°F)\ p_s = 38.79$ psia		
T	*h*	*s*	*T*	*h*	*s*
(°F)	(Btu/lbm)	(Btu/lbm · °R)	(°F)	(Btu/lbm)	(Btu/lbm · °R)
100	121.33	0.2737	100	120.67	0.2600
125	125.44	0.2809	125	124.85	0.2674
150	129.62	0.2879	150	129.09	0.2745
175	133.89	0.2948	175	133.41	0.2814
200	138.23	0.3015	200	137.79	0.2882
250	147.17	0.3145	250	146.80	0.3013
300	156.42	0.3271	300	156.10	0.3140
$(T_s = 25°F)\ p_s = 63.49$ psia			$(T_s = 50°F)\ p_s = 98.74$ psia		
T	*h*	*s*	*T*	*h*	*s*
(°F)	(Btu/lbm)	(Btu/lbm · °R)	(°F)	(Btu/lbm)	(Btu/lbm · °R)
25 (f)	17.66	0.0388	50 (f)	24.69	0.0528
25 (g)	106.81	0.2227	50 (g)	108.92	0.2180
50	111.11	0.2314			
75	115.38	0.2396	75	113.54	0.2269
100	119.65	0.2474	100	118.08	0.2352
125	123.94	0.2549	125	122.58	0.2430
150	128.28	0.2622	150	127.08	0.2506
175	132.68	0.2692	175	131.60	0.2578
200	137.13	0.2761	200	136.16	0.2649
250	146.24	0.2894	250	145.43	0.2784
300	155.63	0.3022	300	154.94	0.2914

TABLE B.10(cont.)   Superheated Refrigerant-22 Properties (Saturation Temperature Table)

$(T_s = 75°F)\ p_s = 146.8$ psia			$(T_s = 100°F)\ p_s = 210.3$ psia		
T	h	s	T	h	s
(°F)	(Btu/lbm)	(Btu/lbm ·°R)	(°F)	(Btu/lbm)	(Btu/lbm · °R)
75 (f)	32.02	0.0666	100 (f)	39.56	0.0801
75 (g)	110.78	0.2139	100 (g)	112.10	0.2097
100	115.79	0.2231			
125	120.63	0.2315	125	117.61	0.2194
150	125.38	0.2395	150	122.82	0.2281
175	130.10	0.2471	175	127.87	0.2362
200	134.82	0.2544	200	132.86	0.2439
250	144.32	0.2682	250	142.76	0.2584
300	153.99	0.2814	300	152.69	0.2719

$(T_s = 125°F)\ p_s = 292.1$ psia			$(T_s = 150°F)\ p_s = 395.8$ psia		
T	h	s	T	h	s
(°F)	(Btu/lbm)	(Btu/lbm · °R)	(°F)	(Btu/lbm)	(Btu/lbm · °R)
125 (f)	47.54	0.0936	150 (f)	55.92	0.1072
125 (g)	112.82	0.2053	150 (g)	112.16	0.1994
150	119.01	0.2156			
175	124.75	0.2249	175	119.77	0.2117
200	130.19	0.2333	200	126.20	0.2216
250	140.67	0.2486	250	137.76	0.2385
300	150.98	0.2626	300	148.67	0.2534

TABLE B.11   *Ideal Gas Properties of Air (Gravimetric)*

T	c_p	h	$s°$	p_r	v_r
(°R)	(Btu/lbm · °R)	(Btu/lbm)	(Btu/lbm · °R)		
536.67	0.240	0	1.599	1	1
550	0.240	3	1.605	1.09	0.9405
600	0.240	15	1.626	1.48	0.7566
700	0.242	39	1.663	2.54	0.5136
800	0.244	64	1.696	4.07	0.3661
900	0.246	88	1.724	6.20	0.2705
1000	0.249	113	1.750	9.07	0.2055
1100	0.252	138	1.774	12.84	0.1596
1200	0.255	163	1.796	17.71	0.1263
1300	0.258	189	1.817	23.89	0.1014
1400	0.261	215	1.836	31.62	0.0825
1500	0.264	241	1.854	41.18	0.0679
1600	0.267	268	1.871	52.88	0.0564
1700	0.270	294	1.888	67.05	0.0472
1800	0.273	322	1.903	84.06	0.0399
1900	0.275	349	1.918	104.32	0.0339
2000	0.277	377	1.932	128.26	0.0291

TABLE B.12 **Ideal Gas Properties of Air (Molar)**

T	$\bar{c}_p$	$\bar{h}$	$\bar{s}°$		
(°R)	(Btu/lbmol · °R)	(Btu/lbmol)	(Btu/lbmol · °R)	p_r	v_r
536.67	6.95	0	46.31	1	1
550	6.95	93	46.49	1.09	0.9405
600	6.96	440	47.09	1.48	0.7566
700	7.00	1138	48.17	2.54	0.5136
800	7.06	1840	49.10	4.07	0.3661
900	7.13	2549	49.94	6.20	0.2705
1000	7.21	3266	50.69	9.07	0.2055
1100	7.29	3991	51.38	12.84	0.1596
1200	7.38	4724	52.02	17.71	0.1263
1300	7.47	5467	52.62	23.89	0.1014
1400	7.56	6218	53.17	31.62	0.0825
1500	7.65	6979	53.70	41.18	0.0679
1600	7.74	7748	54.19	52.88	0.0564
1700	7.82	8526	54.67	67.05	0.0472
1800	7.90	9312	55.12	84.06	0.0399
1900	7.97	10105	55.54	104.32	0.0339
2000	8.03	10905	55.95	128.26	0.0291

TABLE B.13   **Ideal Gas Properties of Methane, CH₄ (Gravimetric)**

T	c_p	h	$s°$	p_r	v_r
(°R)	(Btu/lbm · °R)	(Btu/lbm)	(Btu/lbm · °R)		
536.67	0.531	− 2006	2.772	1	1
550	0.535	− 1999	2.785	1.11	0.9221
600	0.552	− 1972	2.833	1.63	0.6866
700	0.593	− 1914	2.921	3.32	0.3931
800	0.639	− 1853	3.003	6.44	0.2314
900	0.687	− 1787	3.081	12.10	0.1386
1000	0.736	− 1715	3.156	22.17	0.0841
1100	0.784	− 1639	3.228	39.79	0.0515
1200	0.831	− 1559	3.298	70.20	0.0319
1300	0.876	− 1473	3.367	121.93	0.0199
1400	0.919	− 1383	3.433	208.68	0.0125
1500	0.959	− 1289	3.498	352.18	0.0079
1600	0.997	− 1192	3.561	586.39	0.0051
1700	1.032	− 1090	3.623	963.71	0.0033
1800	1.065	− 985	3.683	1564.05	0.0021
1900	1.097	− 877	3.741	2507.91	0.0014
2000	1.127	− 766	3.798	3975.64	0.0009

TABLE B.14   Ideal Gas Properties of Methane, CH₄ (Molar)

T	$\bar{c}_p$	$\bar{h}$	$\bar{s}°$		
(°R)	(Btu/lbmol · °R)	(Btu/lbmol)	(Btu/lbmol · °R)	p_r	v_r
536.67	8.52	− 32179	44.47	1	1
550	8.58	− 32065	44.68	1.11	0.9221
600	8.86	− 31629	45.44	1.63	0.6866
700	9.51	− 30711	46.85	3.32	0.3931
800	10.25	− 29723	48.17	6.44	0.2314
900	11.02	− 28660	49.42	12.10	0.1386
1000	11.81	− 27518	50.62	22.17	0.0841
1100	12.58	− 26299	51.79	39.79	0.0515
1200	13.34	− 25002	52.91	70.20	0.0319
1300	14.06	− 23632	54.01	121.93	0.0199
1400	14.74	− 22192	55.08	208.68	0.0125
1500	15.39	− 20685	56.12	352.18	0.0079
1600	15.99	− 19116	57.13	586.39	0.0051
1700	16.56	− 17488	58.12	963.71	0.0033
1800	17.09	− 15806	59.08	1564.05	0.0021
1900	17.60	− 14071	60.01	2507.91	0.0014
2000	18.08	− 12287	60.93	3975.64	0.0009

TABLE B.15 *Ideal Gas Properties of Carbon Dioxide, CO₂ (Molar/Gravimetric)*

Molar				Gravimetric			
T	$\bar{c}_p$	$\bar{h}$	$\bar{s}°$	T	c_p	h	$s°$
(°R)	(Btu/lbmol · °R)	(Btu/lbmol)	(Btu/lbmol · °R)	(°R)	(Btu/lbm · °R)	(Btu/lbm)	(Btu/lbm · °R)
536.67	8.87	− 169190	51.03	536.67	0.202	− 3845	1.16
550	8.95	− 169071	51.25	550	0.203	− 3843	1.165
600	9.24	− 168616	52.04	600	0.21	− 3832	1.183
700	9.77	− 167665	53.51	700	0.222	− 3811	1.216
800	10.24	− 166663	54.85	800	0.233	− 3788	1.246
900	10.66	− 165618	56.08	900	0.242	− 3764	1.274
1000	11.03	− 164533	57.22	1000	0.251	− 3739	1.3
1100	11.36	− 163414	58.29	1100	0.258	− 3714	1.325
1200	11.66	− 162262	59.29	1200	0.265	− 3688	1.347
1300	11.94	− 161082	60.23	1300	0.271	− 3661	1.369
1400	12.18	− 159876	61.13	1400	0.277	− 3634	1.389
1500	12.40	− 158646	61.97	1500	0.282	− 3606	1.408
1600	12.61	− 157396	62.78	1600	0.287	− 3577	1.427
1700	12.79	− 156125	63.55	1700	0.291	− 3548	1.444
1800	12.96	− 154838	64.29	1800	0.295	− 3519	1.461
1900	13.11	− 153534	64.99	1900	0.298	− 3489	1.477
2000	13.25	− 152215	65.67	2000	0.301	− 3459	1.492
2500	13.79	− 145444	68.69	2500	0.313	− 3306	1.561
3000	14.14	− 138455	71.23	3000	0.321	− 3147	1.619
3500	14.37	− 131323	73.43	3500	0.327	− 2985	1.669
4000	14.54	− 124092	75.36	4000	0.331	− 2820	1.713
4500	14.67	− 116787	77.08	4500	0.333	− 2654	1.752
5000	14.78	− 109423	78.64	5000	0.336	− 2487	1.787
5500	14.88	− 102006	80.05	5500	0.338	− 2318	1.819
6000	14.97	− 94543	81.35	6000	0.340	− 2149	1.849

TABLE B.15(cont.)   **Ideal Gas Properties of Carbon Dioxide, CO$_2$ (Molar/Gravimetric)**

	Molar				Gravimetric		
T	$\bar{c}_p$	$\bar{h}$	$\bar{s}°$	T	c_p	h	$s°$
(°R)	(Btu/lbmol · °R)	(Btu/lbmol)	(Btu/lbmol · °R)	(°R)	(Btu/lbm · °R)	(Btu/lbm)	(Btu/lbm · °R)
6500	15.04	−87039	82.55	6500	0.342	−1978	1.876
7000	15.09	−79505	83.67	7000	0.343	−1807	1.902
7500	15.14	−71950	84.71	7500	0.344	−1635	1.925
8000	15.19	−64369	85.69	8000	0.345	−1463	1.947
8500	15.24	−56761	86.61	8500	0.346	−1290	1.968
9000	15.30	−49126	87.48	9000	0.348	−1117	1.988
9500	15.35	−41465	88.31	9500	0.349	−942	2.007
10000	15.41	−33775	89.10	10000	0.350	−768	2.025
10500	15.47	−26054	89.85	10500	0.352	−592	2.042
11000	15.55	−18300	90.58	11000	0.353	−416	2.059

TABLE B.16   *Ideal Gas Properties of Carbon Monoxide, CO (Molar/ Gravimetric)*

Molar				Gravimetric			
T	$\bar{c}_p$	$\bar{h}$	$\bar{s}°$	T	c_p	h	$s°$
(°R)	(Btu/lbmol · °R)	(Btu/lbmol)	(Btu/lbmol · °R)	(°R)	(Btu/lbm · °R)	(Btu/lbm)	(Btu/lbm · °R)
536.67	6.94	− 47516	47.27	536.67	0.248	− 1696	1.688
550	6.94	− 47424	47.44	550	0.248	− 1693	1.694
600	6.95	− 47077	48.04	600	0.248	− 1681	1.715
700	6.99	− 46380	49.12	700	0.250	− 1656	1.754
800	7.06	− 45678	50.06	800	0.252	− 1631	1.787
900	7.14	− 44968	50.89	900	0.255	− 1605	1.817
1000	7.23	− 44250	51.65	1000	0.258	− 1580	1.844
1100	7.32	− 43523	52.34	1100	0.261	− 1554	1.869
1200	7.41	− 42787	52.98	1200	0.264	− 1528	1.892
1300	7.50	− 42041	53.58	1300	0.268	− 1501	1.913
1400	7.59	− 41287	54.14	1400	0.271	− 1474	1.933
1500	7.67	− 40524	54.66	1500	0.274	− 1447	1.952
1600	7.75	− 39753	55.16	1600	0.277	− 1419	1.969
1700	7.83	− 38973	55.63	1700	0.280	− 1391	1.986
1800	7.91	− 38186	56.08	1800	0.282	− 1363	2.002
1900	7.98	− 37392	56.51	1900	0.285	− 1335	2.018
2000	8.05	− 36590	56.92	2000	0.287	− 1306	2.032
2500	8.32	− 32494	58.75	2500	0.297	− 1160	2.097
3000	8.52	− 28282	60.29	3000	0.304	− 1010	2.152
3500	8.64	− 23990	61.61	3500	0.309	− 856	2.200
4000	8.73	− 19645	62.77	4000	0.312	− 701	2.241
4500	8.79	− 15264	63.80	4500	0.314	− 545	2.278
5000	8.84	− 10856	64.73	5000	0.316	− 388	2.311
5500	8.89	− 6422	65.58	5500	0.317	− 229	2.341
6000	8.94	− 1964	66.35	6000	0.319	− 70	2.369

TABLE B.16(cont.) **Ideal Gas Properties of Carbon Monoxide, CO (Molar/ Gravimetric)**

	Molar				Gravimetric		
T	$\bar{c}_p$	$\bar{h}$	$\bar{s}°$	T	c_p	h	$s°$
(°R)	(Btu/lbmol · °R)	(Btu/lbmol)	(Btu/lbmol · °R)	(°R)	(Btu/lbm · °R)	(Btu/lbm)	(Btu/lbm · °R)
6500	8.98	2518	67.07	6500	0.321	90	2.394
7000	9.00	7016	67.74	7000	0.321	250	2.418
7500	9.02	11520	68.36	7500	0.322	411	2.440
8000	9.05	16037	68.94	8000	0.323	573	2.461
8500	9.07	20566	69.49	8500	0.324	734	2.481
9000	9.09	25107	70.01	9000	0.325	896	2.499
9500	9.11	29659	70.50	9500	0.325	1059	2.517
10000	9.13	34221	70.97	10000	0.326	1222	2.534
10500	9.15	38793	71.41	10500	0.327	1385	2.550
11000	9.17	43375	71.84	11000	0.328	1549	2.565

TABLE B.17   *Ideal Gas Properties of Nitrogen, N₂ (Molar/Gravimetric)*

Molar				Gravimetric			
T	$\bar{c}_p$	$\bar{h}$	$\bar{s}°$	T	c_p	h	$s°$
(°R)	(Btu/lbmol · °R)	(Btu/lbmol)	(Btu/lbmol · °R)	(°R)	(Btu/lbm · °R)	(Btu/lbm)	(Btu/lbm · °R)
536.67	6.95	0	45.74	536.67	0.248	0	1.633
550	6.94	93	45.91	550	0.248	3	1.639
600	6.94	440	46.51	600	0.248	16	1.660
700	6.96	1134	47.58	700	0.248	40	1.698
800	7.01	1832	48.51	800	0.250	65	1.732
900	7.07	2536	49.34	900	0.252	91	1.761
1000	7.15	3247	50.09	1000	0.255	116	1.788
1100	7.23	3966	50.78	1100	0.258	142	1.812
1200	7.31	4693	51.41	1200	0.261	167	1.835
1300	7.40	5428	52.00	1300	0.264	194	1.856
1400	7.48	6172	52.55	1400	0.267	220	1.876
1500	7.56	6924	53.07	1500	0.270	247	1.894
1600	7.65	7685	53.56	1600	0.273	274	1.912
1700	7.72	8453	54.02	1700	0.276	302	1.928
1800	7.80	9229	54.47	1800	0.278	329	1.944
1900	7.87	10013	54.89	1900	0.281	357	1.959
2000	7.94	10803	55.30	2000	0.283	386	1.974
2500	8.23	14848	57.10	2500	0.294	530	2.038
3000	8.43	19016	58.62	3000	0.301	679	2.092
3500	8.58	23270	59.93	3500	0.306	831	2.139
4000	8.67	27584	61.08	4000	0.310	985	2.180
4500	8.74	31939	62.11	4500	0.312	1140	2.217
5000	8.80	36324	63.03	5000	0.314	1297	2.250
5500	8.85	40736	63.87	5500	0.316	1454	2.280
6000	8.90	45174	64.65	6000	0.318	1612	2.307

TABLE B.17(cont.)   **Ideal Gas Properties of Nitrogen, N$_2$ (Molar/ Gravimetric)**

Molar				Gravimetric			
T	$\bar{c}_p$	$\bar{h}$	$\bar{s}°$	T	c_p	h	$s°$
(°R)	(Btu/lbmol · °R)	(Btu/lbmol)	(Btu/lbmol · °R)	(°R)	(Btu/lbm · °R)	(Btu/lbm)	(Btu/lbm · °R)
6500	8.95	49636	65.36	6500	0.319	1772	2.333
7000	8.97	54116	66.02	7000	0.320	1932	2.357
7500	9.00	58607	66.64	7500	0.321	2092	2.379
8000	9.02	63111	67.23	8000	0.322	2253	2.400
8500	9.04	67627	67.77	8500	0.323	2414	2.419
9000	9.07	72155	68.29	9000	0.324	2575	2.438
9500	9.09	76695	68.78	9500	0.325	2738	2.455
10000	9.12	81248	69.25	10000	0.325	2900	2.472
10500	9.14	85814	69.69	10500	0.326	3063	2.488
11000	9.17	90393	70.12	11000	0.327	3226	2.503

TABLE B.18   ***Ideal Gas Properties of Oxygen, O₂ (Molar/ Gravimetric)***

Molar				Gravimetric			
T	$\bar{c}_p$	$\bar{h}$	$\bar{s}°$	T	c_p	h	$s°$
(°R)	(Btu/lbmol · °R)	(Btu/lbmol)	(Btu/lbmol · °R)	(°R)	(Btu/lbm · °R)	(Btu/lbm)	(Btu/lbm · °R)
536.67	7.01	0	48.98	536.67	0.219	0	1.531
550	7.02	94	49.15	550	0.219	3	1.536
600	7.07	446	49.76	600	0.221	14	1.555
700	7.18	1158	50.86	700	0.224	36	1.589
800	7.30	1881	51.83	800	0.228	59	1.620
900	7.43	2618	52.69	900	0.232	82	1.647
1000	7.56	3367	53.48	1000	0.236	105	1.671
1100	7.69	4129	54.21	1100	0.240	129	1.694
1200	7.81	4904	54.88	1200	0.244	153	1.715
1300	7.92	5690	55.51	1300	0.247	178	1.735
1400	8.02	6487	56.10	1400	0.251	203	1.753
1500	8.12	7294	56.66	1500	0.254	228	1.771
1600	8.20	8110	57.19	1600	0.256	253	1.787
1700	8.27	8933	57.68	1700	0.258	279	1.803
1800	8.34	9763	58.16	1800	0.261	305	1.817
1900	8.39	10600	58.61	1900	0.262	331	1.832
2000	8.44	11442	59.04	2000	0.264	358	1.845
2500	8.66	15718	60.95	2500	0.271	491	1.905
3000	8.84	20093	62.54	3000	0.276	628	1.955
3500	9.00	24552	63.92	3500	0.281	767	1.997
4000	9.15	29089	65.13	4000	0.286	909	2.035
4500	9.29	33700	66.22	4500	0.290	1053	2.069
5000	9.43	38381	67.20	5000	0.295	1199	2.100
5500	9.57	43132	68.11	5500	0.299	1348	2.128
6000	9.70	47949	68.95	6000	0.303	1498	2.155

TABLE B.18(cont.) **Ideal Gas Properties of Oxygen, O_2 (Molar/ Gravimetric)**

Molar				Gravimetric			
T	$\bar{c}_p$	$\bar{h}$	$\bar{s}°$	T	c_p	h	$s°$
(°R)	(Btu/lbmol · °R)	(Btu/lbmol)	(Btu/lbmol · °R)	(°R)	(Btu/lbm · °R)	(Btu/lbm)	(Btu/lbm · °R)
6500	9.81	52826	69.73	6500	0.307	1651	2.179
7000	9.90	57754	70.46	7000	0.309	1805	2.202
7500	9.98	62722	71.14	7500	0.312	1960	2.223
8000	10.05	67728	71.79	8000	0.314	2116	2.243
8500	10.10	72765	72.40	8500	0.316	2274	2.263
9000	10.15	77829	72.98	9000	0.317	2432	2.281
9500	10.19	82915	73.53	9500	0.318	2591	2.298
10000	10.22	88019	74.05	10000	0.320	2751	2.314
10500	10.26	93139	74.55	10500	0.320	2911	2.330
11000	10.28	98274	75.03	11000	0.321	3071	2.345

TABLE B.19 *Ideal Gas Properties of Water Vapor, H_2O (Molar/Gravimetric)*

	Molar				Gravimetric		
T	$\bar{c}_p$	$\bar{h}$	$\bar{s}°$	T	c_p	h	$s°$
(°R)	(Btu/lbmol · °R)	(Btu/lbmol)	(Btu/lbmol · °R)	(°R)	(Btu/lbm · °R)	(Btu/lbm)	(Btu/lbm · °R)
536.67	8.02	− 103966	45.07	536.67	0.445	− 5771	2.502
550	8.03	− 103860	45.27	550	0.446	− 5765	2.513
600	8.06	− 103457	45.97	600	0.447	− 5743	2.552
700	8.16	− 102647	47.22	700	0.453	− 5698	2.621
800	8.28	− 101825	48.32	800	0.459	− 5652	2.682
900	8.41	− 100991	49.30	900	0.467	− 5606	2.737
1000	8.55	− 100143	50.19	1000	0.475	− 5559	2.786
1100	8.70	− 99281	51.02	1100	0.483	− 5511	2.832
1200	8.85	− 98403	51.78	1200	0.492	− 5462	2.874
1300	9.01	− 97510	52.49	1300	0.500	− 5413	2.914
1400	9.17	− 96600	53.17	1400	0.509	− 5362	2.951
1500	9.34	− 95675	53.81	1500	0.518	− 5311	2.987
1600	9.50	− 94733	54.41	1600	0.528	− 5259	3.020
1700	9.67	− 93774	55.00	1700	0.537	− 5205	3.053
1800	9.82	− 92798	55.55	1800	0.545	− 5151	3.084
1900	10.01	− 91806	56.09	1900	0.555	− 5096	3.113
2000	10.18	− 90797	56.61	2000	0.565	− 5040	3.142
2500	10.96	− 85506	58.96	2500	0.608	− 4746	3.273
3000	11.59	− 79864	61.02	3000	0.643	− 4433	3.387
3500	12.11	− 73936	62.85	3500	0.672	− 4104	3.489
4000	12.52	− 67774	64.49	4000	0.695	− 3762	3.580
4500	12.86	− 61426	65.99	4500	0.714	− 3410	3.663
5000	13.12	− 54928	67.36	5000	0.728	− 3049	3.739
5500	13.34	− 48312	68.62	5500	0.740	− 2682	3.809
6000	13.51	− 41600	69.79	6000	0.750	− 2309	3.874

TABLE B.19(cont.)   *Ideal Gas Properties of Water Vapor, H₂O (Molar/Gravimetric)*

	Molar				Gravimetric		
T	$\bar{c}_p$	$\bar{h}$	$\bar{s}°$	T	c_p	h	$s°$
(°R)	(Btu/lbmol · °R)	(Btu/lbmol)	(Btu/lbmol · °R)	(°R)	(Btu/lbm · °R)	(Btu/lbm)	(Btu/lbm · °R)
6500	13.66	−34808	70.87	6500	0.758	−1932	3.934
7000	13.79	−27947	71.89	7000	0.766	−1551	3.991
7500	13.91	−21021	72.84	7500	0.772	−1167	4.044
8000	14.00	−14044	73.75	8000	0.777	−780	4.094
8500	14.09	−7021	74.60	8500	0.782	−390	4.141
9000	14.17	43	75.40	9000	0.786	2	4.186
9500	14.24	7143	76.17	9500	0.790	397	4.228
10000	14.31	14279	76.90	10000	0.794	793	4.269
10500	14.37	21450	77.60	10500	0.798	1191	4.308
11000	14.44	28654	78.27	11000	0.802	1591	4.345

TABLE B.20   *Ideal Gas Properties of Hydrogen, H_2 (Molar/Gravimetric)*

Molar				Gravimetric			
T	$\bar{c}_p$	$\bar{h}$	$\bar{s}°$	T	c_p	h	$s°$
(°R)	(Btu/lbmol · °R)	(Btu/lbmol)	(Btu/lbmol · °R)	(°R)	(Btu/lbm · °R)	(Btu/lbm)	(Btu/lbm · °R)
536.67	6.88	0	31.19	536.67	3.415	0	15.472
550	6.90	92	31.36	550	3.420	46	15.556
600	6.93	437	31.96	600	3.436	217	15.854
700	6.97	1132	33.03	700	3.456	562	16.385
800	6.99	1830	33.96	800	3.465	908	16.847
900	6.99	2529	34.79	900	3.469	1255	17.256
1000	7.00	3229	35.52	1000	3.471	1602	17.621
1100	7.01	3929	36.19	1100	3.475	1949	17.952
1200	7.02	4630	36.80	1200	3.482	2297	18.255
1300	7.04	5333	37.36	1300	3.492	2645	18.534
1400	7.07	6038	37.89	1400	3.505	2995	18.793
1500	7.10	6747	38.38	1500	3.522	3346	19.036
1600	7.14	7458	38.84	1600	3.540	3700	19.264
1700	7.18	8174	39.27	1700	3.559	4055	19.479
1800	7.20	8893	39.68	1800	3.571	4411	19.683
1900	7.25	9616	40.07	1900	3.596	4770	19.876
2000	7.30	10343	40.44	2000	3.623	5131	20.062
2500	7.59	14066	42.10	2500	3.765	6977	20.885
3000	7.88	17933	43.51	3000	3.906	8895	21.584
3500	8.14	21938	44.75	3500	4.037	10882	22.196
4000	8.37	26066	45.85	4000	4.152	12930	22.743
4500	8.57	30302	46.85	4500	4.251	15031	23.238
5000	8.74	34631	47.76	5000	4.335	17178	23.690
5500	8.89	39039	48.60	5500	4.408	19364	24.107
6000	9.02	43516	49.38	6000	4.474	21585	24.493

TABLE B.20(cont.)   **Ideal Gas Properties of Hydrogen, H$_2$ (Molar/ Gravimetric)**

Molar				Gravimetric			
T	$\bar{c}_p$	$\bar{h}$	$\bar{s}°$	T	c_p	h	$s°$
(°R)	(Btu/lbmol · °R)	(Btu/lbmol)	(Btu/lbmol · °R)	(°R)	(Btu/lbm · °R)	(Btu/lbm)	(Btu/lbm · °R)
6500	9.15	48057	50.11	6500	4.537	23838	24.854
7000	9.28	52663	50.79	7000	4.603	26123	25.193
7500	9.41	57337	51.43	7500	4.667	28441	25.513
8000	9.52	62070	52.04	8000	4.724	30789	25.816
8500	9.63	66860	52.62	8500	4.779	33165	26.104
9000	9.74	71705	53.18	9000	4.832	35568	26.378
9500	9.85	76602	53.71	9500	4.883	37997	26.641
10000	9.94	81549	54.22	10000	4.932	40451	26.893
10500	10.04	86545	54.70	10500	4.979	42929	27.134
11000	10.13	91586	55.17	11000	5.023	45430	27.367

TABLE B.21 **Specific Heat at Constant Pressure for Some Common Gases at 77°F and 1 atm**

Gas	$\bar{c}_p$ (Btu/lbmol · °R)	Molecular Weight
Air	6.952	28.96
Ammonia° (NH_3)	—	17.03
Argon (AR)	4.964	39.95
Butane (C_4H_{10})	23.74	58.12
Carbon° (C)	—	12.011
Helium (He)	4.961	4.0028
Hydrogen (H_2)	6.892	2.0158
Mercury° (Hg)	—	200.06
Methane (CH_4)	8.52	16.04
Neon (Ne)	4.964	20.18
Nitrogen (N_2)	6.961	28.0134
Oxygen (O_2)	7.021	31.9988
Nitric oxide (NO)	7.133	30.008
Nitrogen dioxide (NO_2)	8.837	46.008
Carbon monoxide (CO)	6.965	28.011
Carbon dioxide (CO_2)	8.874	44.011
Refrigerant-22°	—	86.48
Water vapor° (H_2O)	—	18.0152

Data obtained from the current edition of the *JANAF Thermochemical Data*, Dow Chemical Company, Thermal Research Laboratory, Midland, MI.

°Indicates that substance is not gaseous at 77°F and 1 atm.

TABLE B.22   **Triple Point and Critical Properties of Some Common Substances**

Substance	Critical Properties T_c (°R)	Critical Properties p_c (psia)	Triple Point Temperature (°R)	Triple Point Pressure (psia)
Air	237.60	546.79	—	—
Ammonia (NH_3)	729.65	1639.51	351.88	8.7893×10^{-1}
Argon (A)	271.31	705.61	150.84	$1.0000 \times 10^{+1}$
Butane (C_4H_{10})	765.29	550.58	242.75	9.7175×10^{-5}
Carbon Dioxide (CO_2)	547.60	1070.74	389.86	$7.5124 \times 10^{+1}$
Carbon Monoxide (CO)°	239.69	507.34	—	
Ethane (C_2H_6)	549.59	706.54	162.63	1.6389×10^{-4}
Helium (He)	9.36	33.00	3.20	$4.4135 \times 10^{+2}$
Hydrogen (H_2)	59.74	190.72	25.11	1.0441
Methane (CH_4)	343.00	666.45	163.22	1.6997
Neon (Ne)	80.01	386.38	44.21	6.2916
Nitrogen (N_2)	227.16	493.13	114.30	1.8173
Oxygen (O_2)	278.25	731.41	97.85	2.1176×10^{-2}
Propane (C_3H_8)	665.64	615.25	153.85	4.3511×10^{-6}
Refrigerant-12 (CCl_2F_2)°°	693.27	597.41	210.96	—
Refrigerant-22 ($CHClF_2$)°°	664.74	723.45	208.33	—
Water Vapor (H_2O)°°°	1164.87	3200.11	491.71	8.8724×10^{-2}

Data extracted from ASHRAE Thermodynamic Properties of Refrigerants, R. Stewart, R. Jacobsen, and
   S. Penoncello, Atlanta, Ga., 1986.
°Heat Exchanger Design Handbook, Vol. 5, Physical Properties, D. K. Edwards, etc., Hemisphere
   Publishing, 1984.
°°National Standard Reference Data Service of the U.S.S.R.: A Series of Property Tables, T. Selover, Jr.,
   English Editor, Hemisphere Publishing, 1987.
°°°Handbook of Chemistry and Physics, 72nd Edition, D. Lide, Editor, CRC Press, 1992.

TABLE B.23 (a)   **Enthalpy of Formation and Specific Entropy at Standard Temperature of 77°F and Standard Pressure of 1 atm for Chemical Substances**

Chemical Substance	Chemical Formula	M (lbm/lbmol)	$(\bar{h}_f^\circ)_{537°R}$ (Btu/lbmol)	$(\bar{s}^\circ)_{537°R}$ (Btu/lbmol · °R)
Acetylene (g)	C_2H_2	26.038	97,460	52.78
Benzene (g)	C_6H_6	78.114	35,640	64.30
Butane (g)	C_4H_{10}	58.124	−54,210	74.07
Carbon (graphite) (s) Carbon (Ideal gas) (g)	C	12.011	0 307,825	1.359 37.76
Carbon dioxide (g)	CO_2	44.011	−169,175	51.07
Carbon monoxide (g)	CO	28.011	−47,505	47.21
n-Dodecane (g) n-Dodecane (l)	$C_{12}H_{26}$	170.34	−125,100 −151,300	148.77 117.26
Ethane (g)	C_2H_6	30.068	−36,415	54.81
Ethylene (g)	C_2H_4	28.054	22,525	52.45
Hydrogen (g)	H_2	2.016	0	31.21
Hydroxyl (g)	OH	17.008	16,765	43.92
Methane (g)	CH_4	16.043	−32,200	44.48
Nitrogen (g)	N_2	28.016	0	45.77
n-Octane (g) n-Octane (l)	C_8H_{18}	114.232	−89,595 −107,480	111.48 86.17
Oxygen (g)	O_2	32.00	0	49.00
n-Pentane (g) n-Pentane (l)	C_5H_{12}	72.151	−62,940 −74,375	83.35 62.88
n-Propane (g)	C_3H_8	44.097	−44,625	64.47
Water vapor (g) Water (l)	H_2O	18.016	−103,955 −122,870	45.11 16.75

Data have been computed from the values in Table A.23 (a) using appropriate conversion factors.

TABLE B.23 (b)   **Enthalpy of Combustion and Chemical Exergy (per unit mass) for Common Fuels**

Chemical Substance	Chemical Formula	M (lbm/lbmol)	Combustion in Air $(-\Delta \bar{h}_c)_{537°R}$ (Btu/lbm)		Chemical Exergy χ_c Btu/lbm
			Lower	Higher	
Acetylene (g)	C_2H_2	26.038	20,720	21,450	21,060
Benzene (g)	C_6H_6	78.114	17,410	18,180	18,100
Butane (g)	C_4H_{10}	58.124	19,640	21,280	20,850
Carbon (graphite) (s)	C	12.011	14,100	14,100	14,700
Carbon monoxide (g)	CO	28.011	4,340	4,340	4,210
n-Dodecane (g) n-Dodecane (l)	$C_{12}H_{26}$	170.34	19,100 19,000	20,600 20,400	20,370 20,370
Ethane (g)	C_2H_6	30.068	20,370	22,310	21,450
Ethylene (g)	C_2H_4	28.054	20,250	21,620	20,980
Ethynyl (g)	C_2H	25.03	23,770	24,160	23,940
Gasoline (g)	—	—	20,720	—	22,090 (est)
Hydrogen (g)	H_2	2.016	51,500	60,960	50,120
Methane (g)	CH_4	16.043	21,490	23,860	22,270
n-Octane (g) n-Octane (l)	C_8H_{18}	114.232	19,210 19,080	20,760 20,590	20,460 20,460
n-Pentane (g) n-Pentane (l)	C_5H_{12}	72.151	19,470 19,340	21,060 20,890	20,720 20,800
n-Propane (g)	C_3H_8	44.097	19,900	21,620	21,060
Jet fuel (g)	—	—	18,700	—	19,940 (est)

Data were computed using the values in Table A.23 (b) and appropriate conversion factors from SI to USCS units.

TABLE B.23(c)   **Enthalpy of Reaction of Common Fuels**

Common Combustion Equations	$\Delta \bar{h}_R^\circ = \sum\limits_P^{gas} N_i (\Delta \bar{h}_f^\circ)_i - \sum\limits_R^{gas} N_i (\Delta \bar{h}_f^\circ)_i$ at 537°R and 1 atm (Btu, amount of substance in lbmol)
$H_2 + \frac{1}{2} O_2 \rightarrow H_2O$	$-104{,}000$
$2\,OH + H_2 \rightarrow 2\,H_2O$	$-241{,}400$
$C + H_2 \rightarrow CH_2$	$-140{,}000$
$C + O_2 \rightarrow CO_2$	$-477{,}000$
$H_2O_2 \rightarrow H_2O + \frac{1}{2} O_2$	$-45{,}400$
$CH_4 + 2\,O_2 \rightarrow CO_2 + 2\,H_2O$	$-344{,}900$
$2\,CH + 2\,O_2 \rightarrow CO + CO_2 + H_2O$	$-832{,}900$
$N_2H_4 + O_2 \rightarrow 2\,H_2O + N_2$	$-248{,}900$
$N_2H_4 + N_2O_4 \rightarrow 2\,H_2O + 2\,N_2 + O_2$	$-252{,}900$
$F_2 + H_2 \rightarrow 2\,HF$	$-235{,}000$
$2\,F_2 + N_2H_4 \rightarrow 4\,HF + N_2$	$-511{,}000$

Note: (1) The chemical substances are all assumed to be ideal gases.

(2) Enthalpies of reaction were computed using the values in Table A.23(c) and appropriate conversion factors from SI to USCS units.

TABLE B.24 Equilibrium Constant, ln K_p, for Common Reactions

$$\ln K_p = -\frac{\Delta \overline{G}^\circ (T)}{\overline{R}T}$$

where

$$\Delta \overline{G}^\circ = e\overline{g}_E^\circ + f\overline{g}_F^\circ - a\overline{g}_A^\circ - b\overline{g}_B^\circ$$

for reactions of the type

$$a\,A + b\,B \rightleftarrows e\,E + f\,F$$

with the standard pressure, $p^\circ = 1$ atm

Reaction Equations:

I. $H_2 \rightleftarrows 2\,H$
II. $O_2 \rightleftarrows 2\,O$
III. $N_2 \rightleftarrows 2\,N$
IV. $H_2 + \frac{1}{2}\,O_2 \rightleftarrows H_2O$
V. $2\,H_2O \rightleftarrows 2\,H_2 + 2\,OH$
VI. $N_2 + O_2 \rightleftarrows 2\,NO$
VII. $CO_2 \rightleftarrows CO + \frac{1}{2}\,O_2$
VIII. $CO_2 + H_2 \rightleftarrows CO + H_2O$
IX. $C + H_2 \rightleftarrows CH_2$
X. $N \rightleftarrows N^+ + e^-$

Temp (°R)	Reaction									
	I	II	III	IV	V	VI	VII	VIII	IX	X
360	−250.185	−285.465	−554.57	139.976	−323.019	−105.598	−159.708	−19.729	187.068	—
537	−164.018	−186.979	−367.737	92.214	−212.147	−69.902	−103.824	−11.564	121.912	−565.571
900	−92.844	−105.638	−213.424	52.697	−120.341	−40.45	−57.625	−4.929	67.865	−335.463
1440	−53.148	−60.332	−127.789	30.6	−69.136	−24.145	−32.046	−1.446	37.691	−207.699
1800	−39.82	−45.16	−99.162	23.169	−51.965	−18.708	−23.535	−0.366	27.564	−164.945
2160	−30.889	−35.017	−80.04	18.189	−40.48	−15.081	−17.878	0.312	20.79	−136.357
2520	−24.479	−27.755	−66.36	14.618	−32.257	−12.489	−13.848	0.77	15.941	−115.878
2700	−21.91	−24.845	−60.88	13.187	−28.969	−11.453	−12.241	0.945	14.0	−107.668
2880	−19.65	−22.296	−56.084	11.932	−26.081	−10.546	−10.837	1.095	12.3	−100.473
3240	−15.88	−18.042	−48.08	9.836	−21.27	−9.034	−8.502	1.333	9.466	−88.457
3600	−12.854	−14.633	−41.67	8.157	−17.419	−7.824	−6.642	1.514	7.197	−78.817
3960	−10.369	−11.839	−36.418	6.78	−14.267	−6.835	−5.126	1.654	5.341	−70.906
4320	−8.292	−9.508	−32.038	5.632	−11.639	−6.012	−3.867	1.765	3.793	−64.295
4500	−7.377	−8.483	−30.109	5.121	−10.482	−5.649	−3.315	1.805	3.113	−61.38

TABLE B.24(cont.) *Equilibrium Constant, In K_p, for Common Reactions*

Temp (°R)	I	II	III	IV	V	VI	VII	VIII	IX	X
						Reaction				
4680	−6.531	−7.533	−28.326	4.659	−9.415	−5.316	−2.805	1.853	2.485	−58.685
5040	−5.017	−5.839	−25.142	3.824	−7.51	−4.72	−1.899	1.924	1.362	−53.863
5400	−3.702	−4.369	−22.38	3.099	−5.859	−4.205	−1.117	1.983	0.39	−49.671
5760	−2.549	−3.082	−19.961	2.466	−4.415	−3.765	−0.435	2.031	−0.462	−45.993
6120	−1.531	−1.946	−17.824	1.905	−3.142	−3.36	0.165	2.07	−1.213	−42.739
6300	−1.064	−1.428	−16.846	1.649	−2.56	−3.179	0.438	2.087	−1.556	−41.248
6480	−0.624	−0.9349	−15.922	1.406	−2.01	−3.009	0.696	2.103	−1.88	−39.838
6840	0.189	−0.0299	−14.219	0.96	−0.998	−2.696	1.17	2.13	−2.479	−37.235
7200	0.921	0.785	−12.684	0.557	−0.087	−2.415	1.594	2.152	−3.017	−34.886
7560	1.584	1.524	−11.292	0.193	0.736	−2.162	1.977	2.17	−3.505	−32.755
7920	2.187	2.192	−10.03	−0.138	1.485	−1.933	2.324	2.185	−3.948	−30.812
8100	2.47	2.506	−9.437	−0.294	1.834	−1.827	2.485	2.191	−4.155	−29.904
8280	2.74	2.805	−8.87	−0.442	2.168	−1.726	2.639	2.197	−4.354	−29.034
8640	3.247	3.366	−7.81	−0.721	2.794	−1.536	2.928	2.206	−4.726	−27.4
9000	3.711	3.885	−6.829	−0.978	3.37	−1.361	3.192	2.213	−5.069	−25.892
9360	4.14	4.361	−5.922	−1.216	3.901	−1.201	3.435	2.218	−5.385	−24.497
9720	4.541	4.803	−5.084	−1.437	4.393	−1.054	3.659	2.222	−5.679	−23.201
9900	4.727	5.014	−4.686	−1.541	4.625	−0.985	3.764	2.223	−5.819	−22.588
10080	4.909	5.218	−4.301	−1.642	4.849	−0.918	3.866	2.224	−5.953	−21.996
10440	5.254	5.80	−3.574	−1.835	5.274	−0.792	4.059	2.225	−6.207	−20.871
10800	5.574	5.955	−2.89	−2.014	5.671	−0.675	4.238	2.224	−6.446	−19.818

All constants derived from the current edition of the *JANAF Thermochemical Tables*, Dow Chemical Company, Thermal Research Laboratory, Midland, MI.

Generalized Charts and Psychrometric Charts

The generalized charts for compressibility, enthalpy, and entropy given in this appendix are all based on the original data and initial charts published by A. L. Lyderson, R. A. Greenkorn, and O. A. Hougen, "Generalized Thermodynamic Properties of Pure Fluids," University of Wisconsin College of Engineering, *Engineering Experiment Station Report No. 4* (October 1955). More extensive charts based on this data were published later by Olaf A. Hougen, Kenneth M. Watson, and Roland A. Ragatz, *Chemical Process Principles Charts*, Third Edition, John Wiley & Sons, Inc., New York (1964). The charts presented in this appendix are adapted from the original charts for normal pressures and from the later charts of Hougen, Watson, and Ragatz (1964) for the low pressure range. *Used with permission of the original authors, and of John Wiley & Sons, Inc.*

All parameters in the generalized charts are nondimensional terms. The ordinates of Figures C.1, C.2, and C.3 are given by Equations 4.35, 12.75, and 12.76, respectively.

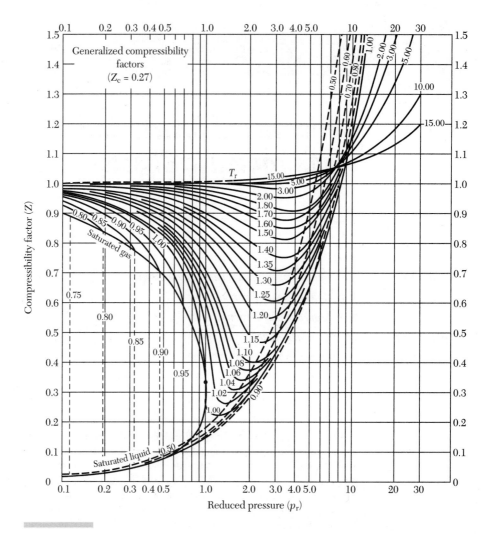

Figure C.1(a) Generalized compressibility chart. (Adapted from Lyderson, A. L., Greenkorn, R. A., and Hougen, O. A. 1955. "Generalized Thermodynamic Properties of Pure Fluids." University of Wisconsin College of Engineering, *Engineering Experiment Station Report No. 4.* With permission from the authors.)

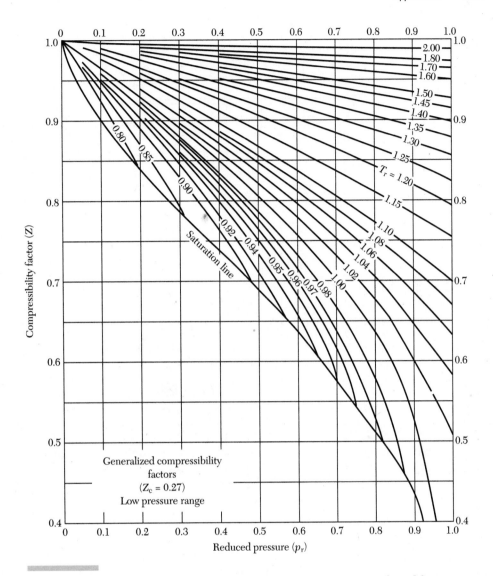

Figure C.1(b) Generalized compressibility chart—low pressure range. (Adapted from Hougen, O. A., Watson, K. M., and Ragatz, R.A. 1964. *Chemical Process Principles Charts.* 3rd ed. New York: John Wiley & Sons, Inc. Figure reprinted with permission.)

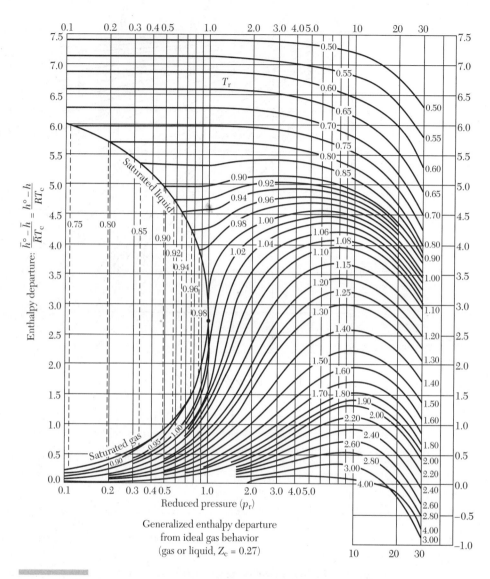

Figure C.2(a) Generalized enthalpy departure chart. (Adapted from Lyderson, A. L., Greenkorn, R. A., and Hougen, O. A. 1955. "Generalized Thermodynamic Properties of Pure Fluids." University of Wisconsin College of Engineering, *Engineering Experiment Station Report No. 4.* With permission from the authors.)

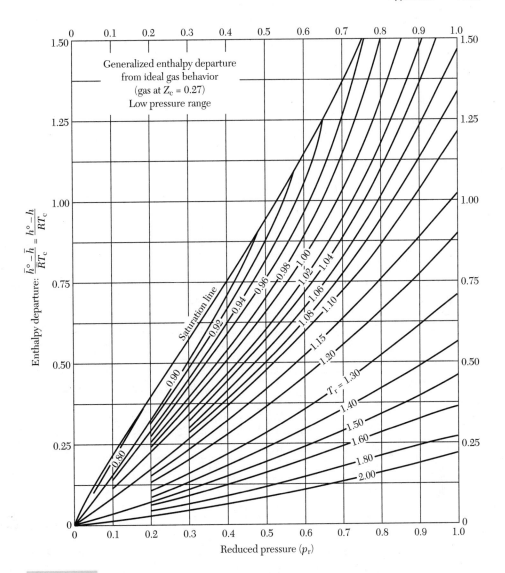

Figure C.2(b) Generalized enthalpy departure chart—low pressure range. (Adapted from Hougen, O. A., Watson, K. M., and Ragatz, R. A. 1964. *Chemical Process Principles Charts.* 3rd ed. New York: John Wiley & Sons, Inc. Figure reprinted with permission.)

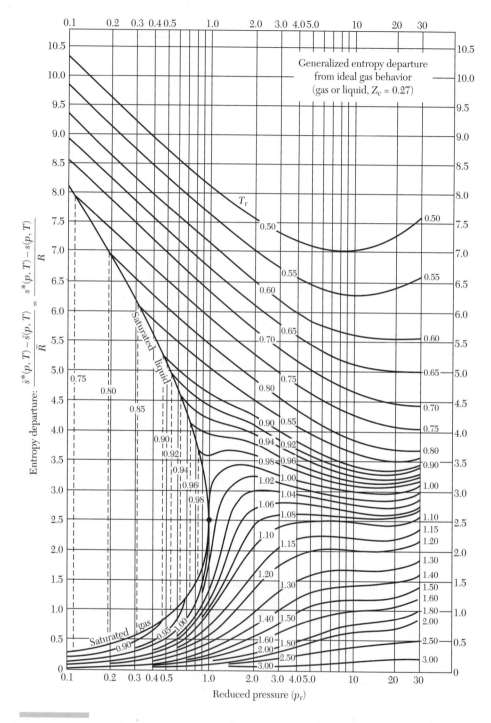

Figure C.3(a) Generalized entropy departure chart. (Adapted from Lyderson, A. L., Greenkorn, R. A., and Hougen, O. A. 1955. "Generalized Thermodynamic Properties of Pure Fluids." University of Wisconsin College of Engineering, *Engineering Experiment Station Report No. 4*. With permission from the authors.)

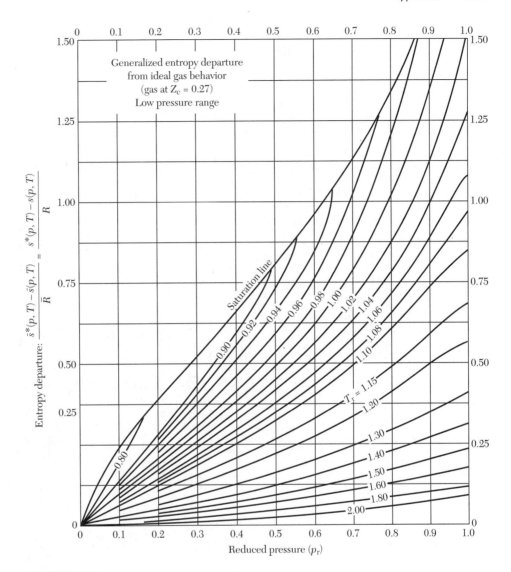

Figure C.3(b) Generalized entropy departure chart—low pressure range. (Adapted from Hougen, O. A., Watson, K. M., and Ragatz, R. A. 1964. *Chemical Process Principles Charts*. 3rd ed. New York: John Wiley & Sons, Inc. Figure reprinted with permission.)

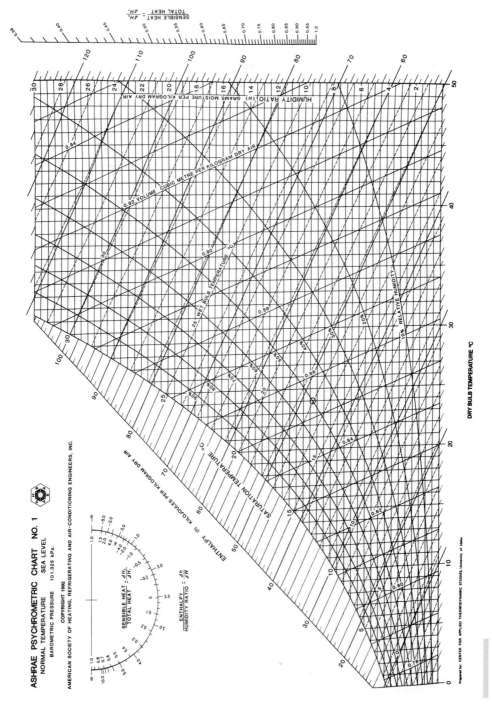

Figure C.4 Psychrometric chart in SI units. (Reprinted with permission of the American Society of Heating, Refrigerating, and Air-Conditioning Engineers, Inc.)

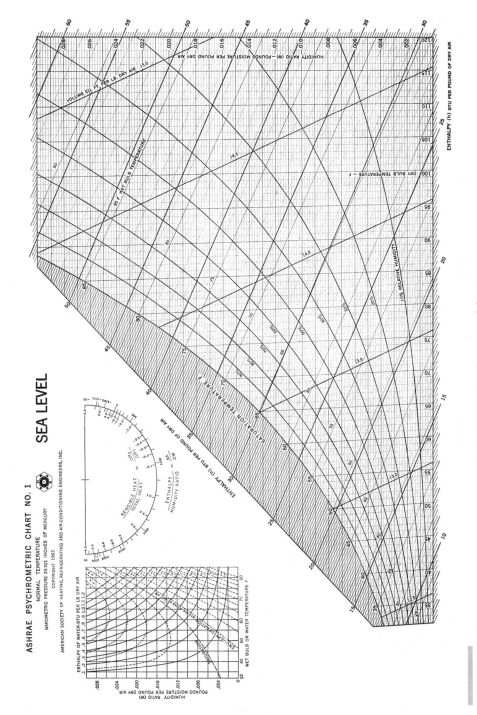

Figure C.5 Psychrometric chart in USCS units. (Reprinted with permission of the American Society of Heating, Refrigerating, and Air-Conditioning Engineers, Inc.)

Computer Codes for the Thermodynamic Properties of Common Substances Encountered in Engineering Applications

Summary

This appendix describes the following four computer codes: **STEAM** (for ice/water/steam properties), **R22** (for refrigerant-22 properties), **GAS** (for ideal gas properties of air, methane, CO_2, CO, O_2, N_2, H_2O, H_2), and **PSY** (for the thermodynamic properties of moist air at near ambient conditions). These codes have been developed by the authors in support of the undergraduate classical thermodynamics course. The range of conditions over which the codes can be used is indicated in Figures D.1 and D.2 for the **STEAM** and **R22** codes, respectively. The **GAS** code gives the specific enthalpy and the specific entropy (at 1 atm) for air and methane in the range of temperatures from 250 K (450°R) to 1200 K (2160°R), while for the other gases, the range is from 250 K (450°R) to slightly over 6000 K (10800°R). Property values can be obtained in either SI or USCS units in all cases except for the **PSY** code, which is presently designed for property values in SI units only. A step-by-step procedure on the use of each computer code is provided in Section D.2 after the general introduction of Section D.1. Each of the codes was written in BASIC and compiled into self-executable programs that can be run on any IBM PC compatible microcomputer.

D.1   Introduction

Thermodynamic analysis requires knowledge of the properties of the substances involved. Most traditional approaches to the teaching of thermodynamics at the undergraduate level accordingly devote time and effort to making the students adept at reading tables of thermodynamic properties. Once the basic table reading skills have been mastered, however, students of thermodynamics should not be encumbered with the burden

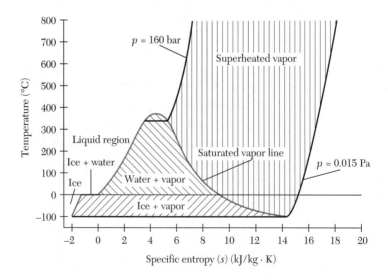

Figure D1 Regions of applicability for values given in steam tables (SI units).

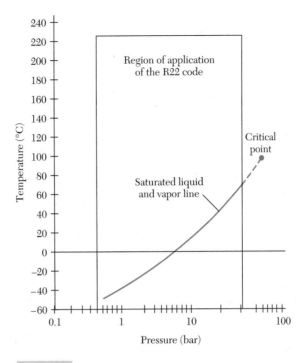

Figure D2 Regions of applicability for values given in R-22 tables.

of looking up property tables whenever thermodynamic applications problems are assigned. This concern was one reason the authors decided to develop computer codes for a number of the widely employed substances in engineering applications. A second reason has to do with the nonlinearity of some of the property relationships that renders inaccurate a linear interpolation often practiced in the computation of properties for states other than those tabulated. A brief description now follows for each of the four codes, including an indication of the range of conditions covered.

D.1.1 **STEAM** *Code*

The **STEAM** code gives the thermodynamic properties p, T, v, u, h, s, and phase (or quality x for wet vapor) for H_2O in the range of states depicted in Figure D.1, which include the ice/water/steam regions. These are

- the ice region from $-100°C$ ($-148°F$) to $0°C$ ($32°F$)
- the saturated water and vapor states all the way to the critical state, which is $p_c = 220.55$ bar (3198 psia) and $T_c = 373.976°C$ ($705.2°F$) for water[1]
- compressed liquid and superheated steam states up to a maximum pressure of 160 bar (2320 psia)

[1]The critical properties assumed for H_2O are those given in the *NBS/NRC Steam Tables* by Haar et al. (1984).

To operate the code, either the pressure or temperature of the H_2O must be specified first, followed by any other intensive property from the above list. The property values can be calculated in either SI units or USCS units. Graphical display of the states is provided and a choice can be made from the following menu of diagrams: *T-s*, *T-v*, *p-v*, *T-p* or *h-s* projections.

D.1.2 **R22** *Code*

A description of the **R22** code and its use in the thermodynamic analysis of refrigeration systems using refrigerant-22 is given in an article by Adebiyi (1991). The range of conditions covered by the code is illustrated in Figure D2, which is a temperature (T) versus pressure (p) diagram drawn to scale. Refrigerant-22 has a critical pressure of 49.86 bar (723 psia) and a critical temperature of 96.13°C (205.03°F).[2] The computer program works for both the compressed liquid and wet vapor states in the temperature range $-60°C$ $(-76°F)$ to 72°C (161°F), while for the superheated vapor states, the pressure should fall between approximately 0.5 bar (7 psia) and 31 bar (450 psia). The list of thermodynamic properties in the program output includes p (and the corresponding saturation temperature, T_s), T, v, u, h, s, and the phase (or quality x when the refrigerant is a wet vapor). The input must be either the saturation pressure or saturation temperature first, followed by any one of the following properties: T, h, s, or x. In an earlier version described in Adebiyi (1991), the output did not include either u or v, but instead the v_f and v_g values were given; now, u and v are given in the output. They are, however, not among the properties that can be specified as input. The graphical display provided in this case is the $\ln p$ versus h diagram. Property values are determined in either SI or USCS units.

D.1.3 **GAS** *Code*

The gases included in the **GAS** code are air, methane, CO_2, CO, O_2, N_2, H_2O, and H_2. For ideal gases, the enthalpy depends only on temperature, while the specific entropy is a function of both pressure and temperature. The standardized entropy (at a pressure of 1 atm) values are calculated by the program, and the list of properties in the output includes this standardized entropy, the specific enthalpy, and the specific heat at constant pressure once the temperature of the gas has been specified. For air and methane, the temperature range covered is from 250 K (450°R) to 1200 K (2160°R), while for the other gases, the range of temperatures is from 250 K (450°R) to slightly over 6000 K (10,800°R). Property values can be determined as molar or gravimetric and in either SI or USCS units.

D.1.4 **PSY** *Code*

Adebiyi (1989) gives a description of the **PSY** code along with illustrations on its use in the computer-aided first- and second-law thermodynamic evaluation and design of air conditioning systems. The graphical display is a scaled version of the psychrometric chart

---

[2]The critical properties indicated for the refrigerant-22 are those given in the refrigerant-22 property tables by Vargaftik (1975).

generated by the code. Computations of thermodynamic properties are made for moist air, liquid water, and pure steam. These substances are those likely to be encountered in air conditioning practice. For moist air, the list of output properties includes the dry bulb temperature (T_{db}), the wet bulb temperature (T_{wb}), the dew point temperature (T_d), the humidity ratio (ω), the relative humidity (ϕ), the enthalpy per unit mass of dry air (h_{ma}), the exergy per unit mass of dry air (χ_{ma}), and the specific entropy (s_{ma}). In the case of liquid water and pure steam, the specific enthalpy, the specific entropy, and the exergy per unit mass are computed. For the computations of thermodynamic properties for moist air, the first property specified must be selected from a list that includes T_{db}, h_{ma}, and ω (or T_d), while that for the second property includes ω (or T_d), T_{wb}, ϕ, and h_{ma}. The first and second properties specified must be different and independent of each other. For computations of exergy, the reference environment is taken as the ambient air whose dry bulb temperature, pressure, and humidity ratio must be specified early in the execution of the program.

D.2   Procedures for Operating the Computer Codes

All the codes are in BASIC and have been compiled as self-executable codes that can be run on any IBM-compatible PC with a CGA compatible monitor.

D.2.1 STEAM Code

Step 1. Boot the PC.

Step 2. Type STEAM and press ENTER (or RETURN).
A logo appears. Strike any key to continue.

Step 3. Type (s)i for SI units or (u)scs for English units and press ENTER.
A scaled *T-s* diagram appears on the bottom right-hand corner of the screen, and a template for the thermodynamic properties of steam/water/ice is displayed in the upper half of the screen. The cursor location is immediately after the following line statement on the screen:

State no? (= 0 to exit)

Step 4. Type any desired state number (1 to 20) and press ENTER. Follow the prompts, which are self-explanatory.

Step 5. To exit (or view other projections), type 0 for the desired state number in step 4 and press ENTER. The following options will be displayed on the screen:

exit,T-s,T-v,T-p,h-s,resume?

Type your choice and press ENTER.

D.2.2 R22 Code

Step 1. Boot the PC.

Step 2. Type R22 and press ENTER (or RETURN).
A logo appears. Strike any key to continue.

Step 3. Type (s)i for SI units or (u)scs for English units and press ENTER. A scaled ln p-h diagram appears on the left-hand side of the screen, and a template for the thermodynamic properties of the refrigerant is displayed to the right of the screen.

Step 4. Use the left ($\leftarrow$) or right ($\rightarrow$) ARROW keys on the keyboard to move the (highlighted) cursor to any desired state from 1 to 20.

Step 5. Once the desired state number has been selected, follow the procedure below to specify two independent properties and determine the other properties:

  a. Use the down ($\downarrow$) or up ($\uparrow$) ARROW key to locate the cursor on either the saturation pressure (p_s) or the saturation temperature (T_s) cell. The code is designed such that either p_s or T_s must be specified as the first of the two independent intensive properties required.

  b. Strike the (/) key for the prompt to enter the value for either p_s or T_s located in step 5a. In the alternative, if the value of p_s (or T_s) is identical to that for a previously defined state (#n), strike the (?) key for a prompting to enter the state number n instead. Enter the appropriate property value (or state number n) and press ENTER.

  c. The cursor moves next to the temperature (T) cell. The down ($\downarrow$)/up ($\uparrow$) ARROW keys can be used to locate the cursor on any of the T, h (specific enthalpy), s (specific entropy), and quality (for the wet vapor condition) cells, after which the procedure outlined in step 5b should be followed to specify the second known independent intensive property for the refrigerant.

Step 6. Properties for other states can be determined by repeating steps 4 and 5 above.

Step 7. Whenever the cursor is located on the State # line, the following additional options are available:

  a. Strike (e)xit for the following prompt options:

  (e)xit   Terminates the program
  (s)i/(u)scs   Allows a fresh start through steps 3 and 4 above
  (r)esume   Restores the current status before (e)xit was typed

  b. Strike (p)rint for the prompt to display (y)es for a hardcopy printout of the complete list of properties or (n)o for screen display of the complete list of properties. Type (y)es or (n)o and press ENTER. Next, the prompting of step 7a appears after the hardcopy printout or the screen display of properties.

D.2.3 **GAS** *Code*

Step 1. Boot the PC.

Step 2. Type GAS and press ENTER (or RETURN). A logo appears. Strike any key to continue.

Step 3. Type (s)i for SI units or (u)scs for English units and press ENTER. The cursor moves promptly to a menu of choices from the list of the following gases and actions: (1) CO_2, (2) CO, (3) O_2, (4) N_2, (5) H_2O, (6) H_2, (7) air, (8) CH_4; (C)HANGE UNITS, (Q)UIT TO DOS.

Step 4. Move the cursor to select a gas or a course of action and press ENTER. Alternatively, type the number (or letter in parenthesis) for a gas (or course of action) and press ENTER. Follow the prompts, which are self-explanatory.

Step 5. To exit, either move the cursor to (Q)UIT TO DOS or type Q in step 4 and press ENTER.

D.2.4 **PSY** *Code*

Step 1. Boot the PC.

Step 2. Type PSY and press ENTER (or RETURN).
A logo appears. Strike any key to continue.

Step 3. Type (y)es if the default state for the reference environment is acceptable, otherwise, type (n)o and press ENTER.

Step 4. If you typed (n)o in step 3, you should now enter the correct values for ω_0, $(T_{db})_0$, and p_0 for the reference environment and press ENTER. You have an opportunity to change your mind about the new entries. If the latest entries are acceptable, type (y)es to proceed to the next step.

Step 5. A scaled psychrometric chart appears at the upper left corner of the screen. Listing of property values for moist air is given on the lower half of the screen. At the upper right-hand corner of the screen, the following prompt appears:

$$\text{State no } (= 0 \text{ for exit, h2o})?$$

Type 0 (zero) to exit or to obtain properties for water/vapor and press ENTER. For properties of moist air, type the desired state number (other than zero) and press ENTER. The pressure of the moist air must be specified, after which menus of options for the second and third independent intensive properties will be presented in succession. Follow the prompts, which are self-explanatory.

D.3   *Formulations for the Thermodynamic Properties of the Substances*

Each of the substances for which a code has been developed can be regarded as a pure substance. One of the two most common approaches used in formulations for the thermodynamic properties of pure substances involves a fundamental (or characteristic) equation of state that is developed for the substance, while the other is basically a "piecemeal" approach. The resulting expression in the first method requires several numerical constants to ensure accuracy. Furthermore, since subsequent determination of the complete list of thermodynamic properties from the characteristic equation of state often requires differentials of the characteristic function, the functional representation must be accurate. Great precision is required if the estimates obtained for these other thermodynamic properties are to be reliable.

The piecemeal approach is the one used for the four computer codes developed by the authors. The method uses the specific heat and pvT data for each phase of the substance for the development of thermodynamic property equations, typically as mathematical integrals of the pertinent functions. The development of expressions for enthalpy and entropy, in particular, involves mathematical integration of specific heat functions. As such, consistent and accurate values for both enthalpy and entropy are achieved even when the specific heat correlations are only approximate. Properties computed by the other method from derivatives of a characteristic function, in contrast, are accurate only to the extent that the fundamental equation of state is exact. This section describes the formulations for thermodynamic properties for each property code developed.

D.3.1 Formulations for the Thermodynamic Properties of Ice/Water/Steam

Wherever possible, existing formulations are used for all of the phases of H_2O. An example of this is the comprehensive formulation by Wexler (1977) for the ice region. Much effort was devoted here to developing correlations for the constant pressure specific heat (c_p) in the superheated vapor region as functions of pressure (p) and temperature (T). The $c_p = f(p, T)$ data used for the correlations are those published in Haar et al. (1984). The form of correlation developed is as follows:

$$\frac{c_p\,(T,\,p)}{R} = \sum_{I=1}^{I1} \sum_{J=1}^{J1} A_{IJ}\, p_R^{(I-1-K1)}\, T_R^{(J-1-L1)} \tag{1.1}$$

where R is the specific gas constant for steam. $T_R\ (= T/T_c)$ and $p_R\ (= p/p_c)$ are the reduced temperature and pressure, respectively. The constants, which include the $I1$, $J1$, $L1$, $K1$, and the $A_{I,J}$'s, are determined in the correlations developed for a range of temperatures and pressures in the superheated vapor region. From such correlations, it is possible to derive analytical expressions for the specific enthalpy and the specific entropy by integrating along constant pressure lines.

Pure substances approximate ideal gas behavior at high temperatures (greater than twice the critical temperature) and low pressures ($<<p_c$). In the case of H_2O, a cursory look at the *NBS/NRC Steam Tables* by Haar et al. (1984), for example, reveals that at a high temperature of 1000°C, the specific enthalpy shows only a weak dependence on pressure right up to a pressure of about 160 bar. Choosing this temperature as the reference temperature (T_{ref}), it is possible to express the specific enthalpy and the specific entropy at T_{ref} as simple functions of the reduced pressure p_R. With these expressions and the above correlation for the constant pressure specific heat, the enthalpy and entropy for superheated steam can be calculated using the following equations:

$$\frac{h\,(T,\,p) - h\,(T_{ref},\,p)}{RT_c} = \int_{T_{R/ref}}^{T_R} \frac{c_p\,(T,\,p)}{R}\, dT_R \tag{1.2}$$

$$\frac{s\,(T,\,p) - s\,(T_{ref},\,p)}{R} = \int_{T_{R/ref}}^{T_R} \frac{c_p\,(T,\,p)}{RT_R}\, dT_R \tag{1.3}$$

A number of other correlations were developed independently, also using the data in the *NBS/NRC Steam Tables*. The more important of these are

1. the saturation pressure (p_s) versus temperature (T_s) equations for the region from the ice point to the critical point for water
2. the latent heat of vaporization (h_{fg}) versus saturation temperature (T_s) relationship from which the s_{fg} $(= h_{fg}/T_s)$ versus T_s relationship can be derived
3. the specific volumes v_f and v_g for the saturated liquid and vapor states, respectively
4. the specific volume v for the superheated vapor state, based on the virial form of the equation of state in powers of pressure

Several of the other properties are then determined from a knowledge of these.

D.3.2   *Formulations for the Thermodynamic Properties of Refrigerant-22*

A detailed write up of the formulations for the thermodynamic properties of refrigerant-22 is given in Adebiyi (1991). Basically, the approach is similar to that described for steam properties. The vapor pressure curve was correlated by equations of the form

$$\ln \frac{p_s}{p_c} = a + \frac{b}{T_R} + \frac{c}{T_R^2} \tag{2.1}$$

where T_R $(= T_s/T_c)$ is the reduced temperature based on the critical temperature (T_c), and a, b, and c are numerical constants determined by using available p_s-T_s data in Vargaftik (1975) for refrigerant-22 and performing a regression analysis.

The enthalpy of vaporization (h_{fg}) was successfully correlated in the following form:

$$\frac{h_{fg}}{RT_c} = \left(\frac{T_s}{T_c}\right)^\lambda \sum_{i=1}^{n} A_i \left(\frac{T_c}{T_s}\right)^{i-1} \tag{2.2}$$

where λ, n, and the A_i values are constants determined using the regression analysis method. As for H_2O, the specific entropy of vaporization s_{fg} can be determined by dividing h_{fg} by the saturation temperature T_s.

The specific volume for the superheated vapor phase was represented accurately by a virial expansion in p with only the second virial coefficient significant. Correlations were also developed for the specific volumes v_f and v_g at the saturated liquid and vapor states, respectively.

The formulations for enthalpy and entropy in the superheated vapor region were made following the procedure outlined for the steam properties. In this case, however, the following correlation was adequate for c_p over the entire region of interest:

$$c_p (T, p) = c_{p0} (T) + \sum_{i=1}^{3} [p \,(\mathrm{MPa})]^{(2i-1)} \left(C_{i,1} + \sum_{j=1}^{4} \frac{C_{i,j+1}}{T_R^{2.6j}}\right) \tag{2.3}$$

The numerical coefficients were determined via a regression analysis using data from Vargaftik (1975). By integrating c_p with respect to dT along a constant pressure line, an analytical expression was obtained for the specific enthalpy of the superheated vapor. Integration of c_p/T with respect to dT along a constant pressure line similarly yielded an analytical equation for the specific entropy in the region.

D.3.3 Formulations for the Thermodynamic Properties of Ideal Gases

The zero pressure specific heat at constant pressure (c_{po}) data given by Rogers and Mayhew (1980) were used for several gases in obtaining correlations having the following form:

$$\frac{c_{po}}{R} = \frac{1}{T_R^{0.5}} \sum_{i=1}^{n} A_i T_R^{i-1} \tag{3.1}$$

where R is the specific gas constant and T_R $(= T/T_c)$ is the reduced temperature. For air, a temperature of 273.16 K was arbitrarily assumed for the correlation in place of the critical temperature T_c, which is ill-defined for the gas mixture. For each ideal gas, the numeric constants n and the A_i's were determined to obtain a close match with the data in Rogers and Mayhew (1980).

For ideal gases, the specific enthalpy h is a function of temperature only, and hence, the following equation is obtained for h by integrating the equation for c_{po} with respect to dT:

$$\frac{h - h_{\text{ref}}}{RT_c} = \left[\sum_{i=1}^{n} \frac{A_i}{i - 0.5} T_R^{i-0.5} \right]_{T_{R/\text{ref}}}^{T_R} \tag{3.2}$$

where $T_{R/\text{ref}}$ is the reduced temperature based on a reference temperature of 25°C and h_{ref} is the standardized enthalpy for the gas at the reference temperature of 25°C. The standardized enthalpy at 25°C is the same as the enthalpy of formation for the substance, and the values used in the present formulation are those given in Hamblin (1971).

The specific entropy s for an ideal gas is a function of both the pressure and temperature. The notation $s°(T)$ is used for the specific entropy at a pressure of 1 atm and temperature T. From the correlation for c_{po}, the following equation was obtained for $s°(T)$:

$$\frac{s°(T) - s_{\text{ref}}}{R} = \left[\sum_{i=1}^{n} \frac{A_i}{i - 1.5} T_R^{i-1.5} \right]_{T_{R/\text{ref}}}^{T_R} \tag{3.3}$$

The standardized entropy s_{ref} for each gas at the reference state of 1 atm and 25°C was also obtained from Hamblin (1971). In the event that the specific entropy at a pressure p other than the reference pressure is desired, the equation to apply is as follows:

$$s(T, p) = s°(T) - R \ln(p/p_{\text{ref}}) \tag{3.4}$$

Equations 3.1 to 3.3 have been coded in the program **GAS**, which gives the ideal gas properties for air, methane, CO_2, CO, O_2, N_2, H_2O, and H_2 over the ranges of temperatures indicated earlier.

D.3.4 Formulations for the Thermodynamic Properties of Moist Air, Water Vapor, and Water

Details on the formulations for the thermodynamic properties of moist air, water vapor, and water for the **PSY** code are given by Adebiyi (1989). The following summary parallels that given in the article.

Formulations for Dry Air

Ideal gas behavior was assumed for dry air, with $R_a = 0.2871$ kJ/kg · K. Thus, the specific volume of the mixture, v_a (in m³/kg da), based on the mass of dry air contained, is given by

$$v_a = R_a(T/p_a) \tag{4.1}$$

where T (in K) is the air temperature, which is the same as the dry bulb temperature of moist air, and p_a (in kPa) is the partial pressure of dry air in the mixture.

For the enthalpy of dry air, the reference condition assumed is 273.16 K and 0.6117 kPa, which is the triple point for H_2O. The choice is that normally used for water vapor in the construction of steam tables. The ideal gas specific heat at constant pressure c_{pa} (= 1.005 kJ/kg · K) was taken as constant for dry air over the anticipated range of temperatures. Thus, the specific enthalpy for dry air is given by

$$h_a(T) = c_{pa}(T - 273.16) \tag{4.2}$$

The computation of entropy, however, is made using the absolute entropy in view of the need for a consistent reference point for all the substances involved. Thus, assuming a reference temperature T_{ref} of 298 K (25°C) and a reference pressure p_{ref} of 1 atm, the specific entropy for dry air can be determined using

$$s_a(T, p_a) = 6.696 + c_{pa} \ln(T/T_{ref}) - R_s \ln(p_a/p_{ref}) \tag{4.3}$$

where p_a is the actual pressure of dry air.

The flow exergy for dry air must be determined relative to ambient (atmospheric) air whose dry bulb temperature is designated T_0 and pressure p_0. The humidity ratio of the ambient air is ω_0, and the partial pressure of dry air in the atmospheric air is p_{a0}. The following equation gives the flow exergy for dry air at a pressure p_a and temperature T:

$$\frac{\chi_a}{c_{pa}T_0} = \left(\frac{T}{T_0} - 1 - \ln\frac{T}{T_0}\right) + \frac{R_a}{c_{pa}} \ln\frac{p_a}{p_{a0}} \tag{4.4}$$

where p_{a0} can be computed using the following equation:

$$p_{a0} = p_0/(1 + 1.608\,\omega_0) \tag{4.5}$$

Formulations for H_2O

The saturation pressure (p_s) versus temperature (T_s) relationship for H_2O in the liquid–vapor region is calculated using the following correlation:

$$\ln p_s = a1 + \tau(a2 + \tau(a3 + \tau(a4 + \tau(a5 + a6 \cdot \tau)))) \tag{4.6}$$

where $\tau = 1000/T$ and $a1$, $a2$, $a3$, $a4$, $a5$, and $a6$ are numerical constants. A similar correlation given by Wexler (1977) was assumed for the ice–vapor region and is of the form

$$\ln p_s = B1/T_s + B2 + B3 \cdot T_s + B4 \cdot \ln T_s \tag{4.7}$$

For the specific enthalpy of H_2O, the triple point of water was assumed as the reference state to be consistent with the values in standard steam tables. The equations for the specific enthalpy of water (h_w) and for water vapor (h_v) are as follows:

$$h_w (T_w) = c_{pw}(T_w - T_t) \tag{4.8}$$

$$h_v (T_v) = 2501 + c_{pv}(T_v - T_t) \tag{4.9}$$

where c_{pw} $(= 4.1868 \text{ kJ/kg} \cdot \text{K})$ and c_{pv} $(= 1.87 \text{ kJ/kg} \cdot \text{K})$ denote the constant pressure specific heat for water and water vapor, respectively. T_w, T_v, and T_t refer to the temperatures of water, water vapor, and the triple point of H_2O, respectively.

The absolute entropy for H_2O in either the liquid or vapor phase must be determined as was the case with dry air. The following equations give the specific entropy for water (s_w) and for water vapor (s_v):

$$s_w (T_w) = 3.882 + c_{pw} \ln(T_w/T_{ref}) \tag{4.10}$$

$$s_v (T_v, p_v) = 10.476 + c_{pv} \ln(T_v/T_{ref}) - R_v \ln(p_v/p_{ref}) \tag{4.11}$$

For flow exergy for water (χ_w) and water vapor (χ_v), the appropriate equations are as follows:

$$\chi_w = h_w (T_w) - h_{v0} (T_0) - T_0[s_w (T_w) - s_{v0} (T_0, p_{v0})] \tag{4.12}$$

$$\frac{\chi_v}{c_{pv}T_0} = \left(\frac{T_v}{T_0} - 1 - \ln \frac{T_v}{T_0}\right) + \frac{R_v}{c_{pv}} \ln \frac{p_v}{p_{v0}} \tag{4.13}$$

The partial pressure of water vapor in the ambient environment may be determined from

$$p_{v0} = (1.608 \, \omega_0)p_0/(1 + 1.608 \, \omega_0) \tag{4.14}$$

Formulations for Moist Air

Equations for the properties of moist air are based on those for dry air and water vapor via the appropriate additive laws. Thus, the enthalpy h_{ma} and entropy s_{ma} for moist air, per unit mass of dry air, are determined using the following equations:

$$h_{ma} (T_{db}) = h_a (T_{db}) + \omega h_v (T_{db}) \tag{4.15}$$

$$s_{ma} (T_{db}, p) = s_a (T_{db}, p_a) + \omega s_v (T_{db}, p_v) \tag{4.16}$$

The partial pressures p_a and p_v for dry air and the moisture in the air are given by

$$p_a = p/(1 + 1.608\omega) \tag{4.17}$$

$$p_v = 1.608 \, \omega[p/(1 + 1.608 \, \omega)] \tag{4.18}$$

The flow exergy χ_{ma} for moist air (on a unit mass of dry air basis) can be computed using the following equation:

$$\frac{\chi_{ma}}{c_{pa}T_0} = \left(1 + \frac{c_{pv}}{c_{pa}} \omega\right)\left(\frac{T_{db}}{T_0} - 1 - \ln \frac{T_{db}}{T_0}\right) \\ + \frac{R_a}{c_{pa}}\left(\ln \frac{p_a}{p_{a0}} + \omega \frac{R_v}{R_a} \ln \frac{p_v}{p_{v0}}\right) \tag{4.19}$$

This equation is the same as that published in the literature (see Wepfer et al. 1979).

The computer code **PSY** written in BASIC uses these equations to enable the thermodynamic properties of moist air, water, and water vapor to be calculated for typical conditions encountered in psychrometric and air conditioning processes.

Adebiyi, G. A. 1989. "Computer-Aided First- and Second-Law Thermodynamic Evaluation and Design of Air Conditioning Systems." *Proceedings of the Third National Conference on Microcomputer Applications in Energy.* Tucson, AZ, Nov. 1–3, 1988. 195–202.

Adebiyi, G. A. 1991. "Thermodynamic Analysis of Refrigeration Systems Using Refrigerant-22." *Mechanical Engineering Systems.* Vol. 1, No. 6. 22–36.

Haar, L., Gallagher, J. S., and Kell, G. S. 1984. *NBS/NRC Steam Tables: Thermodynamic and Transport Properties and Computer Programs for Vapor and Liquid States of Water in SI Units.* Washington, D.C.: Hemisphere Publishing.

Hamblin, F. D. 1971. *Abridged Thermodynamic and Thermochemical Tables: SI Units.* New York: Pergamon Press.

Rogers, G. F. C. and Mayhew, Y. R. 1980. *Thermodynamic Transport Properties of Fluids: SI Units.* 3rd ed. Oxford, U.K.: Basil Blackwell.

Vargaftik, N. B. 1975. *Tables on the Thermophysical Properties of Liquids and Gases.* 2nd ed. Washington, D.C.: Hemisphere Publishing.

Wepfer, W. J., Gaggioli, R. A., and Obert, F. E. 1979. "Proper Evaluation of Available Energy for HVAC." *ASHRAE Transactions.* Vol. 85, Part 1. 214–230.

Wexler, A. 1977. "Vapor Pressure Formulation for Ice." *Journal of Research NBS.* Vol. 81A, No. 1. 5–20.

Answers to Selected Problems

CHAPTER 1

1.1 (a) 4.56×10^{15} kilocalories, (b) 1.91×10^{19} joules (J)

1.3 (a) 5.52 h/day, (b) 0.33 kWh, (c) 0.0945

1.5 2.45×10^6 kg

1.7 14 quads

1.9 2041 A.D.

1.11 1979, 1982, 1986, 1991, 1998

1.13 1.5 kJ (to the heat pump device)

1.15 0.6 kg/m, (a) 25.7 mpg; (b) 19.4 mpg

1.17 16.3%

1.19 (a) (i) 1.19×10^6 ft · lbf (0.447 kWh); (ii) 5.23×10^5 ft · lbf (0.197 kWh), (b) 3.49 kWh; (i) 12.8%; (ii) 5.64%, (c) (i) \$22.40/kWh; (ii) \$50.80/kWh

1.21 111 kWh/million Btu gas; 4.49 ¢/kWh

1.23 (a) 10.4%, (b) 9.88%, (c) 18.7%

1.25 (a) 0.35 kW, (b) −1.8 kW, (c) 7340 Btu/h

CHAPTER 2

2.1 (a) 101.35 kPa (abs), (b) 1.014 bar (abs), (c) 29.9 in. Hg, (d) 33.9 ft water

2.3 10^6 kN

2.5 49.7 psig

2.7 (a) 0.260 psi, (b) 1800 Pa

2.9 18,400 Pa

2.11 (a) 2.5 m^3, (b) 88.3 ft^3

2.13 0.837 m^3

2.15 20 lbm

2.17 466 lbf

2.19 66.7 kg

2.21 19.3 lbf

2.23 49.7 lbm/ft^3

2.25 7.98 lbm

2.27 375 mm^3; 382 mm^3

2.29 93.5°C

2.31 (a) 564°C, (b) 601°C

2.33 −785 J

2.35 30 kJ

2.37 585 Btu

2.39 (a) 0.303 hp, (b) 226 W

2.41 589 J/kg

2.43 93.2 ft

2.45 6.45×10^{-4} m^2

2.47 98.6°F; 60.8°F; 112°F

2.49 0.24 Btu/lbm · °R

2.51 1330 Btu/lbm

2.53 (a) 39.4 in., (b) 4.09 lbf/$in.^2$, (c) 145 lbf/$in.^2$, (d) 1.34×10^4 hp, (e) 1050 N · m, (f) 1330 N, (g) 11.0 in., (h) 96.5 lbm, (i) 29.2 kg, (j) 635 g

2.55 (a) 18 kg, (b) 18 kg

2.57 (a) 9.03×10^{26}, (b) 6.022×10^{26}

2.59 (a) 36.5 m^3, (b) 36.5 m^3

2.61 1.86 kW

2.63 3 kN

2.65 (a) 0.116 lbf/$in.^2$, (b) 1.47 in. Hg

2.67 (a) −1.2 kJ, (b) 1.2 kJ

CHAPTER 3

3.1 (a) Compressed water, (b) Wet steam, (c) Compressed ice, (d) Superheated steam, (e) Wet steam, (f) Compressed water, (g) Wet steam, (h) Wet steam, (i) Superheated steam, (j) Superheated steam

3.5 (a) 318 kJ, (b) 298 kJ, (c) 317 kJ

3.7 3.83 lbm

3.11 (a) 120.2°C, (b) 8.48 kg

3.13 (a) 39.74 bar, (b) 4.40 kg, (c) 0.0907, (d) 5460 kJ

3.15 (a) −25.0°C, (b) 0.321, (c) 0.0338 kJ/kg · K

3.18 (a) 25.0 psia, (b) 136 Btu

3.21 (a)1250 Btu, (b) −839 Btu, (c) −713 Btu
3.25 (a) 22.6 × 10⁻³ m³/kg, (b) 1.95 m³/kmol,
(c) 234.41 kJ/kg, (d) 20,270 kJ/kmol,
(e) 257.96 kJ/kg, (f) 22,310 kJ/kmol,
(g) 0.8942 kJ/kg · K, (h) 77.34 kJ/kmol · K,
(i) −32.06 kJ/kg, (j) −2770 kJ/kmol,
(k) −8.51 kJ/kg, (l) −736 kJ/kmol
3.29 (a) 1.968 kJ/kg · K, (b) 1.968 kJ/kg · K
3.33 (a) 0.625, (b) 29.7 × 10⁻³ m³/kg,
(c) 173.27 kJ/kg, (d) 0.6463 kJ/kg · K
3.35 81.7 psia
3.38 54.5 kg/m³
3.42 0.0119 ft³/lbm; 0.0243 Btu/lbm · °R;
11.8 Btu/lbm
3.46 1.09 × 10⁻³ m³/kg; −374.1 kJ/kg;
−373.9 kJ/kg
3.49 (a) 0.00231, (b) 1.55 × 10⁻³ m³, (c) 754 kJ

CHAPTER 4

4.2 4
4.5 (a) 0.0558 kg, (b) 162 kPa (gage)
4.8 2.50 atm
4.12 35.0 ft³
4.15 44.0
4.19 0.6 ft³
4.22 (a) (i) 8.655 ft³/lbm, (ii) 8.699 ft³/lbm,
(b) 540°F
4.25 125 kPa
4.28 (a) 118 kg, (b) 120 kg
4.31 (a) 0.189 kJ/kg · K (0.0451 Btu/lbm · °R),
(b) 0.819 kJ/kg · K (0.196 Btu/lbm · °R,
(c) 0.630 kJ/kg · K (0.150 Btu/lbm · °R)
4.33 42.3°C
4.38 3.62 × 10⁻³ m³/kg; 10.1 × 10⁻³ m³/kg (ideal
gas)
4.41 0.236 m³/kg (0.233 m³/kg, tables)
4.45 125 kPa (125 kPa, **STEAM** code)

CHAPTER 5

5.1 (a) 0, (b) 90 J, (c) −90 J
5.3 120 N; 360 J
5.6 (a) 600 kJ, (b) −600 kJ
5.9 (a) 0.1 kJ, (b) −0.1 kJ, (c) −0.025 kJ,
(d) 0.125 kJ
5.12 (a) 2.5 J, (b) 52.5 J
5.15 (a) −0.5 J, (b) 1 J, (c) −0.5 J
5.17 4.35 J

5.19 (a) 6.32 × 10⁻² kg, (b) 12.9 × 10⁻² kg,
(c) 0.03 m³ (A), 0.0153 m³ (B), (d) −30 kJ,
(e) −15.3 kJ
5.21 0.35 kJ, −0.603 kJ, −0.501 kJ; 0.754 kJ
5.23 (a) 2.57 × 10⁻² kg, (b) 5 kJ
5.27 203 kJ
5.31 −2990 ft · lbf
5.35 1.225
5.38 19.0 hp
5.41 2160 W, 2.90 hp
5.44 377°F, 156°F
5.47 104°F

CHAPTER 6

6.1 (a) 18 kJ, (b) 38 kJ
6.4 (a) 0 Btu, (b) 13 Btu, (c) −13 Btu, (d) 0 Btu
6.7 0.952°C
6.11 10 J
6.15 135 kWh
6.20 (a) 212 J, (b) 4100 J/kg · K
6.22 19.8°C
6.26 83.7 kWh
6.30 4.59 km
6.33 9.5 kW
6.40 374°C
6.45 (a) 198°F, (b) −5920 ft · lbf, (c) −0.942 Btu
6.49 957 kg/min
6.53 9080 kW
6.55 5860 hp
6.58 38 kW
6.60 508 kW
6.63 857°F
6.66 832°F
6.68 (a) 12.6 bar, (b) 8.67 × 10⁻³ m³/kg
6.73 −17,500 kJ

CHAPTER 7

7.1 (a) 0.596%, (b) 26.5%, (c) 33.1%
7.4 (a) 2.33 × 10⁵ ft · lbf, (b) 30%
7.7 (a) 97.9 kWh, (b) 65.2 kWh
7.11 (a) 66.7%, (c) 4.67 × 10⁵ ft · lbf, (d) 1340°F
7.14 25%
7.18 17.4%
7.20 (a) 490 K, (b) 653 kJ, (c) 155 kJ
7.22 100%
7.26 49.5%
7.28 72.3%

CHAPTER 8

8.1   50 J/K
8.5   (a) -2085 kJ, (b) -7.585 kJ/K
8.9   0.0151 Btu/°R
8.13   1.45 kJ/K
8.17   0.0706 kJ/kg · K
8.19   2.49 atm
8.22   (a) 139°C, (b) 904 kJ/kg
8.24   (a) 11.1 bar, (b) 19.5 kJ/kg
8.27   277°F
8.31   (a) 1.02 atm, (b) 395 kJ/kg
8.34   1036°R, 126 Btu/lbm (T_1 should read 1700°R)
8.37 -186 ft · lbf/lbm
8.39   30 MJ
8.42   (a) 2500 kJ, (b) 7.14
8.46   (a) 300 K, (b) 0.0578 kJ/K, (c) 17.3 kJ
8.49   2.48 kJ/K · kg–steam
8.51   0.0453 kJ/K · kg–refrigerant
8.53   5.24 Btu/h · °R
8.57   98.1 kJ/kg
8.59 3.35×10^{-4} kW/K
8.61   0.909
8.64   (a) -281 kJ/kg, (b) -198 kJ/kg
8.67   2.85 ft/s
8.69   56°C; 0.177 kJ/K; 53.6 kJ
8.72   (a) 74.1°C, (b) -40.2 kJ/kg,
      (c) 5.1 kJ/kg–refrigerant
8.76   33.0 MJ

CHAPTER 9

9.1   (a) 961 kJ, (b) 201 kJ, (c) 200 kJ, (d) 711 kJ
9.3   (a) 21.1 kJ, (b) 21.1 kJ
9.5   (a) 5.25 kJ, (b) 3.30 kJ, (c) 18.4 kJ
9.10   (a) 275 Btu, (b) 195 Btu, (c) 4.64 Btu,
      (d) 17.2 Btu
9.13   (a) 216 kJ, (b) 343 kJ
9.16   496 kJ
9.20   1.44 kWh
9.25   (a) 92.3%, (b) 4.90%
9.30   (a) 37.3 kJ/kg, (b) 3.34
9.33   (a) 15 kg, (b) 1470 kJ, (c) 900 kJ, (d) 61.1%
9.37   (a) 18,300 Btu (bricks); 18,800 Btu (water)
9.41   (a) 1220 kJ/kg, (b) 424 kJ/kg, (c) 3.85 kJ/kg,
      (d) 316 kJ/kg, (e) 232 kJ/kg
9.44   (a) 8.31%, (b) 0%, (c) 58.9%, (d) 86.5%,
      (e) 44.4%
9.47   (a) 78 kW, (b) 7.44 kW

9.49   9.54%
9.52   18.1% (parallel-flow); 39.7% (counter-flow)
9.55   (a) 260 kW, (b) 2.07 kW, (c) 1.25×10^{-3} kW,
      (d) 19.6 kW, (e) 282 kW
9.58   1680 hp

CHAPTER 10

10.1   (a) 40%, (b) 0.667
10.3   4.99 kJ/kg
10.5   (a) 904 kJ/kg; 0.710, (b) 768 kJ/kg; 0.767
10.9   (a) 591 kJ/kg, (b) 827.5 kJ/kg
10.13   (a) (i) 70%, (ii) 70 kJ, (iii) 85 kJ, (iv) 82.4%
10.17   (a) 784 kJ/kg, (b) 30.0%, (c) 0.850
10.20   (a) (i) 413 Btu/lbm, (ii) 0.986, (iii) 39.0%,
      (b) (i) 347 Btu/lbm, (ii) 32.8%, (iii) 0.842
10.24   (a) 1.02×10^5 kg/h, (b) 41.5 MW, (c) 3.14 MW,
      (d) 16.9 MW, (e) 1.40 MW, (f) 48.3%
10.28   34%
10.32   (b) 3520 kJ/kg; 8810 kW, (c) 3700 kW,
      (d) 7050 kW, (e) 52.5%
10.36   (a) 44.1%, (b) 0.997
10.40   46.2%
10.45   (a) 36.9%, (b) 0.572, (c) 18.4×10^6 Btu/h,
      (d) 6.58×10^6 Btu/h
10.49   (a) 25.0%, (b) 0.325, (c) 140 MW, (d) 238 MW,
      (e) 58.6%
10.52   (a) (i) 22.5%, (ii) 93.5; (b) (i) 30.5%, (ii) 127
10.56   (b) 33.4%, (c) 3.86 atm
10.59   (b) 533 kJ/kg, (c) 42.8%, (d) 6.49 atm, (e) 0.790
10.61   (c) (i) 26.7%, (ii) 14.4%, (iii) 31.3%,
      (e) (i) 34.5%, (ii) 15.5%, (iii) 26.5%
10.64   (a) 0.242, (b) 0.835

CHAPTER 11

11.1   41.1 kJ/kg
11.5   207 kJ/kg
11.9   (a) 4.18, (b) 3.5
11.13   1.67
11.17   (a) 1.5 kW, (b) 1.5 kW, (c) 22.5%
11.21   3.01
11.25   (a) 305 W, (b) 101 W, (c) 3.01, (d) 44 minutes,
      (e) 26.3%
11.28   (a) 9.76 kWh, (b) 21.6%
11.31   (a) -40.7°F, (b) 26,600 Btu/h, (c) 4360 Btu/h,
      (d) 23,400 Btu/h, (e) 18.6%
11.36   8.29%

11.38 $COP_C =$

$$\frac{1 - \dfrac{T_3}{T_1} + \eta_{aT}\dfrac{T_3}{T_1}\left(1 - \dfrac{1}{r_p^{(k-1)/k}}\right)}{\dfrac{1}{\eta_{aC}}\left(r_p^{(k-1)/k} - 1\right) - \eta_{aT}\dfrac{T_3}{T_1}\left(1 - \dfrac{1}{r_p^{(k-1)/k}}\right)}$$

11.42 (a) 124 kWh, (b) 62.1 kWh, (c) 10.7 kWh

CHAPTER 12

12.23 251 K
12.25 (a) 158 kJ/kg, (b) 169 kJ/kg
12.27 0.476 kJ/kg · K (chart); 0.428 kJ/kg · K (ideal gas)
12.31 0.141°F/psia (change pressure to 20 atm)
12.33 (a) -9.6×10^{-6} K/kPa

CHAPTER 13

13.1 (a) 52.2% CO, 41.0% CO_2, 6.83% H_2, (b) 0.516 kJ/kg · K, (c) 16.1 kg/kmol
13.5 15.3 psia
13.9 (a) 0.978 atm (air), (b) 0.290 kJ/kg · K, (c) 0.0142, (d) 0.495 kg
13.13 2.43, 4.38, 7.43, 11.4, 16.3, 21.9, 28.2 MJ/kmol
13.18 61.7 kJ/kg mixture
13.22 1.82 kmol
13.25 (a) 71.3°F, (b) 0.0165
13.29 (a) 9.34×10^{-3}, (b) 0.594, (c) 55.5°F, (d) 1070 lbm
13.33 (a) 77.6°F, (b) 39.3% (change water flow rate to 0.02 lbm/s)
13.37 (a) 90.7°F; 0.0163, (b) 2.83%
13.41 (a) 0.0126, (b) 31°C, (c) 49.1%
13.45 (a) 21.3°C, (b) 45.4 kg/h, (c) 26.7%
13.48 (a) 0.0218
13.52 (a) 8.46 kg/h, (b) 17.0 kW, (c) 0.783 kW, (d) 13.8%

CHAPTER 14

14.1 1.38, 3.11
14.5 (a) 11.7% CO_2, (b) 14.1% CO_2, (c) 15.28% CO_2, 84.7% N_2, 0.01% SO_2, (d) 14.7% CO_2, (e) 14.5% CO_2
14.9 (a) 13.5, (b) 86% C, 14% H_2
14.13 (a) 11.99% CO_2, 5.99% O_2, 78.89% N_2, 3.13% H_2O, (b) 12.37% CO_2, 6.19% O_2, 81.44% N_2
14.17 (b) CH_4, (c) 150% theoretical air
14.21 (a) 14.2, (b) 10.8, (c) 30.5% excess air, (d) 14% CO_2, 0.2% CO, 5.1% O_2, 80.7% N_2
14.24 (a) (i) 8.90×10^5 kJ/kmol, (ii) 1.56×10^6 kJ/kmol, (iii) 1.41×10^6 kJ/kmol; (b) (i) 5.56×10^4 kJ/kg, (ii) 5.20×10^4 kJ/kg, (iii) 5.04×10^4 kJ/kg
14.29 (i) 802 MJ/kmol, (ii) 1430 MJ/kmol, (iii) 1320 MJ/kmol
14.33 (a) -3.23×10^6 Btu/lbmol–fuel, (d) 3.23×10^6 Btu/lbmol–fuel, (e) 3.48×10^6 Btu/lbmol–fuel
14.37 -5125 MJ/kmol–fuel; -5125 MJ/kmol–fuel; -5115 MJ/kmol–fuel
14.41 2060°C
14.45 Work—30.8 kWh (32.7%); Exhaust—29.9 kWh (31.7%); Coolant—28.2 kWh (30.0%); Heat loss—5.23 kWh (5.6%)
14.49 49.2 MJ/kg–fuel; 50.0 MJ/kg–fuel
14.54 50.7%
14.57 (a) 5440 MJ, (b) 2755 MJ, (c) 0, (d) 930 MJ

CHAPTER 15

15.1 (a) -2.5 kJ/K, (b) 3.33 kJ/K, (c) 0.833 kJ/K
15.4 385 kW/K
15.6 229 Btu/s · °R
15.9 0.345; 0.327
15.13 36.5% CO, 18.3% O_2, 45.2% CO_2 (molar)
15.17 16.68% C, 1.26% CO_2, 39.07% H_2, 4.11 CH_2, 19.18% H_2O, 19.53% CO, 0.17% O_2 (molar)
15.19 0.0497%
15.21 0.088%
15.25 178 kJ/mol H_2

Index

Note: Page numbers preceded by the letter A refer to the appendices.

SOFTWARE DIRECTIONS

Thermodynamic Properties Data Finder

Four programs are supplied for use with the text. These programs are **STEAM**, **GAS**, **R22**, and **PSY**.

STEAM—gives properties of H_2O in the solid, liquid, and vapor regions

GAS—gives ideal gas properties of CO_2, CO, O_2, N_2, H_2O, H_2, Air, and CH_4

R22—gives properties of refrigerant-22 in the liquid and vapor regions

PSY—gives psychrometric properties for air/water vapor mixtures

To operate any of the above programs you must have an IBM® PC/XT/AT compatible computer with at least 512K of RAM, a 3½ inch disk drive, a CGA or compatible graphics display card and monitor, and DOS 2.0 or higher operating system software.

Operating steps

Step 1. Boot the PC (follow computer/software manufacturers' procedure).

Step 2. Insert the disk into the appropriate 3½ disk drive.

Step 3. Ensure system prompt is for the disk drive (change if required, e.g. a:\).

Step 4. *First option*   Type PROPS and press ENTER to access any of the codes.
Second option   Change directory to \CODES and type the desired program name (i.e. **STEAM**, **GAS**, **R22**, or **PSY**) and return.

Step 5. When the logo appears strike any key to start the program.

Step 6. Follow directions on the screen.

Note: More instructions concerning operation of the programs are given in the Appendix of the text. It may be necessary to refer to these instructions for operating assistance, especially for **R22**.